32nd Center for Chemical Process Safety International Conference 2017 (CCPS)

Topical Conference at the 2017 AIChE Spring Meeting and 13th Global Congress on Process Safety

San Antonio, Texas, USA
26 - 30 March 2017

ISBN: 978-1-5108-4363-9

Printed from e-media with permission by:

Curran Associates, Inc.
57 Morehouse Lane
Red Hook, NY 12571

Printed by Curran Associates, Inc. (2017)

For permission requests, please contact AIChE
at the address below.

AIChE
120 Wall Street, FL 23
New York, NY 10005-4020

Phone: (800) 242-4363
Fax: (203) 775-5177

www.aiche.org

Additional copies of this publication are available from:

Curran Associates, Inc.
57 Morehouse Lane
Red Hook, NY 12571 USA
Phone: 845-758-0400
Fax: 845-758-2633
Email: curran@proceedings.com
Web: www.proceedings.com

TABLE OF CONTENTS

482030 AGILE Award Keynote Address

Abstract:

Factors in Selecting AGILE Honoree

1. Overall sustained excellent industry leadership in businesses that employ chemical engineers

2. Recognized for efforts in innovation and creating strategies that ensure success

3. Value of the keynote to the meeting attendees and the members of the Institute Schedule

Event	Time	Location
Continental Breakfast	7:00AM - 8:00AM	Henry Gonzalez Convention Center, Lila Cockrell Theatre
AGILE Keynote Address	8:00AM - 9:15AM	Henry Gonzalez Convention Center, Lila Cockrell Theatre

The 2017 AIChE Spring Meeting and 13th GCPS AGILE Keynote Address will be presented by AGILE award recipient John Y. Televantos, Partner, Arsenal Capital Partners.

John Televantos is a Partner at Arsenal Capital Partners, a middle market private equity firm focused on the specialty industrial, healthcare and specialty financial service sectors. Before joining Arsenal in 2006, Dr. Televantos was President of Aqualon Company, owned by Hercules, Inc.; President and Founder of Helios Chemical Company; and Chief Executive Officer for Foamex International.

Dr. Televantos began his career with Union Carbide in 1977 in silicones research. There he developed two key processes for the manufacture of Methyl Chloride and Siloxane polymers that were commercialized. These processes not only improved efficiency and reduced cost but also dramatically reduced waste. He moved through a variety of management positions there and later at ARCO Chemical, where he became vice president of strategy and development.

At ARCO John co-invented and led a team that developed and commercialized a breakthrough technology in polyol manufacture (a key raw material for polyurethanes) eliminating a whole process step along with waste and emissions, while at the same time opening the door to the creation of new higher-performance polyurethanes. The technology has since been licensed by most of the major global polyol manufacturers with significant waste reduction worldwide.

At Arsenal Capital, Dr. Televantos serves as co-head of the specialty industrial practice. During his tenure, he has been part of several acquisitions and successful exits in the chemicals and materials sectors. In addition to serving on the boards of several Arsenal-owned companies, he is lead director on the board of Balchem Corp.

Reducing Process Safety Events: An Approach Proven By Sustainable Results

Sheila Van Geffen (SFVanGeffen@dow.com)
The Dow Chemical Company
Houston, Texas

John Champion (JohnWChampion@dow.com)
The Dow Chemical Company
Deer Park, Texas

Lynnette Borrousch (LEBorrousch@dow.com)
The Dow Chemical Company
Midland, Michigan

Keywords: DOW; NEAR MISS; CARDINAL RULES; REDUCING INCIDENTS

Prepared for Presentation at
American Institute of Chemical Engineers
2017 Spring Meeting and 13th Global Congress on Process Safety
San Antonio, TX
March 26 – 29, 2017

Abstract

The Dow Chemical Company has significantly reduced the number of Tier 1 Process Safety events from 69 in 2008 to an average of 10 in 2013-2015. This paper will share some of the key elements that have led to this successful and sustainable reduction in incidents. The structure of the paper is focused around the concept of "Effective Process Safety Programs" published by Klein and Vaughen in Process Safety: Key Concepts and Practical Applications (CRC Press, 2017).

In this paper, we will provide details on Dow's Operational Discipline systems that enable success on a day-to-day basis. These systems include Procedure Use Culture, Process Safety Training programs and Process Safety Collaboration Networks. We will also describe our Process Safety Management Systems that enable long-term success including Dow's Loss Prevention Principles, use of PHA, LOPA and other tools to manage risk, and Dow's Mechanical Integrity program. The success is built on a strong foundation of Safety Culture and Leadership. This paper will illustrate how Dow is structured to support the Process Safety efforts and examples will be shared about how leadership at all levels of the organization is practiced and demonstrated.

1. Introduction

Dow has long emphasized the prevention of Process Safety Incidents (PSIs). In the 1950s and 1960s, the focus was driven primarily by several severe incidents that led to the development of company-wide programs such as Dow's Reactive Chemicals and Risk Management programs, as well as an early version of our Mechanical Integrity program in some locations. On creation of Dow's Reactive Chemicals program in 1966, Dr. Earle B. Barnes, a senior Dow executive, stated "In the future, if any employee is responsible for an incident which occurs because the Reactive Chemicals Program procedures have not been followed, he should remember that his job and his life, as well as the lives of others, are in jeopardy." This statement is still embraced by Dow's current management and accurately reflects the response that occurs when equivalent procedures are not followed and a reactive chemical incident occurs. Other Process Safety tools, management systems, and programs developed over the decades include the Fire & Explosion Index (1964), Technology Centers (1965), Dow's Loss Prevention Principles (1974), and the Chemical Exposure Index (1985).

In 1995, Business and Corporate EH&S Leadership established formal generational goals to significantly reduce incidents over the next 10 years. These goals provided the platform for driving breakthrough performance impacting employees, customers, communities and the environment. This emphasis yielded a more than threefold reduction in PSIs over those 10 years.

In 2005 another 10 year goal was established to further reduce PSIs and their severity by 75% and 90% respectively. This goal was met in 2011 and sustained through 2015.

In late 2015 another generational goal was established through 2025 with the goal of zero Level 1 Process Safety and Containment Events (PSCEs) sustained across the corporation, as well as a significant reduction in lower level events. Note that Level 1 events are equivalent to Tier 1 events as defined in API RP 754 [1], with several notable voluntary inclusions such as laboratories and cross-country pipelines.

Figure 1 shows the trend of PSIs over the last 20 years. Figure 2 shows the Dow incident rate as compared to our American Chemical Council (ACC) peers and normalized by company size. The chart clearly indicates that Dow's breakthrough performance from 2009 to present has taken us from average performance to being amongst the best of our peers.

This paper focuses on some of the key programs that enabled us to break through the plateau we had reached in 2005-2008 and continue to drive us to our 2025 goal of zero.

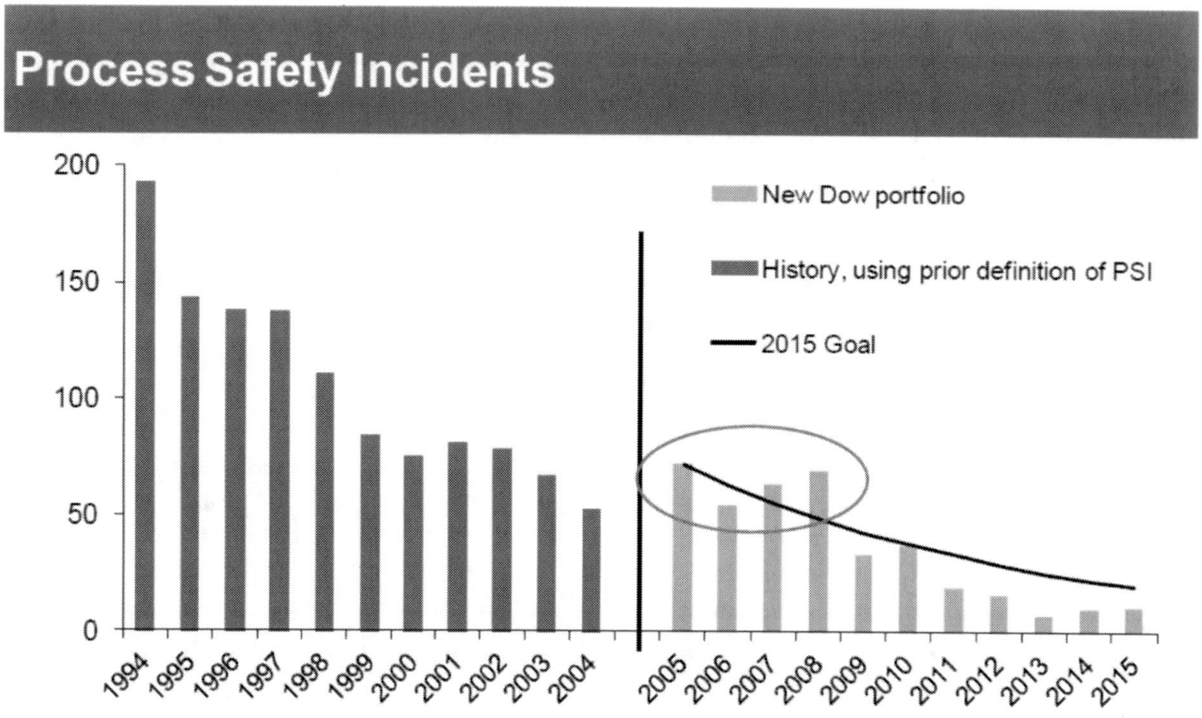

Figure 1 – The Dow Chemical Company Process Safety Incidents by Year

Figure 2 – Dow incident rate vs ACC incident rate [2]

2. Effective Process Safety Programs

Figure 3 illustrates the three foundations of effective Process Safety programs. Each of these three aspects intertwines to form the overall approach to success that we have seen in practice in Dow.

Figure 3 – The Three Foundations of Effective Process Safety Programs [3]

At the base of the triangle, supporting everything we do, is Leadership. Leadership support at all levels of the organization is the foundation that is essential to sustain success in Process Safety. Section 3 covers the structure that enables strong Leadership in support of Process Safety within Dow.

On the left side of the triangle are the Process Safety Systems, or the management systems, that Dow has implemented globally through our Operating Discipline Management Systems (ODMS). These systems include Dow's Process Risk Management Standard (PRMS) and Global Mechanical Integrity Safety Standard (GMISS), as well as our Personal Safety Standards, Pollution Prevention, and many other categories. Compliance with these internal requirements, which typically go beyond regulatory minimum requirements, is what enables long term success. Section 4 provides an overview of Dow's Process Safety Systems and how they are implemented.

On the right side of the triangle is Operational Discipline - what occurs in the plant on a daily basis. Operating Discipline, performed successfully, enables effective implementation of Process Safety Systems. For example, we may have established an inspection protocol under GMISS for a facility's piping circuits. Properly following the inspection protocols allow for effective inspections and a decrease in containment losses resulting from mechanical integrity failures. How the plan is executed will define the outcome – successful or not. Section 5 includes several examples of how Dow executes the Process Safety Strategies.

The authors have observed, both within Dow and in some peer companies, that failures generally do not occur in the design of the Process Safety Systems themselves. Rather, failure tends to occur in the execution of those systems - and accountability for these failures ultimately leads back to failures in Operating Discipline and Leadership. We have concluded that, while we must have strong management systems as a framework for excellence, we must also focus a large portion of our energy on implementation through Leadership and Operating Discipline to reduce incidents year after year.

3. Process Safety Leadership at all levels

Dow's commitment to continuously improve both Personal Safety and Process Safety performance aligns to corporate core values. Dow established EH&S as the top priority of the company, engaging all employees across a multitude of business and functional sectors. Dow has clear expectations that all employees are responsible for safety and measures both business and employee performance against achieving safety expectations. Sponsorship of EH&S begins at the Board of Directors and is a key aspect associated with all leadership roles in the shape of goals, expectations, and resource allocation.

3.1 Leadership Commitment

Operations benefits from leadership support provided by both the Business and Corporate EH&S leadership and aligned to Dow's common Process Safety vision. Business and Corporate EH&S Leadership are responsible for defining the expectations and strategic goals needed to achieve Process Safety targets. They are responsible for allocating the resources necessary to drive Process Safety improvements and sustain performance. A few examples of their direct engagement are provided below:

3.1.1 Leadership Governance

Corporate and Business EH&S Governance is paramount to achieving Process Safety improvements to ensure that on-going support, resources and priorities are managed. The Operations Governance Team, led by the Corporate EH&S Director, is designed to drive business and functional accountability for EH&S performance in Operations facilities. They also provide functional oversight and direction regarding alignment to Dow Corporate EH&S policies, standards and requirements. The Operations Governance Team oversees the corporate risk management of facilities to drive EH&S improvements and ensure that the operating business and regional teams are managed appropriately in accordance with the corporate risk criteria.

3.1.2 Financial Commitment

Substantial corporate capital funding is allocated to supplement business capital to manage overall corporate risk and promote timely completion of Process Safety projects. These projects are identified and prioritized through various Process Safety assessments such as Layer of Protection Analysis and Loss Prevention Principles, which are discussed in section 4. Corporate funded projects are co-managed by the Business Group, Process Safety and EH&S leadership.

These corporate funds are designated as high priority EH&S capital to address targeted risk reduction projects along with high priority personal safety improvement projects.

3.1.3 Process Safety Priority Team

The Process Safety Priority Team, sponsored by Manufacturing, EH&S and Process Safety Leadership, is focused on those facilities in need of additional corporate support due to recent incidents or additional risk factors (called "Challenged Facilities"). Key initiatives of the Challenged Facility program includes leadership support, targeted analysis, and a comparison of facility programs against established best practices for key Process Safety management systems. Corporate support drives the accountability for facilities to implement improvement plans and provides additional support to manage execution plans.

The Process Safety Priority Team is also responsible for leading Root Cause Investigation (RCI) effectiveness reviews. The RCI effectiveness process starts with Manufacturing and EH&S Leadership assigning a neutral RCI facilitator to deliver impartial, high quality investigations. The RCI effectiveness process validates that management system root causes have been identified, corrective actions truly prevent reoccurrence, and leveragble findings are appropriately shared. Key learnings are extracted from the incident review and published in an Learning Event Report (LER) for enterprise wide distribution to drive continuous learning culture. In some cases, significant findings can result in corporate-mandated action plans requiring follow-up actions at all facilities.

3.2 Facility Leadership

At the facility level, the Production Leader manages the resources needed to execute the projects outlined in the strategic Facility Specific Technology plan. The plan is designed to manage the various commitments made to manage risk identified at the facility, business and functional level. The individual facility Production Leaders are expected to demonstrate a competent understanding of the facility's chemistry, technology, and Process Safety risks. One mechanism for validating this knowledge is through the New Leader Reactive Chemicals and Process Hazard Analysis (RC/PHA) work process. This review includes a presentation to the technology knowledgable subject matter experts, Process Safety and EH&S professionals. New Production Leaders are expected to complete this process within the first 90 days of taking responsibility for a facility. The Production Leader is not authorized to approve high risk Management of Change reviews until he/she has successfully completed the New Leader Review. [4]

3.3 Employee Engagement

Employees operate facilities daily with a clear expectation that EH&S is the top priority of their activities. All facility employees, including the operators, are expected to set personal EH&S goals aligned to their facility EH&S goals. These goals align to the Business, Function, Geography and ultimately the Corporate EH&S direction. They are also expected to contribute to a continuous improvement and learning organization. Employees are educated on the technology and the operational Process Safety risks through the Process Safety programs such as Cardinal Rules as well as their involvement in EH&S department and project teams.

Employees participate in programs designed to identify and record near misses both in the Personal Safety and Process Safety. Near miss reporting is a key expectation and an important enabler to ensure that issues are resolved quickly. Proactive resolution of small problems is a desired behavior that is fostered by the corporate and facility cultures. All employees are engaged in behavior based programs focused on intervention and coaching. This real time coaching ensures field practices are occurring in a manner aligned to the standards and procedures designed to manage risk and deliver best in class results.

Key influential leaders focus on driving desired behaviors through continued engagement in field activity and coaching. Proces Safety focal points are in place in all facilities (more in Section 5). All employees, regardless of position, are empowered to report issues and take action to keep themselves and the work environment safe. Operations personnel are are empowered to shut the plant down if they have an EH&S concern, even for a minor leak.

4. Process Safety Systems – Enables Long-Term Success

The Process Safety systems, as well as other supporting management systems, form the framework for a succesful and sustainable program. When properly implemented, they provide the means to prevent the low frequency, high consequence events that can cause great harm to people and the environment. This section provides a basic overview of these systems and how they are practiced within Dow.

4.1 Dow's Organizational Structure for Process Safety

Dow first created Business Technology Centers in 1965. These centers were created around common technologies within Dow. They were tasked with ensuring consistent application of the safest and most effective technology practices. Although the focus of the technology centers has expanded over the years to include responsibilities such as technology specific training and documentation of best operating and design practices, the core focus still remains strongly aligned with Process Safety.

Additionally, there is a Process Safety Technology Center (PSTC) where the corporate expertise in Process Safety practices resides. It is within the PSTC where the work processes for Process Safety are maintained. Process Safety Technology Leaders (PSTLs) are associated with the PSTC and are also aligned to each Business Technology Center. They serve as the liaison between the process technologies and the corporate Process Safety expertise. The PSTLs are responsible for the selection and application of technology, management systems, and hazard assessment tools used to address the risks of unplanned events such as fires, explosions, reactive chemical incidents, and chemical releases.

The Technology Center personnel develop the technology specific manner of meeting the intent of the Process Safety work processes and then disseminate that out to each of the manufacturing facilities. Both the Technology Center personnel and the PSTLs provide oversight and governance to assure that the facilities are implementing the work processes in the most effective manner. Figure 4 illustrates how this flows.

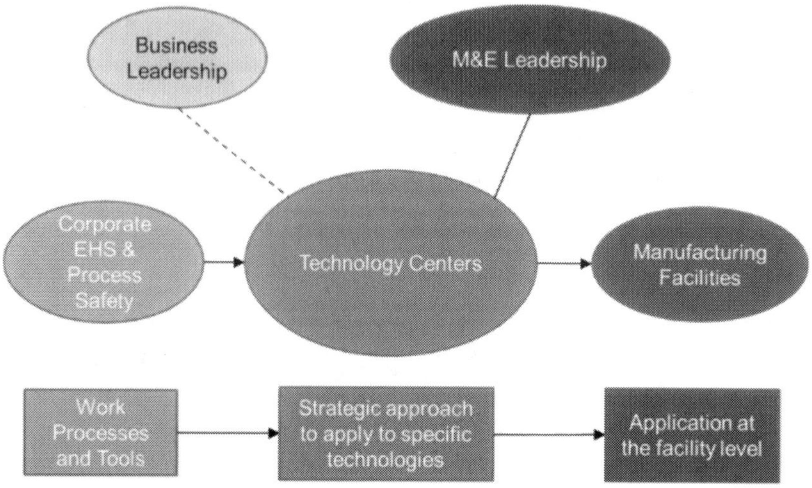

Figure 4 – Technology Center Organizational Alignment

4.2 Dow's Operating Discipline Management Systems (ODMS) and Most Effective Technology (MET)

Operating Discipline Management System (ODMS) is Dow's framework for global requirements in a wide variety of topics, including Employee Health and Safety, Pollution Prevention and Process Safety. All of these are under one umbrella section of ODMS called Responsible Care.

ODMS also has another umbrella section on Common Management Systems, which include Leadership Responsibilities, Communication and Audits. There are also a number of requirements that are related to Process Safety. These include Management of Change (MOC), Procedures, Investigations, and Corrective and Preventative Actions (CAPA). Figure 5 illustrates how these are organized.

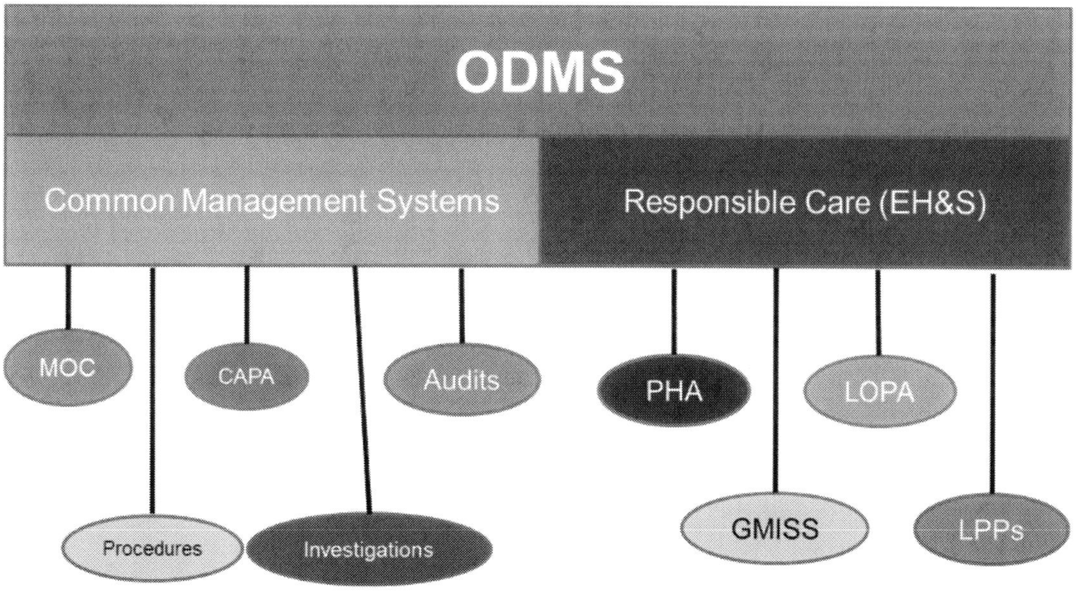

Figure 5 – ODMS Alignment for Process Safety

These ODMS requirements establish the minimum global requirements, many of which are prescriptive. Sites, business units and individual facilities are free to develop additional requirements to suit their needs or to meet specific regulatory requirements, but they cannot deviate from the global requirements without an approved variance.

4.3 The Loss Prevention Principles

The Loss Prevention Principles (LPPs) are the primary collection of Dow's Process Safety requirements. They originated in 1974 as a collection of various safety bulletins, but have evolved over the years to become the foundational minimum requirements for every facility within Dow globally. The LPPs encompass 79 different topics, including Hazardous Area Classification, Storage Tanks, Diking and Containment, Utilities, Flare Safety, Fire Protection Systems, Facility Siting, Relief Systems and Instrumented Systems.

The LPPs collect industry Recognized and Generally Accepted Good Engineering Practices (RAGAGEP). The Subject Matter Experts within Dow develop internal interpretation of the RAGAGEP, and do so with a global perspective by drawing upon standards from around the world. The information is then incorporated into LPP requirements. The benefit of this approach is that the need for each facility or site to own and interpret all of the external standards is eliminated. When RAGAGEP is updated, the LPPs are updated accordingly to reflect those changes.

The LPPs also contain requirements based on learnings from significant incidents, both within and external to Dow. For example, when the learnings from the Buncefield UK terminal explosion were published, Dow added a requirement to the Facility Siting LPP to limit the growth of vegetation in locations where vapor cloud explosion potential existed. [5]

Many of the LPPs are mandatory for compliance for all facilities to which they apply, although some are only required for newly installed equipment. Given the wide variety of technologies and types of operations within Dow, there is no "one size fits all" for some requirements. A variance process can be utilized to demonstrate non-applicability or an equivalenrmeans to manage the hazard that the LPP requirement is intended to address.

4.4 Process Risk Management Standard [3]

The Dow Process Risk Management Standard (PRMS) forms the framework for the hazard identification and risk analysis of chemical processes and supporting operations such as utility systems, waste treatment plants and laboratories. There are four levels to PRMS, which are shown in Figures 5 and 6. Following this stepwise process enables the facilities to be evaluated against Dow's risk tolerance criteria at a level of detail commensurate with the risk of the operation.

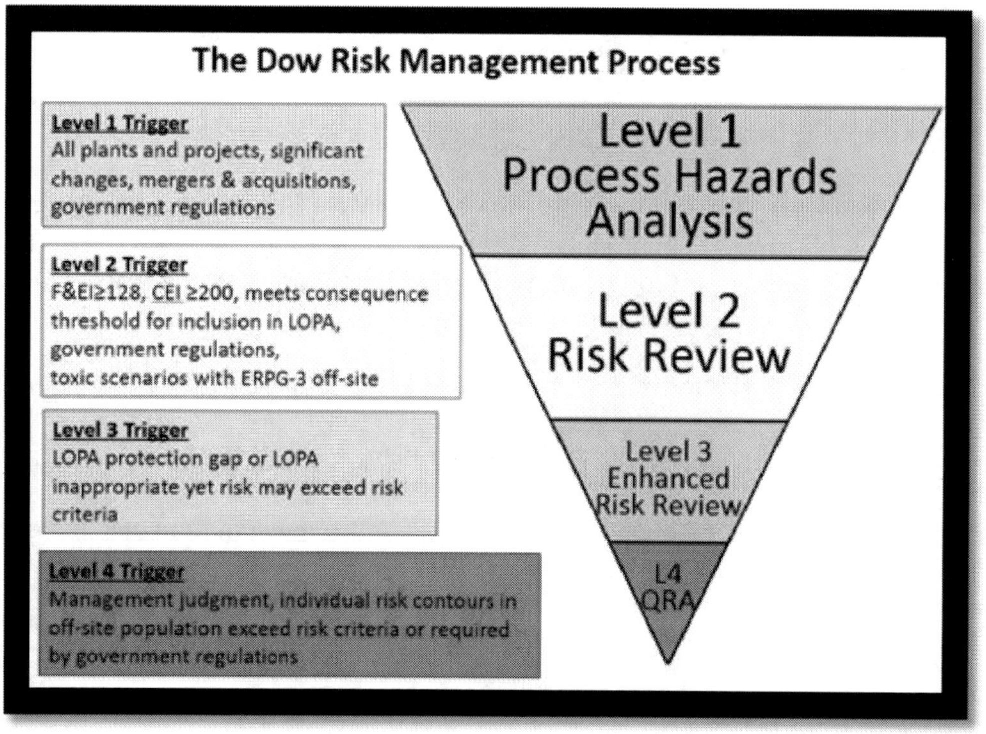

Figure 6 – The Dow Risk Management Process

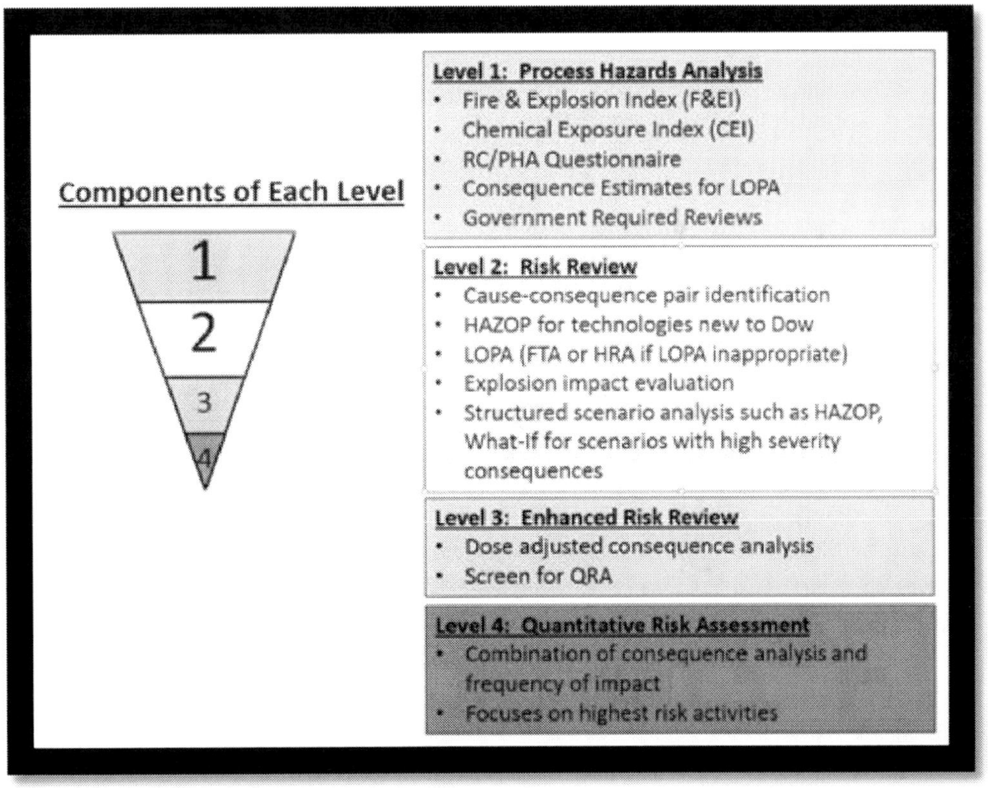

Figure 7 – Components of the Dow Risk Management Process

All Dow manufacturing and support facilities are covered by PRMS and are required to perform at least the first level of the analysis, which is the Reactive Chemicals Process Hazards Analysis (RC/PHA). Those that meet one or more of the trigger criteria specified in Figure 5 must complete a Level 2 analysis, which includes Layer of Protection Analysis (LOPA) and Explosion Impact Evaluation for occupied buildings. Most full scale manufacturing facilities within Dow meet one or more of the Level 2 triggers and perform LOPA.

The purpose of the LOPA is to define the protection layers for identified scenarios and specify the required reliability of those protection layers. If the risk tolerance criteria are satisfied within LOPA, then Level 3 and Level 4 tools are not needed to further evaluate the risk. However, if LOPA cannot define a protection strategy that meets Dow's risk tolerance criteria, then the analysis must proceed to Level 3.

Level 3 analyses are done infrequently and Level 4 analyses are rarely done within Dow unless required by government regulations. Unless legally required, Level 4 studies are typically reserved for high hazard processes for which LOPA is not the appropriate method to analyze and manage risk. When Level 4 studies are required, corporate leadership is engaged in the approval to perform the study and actively drives the risk reduction needed to meet the risk tolerance criteria.

The scenarios that were analyzed, and the protection layers that have been defined in these studies, are critical inputs to some of the operating discipline work processes, such as instrument integrity, operating procedures and the training of plant personnel. While doing a thorough analysis is important, the transition from analysis to implementation into operating discipline is a critical success factor.

4.5 Global Mechanical Integrity Safety Standard

Dow's Global Mechanical Integrity Safety Standard (GMISS) evolved from the first Mechanical Integrity programs that were implemented in all major sites in the 1960s, prior to most regulatory requirements. Sadly, the first Mechanical Integrity program was triggered by a catastrophic incident that occurred within Dow. [6] In 1999, the Mechanical Integrity standard was formalized and unified into the global standard that exists today. The standard is applied at all Dow sites, regardless of whether there are regulatory requirements in a given jurisdiction. When regulatory and GMISS requirements overlap, plants are required to comply with the more conservative requirement.

Similar to regulatory requirements in many countries, equipment and piping is considered to be within the scope of GMISS if it contains hazardous chemicals or is in hazardous service (e.g. high temperature, corrosive). Typical equipment like pressure vessels, storage tanks, relief devices and piping circuits are covered by GMISS. In addition, other components containing hazardous chemicals or in hazardous service are also covered, including expansion joints, flexible hoses, leak repair clamps, line blinds and sight glasses.

All covered equipment is required to be registered, inspected, and repaired/modified as per defined frequencies. Any equipment with deficiencies discovered through the inspection process undergoes a fitness for service evaluation by a trained assessor and is expected to be repaired within the defined timeframe. Any deviations from this protocol require subject matter expert, Manufacturing, Maintenance and Process Safety review, approval and documentation.

A key enabler to implementing a successful GMISS program at the facility level has been the use of Mechanical Integrity Assessment Teams. These teams consist of cross-functional representation from the Inspection Group, the Maintenance Group and Operations (including Plant Leadership). They meet regularly (usually monthly) to jointly discuss the following items:

1. Review all results of completed inspections, especially where there were deficiencies found.
2. Define plans to address deficiencies and other recommendations from the inspections.
3. Review the upcoming planned inspections to ensure that the equipment will be available when needed.
4. Define and review special activities, such as fitness for service evaluations.

The Mechanical Integrity Assessment Teams provide a valuable formal interaction between the experts and equipment owners to define the appropriate strategies, incorporate process knowledge and gain leadership support for the defined plans.

5. Operating Discipline – Enables Success on a Day-To-Day Basis

What is a game plan without execution? We achieve our goals not only through the act of setting a target but also in excuting what is needed to reach and exceed the target. In the three foundations of Effective Process Safety Programs:

- Leadership is the base as setting expectations and goals,
- Process Safety Systems defines the management systems needed to define how to achieve the goals and
- Operating Discipline is the execution.

How the plan is executed will define the outcome – successful or not. Below are descriptions on some of the most impactful ways in which Dow's execution of Process Safety strategies contributed to the sustained success in Process Safety at Dow.

5.1 Process Safety Focal Point and Networks

Each operating facility assigns an individual to the role of Process Safety Focal Point (PSFP). The Process Safety Focal Point (PSFP) is the key resource on technical issues relating to Process Safety for their operating unit. A senior Plant Engineer assumes the role along with their other duties. The role provides focus to the Process Safety management systems and serves as a local first point of contact for Process Safety. The Process Safety Focal Point is the primary technical contact for the identification, assessment and management of Process Safety risks within each Production, Site Logistics and Environmental Operations facility. The PSFP works closely with the Business Technology Center and PSTL assigned to the technology to provide proper interpretation and guidance on Process Safety requirements and practices.

Dow has found that designating a specific PSFP role within an organization, tasked with raising awareness of Process Safety at the facility level, provides a focus and support necessary to ensure Process Safety requirements are met. Core Responsibilities of the Process Safety Focal Point include:

- Working with PSTLs and business Technology Center to provide Process Safety support for the facility, including guidance on implementation of Loss Prevention Principles (LPPs) and compliance with local regulations.
- Performing Process Safety reviews on MOCs for the facility, escalating to a PSTL or other expert for further review where required.
- Acting as a communication focal point, both to and from the PSTL, for the Process Safety related knowledge and issues at the facility.
- Facilitating gap assessments against Dow mandatory requirements and local regulatory standards, and work with plant leadership to gain resources to close identified gaps.
- Participating in, and helping to prepare for, Root Cause Investigations (RCIs), Reactive Chemical Process Hazard Analysises (RCPHAs), Management of Change reviews and other Process Safety requirements as needed.
- Organizing plant resource for Process Safety self-assessments, EH&S audits and LOPA studies.
- Maintaining the Process Safety documentation for the facility.
- Working with Maintenance to review mechanical integrity and instrumented protective system test and inspection results and follow up, as needed.

- Assisting in developing plant, business, and site Process Safety goals and assessing progress and/or defining opportunities to achieve performance goals.
- Actively engaging in geographic and / or Business Process Safety Focal point networks and communication meetings to share best practices and Learning Experience Reports.
- Supporting the development and delivery of Process Safety training in the facility. Ensureing that Process Safety Near Misses (PSNMs) and Process Safety and Containment Events (PSCEs) are reported and investigated

Dow has a number of Process Safety Networks. The purpose of Process Safety Networks in Dow is to share knowledge on Process Safety technology as well as recent learnings. The Process Safety Technology Center has a network composed of PSTLs. Business and Technology Process Safety Networks include technology center representatives and all the Process Safety Focal Points (PSFP) for the business globally. Site and Geographic Process Safety Networks include local representatives and all the PSFPs from that geography.

These network objectives are to:

- Sustain a high level of Process Safety knowledge within Dow manufacturing by sharing best practices and innovative approaches to managing hazards.
- Provide interpretation and guidance on Process Safety requirements and practices.
- Share Process Safety and Containment LERs as well as Near Misses to minimize the probability of occurrence.
- Structured employee development of PSFPs and other network members on the technology of Process Safety.

5.2 Process Safety Near Misses [8] [13]

Major Process Safety incidents in chemical manufacturing are infrequent, but when they do occur, the consequences can be severe. The Deepwater Horizon oil rig explosion and oil spill (April 2010), Imperial Sugar Refinery dust explosion (October 2009), and BP Texas City Refinery vapor cloud explosion (March 2005) are a few examples of industry's major Process Safety incidents. [9] [10] [11] These events had significant impact to human life, the environment and received extensive coverage by the media. Corporate performance goals are often measured in such a way that a single plant could operate for many years without a major process accident and still be operating below an acceptable level of performance. Process Safety Near Miss (PSNM) reporting is intended to be a more sensitive indicator of actual Process Safety performance. The Merriam-Webster dictionary defines the term near miss as "an accident that is just barely avoided." [12] When applied to the discipline of Process Safety, a PSNM can be defined as a narrowly avoided event involving a chemical process. The value of reporting PSNM events is not the counting of events but rather in the ability to recognize the potential for an accident, the opportunity to identify and implement corrective actions and the learning value that may prevent future accidents.

In 2009, Process Safety Near Miss (PSNM) Reporting was emphasized globally in Dow as a key leading indicator of Process Safety performance which presents an opportunity to learn valuable information that may prevent more serious future incidents. The general principles that apply to Process Safety Near Miss (PSNM) reporting are:

- A chemical or chemical process must have been directly involved in the event.
- A failure must occur.
- There was learning value (taking into consideration both the actual and potential severity of the event).

Facilities report and investigate Process Safety near miss incidents to discover and eliminate potential causes for Process Safety incidents. A high number of Process Safety near misses reported is considered to be a positive indicator; the underlying assumption is that investigating and eliminating the causes of Process Safety near misses will also eliminate the underlying causes of potential Process Safety incidents before they occur.

Near misses could be perceived as lack of control of an operation. This negative perception will deter the employees of an organization from reporting and tracking information. It is paramount that all levels of leadership actively support gathering and reporting this valuable information. Leaders are expected to motivate and train their organizations to be skilled at identifying and reporting Process Safety near misses. Senior manufacturing leadership reinforces this expectation regularly.

If we consider the incident pyramid, Figure 8, with a major Process Safety incident being at the top of the pyramid, the Process Safety near misses contribute to the base of the pyramid. There are many more PSNM than major incidents. This ratio of these two should be managed to fit a department's ability to investigate and learn from the near misses. The actual ratio at all of Dow in 2015 was around 260:1, which speaks to how institutionalized the program is. A solid base of the incident pyramid created by the skills to identify, correct and communicate learnings from a PSNM can effectively reduce the number of major Process Safety incidents by preventing them before they occur.

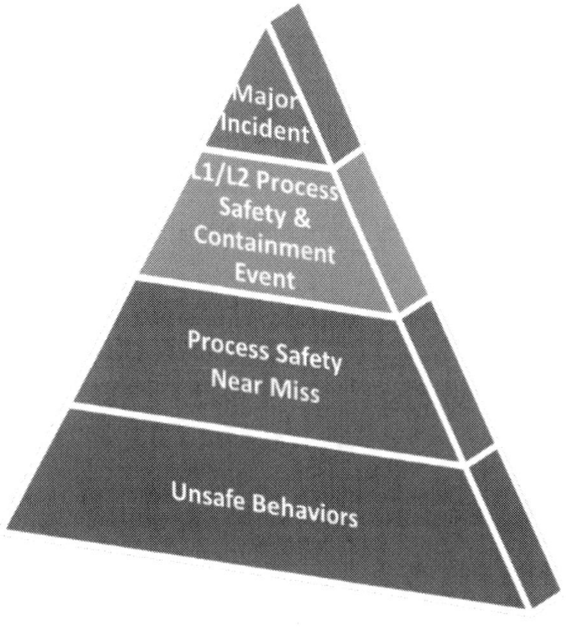

Figure 8 – Process Safety Incident Pyramid

In 2012, a further differentiator was added to Dow's Process Safety Near Miss program. To strengthen the PSNM Reporting program, the opportunity was identified to elevate the learnings from the highest potential PSNM events. The identification, investigation and corporate leveraging of the events within the corporation of the highest potential for major impact can only further strengthen and drive continuous Process Safety performance improvement.

A High Potential Process Safety Near Miss (HP PSNM) is defined as a PSNM event that results in the highest potential for a fatality, numerous Day Away from Work Cases (DAWCs) or most significant community impact if the circumstances had been slightly different. The HP PSNM should provide significant learning value to the corporation or the reinforcement of critical protection layers. The elements of the HP PSNM reporting process include:

- Immediate reporting of event to Business and Technology Leaders.
- Formal Root Cause Analysis conducted, including key roles from Process Safety and the Technology Center.
- Process Safety develops a Learning Experience Report for global distribution across the technologies and sites.
- Positive recognition for the team identifying and reporting a HP PSNM.

Identifying and implementing a High Potential Process Safety Near Miss (HP PSNM) reporting process provides stratification in a data set of events. The value of any near miss is the opportunity to correct and prevent a significant event. The additional focus on the HP PSNM events can leverage that opportunity to prevent future events beyond the impacted facility or technology in which the near miss originated. Effectively leveraging the Management System Opportunities across technologies and across sites within a global corporation has a significant impact to eliminating Process Safety Incidents and protecting the community, personnel, property and the environment.

5.3 Root Cause Investigations and Sharing of Learnings [7] [14]

A Root Cause Investigation (RCI) is the investigation that determines causes that led to a failure. It generally involves people from within the facility, though it could also bring in corporate level expertise. It includes documentation of the event, the investigation, and learning value. The extent of the investigation (number of team members, cause-effect charting, etc.) should be proportional to the event's actual/potential consequences, impact, and complexity. All investigations are not created equal. Every investigation must include the minimum components:

- A root cause analysis,
- Determination of causes that led to each protection layer that failed and associated management system failure,
- Determination of corrective and preventive actions,
- Resolution by emphasizing prevention and
- Documentation of the investigation.

A root cause investigation can range from informal discussions among 2 or more people knowledgeable in the incident to more formal techniques like cause and effect charting, 5 whys, FMEA, fishbone, etc. Effective corrective and preventive actions resolve not only the specific failures, but system and strategy level failures as well to prevent repeat events(or similar events associated with the same management system failures).

Evidence is gathered prior to the scheduled review. Using the collected evidence of process conditions, sequence of events, and operator testimonial, a cause and effect diagram is generated to identify the root cause. This includes the sharp end root cause(s) (such as a failed instrument) and the system root cause(s) (such as ineffective maintenance for the instrument). Corrective actions can then be developed to address all of the identified root causes. The management system root causes and corrective actions are typically the learnings that are most leverageable across a corporation.

With a Root Cause Investigation complete, a Learning Experience Report can be developed. Typically a learning experience report issued from an RCI will include the event description, the root causes, and corrective actions. The impact of the lessons learned depends on the quality and effectiveness of the communication through which it is shared. Summarizing the key lessons learned from any event into a single page learning experience report provides direct and concise information that can be easily communicated throughout an organization, beyond the facility that had the event.

An example of a HP PSNM LER is provided in Figure 9. While it is tempting to include all of the technical details of what occurred during the event, the importance of a learning experience report is to provide enough information on the significant learning value or the reinforcement of critical protection layers. Setting a boundary to a one page learning experience report provides a format that is simple, direct and easily transferable to a range of disciplines. The key content includes a brief description, root cause(s), consequence(s), what should have prevented the event and most importantly, key questions to ask of the recipients of the report. The brief description of the event should provide enough information of what occurred without the full technical details. The intended audience of the report must be considered before including full technical details of the event. If more details are needed, a separate document or presentation can be linked within the one page report.

As with any event, the identified root causes are an important lesson. A picture or schematic of key elements of the process related to the event can provide a visual effect that contributes to an improved understanding of the incident. The LER should include the potential consequences of the event. This could be the potential chemical released, personnel injured or cost of equipment damaged. The corporate standards or industry best practices that could have influenced or prevented the event if effectively in place are highlighted. Finally, challenging questions that someone reading the report could ask of themselves are provided, which allows the reader to internalize the event to their specific situation. The opportunity is to leverage the lesson learned beyond the boundaries of that single event. One or two questions can generate discussion or serve as a reminder when a similar topic arises.

Example

Learning Experience Report

Site Name: Dow Site
Date / Time: *December, 2011*
Presentation: link here to presentation for more details

Brief Event Description

Event Description:

Acetic Acid is fed to the batch reactor. After the acetic acid feeding is complete, a nitrogen pad is used to clear the line into the reactor before closing the downstream block valves on the line. The line uses double block valves to prevent backflow of the reactor contents into the acetic acid feed tank. In this event, the nitrogen pad did not operate properly, resulting in acetic acid trapped in the line liquid full in between the double block valves. The thermal expansion damaged the block valve seating and leaked 20 kg of acetic acid into the reactor.

Project installation recommendation was to install Double Block Valves (ABV) close together and close to the reactor. During construction, the project team assumed that N2 purge would work normally, preventing thermal expansion risks; decided it would be easier to put the upstream ABV closer to acid tank.

Root Cause details

Root Cause of each Protection Layer and its Management System:

- Process control code error in sequence of valve opening and closing did not allow proper nitrogen clearing
- Installation changes during construction:

Picture, Drawing, or Schematic

Changes during construction did not recognize the consequence of the longer distance enlarges the amount of acid that could leak to reactor

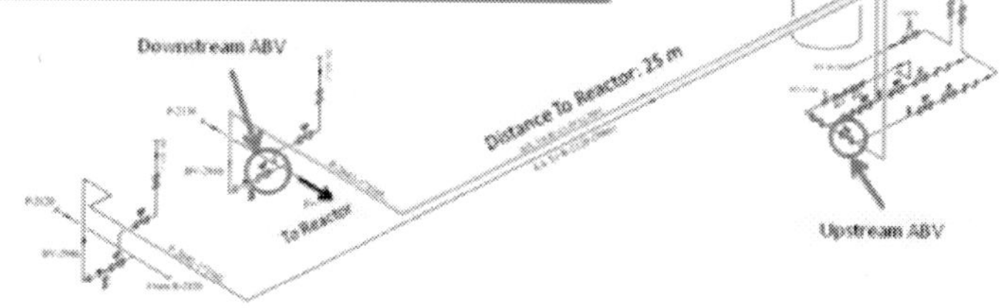

What could have happened?

Potential Consequences:

- Incorrect quantities of raw materials creates potential for an uncontrolled reaction resulting in overpressure and vessel failure

What should have prevented it?

Applicable Process Safety Standards

- Process Hazard Assessment
- Management of Change

Key questions to remember

What can you do? Can this happen in your facility?

(Provide key questions for those not in your facility could ask and identify opportunities for improvement.)

- Are installation details included in project P&IDs or job instructions highlighting the special attention items?
- Knowing the concept of process safety is not enough, being able to implement in the actual process is the most important result.

Figure 9 – HP PSNM Learning Experience Report

5.4 Procedure Use Culture

Within Dow, tasks in a plant must be performed with an approved procedure, under the control of a Safe Work Permit or be specifically exempted as low hazard. It is a core belief that the use of procedures for most tasks is a key contributor to improving Process Safety performance. Proceduralizing work tasks allows for improved planning and collaboration with expertise partners to ensure the best approach is used to define and mitigate hazards while performaing work. Procedure creation also provides an the opportunity to validate that work is being performance in adherence with Process Safety and Personal Safety Standards. It is an expectation that employees will not begin any task unless a procedure exists or the work is exempt.

Many of the events and near misses that still occur within Dow have an element of Procedure Use as a cause, either because the procedure was not followed, was not correct or did not exist. A great deal of focus has been placed on driving improvements in our procedure use culture. A key leadership goal is to foster a culture in which employees truly believe in the need for procedures and utilize them - even when nobody is watching.

The procedure use culture is fostered not only by company and plant leadership reinforcement, but also by identifying the operators that lead by example - those that the other operators look up to and emulate. So while these operators do not have a leadership title, they are effectively "influential leaders." By engaging these influential leaders in the goal of having a healthy procedure use culture, the other operators see their mentors embracing procedures and do so themselves. This has proven to be an effective strategy in driving the culture change.

Operators also own the procedures and are expected to correct them when they find errors. They also conduct more formal periodic reviews of the procedures for accuracy. Metrics are captured on how many procedures are modified when reviewed, with the expectation of a healthy number of changes and improvements be made. Metrics are tracked to validate the correct procedures are submitted following task completion against requirements. Finally, regular field audits are performed to confirm that the procedures are in-hand when required and that they are being adhered to as a task is being performed. This field presence and reinforcement is a critical leadership expectation.

Procedures can be classified several different ways, depending on the risk of the task or the situation when the task must be performed. Each different classification has different requirements, which are illustrated below:

- **Critical Tasks:** "Critical" is defined as having significant potential for exposure to hazardous chemicals or energy, significant potential for a reactive chemicals or other Process Safety incident, or any task that has steps that are defined as a formal protection layer in a hazard analysis. These procedures must generally be in-hand, are signed off stepwise and are adhered to without deviation. These procedures are required to be submitted for document retention following the completion of assigned work.
- **Immediate Response:** These are tasks that must be performed as part of a predefined emergency plan, such as in response to a power or utility failure. In general, these procedures must be readily available (usually hard copy) and routinely trained on. The

training is very important because these tasks need to be ingrained into the operators' ability to respond under stress. The procedures do not need to be in-hand during the task, but are expected to be reviewed once the emergency has subsided to ensure that nothing critical has been missed. Critique of the procedures following the emergency are completed to extract learning value and identify opportunities for improvement.

- **Lower Hazard Tasks, Routine or Non-routine:** These are the more standard tasks that are performed within a plant, such as switching pumps or taking samples. Another example is an infrequent activity like changing a filter in a lower hazard service. These tasks only require the procedure to be in-hand when the operator is in training or needs to refer to the steps, which is likely for non-routine tasks.

5.5 Process Safety Cardinal Rules [15]

The prevention of large scale accidents that have a low frequency of occurring, but high consequences, depends on an acute awareness of worst case scenarios and the assurance that the protection layers are not compromised. Cardinal Rules are used within Dow to maintain a high awareness of these types of scenarios and help prevent major Process Safety incidents.

Process Safety Cardinal Rules are written to be technology specific. These have have been used in some technologies within Dow for many years. In the last few years, an increased emphasis has been placed on ensuring that all technologies have Cardinal Rules developed and disseminated all the way to the shop floor. The Process Safety Cardinal Rules could be compared to the biblical Ten Commandments. They are short statements that are expressed in absolute terms like "never" or "always." Limiting the number of these rules for a given technology emphasizes the importance. A maximum of 10-15 Cardinal Rules is considered effective; any more than that can dilute the emphasis. Most statements are developed as a result of a significant Dow or industry incident within the technology.

All Dow personnel involved in a technology are required to know, understand and abide by the technology's Process Safety Cardinal Rules. Violating a Cardinal Rule may result (and in many cases have resulted) in significant Process Safety incident with potential impact to personnel, property and the community. A violation of a Cardinal Rule without resulting in a Process Safety accident is recorded as a Process Safety Near Miss and appropriately investigated. The learnings from the investigation are leveraged across the technology to help prevent reoccurrence.

Case Study on Implementation of Technology Process Safety Cardinal Rules

The Dow Acrylic Monomers technology owners formalized their list of Cardinal Rules in 2010, based on the internal and external storage and handling standards that had been in existence for many years prior to that. The storage and handling standards are comprised of many pages of detailed rules and guidelines for the safe handling of acrylic monomers. Many of the rules and guidelines are based on best practices and learnings from past events, including the big, small, internal and external events. The key benefit of taking an extensive standard and developing a list of Cardinal Rules is to distill key concepts down into short sound bites that can be more easily hard-wired into people's way of working. Several examples of these rules are shown in Figure 5.

Example Monomer Handling Process Safety Cardinal Rules (Not all-inclusive)
1. **<u>Never</u>** thaw frozen acrylic acid or methacrylic acid with steam 2. **<u>Never</u>** remove liquid from a partially thawed container of acrylic acid or methacrylic acid 3. **<u>Never</u>** contaminate a monomer with a foreign material 4. **<u>Never</u>** sparge nitrogen through aerobic inhibited monomer 5. **<u>Never</u>** store aerobic inhibited monomer beyond the shelf life 6. **<u>Always</u>** know the inhibitor concentration of a monomer

Figure 10 – Example Cardinal Rules for Acrylic Monomers

Behind each of the short statements is the detail to reinforce the importance of the rule. Each rule includes a description of what could happen (or has happened) if the rule is violated, where in the plant the hazard exists and what minimum safeguards shall be in place. Table 1 shows an example for one of the rules from Figure 10.

Table 1 – Details behind an example Cardinal Rule

Cardinal Rule	**What could happen?**	**Where Does the Hazard Exist?**	**Minimum Safeguards**
<u>Never</u> thaw frozen acrylic acid or methacrylic acid with steam	The use of steam can initiate a thermal, uncontrolled polymerization of the monomer, resulting in potential equipment rupture. *Note that a tank truck exploded in 1976 due to this.*	Freezing can occur in pumps, pipe systems and storage vessels that are uninsulated, poorly insulated, have inadequate heat tracing or inadequate temperature control of heating systems	1. Well maintained insulation and heat tracing systems to prevent freezing. 2. Procedures and training on the safe methods that can be used for thawing are critical. 3. Never increase the temperature of the heat tracing system above the maximum allowable temperature 4. The use of a hot water mixing station requires automatic shut-off capability for high temperature.

Once the development of the Cardinal Rules was finalized, they were incorporated into the training program for all personnel who interact with the technology. Operators, engineers, production leaders and others were all trained initially on these rules in instructor-led, interactive training sessions with the technology experts. Periodic formal recurring training is built into the program to ensure retention by those trained initially. Furthermore, they are built into the daily work practices to keep them fresh in everyone's minds. For example, each of the operating plant control rooms has a copy of the rules prominently displayed, as do some of the staff buildings. The rules are frequent topics of safety meetings and even informal discussions.

The concepts contained in the Technology Process Safety Cardinal Rules have been around for decades and have been a part of operator and engineer training as well. The Cardinal Rules provide an emphasis and enough detail to help make correct decisions, including during design, operation, maintenance or unplanned events.

5.6 Process Safety Training Systems

Why is Process Safety corporate memory important? Bad things can, and have, happened in Dow and in industry: uncontrolled chemical reactions, thermal runaway reactions, fires and explosions, mechanical overpressure, building explosions, and more. The consequences can be devastating, from property damage/loss, capital losses, environmental impact and loss of life.

The answer to the question of why is it important is more subtle than just incidents can happen. As we continue to improve Process Safety performance, incidents become less and less frequent and over time we may become less vigilant. Many individuals may never personally experience a serious event in the course of their career. In fact, that is the real goal: to never experience a serious event. In reaching that goal, we must prevent a culture that may result:

We get complacent.

We think it will not happen to me.

We tweak the process or deviate from a standard 'a little bit' with no adverse consequence.

We become more confident and tweak it some more.

Over time we can have normalize of deviations – the wrong way of operating can become the normal way – the abnormal becomes normal.

Institutionalizing Process Safety corporate memory in training can assist in preventing significant events and create a continuum of key knowledge. Dow's Process Safety training includes:

1 *Chemical properties of hazardous materials.*
2 *Hazardous scenarios.*
3 *Main protection systems, design features.*
4 *Preventative and mitigative protection systems.*
5 *Major incident history.*
6 *Process Safety Cardinal Rules.*

1 Chemical properties of hazardous materials

The training program covers the properties of those chemicals used at the facility. It includes a list of chemicals, how to recognize the chemicals, how to respond to exposure, flammability properties, toxic properties, reactivity hazards and properties of those chemicals that operate at conditions that could be hazardous (e.g. high temperatures and pressure).

2 Hazardous scenarios

Process Safety training covers the reactive chemistry and Process Safety hazards associated with systems that could be impacted by those working in the process areas, potentially causing a life-threatening event. The content includes the chemicals involved, typical initiating events, and the extent of the impact (distance / consequence). Hazardous scenarios include potential consequences from normal process scenarios and abnormal modes of operation as well.

3 Main protection systems, design features

The training content includes main protection systems and design features that contribute to safety (e.g. no water in process, layout, building design and no aluminum used in process). One key focus area of this training concerns what aspects of the protection strategy the operations personnel can directly affect, such as procedural and emergency response protection layers.

4 Preventative and mitigative protection systems

It is important that those operating the facilities be able to recognize the systems that are in place to prevent incidents and to mitigate incidents once they begin. Training includes identification of instrumented and mitigative protection systems, their purposes, and the risks posed by impairing or bypassing such systems. Conducting drills and hands-on training for these elements is essential for success.

5 Major incident history

Training on major incident history includes reviews of past serious events incidents / fatalities that have occurred in the facility or in the technology, including the worst toxic or flammable release. Past incidents are described to underline the importance of protection systems and the credibility of the hazards. Included in the training are the worst flammable and toxic releases at the facility as well as any incidents which resulted in fatalities. Photographs are provided if available to illustrate the consequence of the incident. Many trainees will relate more effectively to visual images.

6 Process Safety Cardinal Rules

The technology's Process Safety Cardinal Rules must be incorporated into the training program of all personnel who interact with the technology. Operators, engineers, leaders and others are trained initially on the rules in instructor-led, interactive training sessions with the technology experts. Recurring training is built into the training program to ensure retention.

Each facility must develop the appropriate Process Safety training, with the support of the Technology Center, PSTL and PSFP. The training is delivered as initial training to new employees and on a recurring schedule. A variation of delivery methods can be used to provide the training, from instructor-led training to computer-based, self-paced learning.

6. Conclusions

There is no "silver bullet" to reducing Process Safety Incidents. Strong management systems and constant devotion to Process Safety at all levels of the organization are necessary to drive that reduction. The programs described in this paper were developed and added over many years, some with the hindsight of some serious events that affected people, the environment and company performance. However, anchoring this approach has been the unwavering focus of leadership. This focus has been maintained through countless difficult economic times, mergers and acquisitions and numerous other challenges.

Having company wide requirements defined in the Process Safety and EH&S strategy systems instead of an accumulation of varying location specific practices is key to achieve company wide successful performance, The requirements serve as the framework, but it is also essential to have the expertise and focus by all employees on this common goal.

To achieve our goal of zero incidents, we must continue to drive performance improvement by successful implementation of the established systems described in this paper. In addition, refinements to those established systems and new systems not yet developed will come from innovations, as well as learnings from internal and external events. We are not yet at zero, and we recognize that we have to continue to evolve to get there.

7. References

[1] American Petroleum Institute, *API Recommended Practice 754, Process Safety Performance Indicators for the Refining and Petrochemical Industries, Second Edition*, April 2016

[2] American Chemical Council, *American Chemical Council Responsible Care Fourth Annual Benchmark Report* prepared for Dow Chemical, April 2016.

[3] J. A. Klein and B. K. Vaughen, Process Safety: Key Concepts and Practical Approaches, CRCPress, Boca Rotan, FL, 2017.

[4] K. A. Study, *Training Management of Change Approvers,* presented at the 6th Global Congress on Process Safety, March 2010

[5] United Kingdom Health and Safety Executive, Reports and recommendations arising from the Competent Authority's response to the Buncefield incident, available at http://www.hse.gov.uk/comah/buncefield/response.htm accessed on February 17, 2017.

[6] Amarillo Globe Times, *5 KILLED, 12 HURT AT CHEMICAL PLANT*, April 25, 1960.

[7] J. W. Chastain, P. Delanoy, C. Devlin, P. Mueller and K. A. Study, *Beyond HAZOP And LOPA: Four Different Company Approaches*, presented at the 12th Global Congress on Process Safety, April 11, 2016.

[8] S. F. Van Geffen, *Are Today's Engineering Designs Preventing Tomorrow's Process Safety Incidents?*, presented at the Mary Kay O'Connor Process Safety Center 2013 International Symposium, October 22, 2013.

[9] U.S. Chemical Safety Board, The U.S. Chemical Safety Board's Investigation into the Macondo Disaster Finds Offshore Risk Management and Regulatory Oversight still Inadequate in Gulf of Mexico, available at http://www.csb.gov/the-us-chemical-safety-boards-investigation-into-the-macondo-disaster-finds-offshore-risk-management-and-regulatory-oversight-still-inadequate-in-gulf-of-mexico/ accessed on February 17, 2017.

[10] U. S. Chemical Safety Board, Imperial Sugar Company Dust Explosion and Fire, available at http://www.csb.gov/imperial-sugar-company-dust-explosion-and-fire/ accesed on February 17, 2017.

[11] U.S. Chemical Safety Board, BP America Refinery Explosion, available at http://www.csb.gov/bp-america-refinery-explosion/ accessed on February 17, 2017.

[12] Merriam-Webster Dictionary, available at http://www.merriam-webster.com/dictionary/near%20miss accessed on February 18, 2017

[13] G. Nesmith, J. T. Keating and L. A. Zacharias, *Investigating Process Safety Near Misses To Improve Performance*, Process Safety Progress, Volume 32, Issue 2, pages 170-174, June 2013.

[14] K. P. Stevick, *Next Generation Root Cause Investigation and Analysis - Elimination of Repetitive Incidents through Strengthening Management Systems*, presented at the 11th Global Congress on Process Safety, April 27, 2015.

[15] J. W. Champion, K. P. Stevick, K. A. Study and S. F. Van Geffen, *Dow Learnings And Actions From The Deepwater Horizon Accident*, Process Safety Progress, Volume 34, Issue 4, pages 335-344, December 2015.

Building and maintain a Group Process Safety Management System for a Non-Chemical Industry

Marc Massant
Umicore
Broekstraat 31, 1000 Brussels
marc.massant@umicore.com

Prepared for Presentation at
American Institute of Chemical Engineers
2017 Spring Meeting and 13th Global Congress on Process Safety
San Antonio, TX
March 26 – 29, 2017

Building and maintain a Group Process Safety Management System for a Non-Chemical Industry

Marc Massant
Umicore
Broekstraat 31, 1000 Brussels
marc.massant@umicore.com

Keywords: management system, community, compliance, risk management system, pragmatic, tools, train program, guidance note, standards, integrity, expertise, awareness, non chemical

Abstract

The Umicore organization faced a fatal accident with two fatalities in 2014 due to hydrogen explosion in one of the largest and most advanced sites of the Group (in Olen Belgium).
The accident investigation learned that most of the underlying factors that contributed to this accident could be found in the process safety management approach and the awareness in the organization.

Despite the zero lost-time accident approach of our management the company did not have a systematic 'Process Safety' approach.

This was mainly because the Umicore organization has a lot of different activities, some of them only mechanical and some of them more chemical related.

The accident triggered a more intensified effort in process safety.

A global Umicore project was launched to put more focus on process safety in the whole organization worldwide. The project team started to describe, build and enroll a process safety management system for the group. They started a program to raise awareness and to give training on all levels. They also developed Umicore tools to facilitate the practical work in the Umicore entities.

The program was based on building a process safety community, raise awareness and introduce a process safety culture, perform site visits based on a process safety index of all sites, deploy and train in the whole organization, and develop a software based on the new internal standards and guidance notes. The program was developed and rolled out in a very structured and fast way from April 2014 till now and, while the implementation is in progress, the program is very successful.

This paper gives a short description of every building block of the process safety project that was launched and the approach.

1 Umicore: a non-chemical company?

1.1 History

The history of Umicore goes back more than 200 years. It all started with the coming together of a number of mining and smelting companies, which gradually evolved into the materials technology and recycling company Umicore is today. On 17 December 1805, Napoleon Bonaparte granted Jean Dony control of the Vieille-Montagne mine in Moresnet, on what is today the border between Belgium and Germany.

The story of another Umicore tendril, Union Minière, commenced in 1906. Union Minière du Haut Katanga (UMHK), as it was known at the time, produced copper and other metals in the Congo. After the Zairian government nationalized the company's assets in 1968, UMHK set out to develop new mining and refining activities, eventually becoming a sub-holding of Société Générale de Belgique.

The merger in 1989 of Union Minière with its subsidiaries (Metallurgie Hoboken-Overpelt, Vieille-Montagne and Mechim) transformed Union Minière into an integrated industrial group The acquisition of PMG in 2003 added a new dimension to the company, including a major presence in the automotive catalyst sector. PMG was in fact the former precious metals unit of the German Degussa group – the very company that, in 1887, had been a founding shareholder of Umicore's Hoboken plant. Today Umicore is a global materials technology and recycling group, with more than 10,000 employees and a turnover of €10.4 billion.

1.2 Businesses

Our Businesses are structured in 3 groups

Catalyst

- Automotive Catalysts: A world leader in emission control catalysts for light-duty and heavy-duty vehicles and for all fuel types
- Precious Metals Chemistry: Develops and produces metal-based catalysts used in chemistry, life sciences and pharmaceutical applications.

Energy & Surface Technologies

- Rechargeable Battery Materials: A leading cathode material supplier for lithium-ion rechargeable batteries used in portable electronics and electrified vehicles.
- Cobalt & Specialty Materials: Produces cobalt and nickel specialty chemicals for a wide range of applications, including tires, pigments, catalysts and surface treatment.
- Electroplating: Supplies precious metal electrolytes & processes for technical, functional and decorative applications.
- Electro-Optic Materials: Supplier of products for thermal imaging as well as wafers for space solar cells and high brightness LEDs and chemicals for fibre optics
- Thin Film Products: Producer of evaporation materials and sputtering targets in the fields of optics, wear protection & decorative coatings, microelectronics and semiconductors, and large area coatings.

Recycling

- Jewellery & Industrial Metals: Supplier of precious metals creating products for the jewellery sector and industrial applications as well as recycling old jewellery and production scrap.
- Platinum Engineered Materials: Specialist in the development and manufacturing of PGM based gauzes and components for the special glass and chemical industries.
- Precious Metals Management: Services for hedging, leasing, purchasing and sale of precious metals to internal and external customers
- Precious Metals Refining: Operates the world's most sophisticated precious metals recycling facility and recovers 17 precious and other valuable metals from complex waste streams. Proprietary technology for the recycling of Li-ion rechargeable batteries
- Technical Materials: Supplier of silver and other metal containing products for technical applications in the electric and power, lighting, heating-ventilation-air conditioning (HVAC) and tooling sectors.

1.3 Non-chemical?

A lot of activities are linked to chemical processes based on chemistry with metals.
Other activities are mainly mechanical and can hardly be considered as chemical activities. So the conclusion is mixed.

2 Building and maintain a group safety management system

To start the program we first listed an overview of all activities needed for a successful program. Then we put all these activities into practice with a very tide and strict timeline.

Figure 1 shows the basic scheme of the project program.

In the next section a description of every building block is presented. Where possible the templates and results are discussed.

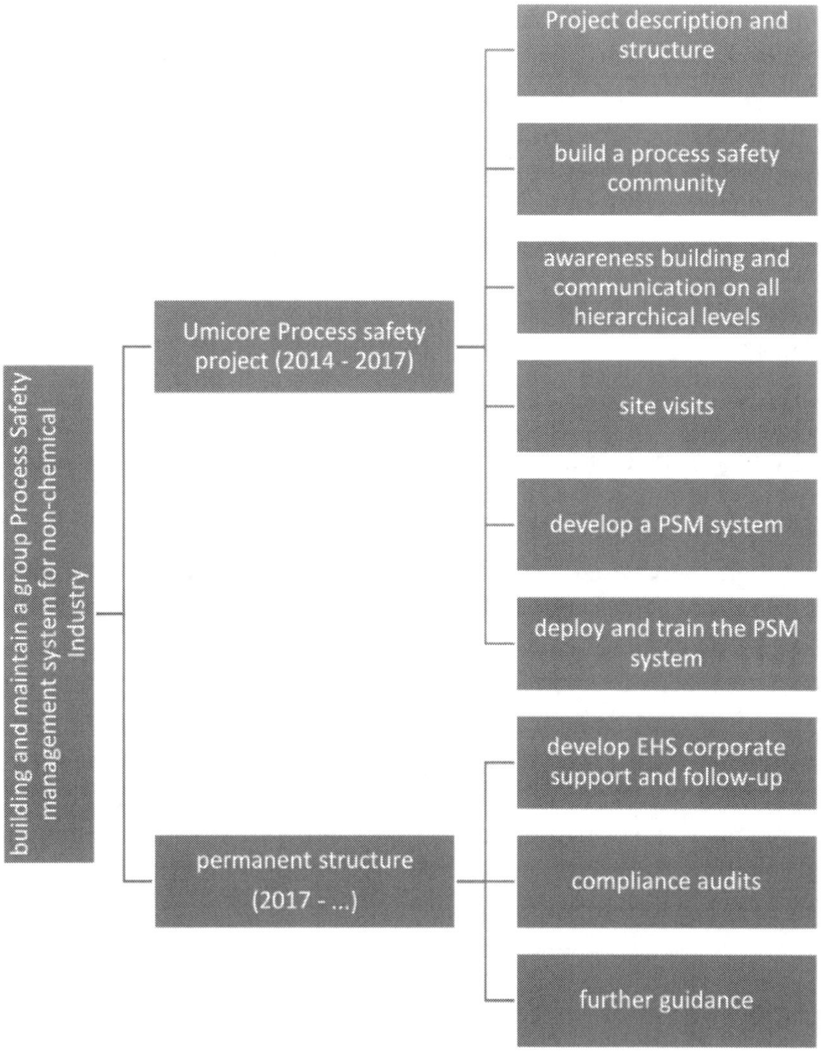

Figure 1: Overview of the project

3 Umicore Process safety project

The project was communicated and organized by the top management of the group (CEO and Executive Committee).

Two internal expert were asked to develop the activities on group level.
One of the experts was more competent in the chemical processes present in our organization. Previous he was based as a senior director in the Group Research department He took up the role of director of the project but was also involved in the project work and the implementation. The second expert was more competent in the implementation of a process safety management systems, the use of PRA (Process Risk Assessment) methods and the necessary risk reduction measures.
Prior to this role, he was the head of the safety department in the factory in Olen at the time of the fatal accident (2014). He was also in charge of the investigation of the fatal accident. He took the first initiatives for an increased effort in process safety for the group in 2013. Previous he was employed as a safety advisor in production factory of monovinyl chloride - MVC.

The project team members had not to start from scratch. A first attempt for a global process safety guidance note for the group was already written in 2013. After the fatal accident in Olen, this first general guidance note on process safety was used as a basis for the process safety project.

This is an overview of the project structure and approach:

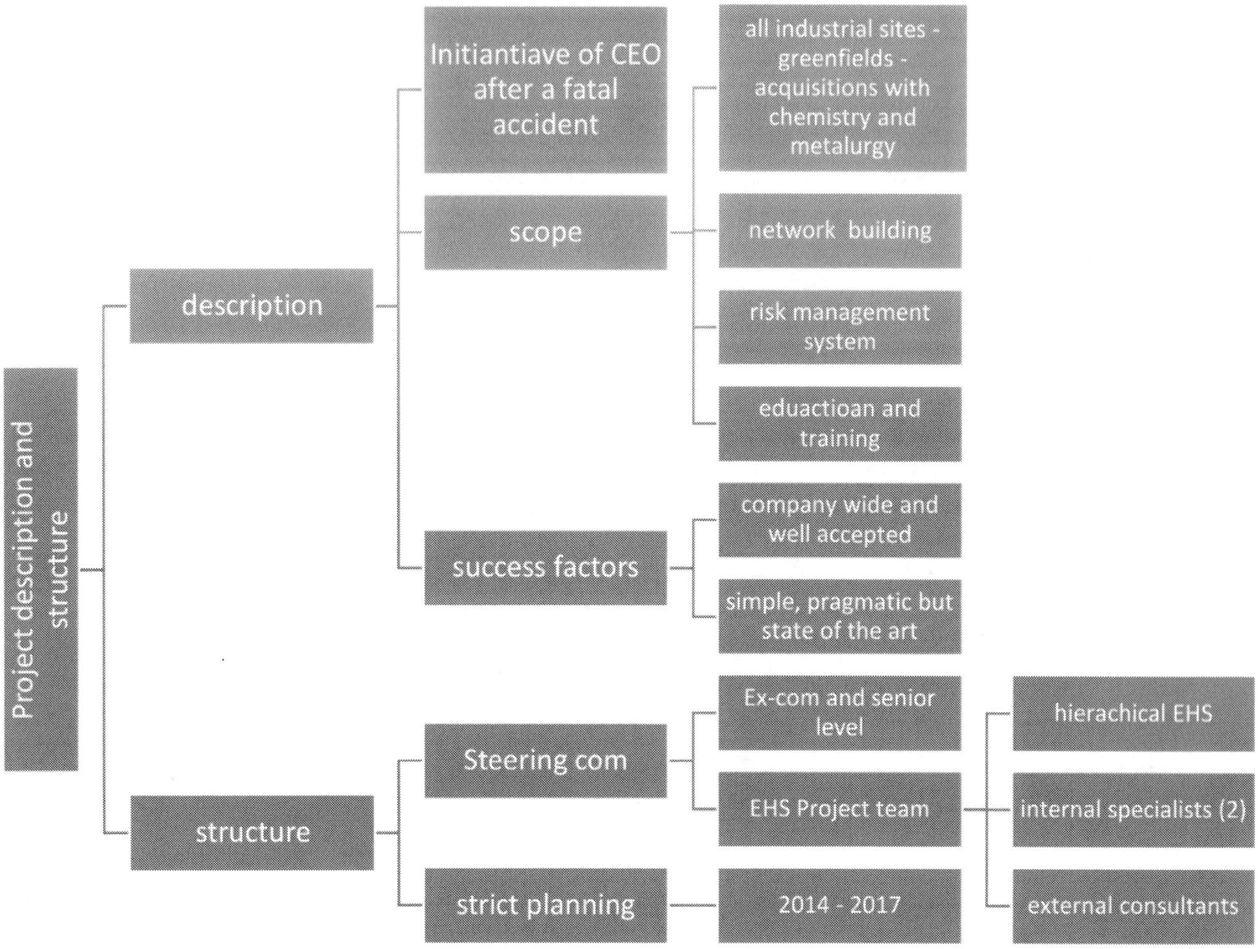

Figure 2: Project description and structure

The structured project approach was a key element in the success of the project. It guaranties continuous attention, full focus and a good structure. It is important to describe 'what you want to do' and even more important, 'what you do not want to do'.

The project was launched and communicated to the whole Umicore community by an internal communication by the top management and an intensive internal communication. The fact that the project was launched and communicated by top management (CEO) contributed to the initial commitment in the whole organization. Of course this was not enough to ensure full participation on all levels. Additional initiatives were taken to ensure full commitment. Describing the roles and responsibilities for process safety for everyone involved was a key element. Besides that we ensured that all relevant functions were part of the steering comity of the project. These functions were (see also next chapter): member of the excom, senior vice president EHS, senior vice presidents of the most hazardous business units, legal counsel, senior vice president of the human resource department responsible for training and development. The contribution and participation was reported in the steering committee and consecutive actions were taken.

The objective of the project was fixed to develop a well-accepted, companywide, state-of-the-art, process safety management approach and implement it in a pragmatic way in all industrial sites. With this objective in mind the project team started to focus on the scope.

The following were the main topics within the overall scope. Each topic is be discussed briefly in subsequent sections.

- all industrial sites
- all greenfield projects
- all acquisitions
- all chemical and metallurgical processes
- building a process safety community
- education and training in Process safety
- building a competence center on group level
- introduction of a risk management system
- building a state of the art process safety management system (but not too complicated)
- KPI's (Key Performance Indicators) process safety
- performance criteria
- assessment program
- start follow-up after project

The project was structured around a steering committee at the senior level and was run by two senior engineers between April 2014 and March 2017.

Steering committee meetings

- 4 steering committee meetings and 1 closing meeting

Project team meetings

- 34 meetings were held till the end of 2016. During these team meetings the progress of the project and the way forward were discussed.
- As part of the approach a separate project was launched to develop a software to document the hazard and risk assessments in a uniform way for the whole group.

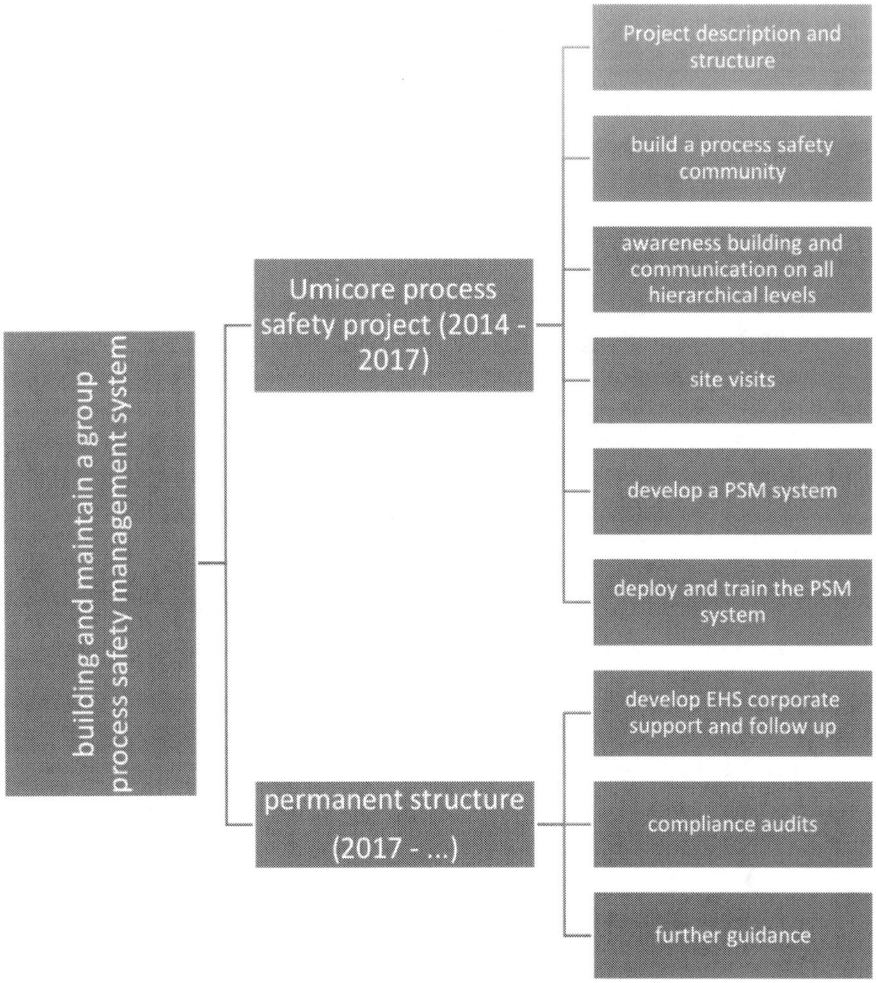

Figure 3: building and maintain a group process safety management system

4 Building and maintain a group process safety management system

4.1 Building a process safety community

Building a process safety community was a key element for the success of project and the certainly for achieving a high acceptance level of the new guidelines and standards. Also, the implementation process was strongly facilitated by this process safety community. Efforts were taken to build an internal and an external network. The internal network was often consulted to review proposals, to explain their needs, to coordinate the rollout in their business units and to follow up on the implementation.

The external network was consulted by the project team to identify best practices in the world and to reflect on ideas for a pragmatic but suitable approach.

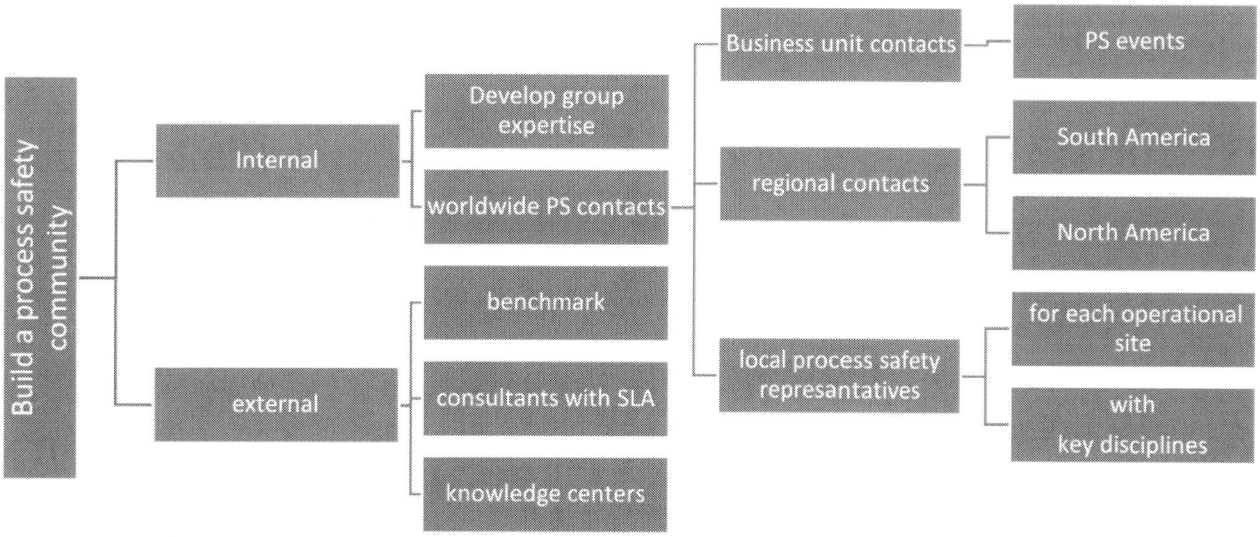

Figure 4: Building a process safety community

4.1.1 Internal process safety community

4.1.1.1 Develop group expertise

As mentioned previously, the knowledge about process safety was rather scattered in the group and certainly not present in every (small) production entity. The knowledge of the internal processes was very good and certainly under control but a lot of the information that could be linked to possible process safety scenarios was not well documented. As a result we concluded that the knowledge management was not well organized yet, especially for the older processes. The availability of information, including process safety information, was not guaranteed during the lifecycle of the processes resulting in a possible wrong decision making process for example during management of change investigations.
The process hazard analysis (please pay attention: this is not the same as a process risk assessment but it is an additional investigation) was implemented to close the gap.

At the group level, the function of process safety manager (or group safety manager) was created to ensure the expertise on management approaches and suitable techniques, and to follow-up on the needs and implementation. The full implementation will continue after the project has closed.

4.1.1.2 Business unit contacts

For every business unit, process safety contacts were appointed by the management in 2014. This group of people also acted as the sounding board for the review of new standards and guidance notes and was consulted to verify if the project plan and schedule was doable and pragmatic. To organize this active participation the project team organized five one-day process safety events; During these events the BU contacts were informed about the status and progress.

4.1.1.3 Local representatives

Local process safety representatives in each operational site (PS community) were appointed in 2014.The list is updated on a yearly base ever since. The list is built around the key disciplines involved in process safety: general management, EHS, research & development, process engineers, operation, maintenance and quality management. Their main focus is the implementation of the project outcomes and the applicable guidance notes and standards.

4.1.1.4 Regional contacts

In the regions of South America, North America and German-speaking sites, a regional contact was appointed to facilitate the communication process. In the other regions, i.e.China, rest of Europe and Asia Pacific, the project team took up this role.

4.1.2 External process safety community

4.1.2.1 Benchmark visits

In 2014 and 2015 the project team representatives performed several benchmark visits in order to get insight into the best practices in the field of process safety management. Also the pragmatic implementation of a process safety management system was discussed. A lot of good ideas and examples were detected during these visits. These practices contributed to the system that was ultimately put in place. The following companies were visited:

- Agfa-Gevaert
- BP
- BASF
- Borealis
- Cytec
- Dupont
- Nikkelverk
- Solvay

4.1.2.2 External specialist and knowledge

To increase the quality of the standards, guidance notes and training packages and to increase the knowledge in every specific field, external partners with specific knowledge and expertise were contacted:

Centre for Chemical Process Safety (CCPS): the complete list of documentation was purchased and used as a guide for the internal standards and guidance notes.

Engineering firm: a service level agreement was written to help out the organization with the knowledge it was lacking. These engineers gave advice on the guidance notes and standards. The firm was already familiar with the Umicore culture before they were asked to help out. Hence, they knew exactly what could work and what not.

Specialists: some of the specific knowledge :electrical energy and industrial gases was covered by suppliers in that field. They prepared training packages in their specific field.

4.2 Awareness building and communication on all hierarchic levels

Awareness building in the whole organization was necessary to change the mind-set for process safety.

To increase process safety awareness and to set the correct priorities several techniques were used:

- calculation of a process safety index
- learning from process safety incidents
- training and learning programs

4.2.1 Process safety index

A risk based approach was necessary to set the correct priorities and to make the management and the operators more aware about hazard and risk ranking of their activities. The technique of risk ranking was chosen on the level of the entity (an activity of a business unit at a site location).

The system to make this risk ranking for the group had to be simple and easy to understand. We decided to include two parameters in the process safety index calculation. We are convinced that the outcome represents the risk level of the activities in the entity

The two parameters chosen were:

- the amount of hazardous chemicals in the entity (weighted according to the GHS system)
- the presence of hazardous operating conditions limited to the Umicore activities

The result was became known as the process safety index for the entity and was calculated by the entity itself.

Figure 5 shows a template that was used. The weighting factors are based on the Seveso regulation in Europe.

PROCESS SAFETY INDEX CALCULATION SHEET

BU :
Site :
Name :
Date :

input | calculated
fixed figures | result

Checklist chemical substances

Nature of Hazard			Unit	Weighting factor	Quantity	Sub result	Correction factor	Outcome	Remarks
Significant inventories of hazardous substances related to	OLD	GHS							
(highly) flammable materials			kg or l or Nm3	1000		0	1	0	
explosive or unstable materials			kg or l or Nm3	500		0	1	0	
(very) toxic materials			kg or l or Nm3	100		0	1	0	
corrosive substances			kg or l or Nm3	1000		0	1	0	
oxidising substances			kg or l or Nm3	100		0	1	0	
harmfull or irritant substances			kg or l or Nm3	10000		0	1	0	
Molten materials			kg or l	1000		0	1	0	
Compressed gasses			Nm3	1000		0	1	0	
pyrophoric materials			kg or l	500		0	1	0	
High reactivity with water			kg or l	500		0	1	0	
Environmentally hazardous substances			kg or l	5000		0	1	0	
							index substance properties	0	

Checklist process hazards and operating conditions

Nature of Hazard	Occurence factor	Weighting factor	Outcome	Remarks
formation or possible release of toxic or hazardous gasses	0	3	0	Occurence factor : 1 = possible, 0 = not possible
formation or possible release of toxic or hazardous liquids	0	2	0	Occurence factor : 1 = possible, 0 = not possible
formation or possible release of toxic or hazardous solid material	0	1	0	Occurence factor : 1 = possible, 0 = not possible
formation of explosive atmospheres	0	3	0	Occurence factor : 1 = possible, 0 = not possible
fire hazard	0	1	0	Occurence factor : 1 = possible, 0 = not possible
use of very high temperatures (>400°C or >752°F)	0	1	0	Occurence factor : 1 = possible, 0 = not possible
use of cryogenic temperatures in processes	0	2	0	Occurence factor : 1 = possible, 0 = not possible
use of high pressure (>2 bar, 200k Pa or 29 PSI)	0	2	0	Occurence factor : 1 = possible, 0 = not possible
use of high voltage/current	0	2	0	Occurence factor : 1 = possible, 0 = not possible
exotherm reaction	0	1	0	Occurence factor : 1 = possible, 0 = not possible
runaway reaction	0	3	0	Occurence factor : 1 = possible, 0 = not possible
		index process conditions	0	

Figure 5: Example of a process safety index calculation sheet

All data was gathered and plotted in a two-dimensional graph with two axes: logarithmic index axe for the amount of chemicals (Y-axe) and a linear axe for the operating conditions (X-axe).

Every point in the figure represents the position of a Umicore entity. The more the point is located in the right upper field of the graph the more hazardous the activity in this activity is. (figure 6 gives only an impression about how this positioning is done. It is not so important that you can read the information in the figure).

After some consideration only two risk areas in the graph were identified:

- high risk (red field and part of yellow) – above the blue line in figure 6
- low risk (all the rest) were identified. – below the blue line in figure 6

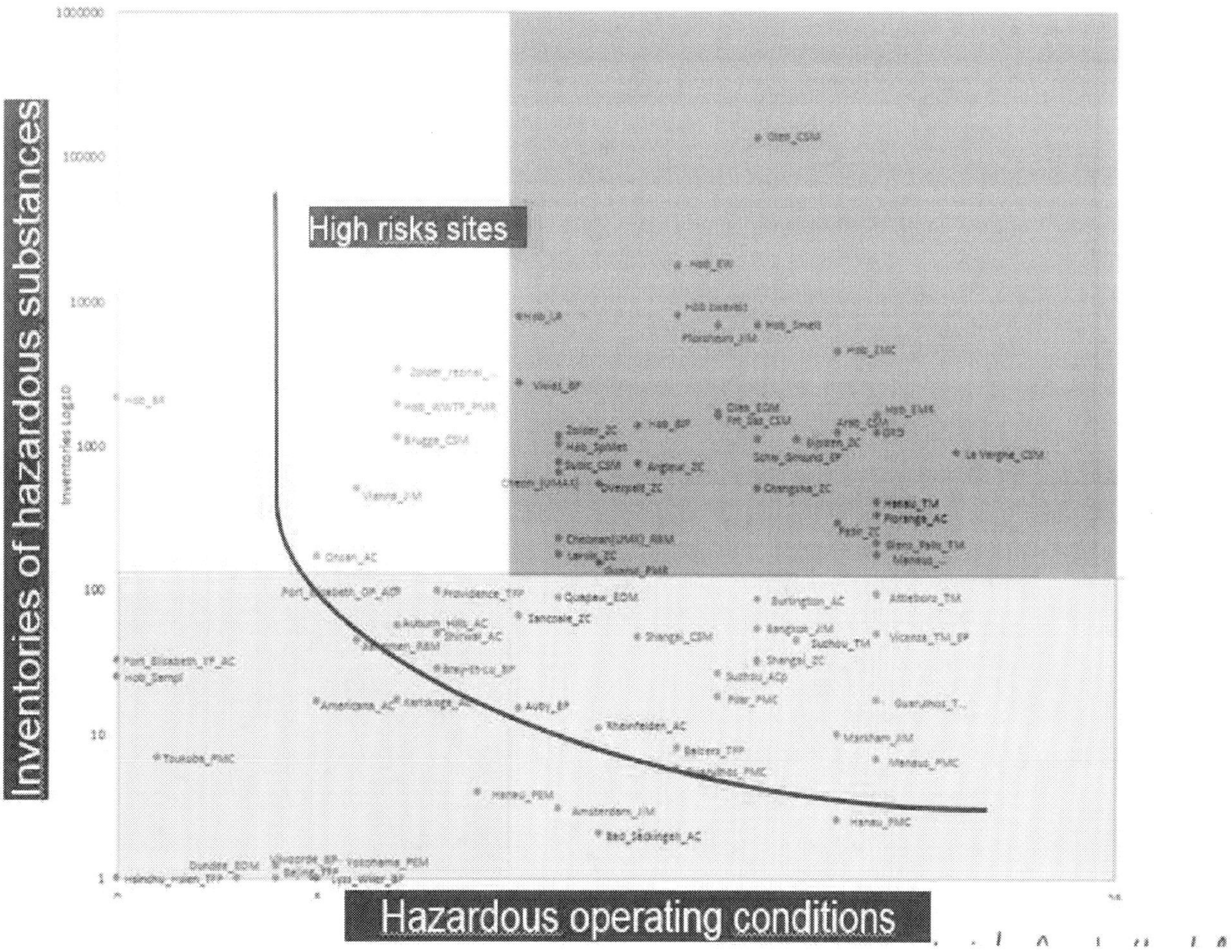

Figure 6: process safety index results

The high risk entities had first priority for

- the gap analysis
- the site visits
- the training programs
- the implementation of tools
- the follow up by the project team.

4.2.2 Learning from incidents

We started to report process accidents. 18 of these cases were reported at the end of 2016. The publication of these accidents (with most of the time no lost time and limited damage) raised the awareness as well because several of these incidents could be more severe.

As part of the guidance note 'incident investigations' additional reporting of important near misses (possible severity level defined as severe or catastrophic) was asked to all Umicore entities. This will significantly increase the number of reported process safety incidents in the future and will ensure that lessons are learned from these incidents.

umicore

Learning from Process Safety Incidents

BU/Site/department: CSM Olen Omnibus
Date/time of the Incident: 5/01/2015
Date of the completion of the form: 2/02/2015

Description Incident	
What has happened?	Picture
When operators tried to start P21290 to add $FeCl_2$ from vessel B27 to a reactor, they noticed that there was no flow. They investigated possible causes and thought maybe manual valve 3036 was broken. They disconnected the line and opened valve 3036 to check if the valve was still working properly. They noticed there was no flow coming out of that valve, meaning it was blocked or broken. They tried to clear the valve by using a water hose. After a while they smelled an "acid vapour". They immediately stopped and checked the atmosphere with a gas detector. The detector indicated there were flammable gasses present although there was still no (liquid) flow. (The	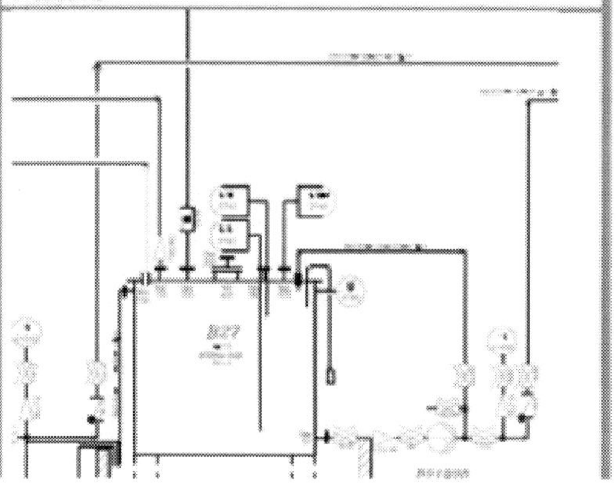

Figure 7: Example depicting learning from incidents

4.2.3 Process safety e-learning

An e-learning 'Process Safety Management (PSM)' and related webinars were organized in the first quarter of 2015. The e-learning was focused on the main elements of the PSM system. In addition, some interactive exercises were included. It became available in 7 languages and is now part of the standard develop and training program.

The program is mandatory for the whole process safety community. These are the key functions as described in chapter 4.1.1. (about 350 people)

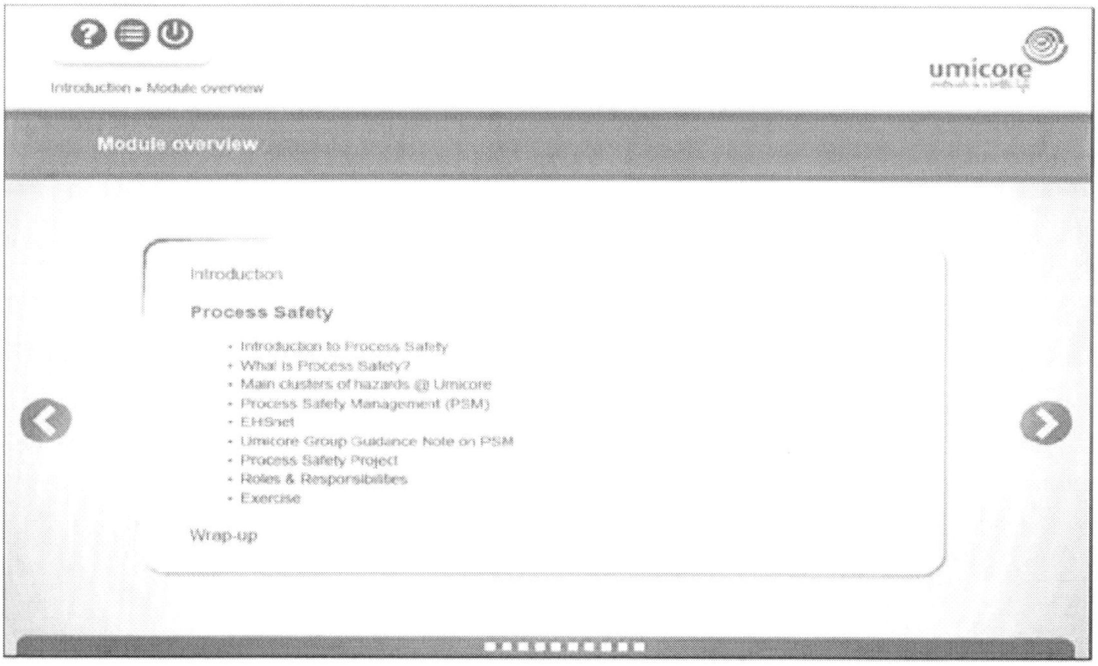

Figure 8: An e-learning slide

4.2.4 Animated on integrity of design

An Umicore animation (120 seconds) on integrity of design was launched in the second quarter of 2016 in 10 languages. It was designed:

- to increase awareness on the work floor; and
- to facilitate all changes (technical and others) on the work floor.

The animation was part of a toolbox and will be used in the future training programs.

Figure 9: Screenshot from the animation

4.2.5 Introduction of the process safety project in Umicore Link

The process safety project was introduced to every employee through the internal communication newsletter named Umicore link.
The project was rolled out from 2014 till 2017. In this timeframe 3 additional communications were done in this internal newsletter (keep the project alive in the whole organization).

4.2.6 Process safety section created on the EHS intranet.

As part of the communication channels of new documents a separate process safety section was created on the Umicore intranet.
All process safety management information is gathered there.
It is a temporary solution to make documents and tools more accessible for all involved. The accessibility of the information is still not perfect because it is still 'hidden' in the intranet structure.

4.3 Assessment - Site visits and reports

Until the end of November 2016 70 process safety assessment visits had been performed. These visits had several objectives:

- raise awareness by personal contacts;
- perform a gap analysis on site;
- perform a full assessment with a self-assessment checklist based on the guidance notes and standards (in the beginning of the project the general guidance note of 2013 was used because the additional guidance note were not available); and
- determine the actual residual risk level of every entity on the site (important for the top management of the business unit and the site manager).

The priority for these visits were based on the process safety index (see 4.2.1.)
Several follow up visits were performed for the sites that had a high residual risk profile and/or an important gap towards the (new) standards and guidance notes.

As a result of these visits process safety risk profiles for every BU were prepared. This summary of the findings of the process safety assessment visits (management summary) on one sheet are especially important for the involvement and the awareness of the management. This summary includes these topics: visited sites, activities, summary, recommendations, hazards, risks, risk management and role of the BU in the implementation of good process safety management practices. These 13 BU risk profiles were reviewed and approved by the respective process safety business unit contacts and the senior vice president of the BU.

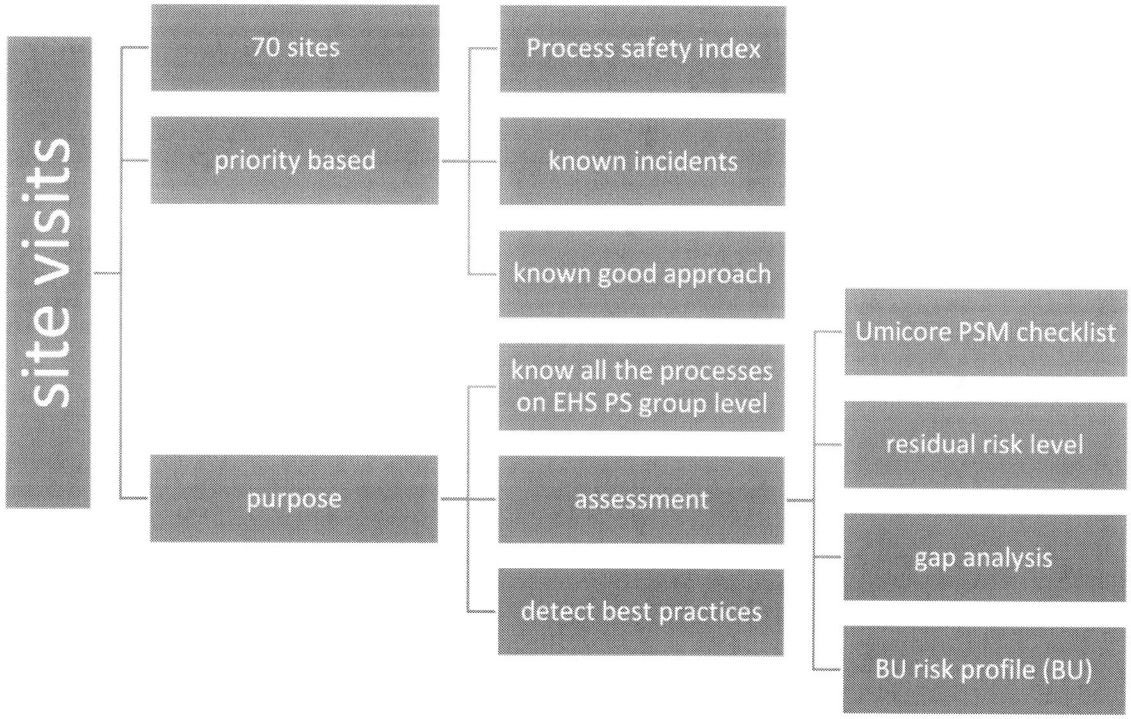

Figure 10: Process safety site visits

4.4 The PSM system

Figure 11: Process safety management

The PSM system was described in guidance notes and standards. The introduction of standards in the field of safety was an important change in culture and habits within Umicore. The general description of the PSM system is an annex to the Group EHS Guidance Note and provides information on process safety requirements and their application. The objective of this general guidance note is to describe a comprehensive PSM system aimed at preventing unwanted releases of hazardous chemicals and energy into locations where employees and others could suffer from severe and catastrophic consequences.

All elements of the system aim at ensuring the integrity of the installations and processes. Each element deals with one specific cause of such unwanted process related events. These elements are introduced in this guidance notes and are the subject of separate guidance notes or standards.

The standards contain mandatory methods that have to be deployed in all operational sites for their hazardous processes.

Description of the Umicore framework

Process safety can only be guaranteed if the integrity of design, the technical integrity and the operational integrity of the installations are guaranteed during the lifecycle of the installation. To ensure that we need a sustained integrity as well.

This approach was summarised in figure 12. The basic elements 'integrity of design', 'operational integrity' and 'technical integrity' are covered in the inner circle. Around that inner circle an additional approach is needed described in the sustained integrity element. A typical example is 'management of change'. Management of change is now part of the sustained integrity element because it includes topics of design, operation and technical integrity.

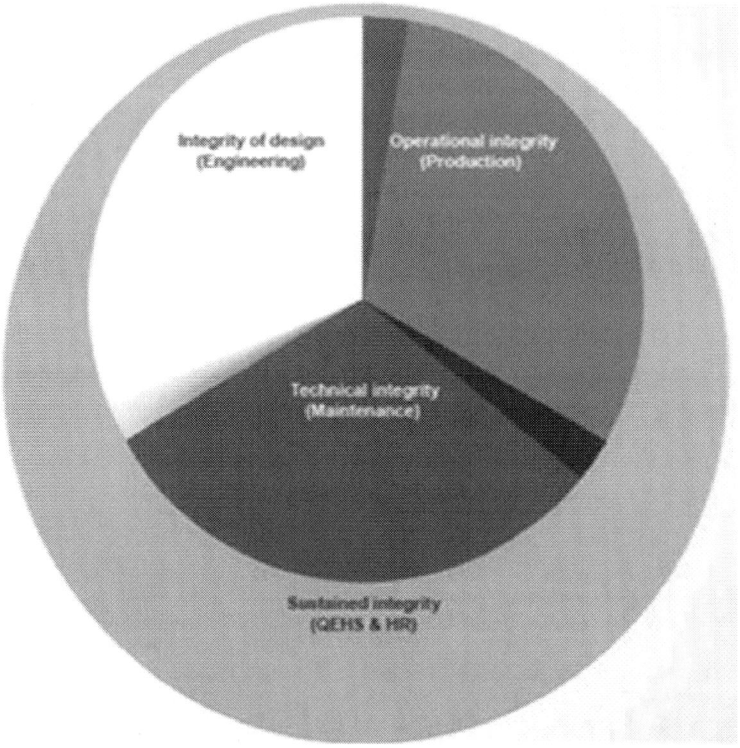

Figure 12: Framework of Umicore PSM system (1)

The standards and guidance notes that are included in these integrity level are listed in figure 13 and 14.

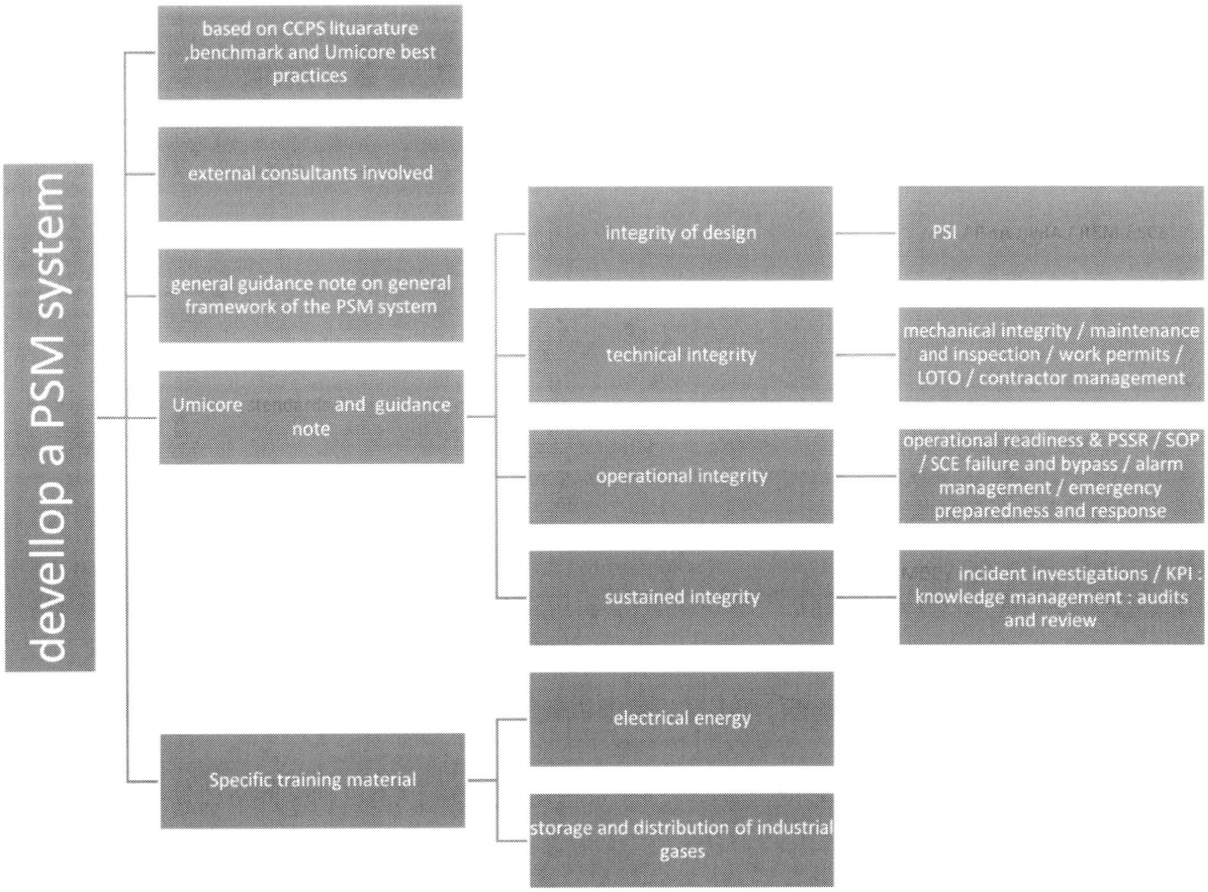

Figure 13: Framework of Umicore PSM system (2)

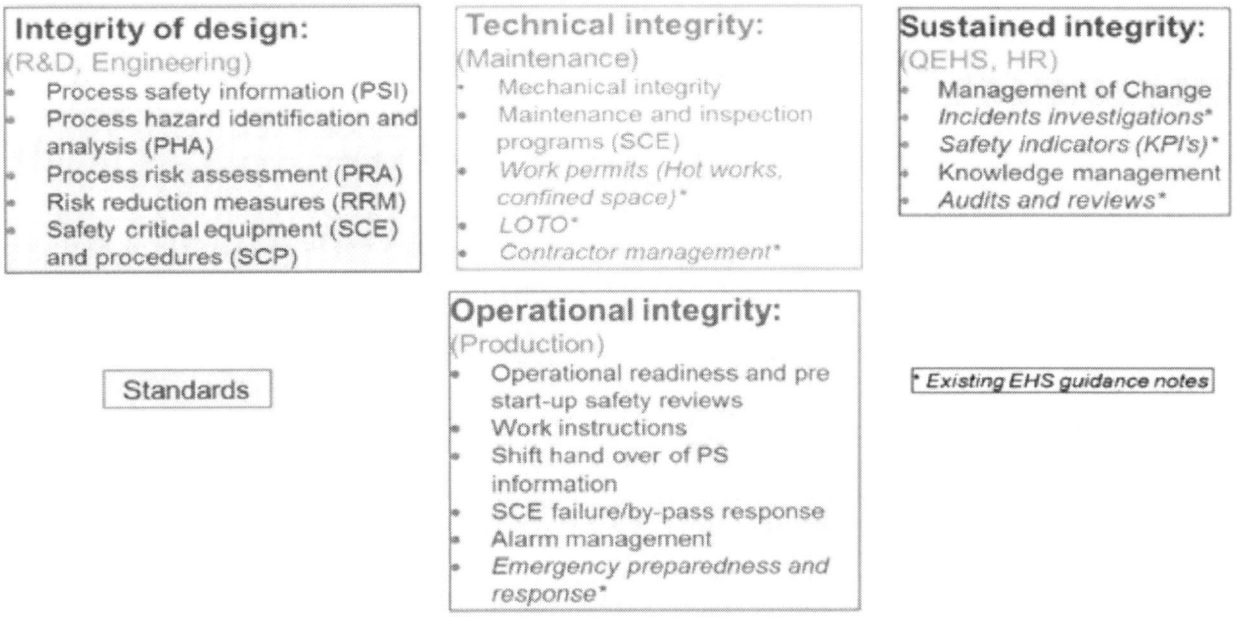

Figure 14: Framework of Umicore PSM system (3)

4.4.1 Integrity of design

There are several phases during the development of a new process technology or plant. Each phase contributes to the management of process safety. The process knowledge grows in extent and level of details as the process development moves through the phases. From a process safety point of view this information is clustered in:

- Process safety information (PSI):Safety related information about the chemical products, the process technology and equipment.
- Process hazard analysis (PHA): Focus on the chemical reaction, its technology and possible hazardous deviations. Hazards associated with intended and unintended reactions.
- Process risk assessment (PRA): Focus on operation of the processing plant. Determination of risks associated with the use of chemicals, reaction hazards and plant equipment.

The identification of relevant risks in the PRA leads to the definition of:

- Risk reduction measures (RRM)
- Safety critical equipment and procedures (SCE and SCP)

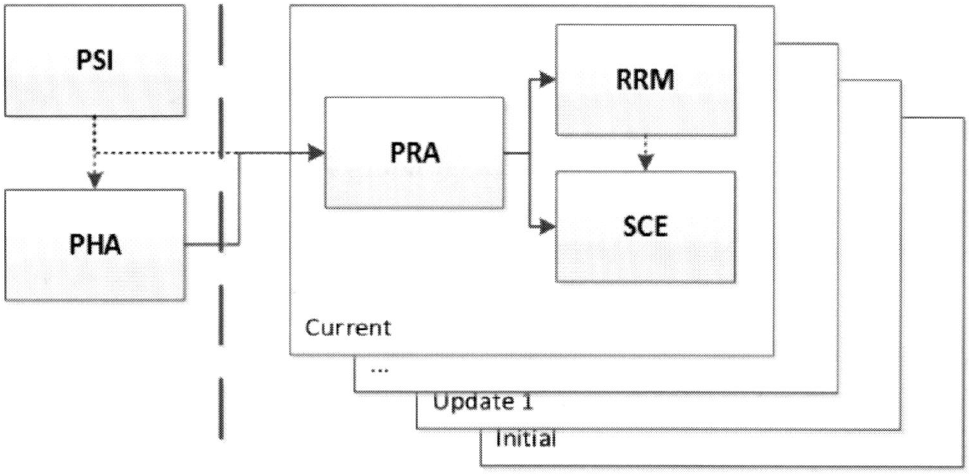

Figure 15: integrity of design structure

4.4.1.1 PHA

A Process Hazards Analysis (PHA) is a study meant to identify the hazards of a chemical process.
The choice of the name PHA can generate some confusion in some areas of the world (North America) because the name PHA is used for the combination of the hazards and risk assessment. At Umicore we decided to explicitly split it into two parts: a hazard assessment (PHA) and a risk assessment (PRA). This split was also part of the recommendations after

the fatal accident in 2014 to ensure knowledge management of hazards over the lifecycle of the process.

The goal of a PHA is the identification and description of hazards that are related to the process. This knowledge is used in the further development of the chemical or physicochemical process in general, performing the process risk assessments, the required risk reduction measures, the safety procedures, the management of change program in later lifecycle phases.

Identification of the hazards of a process is a key step in the process safety management cycle since it helps to map the inherent hazards that can lead to risk(s) (with consequences) in the process. This is in turn essential to take appropriate measures to reduce the risk(s) that will be identified later in the PRA.

A PHA should enable the plant to answer the following questions:

- Which hazards are inevitably associated to the process?
- Can additional process hazards emerge in specific circumstances?
    - What are the conditions in which these additional hazards can occur?
    - What are the safe operating conditions?
    - Can these hazardous conditions be detected?
    - Can the process be steered away from these conditions?
- What combinations of chemical products should be avoided at all times?
- What kind of interaction is possible between products, chemicals and the installation?
- What can influence the lifetime of the equipment?

Mandatory minimum content

The PHA describes:

Equations of main and secondary reactions

- Description of the intended reaction
- Description of the secondary (or unintended) reactions
- Specific reaction heats, clearly specifying exothermic or endothermic behaviour
- Reaction materials and quantities
- Reaction kinetics

Operating conditions (safe and unsafe):

- Operating steps - short description
- Process flow diagram (see PSI)
- Description of the use of special chemicals
- Temperature (or limits for the temperature to ensure a safe process)
- Pressure (or limits for the pressure to ensure a safe process)
- pH

- Special operating conditions (e.g. working with inert gas blanketing, ...)
- Maximum temperature that the reaction mixture can withstand without risk of deterioration
- Concentration
- State of aggregation (solid, liquid, gas, ...)

Process or reaction hazards

- Volume of gas released in the event of decomposition
- Incompatible chemicals - causing a violent reaction upon addition
- Accumulation of unstable impurities
- Delay in the initiation of the reaction
- Reaction interruption (unplanned process or operation shut down)
- Restart of processes after interruption
- Reactive mixtures with runaway potential / influence on reaction kinetics
- Toxic hazard for raw materials, by-products, impurities and finished products
- Contamination/incompatibility (chemical substance compatibility matrix)

Fire or explosion hazard

- Flammable dusts, gases and vapours
- Ignition sources

Corrosion & Erosion

- Specific information on corrosive and erosive behaviour of substances
- Overview of compatible and incompatible construction materials

Notes on anomalies and incidents

- Notes on unexpected conditions and/or incidents occurred during process research, development or operation such as gas, smoke, flame release, foaming, emulsion, overflow, tar formation, sediment, precipitate, etc.
- Reported incidents with similar processes

Attention points for the Process Risk Assessment

- List of special attention points to be discussed in the PRA meetings

4.4.1.2 PRA

The objective of this standard is:

- To introduce the concept of Process Risk Assessment. (PRA)
- To give guidance on the selection of risk assessment methods according to the hazards detected in the PHA
- To introduce the use of a standard risk matrix for Umicore
- To assess the outcome of existing methods towards the preferred ones.

- To share good practices regarding performance and review of a risk assessment in order to guarantee an optimal result
- To clarify how to use the results of the PRA during

Regardless of what technique is used, in general, a PRA addresses:

- The translation of the identified hazards (PHA) into risks in normal and abnormal operating conditions
- Any previous incident which could have led to catastrophic consequences
- Risk reduction measures preventing any harmful consequences
- Consequences of failure of safeguards
- Impact of the facility on its environment
- Human factors
- A qualitative evaluation of the consequences of possible incidents

At Umicore a limited number of mandatory methods have been selected to provide suitable solutions for the assessment of process risks over the lifetime of a process according to the hazards identified in the PHA.

The selected methods for the identification of risks are:

- Checklist
- What if (WI) or structured what if technique (SWIFT)
- Hazard and operability study (HAZOP)

Their application fields are linked to the hazards identified in the PHA.

The Umicore risk matrix is presented below in figure 16. The matrix is developed as a qualitative evaluation method bases on a risk acceptance level of 10^{-5} for catastrophic scenarios

The combination of the evaluation of likelihood and severity, positions the risk in the risk matrix. The risk pertains to one of the following three fields: 'acceptable', 'risk reduction required' and 'unacceptable (do not operate)'.

Severity / Likelihood	Lo w	Important	High	Severe	Catastrophic
Very Often					
Often					
Occasionally					
Seldom					
Rare					

A	Acceptable risk
R	Additional risk reduction is required
DNO	Do not operate without additional risk reduction

Figure 16: Umicore risk matrix

4.4.1.3 RRM

The objectives of this standard are:

- To introduce concepts related to risk reduction for Umicore processes.
- To give guidance on how to use the results of the risk evaluation (PRA) for the selection of risk reduction measures.
- To give guidance for the selection of appropriate reduction measures aimed at reducing the residual risk to an acceptable level.

Risk management systems in the process industry uses often a multilayered approach with independent preventive and mitigating layers (layers of protection).
At Umicore this approach has been transformed into the UPL (Umicore Protection Layers). According to this concept every layer contributes to achieve the risk acceptance level of the Umicore group. Every preventive or mitigating measure is positioned in one of UPL and has a fixed risk reduction factor (RRF). The Umicore acceptable risk level is achieved by the independence of the different protection layers that are applied and the unlikelihood of simultaneous failure of several of these layers.
The UPL describe the different layers in which protective measures can be classified. (see figure 17).
The total risk reduction can be seen as the combination of the measures in different independent layers.

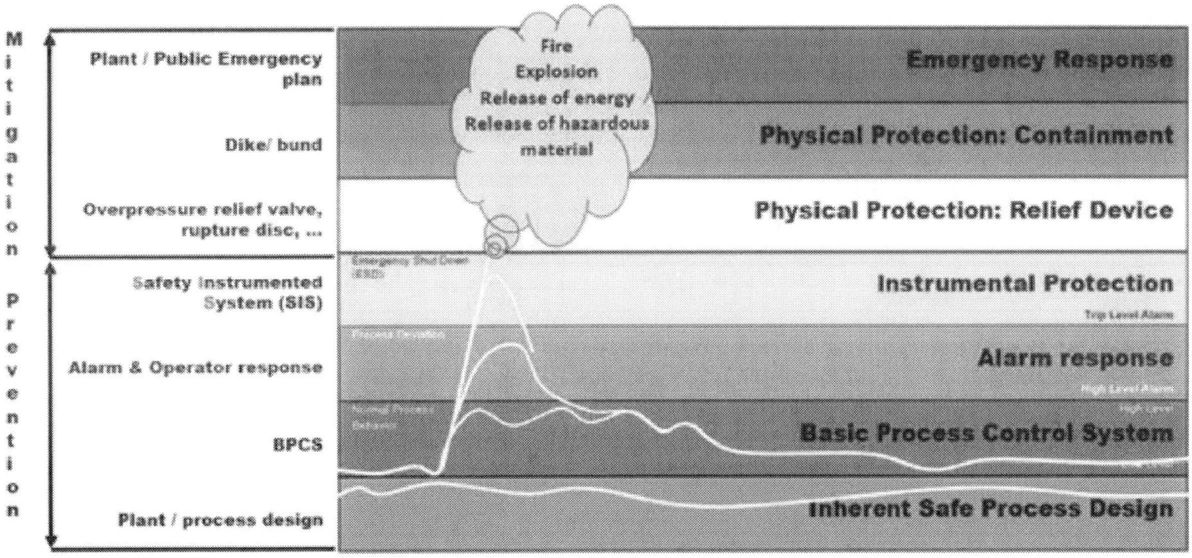

Figure 17: Umicore protection layers

During the Process Risk Assessment (PRA) the unmitigated risk has been positioned in the Umicore risk matrix. This risk matrix can be completed with a required risk reduction factor (RRF)

RRF / Severity / Likelihood	Low	Important	High	Severe	Catastrophic
Very Often	100	1000	10 000	100 000	1000 000
Often	10	100	1000	10 000	100 000
Occasionally		10	100	1000	10 000
Seldom			10	100	1000
Rare				10	100

A	Acceptable risk
R	Additional risk reduction is required
DNO	Do not operate without additional risk reduction;

Figure 16: Risk matrix with Risk Reduction Factor

The fundamental choice underpinning the required risk reduction factor in the matrix is based on the target frequencies per risk scenario:

Severity level	target frequency (1/yr)
Low	0.1
Important	10^{-2}
High	10^{-3}
Severe	10^{-4}
Catastrophic	10^{-5}

Table 1: Umicore target frequency per risk scenario

As a general rule each measure forming an IPL can take into account a maximum risk reduction factor of 10. This means that when after the evaluation with the risk matrix a RRF of 1000 is required, at least three independent measures with a RRF of 10 should be present.

For some specific devices or protection measures such as a properly designed and maintained SIL 2 (Safety integrity level 2) safety instrumented function or a pressure relief valve where design and verification calculations can demonstrate reliability greater than 99%, it is possible to account for a risk reduction factor of maximum 100.

4.4.1.4 SCE/SCP (safety critical equipment and safety critical procedures).

The objective of this standard is

- To give guidance on identifying safety critical equipment and safeguarding its integrity; and
- To give guidance on identifying the safety critical procedures.

The choice of SCE and SCP is based on the position in the UPL(Umicore protection layer) and the severity of the possible scenario.

All equipment, procedures and systems used in the UPL instrumental protection, physical protection and emergency response are considered safety critical for all scenarios of the PRA.
If a scenario is categorized with a severity of severe or catastrophic also the alarm system and the appropriate response becomes safety critical.

This is further visualized in figure 17. You can see that the SCE is selected out of the UPL according to the severity of the scenario.

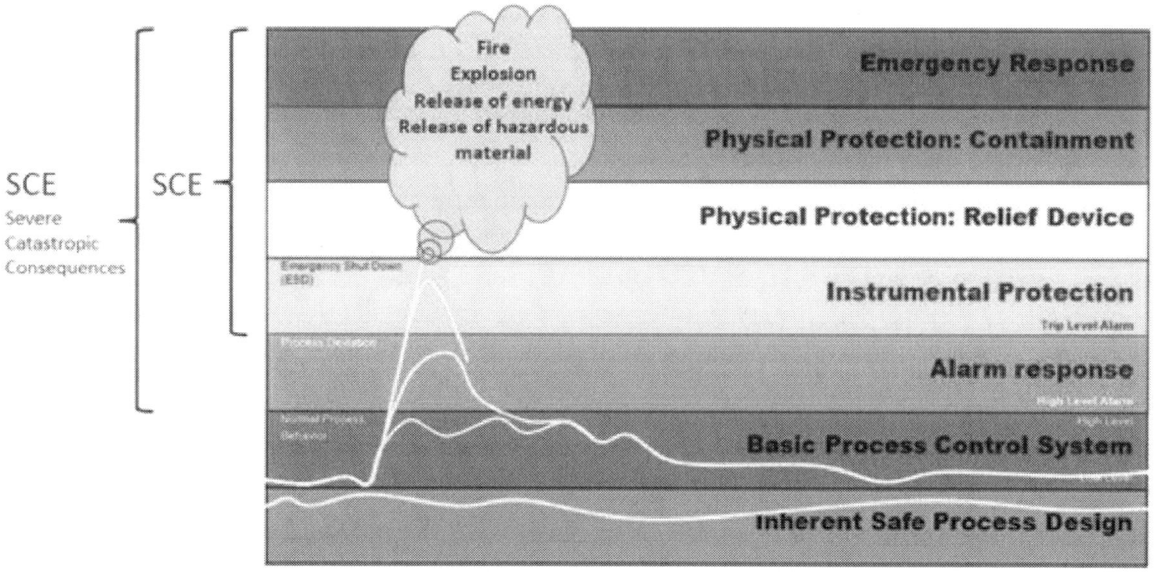

Figure 17 : Selection of SCE and SCP

4.4.2 Technical integrity

Chemical process plants rely on technical equipment for the safe operation of the processes, the containment of the hazardous materials and the reduction of risks to acceptable levels. To achieve this, the technical equipment needs to be in a well-adapted and good working order. In other words the technical integrity of the plant needs to be defined and properly managed. Several aspects and activities can be defined to ensure that.
Each aspect of the program is of vital importance and has to co-exist simultaneously.

The main elements of technical integrity are:

- Mechanical integrity during the lifecycle based on aging and natural events
- Maintenance and inspection programs
- Work permits
- Lock out – tag out (a guidance note for occupational safety already existed but process safety was added based on the isolation of chemicals and energy)
- Contractor management

4.4.3 Operational integrity

The integrity of everyday use – the operational integrity – of a process installation is to be considered as an aspect of process safety management in conjunction with the aspects of integrity of design, technical integrity and sustained integrity.

Operational integrity elements include:

- Instructions and procedures guiding the operational personnel through normal everyday operation (guidance notes, work instructions)
- Instructions and procedures assisting the operational personnel in performing less frequent parts of normal operation, such as process start up or shut down (guidance note work Instructions)
- Instructions and procedures helping the operational personnel to react correctly in alarm and emergency situations (guidance notes work Instructions, emergency response and alarm management)
- The flow of day-to-day information on process status (guidance note shift handover)
- Considerations for evaluation of process plant operational readiness (guidance note operational readiness & PSSR)
- Procedures and control measures for bypassing SCE or handling failure of SCE (guidance note SCE Failure / bypass response)

4.4.4 Sustained integrity

A key element in safe operation of chemical processes lies in long term preservation of the proper control of risks. The safety integrity of the chemical processes needs to be maintained over the operational lifetime of the assets, and taking into account all relevant operational activities, such as normal operation, maintenance, modification, commissioning and decommissioning. In other words the safety integrity of the process needs to be sustained.

The main elements of sustained integrity are:

- Management of change (MOC)
- Audits and reviews
- Safety indicators
- Incident investigations
- Knowledge management

4.5 Deploy and train PSM system in the organization

A lot of new information and tools became available during the project such as new standards and guidance notes with preferred methods and templates, new training packages, new awareness building tools and a new risk documentation software. This information and all these tools had to be deployed to the whole organization with the help of the business units' contacts and the regional contacts.
To deploy the program, we made use of this basic scheme shown in Figure 18 below:

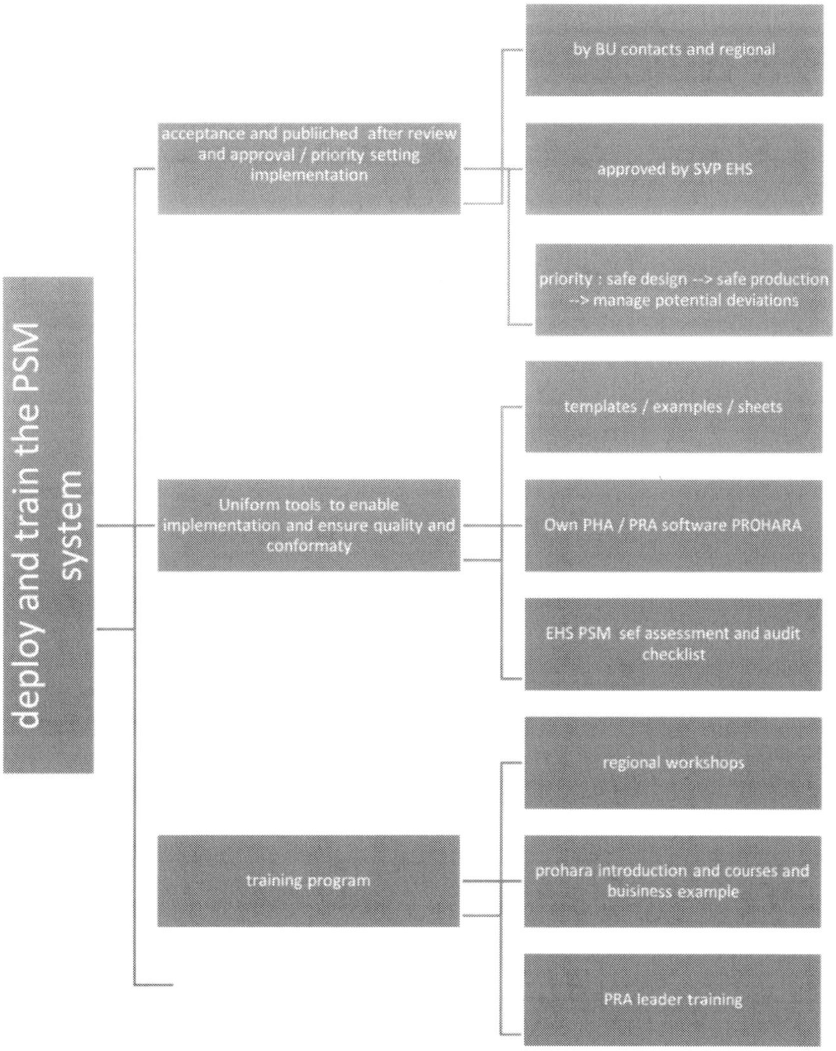

Figure 18 : Deploy and train the PSM system

4.5.1 Review of documents

Before the communication of any new standard and guidance note, all documents were carefully reviewed by the regional contacts and all business contacts. Also the approval of the SVP EHS was asked. This method ensured that:

- the particular process safety language was well understood by people who were not adequately knowledgeable in this field;
- the methods were pragmatic;
- the methods could be implemented in the field; and
- the priorities for new documents were aligned with all people involved. This was organized in tree waves spread over two years.

4.5.2 Tools

The PROHARA (**PRO**cess **HA**zard and **R**isk **A**ssessment) software for the documentation of integrity of design studies was developed in collaboration with the internal software department and an external consultant. The software went live on July 2016. The software is web based and structured in one database for Umicore.

It covers all aspects of process safety information, process hazard analysis, process pisk assessments, the introduction of risk reduction measures according to the Umicore protection layers and facilitates the conformity with the risk acceptance level and the selection of safety critical equipment. This project was originally not foreseen but after the publication of the standards in the field of integrity of design the whole organization was waiting for this tool to start the work in the field.

Figure 19 : PROHARA

4.5.3 Training program and introduction of PROHARA

Regional EHS workshops were organised in 2015 and 2016

- 2015
    - One and a half day workshop dedicated to integrity of design.
    - Locations: North-America, South-America, Asia Pacific, Greater China, Europe, Belgian sites (3 sessions).
- 2016
    - One and a half day workshop dedicated to the introduction of Prohara and the standard/guidance notes related to technical integrity, operational integrity and sustained integrity.
    - Locations: North-America, South-America, Asia, Europe, Belgian sites.

To introduce PROHARA regional test groups were organised in Europe, South America and North America.

5 Permanent structure after the project

To ensure the necessary attention and a continuous improvement program for process safety after the project, a management structure was set up.

5.1 Corporate EHS role

Ensure the deployment of the PSM elements at the sites at the right pace through:

- Appropriate resources allocation
- Interactive communication to ensure continuity of the efforts: conference calls, site visits, workshops
- Supporting actions:
  - Events
  - Trainings
  - IS solutions
- Develop necessary tools to facilitate implementation
- Regular take stock to monitor progress status
  - Requests for progress reporting
    - The degree of introduction of the Process Safety standards at site level
    - The progress made regarding the performance of PHA and PRA studies
  - Compliance and other audits (see next chapter)
    - Integration of process safety in the corporate EHS audits
    - Stimulate the performance of internal process safety audits at site level
    - Selection of process safety elements to complete the scope of due diligences

5.2 Audits

5.2.1 Compliance audits

Compliance audits within the group are already well developed.
The compliance with legal requirements and with internal guidance notes is checked during these audits.
The focus till 2016 was on environment and occupational health and safety. From 2017 on process safety will be integrated in these audits

- Frequency : 1 audit every 5 years for every site or more frequent if needed
- Three teams on site : one for occupational safety, one for environment, and one for process safety
- Help from external auditors foreseen (local regulation covered)

5.2.2 Due diligence audits and integration of the new sites

A lot of the non-compliances at this point of time are a result of the acquisition of small family owned businesses. The level of engineering in these sites was not the same as the level experienced after the acquisition of other lager sites (from lager industrial companies) and also not in line with the engineering level and the documentation level of the Umicore group.

To ensure that this risk is covered in the future process safety will be a topic in all future due diligence audits. So the risk will be known and the effort to bring the site in line with the standards and guidance notes can be calculated in terms of necessary investments and resources needed.

5.3 Further guidance

Further guidance will be needed.

A review of the existing standards will be organized on a regular basis together with the business contacts and the sites.

On top of that a lot of (new) technical guidance noted have to be developed.

6 Abbreviations

- BU: Business Unit
- CEO: chief executive officer
- CTO: chief Technology officer
- EHS: Environment, Health and Safety
- EXCOM: executive comity
- IS: Information Systems
- KPI: Key Performance Indicators
- LOC: Loss Of Containment
- LOTO: Lock Out Tag Out
- MOC: Management Of Change
- MVC: monovinylchloride
- PEM: Platinum Engineered Materials
- PHA: Process Hazard Analysis
- PRA: Process Risks Assessment
- PS(M): Process Safety (Management)
- PSI: Process Safety Information
- PSSR: Pre Startup Safety Review
- PX: Project eXcellence
- RRF: Risk Reduction Factor
- RRM: Risk Reduction Measure
- SCE/P: Safety Critical Equipment/Procedure
- SIL: Safety Integrity Level
- SIF: Safety Integrity Function
- SVP: Senior Vice President
- UPL: Umicore protection layer

How Culture Effects the Implementation of Process Safety Program Elements
13th Global Congress on Process Safety
San Antonio, TX

Michael J. Hazzan, P.E., CCPSC, CPSA
AcuTech Consulting Group Inc.
Vienna, VA

Introduction

Culture underpins everything in a process safety program. Without a sound underlying culture even the most elegantly designed process safety management systems will not be successfully implemented.

Based on a thorough review of major incidents in various industries, several cultural themes have emerged as overarching factors that contributed to those incidents. These themes represent a pattern of behavior that had developed over time and either contributed directly, i.e., they were among the cause(s) of the incidents themselves; or indirectly, i.e., they exacerbated the situation or resulted in inadequate attention to the direct cause(s). These themes are referred to as the *core principles of process safety culture* (Ref CCPS Culture - draft), and consist of:

1. ***Establish an Imperative for Safety:*** Elevating the process safety program to the same level of importance as production/operations in terms of priority and provision of resources.
2. ***Provide Strong Leadership:*** Process safety leaders may or may not be managers but inspire the people, their colleagues as well as those who report to them, if any, to do the right things at the right times. This is often referred to as "Walking the Walk."
3. ***Maintain a Sense of Vulnerability:*** Sustaining a healthy level of respect for the hazards and risk of the facility and its operations, particularly over extended periods of time.
4. ***Understand and Act Upon Hazards/Risks:*** Sustaining complete, objective, consistent, and repeatable hazard and risk analyses that are performed on time by personnel who both understand the process being studied and the analytical methods being used. The risk/hazard analysis process includes the timely and technically valid follow-up of risk reduction action items that result from the analyses.
5. ***Empower Individuals to Successfully Fulfill their Safety Responsibilities:*** Providing personnel with the necessary authority and resources to successfully carry out their assigned roles in the process safety program.
6. ***Defer to Expertise:*** Assigning expert personnel in positions of influence in decision making within the process safety program. These personnel have the requisite knowledge and understanding of the process safety program policies, practices, and procedures (including supporting internal and external codes, standards, and other relevant guidance), how they are applied, how they are interpreted, and how they relate to each other. This core principle also includes seeking out the necessary expertise when it is needed or when it is lacking in the organization.

7. ***Ensure Open and Frank Communications:*** Verbal and written channels of communication are open in the process safety program and also between the program and other parts of the facility and its parent company. An open reporting system where the messenger is never punished has been established.
8. ***Foster Mutual Trust:*** Personnel can trust that their words and actions, if in good faith, will not result in adverse human resource consequences. Process safety programs do not apportion blame.
9. ***Combat the Normalization of Deviance:*** Deviations from approved rules, procedures, codes/standards, or other formal guidance that are not tolerated. When they occur they arouse curiosity, are questioned or challenged, and do not become part of the everyday way of operating the facility, i.e., they do not become normalized. Deviations that do become normalized will likely become latent failures in a process safety incident scenario eventually.
10. ***Learn to Assess and Advance the Culture:*** The application of lessons learned from mistakes are disseminated and then used to improve process safety performance. Curiosity and the determination necessary for the organization to be a learning and advancing culture is encouraged.

These core principles manifest themselves in nearly every activity and practice in a process safety program on a day-to-day basis. Therefore, every element of a process safety program has specific cultural aspects that affect its various procedures, activities, and decision making. These cultural aspects can be categorized by the core principles. Exploring the cultural aspects of each of the twenty Risk-Based Process Safety (RBPS) elements is beyond the scope of this paper, so we will examine the cultural aspects of two important process safety elements:

- Conduct of Operations
- Management of Change

Conduct of Operations

Conduct of Operations (COO), as defined in CCPS's Risk-Based Process Safety (RBPS) program model, is broader than what is contained in process safety regulatory programs, such as OSHA's PSM Standard. COO includes basic operational discipline requirements, such as following approved operating and other procedures verbatim and remaining within approved operating limits, that are part of regulatory and voluntary process safety programs. However, COO also includes other important issues such as: bypassing safety or protective features, housekeeping, the execution of shift turnover, activating an emergency shutdown when warranted, and the granting and use of stop work authority when needed. COO also addresses several other related key issues such as fatigue and its management, alarm management, fitness for duty. "Operations" in this context does not only include activities conducted to produce the products of the facility, and it is not just done by the operators. COO covers activities and practices affecting process safety that are performed by, contributed to, or controlled by other disciplines, such as Maintenance, Engineering/Projects, Safety, Purchasing, and HR.

The process safety culture affects COO in many ways, including the possible hesitance to initiate an emergency shutdown without supervisor concurrence, to confront co-workers and stop work when it is not being performed safely, or to recommend not following an operation because of its potential to cause a process safety incident. The conflict between the fatigue associated with excessive overtime and the extra compensation that it generates is also a cultural issue in Conduct of Operations. As such, COO in a process safety program institutionalizes the pursuit of excellence in the performance of every task and attempts to minimize variations in performance in a process safety program. Workers at all levels are expected to perform their duties with alertness, due thought, full knowledge, sound judgment, and a sense of pride and accountability. Unlike most other process safety program elements, this element is not a distinct set of activities governed by a single stand-alone management system procedure with its own set of distinct records. Rather it is mostly a set of cultural principles by which many other PSM related activities are carried out, particularly but not exclusively by Operations. Several of these principles are summarized as follows:

- Strictly following approved procedures must be a benchmark characteristic of any process safety program and its culture. This includes, but is not limited to the operating procedures. Ad hoc operations of any sort should not be tolerated. One of the most prominent examples of a severe incident occurring because of ad hoc operations was the 1987 Chernobyl accident where an unauthorized test was carried out by facility personnel and this triggered a nuclear accident with a significant release of radioactivity over the western Soviet Union, Eastern Europe, and parts of Scandinavia. This is an example of the effect of the core principle of *Combat the Normalization of Deviance.*
- Written procedures cannot be a substitute for an understanding of the basic principles of safety, scientific knowledge, or plant operation. When the operating procedures are not understood by those who have to use them, or their mere existence becomes a substitute for a thorough understanding of how the facility works or is intended to work, then a cultural problem exists (Ref 4.28 Kletz 2009). For example, during the BP Texas City incident, operators continued to feed the ISOM splitter tower for hours with the outlet to storage closed. Nobody had the presence of mind to ask where all the material was going. It seems obvious that if the input feed flow is constant and the outlet flow is zero, then the process must be accumulating the material somewhere, or it is leaking out, which was not the case in this incident, at least not initially. In this case, several qualified and competent operators failed to ask this simple question over an approximately ten-hour time period. This was not the only cause of this incident, but it was an important contributor. This concept also applies to others who are called upon to provide technical advice or make key process safety decisions, e.g., engineers and managers, who may be offsite when a question or concern is raised. For example, the maximum operating temperature of a process is 400F but the process seems to want to run in a stable manner at a temperature of greater than 400F. Is that allowed? How the engineer or manager responds to this issue is a COO issue. This is an example of the following core principles: *Maintain a Sense of Vulnerability, Understand and Act Upon Hazards/Risks,* and *Empower Individuals to Successfully Fulfill their Safety Responsibilities.*
- The culture of different operating shifts within a given facility sometimes varies widely. What would never be tolerated on one shift might be routine on another. This often represents the difference between staffing levels between day and night/weekend/holiday shifts and the presence and attention of technical staff and management during day shifts on normal weekdays. Most of this difference can be attributed to the attitudes and beliefs of supervisors, foremen, and other mid-level management assigned to off shifts. Having to accomplish the

same level of production with less resources sometimes fosters the attitude that shortcuts and other deviations from approved procedures are acceptable in order to achieve those goals. Sometimes this situation fosters a "can-do" attitude where independently finding any solution to a problem and getting around a production delay/issue quickly is seen as a positive attribute. Yet the process conditions and potential hazards are exactly the same regardless of the time of day or day of the week. One of the definitions of a sound process safety culture is doing the right thing even when nobody else is watching. This definition captures the conditions that exist in some facilities during off shifts. This is an example of the following core principles: *Combat the Normalization of Deviance, Maintain a Sense of Vulnerability*, and *Understand and Act Upon Hazards/Risks.*

- The formality of shift turnover is an important process safety issue. The root causes of some major process safety incidents have included inadequate shift turnover (Ref 4.4 Baker). Some parts of the shift turnover issue are procedural and some are cultural. Personnel need to understand both the importance of this activity and that it must be formally transacted. An analogy is the relieving of the watch in the Navy. This is very formal and includes not only detailed discussions between the oncoming and off-going watchstanders but also a tour of the parts of the ship the watchstander is responsible for, review of log entries, and a precise shift of responsibility that is verbally made, acknowledged, and documented in the log. That level of formality is not required in civilian businesses; however, some aspects of the way Navy watches are turned-over are worthy of consideration for shift turnover in chemical/processing facilities. Following the Texas City incident, there has been significant improvement in this important activity. Many facilities now have written procedures that specify how and where shift turnover is to be accomplished, including its documentation. This is an example of the following core principles: *Combat the Normalization of Deviance, Maintain a Sense of Vulnerability*, and *Ensure Open and Frank Communications.*
- Fitness-for-duty is an important process safety and culture related issue. Fitness-for-duty refers to a range of possible influences on worker performance, such as impairment due to drugs or alcohol, fatigue, illness, distraction due to personal issues, and mental condition (Ref 4.8 CCPS COO). It also includes horseplay, harassment, and other aberrant behaviors. No tolerance for these behaviors goes beyond simply forbidding them in written policies and stating these prohibitions during training/orientation sessions. It means living them at each level of the organization, including self-policing within peer groups and during off shifts. This is an example of the following core principles: *Combat the Normalization of Deviance* and *Provide Strong Leadership.*
- Overtime can be a sensitive subject in many facilities. Sometimes there is pressure to work excessive overtime from facility or company management to help solve more basic staffing problems. Sometimes there is peer pressure in favor of excessive overtime for the financial benefits that result. Regardless of where the pressure originates, when it exists, a culture issue also exists. Therefore, overtime and the fatigue it may cause can be regarded as an issue that is at least related to the process safety culture, particularly when there is pressure present to work excessive overtime. The Baker Panel (Ref 4.4 Baker) noted that the average overtime data at each BP refinery suggested that they were understaffed at the operations and maintenance functions for lengthy time periods and that each refinery had come to rely on excessive amounts of overtime to make up this understaffing. The Panel also noted that at each refinery peer pressure had created pressure to work considerable overtime. This problem was not limited to hourly employees and contractors, as salaried employees, including

supervisors, were working excessive amounts of hours. Industry recognizes that fatigue is an important issue, referencing its contribution to the BP Texas City incident and others. API RP-755 (Ref 4.3 API 755) has been published to provide guidance for facilities to detect and counter fatigue in those parts of the workforce working night shifts, rotating shifts, extended shifts/days, or call-outs and their job requires process safety actions. This reference establishes a fatigue risk management system so that the process of managing fatigue among those positions that may be sensitive to it is a continuous and sustainable process. However, if there are underlying pressures being brought to bear on personnel to work excessive overtime, then there is a process safety culture issue that must be solved first. This is an example of the following core principles: *Maintain a Sense of Vulnerability* and *Understand and Act Upon Hazards/Risks.*

- Another COO issue that is more related to actual operational practices but that contains a cultural component is the periodic operation of installed spare equipment. An example of installed spare equipment is two pumps installed in parallel, either of which can provide the required head/flow. Normally, one of the pumps is operating all of the time, or nearly all of the time, and the other pump receives little or no operating time. Facilities should not wait until an emergency situation to see if the installed spare equipment works as intended or as needed. This is a chronic problem in most facilities. Sometimes the cause is the extra cost, e.g., if the spare pump is steam driven, it costs more to run the steam driven pump. Sometimes it's a chronic problem because of an operating bias against making a change to a unit that is running in a stable and finely adjusted manner. Concern over creating a process transient/upset is a reason for not rotating the operating equipment. Sometimes, the reason is simply that the operators are too busy to perform operations that are seen as unnecessary. In some facilities, operating procedures exist to switch to and operate the spare equipment for fixed amounts of time, but they are not used. These concerns, cost issues, and built-up biases are cultural issues. This is an example of the following core principles: *Maintain a Sense of Vulnerability* and *Combat the Normalization of Deviance.*

Emergency Shutdown

The emergency shutdown of a process or unit (a rare but critical event) is an extremely important aspect of operational discipline. This action can be a controversial activity, or at least regarded as controversial, but should not be. At its core, this is mostly a process safety culture problem rather than an operations problem. The reasons for this are:

- Despite written procedures that say otherwise, some operators do not believe that they really have the authority to activate an emergency shutdown**Error! Bookmark not defined.** without the direct approval of a supervisor.
- Despite operator training that stresses that the individual operators have the authority, by themselves, to activate an emergency shutdown if the conditions warrant, the operators do not believe that they actually have this authority or that they will be second-guessed if they exercise it without the direct order of a supervisor.
- Despite the contents of the standard operating procedures (SOP) or emergency operating procedures (EOP) and the operator training program, some operators are afraid to initiate an emergency shutdown because they believe that they will be punished or reprimanded in some way for it, including fear of termination.

- Despite their training, some operators believe that it simply isn't their responsibility and that such an important decision should be made by supervisors or above.
- In some cases, emergency shutdown procedures are not straightforward to execute, and may have other consequences such as making the process difficult to restart or requiring an extensive cleanout. These productivity factors may discourage operations personnel from making and executing a decision to conduct an emergency shutdown.

Why is there a disconnect between what the SOPs, EOPs, or training says and what the operators actually believe? In some facilities, these important points have been stated but not emphasized adequately. In some facilities the operators have seen punitive actions taken when an emergency shutdown has been initiated and the post-shutdown investigation revealed that alternative actions were possible. This sort of second guessing of a very stressful situation is an indicator of a poor process safety culture. Second guessing in this context means not just investigating the occurrence but using it to punish personnel in some way. Investigating an emergency shutdown is a necessity because these events are near misses at a minimum, and the contributors of the events leading up to the action need to be fully understood. That is an aspect of a sound culture. However, the post-investigatory human relations actions of reprimand or other similar actions are not appropriate in a sound process safety culture. These actions will quickly erode trust and disempower the operators with respect to an extremely important responsibility of their jobs. Emergency shutdown usually requires rapid go/no-go decision making in a very stressful situation. While the operators faced with this very tense situation may have minutes or even just seconds to make crucial, make-or-break decisions, the post-event investigators have the luxury of almost unlimited time to study the time line of the events, consult with experts, perform additional testing or analysis, and deliberate their conclusions. If the post-shutdown investigation reveals that alternative actions were possible, these should be reflected in revised operating or emergency procedures, and in improved training, but not in post-incident second guessing and certainly not in any HR related recriminations. In the end, the organization will have to accept the rare loss of production associated with an unrequired emergency shutdown, and trust that their operators will initiate this action only when it is necessary in accordance with approved operating procedures.

In some cases, positive reinforcement is appropriate to encourage the use of emergency shutdown procedures when necessary. An anecdotal story of such positive reinforcement: An operator exercised the emergency shutdown procedure a few days prior to a visit by a company executive. The visiting executive requested to meet that operator. Everyone thought that the executive wanted to fire the operator for his actions. Instead the executive shook the operator's hand and thanked him for doing what he felt was right. Such action by senior management can only have a positive effect on the process safety culture.

The issue of emergency shutdown authority is affected by many process safety culture core principles: *Establish an Imperative for Safety, Empower Individuals to Successfully Fulfill their Safety Responsibilities, Ensure Open and Frank Communications, Foster Mutual Trust, Combat the Normalization of Deviance,* and *Defer to Expertise.*

<u>Stop Work Authority</u>

A variant of emergency shutdown authority is Stop Work Authority (SWA). A contributing factor in many serious incidents is facility personnel not realizing that they could have or should have stopped themselves or others from continuing with an operation or task that was becoming dangerous or had the potential to become dangerous. In most companies, high level statements of safety or process safety policy encourage personnel to exercise their good judgement and be proactive about stopping an operation or task from proceeding when they recognize an error or other conditions that could make the situation worse. However, many employees are hesitant to actually exercise their judgement and stop work, especially when the work of others is involved. This was a contributor at Piper Alpha in July 1988, where adjoining wells were not shutdown despite a serious fire on Piper Alpha. This was also a contributor to the Deepwater Horizon/Macondo incident in the Gulf of Mexico in April 2010. The platform owner Transocean, as well as BP, had SWA policies that were not followed and this was a contributor to the incident (Ref 4.6 BOEMRE).

As a direct result of the Macondo incident, the Bureau of Ocean Energy Management, Regulation, and Enforcement, which is now the Bureau of Safety and Environmental Enforcement (BSEE), adopted the Safety and Environmental Management Systems (SEMS) regulation in April 2013 (Ref 4.7 SEMS). SEMS is the application of process safety management concepts with very similar elements in the U.S. offshore oil and gas sector. This regulation requires that SEMS programs contain a mandatory Stop Work Authority element. Whether or not the SEMS regulations apply, some organizations have provided written policies that make the exercising of stop work authority a formal requirement. Like emergency shutdown authority, SWA is as much a process safety cultural issue as it is a procedural issue. It will likely be needed or used in very stressful, usually rapidly evolving situations where decisions must be made in seconds or minutes. Personnel must be completely willing to exercise this authority, regardless of potential cost or schedule pressures, loss of production time, loss of product, peer pressure, hesitance to interfere in the work of others, or any other non-safety/non-process safety considerations. Additionally, personnel must be free of any fear of retribution, punishment, criticism by peers or management, or second-guessing as a result of exercising this authority. Otherwise, the culture will not support its use when needed, and the authority has no real meaning.

As with emergency shutdown authority, SWA is related to multiple process safety core principles: *Establish an Imperative for Safety, Empower Individuals to Successfully Fulfill their Safety Responsibilities, Ensure Open and Frank Communications, Foster Mutual Trust,* and *Defer to Expertise.*

Management of Change

Management of Change (MOC) is a classic process safety element that is a mainstay element of both regulatory and voluntary process safety programs. It has also been a contributor in many high-profile process safety incidents over the years, including Flixborough and Bhopal. While several other high risk sectors of industry have had MOC programs in effect for decades, e.g., commercial nuclear power and the airlines, the chemical/processing sector has been slower to adopt this concept as an overarching management system, despite having some operations which exhibit similarly high risks. The success or failure of MOC is directly a function of the underlying culture. It is a difficult program to implement without a significant adjustment in culture among all personnel, and of the organization itself. The cultural influences on MOC are summarized as follows:

- The success of MOC requires, first and foremost, that all personnel believe that MOC is important, and that changes cannot occur, however convenient, simple, or obvious they may seem, without the appropriate review and authorization. This belief is primarily a cultural issue. After initial implementation, continued vigilance is necessary to ensure that the MOC process is being applied when required is crucial, especially in the decision of which changes require MOC and which do not. Over time, particularly when the MOC procedure and forms are complex and are an administrative burden to use, the natural tendency is to find ways to bypass the process rather than use it. Determinations of whether proposed changes are replacement-in-kind (RIK) or not also requires constant vigilance to prevent changes that require MOC from being dismissed as not requiring it. Over time, incorrect thinking patterns can evolve regarding the definition of change and RIK, despite what the MOC procedure says. This can also occur when inexperienced personnel are assigned to make these decisions. This is an example of the following core principles: *Establish an Imperative for Safety, Provide Strong Leadership,* and *Combat the Normalization of Deviance.*
- The review of the impact of prospective changes on safety and process safety is the most important part of the MOC review process. This is a key aspect of the continual process of understanding the risks and hazards presented by the facility. The MOC procedure in each facility should specify a level of hazard review that is commensurate with the prospective risks. A carefully considered, independent, knowledgeable decision on whether the proposed change should be subject to a PHA should be part of the process. If this is not considered necessary, then another form of documented hazard/safety review should be performed. The use of checklists or other approved methods should be incorporated into the process so that this step is consistent and documented. In some facilities this part of the MOC review process has become somewhat pro-forma with little effort beyond routing the MOC to someone in the Safety Department for a routine review. In these cases, the only documentation that this important step has been accomplished is the single signature of the Safety Department reviewer. This is an example of the following core principles: *Understand and Act Upon Hazards/Risks* and *Defer to Expertise.*
- An MOC system where the various MOC reviews, hazard reviews/analysis, and approvals that are performed via an online system by one person at a time in isolation should include provisions for face-to-face meetings when appropriate where facility team members can debate the merits of complex or high risk changes. With the advent of electronic MOC systems the review-in-isolation process has become more prevalent. The MOC review process should not become so insular that outside technical opinions are not solicited when in-house knowledge is not adequate to thoroughly review a particular change. Examples might include complex metallurgical or corrosion issues, or process chemistry issues that are beyond the capability of process engineers to understand completely. This is not to say that all MOC review activity should be undertaken communally, because there are some potential cultural problems associated with such meetings, e.g., the application of peer pressure to get one participant in such a meeting who disagrees with all of the other participants to approve the change. This can also occur during the Pre-Startup Safety review (PSSR) portion of an

MOC (if they are combined which is a very common practice), where a meeting to approve the startup is held very close to the intended startup date and the pressure to approve it can be strong, particularly if the Project Manager or final approver participates in the meeting. Organizations should design the communal portions of their MOC process so it does not place MOC reviewers in such positions. This is an example of the following core principles: *Open and Frank Communications, Understand and Act Upon Hazards/Risks, Empower Individuals to Successfully Fulfill their Safety Responsibilities,* and *Defer to Expertise.*

- Devising different MOC procedures/pathways for different types of changes is a good practice. This customizes the MOC requirements so that they are streamlined for a particular purpose rather than relying one a single, comprehensive, and administratively burdensome "one-size-fits-all" MOC procedure. These all-purpose MOC procedures, which predominate the type of MOC procedures found in most of the PSM community, generally require the same level of review, approval, and recordkeeping regardless of the scope or complexity of the change. The same MOC procedure and forms would be used for very simple straightforward changes, e.g., changing the wording of a caution statement in an operating procedure, as for very complex changes, e.g., a new process train. Often times, people will try to find ways to bypass the "one-size-fits-all" procedure for simple changes, or it becomes more of an administrative exercise more than a thoughtful examination of the ramifications of a change. This is a form of complacency, and an example of the effect of the core principle of *Maintain a Sense of Vulnerability.*
- MOC conflicts of interest should be avoided wherever possible. For example, the initiator of an MOC should not be the only person who reviews the impact of the change on safety or process safety. Also, the person who authorizes the change (i.e., gives final approval for the change to be physically made) should not be the initiator of the MOC, else they are approving their own idea. In very small facilities with few staff this may be difficult to achieve. However, MOC is intended to provide some independence of review. This is an example of the following core principles: *Combat the Normalization of Deviance* and *Defer to Expertise.*
- When done thoroughly, MOC is often a complicated administrative process. A relatively large amount of technical information must be assembled from multiple sources, usually multiple reviewers must take part, separate hazard analyses are sometimes required, sometimes review meetings must be convened, and (usually) a multi-part, multi-page set of forms must be completed. This all must be documented and the status of each MOC tracked. This requires an active and functional system of verbal and written communications. In some facilities the number of active MOCs can be large, further complicating the process. Electronic MOC systems have streamlined the process somewhat but there is still a fair amount of administrative work associated with implementing and maintaining a functional MOC program. If the administrative aspects of MOC begin to overwhelm the purpose of the process itself, then the people involved will tend to lose sight of why they are doing it in the first place and the process becomes an end unto itself. This is an example of the following core principles: *Maintain a Sense of Vulnerability* and *Ensure Open and Frank Communications.*
- The MOC process should require proactive actions in advance of making a change, rather than the change occurring first and the MOC paperwork following merely as a form of documentation. The latter type of MOC process may leave behind a set of records that described what happened, but it obviates the entire purpose of MOC, which is to ensure that a

proposed change receives a thorough and careful review and approval before it is physically made. Establishing an MOC program *only* to comply with a regulation and leave a paper trail for each change does not fulfill the real purpose and intent of MOC. This not only does not follow the requirements of the governing MOC requirements, it is also a form of complacency, where the documentation resulting from a change becomes more important than the original purpose of the activity and people forget why they are doing it. MOC programs can morph into this type of paperwork-first mentality slowly over time, especially when the number of MOCs generated is high and lots of work is required to just keeping up with all of the recordkeeping. Close vigilance by someone who is not immersed in the details of the MOC process every day and is independent from direct responsibility for executing it is required to prevent this from occurring. This is an example of the following core principles: *Maintain a Sense of Vulnerability* and *Combat the Normalization of Deviance.*

- Organizations should not allow temporary changes to become quasi-permanent due to lack of periodic management review and challenge. The MOC procedure of an organization should specify a hard cap on how long a temporary MOC may exist. If additional time is needed and is warranted, then the procedure should contain provisions for waiving the cap following appropriate review and approval. This is a better approach than leaving the expiration date open-ended. This is an example of the effect of the core principle of *Combat the Normalization of Deviance.*
- Emergency/verbal MOCs should be used sparingly. Most MOC procedures allow for the rapid review and approval of a change when the circumstances warrant, e.g., a change that is needed quickly for safety or other reasons during off-hours. The steps covering this type of rapid approval vary, but they usually involve obtaining a verbal approval from one or more designated persons, often via the telephone. The normal MOC forms and process are then completed at the next normal work time availability. Clearly, this is a simpler and quicker way of getting an MOC approved than the normal administrative process and because of its convenience can be abused. At its core, an MOC procedure is a fairly sophisticated and formal communication system where designated persons render an opinion or make a decision about a proposed change. It simultaneously serves as the documentation method for this process. Frequently bypassing the steps that fulfill this core purpose is a strong indicator that the true purpose of MOC is not being achieved. This is an example of the effect of the core principle of *Combat the Normalization of Deviance.*
- MOC should be applied to selected personnel and organizational changes. This is often referred to as Organizational Management of Change (OMOC). Although not originally included in the scope of MOC as adopted in the PSM Standard and other regulations, many organizations have extended the scope of their MOC programs to include certain types of personnel changes. Not all personnel and organizational changes should warrant the use of OMOC and it is currently the choice of each organization what types of personnel changes are included, if any. At a minimum, those positions that are critical to the PSM program within an organization, or personnel actions that can have a significant effect on the PSM program should be included within the scope of OMOC. For example, the change of a PSM Manager would likely be a candidate for the use of OMOC, whereas the change in Human Resources Manager might not. Reducing the number of operators per shift from 5 to 4 should require the use of OMOC, whereas moving an operator from A Shift to B Shift might not. Changing the organizational structure involved in managing process safety elements

should also require an OMOC in order to help ensure that relevant responsibilities are properly aligned. The use of OMOC extends the practice of MOC to an area that can affect process safety but was not originally intended to be part of the scope of MOC. At a minimum, OMOC should include a gap analysis to identify any skill/training gaps that are presented by a different employee. If a change of personnel is followed by a rash of errors, quality problems, or near misses, it may be an indicator that OMOC should have been performed. This is an example of the effect of the core principle of *Maintain a Sense of Vulnerability*.

- Finally, MOC is not supposed to be a roadblock on the way to progress, but it is supposed to place a few speed bumps in front of a proposed change. The culture of MOC, and the overall process safety culture of which it is a part, should recognize this purpose and should govern the execution of the MOC process. This is an example of the effect of the core principle of *Maintain a Sense of Vulnerability*.

Conclusions

In many cases, the cultural core principles are not built into the PSM policies, practices, and procedures themselves (e.g., the MOC procedure) but have been established separately in an unwritten manner. They set the proper overall organizational tone for the PSM program to succeed. Process safety has been a common practice for many years in the chemical/processing industry. Many of the cultural barriers described above for COO and MOC have been overcome in most facilities. This is also true of many of the other process safety program elements. However, this process is not totally complete, and the effort to get all hands to truly believe in the importance of process safety and the management system programs in place to manage it continue. In fact, establishing and nurturing a sound process safety culture is a journey with no definitive end. This is because the influences that affect the culture change as time passes and the culture changes with them. Therefore, this is a continuing process that must be constantly sustained.

References

1. Center for Chemical Process Safety (CCPS), *Guidelines for Process Safety Documentation,* American Institute of Chemical Engineers, 1995.
2. Kletz, T., *What Went Wrong? Case Histories of Process Plant Disasters and How They Could Have Been Avoided*, 5th Ed., 2009.
3. Baker, J. et al., *The Report of BP U.S. Refineries Independent Safety Review Panel*, January 2007 (Baker Panel Report).
4. Center for Chemical Process Safety (CCPS), *Conduct of Operations and Operational Discipline*, American Institute of Chemical Engineers, 2011.
5. American Petroleum Institute (API), Fatigue Risk Management Systems for Personnel in the Refining and Petrochemical Industries, API RP-755, 2010.
6. Bureau of Ocean Energy Management, Regulation, and Enforcement (BOEMRE), *Report Regarding The Causes Of The April 20, 2010 Macondo Well Blowout,* September 14, 2011.
7. Bureau of Safety and Environmental Enforcement (BSEE) 30 CFR §250, *Safety and Environmental Management Systems (SEMS) Final Rule,* 2013.

Ensuring Learning (Not Just Training) – Measuring the Impact of Learning on Business Performance

Lawrence Short
DuPont Sustainable Solutions
500 Studio Drive, Virginia Beach, Virginia
Lawrence.S.Short@dupont.com

Alfonsius Ariawan
DuPont Sustainable Solutions
1919 Minnesota Court, Mississauga, Ontario
Alfonsius.B.Ariawan@dupont.com

Prepared for Presentation at
American Institute of Chemical Engineers
2017 Spring Meeting and 13th Global Congress on Process Safety
San Antonio, TX
March 26 – 29, 2017

Ensuring Learning (Not Just Training) – Measuring the Impact of Learning on Business Performance

Lawrence Short
DuPont Sustainable Solutions
500 Studio Drive, Virginia Beach, Virginia
Lawrence.S.Short@dupont.com

Alfonsius Ariawan
DuPont Sustainable Solutions
1919 Minnesota Court, Mississauga, Ontario
Alfonsius.B.Ariawan@dupont.com

Keywords: Learning, Training, Capability, Competency

Abstract

"How do we know our training is working?" is a frequently asked question by most organizations and it oftentimes asked around budget cycles. Training has often been used as a tool to drive compliance, whether regulatory or organizational, however, many organizations continue to fall short of their performance measures, specifically those related to safety and operational risk management.

For the past 4-5 years, we have been developing a powerful learning model/system that does the following:

- Standardizes learning tools, while allowing for site/individual/organizational flexibility
- Captures millions of data points from all aspects of the learning process, formerly never captured by most organizations
- Pinpoints exactly where improvement is needed to achieve the desired organizational goals

We've worked with many mature organizations that have top-quartile processes and systems in place, yet still see leading indicators of performance continue to drop. After further analysis, we found standards, policies, procedures, adult-focused learning approaches, employee assessments that were great, BUT incidents were still occurring around certain factors (maintenance & reliability, safety, downtime, etc.).

Through our Learning Model and Analysis, we found breakdowns in the following areas and began to immediately mitigate:

- "How" the message was being delivered to employees conveyed a lack of belief in the standard
- Supervisors/Managers did not reinforce the policies through their actions, but words
- Employee assessments were not validated in the field, post workshop
- The learning environment was not appropriate for the desired objective.
- Operations was not engaged in the solution to ensure the upgrade was applicable and owned by the field

Overall, knowing where to target improvements and the ability to quantify the impact learning has on organizational performance are what makes this Learning Effectiveness Model so innovative.

1. Introduction

People drive performance and therefore, the success of an organization depends on its people's capability. Personnel need to be functionally capable to perform their tasks well. But how is this best accomplished?

When it comes to capability building, our first thought is usually that of training. Unfortunately, capability building is not all about training. Rather, it is about learning, and learning happens through both formal interactions and informal experiences. By focusing on learning and on improving learning effectiveness, investments in capability building will be better optimized and will yield more valuable returns.

Consider an example of developing organizational capability in the context of operational risk management. An operator new to a task is usually on-boarded with classroom training sessions. All may be well for a few months until the operator takes a short cut, resulting in a near-miss event or an incident. In this case, the operator has been trained, but there has been a failure in the learning process that results in the event. The organization's response is typically to increase training efforts. However, this may not yield the desired improvement. Instead, learning variables need to be addressed, ranging from training-related variables (trainer's capability, training content and delivery methods, timing of training, etc.) to other variables such as supervisory or peer influence, clarity of procedures as learning tools, awareness of previous incidents related to the task in question, or regular coaching.

Through our work with clients, we developed a Learning Effectiveness model that considers the end-to-end journey of an individual's learning cycle to help ensure improved safety performance and optimal investment.

2. Learning Effectiveness Model

From a capability development perspective, a typical journey for and operator is depicted in Figure 1. The journey usually starts with an on-boarding session and site orientation. This may be followed by a competency assessment and the development of an individual capability building plan. Gaps are then addressed by providing trainings in various formats, including hands-on training. Sometimes, field validations are performed to confirm competency. Once the operator is deemed competent, he/she is allowed to perform the tasks that he/she has been hired to do. Audits and quality assurance are then conducted at some frequency to ensure adherence to training requirements. Some organizations also conduct annual competency assessment and most require their operators to complete refresher training at regular intervals.

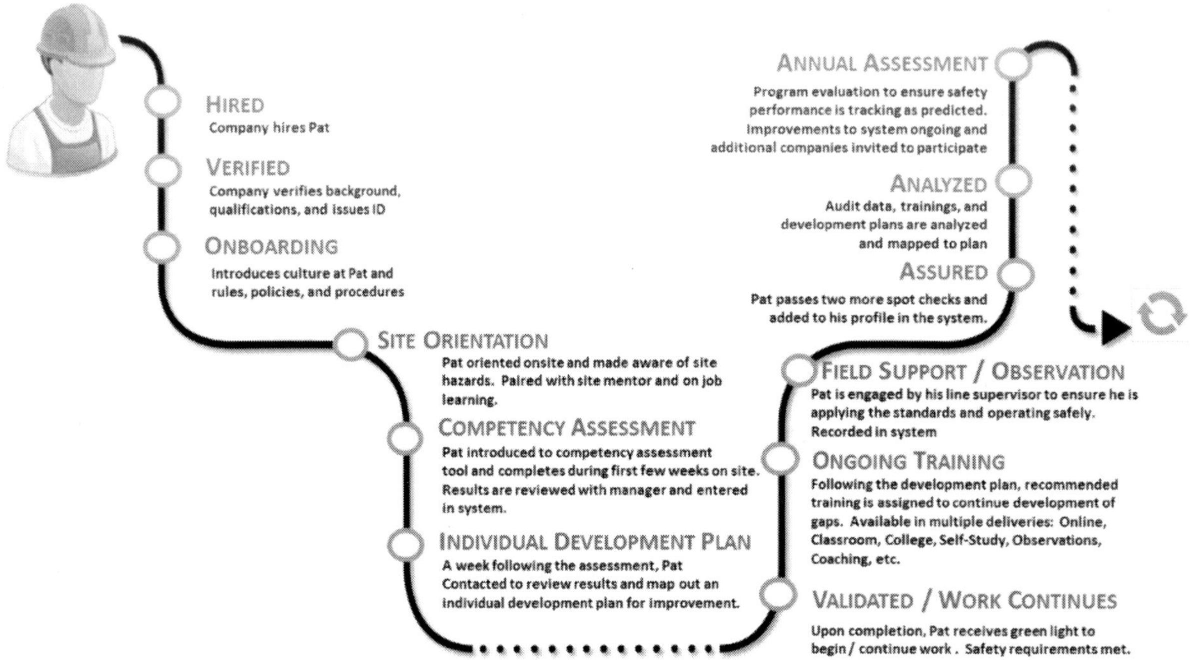

Figure 1. Meet Pat. Pat is an operator. This is Pat's capability development journey, one that is typical for an operator

In our model, we have enhanced an operator's learning journey by incorporating more structures and building in learning assurances throughout the cycle to support continuous improvement. These are depicted in Figure 2. Some of these are not necessarily groundbreaking. These include the more disciplined definition of curriculum for various roles within the organization – who needs to be developed in what areas?

To facilitate the development of curriculum, a curriculum committee needs to be established, with membership made up of personnel closest to the shop floor (e.g. supervisors and operators). In addition to the curriculum, we also include an element around content development that has the primary objective of ensuring the fidelity of course contents to the learning objectives of individual roles. In other words, is the course content supportive and additive to the learning objectives that the participants should achieve?

The design of the content should be deliberately impactful, incorporating adult learning principles and problem-based learnings. What are the most effective media/activities to transfer knowledge? What is the most impactful format? We engage our instructional designers to ensure that these questions are addressed in a structured manner. As an extension of this, the organization should also consider the format and quality of learning providers. How do you ensure quality delivery in class and in the field? How do you strengthen your train-the-trainer/coaching program?

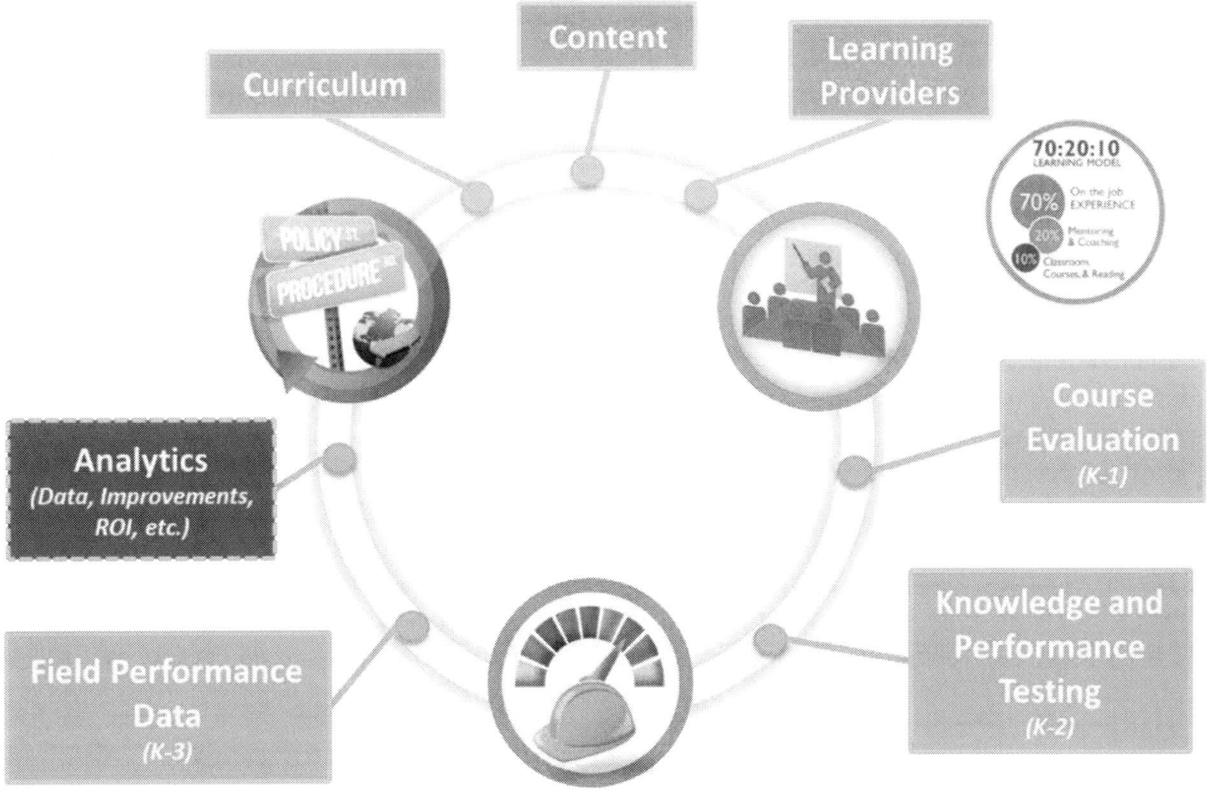

Figure 2. Learning Effectiveness model

Perhaps what differentiates our model is the deliberate learning assurances that we have incorporated throughout an individual's learning cycle. The primary objective here is to ensure that we have confidence in the overall process. How do we know that an individual has learned what is expected of him/her? That is, rather than focusing on the tasks ("Pat has completed his training"), we would like to focus on the outcome ("Pat has learned how to do his job safely"). By focusing on the desired outcome, we can then more readily measure the impact of learning on business performance.

The first assurance is determined by collecting and analyzing the Kirkpatrick K1 measure. The K1 measure is simply the course evaluation data that is distributed to learners following an in-class session. It provides a first level assessment of the effectiveness of the course content and delivery quality. This is then followed by a measure of understanding, or the Kirkpatrick K2 measure. This takes the form of a test, whether in-class or in the field. It provides an indication to the organization on the effectiveness of knowledge transfer. The ultimate assurance of learning, however, is the Kirkpatrick K3 measure. This is a measure that most organizations fail to capture or link to an employee's learning journey. The K3 measure is made up of field performance indicators, such as those related to operational discipline, audit performance, incidents and near-misses, as related to specific tasks. These can be quantified to determine business impacts (for example, in terms of loss avoidance) and with this, Return On Investment can also be estimated.

Our instructional designers combine both the practical experience that we have obtained through client work and instructional theories to design the various assessment protocols in order to determine the right Kirkpatrick measures. This allows us to link the Kirkpatrick measures to the various critical points of employee's learning journey. With some basic analysis, we can then determine where failure modes have happened in the journey. For example, we are able to determine if the way the course content is delivered to employees lacks impact, or if there is a lack of supervisory reinforcement in the field. This then allows the organization to address issues to further enhance learning vs. investing in more training session.

5. Summary

Organizations often associate capability development with training. Training is certainly important. However, it is not sufficient. Capability development is about learning and learning takes place both inside and outside of the classroom. The question organizations should ask is 'How do you measure the impact of your capability development program?' In our work with clients, we have developed a learning effectiveness model that incorporates the critical elements of curriculum and content building, delivery excellence, and several assurance points, using the Kirkpatrick measures. Through thoughtful design of assessment protocols, we can collect the right Kirkpatrick measures that allow us to link these measures back to the various critical points of an employee's learning journey. By understanding this linkage, organizations can address gaps and improve learning effectiveness. Further, the selected measures allow organizations to quantify business impacts and estimate Return on Investment.

Tridimensional Training Matrix improving Process Safety Performance

Presenter/Principal Author
Américo Diniz Carvalho Neto
RSE – RISK MANAGEMENT&SUSTAINABILITY
Address
americodcn@rsem.com.br
americodcn@gmail.com

Other Author(s)
Alex Alves França
Cristina Maretti Dias
Fabiano Fiscina
Isadora Correa
Paula Franco Brandão
Thaiane Duarte
Affiliation(s)
First RSE Post Graduate Course in PSM Address (es)
Email address (es)
atendimento@rsem.com.br

Prepared for Presentation at
American Institute of Chemical Engineers
2017 Spring Meeting and 13th Global Congress on Process Safety
San Antonio, TX
March 26 – 29, 2017

Tridimensional Training Matrix improving Process Safety Performance

Presenter/Principal Author
Américo Diniz Carvalho Neto
RSE – RISK MANAGEMENT&SUSTAINABILITY
Address
americodcn@rsem.com.br
americodcn@gmail.com

Other Author(s)
Alex Alves França
Cristina Maretti Dias
Fabiano Fiscina
Isadora Correa
Paula Franco Brandão
Thaiane Duarte
Affiliation(s)
First RSE Post Graduate Course in PSM Address (es)
Email address (es)
atendimento@rsem.com.br

Keywords: Process Safety training, PSM competence, Training Matrix, PSM Performance

Abstract

This paper outlines how important is to have an efficient Training Matrix in Chemical and Petrochemical companies. To ensure high performance in Process Safety, it is essential to have teams highly trained ready to take action and make great decisions. The high level of human performance is an important aspect of any training program. Process Safety goes beyond of compliance with minimum requirements of legal policies. An efficient Process Safety Management (PSM) can assure both high performances in preventing Process Safety Incidents and saving money for the company. In order to have a PSM program as described, companies have to define what their employees need to know to perform their specific roles successfully and how deep they need to go in each subject. In other words, it is important to have a Tridimensional Training Matrix. Skills development should take place in all parts of an organization, but not everyone needs to know everything deeply.

Focusing on continuous improvement in the processes, companies need to define the proficiency level of their employees in different positions and roles, as well as the applicability of knowledge acquired in training in each position/role. It is also important knowing the difference between positions and roles. A role can be performed by different positions, for instance, a Process Engineer (position) could be the Process Safety Focal Point (role). Knowing this is important in order to set the mandatory competences for positions and critical roles in Process Safety and preventing to overwhelm some

employees with lots of roles decreasing their productivity. A Tridimensional Training Matrix is composed by positions/roles, trainings/courses and proficiency levels. The combination of these three dimensions leads to an effective and lower cost training program. It is also important to define which trainings in which proficiency levels could be taken on the Online Basis to reduce costs with classroom trainings.
Last but not least, it is also important to define when the training should be taken, if it is a new employee training or a refresh one which needs to be taken in a specific time interval.
In conclusion, it is clear that creating a Tridimensional Training Matrix is complex and challenging, but having this tool will improve the performance of the employees and the entire company.

1 Introduction

An important element for Process Safety Management, as discussed in this paper and referenced in Guidelines for Risk Based Process Safety, is the Process Safety Competency element. Related to the RBPS pillar of *committing to process safety*, this element deserves a careful looking, as a main basis for the organizational process safety culture establishment, development and continuous improvement. Associated with the element Training and Performance Assurance, Process Safety Competency element brings the core requirements and concerns for organizations to manage PS knowledge through all functions and levels. Indeed, the PS competency is a so wide subject that CCPS also developed a specific guideline for Defining Process Safety Competency Requirements.

Guidelines for RBPS says that "developing and maintaining process safety competency encompasses three interrelated actions: (1) continuously improving knowledge and competency, (2) ensuring that appropriate information is available to people who need it, and (3) consistently applying what has been learned". The main challenge is to achieve excellence on this management, considering the needs to deliver the right amount of useful knowledge and the adequate charge of information, which means defining which skills might be developed and how deep this development is required to which organizational role or function. The difference between role and function is also a sensitive concept that the company may engage. Information must be available to be consulted whenever necessary, but also shall be formally delivered on specific periods of time and as deep as needed according to the role played by those who are receiving the information. That is why develop a tridimensional training matrix is a very interesting strategy, as this model clearly define who needs the information, in which frequency, and proficiency level, regarding that "only competent people can transform information into knowledge, (...) which helps organizations understand and manage risk and remain competitive." (Guidelines for Risk Based Process Safety).

This paper aims to propose a way to develop tridimensional training matrixes in Chemical and Petrochemical companies as an important tool to pursuit high performance in Process Safety. A methodology is proposed and references and results are discussed to give ideas on how to implement a tridimensional training matrix for development of PS competency and culture.

2 Methodology

In order to build an effective Tridimensional Training Matrix, it is important to do some research and stablish some definitions at the first place. A Training Matrix needs to be objective and fit the company needs. It is important to start the process with the goals and the level of company's Process Safety Culture in mind.

This session will present a process flow, which could be adopted in order of building a Tridimensional Training Matrix for Process Safety Competency (Figure 1).

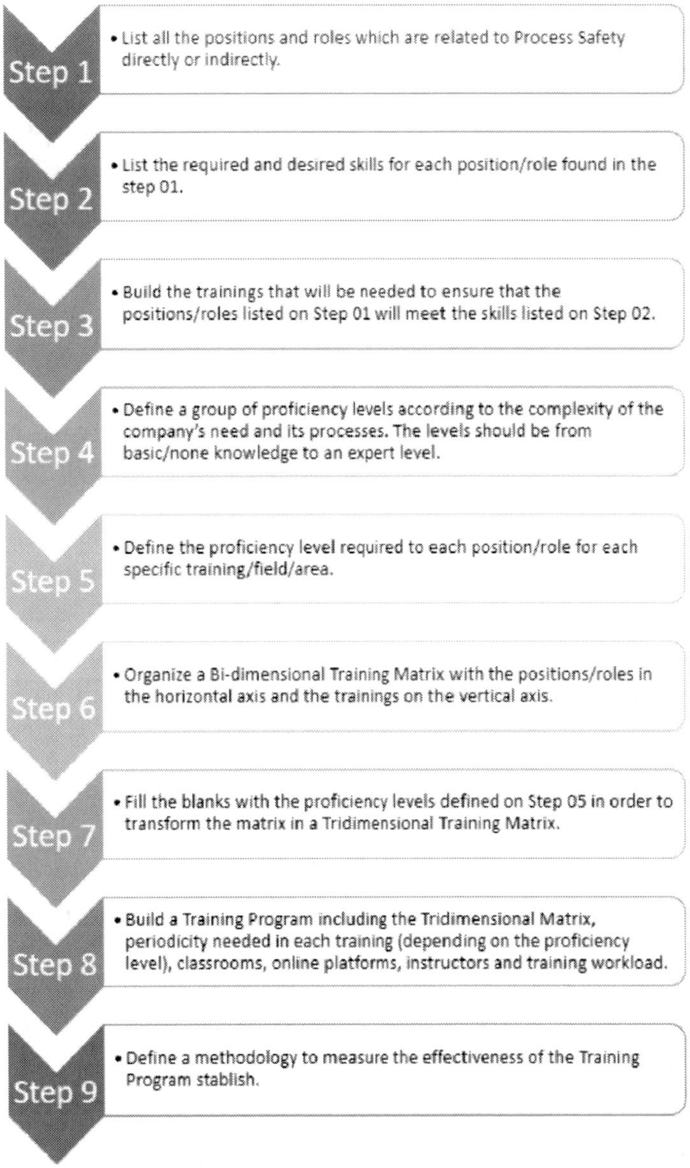

Figure 1 – Building a Tridimensional Training Matrix Step-by-Step

3 Results

In this section, an example of a Tridimensional Training Matrix proposal is presented (Figure 1) and is composed by these three dimensions: trainings/courses, positions/roles and proficiency levels. The Figure 2 represents an initial Tridimensional Training Matrix which can be developed according to the maturity level and needs of the company.

Regarding the dimension of training, this matrix proposal is divided into four different kinds, based on the CCPS pillars: *Commit to process safety*, *Understand hazards and risk*, *Manage risk* and *Learn from experience*.

The benefit of this division is that the company will be able to easily evaluate or focus on the strategic pillar it needs the most. Then the company can improve its process safety culture for further maintenance these four pillars strongest.

For the dimension of functions and roles, this matrix proposal is divided into some examples of areas: Manufacturing; Environment, health and safety; Engineering; Lab; Administrative/ Managers.

For each area, the functions and roles are organized per hierarchy levels. It is important to include all workers involved directly or indirectly to the theme, and also managers and directors who make important decision regarding process safety.

According to CCPS ("Guidelines for Defining Process Safety Competency Requirements"), the dimension of proficiency levels can have five different degrees:

- Awareness: is generally aware of the topic and associated terms. May not know the answer, but knows where to get more information.
- Basic knowledge: has general working knowledge of the topic. Has basic training required to perform general tasks related to the role.
- Practitioner: can execute specific tasks within the topic with minimal direction. Has the experience levels to complete assigned tasks. Is typically an engineer or manager.
- Expert: is a recognized expert with extensive knowledge and skills. Has specialized training or certification which may be required for certain tasks.
- Leader: is responsible for implementation of a PSM element or activity. Knowledgeable of the subject matter, may be a subject matter expert, but that is not required. Coordinates all resources and assigns tasks to ensure effective implementation of the topic.

To exemplify this dimension, this proposed matrix was designed considering trainings in three different levels:

- Level 1: One who takes this kind of training will be on the basic knowledge proficiency level;
- Level 2: One who takes this kind of training will be on the practitioner proficiency level;
- Level 3: One who takes this kind of training will be on the leader proficiency level.

For those functions and roles whose training is not required, the matrix defines it as Not Required (NR).

Depending on the maturity level of the company, it could there be more or less number of training levels.

Example of Tridimensional Training Matrix

CCPS Safety Process Pillars	Course Description Training	Manufacturing: Operator	Manufacturing: Operation Coordinator	Manufacturing: Process Safety Focal Point	Manufacturing: [...]	Maintenance: Maintenance Coordinator	Maintenance: Maintenance Technician	Maintenance: Inspection Technician	Maintenance: [...]	Environment, Health and Safety: EHS Technician	Environment, Health and Safety: EHS Leader	Environment, Health and Safety: Fire Brigade	Environment, Health and Safety: [...]	Engineering: Process Engineer	Engineering: Automation Engineer	Engineering: Instrumentation Engineer	Engineering: [...]	Lab: Lab Coordinator	Lab: Lab Technician	Lab: [...]	Administrative/ Managers: Business Leader	Administrative/ Managers: Industrial Director	Administrative/ Managers: Human Resources Analyst	Administrative/ Managers: [...]
Commit to Process Safety	Process Safety Competency (Culture, Standards...)	L3	L3	L3		L3	L3	L3		L3	L3	L3		L3	L3	L3		L3	L3		L3	L3	L3	
	Loss Prevention and Business Image	L1	L2	L3		L2	L1	L1		L1	L3	L2		L2	L2	L2		L1	L1		L3	L3	L1	
	Process Safety for Stakeholders	L1	L1	L3		L1	L1	L1		L1	L3	L2		L1	L1	L1		L1	L1		L3	L3	L1	
	[...]																							
Understand Hazards and Risk	Qualitative Risk Analysis (PHA, HazId,HazOp...)	NR	NR	L3		NR	NR	NR		NR	L3	NR		L3	L1	L1		NR	NR		NR	L1	NR	
	Quantitative Risk Analysis (LOPA,SIL, QRA...)	NR	NR	L3		L1	NR	NR		NR	L1	NR		L2	NR	L3		NR	NR		NR	L1	NR	
	Facility Siting for Projects and Installations	NR	NR	L3		NR	NR	NR		NR	L2	L1		L2	NR	NR		NR	NR		NR	L1	NR	
	[...]																							
Manage Risk	Critical Activities (Operational procedures)	L3	L3	L2		L2	L3	L3		L3	L3	L3		L1	L1	L1		L1	L1		NR	NR	NR	
	Equipment Mechanical Integrity and Reliability	NR	NR	L2		L3	L3	L3		L1	L1	NR		L2	L3	L3		NR	NR		NR	NR	NR	
	Management of Change and Start-up Analysis	L1	L1	L3		NR	L1	L1		NR	L2	NR		L2	L1	L1		NR	NR		NR	NR	NR	
	[...]																							
arn from perience	Incidents Investigation Methodology (RCA...)	L1	L1	L3		L2	L2	L2		NR	L2	NR		L3	L1	L1		NR	NR		NR	NR	NR	
	Process Safety Indicators Management	L1	L1	L3		NR	NR	NR		NR	L1	NR		L3	NR	NR		NR	NR		L1	L3	NR	
	Safety Barriers Audit	NR	NR	L3		NR	L1	L1		NR	L3	NR		L3	L1	L1		NR	NR		L1	L1	NR	

Figure 2 – Example of Tridimensional Training Matrix Proposal

As the maturity level of the company in process safety increases, the matrix can be developed. It can include information like the kind of training (e-learning or in class), the number of training hours per proficiency level, the recurrence, the level required of the trainer, and the way the efficacy is measured.

4 Discussion

Training and performance assurance, and process safety competency, are the elements of a risk-based process safety management system (RBPSMS) that present themselves as a product delivered through a training matrix. A complete training matrix gathers the expected competency level information for functions and roles, planned trainings to achieve these levels of knowledge, capacity to skills development, and individualized training practices (eg mentoring or coaching) to develop attitudes. The set of "knowledge, skills and attitudes" that forms the competence can be described in levels for each one of the elements of this set; for each element, trainings, capacity and practices can be applied at different levels in order to reach the desired levels of the elements that make up the competences. A three-dimensional matrix makes it easy to visualize these levels of each element of competence and how they are planned to be achieved.

People play functions and roles within organizations. A three-dimensional training matrix ensures that not only the competencies of the functions will be planned to be maintained, but also the competencies of the roles. Depending on the complexity of the roles, the three-dimensional training matrix can help in determining the level of competence that a function must have in order to take on a particular role simply by identifying the competencies envisaged for both a function and a role. In addition, by identifying the persons occupying these functions and roles, the three-dimensional training matrix also enables one to identify whether the distribution of roles is appropriate for the available human resources, and the gaps between the installed competencies and the desired competencies.

The training matrix should provide recycling. However, recycling should consider:

Risk associated with the activity: activities of greater risk or greater impact in the RBPSMS should consider more frequent recycling.

Frequency and complexity with which the activity is performed: activities performed less frequently, or more complex activities, should consider more frequent recycling.

Legal requirements applicable to the activity, for example for recertification.

Recycling strategies (re-training? Workshops? Re-evaluation or recertification?)

Results of efficacy evaluation (as discussed below).

Efficacy assessments should be conducted on a regular basis to identify whether training initiatives, training and practices are meeting their objectives, or whether it is necessary to reapply the same initiatives, or only provide recycling, or even modify form or content (or both) these initiatives. The training matrix will provide only the scheduled recycling. The efficacy assessment may conclude that it is necessary to change the recycling schedule. In addition, it should be borne in mind that recycling is one of the possible actions within the scope of competence training, and that in individual cases it may be necessary to take actions outside the scope of competence training (eg by assigning another person to assume the function or role).

The trainings listed in a training matrix need not be presential. Considering the availability of technology and resources, distance training or electronics can reduce the realization costs of training and capacity building. However, an evaluation of the content should be done in order to identify where e-learning can be employed without compromising the desired results. Since training is only one of the initiatives that can be used to gain competence, it is necessary to assess whether other initiatives are able to cover the gaps in content eventually left by distance or electronic training, so that the goal can be achieved. Again, efficacy assessment is crucial to determining such compliance with objectives.

Finally, it is important to discuss the complexity of the elaboration of such a three-dimensional matrix considering the complexity of the organization. Although the three-dimensional matrix's method of elaboration is simple, if it is applied to an organization with many functions and roles and dealing with significant risks, the matrix will be complex. Such an organization should not decide to construct the matrix of all its functions and roles from the first version. Instead, it should start with the functions and roles of dealing with the most critical risks, and then move on to the others until it get to the full model. On the other hand, people in a company with fewer functions and roles may think that the traditional training matrix model suffices because of the simplicity of their business. But even for simpler business, using a three-dimensional matrix allows the (few) functions and roles of a simpler business to have only the skills they need to have. That is, using three-dimensional training matrices avoids wasting resources while preventing people from being driven to possess skills they will not use because they are exposed to initiatives that far outweigh the needs of the business.

5 Conclusion

Based on what has already been explained so far, it is noted that the challenge of developing, maintaining and strengthening the competencies of Process Safety employees at the most varied levels of an organization is a challenging task, sometimes complex and incessant, since it is necessary to evaluate several factors such as: role played by the employee, workload, proficiency level for each function, applicability, mode of execution, legal requirements, effectiveness and cost.

The execution of the training should follow a planning, so that the development of the employees reaches the level expected by the organization in Process Safety, generating changes in their behavior, making it more qualified and, consequently helping the organization to make better decisions, based on reducing the likelihood of accidents and also reducing the costs associated with their consequences. This planning also ensures that training will not be given indiscriminately, but only for that target audience for which it has been shown there is a competency gap that can be remedied by training (or other initiative in the training of competence).

The development and application of a three-dimensional training matrix allows and facilitates this planning to be put into practice in an organized way, and may even help in the measurement of some indicators, such as percentage of fulfillment in training, management of organizational changes, among others. The application of a three-dimensional matrix also allows optimizing the costs with the formation of competences, structuring the application of initiatives focused on the existing gaps.

However, it is worth emphasizing that depending on the size of the organization, the three-dimensional matrix can become quite complex, being more efficient to subdivide it into departments, making it more feasible.

6 References

[1] CCPS. *Guidelines for Defining Process Safety Competence Requirements.* Center for Chemical Process Safety, American Institute of Chemical Engineers, New York, NY, 2015.

[2] CCPS. *Guidelines for Risk Based Process Safety.* Center for Chemical Process Safety, American Institute of Chemical Engineers, New York, NY, 2007.

[3] RSE Consulting & PSM Center. PSM Post Graduation Course 2016-2017, Module III. Professor Mr. Americo Diniz Carvalho Neto.

479892 Experiences in Development and Deployment of a Process Safety Curriculum in a North American Pipeline Company

Abstract:

A case study will be presented of how a Process Safety curriculum was developed and deployed at a North American Pipeline company. The curriculum included a face-to-face traditional instruction component as well as a computer-based component deployed in 2016. The curriculum was designed to give learners a basic understanding of the principles of process safety management and their application to the company. The face-to-face component of the curriculum employed adult learning theories that bridged individuals whose experiences ranged from new graduates to those with 30+ years. Interactive group exercises and case studies reinforced process safety fundamentals with particular attention paid to a struggle of differentiating between worker or personal safety compared to process safety.

479496 A Proposal for Improving the Efficiency, Effectiveness, Quality, and Usefulness of PHA Identification of Controls

Abstract:

Process Hazard Analysis (PHA) is an important discipline for identification and understanding of hazard causes, consequences, consequence severity, controls / safeguards, end consequence likelihood, risk assessment, recommendations for reducing risk, and prioritization of recommendation resolution.

Controls / safeguard discussion and documentation is fraught with potential problems. If controls identified by members of the PHA team are ***not*** adequate and are ***not*** included in the list, team members may reduce their participation as they become increasingly discouraged and disengaged. If controls that are ***not*** adequate ***are*** included in the list, the sheer number of listed, inadequate controls may result in: a false sense of sufficiency of the listed controls; a risk likelihood and risk rank that are less than actual; and reduced recommendation resolution priority, if a recommendation is made at all.

Accordingly, consistency, discipline, and rigor of controls discussion and documentation is recommended for useful, quality PHAs, whatever methodology is employed. Unfortunately, the time required for useful, quality PHAs employing consistency, discipline, and rigor can also result in tedium, reduced participation of the members, and increasing discouragement and disengagement anyway.

This paper outlines a proposal in which leaders of PHA studies list and annotate controls suggested by members of the PHA Team to be reviewed in the study for quick, instructive, and collaborative disposition. The controls listed for each scenario can be sorted and discussed by the annotations which may be modified in the finalizing discussion.

Using such an approach would encourage participation by all members of the PHA Team and reduce occasion for discouragement and disengagement. Application of the approach supports: quicker, improved evaluation and documentation of the effectiveness of controls; improved quickness and quality of evaluating likelihood and risk rank; better evaluation of the need for recommendations; and proper prioritization of recommendation resolution.

13TH GLOBAL CONGRESS ON PROCESS SAFETY

Effective Human Factors Tools to reduce Plant Costs and the Number and Severity of Injuries

Dennis Attwood, P.Eng., Ph.D.
Senior Consultant, Midland Engineering Limited
Midland, MI
dennis@hf-apps.com

Prepared for Presentation at
American Institute of Chemical Engineers
2017 Spring Meeting and 13th Global Congress on Process Safety
San Antonio, TX
March 26 – 29, 2017

Effective Human Factors Tools to reduce Plant Costs and the Number and Severity of Injuries

Dennis Attwood, P.Eng., Ph.D.
Senior Consultant, Midland Engineering Limited
Midland, MI
dennis@hf-apps.com

Human Factors, ergonomics, valve design, risk analysis, HAZOPs, PHA, Critical tasks, procedures

Abstract

Human Factors is defined as the "systematic process of designing for human use". This process systematically applies our knowledge of the capabilities, limitations, expectations and needs of people to the equipment they use, the environments they operate in, the tasks they perform and the management systems that guide the safe operation of their plants.

This paper describes easy-to-use and cost-effective Human Factors tools for the solution of major issues in process plants. Considering the breadth of the field of human factors and the limited time available for presentation, the paper limits discussion to the following four applications:

- Assessing the physical risk of operating manual valves to reduce maintenance costs.
- Improving the design of operations procedures.
- A method for applying Human Factors to Process Hazard Analyses.
- Critical task identification and analysis.

Each tool is thoroughly explained and its use is demonstrated with a practical example from the process industries with a summary of Benefits. References for papers explaining each tool in more detail are provided.

Introduction

The Human Factors discipline is continually evolving. Many in the field are conducting basic research to determine how humans respond to musculoskeletal, physiological and cognitive stressors. Others are developing tools to analyze plant situations and recommend solutions. This paper focuses on some of the many methods developed for analyzing Human Factors in Process Plants. The methods selected herein, can be used to solve a problem that has been identified, for example, to rank the musculoskeletal risk of manual valve operation. They can be used to determine if a Human Factors problem exists, for example, whether procedures are designed to make users make mistakes. Or they can be used to analyze systems during process design to identify opportunities for human error and to provide a proactive solution, for example, to identify critical operator tasks and analyze them to identify appropriate safeguards to mitigate risks.

The methods on which this paper will focus include:

1. Assessing the physical risk of operating manual valves to reduce maintenance costs. The tool ranks each valve in terms of musculoskeletal risk. Plant management can use this ranked list to lower the long term cost of ownership by scheduling maintenance on the highest risk valves first and reducing human injury
2. Improving the design of operations procedures. Statistics indicate that the majority of safety-related incidents in process plants are the results of not following operating procedures and/or not having complete and easy to follow procedures. This set of tools assesses procedures for inherent risk and provides guidance on redesign. Having improved procedures that are easy to follow and understand reduces the probability of human error/injury.
3. A method for applying Human Factors to Process Hazard Analyses. The paper reviews the development of the API Standard designed to include Human Factors in Process Hazard Analyses and provides an improved format for plant use. The benefit of Including Human Factors in the PHA will make sure the key Human Factor risks are addressed and mitigated
4. Critical task identification and analysis. Critical (high risk) process tasks have the potential for personal injury, equipment or facility loss, environmental release, community exposure or business interruption. The tool provides a systematic method of identifying high-risk tasks, analyzing the tasks for inherent risks and identifying appropriate safeguards to mitigate risks. The benefit is to reduce EHS incidents

Four Human Factors Application Tools

1. Assessing the Physical Risk of Operating Manual Valves to reduce Maintenance costs

Every process plant contains a large number of process valves with a wide diversity of designs that control the flow of feedstock, products and service liquids and gases. The ergonomic issues associated with the design, operation and maintenance of valves include:

- Physical stress to open and close them
- Potential for injuries and subsequent related costs
- Lack of access to process critical valves
- Difficulty to remove and replace them
- Potential for process upsets when valves cannot be operated in the time required

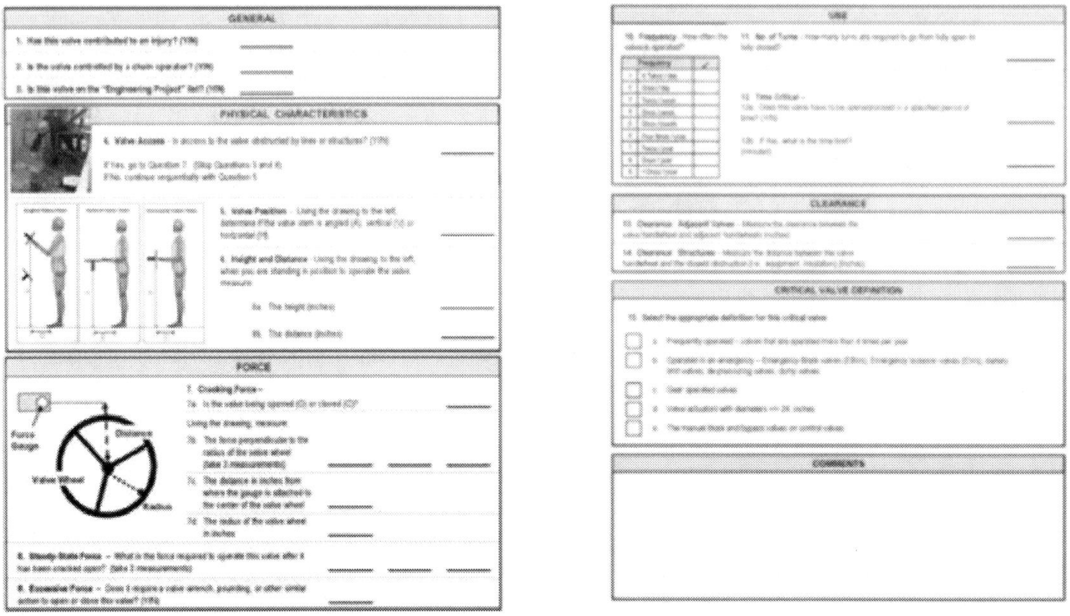

Figure 1: Valve risk assessment forms

As a result of injuries suffered while operating valves, a major U.S. petroleum refinery company initiated a program to identify high risk valves and to modify them to reduce the operating risk of injury.

We developed a systematic approach Attwood and Tys (2013) to identify and assess the operations and maintenance issues associated with process significant valves within the refinery. The approach consisted of eight distinct activities:

1. Develop plant standards for human factors requirements in the design, operation and maintenance of valves.
2. Train a group of refinery personnel, consisting of operators, maintenance techs, plant engineers and safety, health and environment staff, in ergonomic issues associated with valve operations and maintenance.
3. Lead the group in developing the definition of “process significant” valves.
4. Identify the valves that meet the definition in each unit of the refinery.
5. Assist refinery personnel to define the characteristics of valves that determine their risk of personnel injury or their potential for causing process upsets.
6. Develop a quantitative scale for each of the characteristics in (5) above.
7. Field assess each of the “process significant” valves against the characteristics.

8. Calculate the sum of the risk characteristics for each valve and develop a risk prioritization ranking list. This will permit the refinery to repair or replace valves, as business conditions permit, according to their position in the ranking list.

The characteristics that were identified to assess the risk of valve operation are shown in Figure 1. We have used this methodology in two different refineries with excellent results. The data were able to distinguish between high and low musculoskeletal risk valves. Figures 2A and 2B, for example, illustrate and describe both a HIGH and a LOW risk valve and recommends practical improvements for each.

Figure 2A: High risk valve

High Risk Valve

- Risk score = 79 (out of 112 possible)
- Has contributed to injuries
- Third choice location
- Cracking force > 246 lb.
- Steady state force >149 lb.
- Requires 42 turns to open
- It is recommended to increase the height of the actuator to waist height and to increase the size of the actuator to reduce the steady state rim force required

Figure 2B: Low risk valve

Low risk valve

- Risk score = 51 (out of 112 possible)
- Frequently operated
- Cracking force > 181 lb.
- Steady state force > 144 lb.
- Time critical operation (16 turns in 5 minutes)

It is recommended to adjust heights of actuators so pinch points are eliminated and to increase the diameter of the actuator wheels to reduce rim force required

Our valve risk ranking was reviewed by the operations team that provided the data and accepted as a viable risk-ranking technique that can be used to schedule valve maintenance so that the valves with the highest risk are maintained first. Moreover, our recommendations for improving each high risk valve were both practical and economical.

Practical Benefits of this Application

- Reduces the potential for musculoskeletal injuries and days off from work
- Reduces maintenance costs by targeting valves with the highest musculoskeletal risk

2. *Improving the design of operations procedures*

The term "procedure" refers to a written description of the steps that a process worker needs to follow to complete a task. The purpose of a procedure is to guide a process worker, step-by-step, through a task. Not every task requires a procedure (Haas, 1999). If the task has few steps, if it is performed frequently and if the consequence of error is minor, it will not likely require a procedure. However, if the task is complex, if it is performed infrequently and if the consequence of making an error is high, a procedure will likely be required. Failure to follow required procedures or failure to follow them correctly accounts for as much as 70% of the error-related incidents made by process workers (HSE, 2005). Incidents include both injuries and errors that affect process safety.

Evaluation Techniques

The effectiveness of a procedure program can be evaluated in a number of different ways. One of the most effective is a site audit to record any gaps between the existing on-site procedure program and the recommended guidelines. The audit might include:

- Observing multiple operators performing the procedure and comparing their performance against the documented procedure to identify deviations
- Evaluating the written procedure with a checklist. The checklist published in Attwood et al, 2003, contains detailed questions that can be used to evaluate written procedures
- Computing a Readability Score which is an open-source index that includes several well-known indices of readability
- Implementing Simplified English. We are proposing to use a powerful new tool to evaluate and develop your procedures. Simply put, simplified English or simplified technical English (ASD, 2005) is a standard set of English words and vocabulary rules that:
    - Reduces ambiguity
    - Improves comprehension for people whose first language is not English
    - Makes translations cheaper and easier
    - Facilitates automatic checking to ensure that procedures and manuals are written according to the rules and vocabulary
    - Improves understanding among users whose first language is not English

We have evaluated the procedure systems in a number of companies and have concluded that the resulting procedures are easier to understand and use and that the risk of making errors is greatly reduced.

Figure 3 illustrates the some common errors that were identified from one company's procedures based on the evaluation checklist identified above.

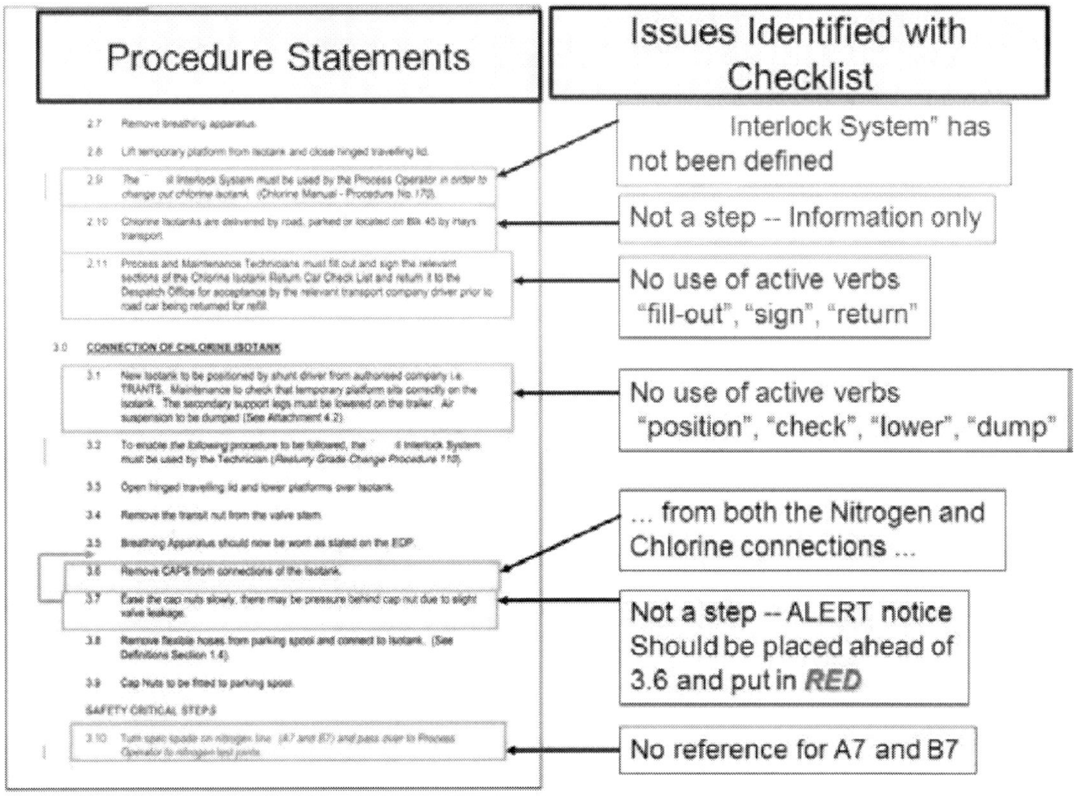

Figure 3: Procedure design issues

<u>Practical Benefits of this application:</u>

Our procedures improvement methodology ensures that clearly defined processes are in place to guide the development and implementation of effective procedures. We need to ensure that process considerations such as those below are considered for any procedures improvement program:

- Jobs that require procedures are identified
- Procedures reflect the way work is done and are technically accurate
- Procedures are written at the reading level of the users
- The detail in the procedure is not so complex that novice employees won't understand it, nor so limited that experienced employees will not read it
- Users can quickly locate the correct procedure for the job
- Users are instructed in the application of each procedure
- Procedures are reviewed and updated when the task, equipment or process changes

- Shared knowledge between operations, maintenance and engineering is routinely incorporated into procedures updates
- Users are advised when a procedure changes and instructed on the changes

3. A method for applying Human Factors to Process Hazard Analyses

The principle objective of the HAZOP is to identify process hazards, especially those that affect containment integrity (Attwood and Froehlich, 2003). Some of these hazards, however, are a result of human factors i.e. either with the process equipment itself or with instrumented process control. Many HAZOP protocols are less systematic than they should be for addressing HF concerns such as Human-Machine-Interface (**HMI**). They rely on the use of a checklist without recommending a process for applying it. Since, Human Factors is not the prime reason to perform a HAZOP review, the HAZOP leader will often elect not to review the checklist during the HAZOP proceedings especially during a HAZOP for major capital projects. More likely, the checklist will be reviewed at the end of the HAZOP study resulting in a less critical and systematic approach.

The objective of this tool is to assist the HAZOP team to identify Human Factors (HF) concerns associated with particular pieces of equipment while conducting the HAZOP.

The HAZOP is a sequential process -- a line by line and equipment by equipment review. It requires that each line and/or process system on the Process and Instrumentation Drawing (P&ID) be followed and checked and that the hazards associated with the equipment that is located at each node be identified and discussed. The HF issues need to be highlighted at these nodes as an **integral** part of the HAZOP discussion and not at the end of the HAZOP when the team has completed each P&ID.

The key to this tool is the use of "Screening Questions" for each of 12 pieces of equipment. The screening questions are a simple checklist that identifies the most important HF characteristics of the targeted equipment and /or system so the team can quickly decide whether a potential problem exists and what action to recommend. If a Human Factors concern is identified, it will be recorded on the HAZOP worksheet. It is possible that sufficient information and knowledge is available to resolve the issue during the HAZOP. However, the resolution of the issue will most likely happen outside the HAZOP.

The following equipment and HF issues are included in the screening lists:

1. Valves: sequential; critical; frequent
2. Blind and Blanks, weight, lifting; location; frequency; clearing, draining
3. Pumps and Compressors; maintaining, draining, start-up; shut-down
4. Field displays; critical; start-up; shut-down

5. Firefighting and deluge systems; critical; accessing; escaping; directing
6. Field instrumentation; rodding; clearing; critical
7. Sample points, exposure; PPE
8. Reactors/Driers, critical; switching/regenerating; clearing; entering; catalyst dumping; lifting
9. Vessels (including exchangers); cleaning; exposure; PPE; lifting; entering
10. Furnaces and Fired Heaters; light-off; decoking, flame controlling; switching; PPE; entering
11. Filters; switching; cleaning; clearing; PPE; weight; lifting; entering
12. Loading and Unloading Facilities; height; space; exposure; weight; clearing; disconnecting

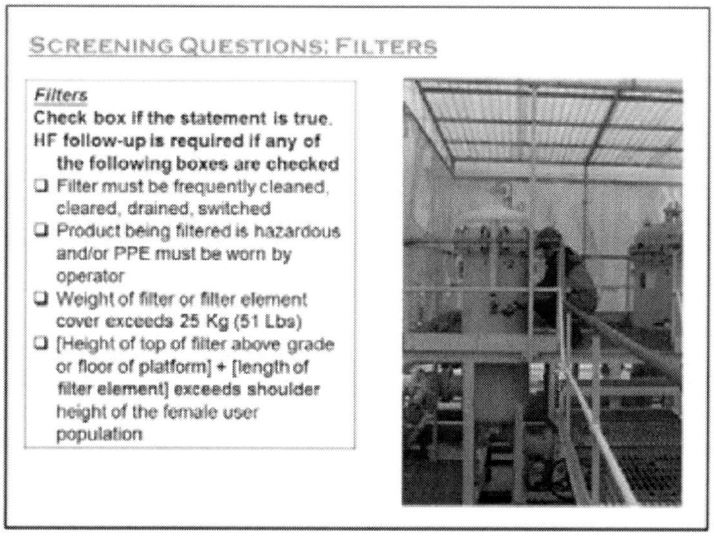

Figure 4: Screening Questions - Filters

Screening questions -- Example Format

An example of a "screening list" for Filters is given in Figure 4. This tool was prepared for the API and was published in 2005 (API, 2005). Our company has revised the original tool to make it easier to use. We have also developed a short introductory awareness program to show HAZOP participants how the tool will work in practice

Practical Benefits of this Application

The major benefit of this application is to ensure that Human Factors Issues are identified and discussed as they are encountered during the HAZOP reducing the probability of an incident.

4. Critical task identification and analysis.

Critical Task Identification Analysis (CTIA) is a systematic method of identifying critical tasks within a process facility, prioritizing their importance, analyzing those tasks that are considered most critical and identifying appropriate interventions to mitigate the risk. Critical in this context means high risk. The tasks that may have the potential for personal injury, equipment or facility loss, environmental release, community exposure, or business interruption are defined as "critical". These tasks represent a hazard and the goal is to identify the hazards and correct them before they turn into incidents.

The phases in the CTIA process are presented in Figure 5. The following paragraphs describe each phase, record examples of the products of each phase and provide insights on the most efficient methods to obtain information.

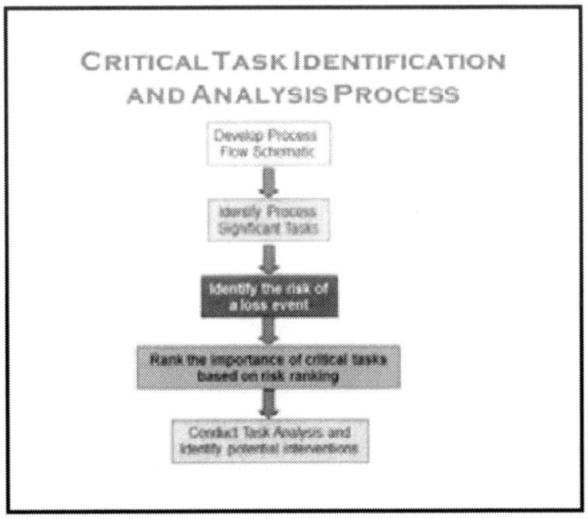

Figure 5: Phases of a CTIA

Develop Process Flow Diagram

The analysis starts by specifying the facility processes and equipment and creating a process flow diagram. A portion of an actual process plant flow diagram is shown in Figure 6.

Identify process significant tasks

The Critical Task Identification process is a 'brainstorming' technique that is designed to encourage the site team to identify the 'critical' tasks that they could be required to perform in each area of the process facility. A critical task could occur during a process upset or other abnormal situation or it could be a task that is difficult to perform or performed frequently. At each block of the PFS, the team is asked to identify the 'critical' tasks that he or she performs by placing them into one of seven different categories including:

1. Safety-critical tasks, e.g., responding to the failure of a pump flange
2. Quality-critical tasks, e.g., responding to an out-of-limits lab analysis
3. Production-critical tasks, e.g., responding to the loss of an exchanger which reduces output
4. Most frequently performed tasks, e.g., filter changes
5. Difficult to perform or complex tasks
6. Time critical tasks, e.g. responding to tasks that, in a short period of time, will cause a larger upset. An example is a compressor shutdown.
7. Environmentally-critical tasks, e.g., responding to an out-of-limit environmental deviation. For example, a high SO_2 reading on the stack of a fired heater.

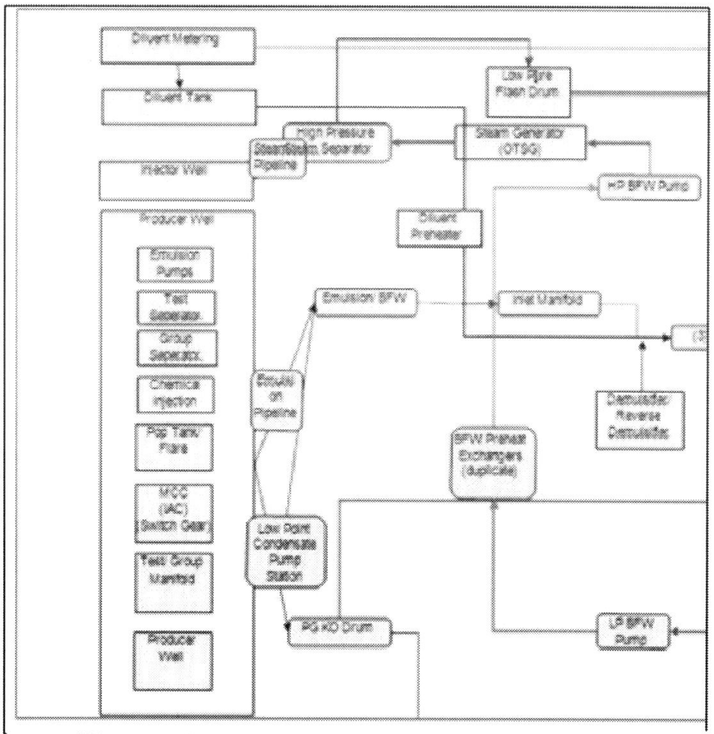

Figure 6: Partial Process Flow Diagram

In many cases the same 'critical' tasks can be categorized in different ways. The purpose of the exercise is to identify the major tasks (it is not important which list they appear in).

Identify the Risk of a Loss Event

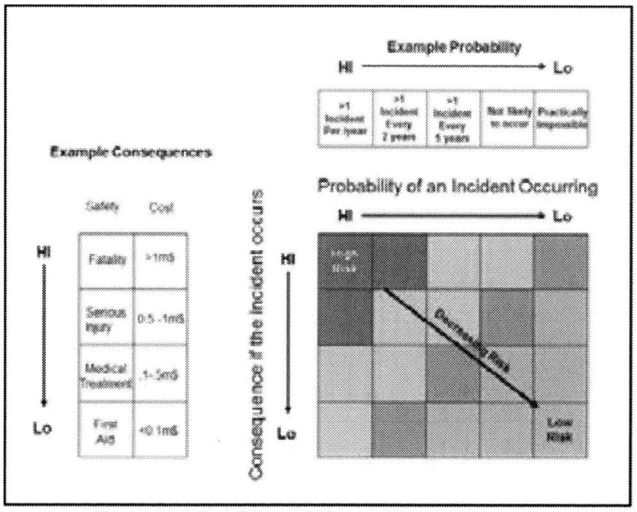

Figure 7: Simplified Risk Matrix

No company can afford to spend money and commit resources on issues that have little benefit to the operation. Thus, it is essential to identify the tasks which, when improved, will improve safety, reduce cost or improve the environment. The approach used at this stage is to estimate the risk associated with each task identified in the above exercise. Many companies have developed their own risk matrix. The next step in this exercise is to prioritize each 'critical' task in terms of its risk to the facility.

Figure 7 illustrates a generic risk matrix that is

designed to help workers prioritize tasks. It is presented for illustration purposes only and is not intended for use in formal risk assessments. Experience has shown that a site team can easily categorize tasks into the simplified consequence/probability risk matrix shown in Figure 7. Alternatively, the team can use the risk matrix developed for their company.

Identify the highest risk tasks

The tasks that most require analysis are those with the highest risk in Figure 7 (shown in RED). The analysis could also be extended to Medium risk tasks shown by the ORANGE sections and the lower risk tasks shown in the GREEN section of Figure 7

Conduct Task Analysis and Identify Potential Interventions

Finally, a task analysis is conducted on the highest risk tasks. Task analyses can take many forms and are explained in several texts and papers. The one that we use is detailed in CCPS (2010).

Table 1 provides, as an example, the activity analysis of a Feed Water Knockout Drum Sampling Task. Column 1 lists the sequential activities performed by the worker, column 2 lists the Human Factors Concerns for each human activity, column 3 lists the potential errors that can occur for each activity, and column 4 lists the potential solutions that can be used as task design recommendations.

Table 1: Activity Analysis for Feed Water Knock-out Drum Sampling

Activity	Ergonomic / Human Factors Concern	Error Possibility and Consequences	Potential Solution / Mitigation
Check sample drain tank for level by looking at overflow at sample position	Inability to check level	Tank overflows	Add level indication on sample drain tank with level controller in loop for tank pump
Approach sample station	Lighting Access	Trips/falls while carrying plastic bottles Poor lighting	Provide adequate lighting Bring sample station to ground Ensure proper head clearance
Ensure ventilation is working	Toxicity Air quality Ventilation area	Injury due to exposure (both inside and by ventilation)	Ensure ventilation is installed Provide safe ventilation area outside

Remove cap from sample point on common header	No human factors concern	Poor quality samples	Design issue
Displace and purge line	Force Fatigue Proximity	Arm strain Body stress	Ensure sample location and valve are located within visual proximity Ensure valve spring tension is low enough for human use Ensure handle is designed for constant pressure
Line up bottle with spout and take sample	Lighting Access Location of sample valves	Poor lighting Take wrong sample	Ensure labeling on valve easily corresponds to bottle (redundant coding with labels/colors) Mount lights outside of
Put bottle into carrier	Bending Lifting	Body stress Dropping samples	Provide space on sampling platform to place carrier
Select next bottle and repeat sampling process	Bending Lifting	Body stress Dropping samples	Provide space on sampling platform to place carrier
Close valve, recap	No new Human Factors	No new hazardous	No recommendations
Take samples to lab	No new Human Factors	No new hazardous	No recommendations

Practical Benefits of this Application

Identifies and analyzes tasks performed by humans that can impact on the overall integrity of the unit, the operating efficiency and reliability of the processes and can influence the safety of humans interacting with the process or facility

Task analysis techniques can be applied to existing tasks or tasks that do not yet exist. The task analysis team is made up of all stakeholders in the plant including engineering, safety and environmental personnel.

Summary and Conclusions

The effectiveness of a Human Factors program in a process plant is enhanced by the use of applications that can identify and solve human-related issues quickly and increase plant profits or reduce expenses or serve to reduce the number and severity of injuries.

The four applications described in this paper have proven, through previous use, to reduce maintenance costs, reduce the time required to launch projects, and make operations more efficient with lower incident rates. Readers who are interested in learning more about Human Factors analysis techniques are invited to read CCPS (2010) and Attwood et al. (2005).

References

1. ASD Simplified English Specification ASD-STE100, Aerospace and Defence Industries Association of Europe, 2005.

2. ATTWOOD, D.A., DEEB, J.M., and DANZ-REECE, M.D. "Ergonomic Solutions for the Process Industries". Gulf Professional Publishing (Elsevier), Burlington, MA, December 2005. ISBN 0-7506-7704-X

3. ATTWOOD, D.A. and TYS, A.T. (2013) "A method for ranking the musculoskeletal risk of operating valves in process industries." Proceedings of the Society of Petroleum Engineers, SPE-165690-MS, Pittsburgh, PA, August 2013

4. ATTWOOD, D.A., and FROEHLICH, B., "A Tool to Identify Human Factors Concerns during the HAZOP Process." ASSE Proceedings, Denver, June 2003

5. ATTWOOD, D.A. "Critical Task Analysis" Report prepared for the Surmont II project, Conoco Phillips Canada, Project No. 06-267, November 2007.

6. [1]CCPS "A Practical Approach to Hazard Identification for Operations and Maintenance Workers." Center for Chemical Process Safety Publication, John Wiley and Sons Inc., (Publishers), New York, 2010. ISBN 978-0-470-63524-7.

7. [2]CCPS "Human Factors Methods for Improving Performance in the Process Industries." Center for Chemical Process Safety (CCPS) publication. John Wiley and Sons Inc. (publishers), New York, 2007. ISBN 13 978-0-470-11754-5.

8. Haas, P. M. (1999). "Human Performance Engineering: A Practical Approach to Application of Human Factors." 1999 National Petrochemical Refiners Association (NRPA) National Safety Conference, Paper no. NSC-99-108.

9. HSE (2005) "HSE Human Factors Briefing Note Number 4: Procedures." United Kingdom, Health and Safety Executive.

[1] D. Attwood, contracted co-author
[2] D. Attwood, contracted co-author

479344 Selection of Cases for Atmospheric Dispersion Modeling

Abstract:

In many industrial facilities, toxic and/or flammable substances are discharged directly to the atmosphere. To ensure compliance with health, safety and environmental regulations, it is important to determine the potential consequences to the public and to personnel working in the facilities. Atmospheric dispersion modeling is a useful tool to provide plant owners the information to make decision in such cases. However, there are numerous potential releases from multiple scenarios, and hence it can be impractical to perform modelling for all cases. Typical industrial standards do not provide sufficient guidance for users about how to effectively choose the representative cases to model. This paper offers the guidance on selecting the relevant cases to perform atmospheric dispersion modeling based on the configuration of relief devices, the nature of discharged fluid, and points of interest in facility.

479160 Dow Laboratory Safety Academy: Improving Awareness of Reactive Chemicals Hazards and Preventing Incidents

Abstract:

University laboratory safety awareness and practices have improved steadily over the past twenty-five years since OSHA issued the Occupational Exposure to Hazardous Chemicals in Laboratories standard (29 CFR 1910.1450). However, developing and sustaining a positive safety culture in academic labs remains a challenge as demonstrated by the major incidents which continue to occur and result in significant injury and loss. In an effort to enhance laboratory safety, The Dow Chemical Company initiated pilot programs with Dow's strategic university partners at U.C. Santa Barbara, University of Minnesota, and Pennsylvania State University. Dow worked closely with graduate students, faculty, and staff to identify areas of improvement and help nurture a sustainable culture of laboratory safety. Content and feedback from the pilot programs were utilized to develop an online platform for safety resources and learning. Subsequently, Dow launched the Dow Lab Safety Academy (LSA) at http://safety.dow.com in a digital learning environment which shares Dow's best-in-class industrial lab safety culture and practices in a quick and accessible format. With over 40,000 visits to date, the LSA is reaching its goal to enhance awareness of safety practices in academic research laboratories and to promote a safety mindset in the future workforce of the chemical community. This paper will provide an overview of the LSA including its modules which provide information on methods for evaluating potential laboratory hazards as well as best practices for mitigating risks. It will also cover Dow's approach to identifying and quantifying potential reactive chemicals hazards, managing change, and sustaining its performance through reactive chemicals/laboratory hazard evaluation (RC/LHA) reviews. Finally, this paper will share best practices at Dow for building and sustaining a strong safe culture in the laboratory through effective mitigation of potential reactive chemicals risks. Our hope is that by broadly sharing what Dow does internally we can help improve the safety of researchers working with hazardous chemicals in any laboratory environment.

Safety and Asset Integrity Excellence – Using Data and Analytics to Control Composite Risk and Add Operating Value

Coen van Driel
Kienbaum Management Consultants
Coen.vandriel@kienbaum.de

Brian Dickerson
Kienbaum Management Consultants
Brian.dickerson@kienbaum.com

Robert Kauer
TÜV SÜD Industrie Service GmbH

Prepared for Presentation at
American Institute of Chemical Engineers
2017 Spring Meeting and 13th Global Congress on Process Safety
San Antonio, TX
March 26 – 29, 2017

Safety and Asset Integrity Excellence – Using Data and Analytics to Control Composite Risk and Add Operating Value

Coen van Driel
Kienbaum Management Consultants
Coen.vandriel@kienbaum.com

Brian Dickerson
Kienbaum Management Consultants

Robert Kauer
TÜV SÜD Industrie Service GmbH

Keywords: risk, asset integrity, process safety, occupational safety, operational performance, data, analytics, culture, transformation

Abstract

Process safety, occupational safety and asset integrity are typically dealt with as separate areas of operating practice and are addressed as such. This creates a limit to the effectiveness of each. All are elements of the same system, and as such should be accounted for in the same system. To change that, a progressive approach is required: a composite risk approach, applied specifically to safety and asset integrity. Viewing them as part of the same system allows for greater risk mitigation and added operating value.

Our transformation model operates on both a strategic and operative level, and maintains a balance between an organizational focus and personnel focus. Model elements are strategy, risk culture, management systems, KPIs and Cockpit, Data and Analytics, Safety and Asset Management, BBS and Training, Compliance, and Change Management.

People and machines create composite risk in a cybernetic (system at the intersection of technology and biological) way. Prescriptive analytics allows understanding of composite risk and proper intervention, which moves operations from cost avoidance to value creation, all while pushing past traditional safety plateaus.

1 Introduction

As the industry we are capable of realizing technically advanced plants and great innovation. But in the approach towards optimizing the performance in process safety as well as asset integrity, the industry tends to keep a silo mentality, approaching from an occupational safety perspective, a process safety (process safety and asset integrity do deal not only with safety but cover health, safety, environment and security issues with the primary goal to avoid loss of containment) perspective or an asset integrity/reliability perspective. Due to recent developments in the world like the IT advancement and cybercrime, security is a new dimension to safety. We believe, however, that these are one and the same. Viewing safety and asset integrity as elements of the same system is an innovative approach that not only manages composite risk but also adds significant operating value. However, these components are both being addressed separately already by many companies without success. To change that, an unconventional approach is required: risk based data and prescriptive analytics, applied specifically to safety and asset integrity.

The essence of why the chemical industry has survived is that the products are produced with good quality, with optimal margin and with minimum risk of an incident happening. People should be able to understand that and convert that into an integrative risk approach using all aspects of process safety, occupational safety and asset integrity (reliability). Striving for excellence in operational performance using a risk management approach is crucial for further success.

How can excellence in operation performance be achieved using a risk management approach? The methodology, seen in Figure 1.1, integrates the existing knowledge of the hazard analysis into the work processes and links it in to performance management. The advantage is that everybody in the organization gets to know their most relevant risk and have that integrated into their daily business. Secondly, the performance management is then built upon the risk meaning the indicators are related to the risk in the plant. The developments with information technology make it possible to use multiple sources of data and analytic tools to make the analysis from descriptive to more prescriptive.

To give a definition of prescriptive analytics: it not only anticipates what will happen and when it will happen, but also why it will happen. Further, prescriptive analytics suggests decision options on how to take advantage of a future opportunity or mitigate a future risk and shows the implication of each decision option[1].

Prescriptive analytics helps leadership at all levels to make decisions in an advanced way.

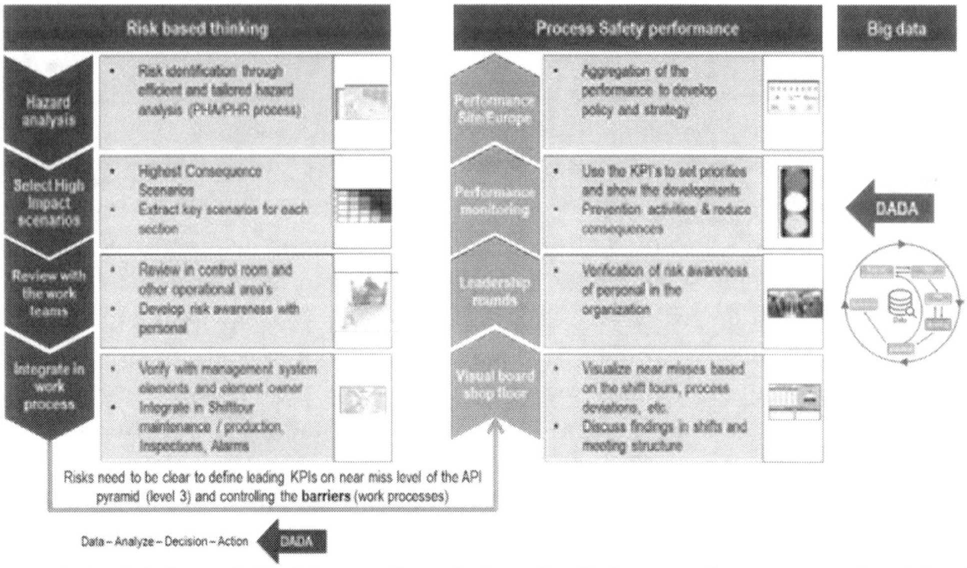

Figure 1.1: risk based thinking as foundation for Safety performance using big data

To that end, an innovative system of Safety and Asset Integrity Excellence is presented. This system addresses composite risk associated with operations using prescriptive analytics instead of lagging, reactive measures. Using risk mitigation strategies, robust data and analytics, key performance indicators, personnel training, workplace design elements, and asset management, organizations have the ability to shift risk mitigation to value creation.

As is known in the industry risk comes from multiple points and all should be proactively managed as major incidents rarely result from a single cause or failure. Rather, multiple failures coincide and collectively result in an exceptional event with severe, and potentially fatal, consequences. The model, which is based on the Swiss cheese model[2], in Figure 1.2, visualizes that when different barriers fail in an organization a major incident can happen. Unfortunately, there are many incidents in the industry where this has happened with catastrophic consequences.

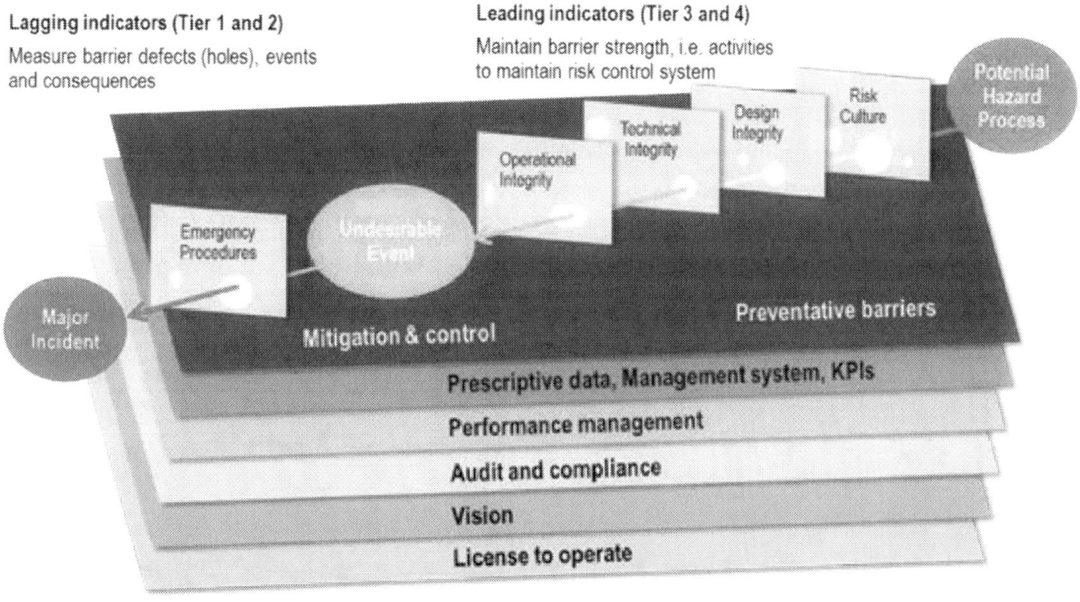

Figure 1.2: process safety control systems

Take, for instance, Three Mile Island. After a thorough investigation of the Three Mile Island event, authorities identified no single action or inaction as the cause. Rather, a complex and interlinked series of mechanical failures, human actions, engineering designs, operational implementations, and team interfaces occurred. The incident was the result of a combination of errors, along with a misaligned safety and asset integrity culture. This incident and others can be avoided by correctly utilizing data and prescriptive analytics. When applied effectively, today's technologies give the ability to gather massive amounts of actionable data, providing the information needed to mitigate this composite risk.

On the one hand, we as an industry are capable of innovating our equipment, plants and hardware, but on the other hand major incidents keep on occurring. Somehow, we as an industry continually run into a problem where we are forced to innovate and change due to major incidents like Three Mile Island. The unconventional approach combined with the risk-based performance management is an innovative step to make an integrative performance improvement possible.

2 Process Safety vs Occupational Safety vs Asset Integrity and Reliability

Process safety, occupational safety and asset integrity are typically seen as separate areas of operating practice and are addressed as such as illustrated in Figure 1.3. A driving factor behind this is the view people take of each, and the differences in consequences for failure and their likelihood. Generally speaking, occupational safety/OSHA incidents have a high likelihood and usually a lower consequence. This is typically the "slips, trips and falls" area. With programs like the 12 Life-Saving Rules[3], driving program, and BBS (behavioral based safety) you create awareness that those incidents are kept to the minimum.

There is also the typical reliability (asset integrity) related area where companies fight to lower the impact of frequent failures on equipment like pumps, compressors, turbines, etc. Here, the target is to either have redundancies or to ensure high reliable operation.

The challenge for process safety is that it is often the blind spot of management and the organization because of the fact that the likelihood of the incidents is fortunately low and it is, therefore, difficult to handle proactively using traditional tools and procedures. It is easier to identify people not wearing helmets than to identify weak points in thousands of kilometers of piping systems. On the other hand, if there is a leak, the consequences might be high on Business and Safety.

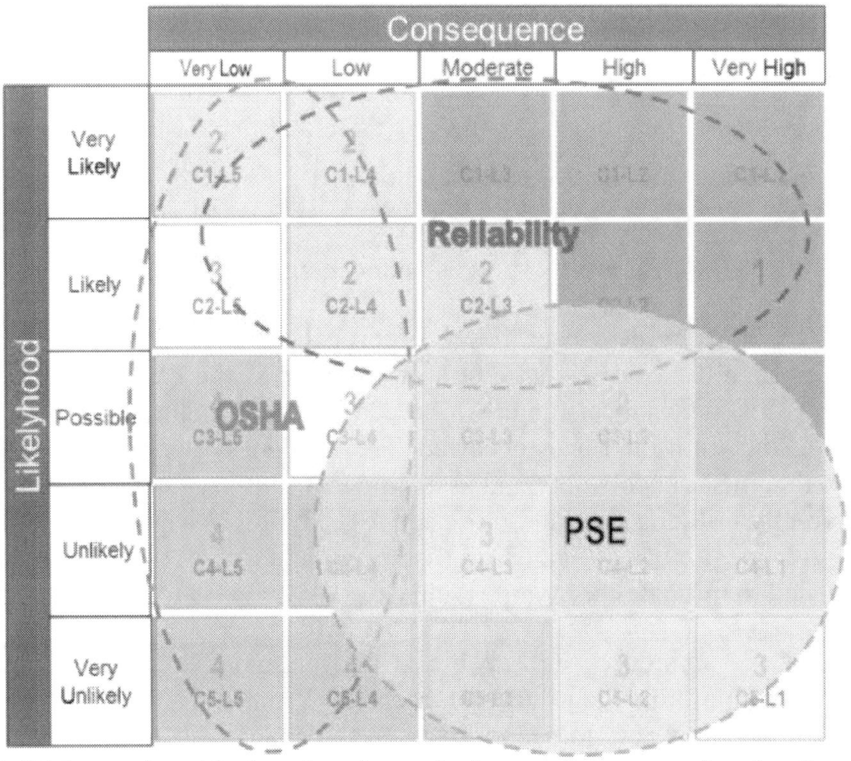

Figure 1.3: Risk matrix with showing the typical areas for occupational safety (OSHA), reliability programs (Reliability) and process safety (PSE)

Another reason is the business model of organization. We tend to organize according to functions with separate departments looking at maintaining each function; even within departments personnel are compartmentalized with small, specific tasks.

A third reason is the compliance-driven behavior within the industry. People look after what they are responsible for and stop there. A larger framework, encouraging people to consider larger-risk influences, is missing which people can identify with when looking after behavior. This separate focus results in increased risk levels, reduced reliability, and increased costs.

There are many reasons why the industry keeps on looking at the three aspects (safety, reliability/ asset integrity and process safety) in a silo way. The "Risk at the Core of the Organization" model, developed by TÜV SÜD/ Kienbaum, enables a focus on all aspects which results into a high performance.

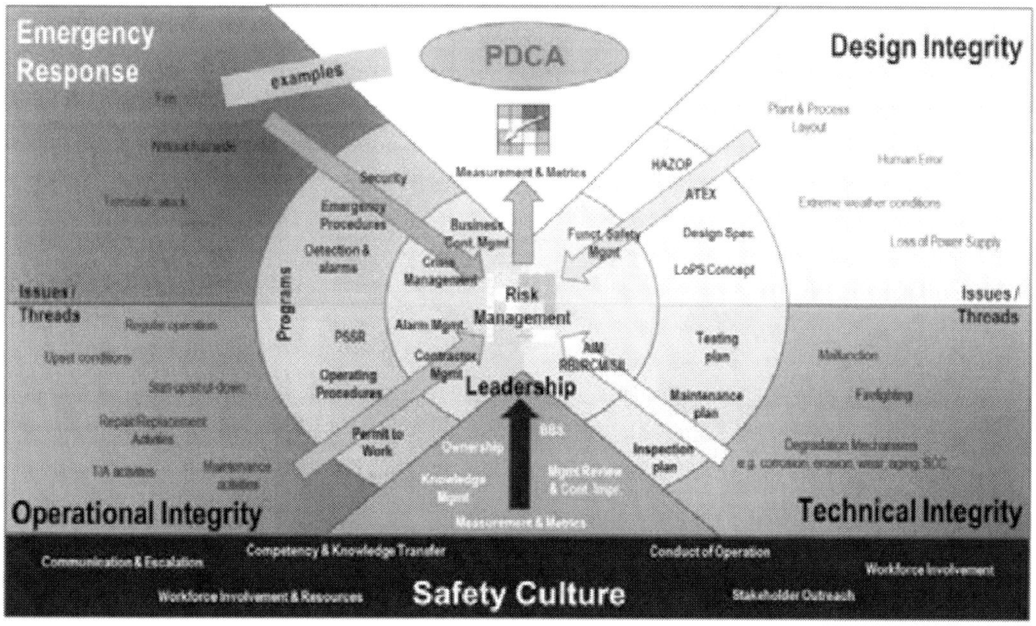

Figure 1.4: Risk management at the core of the organization

The three aspects, occupational safety, reliability/asset integrity and process safety can be managed at the same time with a risk management approach at the core of the organization. In the integrative system (Figure 1.4), there is a strong base of a risk-based safety culture supporting the different processes of design integrity, technical integrity, operational integrity and emergency response. Those processes handle the issues and threats for safety, process safety, and asset integrity using the risk approach. It enables one focus in the organization on continuously managing the issues and threats on the likelihood and consequences, not only the ones that are easily in the view, but also the unlikely ones with high consequence. The management system, discussed in more detail in Section 3.3, based on the Plan – Do – Check – Act (PDCA) circle and using Measurement & Metrics (KPIs) enables a high performance for all three aspects. Viewing them as part of the same system allows for greater risk mitigation and added operating value. In the next section an in-depth description is given on how to realize this integrative way of looking at occupational safety, process safety, and asset integrity.

3 Transformation Approach

All safety and asset integrity excellence elements must be aligned and working together to ensure risk mitigation and value creation. The approach (Figure 1.5) sees an equal weight on the organization of a company with its personnel, split evenly among strategic and operative elements.

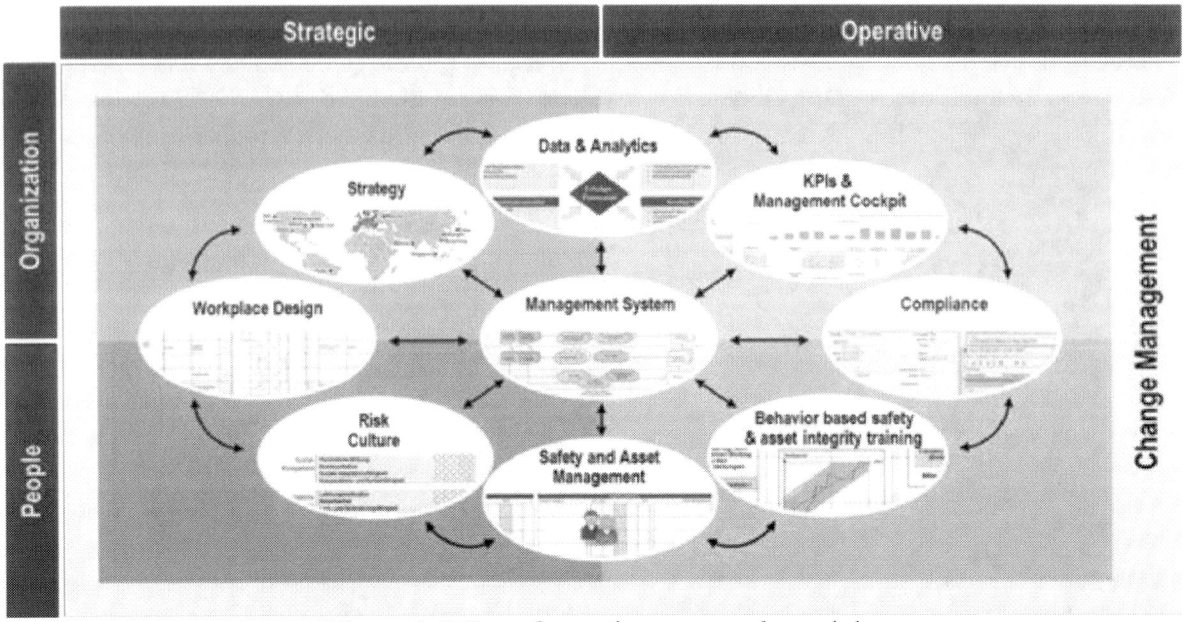

Figure 1.5 Transformation approach model

3.1 Strategy

The first step towards achieving world class Safety and Asset Integrity Excellence is a strategic and systematic composite risk strategy. Assessing composite risk of your organization should be part of the strategic planning process. When components of risk are realized and understood, they can be measured, mitigated, and next steps developed. That is why human, technical, security and environmental issues must be placed in the center of Safety and Asset Integrity Excellence initiatives.

Employees must understand the various components of risk, and feel enabled to perform tasks with a clearly defined and communicated vision of safety and asset integrity excellence. By doing so, employees perform safely when leaders actively & consistently demonstrate that safety is important through their decisions and actions. Similarly, assets perform efficiently when they are proactively managed for risk and prescribed actions implemented for incident avoidance. When organizations implement a strategy of robust risk avoidance, they add value by mitigating composite risk in a data-driven, continuously improving, compliant and holistic way.

3.2 Risk Culture

Risk culture is a culture in which employees are aware of the risks, they know the hazard scenarios for their area and leadership at all levels challenges the employees to search for blind spots in safety. Far too many work incidents continue to take place that could have been avoided with preventable measures in place. Pressure on the organization can have severe consequences. We see that management wants to convey an attitude of 'safety first', but in daily practice this is only of limited help for the front line staff: after all, they have to consider if it is safe enough to perform their jobs many times a day. The mantra 'safety first' is not always helpful for them nor is it a basis for a fruitful discussion between blue collar staff and management.

By implementing a risk culture where employees proactively seek out risk, accidents can be mitigated, or prevented all together. This leads to a safer environment, less monetary damage, while at the same time, empowering employees to feel part of a bigger movement. Perhaps, the even larger benefit is the enablement employees will achieve as they continue to grow and further learn to mitigate risk in a safe and enjoyable work environment. When a proactive culture of risk mitigation and value creation is pursued, the entire organization benefits. It is crucial for organizations to create a company culture where the focus is on "Value Creation" instead of value protection, along with a culture of continuous learning, where all feel responsible for value creation and risk mitigation.

3.3 Management System

An efficient management system is one of the foundations for the successful steering of sustainable business processes.

Management systems serve to link 7 key elements of an organization:

1. Organization
2. RACI
3. Documentation
4. Meetings
5. Reports/Dashboards
6. Indicators
7. Processes

The system supports the direct decision making and action capacity of the company. To this end, the management system must incorporate these 7 elements, and also ensures the specific safety and asset integrity requirements are met. An effective KPI system needs to be established, and this system needs to trigger root cause problem solving and develop remedial actions based on the established KPI system.

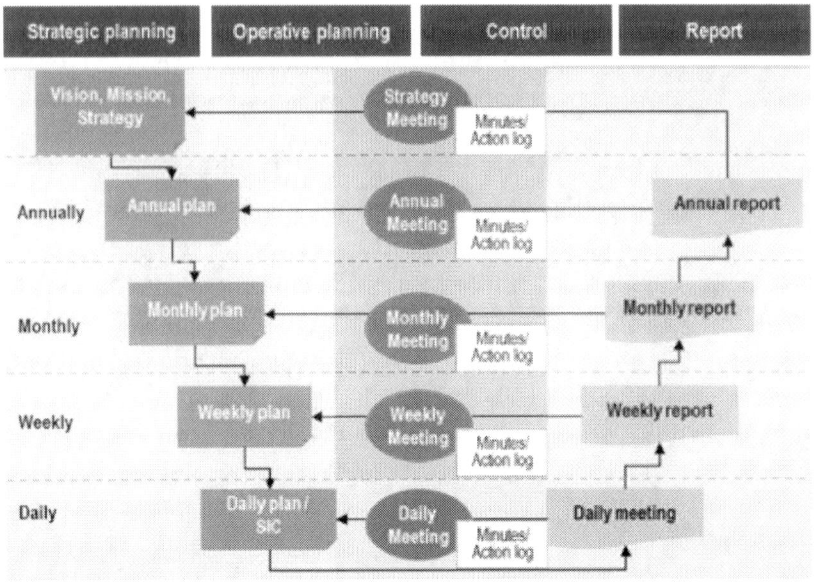

Figure 1.6 Management control system

A well working management system (Figure 1.6) with coordinated elements (KPIs, meetings, reports) stimulates root-cause analyses, following through on decisions, tracking their effectiveness and further development, which adds value while mitigating risk.

3.4 KPIs and Dashboard

When it comes to safety, real-time information and visibility of risk barriers cannot be seen fast enough. A real-time dashboard of KPIs allows for quick decision making and allows further investigation of root-cause analyses of incidents and non-incidents. This allows for proactive elimination of risk and further understanding of non-incidents. Oftentimes, when the right people have visibility of warning signals, major incidents can be prevented. In addition, these signals can reduces the frequency and severity of system failures and their impacts. By implementing and utilizing a real-time risk dashboard, incidents can be reduced and near-misses avoided, leaving organizations to improve their overall safety and profitability.

3.5 Data and Analytics

Data and analytics provide the integral link that holds the entire model together. Before data can be used however, it needs to be collected and aggregated from all available sources. Every source of data should be used, as long as it is proven to be good. If data sources aren't reliable, they should be improved and addressed proactively. Once done, multiple data sets can be transformed into a single data set and allow for algorithms to glean new insights from previously separate data sources.

The idea is to achieve a state of "data and analytics excellence." To do this, there are 5 levels (Figure 1.7) that must be achieved:

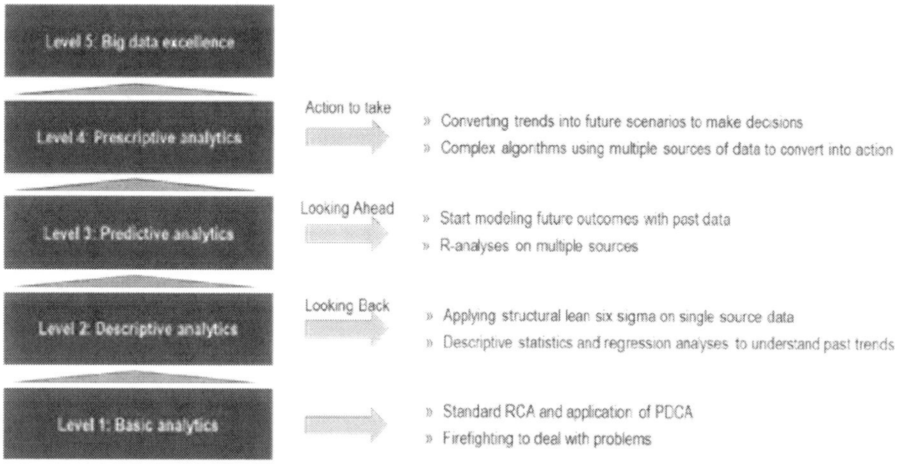

Figure 1.7 The 5 levels of data and analytics excellence

After the data is gathered, a model can be created. This model is updated in real time and gives real time output on risk levels for both personnel and assets. The model output is a real time monitoring of incident risk levels, and it is used to offer mitigation strategies to reduce the likelihood of an incident (Figure 1.8).

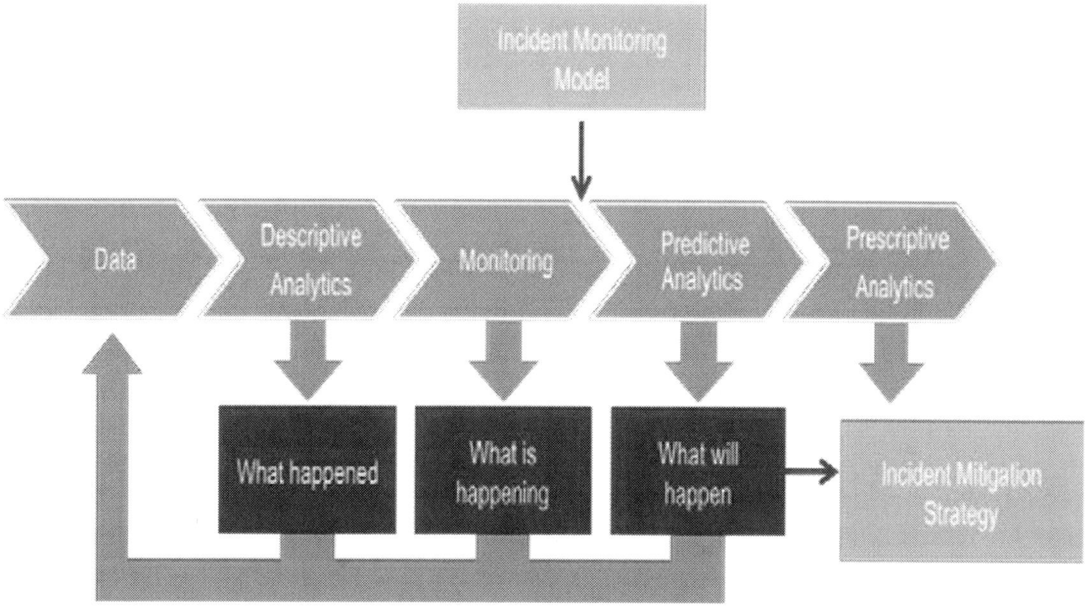

Figure 1.8 Risk mitigation process

3.6 Safety and Asset Management

Safety risks can be identified in workplace designing, and behaviors can be addressed through BBS programs. But, after data is gathered and an analytical model identifies risk and prescribes actions, risk mitigation plans need to be created and executed.

By using analytics, models that take into account dozens of factors can be created to address an issue such as fatigue. Take for example an employee named Thomas from a shipping and logistics industry. Thomas has been with his company for 22 years, working in the northeast United States. On a cold Monday afternoon in Vermont in the middle of January, Thomas has been working for 6 hours. His workload has been extra given the weekend load that came in, and he has already lifted and moved hundreds of pounds. Given his tenure, workload in the past month, week, and even hour, as well as the time/date/location, and dozens of other factors, Thomas is 46% more likely to get hurt in the next 30 minutes. Given all of this information, a pre-defined plan to mitigate the risk can be deployed. This plan includes, for example, actions such as taking mandated breaks, using special equipment and gear designed for these situations, and altering work patterns to avoid the incident.

From an asset standpoint, a similar example exists. Take already available maintenance data in combination with predictive analytics to forecast equipment failures instead of evaluating trends. Current maintenance processes are manual and non-digital, making them prone to errors, while fixed time or planned maintenance is insufficient. Reliability and availability can both be improved with anlaytics. Current TPM (total productive maintenance) activities can go farther to reduce risks and add value to assets with analytics.Current programs use data from all sources such as smart sensors, existing TPM data, and others. With this data, companies can estimate the remaining life of the asset and when a replacement is due, all while scheduling regular preventive maintenance and maintaining the right inventory levels for spare parts. Analytics can go further to aggregate all data, and combine it with other, real-time data such as weather conditions, time, and dozens of other available data sources into one source to use all data to analyze and make new, previously unavailable connections. Then, models compare current, real-time states to normal or ideal states and use advanced algorithms to detect small variances for each asset, which can warn of problems that would have otherwise been missed without aggregated data and associated algorithms. When a problem is detected, operators, supervisors, and maintenance & inspection personnel are notified at the optimal times, mitigating the risk of collateral damage and secondary failure and preventing significant production downtime, maximizing equipment life, improving reliability, and increasing availability.

Take Thomas again. According to the maintenance schedule, his truck is due to have new tires put on the vehicle in two weeks. However, it is the middle of winter in the northeast and a big storm is coming in. Accounting for all of this information, the analytics say to replace the tires early and prevent the potential equipment malfunction. By accounting for not only the past data on tire life, but also the data on weather conditions on equipment, among other factors, an incident is avoided that would have otherwise occurred.

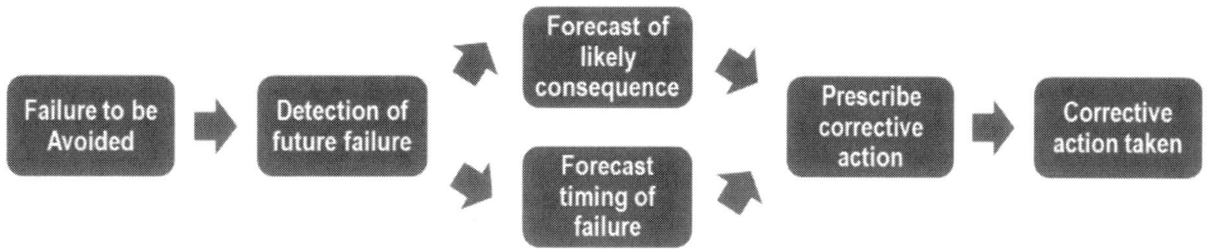

Figure 1.9 Incident avoidance

The real power in the model comes from the aggregation of the data. Once again, consider Thomas, who is fatigued and working in poor weather. However, he has not reached a set threshold for any warning or indication. Also, Thomas's truck is within limits for its condition, but close to needing maintenance. By looking at these data independently, no indication is given of any increased risk or minimum risk level, and an incident is likely (Figure 1.9).

By unifying efforts and taking a composite risk approach, different outcomes occur. A unified system took all this information and gave an alert of a potential situation. The prescriptive actions are prescribed, and preventative measures taken. A composite risk view allows for greater detection of avoidance of incidents.

3.7 Workplace Design

Organizations must put measures into place that optimize worker well-being, while minimizing stress on machines due to controllable factors. For example, by implementing ergonomic workstations, employees will have better posture, less exertion, and fewer motions. The workstation becomes more efficient and the worker more productive. Outside the office area, other crucial elements such as ventilation, machine safeguarding, materials handling, guardrails, or anti-slip floor treatments can be addressed for accident prevention. For example, involve individuals such as engineering, management, operators, and maintenance for ideas or input on safer equipment design and procurement to create more efficient equipment, fewer errors, and easier operation and maintenance. Employees' contribution will make them feel part of the larger picture while helping create a safer workplace for all. Aside from a safer and more enjoyable work area, proper workplace design measures can reduce worker's compensation and health care costs for the organization, including staggering indirect costs associated with injuries. Having all the necessary tools that enable workers to perform their best, improves quality of their work. When a worker feels fatigued, they are less likely to perform their best. This could lead to poor quality in their work and have other, potentially fatal, consequences. When organizations make an effort to ensure safety and health of employees, it does not go unnoticed. It adds value to the overall employee engagement and involvement, while making staff feel empowered to contribute to a culture of safety excellence. Healthy employees are a valuable asset as they create and foster a culture of health and safety, leading to better performance.

3.8 BBS and Asset Integrity Training

Training on behavior based safety (BBS) and asset integrity gives employees the knowledge and skills they need to protect equipment, facilities and, each other. It prepares them to deal with all aspects of the job in order to properly handle any potential risk scenarios. For example, organizations can internalize risk and hazard avoidance strategies and proper use of PPE, while ensuring properly and efficiently functioning assets through training and continued learning. Comprehensive data-driven BBS programs focus on both human and machine factors. They emphasize equipment reliability and integrity by addressing human factors together with smart machines that send warning signals alerting of potential risks. When organizations design and implement a well-supported and managed BBS system and asset training program, they help their employees engage in safe behaviors. This provides opportunities for risk avoidance and improvement in safety for the entire organization, while creating a culture of reinforcement and communication. All employees must be involved, from senior management to front line workers, to achieve a successful BBS and Asset Integrity program.

3.9 Compliance

When organizations implement industry best practices, multiple sources of risk can be minimized significantly reduced. By enforcing compliance with internal and legal guidelines, regulations, and policies with safety and equipment while recognizing best practices, organizations are headed in the direction of a safety culture. At a minimum, organizations need functional standards in place that have been agreed upon by engineering, EH&S, operations, and asset integrity. Lastly, organizations must enforce safe practices from vendors to prevent potential industrial accidents that could disrupt supply chain operations. A key component is disciple. Discipline is following the processes with the possibility to improve performance as the base for compliance. Too much focus on compliance makes it so that people execute the tasks, but stop short or truly thinking, increasing the incident risk.

3.10 Change Management

Shifting to an effective risk mitigation system requires a robust change management program, which helps establish an effective communication strategy and accompanies the change processes on all levels while creating a proactive culture in the organization. Oftentimes, organizations are not ready for change. Change Management acts as an enabler (Figure 1.10), in that it provides a mechanism by which the organization can quickly adapt and respond to changing conditions. Effective Change Management mitigates risk by providing the organization with an adapted workforce and expertise to properly balance the risk and reward change.

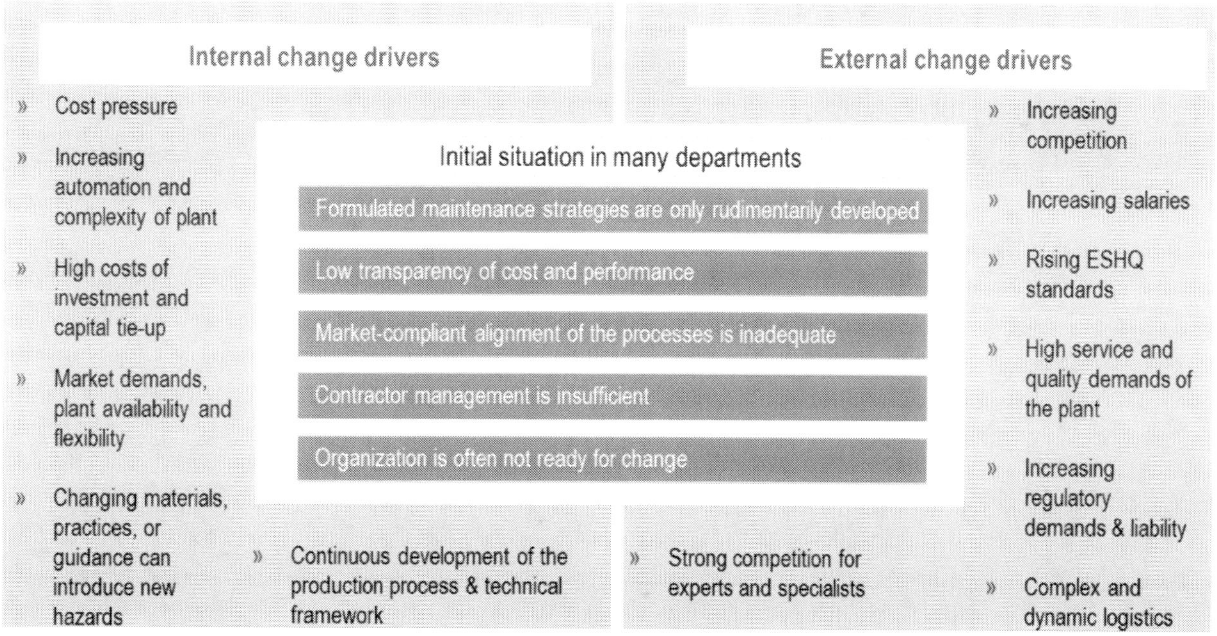

Figure 1.10 Change management drivers

4 Conclusion

The interaction between humans and complex technical systems is a complicated one, but is critical to ensure safe operation and business value. Viewing these as separate elements allows for cost avoidance but stops there, without holistically addressing the system. It is necessary to challenge all processes by looking at all aspects - occupational safety, reliability and process safety – in a holistic way to ensure economic and safe business. The Safety and Asset Integrity Excellence system is an innovative approach that effectively mitigates composite risks using prescriptive analytics, addressing technical, organizational, processes and people related issues allowing organizations to create value in occupational safety, reliability, process safety and asset management. People and machines create composite risk in a cybernetic way. Prescriptive analytics creates understanding of composite risk and proper intervention, which moves operations from cost avoidance to value creation, all while pushing past traditional safety plateaus.

5 References

[1] https://en.wikipedia.org/wiki/Prescriptive_analytics

[2] Dante Orlandella and James T. Reason of the University of Manchester,

[3] Shell: http://www.shell.com/sustainability/safety/personal-safety.html

What if PHAs Included an Analysis of Human Error?

Jeanne M. Curcio
Houston, TX
jmc96@comcast.net

Prepared for Presentation at
American Institute of Chemical Engineers
2017 Spring Meeting and 13th Global Congress on Process Safety
San Antonio, TX
March 26 – 29, 2017

What if PHAs Included an Analysis of Human Error?

Jeanne M. Curcio
Houston, TX
jmc96@comcast.net

Keywords: human error, human factors, PHA, guideword, execution failure, planning failure

Abstract

Process Hazard Analysis (PHA) studies such as What-If, HAZID, and HAZOP are used to evaluate potential risks to personnel, equipment, and facilities in proposed designs in the offshore and onshore oil and gas industry. These studies commonly include a fixed set of guidewords and parameters that identify when human error is a possible cause of process deviations. Further analysis of errors is not typically undertaken during these studies, however. Inputs from a previously conducted HAZOP study were modified to include an analysis of the types and causes of anticipated human error through the addition of specific human factors guidewords. Evaluating the probable cause and type of human error associated with various operator actions or inactions assists in determining when further review through means of a task analysis is advised.

1 Introduction

According to BSEE's Technology Assessment Programs (TAP) Program, "Human error accounts for the vast majority (60 to 80 percent) of failures of marine systems". [REF 1] The UK Health and Safety Executive states that "Human failures are responsible for up to 80% of all types of accidents and figure in almost every major incident." [REF 2] The following list provides typical examples of "human error" cited as the cause of accidents in the oil and gas and nuclear industries:

- Investigators have blamed human error for the oil pipeline explosion that killed at least 55 people in Qingdao [REF 3].
- Death at Pennsylvania gas well blamed on human error, poor supervision [REF 4].
- Air Force officials say human error to blame for damaged nuclear missile [REF 5].
- 'Human Error' blamed for fatal Chevron blast [REF 6].
- Human error blamed for small oil spill at Yokosuka [REF 7].
- Human error is blamed in Bay Area oil spill [REF 8].
- Report Cites Human Error in 2010 Gulf Oil Spill [REF 9].
- Coast Guard: Oil spill probably due to 'human error' [REF 10].

The 2005 accident at the BP Texas City Refinery included the human factors deficiencies of lack of control of worker fatigue, poor human-system-interface design, poor communication by radio/phone, out-of-date and inaccurate operating procedures, and poor (no) communication between workers at shift handover. The settlement agreement between the U.S. Occupational Safety and Health Administration (OSHA) included the following statement [REF 11]:

> BP Products shall retain one or more firms with expertise in the field of process safety management, including pressure relief systems, safety instrumented systems, **human factor analysis** *[emphasis added]* and performing process safety audits and PSM system audits ("PSM Expert") to conduct a comprehensive audit and analysis of the PSM Systems at BP Products' Refinery and assess the robustness of the PSM systems...
>
> An additional requirement is that the safety audit shall be "...comprehensive, conducted in accordance with best practices in the industry, and shall include, but is not limited to... **human factor analysis** *[emphasis added]."*

While all human errors, similar to mechanical errors and equipment breakdowns, cannot be completely eliminated, design approaches and techniques do exist that can reduce the likelihood of these errors PHA studies are useful in identifying potential causes of process deviations that may occur but are not currently structured to address human errors in detail. The purpose of this paper is to present a method that can be added to

PHAs to identify causes of human error and supplement other human factors task analyses.

For purposes of this discussion, the following definitions will be used:

- **Human factors** (also referred to as "ergonomics") is defined by the International Ergonomics Association as the "scientific discipline concerned with the understanding of interactions among humans and other elements of a system, and the profession that applies theory, principles and methods to design in order to optimize human well-being and overall system performance". [REF 12]
- A **Process Hazard Analysis (PHA)** study is an organized and systematic effort to identify and analyze the significance of potential hazards associated with the processing or handling of highly hazardous chemicals. [REF 13]
- **Human error** means that something has been done that was "not intended by the actor; not desired by a set of rules or an external observer; or that led the task or system outside its acceptable limits".[REF 14]

2 Process Hazard Analysis Studies

2.1 Purpose and Types

"Hazard Identification and Risk Analysis (HIRA) is a collective term that encompasses all activities involved in identifying hazards and evaluating risk at facilities, throughout their life cycle, to make certain that risks to employees, the public, or the environment are consistently controlled within the organization's risk tolerance… Hazard analysis identifies possible hazard scenarios for a process or facility." [REF 15] These scenarios commonly include unintended, unplanned, or unexpected "behaviors" or occurrences in a process that could lead to consequences affecting personnel, equipment, or the facility.

Several Process Hazard Analysis (PHA) methods are available for use to meet the needs of the variety of industries, facilities, locations, and scenarios that could require or benefit from a PHA study. The most well-known or frequently used or these are Checklist reviews, What-If analyses, a combination or Checklist and What-If methods, Hazard Identification (HAZID) studies, Hazard and Operability (HAZOP) studies, Failure Mode and Effect Analysis (FMEA), Layers of Protection Analysis (LOPA), and Fault Tree Analysis [REF 16].

Another methodology known as a "Human-HAZOP" or a "procedural HAZOP" involves an evaluation of operating and maintenance procedures that are considered "safety critical", i.e., those 'with the potential to cause or limit the escalation of major accident hazards (MAH)'. The evaluation includes a Hierarchical Task Analysis, which details each step of the procedures; identifies potential human failures; assesses the consequences of failures; assesses the possibilities for recovery; assesses risk reduction measures; and recommends improvements to prevent human failures [REF 17]. A"Human-HAZOP" is a thorough analysis similar to a process HAZOP because of its use of guide words and defined nodes. Conducting a "Human-HAZOP" is suggested for the review of steps completed or omitted during "safety critical" activities determined during the design phase of a project or those in existing operations [REF 17].

2.2 *PHA Guide Words*

While use of the PHA methods listed previously is a standard component of process and facility risk evaluations, analysis of "behaviors" typically is limited to questions of why equipment does not perform as intended and expected.

The Safety and Environmental Management Systems (SEMS), API RP 75 R2013, and OSHA's Process Safety Management (PSM) standards 29 CFR 1910.119 and 40 CFR 68.67 recommend that human factors be evaluated during a PHA. However, discussions of human factors are commonly limited to citing "human error" or "operator error" during PHA studies as causes of potential hazards. Further discussion is limited to a brief mention of "training" or "operating procedures" as recommendations to address the hazards. The OGP "Human Factors Engineering in Projects" [REF 18] provides recommendations for including human factors considerations in HAZOPs and suggests using standard guide words during discussions, each followed by input from the human factors representative regarding issues affecting operations and maintenance personnel.

Examples of fixed guide words and process parameters commonly used in PHAs studies are provided in Table 1 [REF 19]. These terms are used to evaluate effects and causes of chemical, mechanical, and physical changes in the process but cannot be used to evaluate "human error" or to determine the cause of the error. A root cause analysis to identify the type of error is needed.

Table 1. PHA Guide Words and Parameters

		PARAMETER						
		More	**Less**	**None**	**Reverse**	**As well as**	**Part of**	**Other than**
GUIDE WORD	**Flow**							
	Pressure							
	Temperature							
	Level							
	Time							
	Start-up / Shut-down							

3 Human Error

3.1 Categories

Much research into human error has been conducted in the past, and this paper does not attempt to repeat results from those studies. The current discussion is a review of J. Rasmussen's work on HMI and levels of human performance [REF 20]. Human error can be divided into the two main categories of "execution failures" and "planning failures".

Execution failures are classified as either "slips" or "lapses; planning failures are mistakes, either "rule-based" or "knowledge-based". Because the term "slip" is commonly used in process facilities to refer to someone losing his or her footing on a substance on the floor, ground, or deck, the phrase "incorrect action" (IA) will be used. Similarly, the term lapse is commonly used in the phrase "lapse of memory" and may limit understanding of its use in reference to omissions of actions. Thus, the phrase "omitted action" (OA) will be used.

IAs occur when appropriate actions are completed incorrectly and can be attributed to problems with attention (e.g., perceptual confusion, interference, misordering, or mistiming). OAs result from problems with memory, such as when an action is forgotten, omitted, or not completed.

Planning failures are divided into the two categories of "rule-based mistakes" or "knowledge-based mistakes. [REF 21] In this discussion, these terms will be replaced with "procedure mistakes" (PM) and "fact mistakes" (FM), respectively. PMs have one of two causes, either the "misapplication of a good rule" or the "application of a bad rule". FMs involve planning errors that involve limited information about the current conditions or circumstances and result in incorrect decisions or consequences. (Violations are an additional category of mistakes but are not included in the current discussion.) Figure 1 provides a hierarchical tree view of the types of human error.

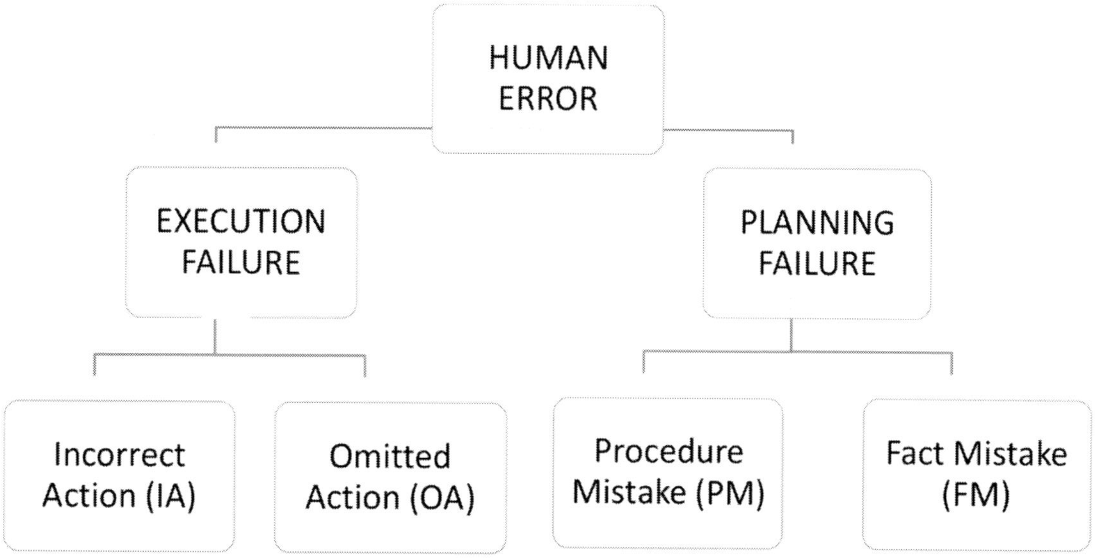

Figure 1. Types of Human Error

Adapted from *http://www.skybrary.aero/index.php/Human_Error_Types*

3.2 "Human Error" as a guide word

Because the focus of PHA studies is commonly on issues with equipment quality and design and because planning and execution failures are not addressed completely in PHA studies, the root causes of "human errors" that may occur can be further evaluated in PHAs by including guide words relevant to the possible types of errors. To assist in identifying these causes, the classifications listed in Table 2 are adapted from the "Inspectors Toolkit: Human Factors in the management of major accident hazards", developed by the Health and Safety Executive. [REF 22]

Table 2. Execution and Planning Failure Classifications [REF 21]

EXECUTION FAILURES	
IAs	**OAs**
Operation too long/short	Operation omitted
Operation mistimed	Check omitted
Operation in wrong direction	Information not obtained
Operation too little/too much	Information not communicated
Operation too fast/too slow	Selection omitted
Misalignment	Plan omitted
Operation incomplete	
Operation too early/too late	
Check incomplete	
Check too early/too late	
Wrong information obtained	
Information retrieval incomplete	
Wrong information communicated	
Information communication incomplete	
Wrong selection made	
Plan incorrect	

PLANNING FAILURES	
PMs	**FMs**
Right operation on wrong object	Information inadequate
Wrong operation on right object	Information interpreted incorrectly
Right check on wrong object	Information communication unclear
Wrong check on right object	

3.3 Example PHA analysis with human factors guide words

Table 3 includes some of the column headings provided in a typical PHA worksheet. Risk rankings, while not shown below, can be assigned to action items as determined by the PHA team.

The example operator errors listed in Table 3 are representative of those which have been identified during actual PHA studies of process facilities.

Table 3. Example PHA Analysis of Operator Error

Guide Word	Deviation	Cause	Consequence	Safeguard	Recommendation
		Operator error	Catalyst spill		

Operator Error	Failure Type	Failure Classification	Recommendation
	Execution	IA: Misalignment (of catalyst drum)	Ensure adequate space for workers to complete tasks. Ensure safe positioning of catalyst drum.

Guide Word	Deviation	Cause	Consequence	Safeguard	Recommendation
		Operator error	Dropped drum cylinder		

Operator Error	Failure Type	Failure Classification	Recommendation
	Execution	IA: Operation incomplete IA: Plan incorrect	Review personnel lifting limits and consider use of smaller drums. Review training documentation. Consider revision of operating procedure.

4 Discussion

4.1 Benefits

Use of the proposed analysis of "operator error" for analysis during PHA studies will result in reduced time and cost and may eliminate the need for the human factors professional(s) to conduct a separate review. Incorporating error analysis during PHA studies provides the additional benefit of further identification of potential operator errors by other engineering disciplines or Subject Matter Experts (SME) present during the HAZOP. Additionally, input from Operations personnel will be immediate rather than if a separate human factors task analysis is conducted, when these individuals may not be available.

4.2 Conclusions

Too often during PHA studies, a cause of a deviation is listed as "human error", but no further analysis is conducted. "Human error" is often used as a reason when the root cause is not identified. Modifications can be made for mechanical, structural, or electrical causes, but "human error" is not typically addressed in depth during discussions of recommendations to mitigate the deviation.

The answer to the question posed in the title of this document will be seen in the benefits of increased involvement of the PHA team in determining the causes of errors, reduced time and cost required for separate evaluations facilitated by human factors professionals, and an increase in information for further human factors task analyses.

5 References

[1] R. G. Bea and W. H. Moore, Management of Human Error in Operations of Offshore Facilities, Available at https://bsee.gov/research-record/tap-167-management-human-error-operations-offshore-facilities Accessed February 2017.

[2] HSE Human Factors Briefing Note No. 1: Introducing Human Factors, Available at http://www.hse.gov.uk/humanfactors/topics/complete.pdf Accessed February 2017.

[3] Investigators have blamed human error for the oil pipeline explosion that killed at least 55 people in Qingdao, Available at http://www.scmp.com/news/china/article/1365685/human-error-blamed-qingdao-oil-pipeline-explosion. Accessed February 2017.

[4] Death at Pennsylvania gas well blamed on human error, poor supervision, Available at http://finance.yahoo.com/news/death-pennsylvania-gas-well-blamed-215541383.html. Accessed February 2017.

[5] Air Force officials say human error to blame for damaged nuclear missile, Available at http://www.foxnews.com/us/2016/01/23/air-force-officials-say-human-error-to-blame-for-damaged-nuclear-missile.html. Accessed February 2017.

[6] 'Human Error' blamed for fatal Chevron blast, Available at

http://www.upstreamonline.com/live/1372020/Human-error-blamed-for-fatal-Chevron-blast. Accessed February 2017
[7] Human error blamed for small oil spill at Yokosuka, Available at http://www.stripes.com/news/human-error-blamed-for-small-oil-spill-at-yokosuka-1.45928. Accessed February 2017.
[8] Human error is blamed in Bay Area oil spill, Available at http://articles.latimes.com/2007/nov/11/local/me-spill11. Accessed February 2017.
[9] Report Cites Human Error in 2010 Gulf Oil Spill http://www.voanews.com/content/report-cites-human-error-in-2010-gulf-oil-spill-113041609/169604. Accessed February 2017.
[10] Coast Guard: Oil spill probably due to 'human error', Available at http://222.cnn.com/2007/US/11/12/california.oil.spill. Accessed February 2017.
[11] OSHA BP Report, Available at https://www.osha.gov/dep/bp/BP_OSHA_Settlement_Agreement.pdf.
[12] International Ergonomics Association, Definition and Domains of Ergonomics, Available at http://www.iea.cc/whats/index.html. Accessed February 2017.
[13] Process Safety Management Guidelines for Compliance, Occupational Safety and Health Standards 1910.119, Occupational Safety & Health Administration (OSHA), Available at https://www.osha.gov/Publications/osha3133.html. Accessed February 2017.
[14] Senders, J.W. and Moray, N.P. (1991) *Human Error: Cause, Prediction, and Reduction*. Lawrence Erlbaum Associates, p.25. ISBN 0-89859-598-3, Lawrence Erlbaum Associates, p.25. ISBN 0-89859-598-3. Accessed February 2017.
[15] Human Factors Methods for Improving Performance in the Process Industries, Center for Chemical Process Safety, Dec. 2006, ISBN 978-0-470-11754-5. Accessed February 2017.
[16] Ibid.
[17] A Practical Application of 'Human-HAZOP' for Critical Procedures, G. R. Ellis and A. Holt, Symposium Series No. 155, p. 434-439.
[18] Human factors engineering in projects, International Association of Oil and Gas Producers (OGP), Report No. 454, August 2011. Available at http://www.ogp.org.uk/pubs/454.pdf. Accessed February 2017.
[19] Hazard and Operability Study, Wikipedia, https://www.en.m.wikpedia.org/wiki/Hazard_and_operability_study. Accessed February 2017.
[20] Information Processing and Human-Machine Interaction: An Approach to Cognitive Engineering (North-Holland Series in System Science and Engineering, 12), J. Rasmussen, Elsevier Science Ltd., 1986.
[21] Human Error Types, SkyBrary, Available at www.skybrary.aero/index.php/human_error_types Accessed February 2017.
[22] Inspectors Toolkit: Human Factors in the management of major accident hazards, Health and Safety Executive, Available at http://www.hse.gov.uk/humanfactors/toolkit.htm Accessed February 2017.

478550 Thermal Decomposition of Mono-Nitrated Toluene (MNT) with Additives

Abstract:

Abstract: On September 21st 1992, an explosion happened in Castleford, UK, at the distillation unit, owned by Hickson & Welch Ltd. The incident killed 5 people and caused nearly 200 injuries. The main reason is due to the decomposition of mono-nitrated toluene (MNT). Nitrated aromatic compounds present a potential run-away reaction threat to the chemical process industry, due to its unstable structure and extensive heat release during upset situations. This study employs the Advanced Reactive System Screening Tool (ARSST) to study the effects of various common contaminants on MNT's thermal decomposition, including sodium nitrate, sodium hydroxide, and sodium carbonate. Experimental parameters and results such as 'onset' temperature, pressure increase rate, temperature increase rate and maximum temperature are present. Sodium hydroxide and sodium nitrate can lower the decomposition onset temperature of MNT and make the consequence more severe. Thermokinetics of the decomposition reaction is shown and SADT prediction of UN25 kg package storage is calculated.

479379 Addressing Challenges of Controlling Change By Implementing Best MOC Practices

Abstract:

Effectively controlling change in a plant environment results in a cumbersome process with an inefficient paper-based Management of Change process or use of a traditional and complex electronic Management of Change (MOC) system. Recent research indicates that the majority of chemical incidents are the outcome of poor implementation of Process Safety Management (PSM) and the lack of efficient MOC processes. As a result, MOC workflow steps are often skipped or prolonged resulting in an ineffective process that does not capture the change fully or safely; approvals are delayed, Process Safety Information (PSI) is not being updated, Pre-Startup Safety Review (PSSR's) are not being performed, and MOCs are not being closed in a timely manner.

The goal of this paper is to address industry challenges with controlling change, and how Siemens implements best practices for an effective MOC process.

479243 Process Change Impact on Relief System Design

Abstract:

Process change impact on relief system design

(Understanding equipment impact and the evaluation warranted)

Various scenarios can cause the pressure within equipment to exceed design pressure. Appropriately sized relief devices are one of the most common methods of protecting equipment and personnel operating this equipment. A key challenge to being in compliance is the dynamic nature of processing plants. Numerous changes can be expected to happen over the lifecycle of a processing plant. Management of Change, one of the Process Safety Management (PSM) elements, requires equipment impact evaluation when these changes occur to ensure adequate pressure relief. This presentation talks about numerous types of changes which will impact equipment and warrant overpressure protection re-evaluation.

479152 Identification and Local Impact Analysis of Projectile Hazard in the LNG Industry

Abstract:

With a strong growth in natural gas production in the United States, a number of liquefied natural gas (LNG) facilities have been proposed and authorized. These LNG facilities have sufficient safety, security, risk assessment and mitigation plans developed in close coordination with local, state, and federal authorities. However, to continue to ensure safe operation, it is vital to continue to improve the scope of risk assessment and management strategies. Recent incidents have highlighted the need to consider the potential hazard and risk from projectiles. This paper studies the sequence of phenomena that occur as part of the local impact of a projectile. The main objectives of this study were to: 1) list projectile (missile/fragment) parameters such as source, type, and characteristics (mass, diameter, velocity) and 2) conduct a local impact analysis for identified potential projectile hazard scenarios at an LNG facility. The paper concludes that the five different calculated thicknesses for concrete penetration, scabbing, and perforation, and steel penetration and perforation, depend directly on the kinetic energy and diameter of the projectile. Furthermore, the authors recommend that operators conduct a projectile hazard analyses for LNG facility projects, which would be useful in making recommendations for the thickness and type of storage tanks and other equipment and for other appropriate mitigation measures.

478563 Thermal Hazard Analysis and Chemical Incompatibility Tests with Custom High-Pressure Crucibles Made from Commonly Used Metals and Alloys

Abstract:

The assessment of reactive hazards and chemical incompatibilities is necessary for safe operations in the chemical and pharmaceutical industries. Commercial ARC (Accelerating Rate Calorimeter), RSST (Reactive System Screening Tools) and high pressure crucibles for Differential Scanning Calorimetry (DSC) can be used for this purpose [1-2]. However, ARC and RSST require significant capital investment and require relatively large amounts of sample (at least 1 g).The alternative, commercial high pressure crucibles for DSC, can also be expensive (e.g., several hundred U.S. dollars for each pair of gold-plated high-pressure crucibles) when multiple samples need to be investigated. These requirements make thermal hazard analysis prohibitively expensive for research laboratories that do not focus on chemical process safety, and where materials under development may be available in limited quantities (e.g., milligrams).

To address these short-comings of current hazard screening methods, inexpensive custom DSC high-pressure crucibles made from different commonly used metals and alloys (stainless steel 304, stainless steel 316, titanium, brass and tantalum) have been constructed in-house using a micro-welding system containing a jewelry spot welder and a microscope. Chemical compounds are tested in different crucibles made from different metal/alloys, which simulates different real-world storage, transportation, and process operations in the chemical and pharmaceutical industry. The cost for each pair of crucibles is less than two U.S. dollars. In a typical experiment, only 1-5 mg of sample is required. Kinetic parameters can be extracted from only a few DSC runs with these crucibles, allowing for prediction of the chemical's thermal behavior at larger scales. This method provides an inexpensive and efficient way to perform thermal hazard analysis and incompatibility tests. The application of these crucibles is demonstrated with case studies of ammonium nitrate (AN) and ionic liquids. Predictions of thermal behavior at larger scales have been verified with ARC tests.

References

1. Oxley, Jimmie C. 2003. "Handbook of Thermal Analysis and Calorimetry. Vol. 2: Applications to Inorganic and Miscellaneous Materials M.E. Brown and P.K. Gallagher, Editors 9 2003 Elsevier B.V. All Right Reserved. 349" 2: 349–69.

2. Han, Zhe, Sonny Sachdeva, Maria I. Papadaki, and M. Sam Mannan. 2014. "Ammonium Nitrate Thermal Decomposition with Additives." Journal of Loss Prevention in the Process Industries. Elsevier Ltd, 1–9. doi:10.1016/j.jlp.2014.10.011.

13TH GLOBAL CONGRESS ON PROCESS SAFETY

Do Dust Hazards Analysis (DHA) Really Differ from a Process Hazards Analysis (PHA) or a Process Analysis (PA)?

Michael E. Robertson

4094 Roslindale Ln
Buford, GA 30519
processesmatter@gmail.com

Prepared for Presentation at
American Institute of Chemical Engineers
2017 Spring Meeting and 13th Global Congress on Process Safety
San Antonio, TX
March 26 – 29, 2017

Do Dust Hazards Analysis (DHA) Really Differ from a Process Hazards Analysis (PHA) or a Process Analysis (PA)?

Michael E. Robertson

4094 Roslindale Ln
Buford, GA 30519
processesmatter@gmail.com

Keywords: NFPA 652, Dust Hazards Analysis, DHA

Abstract

Do Dust Hazards Analysis (DHA) Really Differ From a Process Hazards Analysis (PHA) or a Process Analysis (PA)?

This author reviews the latest NFPA 652 requirement on Dust Hazards Analyses (DHA) and compares it to OSHA PSM's Process Hazards Analyses (PHA) requirement and other process analyses required by industry specific NFPA dust regulations. Though process analyses have been a part of NFPA regulations for several years and in the OSHA PSM regulation for 25 years, the 2016 version of NFPA 652 has clearly defined that a process analysis called Dust Hazards Analysis must be done where there is a combustible dust risk. The subject of this paper is the Dust Hazards Analysis.

The aim of the paper is to find commonality in dust hazards analysis, process hazards analysis and process analysis for tips that may limit the number of analyses a company may have to perform. These tips provided will assist a PHA leader in doing their DHA's. This will not be an exhaustive review of standards or regulations, but an overview and guide for those starting DHA's. It has become obvious that NFPA regulations are modeling a PSM framework and some discussion will be made on the subject in the area of risk assessment and management systems. Some comments will be made concerning the software that can be used, but no software will be sold or marketed. The author is not sharing examples of DHA's from his company due to confidentiality and timing of his DHA's.

Some of the Dust Hazards Analysis differences detailed are the emphasis on dust inside and outside of equipment, the use of NFPA dust thickness calculations, building reviews, and frequency of re-validating the analyses. How does a Dust Hazards Analyses affect the process safety information required, team make-up or report after the review? What are the real requirements for a DHA leader?

1 Introduction

There have been several dust related incidents that resulted in loss of life and severe damage to industrial properties. Imperial Sugar in Savannah, GA (2008) and West Pharmaceutical in North Carolina (2003) are two that top the list. Since the early 2000's, the Chemical Safety Board (CSB) has been recommending the Occupational Safety and Health Administration (OSHA) develop standards surrounding combustible dust. The National Fire Protection Association (NFPA) has had guidelines in several of its standards concerning dust for years.

Table 1. Industry Specific NFPA Guidelines Pertaining to Dust

61 – Agriculture
484 - Metal
654 - General
664 - Wood

NFPA 654, 664, 61, 484, 86 and 499 have dealt with combustible dust for years.

It is 2017 and while OSHA still has not set forth comprehensive dust standards, NFPA has published a comprehensive guideline on the management of combustible dust as 2016 NFPA 652 Standards on the Fundamentals of Combustible Dust. This guideline requires dust hazards analysis (DHA) to be complete by September 7, 2018. long with renewed regulations, risk assessments are now becoming part of every safety performance system. NFPA has required process analysis and hazards analysis in the industry specific guidelines..

A DHA is similar to a Process Hazards Analysis (PHA), but there are some key differences that need to be understood by study leaders. PHA methodology is well- defined especially from the OSHA 1910 PSM regulation, the CCPS book *Guidelines for Hazards Evaluations* **[2]**,and other information from new and vintage PHA courses like Primatech's Process Hazards Analysis Leaders' Training and Process Improvement Institute's PHA Leader's Training.

The similarities between PHA's and DHA's are in the methodology. The methodologies used to conduct a PHA, such as HAZOP, Checklist, What-If, are applicable for identifying and assessing hazards in dust handling processes. Expertise with PHA's can assist with the completion of DHA's. The differences begin when the building has to be evaluated for combustible dust hazards. Process Analysis is defined in NFPA 664-12 – Standard for the Prevention of Fires and Explosions in Wood Processing and Woodworking Facilities. [3]. However, NFPA did not offer much instruction on how to do a Process Analysis (PA).

There are other tools that have to be employed for the building evaluation. Pictures, audits, root cause analysis and cause and effect diagrams can be used for the out of the process review.

Detailed building structure evaluations may require a code inspection or some other engineering tools.

This paper evaluates the elements of dust hazards analysis. Comparisons are made between dust hazards analysis, a process hazards analysis methodology called HAZOP (HAZard and OPerability Study), and a process analysis as described in older NFPA guidelines. These comparisons are made to assist PSM companies in aligning PHA's and DHA's and to assist non-PSM companies on resources to assist them in conducting DHA's.

2 What Is a Dust Hazards Analysis (DHA)

NFPA 652 requires that all facilities handling or creating combustible dust must do a Dust Hazards Analysis. A dust hazards analysis is a systematic review of processes and buildings where dust is created or handled and could create a deflagration.

Per the CCPS process safety glossary: A combustion that propagates by heat and mass transfer through the un-reacted medium at a velocity less than the speed of sound.

This requirement has many businesses scrambling in a similar way to the release of the PSM standard in 1992, which left many oil and gas and chemical companies reeling. The dust hazards analysis is not a PSM requirement nor is it a type of process hazards analysis. While dust hazards analyses are new, analyzing your process for dust hazards has been a part of industrial specific NFPA guidelines for a few years now. Now with NPFA 652, OSHA through Recognized and Generally Accepted Good Engineering Practice(RAGAGEP) has the standards it needs for inspection and enforcement to address the potential for a dust deflagration in a facility.

The dust hazards analysis must be complete by September 2018. The DHA requirement is set forth in Element 7.0 of the 2016 version of NFPA 652. There is a good example of the DHA in Appendix B of NFPA 652. Operating areas are reviewed for the probability of deflagration. Classifications are 'Not A Hazard', 'Maybe a Hazard' and 'Deflagration Hazard'. If the probability exists, then the owner must document how they manage or mitigate that risk. If the owner does not have current controls to prevent the deflagration, steps must be taken to document a hazards management plan with steps to reduce the risk.

Within a process system or building, when there is a sufficient quantity of combustible dust, the following conditions must be considered and documented:

1) Oxidizing atmosphere
2) Credible ignition source
3) Credible suspension mechanism

These are items from the fire triangle used in most fire prevention training. These are key in the dust hazards analysis and must be treated with high importance. Housekeeping also must be properly reviewed and documented to prevent the third item mentioned above – credible suspension mechanism.

3 How Do Dust Hazards Analysis (DHA) Compared to Process Hazards Analysis (PHA)

A major difference between the DHA and a PHA is the consideration of dust risk in a building. PHA's do evaluate the risk to occupied buildings in relation to the potential for releases of highly hazardous chemicals covered by the PSM regulation. This risk is evaluated through facility siting studies and more detailed reviews like a Quantitative Risk Analysis (QRA) or Occupied Building Risk Assessment (OBRA).

Table 2. PHA and DHA Methodology Comparison

Process Hazards Analysis (HAZOP)	**Dust Hazards Analysis**	**Process Analysis**
Source: OSHA 1910.119 (e)	Source: NFPA 652 7.0 and Appendix B – Dust Hazards Analysis	Source: NFPA 664 4.0 and Explanatory Material A4.3
Covers the process system	Covers the process system and the building construction	A methodical review of the facility, each operation housed within, and the identification of where a hazard exists.
Gather Data – Drawings and Process Information	Gather Data – Lab Data, Drawings and Process Information	
Have a Qualified PHA Leader a Team with an Operator and a Process Expert, along with other Subject Matter Experts as needed	Have a Qualified DHA Review Leader and at least one other for a team.	
Set Objectives and Boundaries		
Determine the Study Nodes by marking up the Piping and Instrument Diagrams	Determine the nodes on the drawings and sketches	
Review HAZOP Guidewords	Review DHA questions from Appendix B of NFPA 652	
Describe Process Node	Describe Process Node	
Use all Guidewords in all Nodes	Use all questions in all Nodes	
Review Global Node, Human Factors, Management of Change and Incidents		

With the PSM regulation, PHA's are required in facilities with highly hazardous chemicals. These chemicals are listed in the PSM requirement. PHA's are a group of tools that are used to review processes handling highly hazardous chemicals. PHA's include HAZOP, Failure Mode and Effect Analysis (FMEA), What If Studies, and Checklists.

The DHA requires buildings to be reviewed for the possibility of a fire from dust build-up. Building construction must minimize the effects of a deflagration or explosion from a combustible dust.

From the table above, you can see some differences in PHA's and DHA's. DHA's do not require a review of management of change and incident review history. The team composition is not set in a DHA. P&ID's are the basis, but DHA's do not have a required standard drawing type although P&ID's are applicable for some aspects of dust handling.

4 DHA Toolbox

Similar to PHA's, multiple tools are needed to complete a DHA.A PHA method could be used to evaluate the risk associated with combustible dust in a process. However, beyond typical PHA methods, field audits, pictures and video recordings of the areas may be need of areas where dust can build up.Alternatively, a root cause analysis method like 5 Why method might also be applied in areas where a dust build-up to determine sources of dust leakage.. The key is to locate where dust gets out of a process or accumulates beyond its minimum explosive concentration (MEC) in a process, why the dust gets outside the confines of the process and what can be done to prevent the leakage.

Whereas chemical analysis and laboratory data are helpful in a PHA, parameters like Maximum Explosion Concentration (MEC), Minimum Ignition Temperature (MIT) and Kst are very key in determining dust combustibility and risk for a process and building. NFPA 652 has information in the use of these parameters. There is also good information in NFPA's Guide to Combustible Dust and the AIChE Book on Solids Handling Equipment. This paper will not cover in detail any laboratory or chemical analysis of dust. [4], [5]

5 Documentation for Review

Piping and instrument diagrams (P&ID's) have long been the basis of the PHA techniques, including HAZOP, Checklist, and What-If. Supporting documentation may include SDS's, process descriptions, management of change documents, incident history and standard operating procedures. Plant layouts and general arrangement drawings will help orient the team and provide information for facility siting and items in the global node.

P&ID's will help for the process analysis in the DHA. You may also need general arrangement drawings and assembly drawings where machines are involved. Plant layouts and 3D renderings may help with the building analysis portion of the DHA. Pictures and video of areas where dust buildup can occur will also be helpful.

Lastly, plant conveying, processing and finishing OEM information such as screen sizes, motor power, pneumatic flow rates, and duct sizing are key in understanding out the equipment affects the creating or lack of creation of combustible dust. Information on seals and design pressures must be available for comparison with operating parameters.

The bottom line is DHA documentation is as important to the process as Process Safety Information is for PHA. However, it might not be as standard or available as PSI is for a PHA. Some of the information may also be very old since plant layouts and renderings might come from the original start of the plant.

6 The Dust Hazards Analysis Process

The DHA process can start like a PHA with an overview, process description and agenda. The DHA must cover the fugitive combustible dust that gets out into the building and the building's ability to withstand deflagrations and explosions.

Documentation must be available to follow the flow of the materials that contribute to creation of combustible dust. Equipment like fans and screens are especially sensitive since those components can create dust suspensions and impact the Minimum Explosion Concentration of the system. Critical laboratory must be available for discussion since it will provide the data to determine combustibility.
As with a PHA, a tour of the area would be very critical to see the areas were combustible dust accumulates the most. A start to the process is chosen and each part is reviewed by asking a series of question. Below are the questions from Appendix B of NFPA 652 used in a DHA.

Table 3. DHA Process Questions ([1] – B.4.5.1.1 – B.4.5.1.5)

I.	Is the particulate deflagrable (explosible)?
II.	Is the particulate suspended in air?
III.	Is there sufficient concentration to propagate a flame front?
IV.	Are there competent igniters available?
V.	What hazard management is in place?
VI.	Is there metal detection, spark detection, bonding and grounding, or other hazard management means in place?

These questions are asked for each piece of equipment, piping and ducting. The answers will should lead the team or competent person to the hazard management system measures in place and the gaps that will need attention from site management. The process must continue until the entire process is analyzed and all the outcomes have been reached.

As mentioned previously, Appendix B of NFPA 652 (2016) provides a good example to do a DHA. For those in the wood industry, the example is actually of a wood processing plant.

7 Outcomes

NFPA 652 requires that results from the Dust Hazards Analysis be documented. This includes necessary action items requiring change to the process materials, process itself, operations or facilities. NFPA 652 has requirements for managing the hazards under the heading of Hazards Management Program.

The results from a Process Hazards Analysis are the hazard scenarios, a list of safeguards currently in place, and any action items required to ensure that risk is effectively managed within the process handling highly hazardous materials. This usually involves a risk ranking system as well.

The Dust Hazards Analysis does not require a risk ranking system. Where a hazard exists, identification and evaluation of the fire scenario shall include safe operating ranges, safeguard identification, recommendations for additional safeguards as needed to properly manage the fire event, and lastly a plan for implementation.

8 Leaders

A meeting has a leader and teams have leaders. DHA's and PHA's are no different. Each review must have a leader. Her qualifications are listed in Table 2. There are no certifications or extensive training requirements. Many believe that certification is coming for PHA leaders because the competency of those who lead PSM is being discussed in government and industry especially as we enter the next 25 years of PSM.

Many experts agree that an experienced PHA leader can also make an effective DHA leader. This is a good start, but the DHA leader must also be familiar with the hazards of combustible dust. Additionally, it would be prudent to have a leader who is experienced in the process being reviewed or ensure that a process expert is present as a team member. I would review the qualifications listed in CCPS's book on Hazards Evaluation and information.

Table 4. Comparison of Leader Qualifications

Dust Hazards Analysis Leader	Process Hazards Analysis Leader
Must be able to lead DHA's Must be familiar with the hazards of combustible dust	Must be trained in PHA techniques Must be re-qualified every 5 years

9 Frequency

PHA's are done initially to establish a base evaluation and then every five years after the initial PHA to address changes or revisions to covered processes. . DHA's are done once for the original process and for any major upgrades. There is no renewal period, but the initial reviews have to be done by September 7, 2018.

10 Management Systems

What is a dust hazards management system? How does a dust hazards management system compare to the standard elements of PSM? Chapter 9 of NFPA 652 mentions its elements as listed in Table 6. Both PSM and NFPA 652 have the requirement of risk assessment that has been the subject of this paper.

PSM is embedded in many of the NFPA guidelines. Though the thought of PSM to many firms is frightening, PSM offers a good framework for managing risk whether it is combustible dust or other system or process risks.

Table 5. Management Systems Comparison

Dust Hazards Analysis Management System (Ch. 9)	Elements of Process Safety Management
1. Operating procedures and practices	1. Standard Operating Procedures
2. Inspection testing and maintenance	2. Mechanical Integrity
3. Training and Hazard Awareness	3. Training
4. Emergency Planning and Response	4. Emergency Planning and Response
5. Incident investigations	5. Incident investigations
6. Management of change	6. Management of change
7. Management system review,	7. Compliance Auditing
8. Contractor training	8. Contractor Safety
9. Employee participation	9. Employee Participation
10. Document retention	10. Process Safety Information
11. Ignition Control (Covered in Chapter 8)	11. Hot Work Permits
12. DHA	12. Trade Secrets
	13. Pre-Start Up Safety Review
	14. PHA

11 Conclusions

DHA's must be completed. It is similar but different than a PHA. Expertise with PHA's can assist with the completion of DHA's. The similarity between PHA's and DHA's is in the methodology of reviewing processes for dust. The differences start when the building has to be evaluated for combustible dust. Here other tools have to be employed. These tools as previously discussed may vary from audits and inspections to root cause analysis tools. Detailed building structure evaluations may require a code inspection or some other engineering tools.

12 References

[1] NFPA 652-15 – Standards on the Fundamentals of Combustible Dust, National Fire Protection Association, September 7, 2015.

[2] CCPS. *Guidelines for Hazards Evaluations.* Center for Chemical Process Safety, American Institute of Chemical Engineers, New York, NY, 2007.

[3] NFPA 664-12 – Standards on the Fundamentals of Combustible Dust, National Fire Protection Association, 2011.

[4] NFPA– Guide to Combustible Dusts, National Fire Protection Association, 2012.

[5] CCPS. *Guidelines for Safe Handling of Powders and Bulk Solids.* Center for Chemical Process Safety, American Institute of Chemical Engineers, New York, NY, 2005.

478758 A Leading Indicators Based Decision Support Tool to Predict Blowout Events

Abstract:

Predicting a potential blowout scenario timely and efficiently is often a challenging task due to the complexity of drilling or other well intervention activities. Blowouts are usually preceded by gas kicks and early prediction of gas kick events can provide some precious time window to take control of the well in concern. A leading risk indicators based approach for predicting gas kick or blowout events has been introduced in this work. A causal factors based step-by-step approach for identifying comprehensive sets of leading indicators are presented and discussed. Leading indicators were divided into two broad sections - real-time indicators and long-term organizational safety performance indicators. With the real-time indicators, different decision support algorithms are developed which can be used as training tools. A real-time indicator may not always successfully lead to a well control event but at the same time the situation needs to be assessed carefully considering other factors and indicators. Such uncertainties are taken into account while developing the algorithms. For evaluating the relative importance of different leading indicators Bayesian Network models are developed. The key causal factors for well control barriers failure are identified by conducting fault tree analysis and appropriate indicators are linked with relevant causal factors in a Bayesian model. At the end, a list of efficient leading risk indicators are presented for predicting primary well control barrier failure which can initiate gas kicks.

Consequence Modeling at the Eastman Kingsport Site

David Skelton
Principal Engineer
Eastman Chemical Company
P.O. Box 511, Building 18
Kingsport, TN 37662
davidskelton@eastman.com

Michael James
Senior Chemical Engineer
Eastman Chemical Company
P.O. Box 511, Building 18
Kingsport, TN 37662
mjames@eastman.com

Prepared for Presentation at
13th Global Congress on Process Safety
San Antonio, TX
March 26-29, 2017

UNPUBLISHED

Consequence Modeling at the Eastman Kingsport Site

Michael James
Senior Chemical Engineer
Eastman Chemical Company
P.O. Box 511, Building18
Kingsport, TN 37662
mjames@eastman.com

David Skelton
Principal Engineer
Eastman Chemical Company
P.O. Box 511, Building 18
Kingsport, TN 37662
davidskelton@eastman.com

Keywords: Dispersion modeling, consequence analysis, risk assessment, dosage, shelter-in-place, temporary haven, evacuation, toxicity, toxic chemical release, toxic load, downwind concentration, blast overpressure, vapor cloud explosion, pressure vessel burst

Abstract

The intent of this paper will be to review the methods utilized at the Eastman Kingsport site in determining magnitude of consequences of toxic, thermal and blast overpressure scenarios. This will include discussion on dispersion modeling, dose-response methods for toxic and thermal probits as well as assessing occupant vulnerability through the use of BakerRisk's Explosion Risk Cooperative tools.

1. Potential Consequences of Loss of Containment

Figure 1 is a typical event tree for potential losses of containment [1].

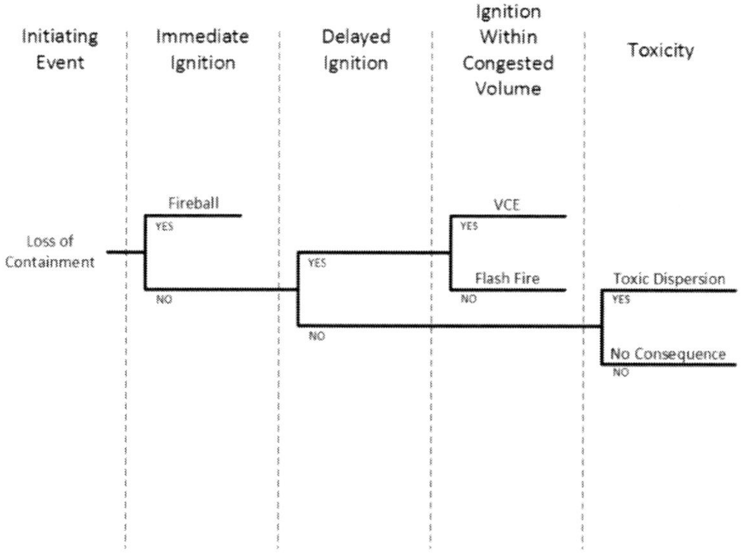

Figure 1. Potential Losses of Containment

The consequences of loss of containment in Figure 1 detail the potential release scenarios that are chiefly of concern. The method for consequence modeling for each type of scenario is described in this paper.

2. Source Terms

Figure 2 details the potential source terms that will need to be developed for each type of scenario.

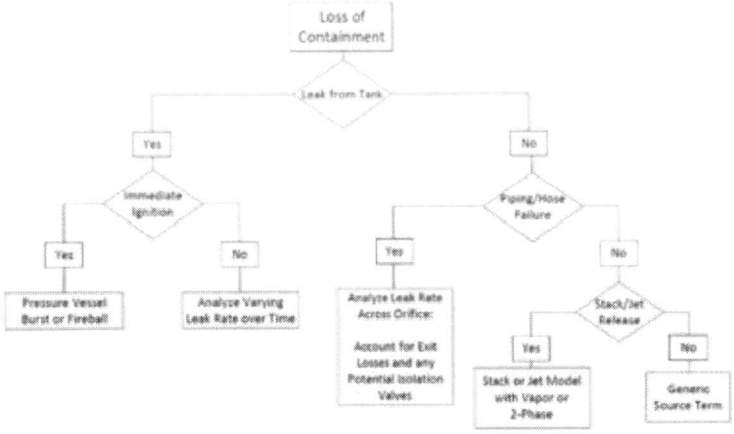

Figure 2. Potential Source Terms.

The source term may have any of the following characteristics:

- Initial momentum
    - Based on the linear velocity across the orifice and other physical properties of the release, the source term may need to account for significant initial momentum of the cloud.
    - In addition, if the linear velocity is sufficiently high, the source of the release may ingest air, resulting in a jetting condition. As a rule of thumb, 10 m/s is a lower limit for potentially jetting releases [2]. There are specific dispersion models to account for jetting conditions during a release.

- Pool Formation
    - Most dispersion models have an integrated pool evaporation model to determine the dynamic evolution of vapor from a liquid spill.
    - If the preferred evaporation model is outside of the dispersion model, the evaporation rate results will need to be analyzed for integration into the dispersion modeling tool.
- Flashing
    - Leaks across an orifice may partially or completely flash from liquid to vapor based on the pressure differential.
    - Modeling tools may report choke conditions at the outlet.
    - Further flash analysis may be required to determine the temperature and liquid/vapor composition from choke to atmospheric conditions.

- Piping Layout between Pressure Source and Leak Point
  - Ensure that all pressure losses are accounted for. This includes entrances, exits, elevation changes, elbows, pipe diameter changes, and other fittings.

- Indoor Release
  - If the release occurs indoors, analyze the concentration changes over time inside the structure based on the source term calculations of the release.
  - Concentration gradients throughout the structure may need to be ignored and perfect mixing assumed.
  - If the vapor portion of the leak will cause overpressuring of the indoor area, determine the amount of vapor required to be vented in order to balance pressure in the room.
  - A flash block can be employed to assess the amount of vapor generated from a leak that will need to be relieved from an indoor area.
  - Flash blocks can also be used to account for vapor suppression systems inside of buildings.
  - This level of modeling is beyond the scope of this paper. However, methods for assessing indoor releases can be referenced from "Consequence Modeling of Dynamic Source Terms" [3].

Presence of the above factors may result in models that require expert judgment and the development of conservative estimates. Most dispersion modeling tools do not have the levels of sophistication required to accurately model these types of source terms. However, the results of a dispersion model employed with conservative estimates may be suitable for site risk assessment.

The following section will discuss the potential leak scenarios and how source terms can be developed, accounting for each of the above factors.

3. Blast Modeling

Simple blast models can be used to predict potential consequences of explosions in lieu of more accurate but complicated methods such as computational fluid dynamic (CFD) models. The TNT equivalency model developed by the U. S. Navy and Air Force was largely used for years, even in industrial settings. Over time it became clear that the TNT model did not predict the behavior of all explosions accurately enough, particularly in regard to vapor cloud explosions more commonly occurring in refining and chemical producing industries. Today several simplified are available to estimate the potential of consequences of explosions, each method with its own strengths and weaknesses.

3.1 Vapor Cloud Explosion (VCE) Modeling

Baker Strehlow Tang (BST) is the simplified method predominantly used at Eastman Kingsport site to model vapor cloud explosions [4][5][7][8]. The BST model is based on the premise that only the portion of a flammable vapor cloud that is confined or contained within a congested area will significantly contribute energy to an explosion. Unconfined or uncongested gas clouds are more likely to result in flash fires or deflagrations. The first step to perform a BST analysis is to identify the confined/congested volume of the plant that has the potential to harbor a flammable cloud of vapor. Examples might include piping and equipment, closely spaced railcars or cars, solid walls, ceilings and floors. A flammable cloud is most often generated from a release of material as a result of an equipment failure, a leak, or a spill. Flammable clouds can drift a considerable distance under the right conditions, so it's important to account for all the equipment within a reasonable radius of the congestion. Once the confined/congested volume is calculated, two parameters contribute the greatest influence on explosion strength using BST methodology; the energy contained within the flammable vapor cloud and flame speed as it burns through the available fuel in the congested region. The first parameter is calculated using the spreadsheet shown in Figure 3. It is a simple tool developed by Eastman to calculate volume based on dimensions of the confined cloud, and then estimate the energy contained with the calculated volume.

ESTIMATION OF VAPOR CLOUD EXPLOSION ENERGY

The user enters dimensions (in ft.) of vapor cloud subject to confinement & obstructions, ambient temperature (F), and the chemical involved in the designated boxes. Thermodynamic data listed in the lookup table to the right is used, or if not available, an approximation is made based on typical hydrocarbons.

The spreadsheet will output the participating energy to be used in V-CLOUD for prediction of pressure and impulse by the Baker-Strehlow method. The methodology is based on the paper "Vapor Cloud Explosion Analysis" by Baker, Tang, and Scheier; Prepared for the AIChE Loss Prevention Symposium, April 1994.

Width (ft)	200	
Height (ft)	50	
Depth (ft)	100	
Volume (ft3)	1000000	
Volume (m3)	2.83E+04	
Ambient Temp (F)	90	
Ambient Temp (C)	32	
Chemical	Acetic Acid	
MW	60.05	
Hc(kJ/mole)	919.73	(heat of combustion in KJ/mole)
Hc(J/m3)	3.67E+07	(heat of combustion in J/m3)
Hc (BTU/lb)	6.59E+03	(heat of combustion BTU/lb for ref.)
Hcs(J/M3)	3.48E+06	(heat of combustion of a stoichiometric mixture with air)
Energy (J)	9.86E+10	
Energy (in-lb.)	8.72E+11	(total available energy in confined / obstructed volume)
% Participation	100	
Equiv Energy in-lb.	8.72E+11	(net energy of explosion in-lb.)

Figure 3. Spreadsheet used to determine energy content of vapor cloud

For a single congested area or multiple areas with little separation distance so they can be effectively considered a single area of congestion, the results from the spreadsheet are used as an input in VCloud®. VCloud® is a simple software tool developed by the BakerRisk Engineering Consultant to produce Pressure/Impulse (P-I) diagrams (this is not to be confused with PI Diagrams representing the response curve for structures and might better be referred to as the side-on blast pressure and impulse as a function of standoff distance). Notice the check box denoted "Plot Using Co-op Energy Factor." Membership in the BakerRisk Engineering Consultant Explosion Research Cooperative (ERC) includes access to experimentally verified energy factors that can reduce conservatism and provide a more accurate prediction of explosion effects. The first page of inputs for VCloud® is illustrated in Figure 4.

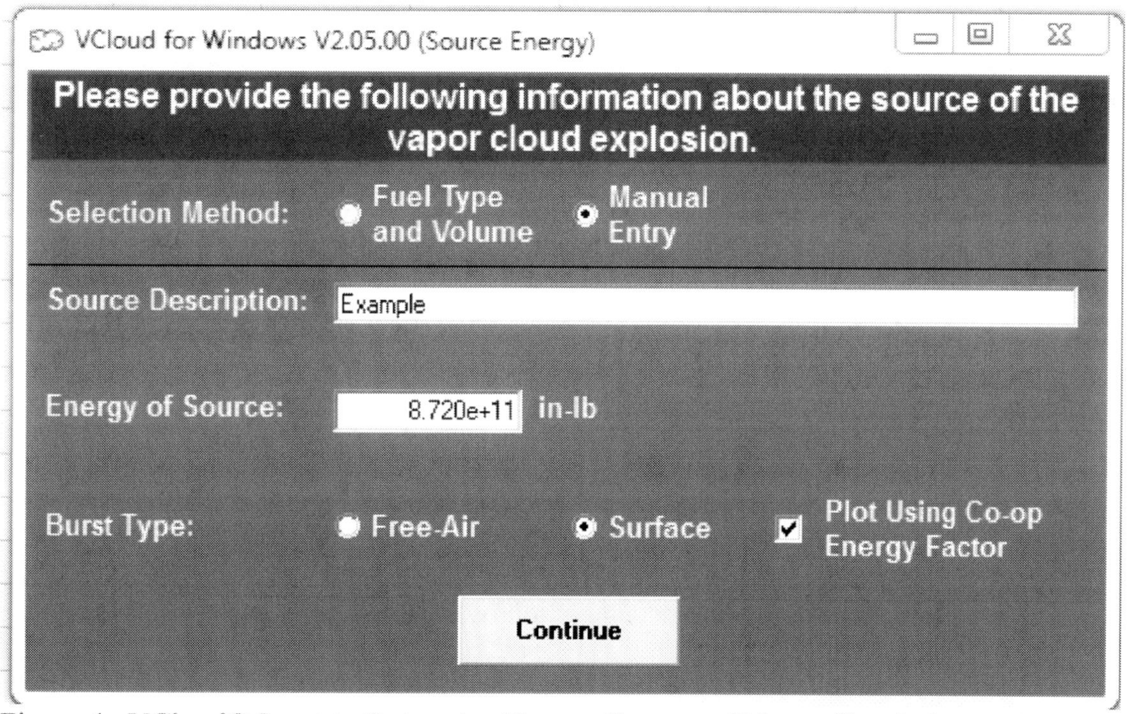

Figure 4. VCloud® Input to Determine Energy Content of Vapor Cloud. Image Courtesy of BakerRisk Engineering Consultant Explosion Research Cooperative.

The second parameter greatly influencing explosive strength when using BST methodology is flame speed, measured as a ratio to the speed of sound or Mach number. Four criterion most influence how fast a flame front burns through fuel: congestion, confinement, fuel reactivity, and the distance to free venting.

- Congestion
    - Burning rate is greatly dependent on the surface area of the flame as it travels through the fuel source. Turbulence causes the flame front to furrow and can greatly increase the surface area which contributes to increasing the speed of the advancing flame. In an industrial setting, turbulence can be generated by piping, equipment, gravel surfaces, vehicles, or any obstacle(s) that inhibits simple linear progression of the burning flame front.
    - The greater the induced turbulence, the faster a flame front will progress increasing the likelihood of detonation. BST method calculations qualify congestion into three categories; high, medium or low congestion. Congestion categorization is the most subjective parameter in BST methodology. Experience is probably the most important tool used to determine degree of congestion.
    - A good rule of thumb is if process equipment occupies an area but can be easily walked through, it is low congestion. If you can manage to slip through an area with difficulty, it's medium congestion. If it is impossible to physically pass through an area, it's high congestion.

- Confinement
    - Any solid surface that prevents the propagation of the flame in a given direction is considered confinement, with the exception of the ground. The BST method models a spherical, free-air vapor cloud explosion, but multiplies the energy of the explosion by a factor of 2 to account for ground reflection of the blast wave from its origin surface.
    - BST calculations categorize confinement in four ways; 3D, 2D, and 2½ D, and 1D.
        - 3D Confinement: A flame front burning through fuel is capable of expanding in all three dimensions with the exception of the ground. It applies to all free-field processing areas not bound by solid walls or ceilings.
        - *2D Confinement:* The propagation of the flame front is free to expand in only two axes because one dimension is blocked by a solid obstacle. A common example is a processing unit located between a solid concrete ceiling with open walls. It also applies if a processing unit is located between two solid walls with no roof. It is important to consider the aspect ratio of 2D confinement, which is the ratio of total width to height of the confined volume. Higher aspect ratios constrict venting capacity and allow additional pressure accumulation over those with a shorter distance to free venting. Typical values are available in a drop down menu in VCloud®, ranging as high as a 25:1 aspect ratio. Anything higher than 25:1 will not provide significantly more pressure accumulation, nor are aspect ratios greater than this commonly found in the field. Any aspect ratio greater than 4:1 should be represented in VCloud as 25:1 to maintain a conservative estimate.
        - 2½ D Confinement: This is a hybrid condition where neither 3D nor 2D accurately describe confinement. For instance, processing equipment contained within a pre-engineered building with lightweight metal walls and roof that will confine the flame front for a time but release at low pressures, before the impending explosion is fully realized. The roof and walls provide some containment, but in short order the blast wave is capable of 3D expansion. Tightly packed equipment on the boundaries of the congested area generating an explosion can also be considered 2½ D confinement, since it provides a small amount of venting, but not enough to consider it 3D expansion.
        - 1 D Confinement: Flame burn in a linear direction, such as inside a pipe.

- Fuel Reactivity
    - The BST method categorizes fuel reactivity in only three ways; low, medium, and high reactivity. Fuel reactivity is based on the laminar burning velocity at a specific concentration, determined by a standard testing procedure in stagnant conditions.
    - Fuels with a laminar burning velocity less than 45 cm/s are considered to have a low reactivity. Fuels with laminar burning velocities above 75 cm/s are considered highly reactive. Everything between is considered to be of medium reactivity.
    - Table 1 illustrates reactivity of a few common chemicals for reference here. NFPA 68 is a good reference for finding laminar burning velocities of many more chemicals so the reactivity category can be determined.

Table 1. Relative Reactivity

Reactivity	Fuel	Group	Relative Reactivity	Formula
Low	Methane	Alkanes	1	CH_4
Low	Formic Acid	Acids	1	HCOOH
Low	Acetic Acid	Acids	1	CH_3COOH
Medium	Ethane	Alkanes	3	C_2H_6
Medium	Propane	Alkanes	3	C_3H_8
Medium	Butane	Alkanes	3	C_4H_{10}
High	Ethylene	Alkenes	6	C_2H_4
High	Acetylene	Alkynes	7	C_2H_2
High	Hydrogen	Inorganics	7	H_2

(Courtesy Eastman Chemical Company)

- Distance to Free Venting
  - Distance to free venting is simply the shortest distance a flame must travel to exit the area of congestion/confinement to find free space.
  - After free venting, turbulence is less likely to accelerate the flame front and under most circumstances it will begin to slow. This reduces the possibility of a transition from deflagration to detonation (DDT).

These four criterion are now input into VCloud® in the format shown in Figure 5. From the provided data, a Mach number (M_f) is assigned to the flame front.

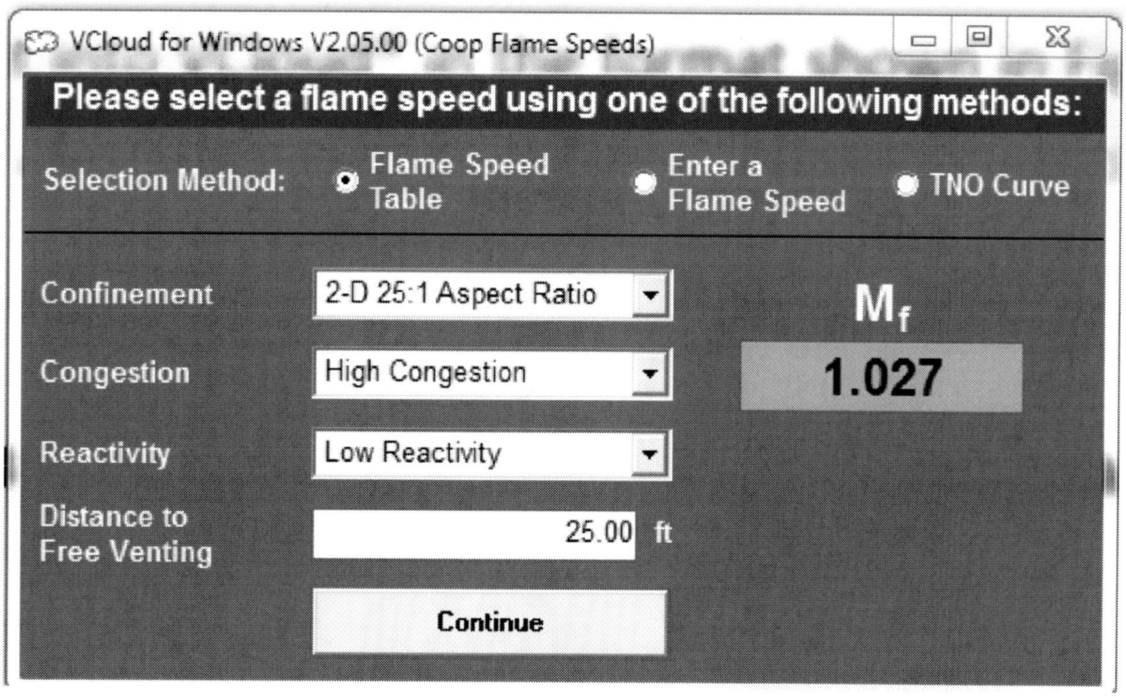

Figure 5. Vcloud® Input to Determine Mach Number. Image Courtesy of BakerRisk Engineering Consultant Explosion Research Cooperative.

A Mach number greater than one indicates the flame front accelerated faster than the speed of sound and beyond, meaning a detonation is more likely to occur. Some typical flame speeds in Mach numbers are provided in Table 2 for reference, with respect to fuel reactivity and degree of congestion.

Table 2. Flame Speeds for Soft Ignition Sources [9]. Provided as a reference but not relevant to the example calculation.

Flame Speed in Mach Numbers (M_w) for Soft Ignition Sources

Flame Expansion	Fuel Reactivity	Obstacle Density: High	Obstacle Density: Medium	Obstacle Density: Low
ID	High	5.2	5.2	5.2
	Medium	2.27	1.77	1.03
	Low	2.27	1.03	0.294
2D	High	1.77	1.03	0.588
	Medium	1.24	0.662	0.118
	Low	0.662	0.471	0.079
3D	High	0.588	0.153	0.071
	Medium	0.206	0.100	0.037
	Low	0.147	0.100	0.037

VCloud® uses the provided inputs to produce a Pressure/Impulse (PI) diagram. The PI diagram is used to determine the overpressure and impulse experienced by a target at a given distance. Figure 6 is an example of a PI diagram produced by VCloud®. The distance between the target and the explosion is manually input to determine the blast effects at that specified distance. The target at 150 feet from the explosion will experience a 10.12 psig side-on overpressure with a 316.05 psi-ms impulse. This is more than enough to completely demolish an unreinforced masonry building. A key attribute of vapor cloud explosions that differs from high explosives is a much higher impulse. Impulse is more threatening to rigid construction than overpressure, because it tends to distort supporting structural members beyond their plastic limits.

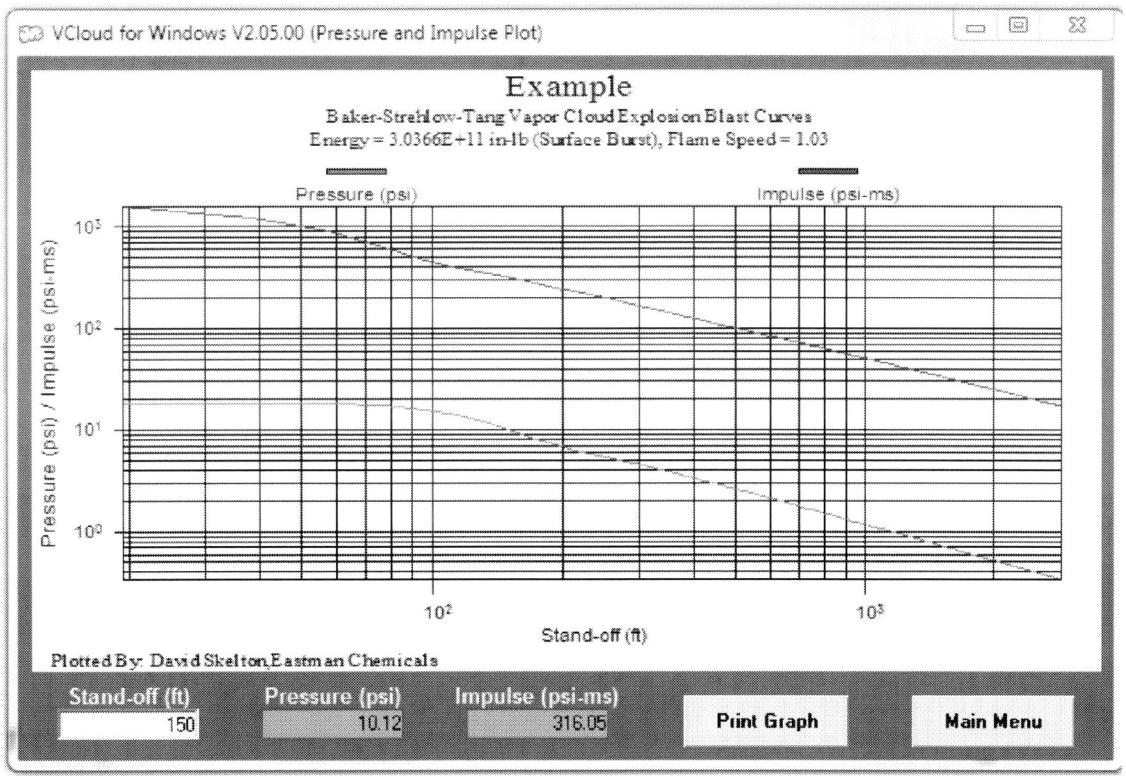

Figure 6. P-I Diagram Output From VCloud®. Image Courtesy of BakerRisk Engineering Consultant Explosion Research Cooperative.

Once a PI diagram is prepared, it can be used to predict the potential damage to conventional industrial buildings from the modeled blast event. Another software tool developed by the ERC can be used for this purpose, the Blast Evaluation and Screening Tool, affectionately known as BEAST®. BEAST® uses empirically based correlations from accidental explosions, shock tube tests, and government sponsored high explosive tests to predict the level of damage imparted by an explosion to a given building or structural component. Originally developed in 1995, BEAST® has been updated over the years to include more building and component damage curves, improve the user interface, and add more building types. The most recent version of BEAST® can evaluate 19 unique building types and many types of individual structural components, such as windows and doors. The first step is to define the building for evaluation. An example using the current BEAST® interface is included in Figures 7a and 7b. Required inputs are type of construction (drop down menu), building dimensions, blast overpressure and impulse (from PI Diagram), and whether wall and roof loading are side-on or reflected.

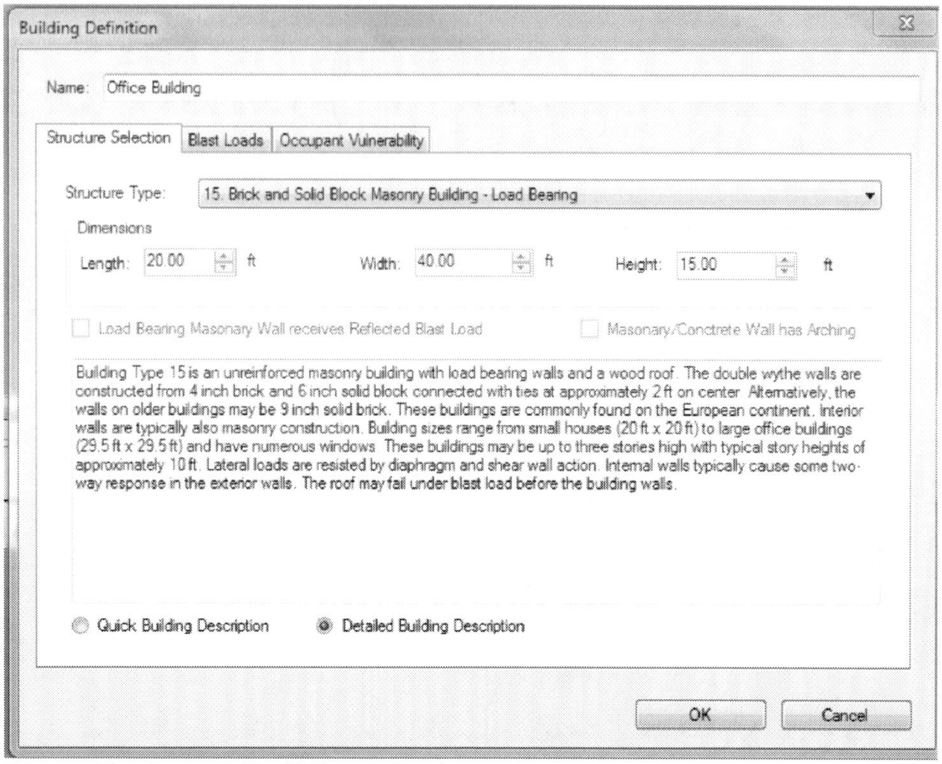

Figure 7a. Beast® Building Selection.

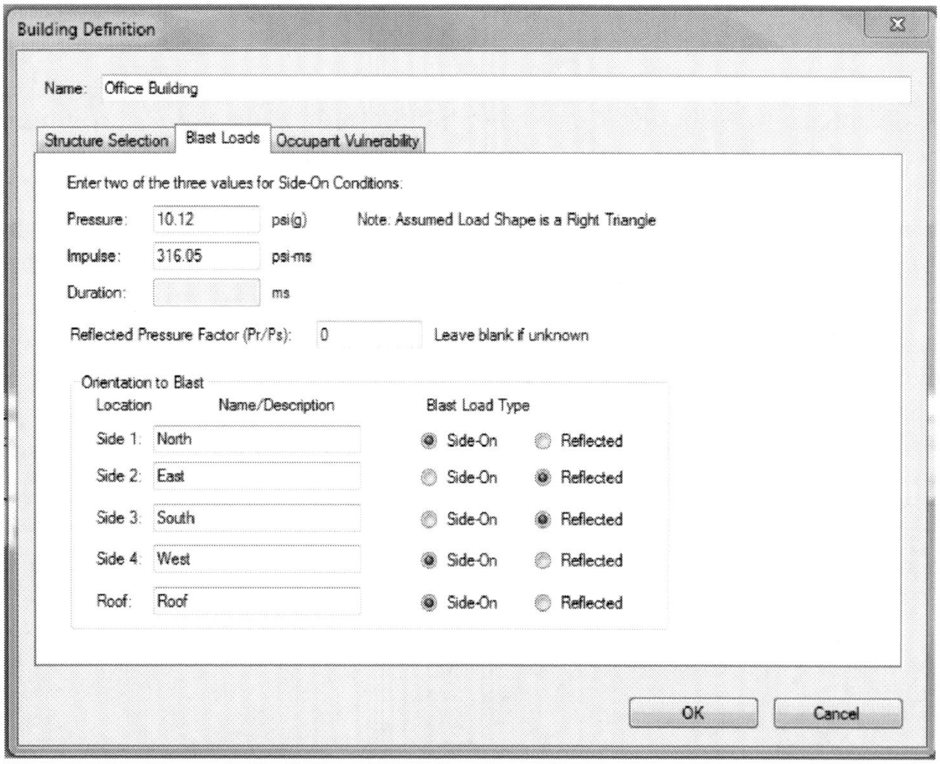

Figure 7b. Beast® Building Selection (cont'd).

Images Courtesy of BakerRisk Engineering Consultant Explosion Research Cooperative.

Given the appropriate inputs, BEAST® predicts a damage level to the building and an occupant vulnerability. Figure 8 illustrates the output provided from BEAST®.

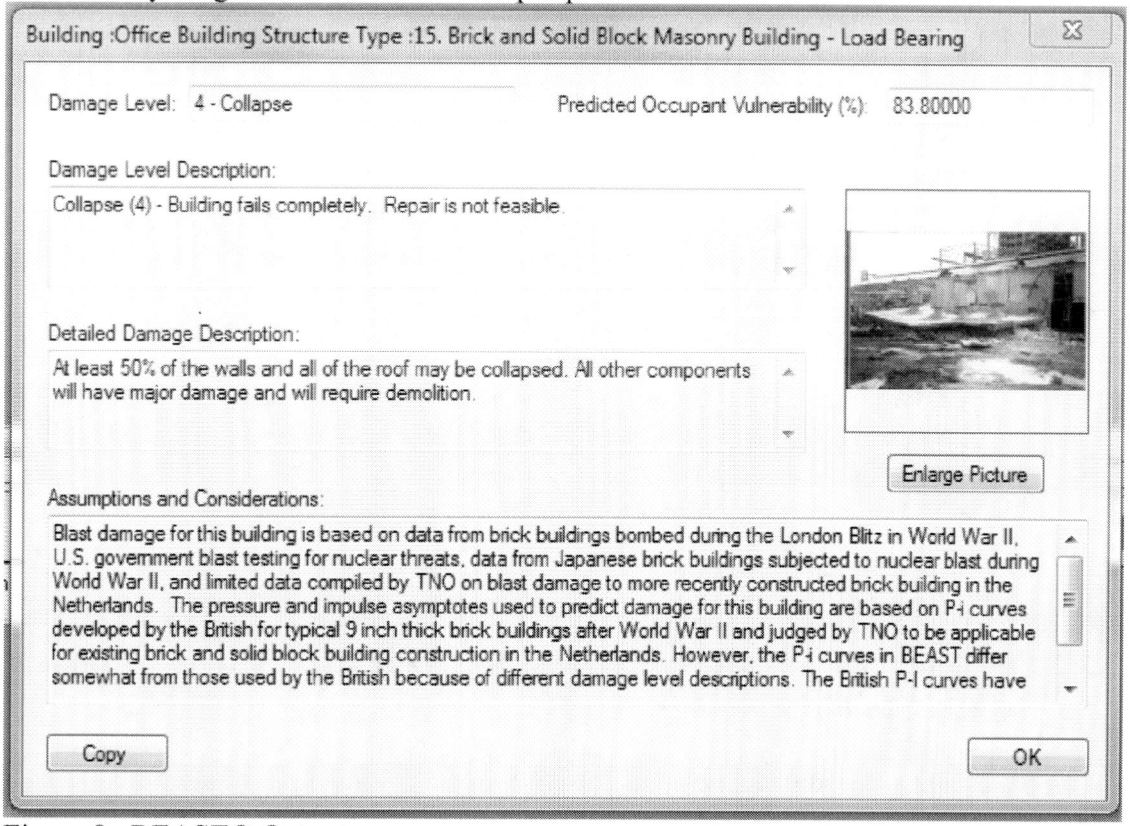

Figure 8. BEAST® Output

Of course, not all explosions result in the complete destruction of a building like the extreme example provided. BEAST® categorizes building damage using the criteria found in Table 3 [6]. A Building Damage Level up to 2B is still survivable by most occupants.

Table 3. BEAST® Building Damage Levels

Building Damage Level	Damage Description
1	Onset of visible damage to building walls facing explosive source. Repairs only needed for cosmetic reasons. Building is reusable following an explosion.
2A	Localized building damage, primarily to components on building walls facing the explosive source. Building performs function and can be used; however, major repairs are required to restore integrity of structural envelope. Total cost of repairs is moderate.
2B	Widespread building damage. Building cannot be used until major repairs are completed. Total cost of repairs is significant, approaching replacement cost of building.
3	Building has lost structural integrity and may collapse due to environmental conditions (i.e. snow or wind). Total repair cost exceeds replacement cost of building.
4	Building fails completely. Repair is not feasible.

Courtesy of BakerRisk Engineering Consultant Explosion Research Cooperative

4. Development of Source Terms for Potential Vapor Dispersion Release Scenarios

Methods for determining the source term for releases from tanks, liquid pools and piping systems are discussed here. All examples provided here do not reflect actual potential release conditions. For all examples, any relation to true conditions is incidental.

Prior to the modeling methods discussion, it is important to review a tool for determining losses due to piping layout.

IoMosaic SuperChems® v 8.21 (2016) allows for development of piping layouts to detail the piping diameters, elbows, and other potential connections. In simple models, as will be discussed in this paper, this can be used as the outlet piping for a pressure source (e.g., tanks) to a release point.

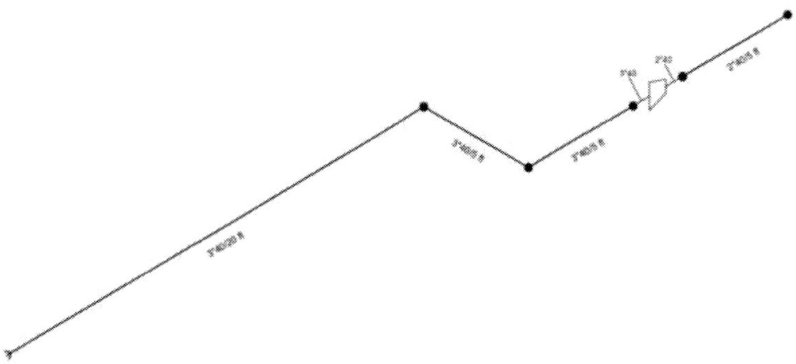

Figure 9. Example Piping Layout.

Figure 3 details a layout from the source at 3 inches with 2 90-degree elbows and a reduction to the 2-inch leak point. The release will be from a guillotine break on the 2-inch line.

4.1 Release from Tank

A reliable model for release from vessels is contained in ioMosaic SuperChems® v8.21 (2016).

This example will model the release of a liquid blanketed by nitrogen. The "Vessels Containing 2-Phases Model" allows for estimation of the vapor and liquid contents of a vessel, accounting for nitrogen blanketing or air in the head space. This same model can be utilized for relief from the vapor phase. There are alternative models for gas-filled vessels.

The overall process for using SuperChems® to model this is as follows:

1. Development of objects for mixture composition, vessel dimensions, piping layouts and other objects. Ensure piping layouts are properly applied.

   The following steps are within the "Vessels Containing 2-Phases" Model.

2. Input the vessel pressure and an initial estimate for overall composition inside of the vessel. The overall composition includes both the vapor and liquid phases. In this example, the modeled vessel is filled to 80% with methyl isocyanate and has a nitrogen blanket.

   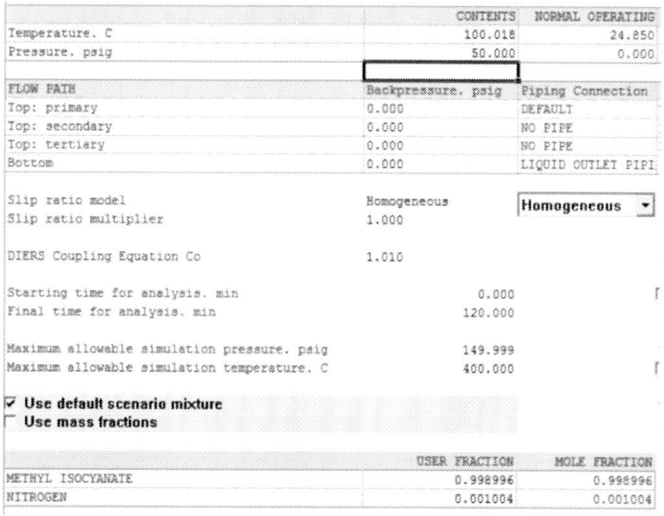

   Figure 10. Vessels Containing 2-Phase Inputs

   Uncheck the "Use default scenario mixture" as defining this model is an iterative process.

3. Use "Specify Fill Fraction" to input the approximate vessel fill level. This example uses an estimate 80% fill level.

   The model it calculates the vessel temperature at the specified overall composition.

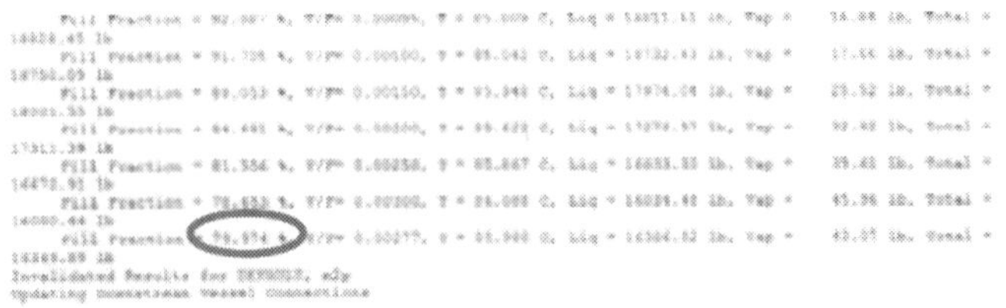

Figure 11. Model Iterations to determine vessel temperature at 80% liquid fill level.

4. Iterate over steps 2 and 3, varying the composition and re-entering the fill fraction until the modeled vessel temperature matches the true conditions. Increasing content of lower boilers will decrease the modeled vessel temperature.

   In the example, the composition was iterated until the storage temperature was approximately 25 deg C.

	CONTENTS	NORMAL OPERATING
Temperature, C	24.918	24.850
Pressure, psig	50.000	0.000

FLOW PATH	Backpressure, psig	Piping Connection
Top: primary	0.000	DEFAULT
Top: secondary	0.000	NO PIPE
Top: tertiary	0.000	NO PIPE
Bottom	0.000	LIQUID OUTLET PIPI

Slip ratio model	Fauske	Fauske
Slip ratio multiplier	1.000	
DIERS Coupling Equation Co	1.010	
Starting time for analysis, min	0.000	
Final time for analysis, min	120.000	
Maximum allowable simulation pressure, psig	149.999	
Maximum allowable simulation temperature, C	400.000	

☐ Use default scenario mixture
☐ Use mass fractions

	USER FRACTION	MOLE FRACTION
METHYL ISOCYANATE	0.991100	0.991100
NITROGEN	0.008900	0.008900

Figure 12. Final composition.

5. Running the model will determine the flow rate through the piping layout and guillotine break.

6. When the run is complete, a report will appear. The "Profiles-Bottom" tab is the most useful for this source term model.

TIME, min	VESSEL CONTENTS TEMPERATURE, C	VESSEL CONTENTS PRESSURE, psig	PRESSURE AT VESSEL BOTTOM, psig	BOTTOM PIPING LIQUID MASS FLOW RATE, lb/min	BOTTOM PIPING GAS/VAPOR MASS FLOW RATE, lb/min	bottom flow reduction factor (Actual flux with pipe/Ideal Flux)	BOTTOM FLOW NOZZLE PRESSURE, psig	BOTTOM FLOW DISCHARGE PRESSURE, psig	BOTTOM FLOW DISCHARGE TEMPERATURE, C
0.0000	24.92	50.000	56.394	2.005E+03	0.000	0.405	56.392	-0.001	25.094
0.0246	24.88	47.142	53.517	1.966E+03	0.000	0.394	53.516	0.000	25.051
0.0506	24.85	44.305	50.662	1.871E+03	0.000	0.383	50.660	-0.001	25.006
0.0782	24.81	41.472	47.809	1.775E+03	0.000	0.372	47.808	-0.001	24.960
0.1077	24.77	38.642	44.961	1.679E+03	0.000	0.360	44.959	-0.001	24.913
0.1394	24.73	35.817	42.116	1.583E+03	0.000	0.347	42.115	-0.001	24.864
0.1735	24.69	32.996	39.275	1.485E+03	0.000	0.334	39.274	-0.001	24.814
0.2104	24.64	30.181	36.440	1.387E+03	0.000	0.321	36.439	0.000	24.761
0.2506	24.60	27.371	33.610	1.289E+03	0.000	0.306	33.609	0.000	24.706
0.2946	24.55	24.568	30.787	1.190E+03	0.000	0.291	30.786	0.000	24.649
0.3433	24.50	21.773	27.971	1.090E+03	0.000	0.276	27.970	0.000	24.588

Figure 13. Vessels Containing 2-Phases Output.

The liquid and vapor release rates out of the piping as well as the discharge temperature and pressure are reported. The discharge pressure will be above atmospheric if there is choked flow inside the piping.

If the choke pressure is significantly above atmospheric, running some select points through the "2-Phase Expansion and Aerosol Model" to determine the final vapor/liquid composition and temperature is recommended.

7. As the release is liquid in this case, evaporation from a liquid forming pool will next be modeled. As stated previously, several dispersion models have integrated pool models. The SuperChems® pool evaporation will be employed in this case, and the output from the model will be input to an external dispersion model.

4.2 Liquid Pools

As with tank leaks, SuperChems® will be utilized for determining the source term of a liquid pool formation.

An important consideration is the variety of weather conditions over which a release could take place. As solar incidence is a major factor in pool evaporation, ensure that the most probable conditions for both daytime and nighttime are employed. Also, to account for solar incidence, ensure that the appropriate time of day and site coordinates are input for the scenario in SuperChems®. In addition, diked area dimensions and substrate composition can be specified.

SuperChems® pool evaporation model allows for a static release rate over a set period of time. As most releases will be dynamic in nature, this may involve expert judgment.

Once a characteristic release rate, duration and temperature have been determined, this information can be entered into the pool evaporation model.

Liquid discharge temperature, C	24.209
Liquid discharge rate, lb/min	583.903
Liquid discharge duration, min	1.949

Use default scenario mixture
Use mass fractions

COMPONENT	USER FRACTION
METHYL ISOCYANATE	1.0000
NITROGEN	0.0000

Average molecular weight 57.0520

Figure 14. Pool Evaporation Model Inputs

The model can perform evaporation on multi-component mixtures. In this case, the evolution of any dissolved nitrogen is not of concern. The liquid rate has been imported from the Vessels Containing 2-Phases model, and the source concentration has been set to pure methyl isocyanate.

Based on the inputs, the model will provide a conservative estimate for the evaporation of methyl isocyanate from the formed pool.

Time. min	OVERALL POOL EVAPORATION RATE. lb/min	Pool diameter. ft	Pool surface temperature. C
0.017	0.000	3.146	4.209
0.017	0.258	3.146	-5.182
0.017	0.257	3.146	-5.235
0.017	0.257	3.147	-5.287
0.017	0.256	3.147	-5.339
0.018	0.257	3.147	-5.389
0.028	0.249	3.171	-5.439
0.038	0.242	3.241	-7.525
0.049	0.252	3.371	-8.514
0.059	0.273	3.553	-8.996
0.070	0.302	3.765	-9.216
0.080	0.335	3.988	-9.306
0.090	0.371	4.212	-9.334
0.101	0.409	4.431	-9.335
0.111	0.447	4.644	-9.322
0.122	0.485	4.849	-9.305

Figure 15. Pool Evaporation Model Output.

Figure 15 details the most critical data points in the pool evaporation model. Understanding these will allow for straightforward input to a dispersion model.

While SuperChems® does have an integrated dispersion model, the author's preference is SAFER Trace® v10.2 (2012). Implementation of pool models into dispersion models is discussed later in this paper.

4.3 Leaks from Piping

For lengths of piping with isolation, it may be appropriate to treat the isolated volume as a vessel and analyze the varying leak rate over time. This section considers the situation where the pressure differential across the orifice may be assumed constant. In addition to the previous source terms, SuperChems® is utilized in determining the source term.

The "Source Term Wizard" can be employed to specify an upstream pressure and temperature to determine the flow rate across a specified piping layout or idealized nozzle. The example here will determine the release rate of methyl isocyanate pressurized to 50 psig and released through a guillotine break in the 2" segment of piping from Figure 9.

OPTIONS	
Liquid Density	Databank
PVT Behavior	Equation of state
VLE Search Direction	Dew point

☐ Use mass basis data to compute fractions
☐ Mixture is soluble in water
☐ Use an existing mixture

COMPOUND	FORMULA
METHYL ISOCYANATE	C2H3NO

Inlet temperature. C	25.00000
Inlet pressure. psig	50.00000
Back pressure. psig	0.00000

☑ Initial state is subcooled liquid

Figure 16. Liquid Source Term Wizard Inputs

The release is treated as a subcooled liquid in order to allow the model to estimate if any portion of the release will flash across the opening. It is not expected to have any flashing due to the release conditions in comparison to the chemical's vapor pressure at 50 psig. However, this model is employed for consistency.

** INLET CONDITIONS	
Inlet temperature. C	25.0
Inlet pressure. psig	50.0
Surroundings/back pressure. psig	-3.974E-06
Slip ratio exponent eta	0.001
Vapor quality	1.E-04
**** EXIT CONDITIONS**	
Exit temperature. C	25.160
Exit pressure. psig	-1.7331E-04
Barometric pressure at site elevation. psig	-3.974E-06
Mixture exit velocity. ft/s	21.213
Exit Mach Number	0.006
Maximum Mach Number	0.006
Exit rho*u*u in kg/m/s2 or in Pascals	38695.90
Maximum rho*u*u in kg/m/s2 or in Pascals	38695.90
Exit steady state impulse. lbf	68.146
Maximum steady state impulse difference. lbf	418.675
Pipe length travelled / total piping length	0.998
Flow rate. lb/min	1714.327

COMPONENT	MW	TOTAL FLOW. lb/min	LIQUID FLOW IN. lb/min	LIQUID FLOW OUT. lb/min	VAPOR FLOW IN. lb/min	VAPOR FLOW OUT. lb/min
METHYL ISOCYANATE	57.052	1713.817	1713.817	1713.817	0	0

Figure 17. Source Term Wizard Outputs.

The results from this model include liquid and vapor release rates as well as exit pressure and temperature. Had flow been choked in the piping, the exit pressure would be above atmospheric. As with previous models, it is recommended to utilize a flash block to determine conditions at atmospheric pressure for releases from systems with choked flow.

For releases from holes or cracks in piping, the piping layout can be developed with an orifice at the end of the layout. The orifice area would be equal to the area of the hole, and the orifice length would be equal to the width of the piping. In addition, due to constraints with the software, a short pipe segment of with the same diameter as the leak should be placed at the exit of the segment.

The results for the above cases can directly be input to a dispersion model, as described in the next section.

5. Application of Source Terms to Dispersion Models

This section will describe the inputs to dispersion models based on the results from the two-phase leak from the sour stream.

Section 5 will review how to utilize the data from the dispersion models in order to assess the hazard posed to a hypothetical shelter 500 ft downwind from the release.

5.1 Test for Jetting

In two-phase or vapor release cases, the linear velocity across an opening may be sufficient to cause jetting. As described previously, there are specific dispersion models for jetting conditions. The rule of thumb for jetting regime is 10 m/s [1]. A simple ideal gas calculation can provide an estimate for linear velocity, given a known release rate, hole size and temperature. This can be expanded to a real gas calculation for highly non-ideal cases. The temperature used in this calculation should be the flashed temperature to atmospheric pressure. The rate employed should be that of the vapor phase for two-phase releases as this will be the largest volumetric component once flashed to atmosphere.

In this example, a 200 psig stream of sour hydrocarbons at 150 deg C is released across a 1-inch hole in 6-inch piping. While several VLE models are available for hydrocarbon mixtures, ideality is assumed in this example.

Piping Isometric / Layout	6-INCH LINE WITH 1-INCH CRACK
Site Name	DEFAULT
Liquid Density Method	Databank
VLE/PVT Method	Equation of state
** INLET CONDITIONS	
Inlet temperature. C	150.691
Inlet pressure. psig	200.0
Surroundings/back pressure. psig	-3.974E-06
Slip ratio exponent eta	0.001
** EXIT CONDITIONS	
Exit temperature. C	133.870
Exit pressure. psig	138.432
Barometric pressure at site elevation. psig	-3.974E-06
Mixture exit velocity. ft/s	190.056
Exit Mach Number	1.0
Maximum Mach Number	1.0
Exit rho*u*u in kg/m/s2 or in Pascals	393239.785
Maximum rho*u*u in kg/m/s2 or in Pascals	393239.785
Exit steady state impulse. lbf	165.061
Maximum steady state impulse difference. lbf	6037.864
Pipe length travelled / total piping length	0.998
Flow rate. lb/min	455.119

COMPONENT	MW	TOTAL FLOW. lb/min	LIQUID FLOW IN. lb/min	LIQUID FLOW OUT. lb/min	VAPOR FLOW IN. lb/min	VAPOR FLOW OUT. lb/min
HYDROGEN SULFIDE	34.082	2.052	2.052	0.866	0.001	1.186
n-PENTANE	72.150	347.584	347.547	268.991	0.037	78.593
HEXANE	86.175	77.840	77.835	66.892	0.005	10.949
n-OCTANE	114.231	27.515	27.515	26.078	0.001	1.438

Figure 18. Flow of stream across 1-inch crack.

The exit conditions show that the flow has choked. As previously discussed, the "2-Phase Expansion and Aerosol Model" will be employed to determine the composition and temperature at atmospheric pressure. In addition, this model will report the linear velocity and cloud diameter when flashed to atmospheric pressure. The linear velocity can be used to determine if jetting is possible.

**TWO-PHASE EXPANSION / DEFAULT

	INITIAL	FINAL
Flow diameter. ft	8.3333E-02	2.3754E-01
Velocity. ft/s	190.42	651.59
Vapor density. lb/gal	0.24	0.02
Liquid density. lb/gal	4.20	5.19
Mixture density. lb/gal	0.98	0.04
Temperature. C	133.87	45.56
Pressure. psig	138.43	0.00
Vapor/Feed (V/F) mole fraction	0.21	0.71
Mass flow rate. lb/min	455.12	455.12
Available energy. BTU/lb	26.45	
Modified Weber number for thermal breakup	392.30	
Critical Weber number for mechanical breakup	12.01	
Viscosity group number	0.01	
Stable droplet diameter. Microns	39.50	
Thermal breakup droplet diameter. Microns	39.50	
Mechanical breakup droplet diameter. Microns	44.18	
Number of droplets	4.9421E+10	
Droplet mass. lb	4.4271E-11	

COMPONENT	TOTAL FLOW	INITIAL LIQ	FINAL LIQ	INITIAL VAP	FINAL VAP
HYDROGEN SULFIDE. lb/min	2.053	0.866	0.038	1.186	2.015
n-PENTANE. lb/min	347.681	269.076	80.158	78.605	267.523
HEXANE. lb/min	77.862	66.912	35.926	10.950	41.936
n-OCTANE. lb/min	27.523	26.085	23.812	1.438	3.711

Figure 19. Sour Stream Contents after Expansion.

Per Figure 19, the temperature of the release cloud will be 46 deg C at atmospheric pressure and will have a linear velocity of 652 ft/s (199 m/s). This is an order of magnitude above the 10 m/s guidance, and should result in significant jetting.

5.2 Input to Jet Model

The dispersion modeling tool utilized for these examples is SAFER Trace® 10.2 (2012).

Figure 20 details the inputs to the jet model based on our example. This is focusing on the vapor portion of the model, ignoring liquid entrainment.

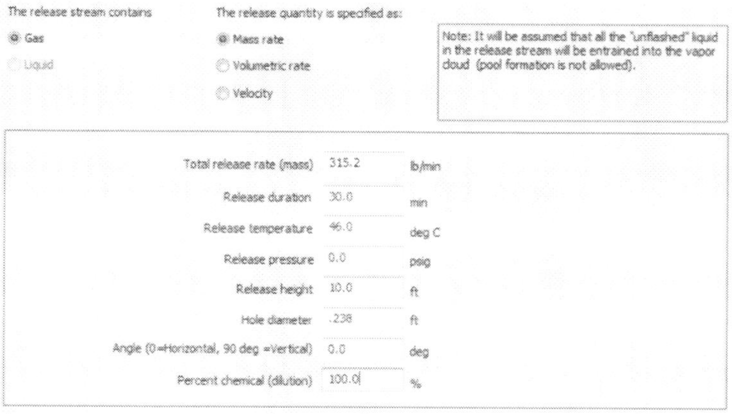

Figure 20. Jet Model Inputs.

The dispersion outputs will report concentrations and plots related to concentration of H2S in air.

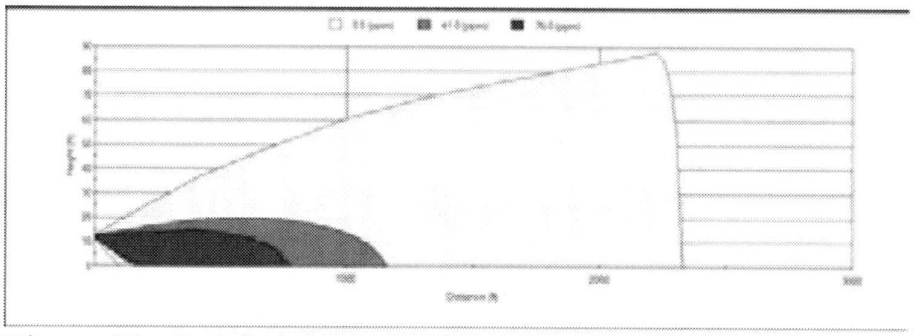

Figure 21. Jet Model Dispersion Output. AEGL concentration limits based on 10-minute exposure to H2S [10].

The H2S concentration profile at the outside of a shelter 500 ft downwind from the release point is shown in Figure 22.

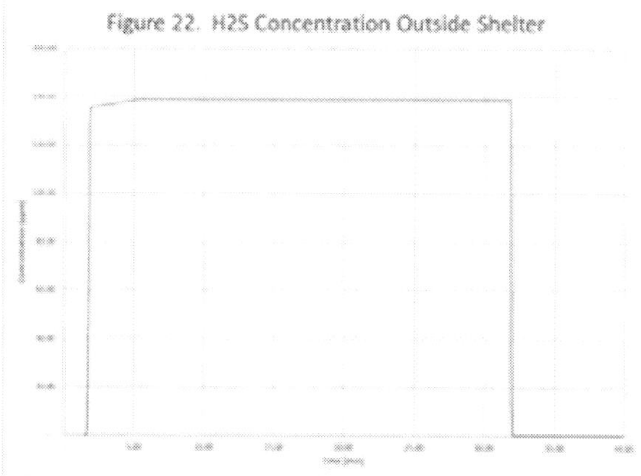

Consequence to personnel in this shelter will be analyzed using a probit function, detailed in Section 5.

5.3 Modeling Pool Evaporation

The liquid portion of the example two-phase release will be treated as the source of a liquid pool.

Spill Surface	DEFAULT
Liquid discharge temperature, C	45.560 <<<
Liquid discharge rate, lb/min	139.934
Liquid discharge duration, min	15.000

☑ **Use default scenario mixture**
☐ **Use mass fractions**

COMPONENT	USER FRACTION
HYDROGEN SULFIDE	0.0006
n-PENTANE	0.6394
HEXANE	0.2399
n-OCTANE	0.1200
Average molecular weight	80.5399

RUN OPTIONS

Figure 23. Liquid Pool Model Inputs

The concentration profile of evaporate is highly dynamic over the studied period of the release.

Figure 24. Pool Evaporate Concentration Profile

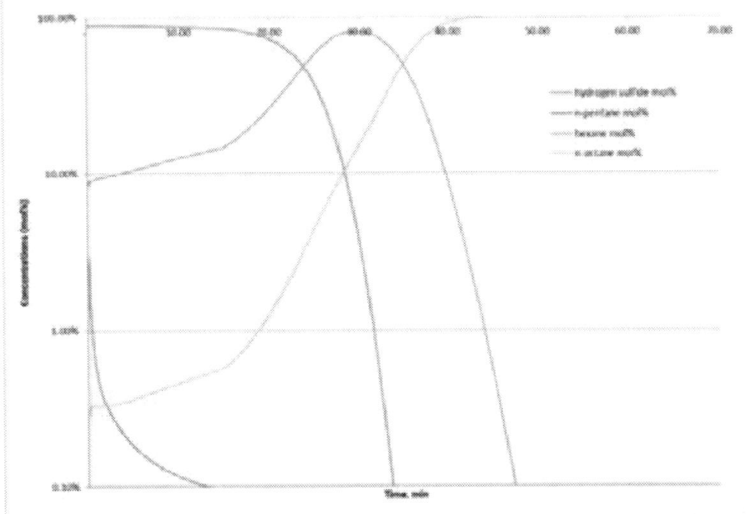

While the concentration is highly dynamic, the main toxicity concern is with H2S. The majority of H2S appears to evaporate in the first 15 minutes. An average rate will be developed based on the total evaporated mass of each component over that time. The effective concentrations and dispersion parameters are shown in Table 4.

Table 4. Molar Concentrations and Dispersion Parameters

Chemical	Mol%
Hydrogen Sulfide	0.1%
Pentane	87.0%
Hexane	12.4%
Octane	0.5%
Average Evaporation Rate	57.0 lb/min
Temperature	2°C
Duration	15 min

The data from this model is input to the SAFER Trace® "Single Phase" steady state model, per Figure 25.

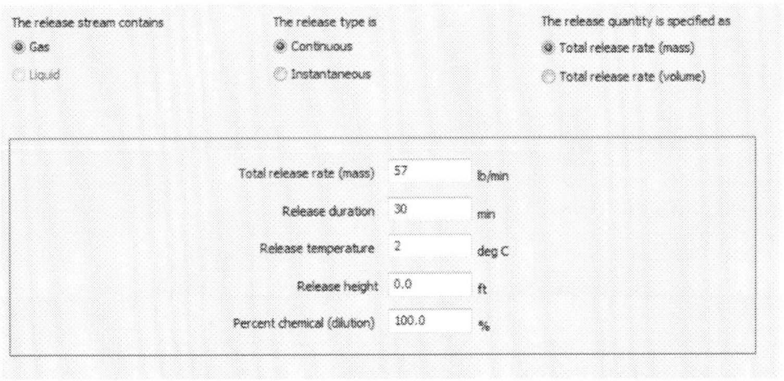

Figure 25. Steady State Model Inputs.

The downwind dispersion of the cloud, focusing on H2S concentrations in air is shown in Figure 26. The concentration is negligible at the 500-ft shelter, so the contribution from evaporation to toxic impact can be ignored.

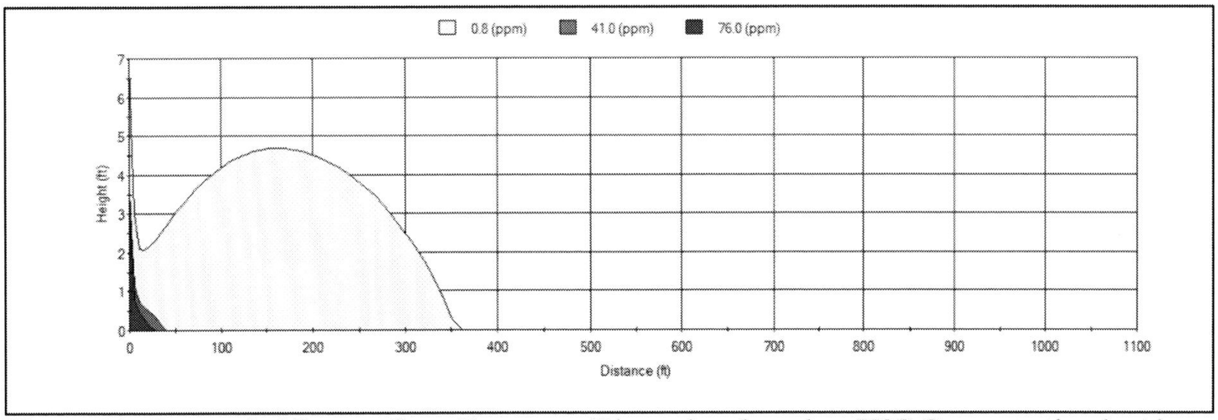

Figure 26. Downwind Impact from Evaporate Dispersion, based on H2S Concentration in Air.

6. Magnitude of Consequence Determination Using Probits

The methods utilized here are described in detail in "Simplified Methods of Using Probit Analysis in Consequence Analysis" [1].

To summarize, the impact to individuals may calculated using dose [5].

$$Dose = \int C^n dt \qquad \text{[Eq. 1]}$$

Dose may be transformed to probability of fatality using a probit transformation [5].

$$Y = A + B\log(Dose) \qquad \text{[Eq. 2]}$$

The constants "A", "B", and "n" are specific to the chemical. The Y value, or probit score is referenced on a probit table to determine probability of studied consequence.

Table 5. Probit table.

%	0	1	2	3	4	5	6	7	8	9
0	—	2.67	2.95	3.12	3.25	3.36	3.45	3.52	3.59	3.66
10	3.72	3.77	3.82	3.87	3.92	3.96	4.01	4.05	4.08	4.12
20	4.16	4.19	4.23	4.26	4.29	4.33	4.36	4.39	4.42	4.45
30	4.48	4.50	4.53	4.56	4.59	4.61	4.64	4.67	4.69	4.72
40	4.75	4.77	4.80	4.82	4.85	4.87	4.90	4.92	4.95	4.97
50	5.00	5.03	5.05	5.08	5.10	5.13	5.15	5.18	5.20	5.23
60	5.25	5.28	5.31	5.33	5.36	5.39	5.41	5.44	5.47	5.50
70	5.52	5.55	5.58	5.61	5.64	5.67	5.71	5.74	5.77	5.81
80	5.84	5.88	5.92	5.95	5.99	6.04	6.08	6.13	6.18	6.23
90	6.28	6.34	6.41	6.48	6.55	6.64	6.75	6.88	7.05	7.33
—	0.0	0.1	0.2	0.3	0.4	0.5	0.6	0.7	0.8	0.9
99	7.33	7.37	7.41	7.46	7.51	7.58	7.65	7.75	7.88	8.09

Source: Finney, 1952 [11]

The Microsoft Excel (2016) function "NORMSDIST" will convert the probit score to a probability.

The Equation 1 is combined with the "NORMSDIST" function to determine impact to occupants of the shelter, per Table 6. This is assuming that the air handling provides 10 air changes per hour to the shelter.

Table 6. Probability of Consequence to Sheltered Occupants.

Time (min)	Dosage ($ppm^{1.43}$-min)	% Probability
1.45		0%
1.46	-	0%
1.47	-	0%
1.48	-	0%
1.49	-	0%
1.50	-	0%
...	...	...
57.00	44,794	0%
58.00	45,354	0%
59.00	45,902	0%
60.00	46,436	0%
61.00	46,957	0%
62.00	47,466	0%
63.00	47,963	0%
64.00	48,448	0%
65.00	48,922	0%

In this example, the occupants can shelter for over an hour with 0% probability of consequence. Probit transformations can be utilized to study probability of fatal exposure from toxicity, thermal radiation, and blast overpressure. In addition, they can analyze probability of burn severity.

7. Conclusion

This paper has summarized the methods utilized in determining impact to personnel from toxic exposure and vapor cloud explosions. The covered models include pipe breach, tank release, and jetting models. These are highly utilized models for consequence analysis. In addition, the paper summarized methods to utilize dispersion modeling data to fully quantify impact to occupants of shelters downwind from releases.

For blast modeling, this paper provided an overview of one simplified method to predict explosion overpressure, impulse, and damage to buildings within the radius affected by the explosion. The method utilized is Baker-Strehlow-Tang (BST), one of several methods engaged by industry for this purpose. Most of the software tools used to perform the analysis were developed by the BakerRisk Engineering Consultant Explosion Research Cooperative, including VCloud® and BEAST®.

For all consequence models, more rigorous methods such as computational fluid dynamic (CFD) modeling will produce more accurate results. However, the methods discussed in this paper are excellent screening tools whose value lies in their ease of use and are suitable for risk analysis.

References

[1] Baybutt, Paul, "Using Layers of Protection Analysis to Evaluate Fire and Gas Systems", Process Safety Progress, Volume 31, Issue 3, 2011

[2] "SAFER Trace® Online Help", SAFER Systems, 2003

[3] James, Michael D., "Consequence Modeling of Dynamic Source Terms", Process Safety Progress, Volume 35, Issue 4, 2016

[4] Baker, Q. A., Tang, M. J., E. A. Scheier, and G. J. Silva, "Vapor Cloud Explosion Analysis," Process Safety Progress, Vol. 15, No. 2, 1996, pp. 106-109

[5] A. Sari, "Comparison of TNO Multienergy and Baker-Strehlow-Tang Models," Process Safety Progress 30 (2011), pp. 23-26

[6] BEAST® Help File, 2000 BEAST® Report, 2000

[7] "VCE Evaluation Method Comparison," The Quest Quarterly, Volume 4, Issue 1, Spring 1999

[8] Baker-Strehlow-Tang Methodology, Guidelines for Evaluating the Characteristics of Vapor Cloud Explosions, Flash Fires, and BLEVEs, CCPS, New York, 1994

[9] Pierorazio, A.J., J.K. Thomas, Q.A. Baker and D.E. Ketchum (2005) "An Update to the Baker-Strehlow-Tang Vapor Cloud Explosion Prediction Methodology Flame Speed Table," Process Safety Progress, 24(1): 59-65

[10] "About Acute Exposure Guideline Levels (AEGLs)", 2016, https://www.epa.gov/aegl/about-acute-exposure-guideline-levels-aegls (Accessed Mar 2, 2017)

[11] D. J. Finney, *Probit Analysis*, Cambridge University Press, New York 1952

Relief Systems Design: The Importance of Accounting for Static Head

Nicholas N. Cristea
Smith & Burgess, LLC
7600 W Tidwell Rd, Ste 600, Houston, TX 77040
Nicholas.cristea@smithburgess.com

Zubin Kumana
Smith & Burgess, LLC
7600 W Tidwell Rd, Ste 600, Houston, TX 77040
Zubin.kumana@smithburgess.com

Guomin Wang
Smith & Burgess, LLC
7600 W Tidwell Rd, Ste 600, Houston, TX 77040
Guomin.wang@smithburgess.com

Prepared for Presentation at
American Institute of Chemical Engineers
2017 Spring Meeting
13th Global Congress on Process Safety
San Antonio, Texas
March 26-29, 2017

Relief Systems Design: The Importance of Accounting for Static Head

Nicholas N. Cristea
Smith & Burgess, LLC
7600 W Tidwell Rd, Ste 600, Houston, TX 77040
Nicholas.cristea@smithburgess.com

Zubin Kumana
Smith & Burgess, LLC

Guomin Wang
Smith & Burgess, LLC

Keywords: Relief Systems Design, Static Head, Upstream, Downstream

Abstract

The process of evaluating an entire facility's relief and flare system design can be time consuming and costly. During the process, there will be systems to be evaluated that have liquid scenarios that may be protecting other assets within the facility. If the static head for the liquid scenarios is not accounted for, then the analysis can become even more costly due to the analysis not being performed correctly the first time. The result could be not meeting code by not adequately protecting system. This paper will look at two different cases where the static head was not accounted for, and how the issue was mitigated in the most cost-effective way. First, when dealing with the liquid heading on an offshore platform, there are many different levels on the platform and ensuring that the static head is accounted for is important to protect the different assets in the liquid handling train. Second, when evaluating a column system, it is important to ensure that the relief device provides protection for every ASME Section VIII Div 1 stamped equipment within the system. These two case studies highlight the necessity of accounting for static head in relief systems design.

1 Introduction

In the process safety industry, relief devices are set such that at least one relief device set pressure is equal to the lowest system MAWP within the protected system, to meet the requirement of ASME Section VIII Div 1 to provide adequate protection against overpressure. ASME Section VIII Div 1 UG-125 states "All pressure vessels within the scope of this Division, irrespective of size or pressure, shall be provided with overpressure protection in accordance with the requirements of UG-125 through UG-138, or with overpressure protection by system design in accordance with requirements of UG-140, or a combination of the two". ASME Section VIII Division VIII UG-125(c) and (1) also state that "when a pressure relief device is provided, it shall prevent the pressure from rising more than 10% or 3 psi (20 kpa), whichever is greater, above maximum allowable working pressure" and "when multiple pressure relief devices are provided and set in accordance with UG-134(a), they shall prevent the pressure from rising more than 16% or 4 psi (30 kpa), whichever is greater, above the maximum allowable working pressure". The vessel is normally hydrotested without accounting for the potential impact of inlet piping to the relief devices. Therefore, when the elevation difference between the top of the vessel and the relief device becomes significant [2], the vessel may experience a pressure higher than its allowable accumulation due to the static head, especially during liquid relief scenarios. The equation below is a conversion from the liquid height to the equivalent liquid head (psig).

$$P = 0.433\ x\ SG\ x\ H$$

P: liquid head in psi

SG: Specific Gravity

H: liquid head in feet

Equation 1: Liquid height to pressure conversion

This paper will look at two different cases where the static head was not accounted for and how the issue was mitigated for each case. The first case will look at a blocked system liquid outlet scenario on a platform where the vessel of interest has the same MAWP as the protecting vessel on an upper level. The second case will look at the liquid relief case (i.e. overfilling) and the mitigation for a column system that contains several vessels. These two cases will show the importance of accounting for static head and the potential mitigation options for these types of concerns.

2 Case 1 – Produced Water Desanders and Accumulators on an Offshore Platform

2.1 System overview

Figure 1 below is an illustration of case study 1. A produced water desander on the upper deck of a platform is fed by a booster pump. The desander then feeds the desander accumulator. The desander and the accumulator both have the same MAWP of 350 psig. The maximum discharge pressure of the booster pump is 415 psig. In the event of a blocked outlet, the relief device PSV-1 (4" P 6") on the desander is intended to provide adequate protection for the desander and the downstream accumulator. The relief device PSV-2 (1" E 2") on the accumulator is intended only for the fire scenario, based on the assumption that the upstream relief device (PSV-1) would limit the maximum deliverable pressure to the accumulator to 350 psig for the blocked outlet scenario. However, the vessels are located on different levels of the platform. The elevation difference was never accounted for in the choice of relief device set pressure. When PSV-1 is relieving at 385 psig, the pressure inside the accumulator would be the relief pressure (385 psig) plus the static head due to the difference in elevation. If the difference in elevation is large, the downstream accumulator may no longer be adequately protected by PSV-1, because the vessel would experience a pressure higher than its allowable accumulation. The elevation difference between the two relief devices illustrated in Figure 1 is roughly 27 feet. The static head from the elevation difference is about 12 psig. This means that when the desander is at the relief pressure, the pressure in the accumulator would be 397 psig, which exceeds the 10% allowable accumulation [2] (385 psig) of the accumulator.

2.2 *Case 1 mitigation*

The common mitigation option to account for the static head is to lower the relief device set pressure on the desander to account for the static head, so that the downstream accumulator would be adequately protected during the liquid relief event. However, for this particular case study, a more detailed engineering analysis revealed that the installation was acceptable as-is.

However, it was determined that there are technically two relief devices protecting the system in the event a blocked outlet on the accumulator. The allowable accumulation for the accumulator can go up to 16% of the system (406 psig) because of the multiple relief devices. Since the manual valve between the two systems will remain open (the initiating event is the closure of the downstream block valve), the accumulator is still protected. If the manual valve between the two systems were to be closed, then the blocked outlet case on the accumulator would no longer be applicable as the source of pressure would be lost as well.

If the static head is greater than 21 psig (~49 feet), the accumulator may experience overpressure in excess of the allowable (a pressure of 406 psig). The potential mitigation would be to lower the set pressure of the desander relief device to account for the static head such that the maximum pressure did not exceed this limit. For this case study, PSV-1 set pressure is not required to be lowered because the accumulator is still within the 16% allowable accumulation after accounting for the static head. Additionally, if the relief device set pressure were lowered, the potential exists that the relief device may be undersized, and additional mitigations may be required to resolve the concerns.

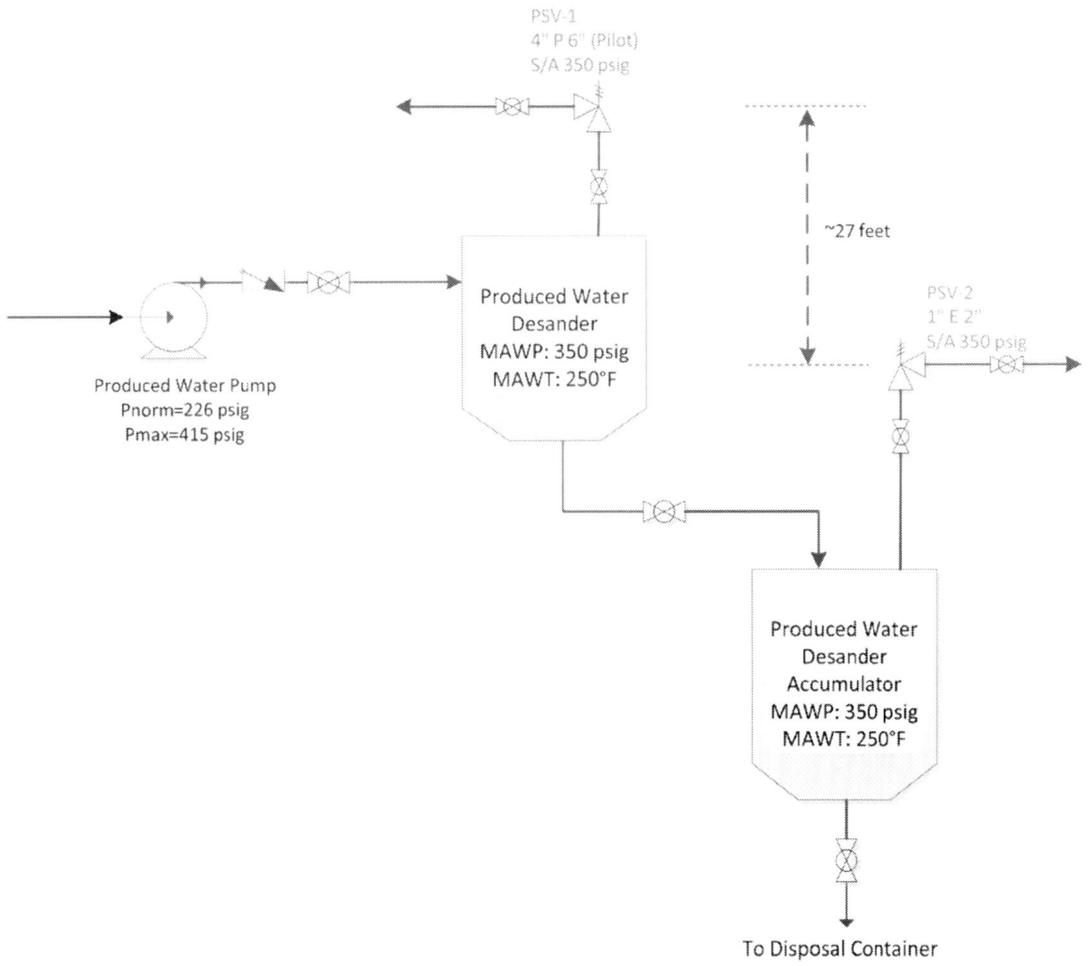

Figure 1, Case 1 – Produced Water System Overview

3 Case 2 – Common Column System

3.1 System Overview

Figure 2 below is a simplified column system that contains several pieces of equipment. The top of column is 50 feet from grade. The reboiler is located at grade. The relief device (PSV-3) is 5 feet above the top of the column and set at 330 psig. The lowest system MAWP is 330 psig based on the MAWPs of the column and the reboiler (shell side). This case study only looks at the "overfilling" scenario for the column. As the lowest system MAWP and the set pressure are the same, it may appear at first glance that the relief device is set correctly to protect the system. However, during the overfilling scenario, there are 55 feet of static head on top of the reboiler (assuming the reboiler is flooded). Therefore, the pressure that the reboiler (shell side) experiences during the liquid relieving event is higher than the allowable accumulation. For this case study, the specific gravity of the fluid is assumed to be 0.7, which would result in roughly 17 psig of static head. This means that during the relief event, the pressure in the reboiler (shell side) is 380 psig (~15% accumulation), which exceeds the 10% allowable accumulation. Therefore, mitigations may be required as the reboiler is not protected due to the static head.

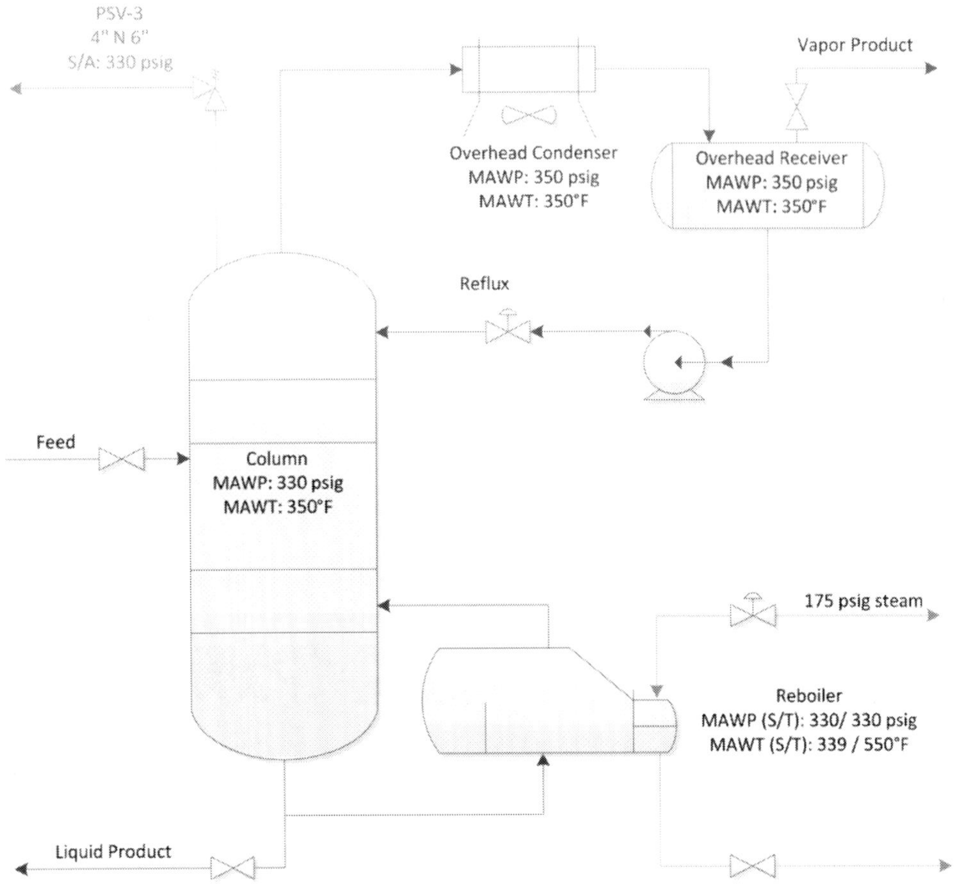

Figure 2, Case 2 - Common Column System Overview

3.2 Case 2 mitigation

There are multiple mitigation options for this type of concern. One mitigation option is to re-rate the reboiler to a higher MAWP to account for the static head. Another mitigation is to lower the set pressure of the relief device to account for the static head. Therefore, the new set pressure for this case study can be 314.5 psig or lower if the reboiler MAWP plus allowable accumulation is not exceeded by the relief pressure plus static head. When the relief device set pressure is lowered to account for the static head during the liquid relief events, engineers must confirm the relief device can still provide adequate capacity at the reduced pressure for all applicable overpressure scenarios. Also, the operating pressure of the column needs to be considered to ensure that the lower set pressure would not result in simmering of the relief device during normal operation. In practice, many plants would also have high independent liquid level alarms and fill time criteria to remove the potential liquid relief scenarios by taking credit for operator responses. Another option is to look at risk assessing the concern as the MAWP is not exceeded by 16% accumulation.

There are many ways to mitigate this concern; this case shows one of the most common mitigation option.

4 Conclusion

The two case studies discussed in this paper show the need to consider the static head when designing the relief system. The issue of static head should be considered during the design phase to ensure the system is code compliant. The cases studies show that the system may not be protected by code, but may be able to be risk accepted based on what the resulting pressure at the vessel is question becomes. It is important during the design phase to be cognizant on the elevation of the systems and the protecting relief devices to ensure the static head has been considered. This can be especially true when designing an offshore platform with multiple levels. Also, when commissioning columns or other vessels, it is important to consider the MAWP of all the vessels in the system to ensure all aspects of the system are protected by the relief device. The case studies give a brief overview of some of the possible mitigation options for static head concerns. It is the responsibility of the engineer reviewing the information to select the option that best fits the facility's needs.

5 References

[1] ASME Section VIII, "Boiler and Pressure Vessel Code," American Society of Mechanical Engineers, 2013.

[2] ANSI/API Standard 520 Part 2, "Sizing, Selection, and Installation of Pressure-relieving Devices," American Petroleum Institute, 6th ed., 2015.

[3] ANSI/API Standard 521, "Pressure-relieving and Depressuring Systems," American Petroleum Institute, 6th ed., 2014.

478548 Study of Factors That Affect Dust Explosion Characteristics of Fibrous Dust Materials

Abstract:

Abstract for 13th Global Congress on Process Safety

Track: Annual Loss Prevention Symposium (LPS)

Study of factors that affect dust explosion characteristics of fibrous dust materials

Bharatvaaj Ravi, Yan-Ru Lin, Chad Mashuga, M. Sam Mannan

Mary Kay O'Connor Process Safety Center

Artie McFerrin Department of Chemical Engineering

Texas A&M University

College Station, TX-77843-3122, U.S.A

+ 1 (774)-262-5215, vaaj26@tamu.edu

Combustible dust hazard is a serious threat to the process industries. Out of the 281 combustible dust incidents reported by the U.S. Chemical Safety and Health Investigation Board (CSB), 66 incidents have been caused by wood dust. In spite of the National Fire Protection Agency's (NFPA) efforts in specifically catering to the wood working and wood-based products manufacturing industries and Occupational Safety and Health Administration's (OSHA) National Emphasis Program on combustible dust hazards, to create awareness and minimize wood dust hazards, the wood dust incidents continue to occur. The fundamental difference between traditional dusts and wood dusts is that the latter are flocculent and fibrous. They are characterized effectively by an aspect ratio (length to diameter ratio) rather than a single dimension, such as diameter, as in the case of traditional - spherical or near spherical - dusts. Various studies in the literature have revealed that the standard dust explosion testing procedures and the categorization of these non-traditional dust based on particle diameter (mesh sizes) might not be valid. This research is concerned with similar line of thought and exploring the effect of aspect ratio on the time of suspension of these fibrous wood dusts. The likelihood of ignition increases with the time of suspension given other conditions to form a combustible dust hazard. The study also deals with characterizing the participating and non-participating portions when a dust explosion occurs through analysis of samples before and after explosion testing. Does the dust concentration give the entire picture of explosion characteristics such as maximum explosion pressure and explosion pressure rise rate? Or is there a better indicator of these characteristics when it comes to flocculent dust? These are the questions that the current research is trying to answer.

Key words: Combustible dust, Fibrous dust, Dust explosion testing, Dust concentration, Dust explosion characteristics, Non-traditional dust.

High Integrity Pressure Protection System (HIPPS) and Implementation Challenges

Janardhanan Kallambettu P.Eng.
Principal Engineer-Control Systems
Bechtel Corporation
3000 Post Oak Blvd, Houston, TX 77459
jkallamb@bechtel.com

Prepared for Presentation at
American Institute of Chemical Engineers
2017 Spring Meeting and 13th Global Congress on Process Safety
San Antonio, TX
March 26 – 29, 2017

High Integrity Pressure Protection System (HIPPS) and Implementation Challenges

Janardhanan Kallambettu P.Eng.
Principal Engineer-Control Systems
Bechtel Corporation
3000 Post Oak Blvd, Houston, TX 77459
jkallamb@bechtel.com

Keywords: PHA, HIPPS, SIS, SIF, SIL, PLC

Abstract

Over the years, the process industries have heavily utilized pressure relief devices (PRDs) to protect the mechanical integrity of system components. While PRDs offer a straightforward and practical way of controlling overpressure in process plants, they do have some inherent limitations such as impacts to flare system design, environmental concerns, and some physical limitations in PRD sizing. In these cases, where the use of PRDs is limited, High Integrity Pressure Protection System (HIPPS) can be utilized as a substitute for PRDs.

While HIPPS offers a solution to address the PRD limitations, there are some challenges associated with a successful HIPPS implementation. The HIPPS are classified as Safety Instrumented Systems (SIS) and require compliance with the functional safety standard IEC 61511[**1**]. The aim of this paper is to guide the reader with the necessary tools to successfully perform all the safety lifecycle activities of HIPPS, per functional safety standards and codes. Challenges in the various phases of the safety lifecycle, from process hazard analysis (PHA) and Safety Integrity Level (SIL) selection to operation and maintenance of HIPPS, are discussed in this paper. Additionally, this article describes many types of HIPPS, their pros and cons, to choose the right system for a given application. The paper concludes with an example HIPPS to achieve high-level risk reductions and demonstrates the appropriate use of quantitative tools, such as fault tree analysis (FTA), during the HIPPS design process.

1 Introduction

HIPPS is an acceptable solution instead of PRDs as a means of overpressure protection for vessels and pipelines. Codes such as ASME Section VIII, Division 1[**2**] and recommended practices and standards such as API Standard 521[**3**], API 17O [**4**] and HSE, U.K [**5**] provide criteria for the usage of HIPPS. The reader is advised to refer to the cited references to understand in detail the selection criteria for HIPPS.

The HIPPS is a safety instrumented function (SIF) implemented in a SIS and, therefore, the codes and the standards referenced above require compliance with the functional safety standards IEC 61508[**6**] and IEC 61511.

The functional safety standards require adherence to a safety lifecycle model. The various safety lifecycle activities are:

- process hazard analysis;
- SIL selection;
- safety requirements specification (SRS);
- SIS design per SRS;
- design verification meeting the target SIL including systematic capability, and architectural constraints (redundancy);
- installation, commissioning, validation, operation, and maintenance per SRS; and
- change management.

The intent of this paper is to expose particularly process engineers and other non-control systems engineers, about specific aspects of the HIPPS in a straightforward language.

2 PHA and SIL Selection

Whenever a HIPPS is indicated instead of PRDs, process and process safety engineers assign an SIL 3 target to the HIPPS as supported by Lawley and Kletz's [7] requirement that the reliability is ten times better than that of a relief device. For clean services, the probability of failure on demand (PFD) of a relief device is typically two orders of magnitude, hence the SIL 3 target for a HIPPS. API Standard 521, while emphasizing the need for an appropriate SIL selection process per established risk tolerance criteria, also suggests SIL 3 target for HIPPS in the absence of a documented risk tolerance criteria.

An assumption, however, that the HIPPS's SIL target is always SIL 3 may not be correct in some circumstances. Applicable codes and standards mandate a requirement for a documented PHA to identify and examine all potential overpressure scenarios. The severity of consequences typically ranges from moderate to catastrophic and, based on the owner's risk tolerance criteria, the HIPPS target can vary from SIL 2 to SIL 4. For example, if the likelihood of an initiating event causing the overpressure is 1 in 10 years and the owner's risk criteria for catastrophic severity is 1in 100,000 years, and in the absence of any other independent protection layers or conditional modifiers, the required risk reduction is 10,000, a SIL 4 target.

Designing a SIL 4 HIPPS is a significant challenge and not recommended. The functional safety standards discourage such high-level risk reduction measures by a single SIF. Every effort should be made to remove or reduce the hazard at the source, introduce other risk reduction measures which are not based on the SIS or minimize the likelihood of the causes of the hazardous event. If the above steps do not help, then use of multiple independent HIPPS, each with a lower integrity, to meet the overall risk reduction requirements should be considered. See section 6 for a hybrid HIPPS design solution meeting an overall risk reduction of 10,000 or more.

3 Types of HIPPS

Many in the process industry believe that HIPPS always includes 2oo3-voted pressure transmitters (sensors), Triple Modular Redundant (TMR) safety PLC (logic solver), and 1oo2-voted ball valves (final elements). In reality, there are many types of HIPPS available to choose from based on the specific application.

HIPPS can be broadly classified as **Electronic**, **Mechanical** or **Hybrid** based on the technology used for the logic solver as shown in **Figure 1**.

A brief description, pros and cons, and applications for each of the above types are given in the following subsections.

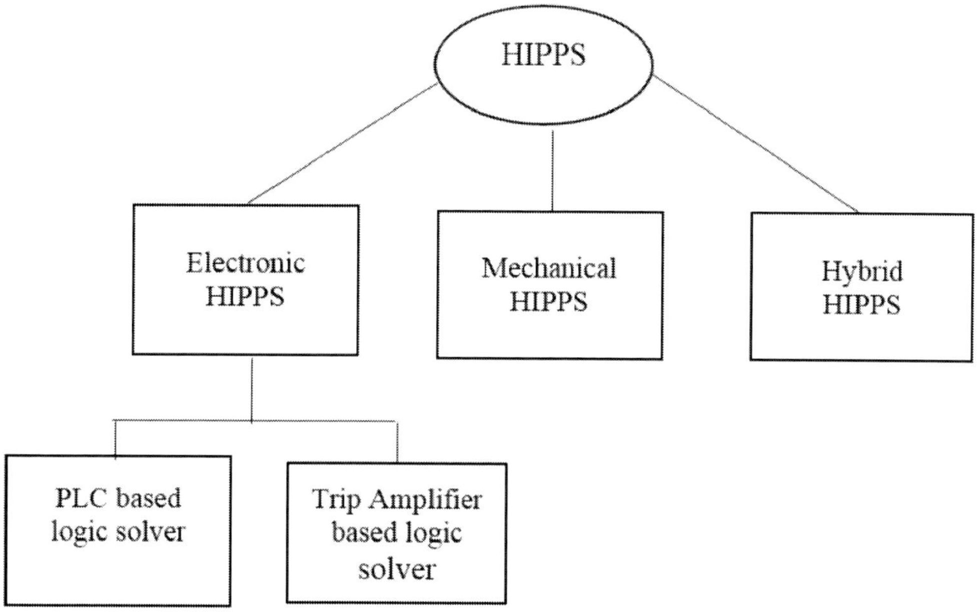

Figure1. Types of HIPPS

3.1 Electronic-HIPPS

An electronic-HIPPS uses either a programmable logic controller (PLC) or a trip amplifier system as a logic solver.

3.1.1 PLC based HIPPS

There are multiple varieties of PLCs equipped with microprocessor or solid state technology in a multitude of architectures such 1oo2, 1oo2D, 2oo3, 2oo4, and 2oo4D. Typically, all these PLCs will meet SIL 3 reliability. These PLCs are further categorized as complex devices and use application programming languages such as ladder logic or functional block diagrams. Refer to **Figure 2**.

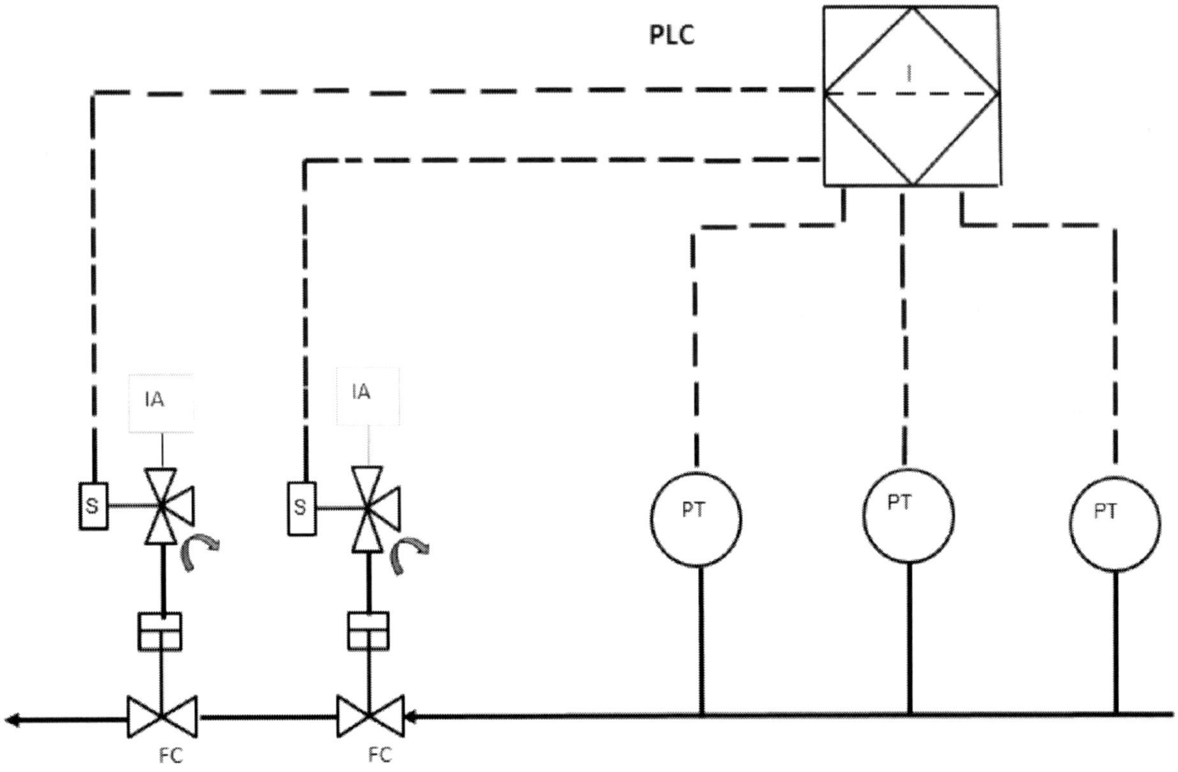

Figure 2. PLC based HIPPS

Pros:

- Flexible in developing the logic program and accommodates higher number of field inputs and outputs (I/Os);
- Equipped with special features for application programming such as variable valve closure speeds while closing the valve to mitigate consequences associated with slamming shut.
- Accommodate multiple HIPPS in a single logic solver within a plant facility;
- Facilitates partial stroke testing to meet target SILs or extend the testing interval, when practicable;
- Easily interfaces with the plant operator stations (human machine interfaces, HMIs).

Cons:

- Often needs specialized application programming expertise, in addition to vendor-specific training for selection of hardware and complex control strategies;
- Validation of the vendor claimed SIL against the safety manual is not a trivial exercise;
- Requires protection against cyber security threats;
- Potential for introducing systematic errors in application program;
- Application program using a low variability language (LVL) calls for special considerations per IEC 61508/IEC 61511 standards including application program safety requirements specification;
- Requires rigorous Factory Acceptance Test (FAT);
- Large footprint is needed -typically 800mmX600mmX1000mm panel;

- Extensive installation requirements such as controlled environment and special grounding;
- Requires complex, and redundant large capacity power supply;
- Voluminous documentation is involved;
- Requires a high level of competence in commissioning, validation, maintenance and operation; and
- Expensive for simple HIPPS applications ($50,000 for a logic solver alone).

Applications:

In a large facility, where many HIPPS are to be implemented for different hazardous event scenarios requiring complex functions or features, a PLC based HIPPS is recommended. For example, over pressure protection of multiple vessels, various piping segments, and process equipment within a facility to avoid loading of the flare system.

3.1.2 Trip-Amplifier based HIPPS

Trip-amplifiers are discrete logic solvers of low complexity. A trip amplifier accepts analog 4-20mA input from a field device such as a pressure transmitter and de-energizes contact output(s) when the process value exceeds the trip set point. Each sensor needs a dedicated trip amplifier. Refer to **Figure 3**.

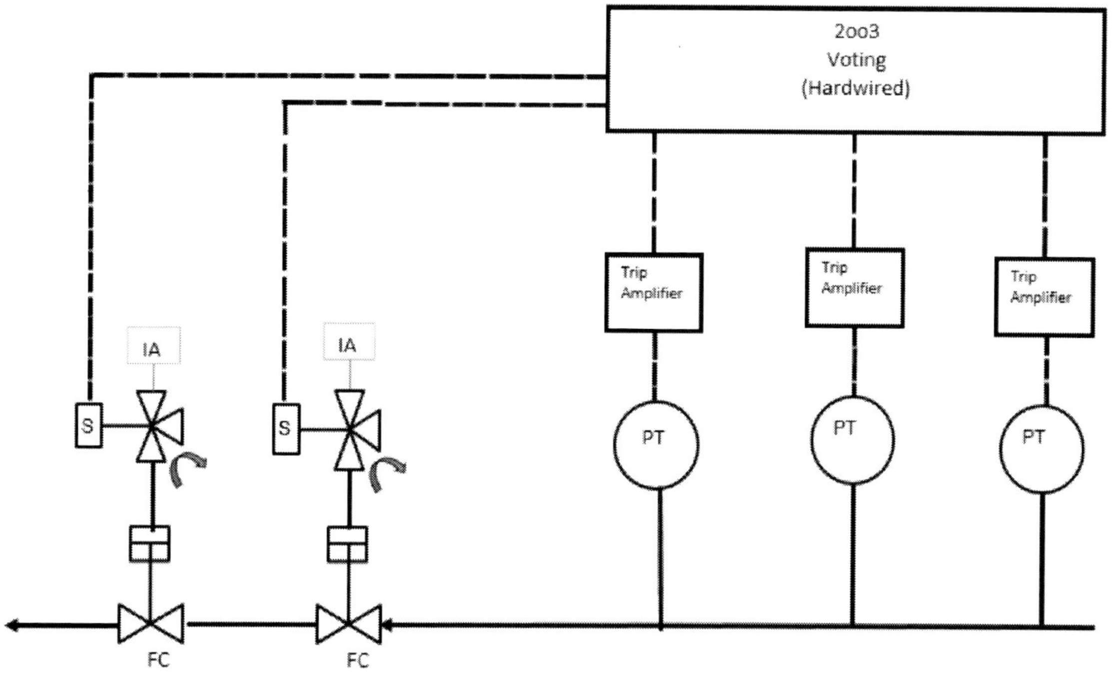

Figure 3. Trip Amplifier based HIPPS

Pros:

- Simple, reliable, with low design complexity;
- Meets SIL 2 reliability without redundancy for Trip Amplifiers and SIL 3 reliability with redundant architecture;
- Does not require application programming; classified as a device with fixed programming language (FPL) per IEC 61508/IEC 61511and, therefore, eliminates the need for application program safety requirements specification;
- No cybersecurity related issues;
- Less potential for systematic errors;
- Well suited for simple logic such as HIPPS logic;
- Low space requirement-typically DIN rail mounted;
- Low power device with no special requirements;
- No need for a controlled environment;
- Being a simple device FAT is not required for the logic solver;
- Easy to install and maintain; and
- Low cost; (logic solver range less than $10000)

Cons:

- Requires external devices for complex logic such as variable speed closing times for valves to avoid slam shut scenarios;
- Interfacing with operator stations requires additional I/Os in the plant DCS;
- Difficult to interface and facilitate partial stroke testing;
- Voting among the trip-amplifiers are external to the trip-amplifiers using the trip contacts and physical wiring which is less straightforward; and
- Implementation of maintenance bypasses is not practical;

Applications:

Well suited for applications involving straightforward, simple logic and for facilities which do not use PLCs. Also, appropriate for applications requiring a small footprint, low lifecycle cost and facilities not having multiple HIPPS. For example, remote pipeline compressor stations or storage tanks.

3.2 Mechanical-HIPPS

A mechanical HIPPS uses mechanical initiators. The mechanical initiator not only senses the over pressure but also serves as a logic solver to detect high pressure and actuate the valve. Mechanical HIPPS do not use any electrical or electronic components and, therefore, are inherently safer. Standalone versions of mechanical HIPPS are available which do not require any energy sources, such as instrument air supplies. Refer to **Figure 4**.

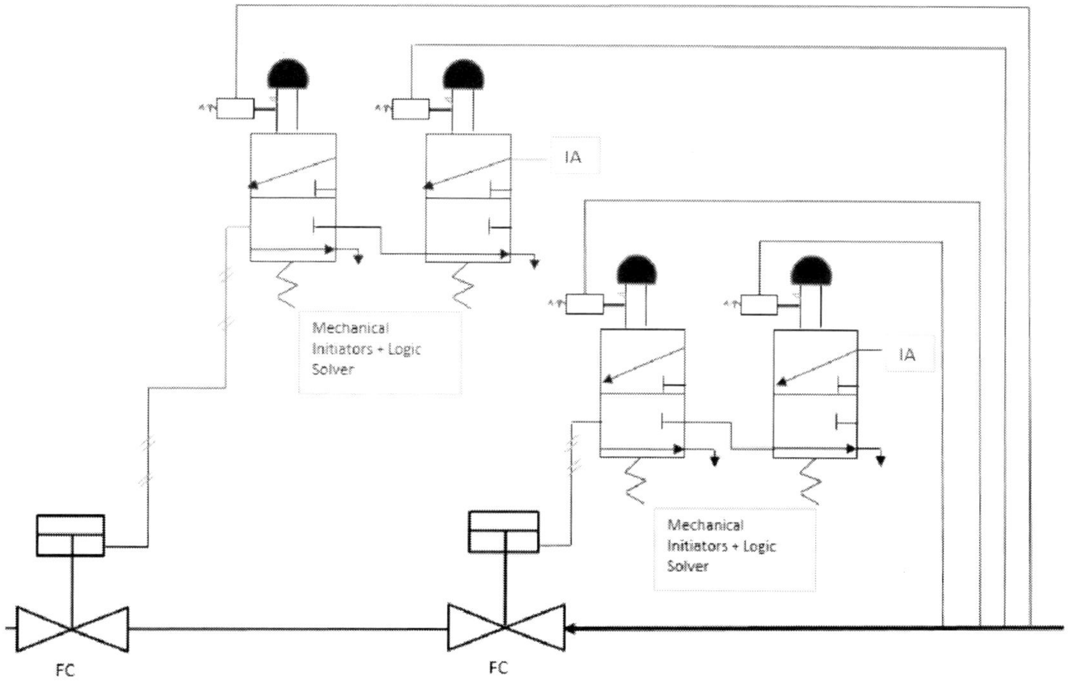

Figure 4. Mechanical HIPPS

Pro:

- Simple, reliable and minimally complex;
- Does not involve wiring and programming;
- Less susceptible to systematic errors;
- Cables, cable trays, conduits, etc. are not required;
- Can be made standalone without external energy sources;
- Single valve system can meet SIL 2 targets, redundant valve configuration can even meet SIL 3;
- Third party agencies certify an entire system in contrast to individual elements/subsystems of electronic HIPPS;
- Fewer components to fail;
- Easy to install, commission, validate, maintain and operate;
- Minimal documentation: no wiring diagrams or, installation drawings for logic solvers and initiators, application program SRS required;
- Competence in programming or maintaining a complex electronic system is not required;
- There are no EMC interference issues;
- There are no impacts due to cyber security threats;
- Suitable for harsh environments-no environmental climate control is needed;
- Easy to comprehend by operators and maintenance mechanics;
- Lowest life cycle cost among all the types of HIPPS;

Cons:

- Limited distance between the pressure sensing points, mechanical initiators and the valve- all components must be in the close proximities to avoid transportation time lags impacting the response times;
- Mechanical initiators need to be carefully selected as they are sensitive to setpoints and their accuracies;
- Interfacing with operator stations requires additional transmitters in the field and Inputs/outputs (I/Os) in the plant DCS;
- Requires external devices to facilitate partial stroke testing;
- Implementing maintenance bypasses consistent with the overall facility philosophy is not practical;

Applications:

Mechanical HIPPS are attractive for standalone applications in remote plant locations, for areas where electrical cabling is impractical and in facilities, like pipeline pump/compressor stations, where PLCs are not commonly used.

3.3 Hybrid-HIPPS

Hybrid-HIPPS are a combination of electronic and mechanical HIPPS.

Pros:

- High level of risk reduction- 10,000 (SIL 4) or higher can be obtained;
- Significant reduction of common cause and common mode failures due to diversity (mechanical and electronic);
- Low-level of systematic errors;

Cons:

- Complex system;
- Requires high level of competence in the installation, commissioning, validation, maintenance and operation;
- Requires a sizeable spare parts inventory;
- Many of the Cons of electronic and mechanical HIPPSs described above are applicable for the hybrid HIPPS as well.

Application:

High-pressure feed gas to a facility typically requires a high-risk reduction in the absence of non-instrumented IPLs. When a high level of risk reduction (greater than 10,000) is required, hybrid systems are well-suited as they are diverse systems with minimum common cause and common mode failures.

4 Safety requirements specification (SRS)

Functional safety standard IEC 61511 requires the development of a SIS safety requirements specification (SRS), including application program safety requirements, when applicable. HIPPS as a functional SIS, requires an SRS. There are a minimum of 29 SRS requirements specified by the standard. While every element is important, a few present unique challenges that can significantly impact the selection, design, and quantification of random hardware failures of the HIPPS.

4.1 Process safety time (PST)

Process safety time, per clause 3.2.52.1, IEC 61511, is the period between a failure occurring in the process or the basic process control system and the occurrence of a hazardous event, if the SIF is not performed. The SIF has to detect the failure and complete its action soon enough to prevent the hazardous event taking into account any lag in the process.

HIPPS that implements the overpressure protection function must complete the action to bring the process to a safe state within the process safety time. Quite often, the process safety times for HIPPS are less than few seconds. The valve subsystems may need quick exhaust valves, larger size air supply tubing, or solenoid valves to meet the response times. Therefore, the PST determination is an important specification which in turn determines various components within the HIPPS to achieve the response time.

4.2 HIPPS output action success criteria

Output actions of a SIF will prevent a particular consequence of concern only when the action is completed with certain performance criteria. For example, stating an action as to close a valve is inadequate. The criteria for the successful operation should be specified. Many HIPPS applications do not require tight shut-off requirements for valves. However, there are situations where larger leakage rates may impact the process safe states. In such scenarios, the process engineer must specify an acceptable leakage rate for the valves, and ensure the design shall have means to achieve such low leakage rates. The design may require unique valve types and trims. Also, additional pressure transmitters and isolation/bypass valves are required for measuring the extent of leakages.

Low leakage rate specification may impact the proof test coverages for the final elements in the quantification of random hardware failures (PFD_{avg}). Achieving a SIL 3 target with tight shut-off requirements is a significant challenge.

4.3 Proof test interval and overhaul interval (mission time)

IEC 61511 defines the proof test as a periodic test performed to detect dangerous (hidden) failures in a SIS so that, if necessary, a repair can restore the system to an "as new" condition or as close as practical to this condition.

The frequency of testing plays a significant role in determining the HIPPS's availability to perform intended safety function on demand. However, frequent testing may affect the production, introduce spurious shutdowns due to human error or affect the life of a device. If only a full stroke test is performed for a valve, then the test is imperfect since it will not detect all the covert failures. The undetected, hidden failures are detected and repaired during overhaul periods (mission time). The longer the mission time, the higher the PFDavg; therefore, it will be difficult to achieve SIL 3 targets with longer mission times.

Bypass valves and isolation valves are required for proof testing intervals shorter than the overhaul periods to avoid process interruption. Provision of bypass and isolation valves increases the complexity of the design, footprint, capital and the life cycle costs of the facilities.

4.4 Survivability

As per IEC 61511, the SRS shall specify any requirements for a SIF to survive a major accident event, e.g. the time required for the HIPPS valve to remain in operation in the event of fire shall be specified. Survivability specification impacts the selection of valve trims and additional fireproofing requirements.

4.5 Common cause failures

The SRS must specify the measures to identify and minimize common cause failures. Common cause failures will significantly impact PFD_{avg} for high SILs, such as SIL 3 and SIL 4. Common cause failures can be due to process conditions or similar redundant equipment configurations or systematic failures.

Physical and electrical separation and use of diverse devices which respond differently to the common stresses will minimize common cause failures. An effective functional safety management plan and verification processes will reduce systematic failures.

5 Quantification of random failure of HIPPS

The HIPPS design must be evaluated to ensure that the calculated failure measure (PFDavg) is less than the target related to the SIL as specified in the SRS.

The functional safety standard mandates assessment of the following as part of the SIL verification:

- Systematic capability of all the elements associated with a HIPPS meets the specified SIL target
- Calculated PFD_{avg} is less than the target PFD_{avg} of the selected SIL
- Hardware fault tolerance (HFT) complies with the required SIL target

5.1 Systematic capability

IEC 61511 defines the "systematic capability" as a measure (expressed on a scale of SC 1 to SC 4) of the confidence that the systematic safety integrity of a device meets the requirements of the specified SIL, in respect of the specified safety function, when the device is applied in accordance with the instructions specified in the device safety manual.

Systematic capability assessment is for addressing systematic failures which are not quantified and hence not included in the SIL verification calculations. IEC 61508 and IEC 61511explicitly identify a systematic capability parameter (SC). A systematic capability of "SC N" for a device means that the systematic safety integrity of "SC N" has been met when the device is applied following the instructions specified in the instrument safety manual. If a device is to be used in an SIL 3 application, then it must be assessed for the systematic capability of SC 3.

The systematic capability of a device depends on the nature of the device. If a device has no software, then only hardware systematic failure mechanisms are considered. If a device has both hardware and software, then both hardware and software failure mechanisms shall be considered.

There are agencies, such as TUV, Exida, and SIRA that perform systematic capability assessment of a product as per IEC 61508 and provide a certificate with a report specifying an SC parameter. Following a well-established functional safety management plan during detailed engineering, installation, operation and maintenance phases ensures that the systematic safety integrity of a device meets the requirements of the specified SIL.

5.2 PFD_{avg}

Many factors influence the PFD_{avg} of HIPPS function. A few of the factors that pose unique challenges are considered in this section.

5.2.1 Failure rates

There is always an uncertainty in the quantification of random failures. The failure rate is, probably, the least precise parameter used in performing SIL verifications [**8**]. Therefore, one must acknowledge that the results of the quantification are not real values, but rather a basis for comparing different design options and for monitoring the reliability performance during the operational phase of the safety lifecycle. The analysis is only a prediction. Historical performance is not the same as future performance.

If the failure rates are conservative, then achieving target SIL will be demanding. However, optimistic failure rates will result in inadequate protection integrity. Therefore, the failure rates of components used in the analysis should be sufficiently realistic.

There are many sources of failure rate data, such as application-specific, site-specific, company-specific, manufacturer-specific, industry-specific, and generic data. These failure rates can vary significantly by several orders of magnitude. While application specific or site specific data is the best option, but it can often be difficult to obtain. A detailed discussion on the selection of failure rates to use in the analysis is beyond the scope of this paper, and the reader is advised to refer to other sources [**9**].

5.2.2 Proof test interval, test coverage and mission time

As explained in section 4.3, a proof test may be imperfect based on the following:

- The extent of tests that could be performed to reveal the hidden failures. For example, if the valve is only full stroke tested without examining the internals then the proof test is imperfect;
- The conditions under which the proof test is carried out may differ from a real demand;
- Lack of competence of testing crew;
- Inadequate procedures.

Perfect testing is not a realistic assumption. IEC 61511/IEC 61508 requires consideration of the impact of imperfect testing in the analysis. In the SIL verification calculations, a proof test coverage (PTC) is considered for incomplete testing. The incomplete testing will increase the PFDavg. The lower the proof test coverage, the higher the PFD_{avg}. Undetected dangerous failures due to imperfect proof testing will be detected either during the complete overhaul period (when repaired or refurbished) or when the component is replaced or during an actual demand. Mission time is the terminology used by many reliability specialists for the above overhaul period.

For example, if an overhaul period or demand interval for HIPPS is ten years and the imperfect proof test interval for a valve is one year with a proof test coverage of 90%, then achieving an SIL 3 target for HIPPS is an arduous task. If tight shut-off requirement is specified, then the proof test coverage could be even less than 30%.

From the above, it is evident, that the significance of proof test interval, the effectiveness of proof test, and the mission time cannot be ignored and must be taken into account in the analysis.

5.2.3 Partial stroke testing

Partial stroke test will help to improve the PFD_{avg}, extend the proof test interval, or increase the safe failure fraction. A single valve with partial stroke test may achieve an SIL 2 integrity. However, for many HIPPS with short process safety times, partial stroke test may not be useful. Moreover, partial stroke tests may result in spurious trips and cause potential damage/wear to the valve seat due to the frequent testing.

A perusal of literature reveals that the partial stroke testing coverage is in the range of 40-80% [10]. However, the above-stated coverage is widely debated.

If the partial stroke testing is not automated, then it is not considered as the diagnostic testing per functional safety standards. Also, the analyst shall take into consideration the partial stroke testing interval with the repair time in the PFD calculations.

5.2.4 Common cause failures

Common cause failures impact the redundant components such as valves and sensors. The common cause failures occur simultaneously or, at least, within a short interval as to be perceived as simultaneous.

The following are the defenses against the common cause failures:

- Diversity (different types of valves and systems);
- Separation (physically and electrically);
- Staggered testing;
- Use of low complexity devices- simpler equipment less prone to common cause failures;
- Well administered and complied procedures;
- Environmental controls;

When diverse components are used, the common cause factor (Beta factor) is minimal. Using IEC 61508, Part 6 [5], Annex D, Tables, D.1, D.2 and D.3, the Beta factor can be estimated to be as low as 2% with diverse components. However, the diverse components must be suitable for the application with comparable performance without any degradation.

Also, staggered testing reduces the overall PFD_{avg} by a factor of 30 to 50% depending upon the voting configurations [**10**].

5.2.5 Modeling of components

Modeling of HIPPS elements in the analysis must include all the components that are involved in bringing the process to a safe state within the process safety time.

Many HIPPS valves are equipped with quick exhaust valves (QEVs) to reduce the valve closing time to meet the process safety time. Sometimes, redundant QEVs are provided in various voting configurations such 1oo2 or 3oo4, etc. The contribution of QEV failures to the PFD_{avg} is significant and cannot be ignored. Therefore, the reliability analyst must consider the QEVs with their appropriate voting architecture in the PFD_{avg} calculations. Similarly, the likelihood of plugging of sensing lines must be taken into account in freezing applications.

Any interfacing components such as barriers, isolators, interposing relays, and poppet valves, which are part of the safety instrumented function to achieve or maintain the process to a safe state must be modeled and analyzed.

5.3 Hardware fault tolerance (HFT)

Irrespective of the hardware reliability calculated (PFD_{avg}) for the design, the functional safety standard IEC 61511 specifies minimum levels of redundancy for each safety integrity level.

There are two approaches:

- Route 1_H: Determination of HFT based on the calculated safe failure fraction (SFF);
- Route 2_H: Determination of HFT based on field experience and field data (proven-in-use)

5.3.1 Route1$_H$

The Route1$_H$ approach uses SFF and the SIL target to determine the hardware fault tolerance. The SFF is defined as the sum of the potentially dangerous failures revealed by auto-test and those failures which result in a safe state, as a fraction of a total number of failures.

$$SFF = \frac{[dangerous\ detected\ failures(self\ test) + safe\ failures]}{Total\ failures}$$

The above is calculated for each element (sensor, complete valve assembly, logic solver). Two tables, Table 2 and Table 3, are provided in IEC 61508- Part 2. Table 2 is for "Type A" devices which are simple devices such as valves (no microprocessors and programming are involved), and Table 3 for "Type B" devices which are complex devices such as electronic transmitters, trip amplifiers, and PLCs. Based on the SFF, type of the device, and the SIL target, the hardware fault tolerance is determined using Table 2 and Table 3.

5.3.2 Route2$_H$

In Route 2$_H$, the hardware fault tolerance is specified by the functional safety standard for each SIL target if the device is selected based on well documented and verified field based failure rate data with 90% statistical confidence. For low demand application with SIL 1 or SIL 2 target, there is no need for any redundancies (HFT=0), and SIL 3 requires one hardware fault tolerance.

Unless the field failure data with 90% confidence levels are available, the Route 2$_H$ cannot be justified. There are third party agencies who provide Route 2$_H$ certifications and should be evaluated carefully before adopting the Route 2$_H$ approach based on those certifications.

6 High level risk reductions (RRF> 10,000)

As discussed in section 2, sometimes, PHA/SIL selection may indicate a required risk reduction greater than 10,000 in the absence of any non-instrumented protection layer or, human intervention is impractical due to process safety time considerations or safety reasons.

If there are no options other than providing HIPPS with a risk reduction greater than 10,000, then the functional safety standard IEC 61511 mandates certain requirements. The following are a few of the clauses and their requirements:

Clause 9.2.6: *If further consideration of application and confirmation that a risk reduction requirement >10,000....is still required, then consideration should be given to achieving the safety integrity requirement using a* <u>*number of protection layers with lower risk reduction requirements*</u>. *If the risk reduction is allocated to multiple protection layers, then such protection layers shall be independent from each other or the lack of independence shall be assessed and shown to be sufficiently low compared to the risk reduction requirements.*

Clause 9.2.7: *If a risk reduction greater than 10,000....is to be implemented, whether allocated to a single SIS or multiple SIS or SIS in conjunction with a BPCS protection layer, then a further risk assessment shall be carried out using a quantitative methodology to confirm that the safety integrity requirements are achieved.*

Clause 9.2.8: *If the risk reduction required for a hazardous event is allocated to multiple SIFs in a single SIS, then the SIS shall meet the overall risk reduction requirement.*

A hybrid HIPPS using the combination of electronic and mechanical HIPPS is a solution to achieve risk reductions greater than 10,000. Each HIPPS can be designed to be better than SIL 2, and the overall risk reduction will be better than 10,000. The performance of electronic and mechanical HIPPS are comparable.

High level of risk reduction is achieved due to the diversity and very low common cause failures between the electronic and mechanical HIPPS. All the subsystems of the HIPPSs are diverse with different technologies. Also, systematic failures are minimal. The integrity of such multiple HIPPS is best analyzed using a fault tree analysis (FTA).

An example fault tree analysis for a hybrid HIPPS using ISOGRAPH's Reliability Workbench™ is shown in **Figure 5**.

The example hybrid HIPPS achieved a risk reduction of 12,815(PFD_{avg} = 7.80E-05).

The FTA shows an approach for modeling of various components including the common cause failures. The failure rates assumed are generic for the purpose of demonstration. The actual failure rates and other factors are application specific and cannot be generalized.

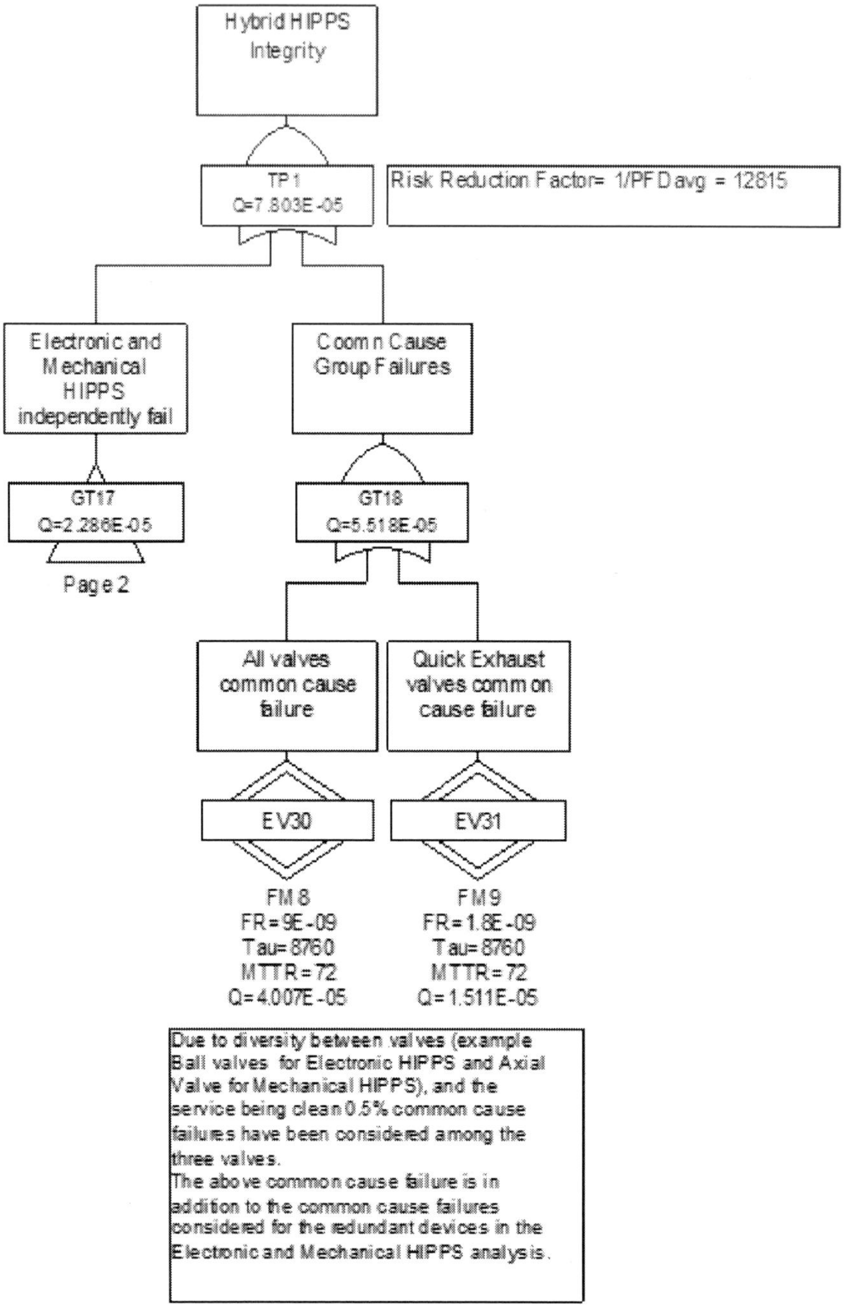

Figure 5. FTA for multiple HIPPS to achieve high risk reduction

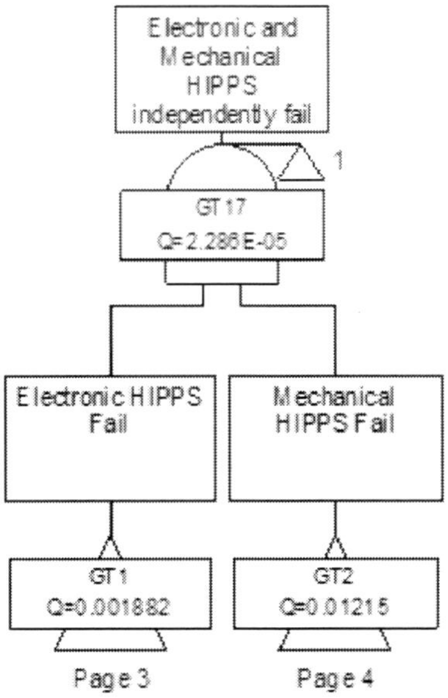

Figure 5. FTA for multiple HIPPS to achieve high risk reduction (contd..)

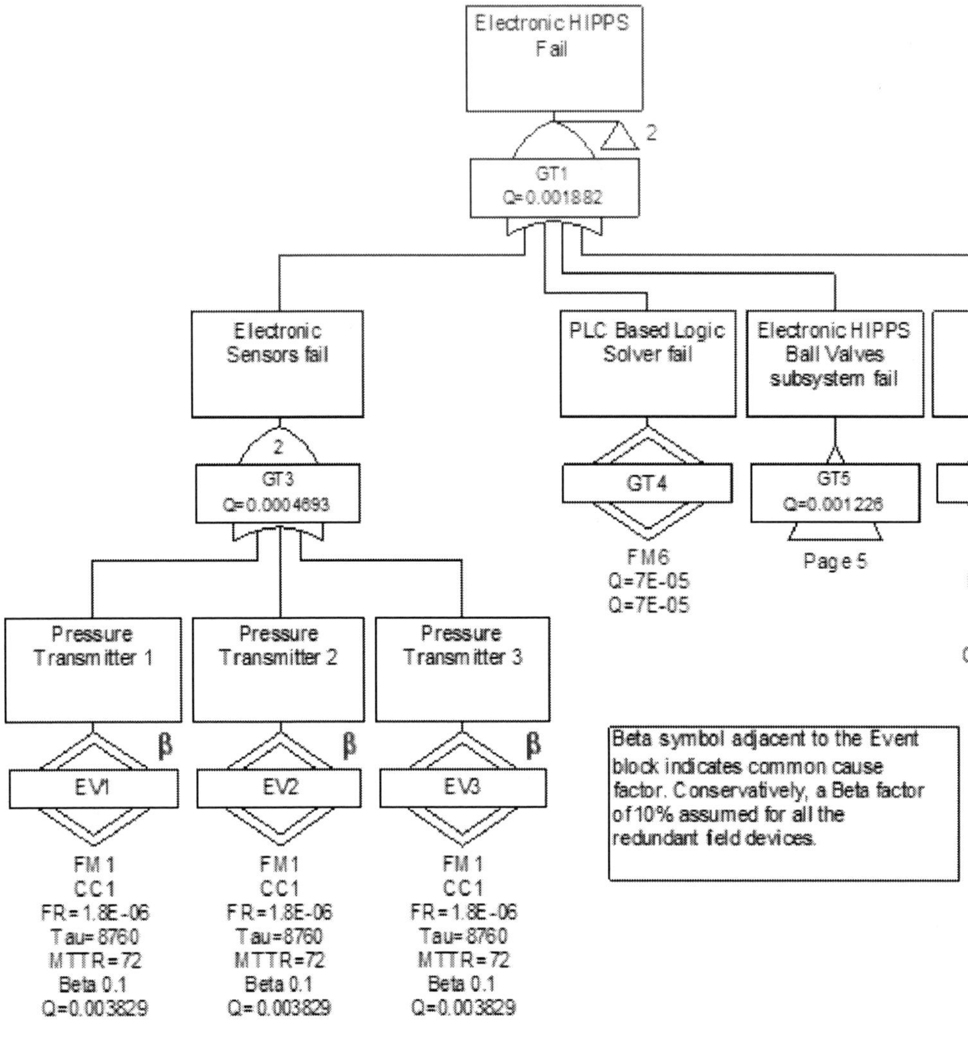

Figure 5. FTA for multiple HIPPS to achieve high risk reduction (contd..)

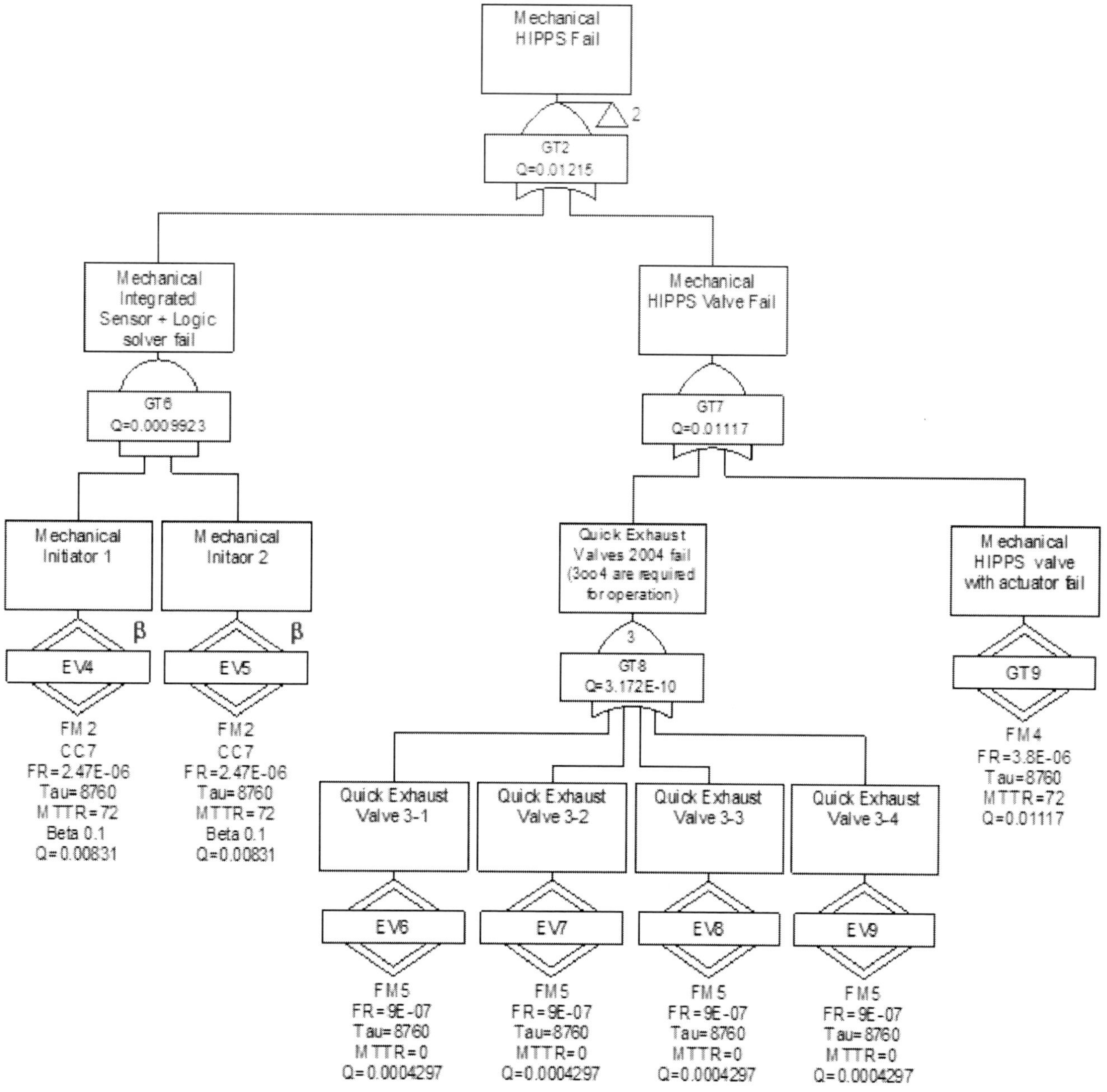

Figure 5. FTA for multiple HIPPS to achieve high risk reduction (contd..)

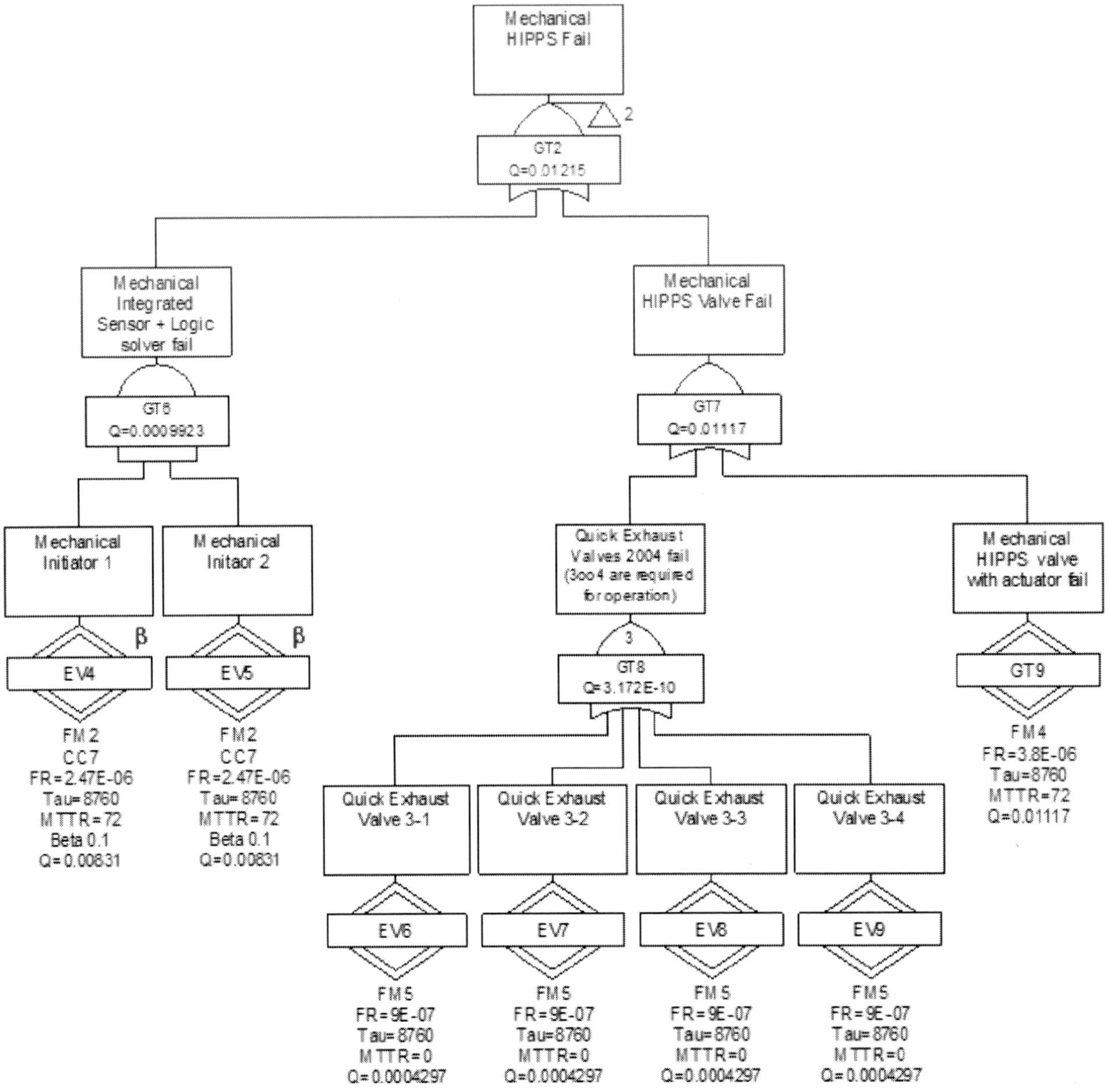

Figure 5. FTA for multiple HIPPS to achieve high risk reduction (contd..)

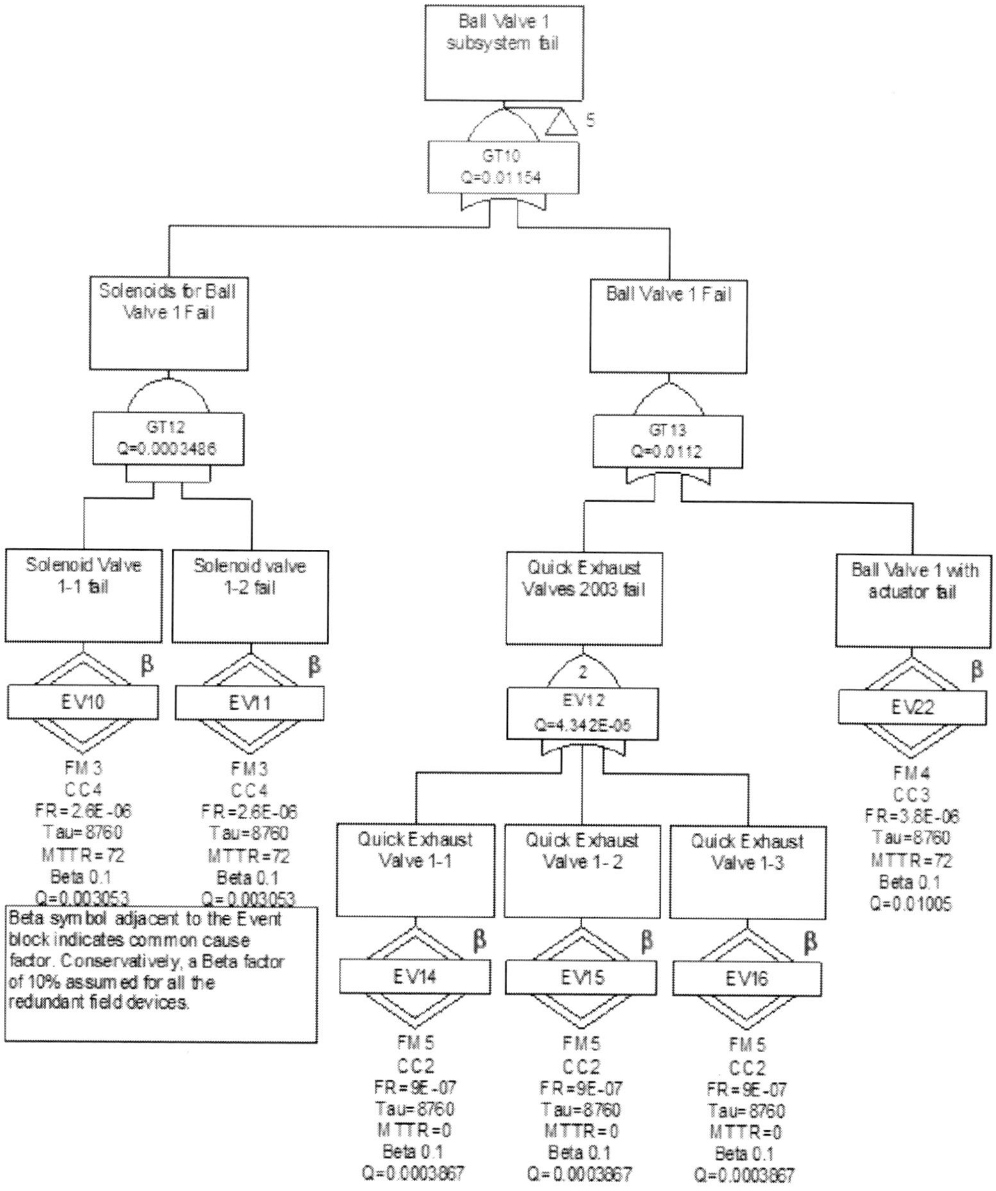

Figure 5. FTA for multiple HIPPS to achieve high risk reduction (contd..)

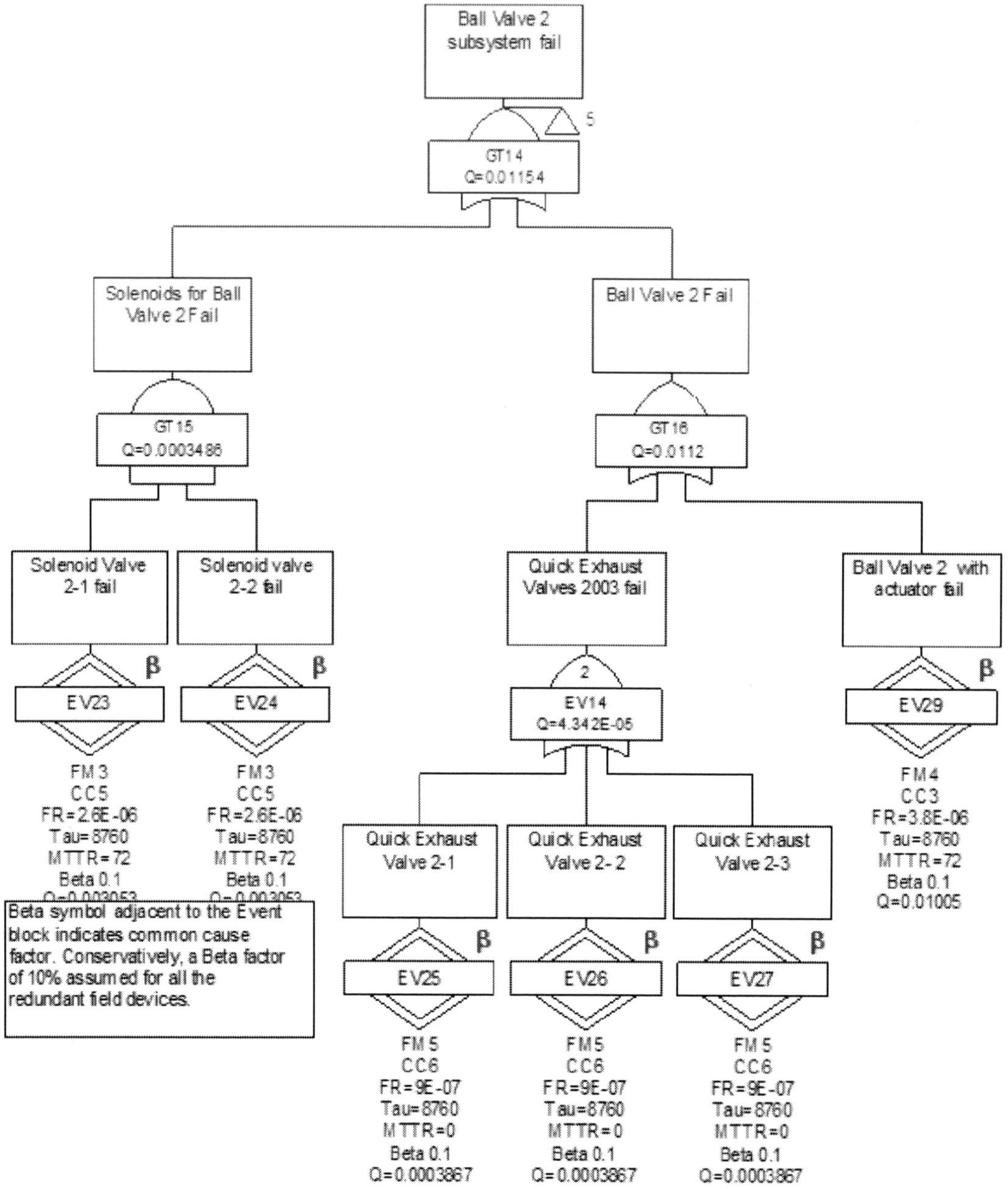

Figure 5. FTA for multiple HIPPS to achieve high risk reduction (contd..)

7 Installation, commissioning, operation and maintenance

Installation and commissioning of the HIPPS must be completed in accordance with the installation drawings, safety manuals, and the SRS.

Operation and maintenance is the longest period in the safety life cycle. The functional safety standards mandate continuous performance monitoring, including demand rates, failure modes and failure rates of various components. The actual performance must be compared with the predicted performance to identify any gaps so that remedial measures can be provided with appropriate safety review.

A robust change management process must be in place to ensure HIPPS integrity throughout the operation and maintenance phase.

8 Conclusion

Assignment of an SIL target and choosing an appropriate technology for a HIPPS is not trivial. In this paper, the significance of PHA and SIL selection for HIPPS has been re-emphasized and a brief account of available options for HIPPS to choose from, their pros and cons, and the selection criteria with examples are provided.

The design, installation, operation, and maintenance of HIPPS shall comply with the requirements of functional safety standards IEC 61508 and IEC 61511. In this article, the challenges in each of the above activities are discussed and appropriate solutions to mitigate the above challenges are provided.

In applications where a high risk reduction greater than 10,000 is required a hybrid HIPPS design approach is proposed. Using FTA demonstrated the modeling of complex hybrid HIPPS to achieve high level risk reduction.

9 References

[1] IEC. IEC 61511 Functional Safety: Safety Instrumented Systems for the Process Industry Sector- Part 1-3, Edition 2, Geneva, 2016.

[2] American Society of Mechanical Engineers (ASME), Boiler and Pressure Vessel Code, Section VIII-Pressure Vessels, New York, 2010.

[3] API, Standard 521, Pressure-relieving and Depressuring Systems, 6th Edition, Washington, DC, 2014.

[4] API, RP17O, Recommended Practice for Subsea High Integrity Pressure Protection Systems (HIPPS), 1 Edition, Washington, DC 2009.

[5] HSE, UK. SPC/TECH/OSD/31, Safety instrumented systems for the overpressure protection of pipeline risers, Available at

http://www.hse.gov.uk/foi/internalops/hid_circs/technical_osd/spc_tech_osd_31.htm, Accessed on Jan 29, 2017.

[6] IEC. IEC 61508 Functional safety of electrical/electronic/programmable electronic safety related systems- Part 1-7, Edition 2, Geneva, 2010.

[7] H.G. Lawley and T.A. Kletz. High-pressure-trip systems for vessel protection, Chemical Engineering, Vol 82, (1975).

[8] Dr. D. J. Smith, Reliability, Maintainability and Risk, 8th Edition, Butterworth-Heinemann, Waltham, MA, 2011.

[9] S. Hauge, S. Habrekke, M.A. Lundeteigen. Reliability Prediction Method for Safety Instrumented Systems- PDS Example Collection, SINTEF Technology and Society, Safety Research, 2010.

[10] Dr. D.J. Smith, and K.G.L. Simpson, The Safety Critical Systems Handbook, 4th Edition, Butterworth-Heinemann, Cambridge, MA, 2016.

475693 Managing Risk Associated w/ Electrical Energy Ignition Sources

Abstract:

In facilities that handle flammable hydrocarbons it is extremely important that all forms of potential ignition sources be carefully managed. This presentation will focus on key points associated with process electrical safety, open flame process heaters, and mechanical forms of ignition sources. OSHA 1910.119 has for many years required that hazardous area classification documentation remain in an evergreen state that requires re-certification at least once every five years. For most process facilities this OSHA mandate presents many problems. In most cases documentation has not been maintained in a current state, the facility perhaps has changed owners if not once many times over the years. The presentation attendee will be exposed to a common sense methodology that focuses primarily on minimizing electrical energy from becoming a source of ignition. A process for updating hazardous area classification documentation will be presented with tips on providing the end-user with an audit proof set of documentation. Once the hazardous area map has been developed then methods of properly documenting and conducting a site compliance assessment will be explained. This compliance assessment is a gap analysis that compares the installed electrical systems against the requirements of the National Electrical Code (NFPA 70 Article 500). Methods of presenting this documentation will be explained so the end-user has a clear understanding of the proper steps for remediation.

The presentation attendee will be exposed to the following:

- Managing risk associated with roadways that are impinged by hazardous classified boundaries
- Open Flame Process heaters encroachment by hazardous classified locations
- Methodology for area classification assessment *(it's determined by the process conditions)*
- Hazard's compliance for instrumentation and electrical system installations in hazardous process locations
- How to interpret documentation in support of the area classification assessments which include:
    - Plan Drawings (maps), Elevations, Site Specific Details. Legends, and notes
    - Assessment report documentation that addresses General Considerations, Identification of Point Sources of Emissions, Physical properties of Materials being Handled, Processed, or Stored, and Gas Dispersion Modeling Graphs and plots
- Methods for mathematical determination of hazard radius thru the concept of commercially available computer assisted gas/vapor dispersion modeling software.
- Application of common standards, codes and installation practices i.e. (NFPA, ISA, API and OSHA)

Detailed handouts as well as a complete audio-visual presentation are used to supplement the presentation.

479320 Review and Analysis of Recent Cyber Security Attacks

Abstract:

Process plants need to ensure that adequate safeguards are in place for a variety of upset conditions, including cyber security attacks, which can result in loss of product, damage to equipment, downtime, and potential safety and environmental disasters. Select cyber security incidents across a variety of industries will be reviewed to determine the modes of cyber security attack and their subsequent impacts. Specific areas of vulnerability will be identified with corresponding mitigation strategies. Possible solutions will be presented for incorporating cyber security into an existing process safety program to ensure a tolerable level of protection for high-consequence events.

478316 Resilience-Based Survival Analysis in the Event of a Process Upset: A Case Study of Utility Plant

Abstract:

Abstract: In a process plant system, safe and reliable operations are highly sensitive to the power or steam failure, loss of cooling water or nitrogen or instrument air. Most of these products and services from utilities plant play an important role as safety barriers. The main focus of this paper is to propose a model for survival evaluation of a process system under an upset condition. This model is based on the holistic resilience-based method that integrates both technical (process parameters variations) and social (policy/regulations, human and organizational) factors. The resilience methodology emphasizes non-linear dynamics, new types of threats, uncertainty, survival through upset conditions and recovery from catastrophic situations [1], [2].

The purpose of this paper is to maximize the performance of safety barriers that will be utilized in case of a process upset condition to prevent the escalation of the event into a catastrophic incident. In this work, a methodology is presented to select the best maintenance strategy [3] while considering the statutory safety regulations for maintenance of equipment like boilers, generators and more. Three different types of maintenance alternatives considered as part of this analysis are: predictive, preventative and condition-based. The key ideas of survival evaluation are presented using a utility plant as the case study.

References

1. Dinh, L. T., Pasman, H., Gao, X., & Mannan, M. S. (2012). Resilience engineering of industrial processes: principles and contributing factors. Journal of Loss Prevention in the Process Industries, 25(2), 233-241.
2. Jackson, S. (2009). Architecting resilient systems: Accident avoidance and survival and recovery from disruptions (Vol. 66). John Wiley & Sons.
3. Vassiliadis, C. G., & Pistikopoulos, E. N. (2001). Maintenance scheduling and process optimization under uncertainty. *Computers & Chemical Engineering*, *25*(2), 217-236.

478816 Model-Based Pressure Relief Design for Reactive Systems

Abstract:

Abstract

Reactive systems behave unpredictably under overpressure conditions. Varying composition impacts the physical properties and response of the mixture to temperature and time. Pressure generation in a reactive system is often a function of both vapor pressure and gas generation rate. Standard methodologies applied in non-reactive pressure relief system design cannot be relied upon to give appropriate sizing for reactive systems. Consistent with DIERS methodologies, model-based design provides accurate, reliable vent sizing. A comprehensive approach to modeling is presented which incorporates vapor-liquid equilibrium, reaction stoichiometry, and reaction rate.

Dynamic Modeling of Inbreathing Requirements for Low-Pressure Storage Tanks

Dona Abou-Chakra
ioMosaic Corporation
93 Stiles Road | Salem, NH 03079
chakra.d.nh@iomosaic.com

Georges A. Melhem, PhD, FAIChE
ioMosaic Corporation
93 Stiles Road | Salem, NH 03079
melhem@iomosaic.com

Ronald J. Willey, PhD
Northeastern University
360 Huntington Avenue | Boston, MA 02115
r.willey@neu.edu

Prepared for Presentation at
American Institute of Chemical Engineers
2017 Spring Meeting and 13th Global Congress on Process Safety
San Antonio, TX
March 26 – 29, 2017

Dynamic Modeling of Inbreathing Requirements for Low-Pressure Storage Tanks

Dona Abou-Chakra
ioMosaic Corporation
93 Stiles Road | Salem, NH 03079
chakra.d.nh@iomosaic.com

Georges A. Melhem, PhD, FAIChE
ioMosaic Corporation
93 Stiles Road | Salem, NH 03079
melhem@iomosaic.com

Ronald J. Willey, PhD
Northeastern University
360 Huntington Avenue | Boston, MA 02115
r.willey@neu.edu

Keywords: Thermal Inbreathing, API 2000, Storage Tanks, Venting.

Abstract

A Process Safety Office™ - SuperChems™ component based model is proposed to study the dynamic behavior of the thermal inbreathing of low-pressure storage tanks upon the occurrence of a sudden cooling rainstorm. The objective here is to establish a model that is capable of properly calculating the design relief capacity of the relief device to be installed on such tanks. The impact of phase-change, particularly vapor condensation, on the thermal inbreathing of the tank assessed. The model was validated against two large-scale experiment datasets and some other published theoretical models. The results were then discussed and compared to the relief requirements obtained using the API 2000 7th Edition method, which is the current standard for venting atmospheric tanks.

1 Introduction

Low-pressure storage tanks are protected against extreme weather changes following the API 2000 (7th edition) standard. The threat of creating a vacuum occurs when the contents of the tank vapor space are suddenly cooled by an intense rainfall causing the vapor space to contract and condense, hence, rapidly decreasing the internal pressure. This pressure drop will pull outside air into the tank's vapor space and acts as the driving force for thermal inbreathing.

The current limitation with the API standard is that it only addresses tanks storing non-condensable gases and have no liquid residue, where in fact most storage tanks contain condensable vapors such as water vapor which will have a big impact on the thermal inbreathing of the tank. This paper assesses the API 2000 relief requirement calculations for low-pressure storage tanks by comparing them to (1) the results of the dynamic simulations from a proposed SuperChems™ based model and (2) to several methods found in the literature (PTB W-22, Salatino, etc.) for both condensable and non-condensable vapor services.

2 API 2000 Methods

The tank contents are affected by the ambient conditions especially when the tank is uninsulated, which is the case for most storage tanks. This method uses air at normal conditions (0°C and 101.325 kPa) as the inbreathed gas. The maximum inbreathing rate, $\dot{V}_{IT}$, is primarily a function of tank volume, insulation, and a C-factor (found in Table 1 below) as shown in Equation (1) below:

$$\dot{V}_{IT} = C\, V_{tk}^{0.7} R_i \tag{1}$$

where $\dot{V}_{IT}$ is the maximum inbreathing volumetric flow rate of air at normal conditions (Nm^3/h), C is the dimensionless factor that depends on vapor pressure, storage temperature and latitude, V_{tk} is the tank volume (m^3), and R_i is the reduction factor for insulation.

Table 1. API 2000 *C*-factor for Thermal Inbreathing

Latitude, φ	C-factor			
	Vapor pressure similar or below hexane		Vapor pressure higher than hexane or unknown	
	Average storage temperature (°C)			
	< 25	≥ 25	< 25	≥ 25
φ < 42°	4	6.5	6.5	6.5
42° ≤ φ ≤ 58°	3	5	5	5
φ > 58°	2.5	4	4	4

The assumptions used for the scenario of "tank cooling by rain" are found in Annex E of API 2000 [1] and are listed below:

- Tank is filled with air only.
- Initial vapor space and shell wall temperature are in thermal equilibrium at 55 °C.
- Rain temperature and intensity are constant at 15°C and 225 kg/m^{2}h (This rainstorm has a probability of 1/100 year) respectively; rain angle is at 30° from the horizontal.
- Inside free convection heat-transfer coefficient is 5 W/m^{2}K.
- Outside film cooling heat-transfer coefficient is 5,000 W/m^{2}K.

2.1 API 2000 Annex A Method

The method found in Annex A was used prior to 7th Edition (2008) and is quite different than the current sizing method for venting devices. In order to quantify the thermal inbreathing for storage tanks volumes smaller than 3180 m^3, Annex A [1] assumes that the tank starts cooling from an initial temperature of 48.9°C and at a constant temperature rate of -56 K/h. The thermal inbreathing rate, $\dot{V}_{IT}$ (Nm^3/h) is as follows:

$$\dot{V}_{IT} = 0.169\,V_{tk} \quad (2)$$

where V_{tk} is the tank volume (m^3). For tank volumes equal or greater than 3180 m^3, the thermal inbreathing is assumed to be limited by a constant heat flux, $h\Delta T$, equal to 63 W/m^2. The thermal inbreathing rate in this case becomes:

$$\dot{V}_{IT} = 0.577\,A_{exp} \quad (3)$$

where A_{exp} is the external tank area exposed to rain (m^2). Table 2 below summarizes the thermal normal venting requirements as found in Annex A of API 2000. Volumes greater than 30,000 m^3 are outside the scope of this method. However, interpolation is allowed for Table 2 as long as the tank volumes lie between the specified volume limits of 10 m^3 to 30,000 m^3. [1]

Table 2. API 2000 Annex-A Normal Venting Requirements Due to Thermal Effects [1]

Tank Capacity	Inbreathing	Out-breathing	
Column 1 [a]	Column 2 [b]	Column 3 [c]	Column 4 [d]
		Flash Point ≥ 37.8 °C or Normal Boiling Point ≥ 149 °C	Flash Point < 37.8 °C or Normal Boiling Point < 149 °C
m^3	Nm^3/h of air	Nm^3/h of air	Nm^3/h of air
10	1.69	1.01	1.69
20	3.38	2.02	3.38
100	16.9	10.1	16.9
200	33.8	20.3	33.8
300	50.4	30.4	50.4
500	84.5	50.7	84.5
700	118	71.0	118
1,000	169	101	169
1,500	254	152	254
2,000	338	203	338
3,000	507	304	507
3,180	537	322	537
4,000	647	388	647
5,000	787	472	787
6,000	896	538	896
7,000	1,003	602	1,003
8,000	1,077	646	1,077
9,000	1,136	682	1,136
10,000	1,210	726	1,210
12,000	1,345	807	1,345
14,000	1,480	888	1,480
16,000	1,615	969	1,615
18,000	1,750	1,047	1,750
20,000	1,877	1,126	1,877
25,000	2,179	1,307	2,179
30,000	2,495	1,497	2,495

3 Thermal Inbreathing of Non-Condensable Vapors

One of the major assumptions of API 2000 is that the tank experiencing inbreathing is empty (i.e. only filled with dry air) and the vapor space is considered non-condensable. This section will review recent work on inbreathing dynamics for non-condensable dry air services only.

3.1 Salatino's Model

Salatino et al. [2] compare their tank inbreathing model to the older version of API 2000 (1992) that utilized the method of Annex A: a method based on the maximum estimated heat loss rate of 63 W/m^2 upon sudden cooling by a rainstorm of 225 kg/m^{2}h rain density. They recognized that all existing thermal inbreathing models assume a uniform vapor-space temperature, thus his team's contribution was accounting for the temperature non-uniformity inside the tank by dividing the tank into two individual volumes: the shell and the roof.

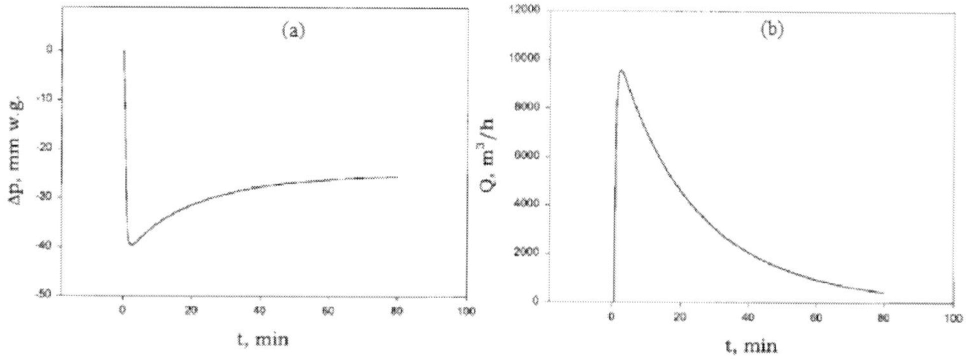

Figure 1. Pressure drop (a) and thermal inbreathing (b) profiles of Salatino's Model [2]

Salatino derived an equation for the maximum thermal inbreathing (Equation 4), stating that the maximum occurs at time t=0 (Figure 1) when the sudden weather change causes (1) vanishing of the radiative flux absorbed by the tank, (2) abrupt drop in ambient temperature and finally (3) rainfall.

$$\dot{Q}_{max} = -\frac{V}{T_G}\frac{dT_G}{dt}\bigg|_{t=0} = \frac{V}{T_{G,1}}\frac{1}{\tau}\left(T_{G,1} - T_{G,2}\right) \tag{4}$$

$$\tau = \frac{pV}{RT_G}\frac{c}{\sum A_i h_{iG}} \tag{4a}$$

where $\dot{Q}_{max}$ is the maximum thermal inbreathing rate (m^3/h), V is the tank volume (m^3), $T_{G,1}$ and $T_{G,2}$ are the gas temperatures (K) before and after the rainstorm, respectively. τ is the time constant that sums the effects of pressure, volume, heat transfer parameters (Equation 4a).

Salatino et al. further simplified the model by assuming an average value for the heat transfer coefficient of 4 W/m^{2}K and a gas temperature drop (after rainfall) of T_{G1}-T_{G2} ≅ 25°C claiming that these values will result in a conservative approximation of the inbreathing rate. Replacing these values in Equation (4) yields the simplified maximum inbreathing rate, $\dot{Q}_{max}$, which is a function of the tank volume, V, and the height-to-diameter ratio, f:

$$\dot{Q}_{max} = 2.6\, V^{2/3} \left(\frac{1 + 2f}{\sqrt[3]{f}} \right) \tag{5}$$

For a 63,000 m^3 cone-roof tank, Salatino predicted a maximum thermal inbreathing value of 9,900 m^3/h, using Equation 5. However, using the API 2000 Annex A method, the maximum inbreathing for the same tank was found to be 4,100 m^3/h. Salatino proved that his method is more conservative than the API 2000 equations and advised that an oversized thermal relief device is considered a safer design in this case.

3.2 Lloyd's Register Energy Americas, Inc. (LR) Model

Lloyd's Register (LR) Energy Americas, Inc. analytically solved the heat balances of a storage tank that is suddenly cooled by rain [3] and compared their results against ISO 28300, which is an equivalent of API 2000 6th edition (2009). The model results are summarized in Figure 2 below.

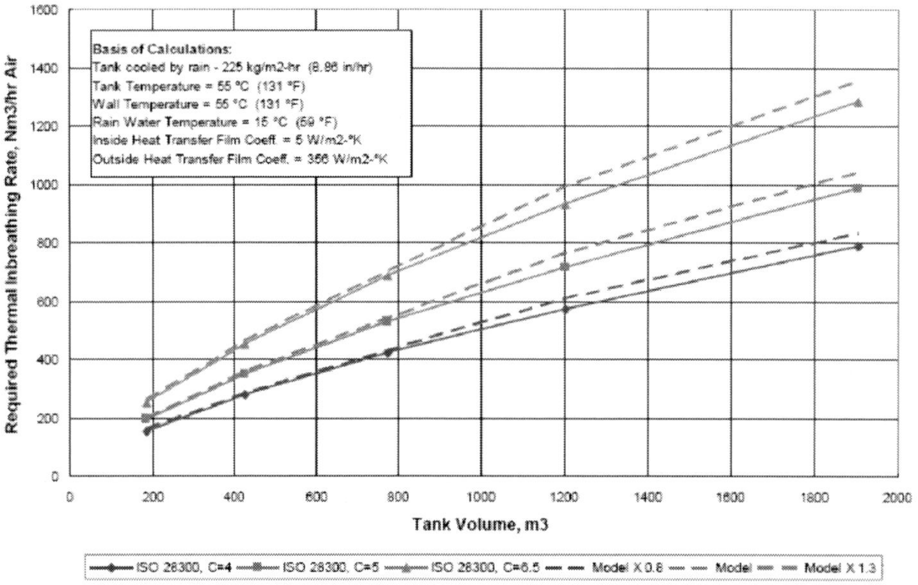

Figure 2. LR Model versus API 2000 6th ed. (ISO 28300)

The LR model was slightly more conservative than the API 2000 thermal inbreathing predictions by factors ranging between 0.5 and 1.3. Note that the C value in Figure 2 refers to the C-factor in the API 2000 thermal inbreathing in Equation (1) of this paper.

The model was then used to study the impact of several variables, some of which had minimal impact on the predicted thermal inbreathing (less than 5% error) such as the tank weight and initial gas temperature. Other variables, like the outside heat-transfer coefficient (Table 3) and rain intensity, had substantial impact on the calculated thermal inbreathing rates.

Table 3. LR Model - Impact of Outside Heat-Transfer Coefficient, h_o [3]

Case No.	Tank Volume m3	Predicted Thermal Inbreathing Rate, Nm3/hr		
		ho = 200 W/m2- K	ho = 300 W/m2- K	ho = 400 W/m2- K
1	186	169.6	191.2	205.1
2	427	308.3	341.4	362.7
3	775	475.1	520.7	548.6
4	1202	678.9	740.5	778.6
5	1903	935.2	1011.6	1058.5

The LR model was used to derive a relationship between the rain intensity (mm/h) and the heat transfer coefficient between the outer tank wall and rain (W/m^2K) as shown in Figure 3 below. This linear dependence relationship was implemented in SuperChems™ for this study.

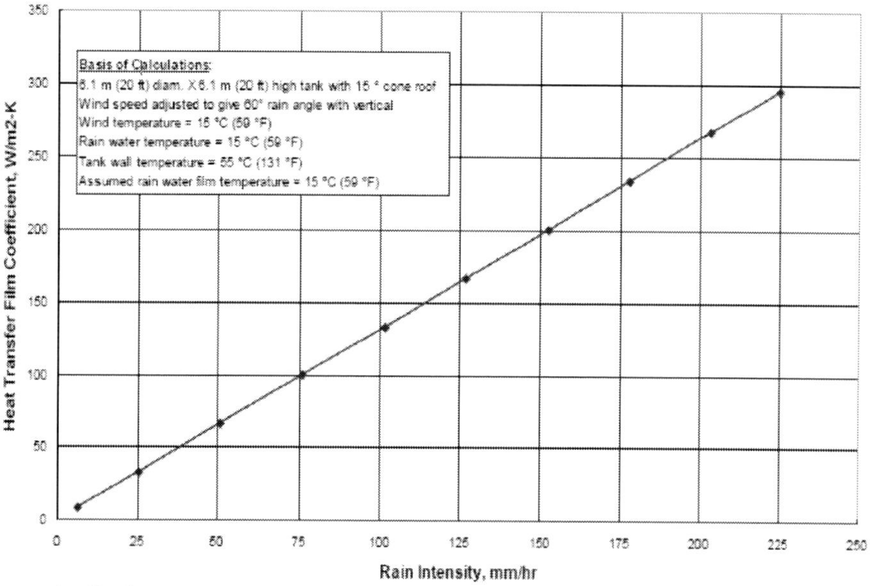

Figure 3. Rain/wall heat transfer coefficient versus of rain intensity

3.3 PROTEGO© Model

Scholz and Weber [4] from PROTEGO© derived the API 2000 7th Edition thermal inbreathing formula (Equation 1 of this paper) and determined that the maximum thermal inbreathing was a function of primarily tank volume with a coefficient that is less than or equal to 9.99:

$$\dot{V}_{max} = Cofficient \cdot V^{2/3} \quad (6)$$

The above equation is applicable to both flat-roof and conical-roof storage tanks. The derivation of Equation (6) was obtained by performing energy balances on the stored gas, the tank walls, and on the rainwater film. The differential equation and its solution are not included in this document, but can be found in Scholz and Weber's publication [4].

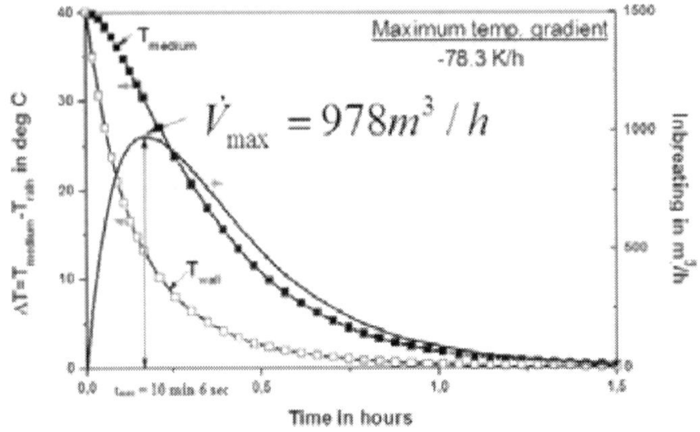

Figure 4. PROTEGO© Model Results

PROTEGO© then compared their results with the Foerster's model [5], whose main assumption was a fully immersed tank, which proved to be in good agreement with the published data. Figure 4 contains the results of a simulation using a conical-roof storage tank of 3,978 m^3 volume and an aspect ratio f = 0.6. Light rain was assumed (75 kg/m^2h) at 15°C. The inside and outside convective heat-transfer coefficients were set to 5 and 37 W/m^2K respectively.

PROTEGO© used their model to study 5 different types of condensable vapors and their effect on the maximum thermal inbreathing rate, however, the results will be discussed in Section 4 of this paper.

4 Thermal Inbreathing of Condensable Vapors

API 2000 neglects the possibility of vapor condensation for the thermal inbreathing/out-breathing relief calculation since only dry air is present in the tank i.e. the tank is empty. One reason to consider the effects of having condensable vapors is the steam cleaning that storage tanks undergo. Hot steam will occupy the tank after the cleaning process is complete and if a rainstorm occurs, the water vapor will condense/contract and pull a vacuum inside the tank.

4.1 Fullarton Model – Water Vapor

Fullarton [6] introduced two dimensionless factors in his model: (1) The reduction factor "η" that accounts for the tank wall thickness by factoring in the metal heat capacity of the walls and (2) the condensation factor "χ" that accounts for the stored-vapor condensation i.e. the additional condensation mass flux and the additional heat of condensation. The formulas of each "η" and "χ" can be found in Fullarton's paper [6].

It is expected that the tank wall temperature drops much faster than the temperature of the vapor stored inside the storage tank because of the metal's higher heat capacity. However, a thicker metal wall acts as a resistance to heat transfer. This means that the tank walls cool at a slower rate than when the wall thickness is neglected. The maximum inbreathing in this case will be lower, hence the reduction factor "η" which "reduces" the maximum thermal inbreathing of a storage tank with thicker walls.

Table 4. Range of χ-factor for Different Condensable Vapors

Condensable Vapor	χ-factor
Water vapor Heptane Hexane	1.5-2
Ethanol	2-2.5
Methanol	3-4

The condensation factor "χ", on the other hand, is primarily a function of the volatility of the condensable vapor being studied. For that reason, Fullarton established a range for the "χ" factor for different vapors as seen in Table 4 above [6].

Therefore, the maximum thermal inbreathing for a condensable vapor space per Fullarton can be obtained by using each of the Naumann approximation, $\dot{V}_N$, the reduction factor, η, and the condensation factor, χ.

$$\dot{V}_{K,max} = \dot{V}_N \eta \chi \quad (7)$$

One must keep in mind that properties of air at 288K were used to derive this equation in addition to the assumption of the initial temperature gradient ($T_{G,0} - T_A$) equal to 40°C. If other values are to be used, the above equation must be re-derived to account for the appropriate conditions.

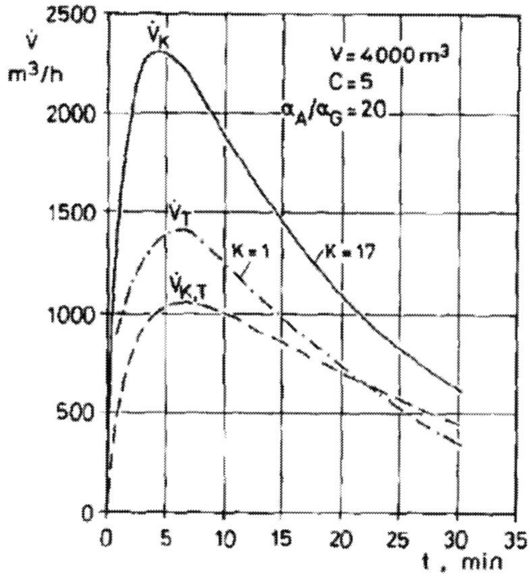

Figure 5. $\dot{V}_{inbreathing}$ **profile – Fullarton Model** [6]

Fullarton performed a dynamic simulation on a 4,000 m^3 empty tank intended for the storage of ethanol. The initial gas temperature was 55°C and was suddenly cooled by rain at 15°C. The rain fall was assumed to be an extremely heavy downpour (probability of 1 every 100 years) with $\alpha_A/\alpha_G = 20$ and $\alpha_G = 5$ W/m^2K. Figure 5 above shows the resultant inbreathing profile for each of dry air, K=1, and saturated water vapor, K=17. The maximum thermal inbreathing occurs at time t = 6 min where the inbreathing of the condensable vapor space ($\dot{V}_{K=17}$) is *1.6 times larger* than that of dry air ($\dot{V}_{K=1}$).

4.2 PROTEGO© Model – Condensable Vapors

PROTEGO© used their model [4] to obtain the inbreathing curves of five different vapors: methanol, ethanol, saturated water vapor, heptane and decane (Figure 6) and comparing them to the inbreathing curve of dry air in the same tank.

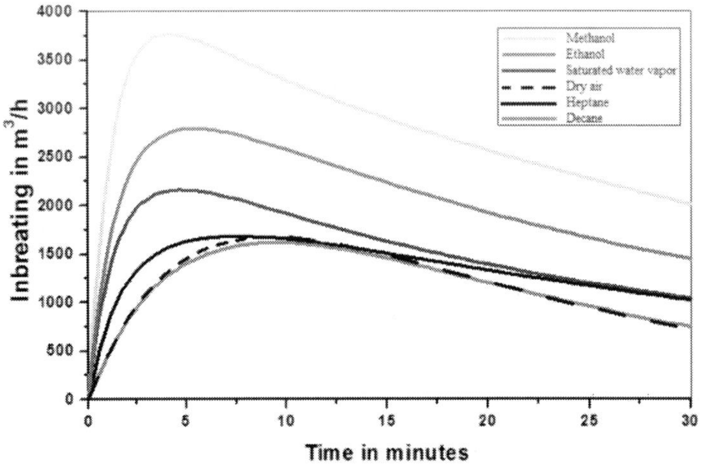

Figure 6. $\dot{V}_{inbreathing}$ profile of different vapors – PROTEGO© Model [4]

It can be seen that the condensation of volatile vapors has a significant impact on the inbreathing rate causing it to have a twofold increase. On the other hand, non-volatiles such as decane do not affect the inbreathing rate and yields the same inbreathing profile as with dry air.

Table 5. Condensation Factor "χ" Results

Vapor	$\chi_{PROTEGO}$	$\chi_{Fullarton}$
Methanol	4.76	3-4
Ethanol	2.23	2-2.5
Heptane	1.61	1.5-2
Water	1.29	1.5-2
Decane	1.01	N/A

PROTEGO© calculated the condensation factor "χ" for each of the simulated vapors and compared their values to that of Fullarton's (Table 4) and summarized the results in Table 5 above. The condensation factor values from this model are all in agreement with the published values [6] except the $\chi_{methanol}$ which is greater than the range specified by Fullarton (1986). A possible explanation for the latter is that PROTEGO© may have been more conservative (higher safety margin) when dealing with extremely volatile vapors like methanol.

5 Large-Scale Experimental Data

A thorough literature review yielded two sets of large scale data from actual experiments to study the thermal inbreathing of low-pressure storage tanks. Large scale datasets are preferred for model validation. The first dataset is for non-condensable gases (air) by *Sigel et al.* [7] and the second dataset is for condensable gases by *Holtkoetter et al.* [8].

5.1 Sigel's Experiment

Sigel performed actual data collection for the thermal inbreathing rate from a large tank of 617.5 m^3 capacity exposed to high solar radiation. The tank was empty without any trace of product vapor or liquid residue.

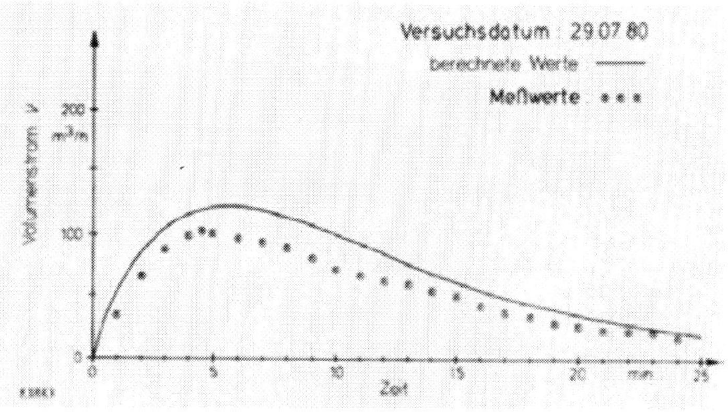

Figure 7. Sigel Experimental Inbreathing Profile

Sigel et al. [7] developed their own thermal inbreathing model to compare it with their experimental results. Figure 7 above displays the measured values of inbreathing (markers on the graph) versus the model (solid line). The inbreathing rates predicted by his model proved to be more conservative than the measured rates.

5.2 Holkoetter's Experiment

Holtkoetter performed an experiment on a 1.18 m^3 tank to measure the actual inbreathing rates and compare the results to his proposed model. This experimental investigation aimed at studying the effects of condensation on thermal inbreathing for three different vapors: water, isopropanol, and methanol.

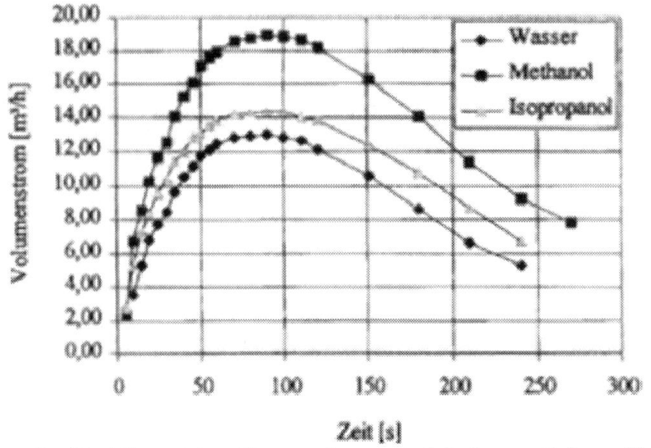

Figure 8. Holtkoetter Experimental Inbreathing Profile

The maximum thermal inbreathing is recorded after almost 95 seconds for all three gases. The maximum thermal inbreathing of methanol $\dot{V}_{max} \approx 19$ m^3/h is significantly higher than that for both water ($\dot{V}_{max} \approx 13$ m^3/h) and isopropanol ($\dot{V}_{max} \approx 14$ m^3/h) due to the fact that methanol is the most volatile vapor.

6 SuperChems™ proposed Model

The inbreathing profiles were generated using SuperChems™ dynamic simulation of storage tanks experiencing a sudden weather change. Dynamic models continually generate results that are used as the initial conditions for the next time interval calculation which perpetually drives the model to completion.

The proposed model first starts by defining a variety of objects to represent components of the system such as the storage tank vessel, the mixture that is stored, the piping segment and relief device, and the water sprays that are used to simulate a rainstorm. A built-in thermodynamics package is selected (modified Peng-Robinson EOS [9]) and scenarios are then created to collect the objects. Once the scenarios are created, single and multi-phase flow dynamics models are associated with the individual scenarios and the dynamic solutions for the inbreathing profiles are produced.

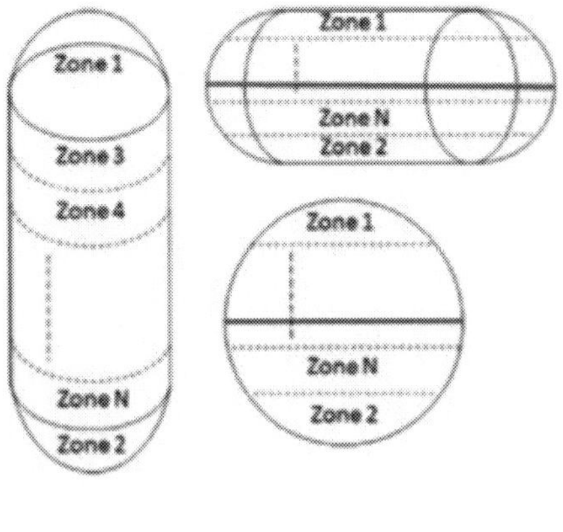

Figure 9. SuperChems™ Segmentation Scheme

An important feature of the software is wall dynamics which segments the vessel into multiple zones (Figure 9) allowing the user to account for heat conduction across the wall and between the zones using the effective conduction length method.

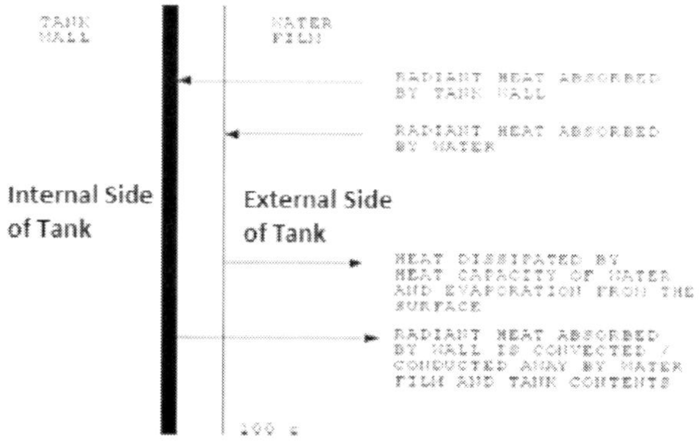

Figure 10: Heat Transfer Mechanisms for a Tank during a Rainstorm

Once segmentation is defined, heat transfer analysis is applied per wall segment (Figure 10) where the impact of rain, solar radiation, wind, and insulation is considered. Inside the tank, the wall-to-fluid heat transfer includes radiation and natural convection for the vapor phase; and nucleate, film and pool boiling for the liquid phase.

It should be noted that many parameters from the existing models and large-scale datasets were missing (vessel, dimensions, wall thickness, mixture composition, heat transfer coefficients, rain intensity, etc.) and had to be scientifically estimated in our dynamic simulations.

6.1 SuperChems™ Model Prediction for Non-Condensable Vapors (Large-scale Data)

The dimensions of the 617.5 m^3 empty tank used in Sigel's experiment [7] were introduced as input parameters to the SuperChems™ vessel object. The rainstorm used in this simulation had an intensity of 0.38 inch/h and a duration of 30 minutes with a temperature of 15°C. The inside heat transfer coefficient value was set to 5 W/m^2K as specified by API 2000 [1].

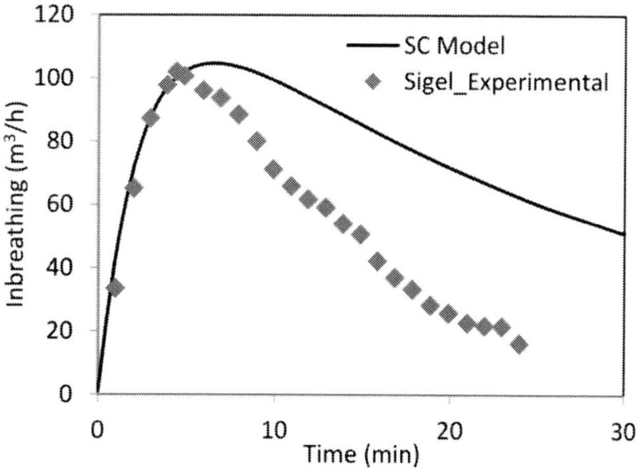

Figure 11. Simulation of Sigel's Tank using SuperChems™

The maximum value for inbreathing predicted by SuperChems was 104 m^3/h at time t = 6.6 min, as seen in Figure 11 above. This value is at a 2% deviation from the actual recorded maximum. The model's slope matches the first part of the inbreathing curve up until the maximum. Although the slopes of the decay of the inbreathing are different, they do not contribute to the design of the relief device because the most important value extracted from such studies is the maximum inbreathing that will act as the design relief capacity of the venting device.

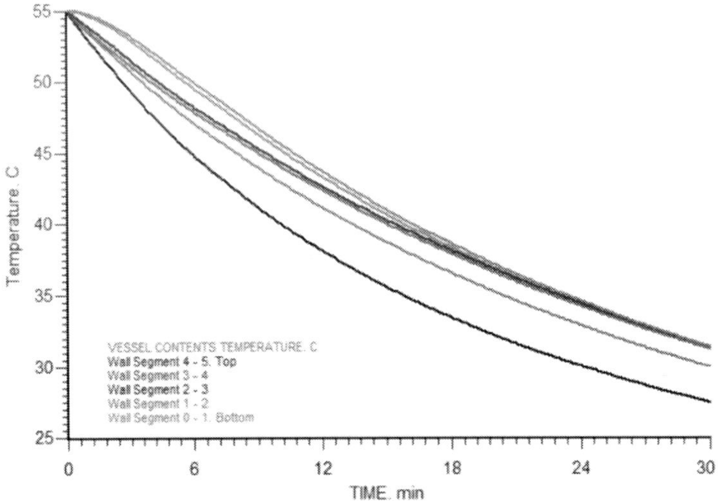

Figure 12: Temperature Profiles of Sigel's Tank using SuperChems™

Figure 12 above shows the temperature profiles of the wall segments and air inside the tank from the SuperChems™ simulation. Since Sigel [7] did not disclose the actual temperature profiles, therefore, we could not compare the computed temperature profile from SuperChems™ to the actual readings from Sigel's experiment.

6.2 SuperChems™ Model Prediction for Condensable Vapors (Large-scale Data)

Holtkoetter's experimental setup [8] consisted of a 1.18m^3 tank with walls that are 1.2 mm wall thick. The three vapors Holtkoetter studied were: methanol, isopropanol and water. For each type of vapor, the tank was filled with a small amount of liquid which was then vaporized using three infrared radiators (2.5-4.2 kW). After vaporization step was complete, water sprays were activated to cool the tank and the thermal inbreathing was recorded (Figure 8).

Using SuperChems™, the vessel was defined, then the rain-wall heat transfer coefficient, h_{rain}, was fixed at 600W/m^2K that corresponds to a rain intensity of 0.1262 kg/m^2s as specified by Holtkoetter. The duration of the rain for this simulation was 300 seconds. The h_{gas} was initially set to the API 2000 value of 5 W/m^2K, but was then adjusted to compensate for the vapor condensation inside the tank.

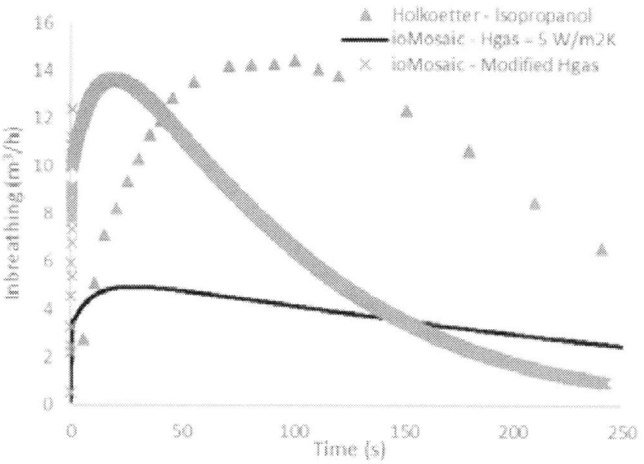

Figure 13. SuperChems™ Simulation of Holtkoetter's Tank with Isopropanol

Figure 13 illustrates the inbreathing profile for two different values of the free-convection heat transfer coefficient of the isopropanol vapor, h_{gas}, inside the tank. It can be clearly seen that using the API 2000 value $h_{gas} = 5W/m^2K$ underpredicts the thermal inbreathing by almost 62%. After numerous iterations, a modified value for h_{gas} was found to be 30 W/m^2K. This higher value makes physical sense since the condensation of isopropanol occurring on the inner walls of the tank enhances the heat transfer between the vapor and the wall.

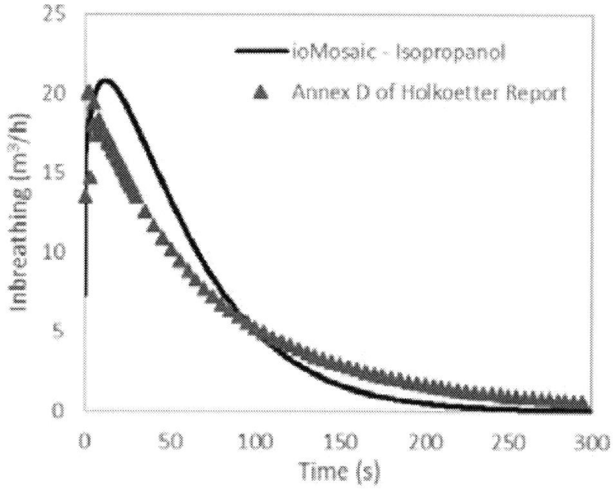

Figure 14. SuperChems™ Reproduction of Holtkoetter's Inbreathing Model (Isopropanol)

Although *Holtkoetter et al.* [8] performed experimental work to measure the maximum inbreathing, they did not, however, compare their experimental results to their proposed model. Hence, the SuperChems™ results were compared to Holtkoetter's model found in "Annex D" of his publication [8].

Figure 14 above proves that the SuperChems™ based model shows good agreement with the Holtkoetter model that considers spontaneous condensation. The predicted value for the $\dot{V}_{max}$ using SuperChems™ was 20.7 m^3/h and is at a +3% error from the value reported by Holtkoetter [8].

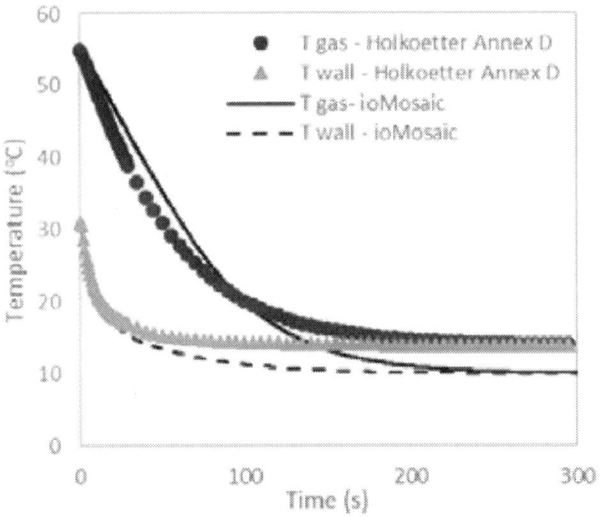

Figure 15. SuperChems™ Reproduction of Holtkoetter's Temperature Model (Isopropanol)

The gas and wall temperature profiles were almost accurately reproduced using the SuperChems™ model and are shown in Figure 15. Holtkoetter attributes the discrepancy between his model and his experimental result to the fact that the experiment was done in the month of December where the available water had a temperature of 8°C instead of 15°C. Colder rain will eventually lead to higher inbreathing rate values, therefore, the inbreathing should be inspected as a function of temperature to account for this 7°C difference in the water temperature. For more information about this method, please refer to the actual publication [8].

6.3 SuperChems™ Model vs. Published Models

In addition to validating our proposed model against large-scale experimental data, the SuperChems™ based model was also compared to the published models listed in Section 3 and 4 of this paper [10].

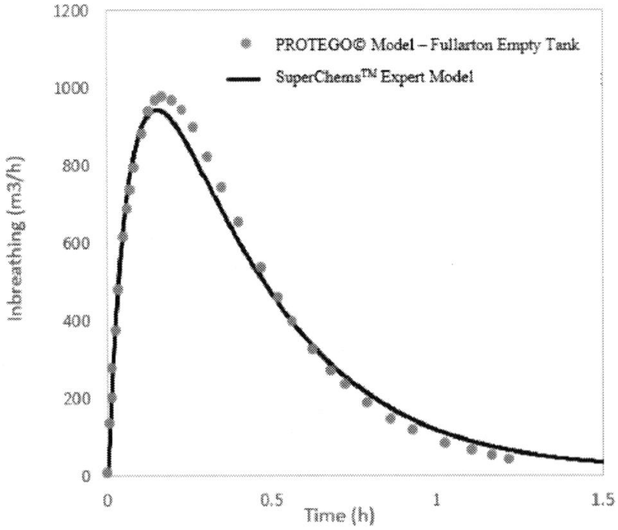

Figure 16. SuperChems™ Inbreathing Profile for Air (Fullarton's tank)

Figure 16 shows the predicted inbreathing profile of dry air using the SuperChems™ model and compared against the PROTEGO© model. The maximum recorded inbreathing from the SuperChems™ Expert simulation is around 940 Nm^3/h which has a -4% deviation from $\dot{V}_{max}$ predicted by PROTEGO©. The temperature and temperature-rate profiles were also calculated and can be found in the actual publication along with the fits for the different models and vapors using SuperChems™ Expert [10].

6.4 Modified Heat Transfer Coefficient

The heat transfer coefficient between the stored vapor and the inner-tank walls, h_{gas}, was modified for each type of volatile condensable vapor [10]. Table 6 below summarizes the values of h_{gas} used in the two-phase dynamic model for a tank-wall thickness of 5 mm. These results show that condensation "enhances" heat transfer thus providing evidence that use of the API 2000 value heat transfer coefficients h_{gas} will underpredict the maximum thermal inbreathing (Figure 13). Also, note that h_{gas} is directly proportional with the vapor's volatility.

Table 6. Modified h_{gas} Values for SuperChems™ Two-Phase Model

h_{gas} (W/m^2K)	Wall Thickness = 5 mm
Methanol	40
Ethanol	30
Water	20
Heptane	5
Decane	7

7 API 2000 Conservativeness

The maximum thermal inbreathing rates, V_{max}, obtained from the SuperChems™ model were compared to those obtained using the API 2000 thermal inbreathing equation, Equation (1), to assess the performance of the API 2000 inbreathing method.

In order to determine the appropriate value for the C-factor in Equation (1), the vapor pressure and storage temperature of the stored gas are needed.

- Since all the investigated vapors (methanol, ethanol, water vapor, etc.) do not have vapor pressures higher than hexane, the vapor pressure was classified as "Similar to Hexane".
- The storage temperature of the gas is 55°C, which is greater than 25°C, so according to Table 1, the C-factor has a *value of 5.*

The storage tanks in this study were all uninsulated, hence, the insulation reduction factor R_i has a value of one. The API 2000 inbreathing equation used for this case now becomes:

$$\dot{V}_{IT} = 5\, V_{tk}^{0.7} \quad (8)$$

To determine whether the API 2000 inbreathing relief requirement is adequate, a deviation percentage of the simulated (or experimental) results the API 2000 $\dot{V}_{max}$ method is computed as follows:

$$\% \, Deviation = \frac{\dot{V}_{max,SC} - \dot{V}_{max,\ API\ 2000}}{\dot{V}_{max,\ API\ 2000}} \times 100 \tag{9}$$

where $\dot{V}_{max,\ API\ 2000}$ is the maximum thermal inbreathing obtained from Equation (8) (m^3/h) and $\dot{V}_{max,SC}$ is the maximum inbreathing from SuperChems™ model simulation (m^3/h).

7.1 API 2000 performance – Large-scale Experiments

Sigel et al. [7] used a tank volume of 600 m^3, which when replaced in Equation (8), results in a maximum inbreathing value of 440 m^3/h. Figure 17 compares $\dot{V}_{max}$ using the API 2000 method to the actual recorded $\dot{V}_{max}$ values during Sigel's experiment. The API 2000 inbreathing prediction is conservative for Sigel's empty tank where the relief device is designed to handle an inbreathing rate 4.2 times larger than the recorded $\dot{V}_{max}$.

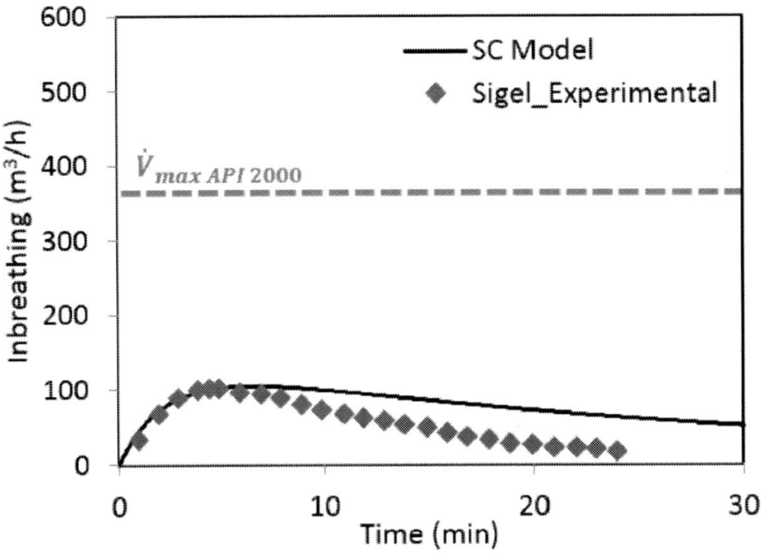

Figure 17. API 2000 Inbreathing Requirements – Sigel's Tank with Dry Air

On the other hand, *Holtkoetter et al.* [8] used a smaller tank with a capacity equal to 1.18 m^3 which results in a maximum inbreathing rate of 5.6 m^3/h according to the API 2000 method – Equation (8). Figure 18 shows that the API 2000 inbreathing relief requirement is insufficient for water, isopropanol, and methanol vapors in Holtkoetter's tank.

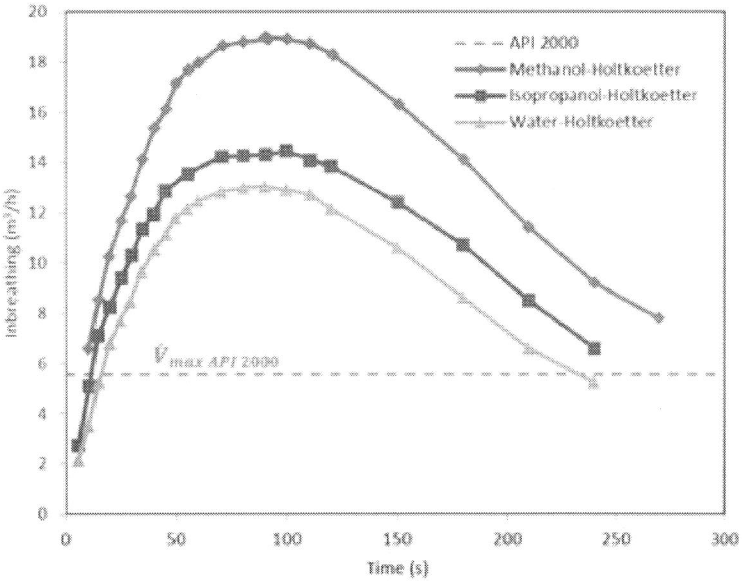

Figure 18. API 2000 Inbreathing Requirements – Holtkoetter's Tank

The recorded inbreathing rates are 2.3 to 3.7 times larger than the rate calculated by API 2000; this leads to an undersized relief device that puts the storage tank at a high risk of failure if a sudden change of weather occurs.

7.2 API 2000 performance – Computer Model

The tank size used in Scholz and Weber's model [4] was 6,283 m^3 to study the condensable vapors. Replacing the tank volumes in Equation (8), results in a maximum inbreathing, $\dot{V}_{max}$ = 2,278 m^3/h. Figure 19 graphically shows the API 2000 $\dot{V}_{max}$ in comparison to the dynamically simulated inbreathing profiles for the different types of vapors, including dry air.

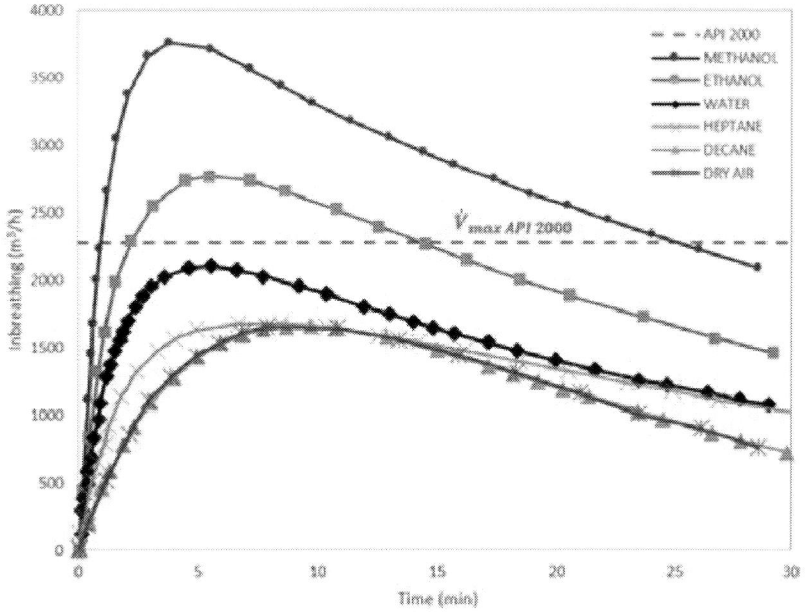

Figure 19. API 2000 Inbreathing Requirements – Different Types of Vapors

It can be seen that API 2000 method has a conservative prediction for the maximum thermal inbreathing of: (1) dry air, (2) heptane, (3) decane, and (4) water services. However, the API 2000 method underpredicts the relief requirement for the thermal inbreathing of a tank saturated with *methanol* and *ethanol*. The relieving capacity, for this case, has to be 1.7 times higher for methanol and 1.2 times higher for ethanol.

8 Conclusion

A dynamic simulation of the thermal inbreathing of low pressure storage tanks experiencing sudden change of weather conditions was performed using a SuperChems™ model. This model was validated by successfully reproducing the thermal inbreathing and temperature profiles of large scale experimental datasets and other available inbreathing models.

The thermodynamic variables specified in the API 2000 7th Edition standard have been tested for adequacy. It was found that the gas/wall free-convection heat transfer coefficient, $h_{gas} = 5W/m^2K$, specified by API 2000 was adequate when the tank was filled with non-condensable vapors like air. However, higher values were suggested for condensing vapors with a range of 10 W/m^2K to 45 W/m^2K depending on the type and volatility of the vapor. As for the rain/wall heat transfer coefficient, the API 2000 value of 5,000 W/m^2K proved to be too conservative, and even unrealistic, therefore leading to an oversized relief device. Suggested values for the rain/wall heat transfer coefficient were found to be in the range of 12 to 600 W/m^2K depending on the intensity and angle of the rainfall.

Finally, the API 2000 7th Edition method for calculating thermal inbreathing was compared with several tanks of different vapor services. The method proved to calculate an adequate relief requirement for tanks storing air, heptane, decane, and even water vapor in one case. However, the application of the API 2000 thermal inbreathing method was not conservative enough when the vapor services became more volatile, like in the case of isopropanol, ethanol, and methanol, that resulted in high risk of failure for these types of storage tanks.

9 References

[1] API Standard 2000, Venting Atmospheric and Low-Pressure Storage Tanks, 7th Edition, March 2014

[2] P. Salatino, G. Volpicelli and P. Volpe. "On the design of thermal breathing devices for liquid storage tanks", Process Safety and Environmental Protection, 77.6, pp. 354-359, 1999.

[3] D. Brooks, "Thermal Inbreathing Requirements of Low Pressure Storage Tanks at Elevated Temperatures". Lloyd's Register Energy Americas, Inc. Spring DIERS Users Group Meeting, Presentation, May 2012.

[4] R. Scholz and R. Weber, "Cooling Down Industrial Tanks". Prepared for PROTEGO Braunschweiger Flammenfilter at API 2000 Meeting Presentation, Fall 2010.

[5] H. Foerster, K. Schampel, and H. Steen, "Atmungsvorgänge infolge von Witterungseinflüssen an Lagertanks für brennbare Flüssigkeiten", Physikalisch Technische Bundesanstalt (PTB) W-22. Braunschweig Univ. Phys.-Tech. Bundesanst, 1984.

[6] D. Fullarton, J. Evripidis and E.U. Shclünder, "*Influence of Product Vapour Condensation on Venting of Storage Tanks*", Institut für Thermische Verfahrenstechnik, Universität Karlsruhe (TH), 1986.

[7] R. Sigel, B. Kuxdorf, R. Meiss and H. Schwarz, "Praktische und theoretische Untersuchung der Beatmung von Festdachtanks". Chemie Ingenieur Technik, 55.5, pp. 377-379, 1983.

[8] T. Holtkötter, J. Shang and H.G. Scheckerm "Behälteratmung - Experimentelle Untersuchungen und Entwicklung eines Vorhersagemodells" (Master Thesis), 1995.

[9] G.A. Melhem, R. Saini, and B. M. Goodwin, "A Modified Peng-Robinson Equation of State," *Fluid Phase Equilibria,* 47, pp. 189-237, 1991.

[10] D.D. Abou-Chakra, G.A. Melhem, and R.J. Willey, "Dynamic Modeling of Inbreathing Requirements for Low-Pressure Storage Tanks" (Master's thesis), 2016. Retrieved from **http://hdl.handle.net/2047/D20209806**

478223 Studying the Relationship Between Inherently Safer Design and Equipment Reliability

Abstract:

Risk associated with a process depends on both likelihood and consequence of an incident. Inherently safer design is based on the use of technologies and chemicals with intrinsic properties that reduce or eliminate hazards. This philosophy is mainly focused on reducing the consequences associated with an incident. However, recent research has revealed that applying principles of inherent safety can have an effect on the likelihood element of risk as well. One of the major drawbacks of implementation of an inherently safer design (ISD) is that it can lead to risk migration, Thus, attempting risk reduction through implementation of ISD by mitigating the consequence element of risk can undesirably lead to an increase in risk by affecting the likelihood element of risk. This likelihood element of risk is the mathematical complement of system reliability. This study mainly focuses on assessing the effect of implementation of ISD principles on the reliability of a chemical process as it progresses along different stages of process design and manipulating this relationship to achieve maximum economic benefits while simultaneously minimizing risk associated with the process.

In this study, safety indices are used for quantification of inherent safety of a process. The choice of safety index depends on the stage of chemical project under consideration. Reliability of an equipment involved in a process is determined through the use of offshore and onshore reliability data (OREDA) and center for chemical process safety (CCPS) reliability databases for process equipment. The reliability of process system is then determined from the reliability of the individual equipment involved in the process. Lastly knowing values of safety indices and reliability of systems for competing designs during different stages of a chemical project (such as process selection stage, conceptual stage etc.), the required relationship between ISD and reliability is established for that particular stage.

Applying this methodology to process selection stage reveals that selecting a process that's inherently safer in terms of the involved reactions, chemicals and equipment might lead to risk migration, since the selected process might have lesser system reliability as compared to other competing process. When this methodology is applied to conceptual stage, where-in the competing processes are similar in terms of reactions and chemicals and vary only in terms of the equipment involved, it is observed that a process that has lesser number of equipment might be more reliable and inherently safer as compared to other competing processes. Lastly, when this methodology is applied for a specific equipment such as a reactor where the reliability is modeled as a function of design parameters of the reactor, it has been observed that within a certain range of design parameters, the reactor becomes sufficiently reliable and inherently safer.

Thus, this study provides a framework for designing of process systems and equipment that are reliable, economically profitable and inherently safer during different stages of project.

479377 Pressure Relief Analysis - It's about More Than Just Compliance

Abstract:

Pressure relief valves and disposal systems are the last layer of defense in petrochemical plants and refineries. Therefore, there are many codes and standards that must be followed to design those systems. Failure to ensure safe plant operations can result in fatalities, injuries, and environmental releases which can result in significant cost and poor public relations.

OSHA's CFR 1910.119 - Process Safety Management of Highly Hazardous Chemicals is designed to manage the hazards associated with facilities which handle highly hazardous chemicals. Many of the requirements in PSM are also valuable for facilities which are not governed by the requirement and application of these principals may improve safety for all facilities.

This paper will present the proposal that Process Hazard Analysis (PHA) be the heart of process for ensuring safe operations to prevent negative impact on companies' image and goodwill with the community.

480249 How to Leverage Limted Design Data Along with Historical Incident Data to Assess Inherent Process Safety During Early Stage of Design?

Abstract:

Inherent safety level quantification remains as one of the challenges for the inherent safety concept to gain industrial acceptance comprehensively. In current practice, most of the time safety assessment is conducted once the process design has been completed. At this stage of design implemented strategies look to control the hazard level. Any design modification necessary as a result of unfavorable safety performance at this stage of design, would be costly when compared to alteration during conceptual design stage [1]. It is beneficial to design industrial processes that are user friendly, inherently safer and have a level of tolerance for the deviations from ideal performance, which will eventually reduce the number of hazardous accidents. Therefore, both industrial experts and researchers are interested in safety indices that can be applied during the early design stage of a process, where the extent of knowledge regarding the process is very limited. Again, the available indices to quantify inherent safety still leave room for enhancement to represent process stream condition more accurately as a mixture rather than pure component(s) and to account for the interaction of all process parameters in determining the level of inherent safety [2].

In this work, a property based methodology to quantify inherent safety level has been proposed by assessing chemical and process unit risk at earlier stage of design. This will assist process designers in producing safer designs by the application of inherent safety principles in a more efficient and cost effective manner. It would offer process/design engineers an alternative to include safety without waiting until detailed design stage, where safety costs increase. Degrees of freedom at this stage would include alternative route selection, equipment sizing, and selection of operating conditions. The additional benefit of safety consideration in the early design stages is that it decreases the level of dependence on safety systems. The proposed safety assessment method here considers the perspective of five main safety factors associated with chemicals such as flammability, explosiveness, corrosiveness, toxicity & reactivity. These factors were linked to the process operating conditions (temperature and pressure). The process unit risk are calculated by classifying them into six main categories (i.e., pump, fired heat exchanger, pressure vessels, etc.) based on API 14C standard and identifying the undesirable events associated with each category. Major incidents database and the probability of unwanted incidents were used to link the undesirable events to operating conditions; this allows for the calculation of unit safety factors using the limited available information. Lastly, an enhancement technique to reflect the contribution of the individual components in the mixture is introduced based on the mixture properties, where the severity score was calculated for the mixture of chemicals in each unit at the actual operating conditions (whenever applicable) and multiplied by the flow rate of the mixture.

This proposed methodology of safety assessment gives a more intuitive, easy-to-understand way to represent the chemical and equipment safety factors posed by each individual unit. The steps in the methodology calculate, compare and prioritize the level of inherent safety of process streams during simulation work that influences the abovementioned safety factors in order to offer the user an insight into the risk associated with alternative design decisions. By prioritizing the streams based on the potential for chemical and equipment safety factors, the design engineers can easily identify the critical streams/units to be considered for improvement in order to avoid or minimize the risk. This methodology of quantifying inherent safety level will provide users another dimension to consider when selecting optimal design with multi objectives (e.g. techno, economic, and environmental).

The assessment of inherent safety level using the proposed method is demonstrated through a case study. HYSYS simulation is used to design two alternative designs for ammonia production.

1. Thiruvenkataswamy, P., et al., *Safety and techno-economic analysis of ethylene technologies.* Journal of Loss Prevention in the Process Industries, 2016. **39**: p. 74-84.

2. Leong, C.T. and A.M. Shariff, *Process route index (PRI) to assess level of explosiveness for inherent safety quantification.* Journal of Loss Prevention in the Process Industries, 2009. **22**(2): p. 216-221.

Inherently Safer Design from Overpressure Protection Viewpoint

Chris Ng
Sachin Kanade
Nancy Faulk
Siemens Energy, Inc.
4615 Southwest Freeway, Suite 900
Houston, TX 77027
chris.ng@siemens.com

Prepared for Presentation at
American Institute of Chemical Engineers
2017 Spring Meeting and 13th Global Congress on Process Safety
San Antonio, TX
March 26 – 29, 2017

Inherently Safer Design from Overpressure Protection Viewpoint

Chris Ng, Sachin Kanade, and Nancy Faulk
Siemens Energy, Inc.
4615 Southwest Freeway, Suite 900
Houston, TX 77027
chris.ng@siemens.com

Keywords: pressure relief, process safety, inherent safety

Abstract

Pressure relief and flare system design is mostly evolving or iterative throughout the various engineering process design stages. This iterative nature stems from the need to capture the impact of engineering design changes on this critical layer of protection, which often is designed as the last line of defense for overpressure events.

The systematic review of pressure relief and flare system design basis at various engineering design and project execution phases will not only help to ensure adequate overpressure protection, but can also highlight opportunities for inherently safer processes and project cost savings. Often, evaluating trade-offs between process safety design improvements against project economics may be challenging.

In this paper, several case studies are presented to demonstrate how inherently safer design options and cost-effective solutions could be implemented across the various engineering design and project execution phases to minimize gaps in terms of overpressure protection between early and as-built designs.

1 Introduction

Figure 1 depicts the stages of a typical engineering project lifecycle, excluding decommissioning. Each stage has associated activities in terms of selection and refinement of design details which contribute to the overall integrity of the facility and economic feasibility of the project. The general approach involving timely review of pressure relief and flare system design could be applied to both large projects involving major Capital Expenditures (CAPEX) as well as smaller projects with limited scope.

Figure 1: Typical CAPEX Project Life Cycle.

During the early phases of design such as conceptual design and Front End Engineering Design (FEED) phases, the scope, objectives, and requirements are first defined, followed by development of a process basis (i.e., heat and material balance), Process Flow Diagrams (PFDs), and preliminary Piping and Instrumentation Diagrams (P&IDs). Equipment, line and instrument lists are then prepared, and preliminary equipment sizing and specifications for process and utility systems are generated. Design options are selected, and design drawings such as equipment layout plans are completed. A budget is developed and costs are estimated.

During later phases such as Detailed Engineering and Procurement, the remaining design information is added, such as vendor data, drains and vents, and piping details. The actual piping layout is verified and updates are incorporated. In addition, electrical, instrumentation, structural and building requirements to support the project are specified. Often Hazard and Operability (HAZOP) reviews are carried out. Typically procurement of the main equipment and auxiliaries takes place followed by construction, installation, commissioning, and Pre-Startup Safety Reviews (PSSRs), in preparation for start-up, operation and maintenance.

However, what is often overlooked during the early phases is input from pressure relief and effluent handling system design engineers. This paper will describe how early involvement by these engineers could help validate relief system design and reduce overall costs and delays later. The case studies described in Section 3 below offer examples of the potential savings.

Relief systems engineering analysis should be performed during various process design and project implemented phases, which allows identification of potential gaps as well as opportunities for improvement. Typical activities include:

- Overpressure scenario analysis
- Determination of relieving loads
- Selection of relief technology
- Sizing of relief systems
- Evaluation of relief device inlet loss and backpressure effects
- Depressuring analysis
- Hydraulic evaluation of common relief headers
- Flare radiation calculations
- Dispersion analysis, including suitability of atmospheric release
- Acoustic induced vibration (AIV) analysis
- Presenting findings and developing mitigation
- Developing comprehensive documentation
- Reviewing impacts of design changes and updating documentation to capture changes are they are implemented.

It is recommended that the reviewer(s) be well-versed in the appropriate industry standards, e.g., American Petroleum Institute (API) Standards 520 Parts I and II, 521, and 2000 [1][2] [3][4]; American Society of Mechanical Engineers Boiler and Pressure Vessel Code (ASME BPVC) Sections I and VIII [5][6]; as well as corporate guidelines and site-specific requirements. In addition, it is recommended that the reviewer establish and utilize a consistent approach for review of all designs, irrespective of the Engineering, Procurement, and Construction (EPC) companies working on the project.

2 Benefits of Early Review of Relief Systems Requirements and Design

2.1 Early Involvement Allows Critical Input

Early involvement of relief system engineers can allow critical input to be incorporated which influences selected design options from an overpressure protection point of view, thus promoting an inherently safer design. By careful consideration of a number of factors, the design team has the potential to reduce overpressure relief and flaring requirements which in turn can reduce sizing of relief devices and associated piping, as well as flare components such as headers, knockout drums, flare stack, and flare tip. Potential reduction of environmental emissions is also realized. The examples below demonstrate some ways inherently safer design options and cost-effective solutions were implemented to minimize gaps in terms of overpressure protection between early and as-built design. In addition, early review can help to reduce or minimize the need for later design changes or mitigating measures, which could add significant costs and delay completion schedule for the project.

2.2 Identify Opportunities to Increase Inherent Safety

2.2.1 Centrifugal Pump Design

Consider a centrifugal pump as a pressure source for downstream equipment, as in Figure 2. As it stands, P-1 is capable of overpressuring the downstream vessel, V-1, in the event of a blocked outlet or an inlet control valve failure.

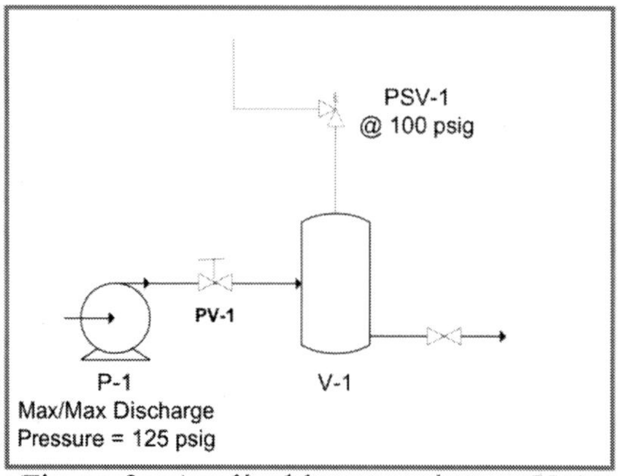

Figure 2: Applicable scenarios on V-1.

By properly sizing this pump during the design phase, however, it may be possible to minimize the relief rates for blocked outlet and control valve failure-open overpressure scenarios of the downstream equipment or even to eliminate the applicability of such scenarios altogether. Working with the process and mechanical engineers as well as rotating equipment designers, the maximum pump impeller size or speed and/ or maximum pump suction pressure could be carefully specified within process operation limits, which in turn limits the maximum pump discharge pressure. By reducing the maximum discharge pressure of the pump to a value less than the downstream equipment Maximum Allowable Working Pressure (MAWP), the above-noted scenarios may not be applicable, as in Figure 3a. An alternative to this, as shown in Figure 3b, is to ensure that the downstream equipment MAWP would be higher than the maximum discharge pressure of the upstream pump. These engineering design choices, which would result in minimizing relief loads or eliminating applicable scenarios, helps make the process inherently safer.

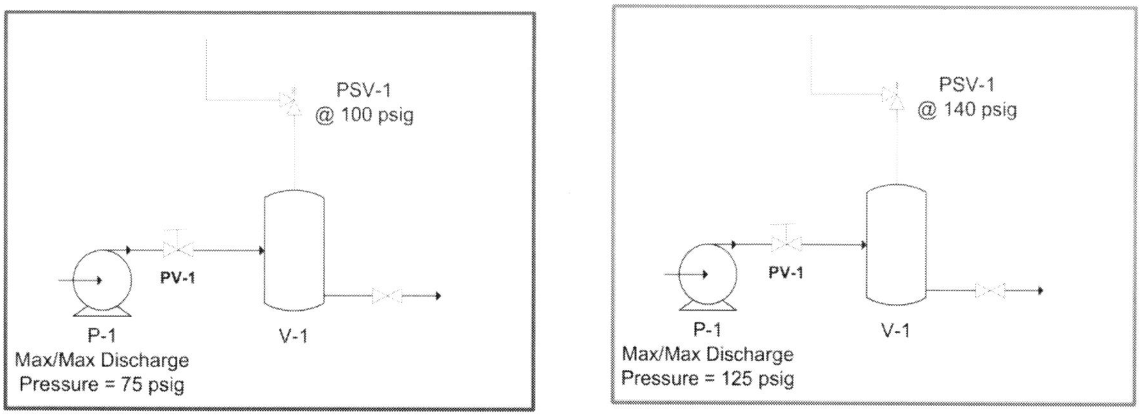

Figure 3: a) Limiting pump discharge pressure. b) Increasing vessel design pressure.

2.2.2 Pump and Compressor Discharge Check Valve Design

A related example involving pumps (and compressors, for that matter) is to ensure appropriate discharge check valve(s) are allocated to mitigate potential for reverse flow, impacting upstream equipment. In some instances, such as in Figure 4, feed could continue into an upstream vessel such as a feed surge drum if the liquid outlet pump is stopped with a discharge check valve stuck open. If this relief scenario is not identified by the relief designer, the relief valve on the feed surge drum may not provide adequate relief capacity.

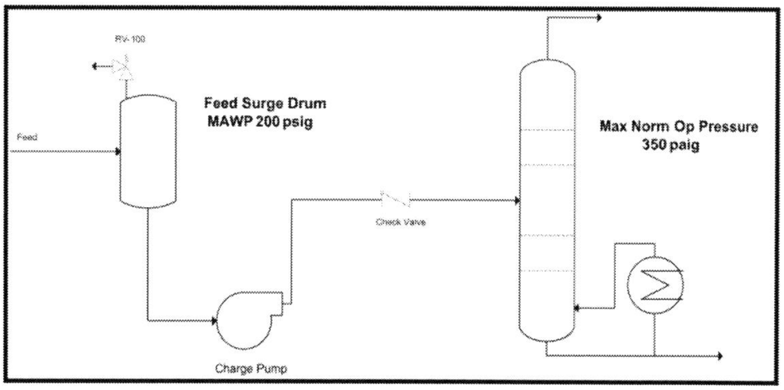

Figure 4: Applicable scenario on Feed Surge Drum

Allocating dual check valves in series of similar (or diverse) design may be required to minimize the backflow rate in the event of a check valve failure; if they meet the requirements set forth in API STD 521, the backflow rate may be limited to check valve leakage instead of check valve failure [7]. Careful specification of the MAWP of the upstream vessel may also help mitigate the overall relief requirement. Consideration of such issues early on is critical to avoid potentially costly design changes later as well as to increase the safety of the process; however, this is commonly missed due to factors such as improper management of change or poor communication between different groups designing upstream and downstream systems.

In a similar fashion, an early design review of several capital projects was performed, focusing on inherently safer design. One recommendation was proposed and accepted to raise the design pressure of the feed surge drum to mitigate the scenario in which a shutdown of a downstream charge pump results in reverse flow to the surge drum, while normal feed to the surge drum continues. By doing so, the feed and reverse flow sources would no longer be capable to overpressure the drum. Early review made the design inherently safer as the scenario was no longer applicable. Without this recommendation, installation of either a safety-instrumented system (SIS) or three 8T10 relief devices would have been necessary.

2.2.3 Control and Bypass Valve Design

Early input is critical in the selection and sizing of a control valve and its bypass valve, since the relief designer must account for the scenario in which *both* the control valve and its bypass valve fail open, as discussed in API STD 521 on "Pressure-relieving and Depressuring Systems". Early engagement of a pressure relief systems expert will help optimize the control valve/bypass and pressure relief requirements to provide adequate overpressure protection within the process operation envelope and requirements, while maintaining industry design and safety best practices. In one instance, a 2" gate valve was specified as the bypass for a 1" control valve, as shown in Figure 5. During a later design stage, it was discovered that this installation resulted in a very undersized relief device for the scenario described above, and thus the need to replace the bypass valve with a different and smaller-capacity type of valve, such as a globe valve. As in this instance, if the nominal size of the installed bypass valve is bigger than the control valve size, it necessitates careful review and possible re-design.

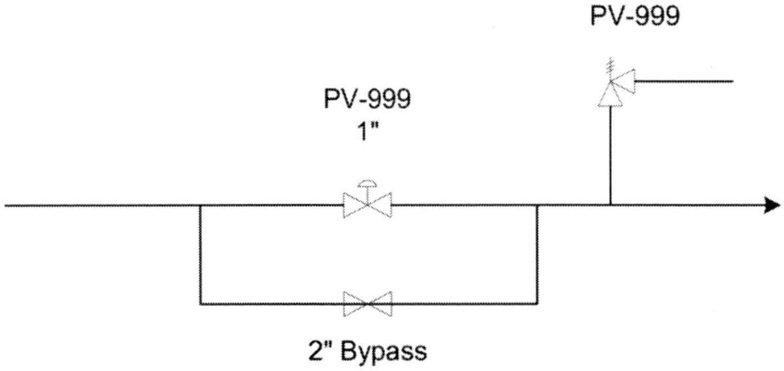

Figure 5: Bypass valve initially specified results in undersized relief device

2.2.4 Static Head

Static head is often overlooked in the preliminary design phases as equipment layouts may not yet be available. Relief designers should encourage vessel designers to specify vessels to be designed to be liquid full at the requested MAWP. Alternatively, relief device set pressures can be reduced to compensate for static head effects provided that the vessels can be filled above the top weld seam (into the top head). In another instance, the effects of static head on the set pressure of multiple relief valves had not been accounted for by an EPC company during detailed engineering. The potential for the internal pressure of affected equipment to exceed MAWP plus allowable accumulation per code was identified, allowing the relief valve set pressures to be adjusted prior to field installation.

2.2.5 Detailed Design Reviews

An independent pressure relief design audit was conducted for a large operating company. Relief design analysis and calculations had been performed by an EPC company as part of an expansion project. A review was conducted during the detailed engineering phase, during which it was found that most of the relief valve and associated piping upgrades which were being proposed by the EPC company were not required. This review included one application for which the EPC company had recommended upgrades to 6 relief valves (including recommendation to install two additional 8T10 relief valves), based on a recirculation flow rate which had been incorrectly included in a blocked outlet scenario relief rate estimate. Upon review, it was concluded that the existing relief valves were adequate and did not need to be upgraded. This reduced the potential for oversized relief capacity.

2.2.6 Design Gaps

In another example of early design review, engineers were able to identify a proposed design gap which would have resulted in costly and unnecessary modifications to a flare system. The proposed design project involved tie-ins of multiple large atmospheric relief valves to the associated flare for a process unit, resulting in inadequate flare capacity and the necessity of making expensive upgrades to this flare system. By performing detailed flare hydraulic modeling, relief engineers were able to show that re-routing the additional relief loads to an alternate adjacent flare having some spare capacity would be more practical.

2.2.7 Design of Feed Surge Drum

During a design review of a feed surge drum, additional examples of recommendations were developed to further increase the inherent safety, such as:

- By limiting the maximum discharge pressure of a feed pump to the proposed design pressure of the downstream equipment, a blocked outlet overpressure scenario was no longer credible for that equipment.
- Rerating the feed/bottom exchangers to account for static head in the event of overfilling in the associated distillation column.
- Car-sealing open (CSO) appropriate valves to ensure that various overpressure scenarios would no longer be considered as design contingencies and relief paths would be available for impacted equipment.

2.2.8 Car-Sealing Strategic Valves

In a final example, several recommendations were developed and adopted prior to start-up for a Coker Vapor Recovery Unit (VRU) revamp project. These recommendations include the following:

- Re-rating equipment
- Car-sealing closed (CSC) the control valve bypass valves
- Reducing the control valve flow coefficient (C_v) to limit maximum flow through the valve in the event of failure open
- Increasing the vessel design pressure to account for static head
- Adding dual check valves to mitigate reverse flow

These changes helped make the overall design inherently safer, so fewer safeguards were required to reduce the risks associated with overpressure scenarios for multiple systems.

2.3 Relief Systems Engineer's Role

Numerous other examples of inadequate design issues as well as missed opportunities to increase the inherent safety of the design are routinely identified by the relief systems engineers after the design has already progressed to a late stage or even after field implementation. Design review can begin early in the process, when preliminary P&IDs are issued for HAZOP or other types of Process Hazard Analysis (PHA). The relief systems engineer should also validate the design once piping isometrics are issued for approval and construction or after other design changes are proposed later. Taking these steps can help to ensure that design shortfalls can be resolved at the earliest opportunity. Making design changes early on is easier and less expensive than during later design stages and even more so than after commissioning and startup, thereby preventing subsequent shutdown, delay or major re-design work.

3 Critical Factors for Success

All pressure relief and effluent handling design documentation and supporting data should be stored in a central repository, for which access can easily be provided to all who have a need for such information. This access can be controlled, so that most viewers will have "read-only" access, and only authorized personnel can make required changes associated with Management of Change (MOC) items, thus keeping overall content up-to-date (evergreen) throughout the life of each facility as MOC activities, major expansion projects, or process modifications are implemented. These activities should trigger design review by relief systems engineers to ensure that relief design continues to be designed to provide adequate relief for all credible overpressure scenarios and that associated documentation is kept up-to-date.

In order for the review process to achieve the desired results, the three parties which are commonly involved (relief systems team, operating company, and EPC company) should work together to understand the goals of each party and how the design inputs from each party should be integrated in the various stages of engineering design and project execution. The following factors have been identified as critical for success:

- Conducting independent relief system design review at the right time in the project scope

- Communication among all parties
- Sharing of critical design data
- Effective management of change
- Discussing and resolving identified issues promptly
- Integrating relief system design review team with project design and execution teams

4 Conclusion

As shown in several examples in this paper, the systematic review of pressure relief and flare system calculations at various process design phases will not only help to ensure adequate overpressure protection, but can also highlight opportunities for inherently safer processes and promote overall project lifecycle cost savings. In addition, such a review process can easily be extended to resolve similar design issues commonly addressed during HAZOP reviews and integrated with the MOC workflow.

5 References

[1] American Petroleum Institute, *API Standard* 520, *Sizing, Selection, and Installation of Pressure-Relieving Devices, Part I-Sizing and Selection*, 9th Edition, 2014.

[2] American Petroleum Institute, *API Standard* 520, *Sizing, Selection, and Installation of Pressure-Relieving Devices, Part II-Installation*, 6th Edition, 2015.

[3] American Petroleum Institute, *API Standard 521, Pressure-relieving and Depressuring Systems,* 6th Edition, 2014.

[4] American Petroleum Institute, *API Standard 2000, Venting Atmospheric and Low-pressure Tanks*, 7th Edition, 2014.

[5] American Society of Mechanical Engineers, *ASME BPVC Sec I, Rules for Construction of Power Boilers*, 2015 Edition.

[6] American Society of Mechanical Engineers, *ASME BPVC Sec VIII-1, Rules for Construction of Pressure Vessels*, 2015 Edition.

[7] API STD 521, §4.4.9.3

exida

Three Reasons Major Event Risk Management Fails on Capital Projects

Denise Chastain-Knight

P.E., CFSE

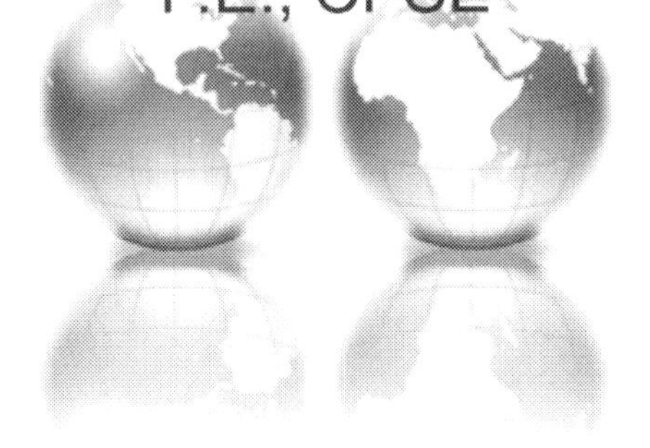

exida

Make sure all deliverables are in the project plan and reflected in the integrated project schedule

2

About the Presenter

- **Denise Chastain-Knight, PE, CFSE, CCPSP**
  - BS ChE, Georgia Institute of Technology
  - Senior Functional Safety Engineer
  - More than 31 years experience in engineering design, process improvements, process risk analysis and safety instrumented system implementation
  - Experienced in the chemicals, resins, coatings, monomers, polymers, pharmaceutical, silicate, bio-fuels, petroleum, power…
- **Affiliations**
  - CCPS Technical Steering Committee
  - Past Chair NCEES PE (ChE) exam development committee
  - Fellow AIChE

3

Familiar Conversation?

How is the project going?

Great we are commissioning now and startup will begin in a couple weeks

I'll need to review the final SIL Verification and FSA so I can sign off on PSSR

SIL Verification? FSA? What are those?

4

Familiar Conversation?

Similar conversations occur all to often as capital projects are nearing completion.

Often the topic of the conversation involves an aspect of process safety that is critical for management of high consequence events that pose a significant risk.

Why does this keep happening?

Missed Opportunity

- Capital projects offer a huge opportunity to make step changes in process safety
- Best practices can be implemented before capital is spent
- Clear guidance will assure that performance standards effecting process safety be implemented in the way the owner/operator desires.

Operating Company

- Extensive experience managing risk in an operating environment
- Little practice managing process risk within a capital project
- Process Safety model relies heavily on culture, procedures and training
- Assumes the contractor knows how they want to approach complaince

 7

Engineering Design Firm

- Experienced at project design and execution
- Traditional project execution models are focused on delivering code compliant scope on schedule and budget
- Process safety deliverables are often excluded from project scope
- Knows what performance standards require, but clients approach compliance in different ways

 8

PS Activities that Usually Happen

- Hazards Identification and Risk Analysis
- Relief system design
- Compliance with code mandated standards (ASME VIII, NFPA 30, NFPA 70)

 9

PS Activities that Often Fail

- Compliance with RAGAGEP standards (ISA 18.2, ISA 84, ANSI, ASHRAE, Chlorine Institute)
- Inherently Safer Design
- Process Risk Mitigation
- Asset Integrity and Reliability
- Preparation for operation & maintenance and beyond

 10

exida

Why We Fail

- No or inadequate overall vision
- Insufficient preparation
- Deficient project execution

Communication

exida

Lack of Overall Vision

- What is the risk reduction required and what are the tradeoff priorities?
- How will responsibilities be divided?
- What tools and standards to be used?

Insufficient Preparation

- Philosophy and base procedures not defined by operator owner
- Capital project scope (deliverables) and schedule do not support process safety objectives
- Key role and staff competencies undefined

Deficient Project Execution

- Underestimate of the workload required and dependencies with other scope objectives
- Process Safety /Risk Management not treated as a discipline
- Reactive not proactive

How Project work

- Stage gate process
- Scope identified, schedule and budget developed to support scope
- Detail built in layers. Early definition important to minimize rework
- Often involve many contractor teams working on a subset of the scope

 15

Establish Vision

- Define Process Safety – it is more than a PHA
- Prepare Philosophy and Design basis documents
- Define Competency Requirements
- Define Performance Metrics

 16

Prepare

- Identify process safety deliverables and expectation for contents of the deliverables
- Set milestones for project stage gates
- Understand the inputs and outputs for PS deliverables and reflect dependencies in the project schedule
- Assess competency and close gaps

 17

Execute

- Execute PS scope as a project discipline
- Include PS requirements in specifications
- Complete technical review of bid packages

 18

Pre-Project Documents

- List of Standards for Compliance
- Risk Management Plan & Procedures
- Hazards Assessment Procedures
- Cybersecurity Policy & Procedure
- Functional Safety Management Plan (organization)
- Functional Safety Management Procedures
- Approved vendors list for critical/safety equipment and instrumentation
- Alarm Philosophy
- Inherently Safer Design Philosophy
- Major risk management philosophies (fire, relief, etc)
- Contractor Competency Requirements
- Procedures for required studies (dropped object, etc)
- Maintenance Philosophy (isolation, clearing requirements etc)
- Measurement & Metrics philosophy
- Communications Philosophy
- Spares philosophy

19

Preliminary Engineering

- Functional Safety Management Plan (project)
- Chemical Reactivity Matrix
- Hazards identification
- Conceptual SIF list
- Hazards Identification Study
- Define credible accident scenarios
- Major and long lead equipment specifications
- Controls Description
- Cyber Security Risk Assessment & Gap Analysis
- Maintenance Plan

20

Basic Engineering

- Project phase HAZOP
- CHAZOP
- HIPPS Studies
- Fire and Explosion Analysis
- Chemical release dispersion modeling
- SIL Determination
- SIS Safety Requirements Specification
- SIF Safety Requirements Specification
- Stage 1 Functional Safety Assessment (FSA1)
- Cyber Security Requirements Specification
- Countermeasure Inspection & Test Procedure
- Penetration Testing
- Cyber Security Assessment (CSA-1)
- Emergency notification, evacuation & response review
- Reliability, Availability, and Maintainability (RAM) analysis

 21

Detailed Design

- Balance of specifications
- Detail Design documents
- Alarm Rationalization
- SIL Verification
- SIS Factory Acceptance Test (FAT)
- Stage 2 Functional Safety Assessment (FSA2)
- Cybersecurity Factory Acceptance Testing (CFAT)
- Cyber Security Assessment (CSA-1)

 22

Procure and Construct

- Technical and financial bid reviews
- Operating & Maintenance Procedures
- SIS/SIF Proof Test Procedures
- Stage 3 Functional Safety Assessment (FSA3)
- SIS Site Acceptance and Integration Test (SAT, SIT)
- Pre-startup SIF Validation
- Pre-startup Cybersecurity Review (CSA 3)
- Training
- Operational Readiness Review
- Configure maintenance management system
- Emergency management plan

 23

Operate & Maintain

- MOC for ALL changes
- SIS/SIF Proof tests & results analysis
- Revalidation of Detailed Risk Assessment
- Cybersecurity Vulnerability Assessment (CVA)
- Cybersecurity Assessment - CSA 4 & 5
- Periodic Revalidation of PHA

 24

25

A Comparative Analysis of Safety Rating System

Linlin Wang[a]

[a]Mechanical and electrical engineering college, China University of Petroleum (East China)

Abstract: With the development of petrochemical enterprises, the scale and petrochemical enterprises frequent accidents, the high risk of petrochemical enterprises has been more and more extensive attention. It's particularly important to establish a sound safety rating system for the timely detection and elimination of risk factors, reduce the probability of accidents appears. However, in China there is less research and application on safety rating system. This paper summarizes the similarities and differences between the various systems and puts forward the development of China's safety rating system.

Keyword: Safety Rating System; Petrochemical enterprises

1 Introduction

Petrochemical enterprises have a lot of flammable and explosive, toxic substances, and the installation of large-scale and continuous production, can easily cause major accidents. In 2013 Sinopec Huangdao Oil pipeline explosion, 62 people were killed, 136 people were injured. With the occurrence of petrochemical enterprises accidents, the safety production of petrochemical enterprises has become one of the issues that government and national focus on. Evaluation and rating of the status quo of enterprise security can promote the enterprises to find their own problems, develop improvement goals and implement continuous improvement. At the same time, it can also improve the government safety supervision departments to find out the status quo of enterprises, determine the key supervision enterprises, and optimize the allocation of resources, enhance the safe production level of enterprises in China, it is also an important means to improve business efficiency. The Chinese government's security supervision department attaches great importance to the assessment of enterprise safety, since 2006, has issued a "Basic norms of enterprise safety production standards", " General specification for safety standardization of hazardous chemicals", "Guidelines for oil industry safety production standardization "etc., The government promotes the implementation of the safety production standardization assessment throughout the country, take the "enterprise self-evaluation, government review" approach to the enterprise rating, and achieved important work results. But in the implementation a lot of problems were also found. Based on the status quo of safety rating of petrochemical enterprises in China, this paper analyzes and compares the implementation process of enterprise safety rating system such as ISRS (International Safety Rating System) ,NOSA five-star comprehensive evaluation system and VPP project implemented by OSHA. It is expected to provide support for the selection and development of our country's petrochemical enterprise safety rating tool.

2 Status Quo of Safety Rating of Petrochemical Enterprises in China

Safety standardization rating is developed, implemented and reviewed by China's relevant laws and regulations, industry standards. It is led by the State Administration of Safety Supervision and implemented on a national scale. It is the most extensive and highest commitment safety rating system in China. In addition, some petrochemical enterprises received the support of government, and independently introduced tools such as ISRS and NOSA. For example, 2014 China Petroleum West Pipeline Company reached the rate 7 of ISRS, and became first enterprise which reached this level in China.

Based on the safety standardization assessment of petrochemical enterprises in China, it is found that there are three main problems in the process of implementing safety standardization assessment in our country:

(1) Safety standardization system and some other implementation systems, such as HSE system, have incompatible problems, resulting in enterprises to repeat engaging in several sets of systems, repeat records, repeat systems. Companies not only feel cumbersome, but also waste human and material resources;

(2) Enterprises have low motivation and a bad attitude to practice safety standardization. Standardization of safety production is a measure imposed by the government to improve the safety level of the enterprise, and enterprises do not really realize its importance. So that the work is easy to become formalize and surface, and did not achieve the purpose of improving the level of safety, but only to let the enterprises to meet the rating and inspection;

(3) China's safety standardization assessment consists of three grades. The grade of the score range is quite large. And the flexibility of the assessment is not sufficient. Besides, in the absence of experimental circumstances, the assessment has a low recognition for some enterprises which take a new process or a new management approach.

3 System Introduction

3.1 ISRS（International Safety Rating System）

ISRS（International Safety Rating System）is a rating system of international security measure, and improving and showing the level of health, safety, environmental and business performance and other aspects of the organization, designed to help organizations manage risk and promote continuous improvement of the evaluation system. Det Norske Veritas (DNV) international safety rating system (ISRS) eighth edition published in 2009 after decades of practicing and improving.

ISRS ^{7} international safety rating system has 15 elements, each element is divided into several sub-elements of a total of 119 sub-elements, and list the number of points for each sub-element of the assessment, the total number is 661. 15 elements are: leadership, planning and administration, risk assessment, human resources, compliance assurance, project management, training and skills, communication and marketing, risk control, asset management, procurement and contractor management, emergency preparedness , the event learning, risk monitoring, and review the results, in which risk management (including 3 risk assessment,9 risk control, 14 risk monitoring) system for the trunk. During the assessment process, the customer selects the subroutine according to the grade to be evaluated, as follows::

Grade	1	2	3	4	5	6	7	8	9	10
Level range	1-4				5-6		7-8		9-10	
Required number of subroutines	30				60		90		119	
The number of customer chooses	20				15		10		0	
The total number of subroutines	50				75		100		119	
The lowest score of each program%	10	15	20	25	30	35	40	50	60	70
Average minimum score%	20	30	40	60	60	70	70	80	80	90
Live tour lowest score %	65	65	65	65	70	75	80	85	90	95

3.2 NOSA integrated five-star management system

In 1951, the International Occupational Safety Association (NOSA) created a five-star safety management system by conducting a comprehensive hazardous and harmful factors identification and risk assessment to identify all possible hazard identification and develop appropriate management for reduce hazard and risk. NOSA management system regards hazard identification and risk management as the core, Heinrich's "iceberg theory" as the basis and PDCA manner as its operation mode. System consists of five main elements, 72 elements, more than 300 minimum standards, the implementation of 1200 rules and standard instructions (against NOSA Integrated Five Star system revision and reorganization, the system consists of 75 elements). System evaluation content mainly in the following two aspects:

(1) Performance evaluation

Evaluation content of NOSA five-star rating system mainly consists of five areas, particularly in the following table:

NOSA integrated five-star rating system

project	Unit content	The number of elements	score	Share of the share
1	Building and housekeeping (factory) management	11	450	15
2	Mechanical electrical and personal protection	17	600	20
3	Fire and other emergency risk	8	400	13.3

	management			
4	HSE accident recording and investigation	5	400	13.3
5	HSE management organization	31	1150	38.4
total		72	3000	100

Through five major categories of 72 elements of qualitative evaluation, the project almost involved all activities of production and operation. It evaluates each element from the "four levels, three aspects, five grades" to carry out a comprehensive calculation to measure whether the risk of the element has been effectively controlled. "Four levels" means: risk awareness, the file system, compliance, implementation effect. "Three areas" means: safety, health, environment. "Five grades" means: 0, 25, 50, 75, 100 five scoring levels。

（2） Experience assessment

Introducing DIIR (Disabling Injury Incidence Rate) rate of accidents concept of enterprise, and performance assessment combined score, drawn 1 to 5-star level. DIIR shows annual percentage of the total number of employees were injured, according to NOSA claims, statistics should include employees of the enterprise, the tertiary industry and outsourcing construction team personnel. DIIR calculated as follows：

$$\text{DIIR} = \frac{The\ number\ of\ accidents\ caused\ by\ work \times 200000}{Number\ of\ employees\ working\ hours\ per\ year}$$

Where：DIIR——Disabling injury incidence rate；

The number of accidents caused by work - refers to the work caused by or caused by the injury, so that the injured cannot or cannot fully perform the daily designated job work up to one or more working days；

200000-- base by 100 people work for a year (50 weeks), 40 hours a week calculated；

Number of employees working hours per year - Total number of employees × Actual hours per person per year

The final combination of the above two aspects of the results given the final star rating results, the corresponding assessment of the star method as shown in the following table：

NOSA integrated five - star rating system assessment standard table

Star rating	**score（%）**	**DIIR（%）**
NOSCAR	≥95	≤0.8
Five star	≥91	≤1.0
Four star	≥75	≤2.0
Three star	≥61	≤3.0
Two star	≥51	≤4.0
One star	≥40	≤5.0

NOTE: DIIR = the number of industrial accidents × 200000 (base) ÷ annual working hours of employees

3.3 VPP（The Voluntary Protection Programs）

The Voluntary Protection Programs (VPP) is a safety and health management system to prevent workplace accidents and death, by allowing enterprises to implement additional higher than the US Occupational Safety and Health Administration (OSHA) Voluntary standards measures to achieve excellent safety and health management goals. OSHA is through a detailed site assessment and comprehensive site accident rate / illness rate to assess whether the corporate site has a good protection system。

The major elements of the program: management leadership and employee involvement; worksite analysis; hazard prevention and control and safety and health training. To assess something like this, there are three main forms of evaluation methods, including the inspection of documents, on-site inspection and interviews。

Leadership and employee participation	Written safety and health management system, management commitment and leadership, Authority and Line Accountability (authority and responsibility), contractor employees, employees' participation
Workplace analysis	Baseline Hazard Analysis baseline risk analysis, hazard analysis regular activities, regular inspection, hazard reporting systems, accident / incident investigation, trend analysis
Hazard Prevention and Control	Whether there are effective measures to combat control and control, as well as written contingency plans safety and health training

In addition, team members also need to The Total Case Incidence Rate (TCIR)以及The Days Away from Work, Restricted Activity and Job Transfer (DART) Case Incidence Rate evaluated to confirm the company level must be below the national average. After a series of reviews, OSHA decides that the applicant's qualifications are Star, Merit, Demonstration program, if the applicant fails to meet the requirements, return the application, and VPP level is relatively flexible, for those who do not meet the Star designing 1-Year Conditional Star Participants and Star Demonstration Programs.

3.4 Safety standardization assessment

Safety production standardization refers to the establishment of safety production responsibility system, the development of safety management systems and operating procedures, troubleshooting and monitoring of major hazards, the establishment of preventive mechanisms to regulate production behavior, so that the production links in line with relevant safety laws and regulations and standards Standard requirements, people, machines, objects, ring in a good state of production, and continuous improvement, and constantly strengthen the standardization of enterprise safety construction.

Safety standardization evaluation mainly taken independent assessment, external review of the way, self-assessed score of 60 points or more companies can apply for formal standardization rating, the final assessment is divided into one, two, three, one is the highest, from the security Production supervision and management departments to review the classification of supervision and management. During the review process, the following aspects are mainly reviewed: objectives, organization and responsibilities, safety production, laws and regulations and safety management system, education and training, production equipment and facilities, operation safety, hidden trouble investigation and management, major hazard monitoring , Occupational health, emergency rescue, accident learning, performance assessment and continuous improvement of the thirteen major aspects of compliance assessment, the final score to score how much. The enterprises that need to be graded need to carry out self-evaluation on a yearly basis to determine the safety standardization status of the enterprises and to re-order them every three years to urge enterprises to continuously improve the standardization work level.

4 System comparison results

Through the understanding of the information, compared the above systems, found the following the same and different:

1. Purpose

In essence, the above mentioned rating methods are all aimed at checking and enhancing the enterprise to achieve a certain level of security. However, the basic purpose of the rating system is different because of the different backgrounds of the development period. For example, the NOSA five-star rating system was developed in the 20th century, the rapid development of industrial and mining in South Africa and the quality of personnel and other issues caused the high rate of injury accidents, so its basic purpose is to reduce the high incidence of work accidents. In the final analysis, the rating is only a means and security is the purpose.

2. Applicable scope

ISRS is available in 11 languages and has been implemented on thousands of sites worldwide. The industries involved include marine, oil and gas, energy, nuclear, etc., and ISRS is introduced into the oil industry in China, such as China Petroleum West Pipeline Company reached the international security rating of 7.

NOSA five-star rating also can evaluate several kinds of industries including mining, forestry, railways, health care and natural gas, but its vast application in China is applied to the mining industry.

The VPP assessment program is applicable to all companies that use OSHA standards. VPP's participation companies include construction, agriculture, marine, oil and gas, health care and other industries.

Since the implementation of the Safety Production Standardization, under the guidance of the government, Safety Production Standardization has been applied in most enterprises in China, including industry and trade, power industry, transportation, hazardous industries, non-coal mines, construction industry, fireworks, coal industry, Its application covers most of the production enterprises.

3. Evaluators

Safety Production Standardization rating system take the way of "enterprise independent assessment then external assessment " to carry out the rating, Safety Supervision and Management Departments supervises the classification. The external review is conducted by the relevant experts. Similarly, the NOSA five-star rating system is evaluated by the company's team of experts. In the composition of the assessment team, VPP projects and ISRS systems have made innovations.

The VPP site assessment team is composed of experts from all areas of OSHA and SGE (The Special Government Employee), which has performed on-site evaluation. The implementation of the SGE project is intended for enterprises employees to participate in the process of VPP's on-site assessment. The implementation of this innovative project is more conducive to government and corporate information and ideas sharing.

ISRS needs to form an evaluation team consisting of two or more persons prior to evaluation, and to ensure the depth of the HSEQ field in evaluation, the team members contain relevant experts and do not restrict internal / external members, and All evaluation members will receive DNV training courses.

4. Rating method

Whether it is ISRS system, NOSA, VPP or safety standardization rating system are all required to compare the relevant standards and scoring rules with the actual situation of enterprises, through the document inspection, on-site visits and consultation to decide the company's compliance to standards.

However, NOSA and VPP projects need to combine other indicators before the final results. VPP not only requires companies to meet or exceed OSHA standards, but also requires companies to maintain below the average rate of disease released by National Bureau of Labor Statistics. And NOSA combines the evaluation rate of qualified with DIIR ratio to determine the stars should be given.

5. Follow up later

The "PDCA" model has been adequately applied in several rating systems, "plan, implement, check, action", adopt the infinite loop method of " find problems then solve problems, and then find the problem, and then solve the problem" to promote the continuous improvement of enterprise safety management system and continuously improve the level of enterprise security.

But for the rating implementation unit, the follow-up degree is different from others. Among them, the VPP project's follow-up work is very good. VPP participants need to submit an annual self-assessment report, OSHA's site assessment team will confirm whether the enterprise has made a corresponding improvement, and re-examination time interval varies by the enterprise level, such as star business, every 3-5 years conduct a reassessment. For companies with a one-year correction, if they do not meet the requirements after one year, they will withdraw the VPP application.

System comparison table

	ISRS	NOSA	VPP	Safety Standardization
Basic Purpose	get management	Reduce the	prevent	Promote

	system to work more effectively	incidence of industrial accidents	workplace accidents and death	enterprises to achieve standardization and standardization
Applicable scope	ISRS is available in 11 languages. The industries involved include marine, oil and gas, energy, nuclear, etc.,	Including mining, forestry, railways, health care and natural gas, but its vast application in China is applied to the mining industry	all companies that use OSHA standards	Covers most of the production enterprises, such as industry and trade, power industry, transportation, hazardous industries, non-coal mines, construction industry, fireworks, coal industry
Evaluators	Two or more people (DNV training)	Company experts team	OSHA experts in various fields and SGEs who have conducted on-site evaluations	Independent assessment, external review The external review is conducted by the relevant experts
Rating method	Performance evaluation	Performance evaluation and DIIR	companies to meet or exceed OSHA standards, low disease rate	Performance evaluation
Assessment Frequency		Self-evaluati on Twice a year Re-rating depend on clients	Self-evaluation once a year re-examination time interval varies by the enterprise level	Self-evaluation once a year Every three years have re-examination
Grades	10	6	Have Merit、Star、Demonstratio n program etc., grade flexibility is high	3
Voluntary	voluntary	voluntary	voluntary	Forced

5 Conclusion

Combine comparison and thinking above, for the Chinese petrochemical enterprise safety rating system selection, provide the following suggestions:

1. For small private enterprises, Safety Standardization has been fully compatible with security requirements, for other state-owned enterprises, large-scale Sino-foreign joint ventures can choose to introduce foreign security rating system.
2. ISRS system covers the ISO standard and OSHA standard requirements, and also developed a special version of ISRS-Pipe, the rating system also covers the majority of indicators in China oil safety management system, ISRS has been applied in a lot of petrochemical enterprises in China But the ISRS system uses the lowest score of the total score in the first-level elements as the criterion for determining the final grade, so it is not easy to reflect the advantages of the enterprise in the final level judgment.
3. NOSA five-star rating system considers the performance evaluation and DIIR value together in the final evaluation to get the overall level, but when the enterprise finally decided to use the NOSA system, NOSA was originally used for mining assessment should be considered; For the company implementing OSHA standards can apply for the VPP program because VPP determines the enterprise level by checking whether the company is performing a higher-than-OSHA standard.

Acknowledgement

This work is supported by the National Natural Science Foundation of China (21406271) and the China Scholarship Council "Innovative Talents International Cooperation Program" (Liu Jin Mei [2016] 7988). Thank you very much.

Reference

1. https://www.dnvgl.com/services/isrs-for-the-health-of-your-business-2458
2. https://www.nosa.co.za/

478942 The Difficulty and Prospect of the Development of Quantitative Assessment Technology in the Field of Chemical Safety in China

Abstract:

Abstract: With the increasingly high demand for safety of chemical industry quantitative assessment technology has been widely recognized in chemical industry safety field because of its more accurate assessment results. Nowadays many existing quantitative assessment techniques,(such as quantitative risk assessment (QRA), reliability assessment, safety integrity level assessment (SIL) .etc)have been gradually introduced into the filed of China chemical safety to provide more assistance for the risk management of chinese chemical enterprises. However, some basic data of the quantitative assessment on chemical industry in China are lost due to the lack of systematic statistical analysis of data(such as equipment failure data, accident event data .etc), which leads to the situation that our country often adopt foreign related data when conducting a quantitative assessment. In result, results of quantitative assessment are out of their authenticity. In addition, because there is no more specific and more operational assessment guidelines for the part of quantitative assessment techniques issued in our country, it also affects the accuracy and scientific of the results. According to analyzing some misunderstandings and difficulties in chinese chemical safety field when carrying out the quantitative assessment technology, the paper puts forward some suggestions on the change and improvement of related industries to make quantitative assessment technology to provide greater help for the development of chemical safety field in China.

Recipe for a *Sizzling* HAZOP

Poster Presentation to 13th GCPS

Angus Keddie, Process Safety Matters

Benefits of Running an Effective HAZOP

- Commercial: Meeting schedule; Risk optimisation of facility
- Safety: Pick up all significant potential hazards
- Psychological: Limiting stress on participants
- Chairman: High probability that the chairman will be invited back

PROCESS
SAFETY MATTERS

Recipe Success? – Key Components

- Recipe Disasters……
- Provenance
- Preparation
- Supporting Team
- Kitchen Knives
- Ingredients
- Spice and Flavour
- Oven Conditions
- ….and How to Avoid Them
- Ta Dah!

PROCESS
SAFETY MATTERS

Recipe Disasters

Recipe Disasters

- Behaviour to avoid if Effective HAZOP to be achieved:
  - Falling asleep
  - Fighting with team members
  - Allowing fighting between team members
  - Silence of participants
  - Permitting interminable discussions on minor issues
  - Scribing yourself
  - Working well in excess of 7 hours per day

HAZOP Chair Provenance

To paraphrase the Bard "Some or born to HAZOPness, some achieve HAZOPness and some have HAZOPness thrust upon them"

There are around 5000 HAZOP chairmen worldwide all of them being equal, but some are more equal than others.

In order to optimise process safety for existing and new facilities worldwide, excellent PHSERs need to be carried out. A key ingredient of this is the performance of the facilitator. What personal characteristics combine to create this performance?

HAZOP Chair Provenance

- Undergraduate degree in Chemical Engineering
- Professional exposure to relevant aspects of industry:
  - Process Design
  - Operations/Commissioning
  - Team Leadership/Management
  - Process Safety
- Top up with Masters in Process Safety
- Modal age range: 40 to 60
- CPD of IChemE Training Course
- Observation of Other respected HAZOP chairs to evolve best practice
- But like all great recipes, a little variety doesn't do any harm and may enhance the final product.

HAZOP Chair Characteristics

Desirable attributes for the leader include:

- wide experience of all the stages of process hazard studies, including quantitative risk assessment;
- extensive experience as a HAZOP study member and, preferably, some as a HAZOP study scribe;
- training in the leadership of HAZOP studies;
- technical competence and the ability to quickly understand the system and its operation;
- meticulous attention to relevant detail;
- good analytical thinking;
- motivational skills including the encouragement of creativity and open speaking;
- independence from the project itself with no other direct responsibility to the project manager other than completing the scope of the HAZOP study.

PROCESS
SAFETY MATTERS

Fail to Prepare......

- Hold a Pre-meeting with key project stakeholders to establish significant draft tenets:
  - Project Background and Scope
  - Stakeholder Drivers
  - Optimal Methodology
  - Key Hazards
  - Putative Assumptions
  - Principal Personalities
- Outcomes and Terms of Reference inform Intro presentation and initial approach

PROCESS
SAFETY MATTERS

The Scribe - Sous chef or trusty side kick?

PROCESS SAFETY MATTERS

LinkedIn HAZOP Discussion:

- **Does a PHA scribe need to have a technical background? – Ron Brachten, Nov 2014**
- **AK contribution: scribal attributes -**
  - Fast and accurate typing skills
  - Native or excellent English language skills
  - Sufficient *confidence* to capture the consensus and to overlook the inevitable spelling and grammar gremlins
  - *Stamina*
  - Good humour
  - Interest in the process
  - Threshold technical knowledge

Technical knowledge is handy but....

PROCESS SAFETY MATTERS

HAZOP Tools – Sharpen those knives

- Climate controlled room, good natural light
- A1 PIDs on walls or second screen
- Availability of Flip charts or White boards
- Wifi/LAN access to upload project tech info and wider references
- Good food during breaks and lunch
- Effective HAZOP recording software

Ingredients

- Meat and 2 Veg:
  - Process Description
  - Noded P&IDs
  - Intro Presentatiion
- Side Dishes
  - PFDs
  - Cause and Effect Diagram
  - Report from any previous HAZOP(s), HAZID
  - HAZOP Terms of Reference
  - Facility Design Standards
  - Plot Plan
  - Line List
  - Piping Classification Overview
  - Toxicity/flammability data for materials being handled
  - Relevant National/International Codes/Laws/Standards

The texture and flavour – the local element!

The experience of HAZOP can be vary significantly depending on technical, cultural and stakeholder perspectives:

Stakeholders

- Operators focus on operability and consistency of design. (gooey and sweet)
- The EPC contractor's focus is on schedule and no changes (hard and dry)
- The operator's safety team focus on strict code compliance (tart and bitter)

Technical

- Upstream – no steady state; little standardisation
- Downstream – Large through puts/inventories; replication

Cultural

- "Many nations separated by a common language – HAZOP!"

Oven Conditions – HAZOP Ambience

PROCESS SAFETY MATTERS

The team works better if relaxed and motivated – try to make this happen early on:

- Introduce yourself to team members
- Get everyone to introduce themselves round the table, including what contribution they expect to make
- Directly engage all team members in the first 30 mins
- Allocate responsibilities based on member skills and attributes
- If team is flagging, call a break
- It two team members become adversarial, take a time out
- Smile

Recipe Disasters – and their remedies

PROCESS SAFETY MATTERS

- Behaviour to avoid if Effective HAZOP to be achieved:
  - Falling asleep – stand up and walk around when facilitating
  - Fighting with team members – develop excellent rapport with team from the off;
  - Allowing fighting between team members – call time out; verbal warnings; firing culprit(s)
  - Silence of participants – engage in meetings (and in breaks).
  - Permitting interminable discussions on minor issues – set time limits for discussion; "is there anything we need to record at this point?"
  - Self Scribing – refuse!
  - Working well in excess of 7 hours per day – planning; negotiate schedule; identify key risks up front; prioritise nodes (Shell approach – 6 working hours per day, 4 days per week, 4 weeks out of 5)

Ta Dah!

Angus Keddie
Director
Process Safety Matters
+44 (0) 7515897869
+44 (0) 1344 455800
info@processsafetymatters.com

477894 Bridging the Divide - OHS and Process Safety

Abstract:

The occupational health and safety field is well established, with a number of professional bodies, undergraduate programs and a comprehensive body of knowledge. Conversely, the process safety field has evolved largely over the past 40 years as an engineering discipline, with a focus on engineering solutions. Both of these fields utilise a risk management basis for decisions and actions. But both fields still largely operate independently of each other. There is much to be learned from sharing knowledge and responsible collaboration to deliver safer plants and operations from an OHS and a process safety perspective. This paper will explore how we can bridge the divide between OHS and process safety to achieve better safety outcomes. This includes better education and information flow between both professions as well as leveraging off existing systems to advance process safety.

20 Years Of Personal Process Safety Lessons

Steve Streblow
ACS Engineering
10914 Milano Ct.
Richmond, TX 77406
streblowsr@yahoo.com

Prepared for Presentation at
American Institute of Chemical Engineers
2017 Spring Meeting and 13th Global Congress on Process Safety
San Antonio, TX
March 26 – 29, 2017

20 Years of Personal Process Safety Lessons

Steve Streblow
ACS Engineering
10914 Milano Ct.
Richmond, TX 77406
streblowsr@yahoo.com

Keywords:

Amine tower; carbon dioxide; communication and restriction; competency; compliance; culture; documentation; failure analysis; intimidation; investigation; lessons learned; material of construction; mechanical integrity; pressure; stress; risk assessment; vessel repair.

Abstract

This paper presents in detail one of the many process safety issues encountered by the author in his 20 years of industrial experience. This issue deals with an amine tower, but touches upon the following topics: carbon dioxide; communication and restriction; competency; compliance; culture; documentation; failure analysis; intimidation; investigation; lessons learned; material of construction; mechanical integrity; pressure; stress, risk assessment and vessel repair.

It describes better ways to address the situation and applicable RAGAGEP. It is hoped that sharing this information will enable others to learn from the past and prevent similar issues in the future.

1 Introduction

This paper presents in detail one of the many process safety issues encountered by the author in his 20 years of industrial experience. This issue deals with an amine tower, but touches upon the following topics: carbon dioxide; communication and restriction; competency; compliance; culture; documentation; drawings; failure analysis; intimidation; investigation; lessons learned; material of construction; mechanical integrity; pressure; stress, risk assessment and vessel repair.

It describes better ways to address the situation and applicable RAGAGEP. It is hoped that sharing this information will enable others to learn from the past and prevent similar issues in the future.

2 Amine Tower Leak

In one of the operating companies I worked for, I was supposed to be the technical support for many gas processing facilities, which included amine systems for removal of carbon dioxide. I was verbally informed of a small leak in one of the facilities amine regenerator towers which was stopped by placing a screw in the side of the tower, as shown in the photo (taken later) immediately below.

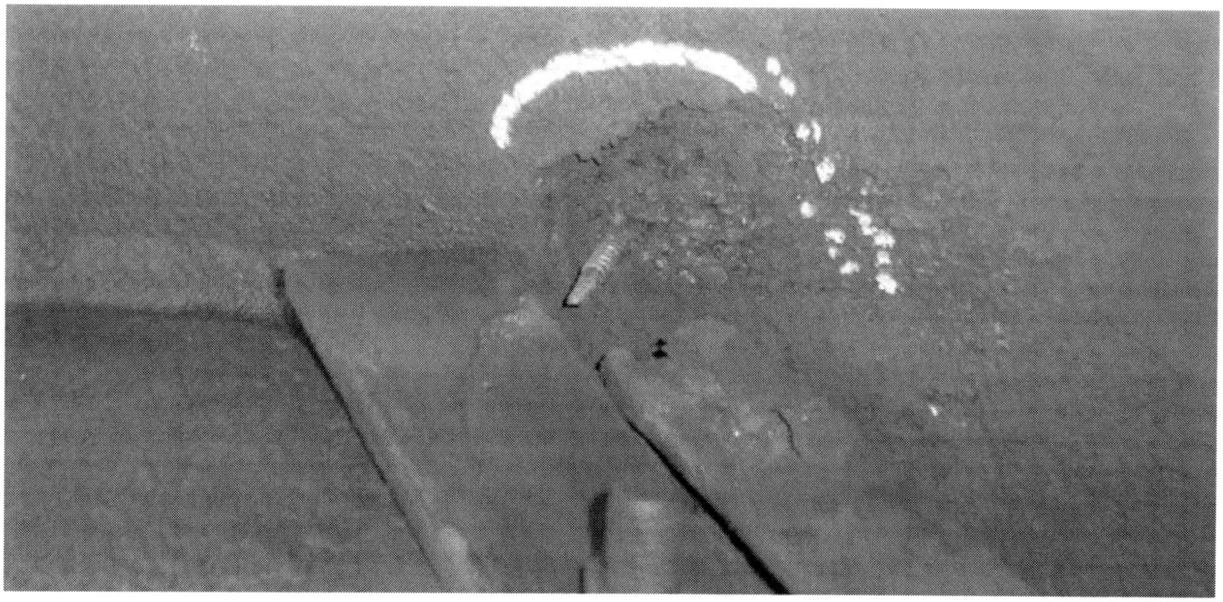

I was shocked for several reasons. First, I was not consulted for my technical opinion on how to proceed. Second, because after I communicated with the plant, they stated they were told by a service provider that they normally use that this type of repair was proper.

So much of this situation seemed wrong to me, so I provided the plant suggestions to restrict access to the area nearby the tower, while I determined how to proceed. The plant did put this area restriction in place.

I started to research what to do. First, I verified the pressure vessel law, as one can do by checking NB-370 at:
https://www.nationalboard.org/PrintAllSynopsis.aspx?Jurisdiction=Select
The location was in the state of Texas, which did not have a pressure vessel law. I next tried to find RAGAGEP for this. I called some pressure vessel manufacturers and had some brief conversations, which helped me focus my search. I found out that to keep the ASME code status, the repair should have been performed through a company which has an R stamp and filled out the appropriate paperwork. The R stamp status of a company can be confirmed at:
http://www.nationalboard.org/ManufacturerDirectory.aspx
The RAGAGEP to be used to maintain ASME code compliance is NBBI NB 23 (NBIC), National Board Inspection Code, Parts 1-3, reference 1.

If the vessel is a non-code vessel or if ASME code status does not need to be maintained, then one can refer to RAGAGEP API 510, Pressure Vessel Inspection Code: In-service Inspection, Rating, Repair, and Alteration, Tenth Edition, May 2014, reference 2.

After seeing that there was much technical knowledge needed for this activity that I did not have, I checked with management to see if there was someone better qualified than I inside the company, and I was disappointed to learn that there was not. Management did not seem concerned about the lack of expertise or the fact that I was being asked to work outside my area of expertise. It would be good to have a dedicated mechanical integrity expert in the company, since there is a great breadth and depth of technical knowledge needed.

I then tried to get the results of the previous mechanical integrity inspections for the tower, which I was told existed, but I was not provided. This withholding of information without any explanation seemed suspicious to me. This is a weakness in the culture. So, I tried to determine what the inspections should be for this item.

I tried to get the paperwork on the tower screw "repair", but none was provided, again showing a cultural problem.

I informed the plant that the company they were using did not have an R stamp and the screw action was not a proper repair.

I suggested contacting an inspection company to perform some external vessel wall thickness inspections, to see if additional thin spots were present. The external inspection was performed, which showed some additional thin areas.

I was verbally informed of a second small leak in the subject amine regenerator tower which was stopped by placing a screw in the side of the tower, as shown in the photo (taken later) immediately below.

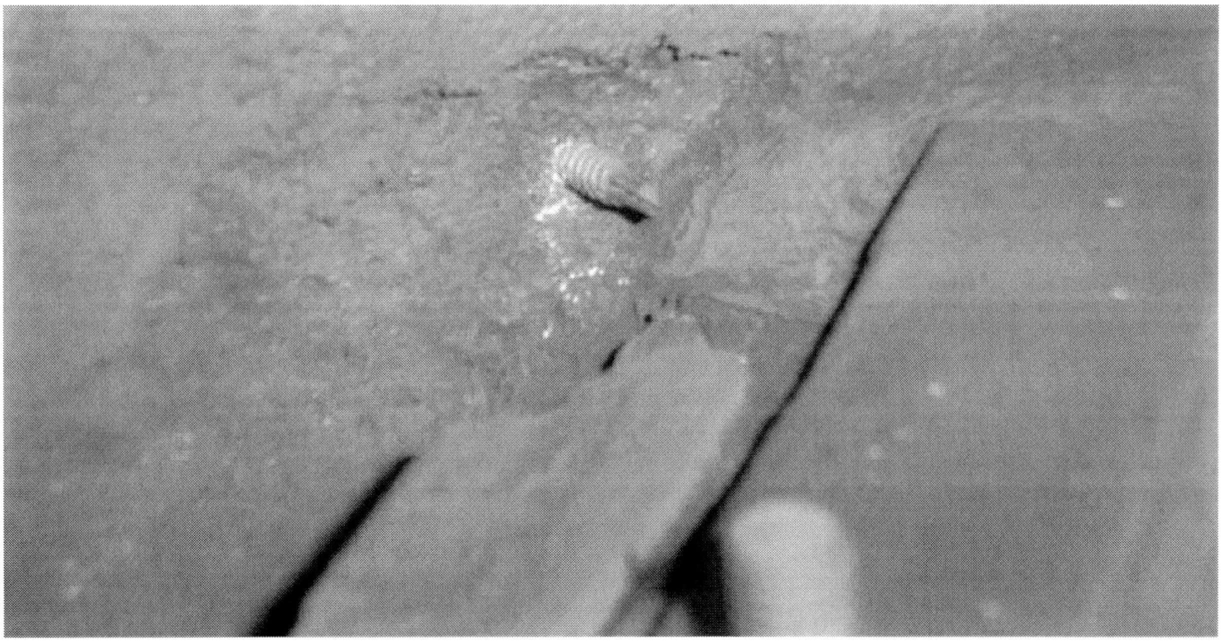

It was disappointing that the facility took the same actions as last time, when they had been advised otherwise. I notified my management, and they did not seem concerned. I again reminded the plant that the company they were using did not have an R stamp and the screw action was not a proper repair.

I was verbally informed of a third small leak in the subject amine regenerator tower which was stopped by placing a screw in the side of the tower, as shown in the photo (taken later) immediately below.

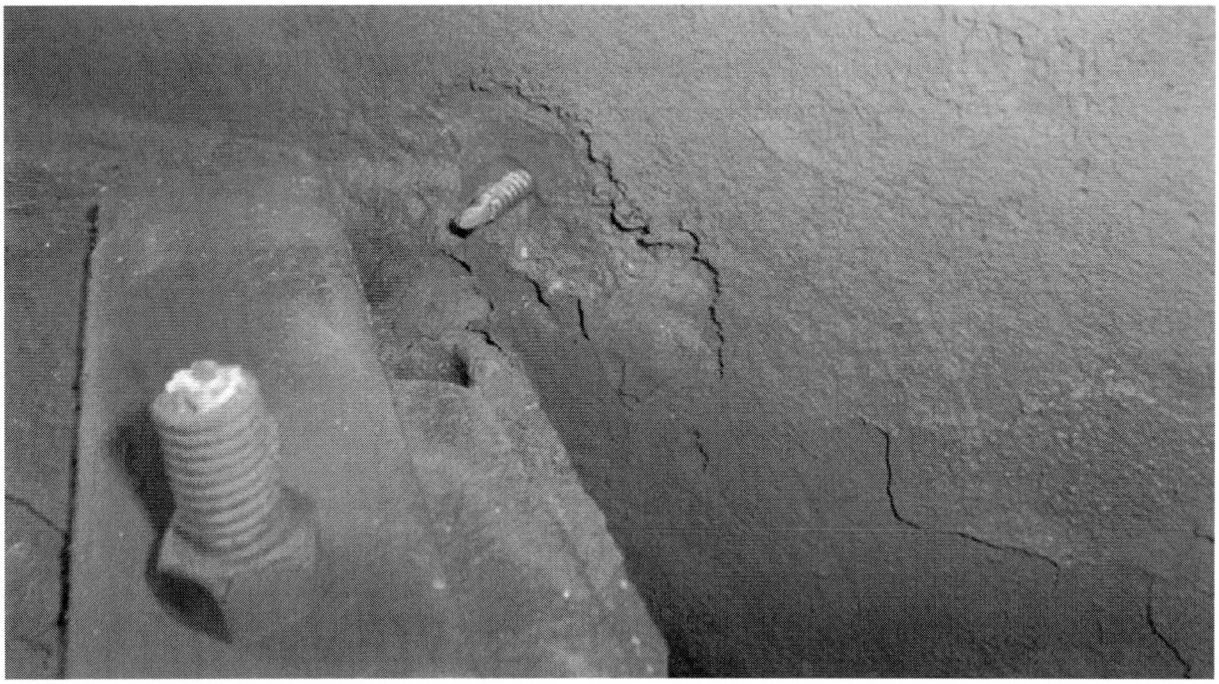

It was disappointing that the facility took the same actions as the last two times, when they had been advised otherwise. I again reminded the plant that the company they were using did not have an R stamp and the screw action was not a proper repair.

This time when I notified my management, they may have over reacted, since they now changed their direction drastically in stating that the right thing to do was shut down the facility, again without trying to get input from me or the plant manager. I was concerned about this drastic change in direction. When I spoke with management about this, they said this was a great success, since the facility was shut down in response to the latest leak. I informed management that I did not view it as a success to take this action in response to the third instance, but would have viewed it more positively if more action was taken on the first issue. A consequence analysis could have been conducted to determine the extent of the consequences. Then the likelihood could have been determined. Combining the consequence and the likelihood results in the risk, which could be used to help make a quantitative risk based determination, rather than a subjective decision.

I then asked management if we could share the learnings from this one site with the other 30+ sites in the company, and they said no, they would not do so until the incident happened at 3 or more facilities. It greatly disturbed me that management would wait for more issues to happen rather than sharing knowledge to prevent future incidents. This shows another cultural weakness.

Since the facility was now shutdown, it presented an opportunity to do some internal inspection, which was done. This effort resulted in many pictures, including the ones in this paper, as well as many wall thickness readings. The wall thickness readings were as thin as 0.110".

The original vessel wall thickness was supposed to be 0.375", with 0.125" corrosion allowance, which shows that the end of life wall thickness should be 0.250", well above the 0.110" measured. The vessel had a 50 psig design pressure, was 78" ID, with an 18 Ft skirt, 64 feet shell height and was made from SA-516-70.

I put together a spreadsheet, based on ASME Boiler and Pressure Vessel Code, Section VIII, Division I, 1980, reference 9, to evaluate what wall thickness was needed to handle different scenarios. Joint efficiencies were taken from reference 9, UW-12. Allowable stresses and overstress for combined loads were from reference 9, UG-23.

Seismic and wind speed were considered by calculating the moment at the bottom head to shell weld, and then converting to metal thickness.

I got the original calculations for the tower, and found that the original seismic calculations were in error. Seismic tension resulted in a required wall thickness of 0.048", which is less than the 0.110" measured wall thickness, so the wall thickness was acceptable. Seismic compression resulted in a required wall thickness of 0.113", which is just above the 0.110" measured, meaning the wall is not thick enough. A useful reference for seismic design is
Guidelines for Seismic Evaluation and Design of Petrochemical Facilities, Second Edition, ASCE, Reston, Virginia, 2011, reference 10.

For the wind loading, the tower was divided into 4 different zones. Wind tension resulted in a required wall thickness of 0.072", which is less than the 0.110" measured wall thickness, so the wall thickness is acceptable. Seismic compression resulted in a required wall thickness of 0.169", which is above the 0.110 measured, meaning the wall is not thick enough. A useful reference for wind loading is ASCE 7-05, Chapter 6, Figure 1A Basic Wind Speed – Western Gulf of Mexico Hurricane Coastline, reference 11. A second useful reference is Wind Loads for Petrochemical Facilities and Other Industrial Facilities, ASCE, Reston, Virginia, 2011, reference 12.

Reference 9, UG-27 was used to determine the minimum thickness for internal pressure and liquid height. The thickness needed for circumferential stress is 0.176", which is above the 0.110 measured, meaning the wall is not thick enough. The thickness needed for longitudinal stress is 0.052", which is less than the 0.110" measured wall thickness, so the wall thickness is acceptable. The thickness needed for compressive stress is 0.085", which is less than the 0.110" measured wall thickness, so the wall thickness is acceptable.

For the combined load temperature, pressure, level, wind and seismic, the thickness needed for longitudinal stress is 0.144", which is above the 0.110 measured, meaning the wall is not thick enough. The thickness needed for compressive stress is 0.306", which is above the 0.110 measured, meaning the wall is not thick enough.

Management tried to direct me to just solve the issue by reducing design pressure until the wall was thick enough. However, this was not a reasonable path forward, for several reasons, such as that if the same corrosion allowance was used, the remaining life (0.110" measurement minus 0.125" corrosion allowance) was 0. Also, there was no mention of the other aspects that would need to be considered, such as increased relief valve sizing at lower pressure. Also, the wall was not thick enough for several other cases listed above, such as seismic and wind, which are independent of the design pressure. Reducing the design pressure from 50 to 15 psig, meant that the wall thickness was still not adequate for the combined loading.

I also did some process simulation to determine if there could be changes made to the process and/or equipment to reduce the corrosion, such as:

- Changing type of amine
- Changing flowrate of amine
- Changing tower diameter
- Changing tower materials of construction
- Changing heat input to the reboiler

I was told to stop on this effort, which meant no better design would be in place when the next tower has problems. Production again won out over process safety when I was told to get a new tower the same as the last one.

A fitness for service assessment could have been conducted, however, production again won out over process safety when I was told to get a new tower the same as the last one.

A risk based mechanical integrity program could have been developed and put into place, but it wasn't. The emphasis was on production over process safety.

A root cause analysis could have been conducted, but the emphasis was on production and appearance over process safety. This is demonstrated by the operations direction to rapidly take the tower offsite to keep the plant appearance attractive. There was not an effort to engage in dialogue to discuss the issue. A better approach is to engage in dialogue with employees and seek their input and comments on suggestions. When management eventually realized the value of the investigation process, the tower was already destroyed. This was unfortunate, because the same type of failures starting appearing at other sites.

I told management that they needed to institute a mechanical integrity group. They eventually did, however, when I was informed of the head of the mechanical integrity group and the recent hire for our area, I could see that they were not qualified. I informed management of this, and I was told that was not my concern. This again shows a poor process safety culture. No attempt was made to resolve my valid concern, just to dismiss it.

I finally put many items in writing to my management, and was provided an individual visit by HR from headquarters. In this closed-door meeting, I was told many unacceptable things, including that I should not write about process safety, but management is glad to discuss orally. This shows a poor process safety culture. I resorted to writing when it was clear that oral was not effective. I did not agree with HR. I was then sent to an individual training session with the head of compliance, which was titled "Write Now, Pay Later." It is clear that the company tried to restrict internal written communication on safety and health, when it is an employee right to write written complaints to external agency, OSHA. This does not show a safety culture.

3 Additional RAGAGEP

One good resource is Guidelines for Mechanical Integrity Systems, AIChE CCPS, 2006, Wiley, reference 3. For pressure vessels, a useful item is "Table 9-1 RAGAGEPs for Pressure Vessels".

Another useful item in reference 3 is "Table 9-13 Mechanical Integrity Activities for Pressure Vessels". It has example activities and typical frequencies for:

- New Equipment Design, Fabrication, and Installation
- Inspection and Testing
- Preventive Maintenance
- Repair

Reference 3 also has Technical Basis for Activity and Frequency, Sources of Acceptance Criteria, Typical Failures of Interest, Personnel Qualifications, Procedure Requirements, Documentation Requirements for the following categories:

- New Equipment Design, Fabrication, and Installation
- Inspection and Testing
- Preventive Maintenance
- Repair

Reference 3, Chapter 11 contains an introduction to a risk based approach for mechanical integrity.

Reference 3, Section "12.4 Equipment Failure and Root Cause Analyses" provides useful information such as:

- "Step 1 - Evaluate Conditions at the Site."
- "Step 2 - Perform Preliminary Component Assessment."
- "Step 3 - Preserve "Fragile" Data Sources."
- "Step 4 - Perform Macroscopic Examination."
- "Step 5 - Perform Microscopic Examination (if warranted)."
- "Step 6 - Perform More Detailed Testing (if warranted)."
- "Step 7 - Test Under Simulated Conditions (if warranted)."
- "Step 8 - Make the Failure Mechanism Determination and Develop Corrective Actions."

Listed below are several additional API documents that contain excellent information related to mechanical integrity, with specific applicable sections listed.

API Recommended Practice 571, Damage Mechanisms Affecting Fixed Equipment in the Refining Industry, First Edition, December 2003, is reference 4. See sections:

- "4.2.14 Erosion / Erosion-Corrosion"
- "4.3.6 CO2 Corrosion"
- "5.1.1.1 Amine Corrosion"
- "5.1.2.2 Amine Stress Corrosion Cracking"

API Recommended Practice 572, Inspection of Pressure Vessels (Towers, Drums, Reactors, Heat Exchangers, and Condensers), Second Edition, February 2001, is reference 5. See sections:

- 7 for reasons for inspection
- 8 for causes of deterioration
- 9 for frequency and time of inspection
- 10 for inspection methods and limitations
- 11 for methods of repair

API Recommended Practice 579, Fitness for Service, First Edition, March 2000 is reference 6. Reference 6 states "The ASME and API design codes and standards for pressurized equipment provide rules for the design, fabrication, inspection and testing of new pressure vessels, piping systems, and storage tanks. These codes do not address the fact that equipment degrades while in-service and that deficiencies due to degradation or from original fabrication may be found during subsequent inspections. Fitness-For-Service (FFS) assessments are quantitative engineering evaluations which are performed to demonstrate the structural integrity of an in-service component containing a flaw or damage. This Recommended Practice provides guidance for conducting FFS assessments using methodologies specifically prepared for equipment in the refining and petrochemical industry. The guidelines provided in this recommended practice can be used to make run-repair-replace decisions to help ensure that pressurized equipment containing flaws which have been identified by inspection can continue to operate safely."

API Recommended Practice 580, Risk-based Inspection, First Edition, May 2002 is reference 7. Reference 7 states "1.1 PURPOSE -The purpose of this document is to provide users with the basic elements for developing and implementing a risk-based inspection (RBI) program. The methodology is presented in a step-by-step manner to the maximum extent practicable."

API Recommended Practice 581, Risk Based Inspection Technology, Second Edition, September 2008 is reference 8. In reference 8, "PART 1- INSPECTION PLANNING USING API RBI TECHNOLOGY", see sections:

- 4 for RBI Concepts
- 5 for Pressure Vessels

In reference 8, "PART 2 - DETERMINATION OF PROBABILITY OF FAILURE IN AN API RBI ASSESSMENT", see section 8 for "Damage Factor – Amine Cracking".

Reference 8, "PART 3 - CONSEQUENCE ANALYSIS IN AN API RBI ASSESSMENT" covers hole size, release rate calculation, release inventory, impact of detection systems, flammable and explosive consequences, toxic consequences, non-flammable non-toxic consequences, component damage and personnel injury consequence areas, financial consequences.

4 Conclusions

This paper presented in detail one of the many process safety issues encountered by the author in his 20 years of industrial experience. This issue deals with an amine tower, but touches upon the following topics: carbon dioxide; communication and restriction; competency; compliance; culture; documentation; failure analysis; intimidation; investigation; lessons learned; material of construction; mechanical integrity; pressure; stress, risk assessment and vessel repair.

It describes better ways to address the situation and applicable RAGAGEP. It is hoped that sharing this information will enable others to learn from the past and prevent similar issues in the future.

5 References

[1] NBBI, NB 23 (NBIC), National Board Inspection Code, Parts 1-3.

[2] API 510, Pressure Vessel Inspection Code: In-service Inspection, Rating, Repair, and Alteration, Tenth Edition, May 2014.

[3] Guidelines for Mechanical Integrity Systems, AIChE CCPS, 2006, Wiley.

[4] API Recommended Practice 571, Damage Mechanisms Affecting Fixed Equipment in the Refining Industry, First Edition, December 2003.

[5] API Recommended Practice 572, Inspection of Pressure Vessels (Towers, Drums, Reactors, Heat Exchangers, and Condensers), Second Edition, February 2001.

[6] API Recommended Practice 579, Fitness for Service, First Edition, March 2000.

[7] API Recommended Practice 580, Risk-based Inspection, First Edition, May 2002.

[8] API Recommended Practice 581, Risk Based Inspection Technology, Second Edition, September 2008.

[9] ASME Boiler and Pressure Vessel Code, Section VIII, Division I, 1980.

[10] Guidelines for Seismic Evaluation and Design of Petrochemical Facilities, Second Edition, ASCE, Reston, Virginia, 2011.

[11] ASCE 7-05, Chapter 6, Figure 1A Basic Wind Speed – Western Gulf of Mexico Hurricane Coastline.

[12] Wind Loads for Petrochemical Facilities and Other Industrial Facilities, ASCE, Reston, Virginia, 2011.

Identifying Opportunities of Enhancing Safeguard Stewardship through IPL Rationalization

SreeRaj R Nair
Technical Safety, Chevron U.S.A Inc.
Houston, USA
snair@chevron.com

Bernadette Thornton
Safety Systems, Chevron
ETC, Houston, USA
bernadette.thornton@chevron.com

[Rev2]

Prepared for Presentation at
American Institute of Chemical Engineers
2017 Spring Meeting and 13th Global Congress on Process Safety
San Antonio, TX
March 26 – 29, 2017

Identifying Opportunities of Enhancing Safeguard Stewardship through IPL Rationalization

SreeRaj R Nair
Technical Safety, Chevron U.S.A Inc.
Houston, USA
snair@chevron.com

Bernadette Thornton
Safety Systems, Chevron
bernadette.thornton@chevron.com

Keywords: Process Hazard Analysis, Layer of Protection Analysis, Independent Protection Layers, Rationalization

Abstract

Ensuring effective safeguard stewardship is an integral part of the journey to address the challenge of moving from 'I think" to "I know" that a major accident hazard is managed. One of the steps in achieving effectiveness is to ensure that the process design is robust and that the safeguards provide the necessary protection to reduce the process risk.

Process Hazard Analysis and Layer of Protection Analysis are carried out to identify hazards, determine the potential consequence and evaluate the adequacy of the safeguards. Opportunities for further risk reduction are identified and risks are eliminated or reduced through established risk management programs. Though the study methodologies have been used for decades, there are opportunities for improvement.

This paper specifically addresses the opportunities in the evaluation of safeguards during the design of major capital projects. It explores some of the key concerns including lack of independence and variations in modifier application that could lead to over or under engineering; this could in turn may result in increased risk exposure. Additionally, this paper explains how the use of a defined methodology for the rationalization of Independent Protection Layers can assist in overcoming these challenges. The significance of performing IPL rationalization at appropriate stage of engineering progress is key in ensuring effective safeguards through design. A comparison of the advantages of rationalization exercise performed at various project phases is also discussed in the paper.

1 Introduction

A critical starting point in process safety management lies with the identification of facilities with potential for high impact (Safety and Environment) events and an understanding of the high consequence scenarios. For the high impact facilities, a detailed Process Hazard Analysis is conducted to determine the scenarios of concern. The potential impact from realizing those scenarios and the likelihood considering the safeguards (prevention and mitigation) is evaluated. A process safety management system is established with a system of "controls" (including critical safeguards verifications, and validation) to prevent the occurrence of these incidents or reduce their severity.

Layers of Protection Analysis (LOPA) was introduced as a detailed frequency evaluation method compared to Hazard and Operability Studies and as a simpler version compared to Quantitative Risk Assessment. This method is used to assess the adequacy of independent protection layers (IPL) and to ensure that required risk reduction is achievable (i.e., there will be no gaps post implementation of the risk reduction plan). LOPA is a systematic approach for applying IPL which follows guided principles and rulesets as prescribed in the industrial standards (AIChE 2001). CCPS publication (AIChE, 2007) provides guidance on choosing and evaluating initiating event frequency and independent layers of protection.

An independent protection layer (IPL) is a device, system, or action that is capable of preventing a scenario from proceeding to its undesired consequence independent of the initiating event or the action of any other layer of protection associated with the scenario (AIChE, 2015).

An effective IPL should be independent of the initiating causes, survive the initial event, and available to perform at the expected level for the intended duration. In order to determine the effectiveness, each safeguard which has been identified as a potential IPL needs to have the entire core attributes; the attributes are independence, specificity, integrity, reliability, auditability, access security, management of change. IPL can be either 'Active' or 'Passive' with Active IPL being those with the ability to detect (sense a condition or problem), decide (whether an action or intervention is necessary or not) and deflect (action capable of preventing consequence in a timely manner) and Passive IPL being those which primarily deflect to reduce the impact.

The following sections review some of the challenges in determining the effectiveness of active IPL and how to overcome the challenges through a structured approach known as IPL rationalization. It is followed by discussion on the inherently safer approach and its significance in improving the performance of the LOPA results. The potential economic benefits of IPL rationalization (Capital and Operational) are also evaluated with relevant project case studies.

2 LOPA Process Concerns

After evaluation of multiple projects, we have determined that the major LOPA concerns tend to fall into three categories: lack of independence, variations in modifier application, and appropriate documentation. The challenges in IPL related documentation can be addressed effectively by implementation of Process Safety Information management plan.

2.1 Lack of Independence

The first issue, lack of independence, relatively an easy one to eliminate, yet it continues to appear in reviews. Most of this is due to a failure to understand packaged equipment controls and a failure to identify the critical actions required by a safety function. Packaged equipment controls typically are not well defined in early Process Hazard Analysis (PHA) / Hazard and Operability (HAZOP) studies. Assumptions concerning the independence of systems and poorly defined functionality tend to lead to IPL credit being given to these systems in error. For example, vibration monitoring systems tend to send requests to other systems, such as the Safety Instrumented System (SIS), to complete critical actions such as the closure of valves or motor stops. If a Safety Instrumented Function (SIF) exists with the same critical actions, then credit cannot be taken for both. The critical actions of each would have to be defined to remove the conflict and the system common cause would need to be reviewed to ensure that the combination will provide the appropriate layers of protection. Failure to validate packaged equipment independence as the project progresses leads to discovery late in the project that the required number of IPL is unavailable.

2.2 Modifier Application

Differences in the team make-up and the facilitation of the LOPA studies leading to differences in the application of enabling events and conditional modifiers and therefore poor application of IPLs. Enabling events and conditional modifiers are applied to change the likelihood of a consequence or severity of a consequence, so these take into account factors such as the likelihood of personnel being in the area and the phase of operation. Failing to apply enabling events and conditional modifiers properly can lead to major operational issues, including alarm floods and excessive instrumentation. For example, if one facilitator applies a conditional modifier to a scenario and another facilitator does not, the required risk reduction is greater for the second facilitator and will require the application of additional IPL. This inconsistency in the application of enabling events and conditional modifiers leads to under-design (with gaps) and over-design (most-often) of instrumented systems and, potentially requiring additional maintenance and testing requirements.

Effective IPL rationalization can reduce the concerns of lack of independence and modifier application by providing an opportunity to look at the application of IPL on a facility basis rather than a scenario by scenario basis.

3 IPL Implementation Challenges

Poor application of IPL in the processes could result in facilities operated at elevated risk levels. The key challenges in the implementation of LOPA findings include the following:

- Risk reduction during design primarily relies on engineered systems, leaning to the route of 'Add-on' safety;
- Lack of credits for non-engineered / soft barriers as additional layers of protection;
- Operational inefficiency due to overly engineered protection system (multiple interlocks, trips for dedicated systems designed without considering overall facility reliability requirements);

- Regulatory requirements – prescriptive versus performance based approach;

Kletz and Amyotte [Ref. 3] have said, "Intensification, substitution, attenuation, and limitation of effects produce inherently safer design because they avoid hazards instead of controlling them by adding protective equipment. The term 'inherently safer' implies that the process is safer because of its very nature and not because equipment has been added to make it safer". However, this very point is often overlooked during design and often adding additional layers (mainly instrumented and hardware) are considered as improvement to safety. Industries' traditional solution to increased levels of hazard has been through more controls. For example, an offshore facility designed to be operated as 'normally unmanned' could become 'normally manned' just because of the manning distribution required to ensure the effectiveness (through inspection and maintenance) of the 'added on' engineered safeguards.

Projects have become more complex and personnel involved in projects rarely see the effect of their designs, so design complexity is less of a consideration and the more cost effective solution, typically an instrumented one, is implemented to ensure enough safeguards are present to manage the risk. Unfortunately, this leads to the massive problem of facilities having to manage large numbers of safeguards and IPL.

While industry is moving towards performance based approaches to define risk reduction, prescriptive regulatory environments still exist. Prescriptive environments ensure a minimal level of risk reduction is implemented, but they also can add challenges to the implementation of a performance based approach. Prescribed testing and testing procedures may provide some reduction in the overall risk of a facility, the tests rarely are sufficient to provide assurance of the effectiveness of the prescribed functions as IPL; therefore, additional testing must be performed on top of the regulatory required testing potentially requiring more personnel to complete ensure the functions meet their risk reduction target.

4 How to Overcome the Challenge

Eliminating hazards using the inherently safer design route is the most effective way to overcome the challenge of safeguard stewardship simply because there will be only fewer safeguards as IPL (Kletz and Amyotte, 2010). Time and effort can be spent on managing the asset integrity of pipe, vessels, and major equipment rather than managing hundreds of safeguards. In such cases, the projects compare options for managing risk. If elimination is not a viable option considering safety and operability factors, decisions are made to proceed with acceptable levels of risk using adequate safeguards/IPL. The following words of wisdom shall be considered as part of ISD approach (Nair 2009, Nair 2017, Stuart, et. al 2003, Kletz and Anyotte, 2010):

- More is not always safer
- Identify all hazards (you can't eliminate or manage a hazard which have not been identified)
- Design against accidental loads
- Design plants for easy and safe maintenance (*reliable equipment which requires less maintenance*)
- Simplify the process and minimizing the requirement to rely on human intervention

While companies acknowledge that management of IPL is necessary, the number of IPL being identified prioritizes finding ways to reduce the load on maintenance and testing personnel. One way to reduce the load is by using IPL rationalization. IPL rationalization can be defined as a process of defining the most effective set of IPL to provide the appropriate risk reduction while also reducing the number of management activities and operator tasks and the maintenance and testing load a facility must endure. So the question of how to define the most effective set of IPL must be answered.

IPL rationalization uses the results of a LOPA to identify IPL which are unnecessary to provide the facility with the risk reduction required. The process requires a functional safety Subject Matter Expert (SME), a process engineer and an operations representative to review the application of enabling events and conditional modifiers, identify SIF with multiple Safety Integrity Level (SIL) and ensure that SIF, critical alarms and basic process control system (BPCS) loops identified as IPL are, in fact, effective. During a LOPA, many of these items may be addressed, but as projects require more and more facilitators to make large HAZOP more manageable, more and more issues creep into the design. The smaller group allows for a more in-depth review of independence, consistency in identification of the SIF and its actions, and a way to compare the results of one study with another study to ensure IPL chosen between similar units and similar scenarios provide the appropriate risk reduction while applying the same enabling events and conditional modifiers. Most importantly, IPL rationalization provides a way for the project to review the IPL from a bird's eye view as well as scenario by scenario to identify whether IPL are appropriate to mitigate the risk for the overall facility as well as the individual scenarios. The most surprising element of the process is that it requires very little time to complete, especially with the correct pre-work completed.

The process of IPL rationalization is simple. First, the team must first identify the order of preference for implementation of IPL. This preference will take into account the facility's maintenance staff size, the competency of the staff and the operating basis of the facility. For example, some owners may prefer to take more SIFs as they have a good management program for them and no management program for other engineering IPL. Other owner may prefer other instrumented IPL over SIF as they do not expect to have the competency they would prefer to manage the IPL.

Secondly, the team reviews the application of enabling events and conditional modifiers to guarantee their effectiveness or applicability. Next, the IPL are reviewed in order of preference and where different IPL have been provided for scenarios with the same consequence, the use of different IPL is either rationalized or the IPL are revised to provide the most effective set. Unused IPL are either removed or removed from the IPL list.

Finally, the team reviews any recommendations to identify if there are LOPA gaps that may be addressed using existing IPL or whether new IPL must be created.

Experience has demonstrated that the process can be effectively used for brownfield facilities as well. However, it need to be noted that there could be differences in the IPL preferences and that unused IPL are less commonly found as facilities choose to use existing IPL rather than add new ones, even if the final result may be less efficient.

5 A Representation of the Benefits

The benefits of completing IPL rationalization depends on three things:

1. The objective of the project
2. The timing of the study
3. The regulatory environment of the facility

The objective of the project may affect IPL rationalization results as looking for ways to eliminate or reduce the numbers of a particular IPL type or may require the addition of IPL in other categories. This activity may not be a reduction of IPL, but rather a reallocation of IPL. Alternatively, the project may be looking to reduce the number of IPL overall by ensuring the application of IPL is streamlined. Illustrations of IPL rationalization benefits in terms of IPL and potential economic saving are given in Figure 1 and Figure 2. A recent project looking to reduce the number of alarms was able to achieve a 90% reduction in alarms with little effect to the number in other IPL categories due to the proper application of conditional modifiers. If one assumes that maintenance of these alarms requires approximately \$100 per year and that the number of alarms was reduced from 1000 to 100, then it can be found that the facility saves approximately \$90000 per year in alarm maintenance alone (*representative figures only*).

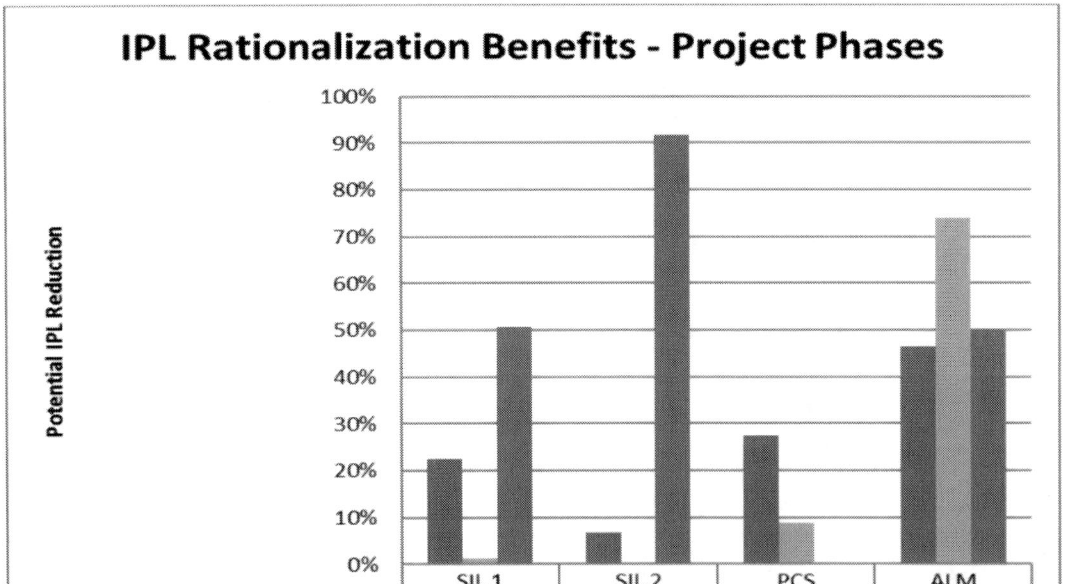

	SIL 1	SIL 2	PCS	ALM
Project 1 - Basic engineering	23%	7%	27%	46%
Project 2 - Detailed design	1%	0%	9%	74%
Project 3 - Post commissioning	51%	92%	0%	50%

Figure 1: IPL Rationalization Benefits – At Different Phases of Project

The timing of the study has a significant effect on the outcome of the IPL rationalization. Performing rationalization before the design is complete (i.e., before construction) provides opportunity to eliminate IPL, such as SIF, that are found to be unnecessary (Figure 1). A project may choose to keep the functions or alarms for operational or reasonable risk reduction justification/demonstration; but take the opportunity to reduce the testing or maintenance associated with these. In addition, the team may be able to identify designs which could be modified to provide a more inherently safer facility. A study performed late in the project will not achieve the same benefits. They may be able to make reductions in some areas, but if the design or installation is almost complete, there is only limited opportunity to remove unnecessary IPL (Figure 1). Alarm priorities may be reduced and maintenance and testing requirements may be relaxed, but the facility will likely be forced to retain most IPL. Performance of IPL rationalization post-commissioning provides even less opportunity for savings. IPL have not only been designed and installed, but they have already been commissioned and, in many cases, changing maintenance and testing requirements will require changes to maintenance managements systems that likely have already been completed, potentially adding cost short-term. A review of projects showed potential IPL reduction of 25-30% for projects in Basic Design / Front End Engineering Design (FEED) and 10% reduction for those in Detailed Design or later.

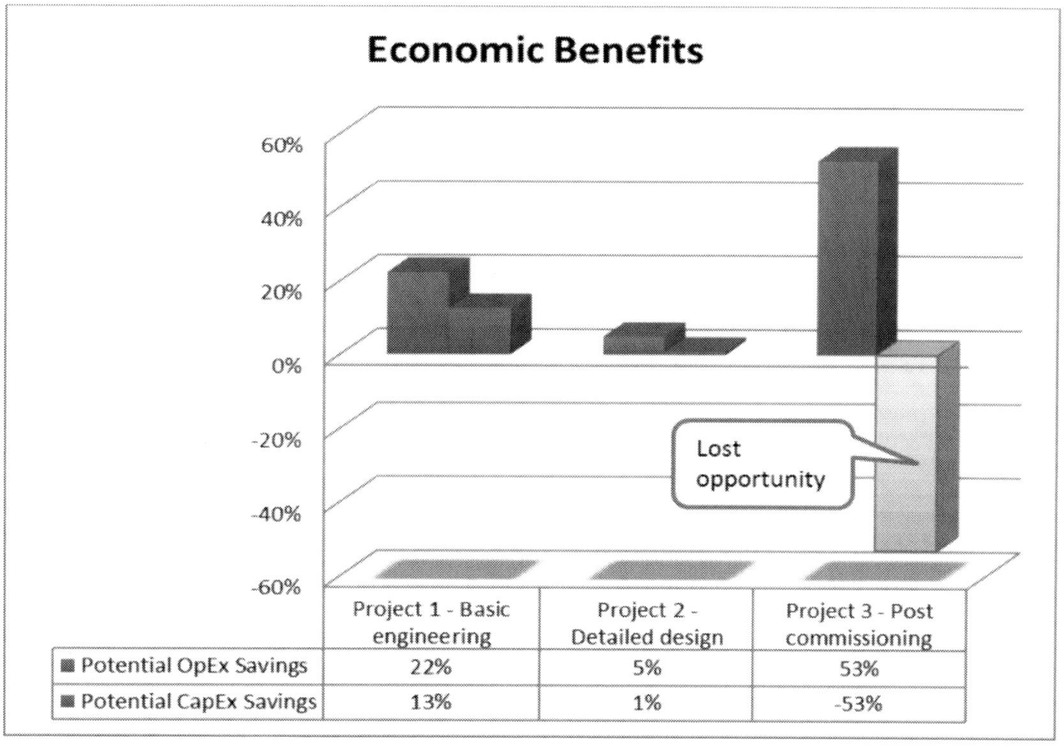

Figure 2: IPL Rationalization and Potential Economic Savings

When taking into account both Capital Expenditure (engineering design, construction, installation) and Operational Expenditure (inspection, testing and maintenance), the savings to a facility can be significant if the appropriate evaluation is carried out in early project stages; even in later stages of the project, the suggested approach can reduce IPL requirement and remove some of the maintenance and testing cost of the so-called "gold plating" that occurs on projects (Figure 2).

Finally, the regulatory environment can have an impact on the study. Where regulators follow a prescriptive regimen for safety instrumented systems (e.g., U.S. Gulf of Mexico), then the benefits of IPL rationalization are realized through the removal of extensive testing and maintenance related to IPL, but not through removal of IPL or through the removal of regulatory required testing. The potential to gain savings still exists though, so the exercise provides benefit by removing the "gold plating" of the facility.

6 Conclusion

Critical safeguards identified, performance requirements established and an evergreen process to ensure safeguards remain in place are key to keep high consequence scenarios from occurring. Through safety studies and by following industrial codes and standards, multiple layers of protection are factored in to prevent or mitigate potential events. However, it is experienced that often the designs and facility become complicated due to 'add-on' safety. Such facilities could experience operational inefficiency, high cost of maintenance and potentially higher risk exposure to personnel.

IPL rationalization should be part of the design engineering process to overcome the challenges and a systematic approach will help in:

- establishing 'effective' safeguard
- optimizing IPLs
- lifecycle benefits

The paper reviewed some of the challenges in determining the effectiveness of IPL and how to overcome the challenges through a IPL rationalization exercise. The paper discussed the inherently safer approach and its significance in improving the performance of the LOPA results. The timing of the exercise has a significant effect on the outcome of the IPL rationalization. The potential economic benefits (Capital and Operational) are also evaluated with relevant project case studies.

7 References

1. CCPS. "Guidelines for Independent Protection Layers and Initiating Events", American Institute of Chemical Engineers, New York, NY, 2007
2. CCPS, "Guidelines for Initiating Events and Independent Protection Layers in Layer of Protection Analysis", American Institute of Chemical Engineers, New York, NY, 2015
3. CCPS, "Layer of Protection Analysis – Simplified Process Risk Assessment", American Institute of Chemical Engineers, New York, NY, 2001
4. Nair SreeRaj, "Determining the criteria for evaluation of toxic hazards, Journal of HSE & Fire Engineering", ASFE, Issue 2 March 2009
5. Nair SreeRaj, "'Inherently Safer' – the Concept and its Application in Oil & Gas Sector," International Conference on Safety and Fire Engineering, SAFE-'17, 2017
6. Stuart Chia, Kieran Walshe and Ed Corpuz, "Application of inherent safety challenge to an offshore platform design for a new gas field development – approaches and experiences," Symposium Series No. 149, IChemE, 2003
7. Trevor Kletz and Paul Amyotte, "Process Plants: A Handbook for Inherently Safer Design," Second Edition, 2010

Process Safety Site Assessment Program (PSSAP) – 5 years of Assessments

Andrew P. Broadbent
American Petroleum Institute (API)
1800 West Loop South, Suite 475
Houston, TX 77027
broadbenta@api.org

Prepared for Presentation at
American Institute of Chemical Engineers
2017 Spring Meeting and 13th Global Congress on Process Safety
San Antonio, TX
March 26 – 29, 2017

Process Safety Site Assessment Program (PSSAP) – 5 years of Assessments

Andrew P. Broadbent
American Petroleum Institute (API)
1800 West Loop South, Suite 475
Houston, TX 77027
broadbenta@api.org

Keywords: Process safety, assessments, benchmarking, protocols, scoring, rating, weighting

Abstract

The API Process Safety Site Assessment Program was developed in 2011 as part of the AFPM and API Advancing Process Safety Programs. PSSAP began conducting assessments in 2012. By the end of 2016, the API Site Assessment program will have conducted 52 General (7 protocols) and 27 HF Alkylation/RP 751 assessments. These assessments have been conducted 59 different refineries and petrochemical facilities. The areas assessed are: Process Safety Leadership, Management of Change, Mechanical Integrity, Safe Work Practices, Operating Practices, Facility Siting, Process Hazards Analysis, and HF Alkylation/RP 751.

This presentation will provide a brief overview of the program and present some the learnings, trends and benchmarking developed from the assessments based on the five-year history of the program.

In 2010, the American Petroleum Institute (API), American Fuel & Petrochemicals Manufacturers (AFPM) and industry joined together to address improving process safety performance with the goal of preventing process safety events at refineries and petrochemical plants. API, AFPM and the industry wanted this effort, known as Advancing Process Safety, to be:

- Focused on higher risk process safety activities;
- Refining & petrochemical focused;
- Voluntary; and
- An avenue to communicate and share experiences and knowledge.

Advancing Process Safety (APS) consists of six different programs which are:

- Process Safety Metrics and Analysis (RP 754)
- Process Safety Hazard Identification
- Event Sharing
- Focused Improvement
- Regional Meetings
- Process Safety Site Assessments Program (PSSAP).

The intent of all the programs is to raise the level of process safety performance at the sites to prevent process safety events.

The management of these programs is split between API and AFPM. API manages the Process Safety Site Assessment Program, Process Safety Metrics and Analysis, and the Focused Improvements effort. AFPM manages Process Safety Hazard Identification, Event Sharing and Regional Networks.

Benefits/Opportunities

The Process Safety Site Assessment Program is perhaps the most hands-on of the six programs since it involves a team of assessors being onsite at a refinery or petrochemical plant for an entire week. The program uses credible, third-party assessors who have specific expertise in one or more of the eight areas being assessed. The team utilizes a set of protocols that are applicable to both refineries and chemical plants. These protocols were developed through a collaborative effort of subject matter experts from a wide range of companies.

Through these site assessments, the program provides the sites with feedback regarding what they are doing well and what improvements they can make to their process safety programs. Additionally, the program provides the sites with benchmarking data so the sites can compare how they are doing versus their industry peers.

The assessments are not process safety management compliance audits, and so we refer to them as assessments and not audits. We speak to process safety rather than process safety management so people will understand that these assessments are not the three-year process safety management (i.e., 29 CFR 1910.119) compliance audits, rather they are assessments of how sites are performing relative to 'what good looks like" based on the experience of industry and the expert assessors. As you will see, the assessments only cover seven areas of process safety (compared with the 14 areas in OSHA 1910.119) as well as HF alkylation/API RP 751.

Protocols

The eight protocols address:

- Process Safety Leadership
- Operating Practices
    - Procedures
    - Shift Handover
    - Operator Rounds
    - Alarm Management, and
    - Workplace Conditions
- Mechanical Integrity – Fixed Equipment
- Safe Work Practices
    - Procedures
    - Work Permit Procedures
    - Utilities, etc.
- Management Of Change
- Process Hazards Analysis
- Facility Siting
    - Permanent Buildings
    - Portable Buildings, and
    - Tents
- HF Alkylation/API Recommended Practice (RP) 751

The first seven elements are areas that could prevent a process safety event from occurring or reduce the severity of an event (i.e., Facility Siting). The eighth area – HF alkylation – deals with process safety in this specialized operation. Assessments involving the first 7 protocols are referred to as "General" Assessments and the HF Alkylation/RP 751 ones are known as "HF In-Depth RP 751" Assessments.

Note: Since the HF Alkylation protocol specifically follows API RP 751, the HF-In-Depth RP 751 Assessment does meet the RP 751 requirement to conduct an audit once every three years.

Sites may also choose to do just some of the protocols (for example, MOC and Facility Siting) rather than all seven at a time. These are known as "Partial" assessments.

These protocols were developed by industry for industry utilizing 54 subject matter experts from a wide range of companies. The protocols were developed, balloted using the API standards development consensus process, and then piloted at three sites. The protocols were then reviewed and modified based on learnings from the pilot assessments, including input from the assessors.

Keys to the Program

There are three very important components the PSSAP program, which are:

- Benchmarking
- Mentorship
- Expertise of the Assessors

Benchmarking

Refineries and petrochemical plants always like to see how they compare to their peers in the industry. In the PSSAP program this is done by providing a score for each protocol and each protocol question.

The 7 General Assessment protocols have a total of 577 questions and the HF In-Depth protocol has 394 questions.

When developed, each protocol question was given a weighting indicating the question's importance to preventing a process safety event from occurring. The question's weighting is a fixed number between 1 and 4 in half increments (i.e., 0, 0.5, 1, 1.5, etc.) with 4 indicating the most importance. During the assessment the assessor will rate each question based on the information and evidence the site provides. This is done on a scale of 0 to 4, also in half increments. The score is a simple calculation of the weighting times the rating.

Keeping the scoring objective is done through ensuring the protocols are well written so that different assessors will understand and interpret the question the same way. Consistency is also addressed through the training provided and calibration sessions with the assessors at the start of the assessments. Additionally, the PSSAP Program Manager provides oversight so that a consistent approach is used by the assessors.

Mentorship

One of the most important pieces of the program is the information exchange the sites get with the assessors. As mentioned, the assessors have considerable experience and expertise based on their years of service in the industry and through performing PSSAP assessments.

We inform the sites (we typically visit the site six months prior to the assessment to set up all the logistics), to get the most value out of the assessment, that their employees should be interviewing the assessors as much as the assessor is interviewing the employee.

As an example, on one assessment the assessor sat down with three site employees for an hour and discussed various ways to manage car seals.

We encourage sites to take advantage of the time with the assessors throughout the week, including the daily debriefs, lunch time discussions, and the closing conference of Friday.

Assessors

As mentioned, the assessors are major piece of the PSSAP program. They are handpicked based on their expertise and industry background/knowledge. Typically they are retired from industry companies, though some work for consulting firms (or have their own firm) but worked in the industry at some point in time.

An assessor candidate with appropriate qualifications and references is selected to attend a 1-1/2 day training course which explains the API Process Safety Site Assessment Program to them. The candidate must undergo an interview with a three-person panel – the PSSAP Manager and two industry representatives before being selected to be an assessor. Not all candidates that attend the training get selected. Also, assessors are picked for specific protocols (i.e., we do not look for Mechanical Integrity assessors to be Operating Practices experts and vice versa). Currently, over 70 people have inquired about being an assessor and we have around 20 in our assessor pool.

To date, a typical assessment team is averaging 39 years of industry experience. Typically the team has 7 assessors.

An in-depth HF alkylation unit assessment follows the API RP 751 Recommended Practice in detail. This assessment can be done at the same time as a "basic" assessment or completely separate and consists of 5 assessors. As noted, an In-Depth RP 751 assessment can be used by the site as their 3-year RP 751 audit.

Assessments

An assessment gets started when the site contacts the PSSAP Program Manager to schedule a week for the assessment. It is important that the site assessment date, be selected well in advance – at least 6 months – so appropriate time is available for planning, including scheduling the assessors.

A pre-assessment visit is then scheduled and conducted at the site by the PSSAP Program Manager. This meeting serves two purposes. First, it allows the site management team to learn more about the program and ask any questions. The second purpose is to meet with the site contact (the person who will work with the PSSAP Program Manager to ensure all arrangements are made prior to the start of the assessment) to go over the logistical needs for the assessment - including conference rooms, scheduling interviews, documents needed prior to the assessment, etc.

Conducting the Assessment

On the week of the assessment the team arrives on Sunday afternoon/evening ready to go. The assessors will have already spent about 8 hours reviewing site documents, provided by the site through an FTP file share site, so they will be well versed on the site's programs relative to the protocols they are assessing.

Monday begins with an opening conference to review the ground rules and usually a quick tour of the site. After the tour, each assessor interviews key people to learn more about the site's management systems, programs and procedures. These interviews have been pre-scheduled by the site using a template provided to them by the PSSAP Program Manager.

The assessors also work to spend as much time in the field as possible. The intent is to not only review whether the site has good management systems, but to provide them with insight on how well their systems are being implemented in the field.

On Tuesday, Wednesday, and Thursday, the assessment team meets with site staff, spends time in the field, and holds a debrief with the site personnel, typically around 4:00 pm. These daily debriefs have been extremely beneficial as they are excellent opportunities for site personnel to continue to interact with the assessors. Often when the debrief has been completed, there are follow-up conversations with the assessors so site staff can gain more insights.

Thursday night and Friday morning is when the assessors compile all the information and the protocol ratings. The assessors enter observations and ratings for each line item in each protocol, thus the facility is provided with a considerable amount of information.

On Friday, an hour to an hour and a half closing conference is held with the site personnel, including the site management team. The site receives a PowerPoint overview, the completed protocol spreadsheets, and some preliminary benchmarking data. The draft report is provided to the site the following week. Distribution of the reports is as designated by the site.

When the assessment team leaves the site on Friday, the assessors return all hard copy documents and delete any electronic documents that were provided to them by the site.

<u>Handling Report & Data</u>

The site will typically be given two to three weeks to comment on the draft final report. This review period is not a discussion of the scores, but whether the wording of the Observations and Good Practices is clear. A final report is then issued by the program manager to the site.

Once the final report is issued, the data and information including the good practices and observations, with the permission of the site, are blinded and entered into the database for trending and benchmarking purposes. API does not ask for corrective action plans or updates.

<u>Program Status, Experience & Keys</u>

As of year-end 2016, we have conducted 53 General site assessments, 27 HF alkylation/API RP 751 in-depth assessments and 5 Partial assessments. The program has been to 28 different companies at 60 different locations (note: several companies have had a site assessment at more than one site.)

Item	Total	Average per Assessment
Protocol Questions	> 41,200	577 Basic 394 HF
Interviews	> 8,000	>100
Observations	> 2,800	~35
Good Practices	> 2,400	~30

Below are a couple of comments received from sites that have been assessed:

- "You cannot fool these Assessors — they have

done *your* job for 30 years"

- *"In one week the assessment team found issues that our internal teams could not find in 3+ weeks, and it was stuff we needed to know and address. Well focused, deep dive process."*

Benchmarking Data

The benchmarking data is provided in several formats. For example, the data is analyzed and presented by:

- Protocol
- Sub-protocol (some protocols have multiple areas of review – Shift Handover, Operator Rounds, etc.)
- Specific protocol sections (groupings of questions for specific areas)
- Management Systems vs. Implementation (having the procedure vs. how is it being done in the field)
- Size of site
- Size of company
- Quartile

Benchmarking data updates are provided generally every six months to sites that have had an assessment.

Data Review – Good Practices and Observations

Below are some examples of the types of Good Practices and Observations that the assessors have been identifying through the site assessments. If you note the last bullet on both slides this is an example of exactly what we are after. At some sites it's an item that needs to be addressed while at other sites we have seen them doing it well.

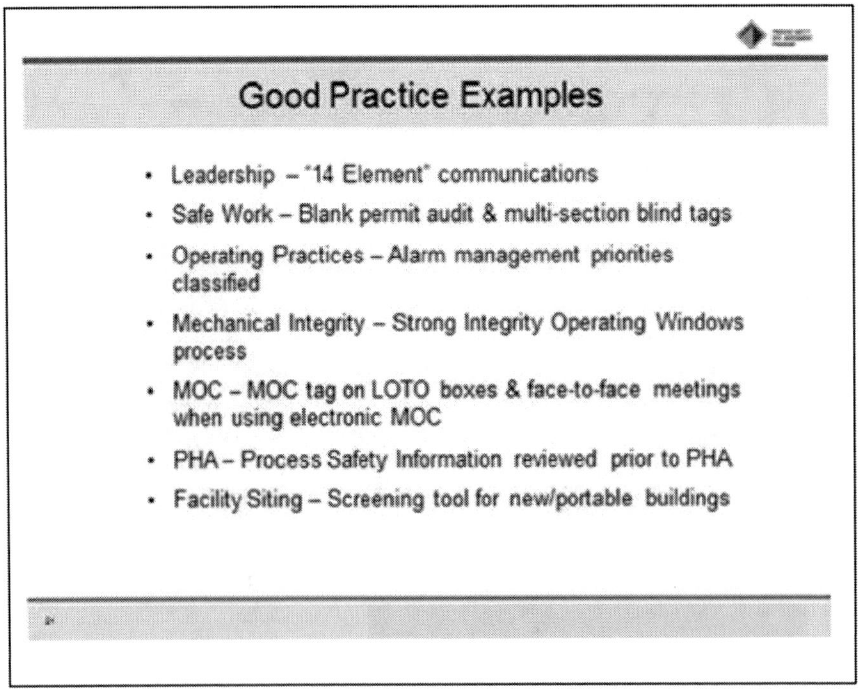

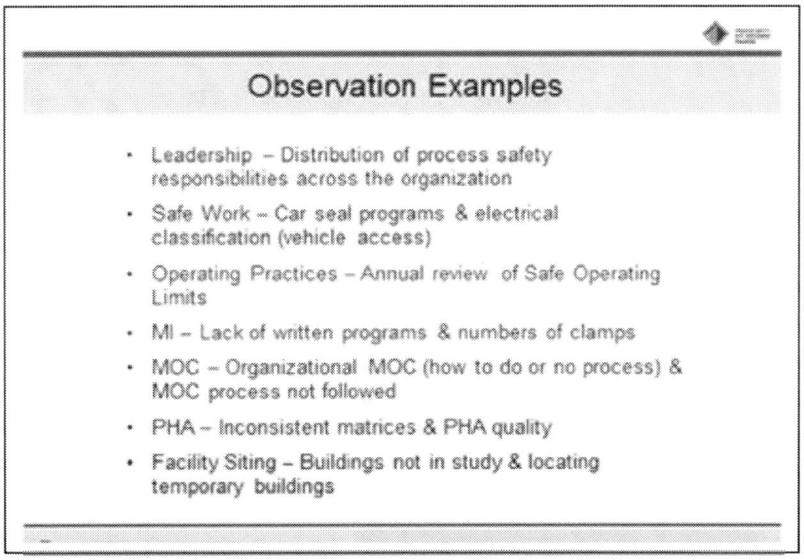

This points to a key element of the site assessment program - sharing the information from sites doing it well with others that are working to put systems/programs/procedures in place at their site. A variety of methods are used for sharing the practices learned through the site assessment program including:

- The Advancing Process Safety Program
  - Regional networks
  - Hazard Identification program
- Industry Conferences and Seminars
- API standards

The data, information, and learnings from the site assessments have been used to focus resources to develop informational and guidance documents regarding areas identified as improvement opportunities. For example, an informational document for mechanical integrity regarding small bore piping was developed based on observations during PSSAP site assessments and data from the 754 metrics and Event Sharing database..

Conclusion

In conclusion, the "boots on the ground" performance of site assessments began April of 2012. Feedback from these assessments has told us that the process of preparing for an assessment (going through the protocols prior to the assessment) and the mentorship/interaction with the assessors has been very beneficial to the sites. And now that we have done enough sites to allow the sharing of benchmarking data the sites are getting even more value from the program.

Alarm Management Meets SIS

SIS Alarm Requirements

How many alarms must be configured per the SIS Standard (ISA-84 Part 1) for an SIF?

Zero

How many alarms must be configured per the AM Standard (ISA-18.2) for an SIF?

Zero

How many alarms are necessary for an SIF to correctly perform its function?

Zero

Pre-alarm – What makes sense?

Pre-alarm before an SIS impending trip?

True – This gives the operator an opportunity to prevent the trip, keeping the plant running and reducing the demand frequency on the SIF.

Pre-alarm should always be on the BPCS control or indicator tag rather than the SIS transmitters?

False – Choice of where to put the pre-alarm is up to the facility, and it's a good idea to document this in the Alarm Philosophy.

Each transmitter in a voting arrangement should alarm individually?

False – Are we trying to keep our plants operating safely or do we just throw multiple alarms at the operator?

SIS and BPCS transmitter disagreements among redundant SIS transmitters should be alarmed?

True – With this alarm in place, there is no need for a pre-alarm on more than one transmitter.

Introduction

Most plants focus on meeting the alarm management requirements laid out in ISA 18.2. Even if the proper approach is taken in configuring the BPCS, there is one area that is sorely neglected, the SIS system. SIS is addressed based on the requirements laid out in ISA 84. Unfortunately, this standard does not address alarming related to SIS events. ProSys contends that a successfully implemented alarm management philosophy is only effective when taken as a holistic view of the BPCS and SIS systems.

Unified Alarming Scheme

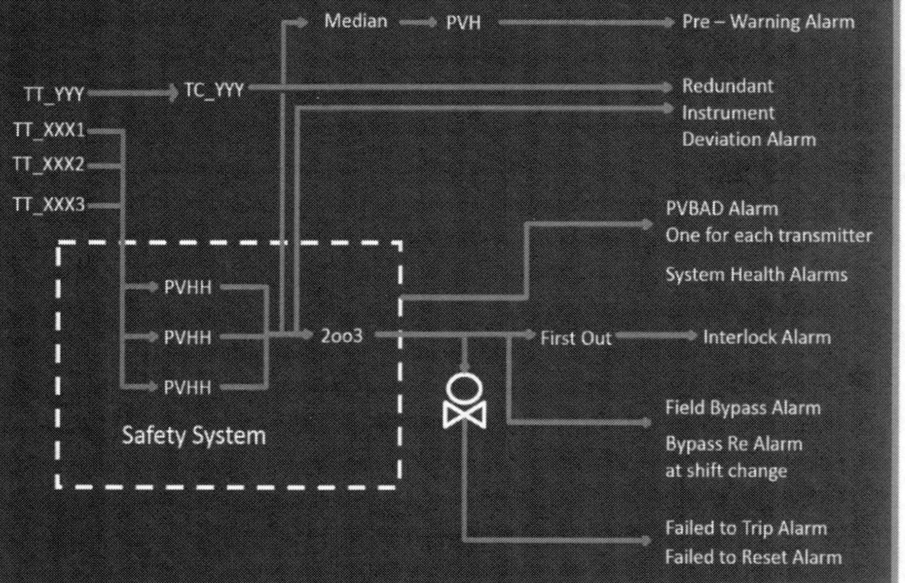

Trip – What makes sense?

First out alarm indicating the reason for a trip?

True – There is no better way to announce to the operator that a trip has been initiated.

Alarm SIS valve moving to trip position?

False – Going to the trip position is the expected response to an SIS trip – better, provide a failed to trip alarm, which makes noise if the valve doesn't work as expected.

Bypass – What makes sense?

Alarm board operator initiated SIS bypass?

False – Although a common practice, this alarm adds little or nothing to the integrity of the system: "Hey Joe/Mary/Sam, you just pressed a button!"

Alarm field operator initiated SIS bypass?

True – This may come as a surprise to the console operator, so it's not a bad thing to annunciate.

If a bypass is in place, trip point alarm on the SIS transmitter?

False – Why give the operator an alarm that isn't reliable or won't be believed? Find an alternative monitoring plan.

Re-alarm on SIS Bypass?

Neutral – Bypass and shift changeover procedures exist, but do we follow them 100%? This is a facility decision. If used, re-alarm shortly after shift change.

PROSYS Maximize Your Plant™ www.ProSys.com

ENHANCING PROCESS SAFETY MANAGEMENT INTEGRATION TO IMPROVE PROCESS SAFETY AND PROFITABILITY IN THE FUELS AND PETROCHEMICALS INDUSTRY

Thomas D. Watson, P.E.
TDWatson Consulting LLC
Pittsburgh, PA USA
twatson@tdwatson.com

Ronald D. Fuchs, P.E.
RD Fuchs Engineering LLC
Lake Jackson, TX USA
rdfuchs13@aol.com

Harold Phillippi, P.E.
Independent Process Safety Consultant
Charleston, SC USA
butch.phillippi@gmail.com

Keywords: PSM, Process Safety Management, Relief Systems, Incident Investigation, Mechanical Integrity.

Abstract

Process Safety Management (PSM) in the Fuels and Petrochemicals industry has the paradox of being well established and yet constantly challenged to improve its effectiveness and impact on process safety and profitability. Such a paradox is a natural consequence of an industry exhibiting a positive process safety culture that maintains a sense of vulnerability and sees the imperative to improve process safety. This paper addresses the frequently asked question, "How do we continue to improve our organization's process safety culture?" and presents opportunities and examples for organizations to enhance their existing PSM systems and culture by further integrating PSM elements both within an organization's established PSM systems and its other systems, processes, and procedures. The paper illustrates integration and enhancement opportunities using Relief System Design Basis (RSDB) documentation, a component of the PSM element - Process Safety Information (PSI). Illustrations show RSDB integration with the PSM elements of Mechanical Integrity (MI), Process Hazard Analysis (PHA), and Incident Investigation (II). The paper then illustrates opportunities to increase profitability by highlighting RSDB integration with partial downtime throughput opportunities and improved project definition. The illustrations challenge organizations that truly "own" a process safety culture to look for additional opportunities to further enhance and integrate established PSM systems into their organization to further improve safety and profitability.

1. Introduction

The Fuels and Petrochemicals Industry in partnership with governing regulatory agencies have established effective Process Safety Management (PSM) systems to handle hazardous materials and energies over the past several decades. Examples include the U.S. Occupational Health and Safety Administration (OSHA) PSM Standard and the European Commission Seveso II Directive. The Center for Chemical Process Safety (CCPS) offers an extensive reference list of regulatory/industry/system requirements.[1] The catalyst for these industry systems have often been the tragic circumstances around industry incidents and catastrophes. Today, PSM in the Fuels and Petrochemicals Industry has the paradox of being well established and yet constantly challenged to improve its effectiveness and impact on process safety and profitability. Such a paradox is a natural consequence of an industry exhibiting a positive process safety culture that maintains a sense of vulnerability and sees the imperative to improve process safety.[2] The vulnerability and sense of urgency are well founded as the industry continues to experience preventable incidents. The Marsh Report states the 100 largest property damage losses in the hydrocarbon industry from 1974 to 2013 had a combined property damage cost of approximately US$34 billion.[3] Organizations demonstrating a committed process safety culture continually challenge and address the question, "How do we improve our organization's process safety culture?"

Organizations that have successfully managed the PSM implementation and sustainment phases have already experienced and moved past the common misconception that process safety and profitability have conflicting objectives. These organizations further realize they must move beyond a "separate but equal" PSM system established among their current business management systems, and fully integrate PSM into their systems to further improve their organization's impact on process safety and profitability. Recent efforts in the Fuels and Petrochemicals Industry to improve process safety through management system integration include industry efforts such as:

- The American Fuel and Petrochemical Manufacturers (AFPM) "Advancing Process Safety Programs," which promotes sharing PSM learnings across member organizations of AFPM
- The CCPS publication containing guidelines for integrating management systems and metrics to improve process safety performance.[4]

This paper identifies and presents specific examples and opportunities for organizations to continue to enhance their existing PSM systems and process safety culture by further integrating PSM elements both within an organization's established PSM systems and its other business systems, processes, and procedures. The paper illustrates integration and enhancement opportunities using Relief System Design Basis (RSDB) documentation, a component of the PSM Process Safety Information (PSI) element. Specifically, the examples illustrate RSDB integration with the PSM elements of:

- Mechanical Integrity (MI)
- Process Hazard Analysis (PHA)
- Incident Investigation (II)

RSDB integration with other business processes also present opportunities to directly improve profitability. Two examples presented here include RSDB integration with:

- Production planning (partial downtime throughput opportunities)
- Improved project definition

The following sections provide a general introduction to RSDB and illustrate the five integration opportunities in more detail.

2. An Introduction to RSDB and its Relationship to PSM

Within the Fuels and Petrochemicals Industry, many processes involve highly hazardous chemicals. To adequately assess and address a process' hazards an organization must adequately define the Process Safety Information of the process. Process Safety Information (PSI), includes information regarding the process' hazards, technology, and equipment.[5] Examples of hazard information include toxicity and permissible exposure limits. Examples of technology information include a process' chemistry and operating limits. Finally, examples of equipment information include materials of construction and the Relief Systems Design Basis (RSDB). There are many references for what constitutes RSDB and how to effectively manage RSDB documentation.[6] The American Petroleum Institute states RSDB documentation should demonstrate that all pressure-containing equipment has been identified and the relief requirements adequately analyzed. Documentation should include specific engineering guidance, methodology, analysis, and reference information that collectively reflects Recognized and Generally Accepted Good Engineering Practices (RAGAGEP). [7]

RSDB is a key component of PSI, but RSDB is also an important component in maintaining other Process Safety Management (PSM) elements such as Mechanical Integrity (MI), Process Hazard Analysis (PHA) and an Incident Investigation (II). Furthermore, RSDB can be a significant aspect of process production constraints and project opportunities associated with processes. The next sections discuss integrating RSDB into the aforementioned PSM elements, production planning, and project development processes.

3. RSDB Integration with Mechanical Integrity (MI)

Integrating Relief System Design Basis (RSDB) analysis with Mechanical Integrity's (MI's) Risk Based Inspection (RBI) through a combined risk assessment can identify acute process equipment vulnerabilities. Mechanical Integrity (MI) activities ensure the proper design, installation, operation, and maintenance of process equipment. Risk Based Inspection (RBI) is one component of an effective MI program that links risk assessment and mitigation activities with appropriate inspection and process control activities. RSDB analysis, on the other hand, identifies the applicable relieving scenarios and required relieving rates affecting pressure-containing process equipment to ensure the relieving scenarios do not expose the equipment to operating conditions outside of design limits. The American Petroleum Institute's Recommended Practice (RP) 581 identifies the interdependency of the Pressure Relief Device (PRD) with the fixed equipment the PRD is protecting.[8] RP 581 evaluates PRD criticality to set mechanical integrity activities such as inspection frequency. The RBI methodology combines the PRD's anticipated demand rate with the probability of failure on demand of the PRD to determine an overall risk for the PRD and the process equipment the PRD protects. While the API RP 581 methodology is sound, when the RSDB documentation is not available or the RSDB analysis has identified deficiencies in the PRD design, the RBI methodology develops shortcomings that should be addressed.

As an example, consider a process vessel containing a flammable hydrocarbon liquid. An RBI of the vessel defines a risk associated with the vessel, the vessel's PRD and the anticipated overpressure scenarios. The RBI analysis is predicated in part, however, on the PRD adequately protecting the vessel in the anticipated overpressure scenarios should the PRD open. The RSDB documentation identifies fire as the only applicable overpressure scenario but concludes the PRD is undersized. The RSDB can provide additional insight into the level of overpressure that would occur (and the anticipated consequence). The additional insights would significantly impact the RBI risk assessment and potentially warrant a reassessment of inspection and mitigation actions for the PRD and the protected vessel.

The example illustrates how integrating RSDB with MI's RBI can identify acute equipment vulnerabilities and aid in further prioritizing MI efforts and RSDB deficiency resolution, including risk mitigation efforts. Consequently, integrating both RSDB and MI can further enhance PSM integration and improve process safety and profitability.

4. RSDB Integration with Process Hazard Analysis (PHA)

Integrating RSDB with PHA can reduce costs and improve the effectiveness of both the PHA and the RSDB. As most readers are aware, a Process Hazard Analysis (PHA) in the United States and many other countries, is a requirement of a Process Safety Management (PSM) program, (for example, 29 CFR 1910.119 requires PHAs and periodic revalidations). No matter which methodology a PHA employs [Hazard and Operability study (HAZOP), "what-if," Failure Modes and Effects Analysis (FMEA), etc.], capable teams must perform the analysis. The teams typically range from 4 to 8 (or more) persons, consisting of owner's personnel and often outside consultants/specialists. No matter who is present, the participants represent a necessary and significant staff resourcing cost. Furthermore, each PHA finding should be well documented, especially those requiring additional evaluation and stewardship before the findings are declared either a confirmed issue or not. The PHA finding documentation and corresponding follow-up/stewardship also represent a significant staff resourcing cost. Finally, actions required to address the confirmed issues, whether facilities changes, work process changes, or administrative changes represent significant costs.

With the PHA costs sufficiently identified, now consider how integrating RSDB with PHA reduces costs and improves PHA and RSDB effectiveness. PHAs always identify questions and concerns about the adequacy of the over (or under) pressure protection systems as well as items such as Maximum Allowable Working Pressure (MAWP), hydrotest pressure, materials of construction, etc. Furthermore, PHAs require the most recent drawings, heat and material balances, and other Process Safety Information (PSI). Similarly, the most recent RSDB should have many (if not all) of these items assembled and documented, with any changes since the most recent RSDB tracked and provided as a part of the facility's Management of Change (MOC) program. The overlap of PSI and resources between RSDB and PHA is substantial. Consequently, completing RSDB analysis prior to and integrating it with the PHA process and follow-up activities introduces cost savings. Common savings occur relating to over (or under) pressure findings. If properly completed and documented, the RSDB can provide the PHA participants with immediate feedback regarding the adequacy of pressure protection devices, which in most cases are code items and must be addressed. A benefit of having a quality RSDB done prior to the PHA is that the over or under pressure concerns can be quickly resolved. On many occasions, no action item follow-up work is necessary. This reduces costs both during the PHA and the PHA follow-up. Conversely, the PHA serves as an audit check of the RSDB as a capable team reviews the RSDB basis as needed.

As a quick example, a PHA team questioned the adequacy of an exchanger shell's design pressure during a blocked-in condition. The maximum pressure generated from a blocked-in condition from a manual valve closure was greater than the design pressure by approximately 10 percent. The PHA team referenced the RSDB. The RSDB previously identified the same issue and investigated the situation in more detail. The RSDB identified the exchanger's MAWP to be approximately 20 percent more than the noted design pressure, and the hydrotest pressure was based on 1.5 times the MAWP. The U-1 form that was part of the RSDB package clearly indicated this. The RSDB package also noted the exchanger shell was adequate based on ASME VIII, Appendix M, M-5.7, in this situation, because the pressure was below 116% of MAWP and it would be immediately noted by operators, the valve closure was not identified as an issue and the PHA follow-up item was closed. Because the PHA utilized the RSDB, the PHA saved costs on discussion and documentation of an action item and the necessary follow-up and tracking of a code related item. Conversely, the PHA team validated the RSDB, enhancing PSM integration and improving process safety and profitability.

5. RSDB Integration with Incident Investigation (II)

Incident Investigation (II) is also a mainstay of any effective personnel or Process Safety Management (PSM) program. II is an opportunity to learn from mistakes, system failures, equipment failures, and/or overlooked vulnerabilities. II is also an opportunity to evaluate the effectiveness of preventive and control measures installed to prevent or mitigate an incident. In the CCPS Vision 20/20 Characteristics of Companies with Great Process Safety Performance the authors write: "To reduce incidents, employers and employees must enthusiastically support a culture that is driven to learn from many sources including benchmarking, near misses and incidents, and jobs done well."[9] For incidents involving release from a relief device and/or a flaring event, II intersects Relief System Design Basis (RSDB) and together they provide a learning opportunity that enhances PSM and improves process safety.

Yogi Berra said; "When you come to a fork in the road, take it." When an incident occurs investigate it. Definitions may vary for what constitutes an incident. Indeed, there are legal requirements for investigation of safety and environmental incidents. These form the baseline for what organizations must do but do not limit organizations from doing more. For instance, some organizations with very effective PSM programs investigate all flaring events. The organization calling the event an incident, a near miss, or flaring event is least important, while the organization's learning opportunity from the event is most important.

By integrating RSDB with II organizations can do real case study evaluation of the RSDB's assumptions and technical predictions. Integration can help answer such questions as:

- Did the scenarios of the RSDB represent the incident that occurred?
- Did the relief valve effectively handle a release rate lower than the controlling case without issues such as chattering?
- Were the release rate, pressure limitation, and system temperature change what the calculations predicted they would be?
- Did the flare handle the relieving event without an environmental release or safety issue? For example: flare smoking, toxic release, or high temperature exposure of personnel and adjacent equipment.

RSDB integration with II may facilitate getting to the root cause of the incident. The RSDB contains valuable information about the relief system response and may help identify the original process design intent. If the relief system response was not as intended, the relief system may be the cause of the incident. Additionally, if there are layers of protection designed to make flaring or release valve opening an infrequent event but flaring is occurring often, the preventing layers of protection may need further evaluation.

The RSDB and II integration may also help identify additional inspections and corrective actions. In this case II also integrates Mechanical Integrity (MI). For example, damage to the relief system may have occurred from too-high or too-low temperature or pressure. Mechanical damage resulting from corrosive chemicals or relief valve chattering may have occurred. Additionally, the RSDB may shed light on damage that the incident caused to other process equipment. Finally, if the RSBD does not accurately predict what will occur during such an incident there should be a corrective action to change the RSDB and possibly modify the relief system. Consequently, the II serves to further validate the RSDB and the RSDB enhances the II. An important point to note: incorporating RSDB into II requires someone knowledgeable in RSDB be on the II team.

In conclusion, a relief device release or flaring event II that does not consider the RSDB leaves a valuable tool unused. Integrating RSDB with II enhances PSM integration and improves process safety and profitability by gaining additional benefit from both RSDB and II.

6. RSDB Integration with Production Planning

Opportunities often present themselves in complex facilities to ensure process safety and increase profitability by integrating Relief System Design Basis (RSDB) with production planning. Many facilities integrate processing units in a complex manner. Such facilities often have multiple production constraints preventing individual processing units from running maximum capacities simultaneously. Production constraints may include downstream processing capacity, feed qualities, processing unit downtimes, permitting limits, or even facility RSDB constraints. When a facility RSDB constraint limits production, understanding the relationship between other production constraints and the facility RSDB constraint becomes critical to ensure process safety and optimal profitability. Hence the need and opportunity to integrate RSDB with production planning.

Before presenting RSDB and production planning integration examples, there are a few points to note. First, RSDB analysis for an individual processing unit typically establishes the unit's safe operating window at a maximum capacity. A facility RSDB, however, reflects multiple processing units operating simultaneously. Second, due to the myriad of production constraints a complex facility may have, the facility RSDB often reflects a limited set of facility production constraints (often with a very conservative basis). Third, in some instances, the facility RSDB may constrain various processing units to a fraction of their maximum production capacity. Finally, when production constraints setting the facility RSDB change, production planning needs to evaluate alternative production options that the facility RSDB does not adequately reflect.

In one example, a complex facility relief system constraint of excessive relief system backpressure for low set point relief devices exists. The excessive backpressures result from a potential utility failure scenario impacting multiple processing units. This facility RSDB constraint limits the facility's production. When the processing unit with the low set point relief devices is down for maintenance, however, the facility RSDB constraint relaxes. Thus, opportunities present themselves for other processing units to increase production. In another example, a complex facility schedules a portion of the facility relief system maintenance downtime with processing units having a substantial impact on the facility relief system constraint. During the downtime, however, the startup timing for the relief system is delayed due to discovery work and opportunities arise to increase profitability by minimizing production cuts within a newly defined facility RSDB constraint.

7. RSDB Integration with Project Development

Integrating Relief System Design Basis (RSDB) with projects presents opportunities to improve project definition, increase project profitability (through project cost savings), and enhance Process Safety Management (PSM). Having a preliminary well documented RSDB at the on-set of a project — often called the Front End Engineering Design (FEED) stage — will identify the various scenarios that should be a part of the sizing of not only the individual relief devices, but also the relief device collection system functional requirements. The potential costs of the project's impact on the relief devices and systems can then be reviewed and optimized in coordination with the project's functional equipment. The following examples best illustrate these points.

In a larger project a liquid feed drum was required. The feed drum's initial design was 50 psig, and the feed sources ranged from 35 psig to 100 psig. Because the feeds to the drum were greater than the drum's proposed design pressure, the overpressure device and corresponding system had to be designed for liquid overfill. All the sources that were greater than 50 psig represented an overfill potential of around 70,000 barrels a day, and consequently the relief device and the corresponding collection system size and costs were unattractive. Upon further investigation and review of the options, the project team increased the design pressure of the feed drum to 110 psig, which had minimal cost differential when compared to the necessary relief system design. Thus, the 110 psig design was chosen and the relief device was then sized for fire and some potential back-flow on stoppage of the feed pumps. Cost savings were in the millions of dollars.

A second project involved design and installation of equipment that had to be tied into the flare system. A preliminary RSDB was done early in the project. Upon analysis of some typical global scenarios such as loss of power, loss of cooling water, etc., the new project's primary relief device(s) would require increasing the collection system's piping and flare size. Furthermore, the sizing of the relief device and piping alone had a large estimated cost.

The relief device's size was driven by the main tower's selected design pressure and steam reboiler pressure available. The project team, in coordination with the RSDB personnel, were tasked with optimizing the reboiler and the relief system. During the optimization phase, the team identified 2 key items: 1) the reboiler steam had only one pressure available, but it could be let down to a lower pressure, and 2) at a slightly lower pressure the steam could easily provide the available heat necessary for the process.

Additional investigation by the team revealed that the tower material and thickness had not initially taken advantage of the potential MAWP. The material thickness needed was ~0.82 inches, and 0.875 inches was to be selected. This allowed for a MAWP higher than design. However, because the steam pressure was reduced, good engineering practice would require that the design consider failure of the steam pressure control valve.

Failure of the steam pressure control valve would provide sufficient heat to overpressure the tower based on the new design pressure and the relief device to the flare was still needed. To address this failure, the steam system (downstream of the pressure let-down valve) was provided with its own pressure relief valve which could be vented to atmosphere.

With the team's recommended changes noted above, the only scenario for the main tower relief device was fire, which had a lower cost of installation and did not require an update of the flare system. Having the RSDB done early in the project allowed the designers and project team to make some easy (paper) cost effective changes, and did not require any modifications to the existing collection and flare systems.

The examples above demonstrate organizations integrating RSDB during project development to improve project definition, reduce project cost, and even make processes inherently safer. Organizations realize further PSM enhancements by integrating a process unit's existing RSDB documentation with project development activities. In addition to the benefits illustrated above, organizations incorporate project impacts on a process unit's RSDB with minimal duplication of effort (minimal cost), and enhance the process unit's RSDB integrity. From an overall PSM perspective, the project development integration enhances both the PSM and project development system.

8. Conclusion

This paper presents opportunities for organizations to enhance their PSM systems and process safety culture by integrating Relief System Design Basis (RSDB) with the following PSM elements and business processes:

- Mechanical Integrity (MI)
- Process Hazard Analysis (PHA)
- Incident Investigation (II)
- Production planning
- Project development.

Organizations that truly "own" a process safety culture should use each successful opportunity implementation as a spring board to further enhance PSM integration and improve process safety and profitability.

9. References

1. CCPS, "Guidelines for Integrating Management Systems and Metrics to Improve Process Safety Performance", Appendix A, Center for Chemical Process Safety, American Institute of Chemical Engineers, New York, NY 2016.

2. CCPS, "Vision 20/20 Booklet – Committed Culture", Center for Chemical Process Safety, American Institute of Chemical Engineers, New York, NY 2016.

3. Marsh and McLennan Companies, "The 100 Largest Losses", 23rd edition (2014).

4. CCPS, "Guidelines for Integrating Management Systems and Metrics to Improve Process Safety Performance", Center for Chemical Process Safety, American Institute of Chemical Engineers, New York, NY 2016.

5. Occupational Health and Safety (OSHA) "Process Safety management of Highly Hazards Chemicals" 29CFR1910.119(d).

6. Thomas Watson, "AM-13-24 Considerations for Effectively Managing Emergency Relief Systems Design," American Fuels & Petrochemicals Manufacturers Annual Meeting, 2013.

7. American Petroleum Institute, "API Standard 521 – Pressure-relieving and Depressuring Systems", section 4.7.3, sixth edition (2014).

8. American Petroleum Institute, "API RP 581 – Risk-Based Inspection Technology, Part 1", Section 7, second edition (2008).

9. Center for Chemical Process Safety's Vision 20/20: "Characteristics of Companies with Great Process Safety Performance" (2013, McCavit, Berger, and Nara).

479098 Reactive Hazards Management: Lessons Learnt from Major Incidents

Abstract:

Chemical process industry has long suffered from major incidents due to lack of proper understanding of reactive hazards. The methyl-isocyanate leak in Bhopal, India, has been by far the worst industrial incident in history which led to the development of many process safety regulations. However, the formation and implementation of regulations has not prevented the occurrence of these incidents. Many incidents such as Concept Sciences Inc. where a hydroxylamine manufacturing unit was destroyed, the explosion at the Sterigenics International Inc., ethylene oxide sterilization facility, Bayer CropScience Pesticide Chemical Runaway Reaction Explosion and West Fertilizer Explosion have highlighted the importance of having a better understanding of the reactive hazards. The focus of this presentation is to identify different types of reactive hazards through examples of chemicals incidents and analyze how a better understanding of these hazards could have avoided these incidents.

479245 The Study of Safety Evaluation Technique for High Temperature Process

Abstract:

There are many accidents of a firing furnace and kneading machine by overheating and the abnormal injection of the product. When the gas which suddenly expanded by thermal decomposition destroys plant equipment and enters the explosion concentration range, a gas explosion is caused and damage spreads. The material which it is easy to decompose is invested by mistake by a firing furnace and has the trouble that the pressure of the furnace abnormally rises to in our factory. Also, troubles such as an explosion of the plumbing by the thermal decomposition of the polymer by the overheat, vent explosion by the thermal decomposition by the false injection of an additive having low heat stability has broken out with kneading machine

These troubles are caused by the thermal decomposition by a sudden temperature rise. The temperature program of the analytical instrument which is generally used by a thermal analysis is 10-20 /min. The conventional thermal analysis equipment cannot evaluate the temperature rise rate at the time of the trouble exactly to become than 20 /min. It becomes the problem that the slow temperature rise may be different from rapid heating up in a decomposing reaction.

We produced a device capable of quickly raising temperature to solve these problems.

This device is comprised of the vertical plumbing which it sealed and heating furnace, and it is very simple structure. The sample holder is fixed to the upper part in the plumbing by a magnet. And the holder drops in the heating zone by keeping a magnet away from the plumbing and is heated rapidly. The materials of the plumbing are SUS316L, and, as for the use upper limit temperature, 500 , the use upper limit pressure are 1MPa. Because the deterioration of SUS materials by the heat is intense when it is raised the temperature of the device to 750 , the plumbing is disposable.

The inside of the plumbing is replaced with inert gases such as nitrogen, argon. The maximum temperature rising rate is 150 /min (furnace temperature: 750).

We introduce an example of the trouble analysis of a firing furnace and kneading machine using this device.

[Case example 1] Study of firing furnace trouble

After burning with a new additive at 720 , abnormal pressure rise happened in a furnace, and a plant stopped. After measuring quantity of outbreak gas using the rapid heating device, it was recognized that gas of 7 times was generated and a gasification rate tripled in comparison with a conventional additive.

[Case example 2] Study of kneading machine trouble by overheating

The additive which had low heat stability was spent mistakenly in large quantities to a kneading machine, and a vent board exploded. After performing a rapid temperature rise examination under a kneading temperature condition, as for the pressure raising rate when an additive disintegrated, 1.8MPa/min, the quantity of gasification became 190L/kg. The additive concentration became (mixing ratio with the resin) more than at least 75%, and it was recognized that a large amount of additives were invested mistakenly.

Fortunately, this example does not become the big accident with resulting in injury or death. In this explosion example, a vent board penetrated a checker plate. It might become the casualty when there was a person near a kneading machine.

In comparison with the plant of the petrochemical and fine chemistry that heat risk analytical technique establishes, the firing furnace and kneading machine is often run without detailed risk assessment. The rapid heating device becomes the powerful tool to do the heat risk evaluation of these high temperature processes.

In future, it is going to be improved the rapid heating device changes the plumbing materials, and to use it to the temperature ranges more than 1,000 .

Handling uncertainty in LOPA and SIF design

Akshat Khirwal
IRESC
Unit 2801, Enterprise Square 3
39 Wang Chiu Road
Kowloon Bay
Hong Kong
Akshat.khirwal@irescglobal.com

Arjan Abeynaike, Venkatesh Sourirajan
IRESC
Arjan.abeynaike@irescglobal.com, Venkatesh.s@irescglobal.com

Keywords: Uncertainties, LOPA, SIL, SIF design, reliability

Abstract

Layer of protection analysis (LOPA) is a widely used semi-quantitative method for determining the reliability requirement for safety instrumented functions (SIFs). An advantage of LOPA is that it specifies a numerical reliability target - the average probability of failure on demand (PFD_{avg}). However, uncertainties in the input parameters and simplifications within the LOPA method lead to a margin of error in the target PFD_{avg} that is often not accounted for in the design of the SIF. Further uncertainties are introduced when carrying out a reliability analysis of the SIF design to ensure that the target PFD_{avg} is met. The result of overlooking such uncertainties is that the target PFD_{avg} may be set low and the calculation of the actual PFD_{avg} achieved by the design may be optimistic, leading to a SIF that is under-designed. The need to account for uncertainty is emphasized in the latest version of IEC 61511, the international standard for SIF design.

This paper investigates and highlights the sources of such uncertainties in LOPA and SIF reliability analysis. In particular, the quality of the data used as input parameters to LOPA and reliability models has a large effect on the accuracy of the results. These parameters include failure rates, conditional modifiers, beta factors for common cause failures, and proof test coverage factors. Available data for these parameters can be sparse and may not be representative of the conditions in which the SIF is applied. Where the confidence intervals of the data are available, these can be incorporated into the calculations to reach a more conservative result. Alternatively, a safety factor can be applied to try and capture all the uncertainties. In this approach, the target PFD_{avg} obtained from LOPA is reduced by the safety factor to give a more stringent target PFD_{avg} for the SIF design to meet. The methods for handling uncertainty are reviewed in this paper.

1 Introduction

SIL Determination is the process of assigning a Safety Integrity Level (SIL) to Safety Instrumented Functions (SIFs) following any of the methods i.e. Risk Graph, Risk Matrix or Layer of Protection Analysis (LOPA); as laid down in IEC 61511 and IEC 61508 [1] [2]. Of these methods, the semi-quantitative LOPA approach is increasingly being adopted for SIL Determination as compared to other qualitative approaches such as Risk Graph or Risk Matrix. An advantage of LOPA is that it specifies a numerical reliability target for a safety instrumented functions (SIF), which is expressed in terms of the target average probability of failure on demand (target PFD_{avg}), or target risk reduction factor (RRF, where RRF = 1/ PFD_{avg}). This target PFD_{avg} then forms the basis for design of the SIF wherein a reliability analysis is carried out to ensure that the target PFD_{avg} is met. However, uncertainties in the input parameters and simplifications within the LOPA method lead to a margin of error in the target PFD_{avg} that is often not accounted for in the design of the SIF. Further uncertainties are introduced when carrying out a reliability analysis of the SIF design. Without accounting for these uncertainties in the assessment, the SIF may be under-designed since LOPA may result in under prediction of the target PFD_{avg} or the reliability analysis may show that the target PFD_{avg} has been achieved although this may not be the case. This paper has reviewed the sources of uncertainties in the LOPA method and design of the SIF (SIF reliability analysis) and provides a way forward for handling such uncertainties.

2 Uncertainties in LOPA

Initiating event frequency, consequence severity, and likelihood of failure of Independent Protection Layers (IPLs) are the key parameters used in LOPA to approximate the risk of a scenario in case the SIF fails to act when the demand arises. In addition to the above, conditional modifiers [3] may also be used to factor in specific parameters pertaining to the demand on the SIF. The simplicity of LOPA allows it to be easily understood by engineers of all disciplines, leading to its widespread use. The drawback of this simplicity is the greater uncertainty in the results, as described in the following sub-sections. The governing equations and the LOPA worksheet used for assessing the scenarios discussed in this paper are referenced from IEC 61511-3 [1] and CCPS, Simplified Process Risk Assessment [4].

2.1 Initiating event frequency

The demand on any SIF may arise due to one or more events, each with a frequency in the range of once per year to once in plant life (for example, 30 years). There is some approximation involved in the selected initiating event frequency values especially when using generic databases. For instance, the source data in the generic database may not be appropriate for application to the facility under review or there may not be enough confidence in the dataset. Power failure or failure of other utilities such as cooling water or steam is a good example. The frequency of these failures from generic databases is typically $1x\ 10^{-1}$ per year (once in 10 years) [4]. However, in the author's experience, based on feedback from many operators, these values are likely to be somewhere between once per year to once in ten years. Similarly, the frequency of pump trip based on generic databases is $1x\ 10^{-1}$ per year, however operational experience suggests that a value in the range of once in 2 years to once in 3 years may be more appropriate. An analysis of the OREDA [5] data also provides a value close to once in 3 years. Additionally, when there are multiple demand scenarios for a particular SIF, failure to consider the cumulative frequency of all the demand scenarios can also lead to under-estimation of the overall initiating frequency. A simple approach is to consider a suitable approximation for the cumulative demand frequency [6] [7]. In addition, common cause events specific to the facility is also a contributor which needs to be identified. Without such considerations, the required target PFD_{avg} may be under-predicted. All of the above introduce uncertainties in the LOPA assessment noting that the initiating event frequency is the starting point of a LOPA assessment. Therefore, care must be taken in selecting an appropriate value for this parameter.

To better understand the dependence on the initiating event frequency, consider a typical distillation column as shown in Figure 1. There are many scenarios such as overhead condenser failure, reflux pump failure, reboiler control failure, etc. which may lead to a demand on the high high pressure SIF provided on the column. For the sake of brevity, only the case of Reflux Pump trip is used further in this example.

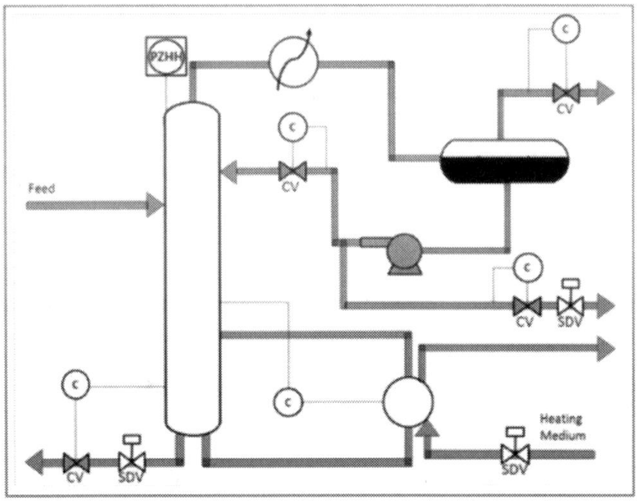

Figure 1. Typical Distillation Column

Assuming an initiating event frequency of 1 x 10^{-1} per year for pump trip and considering a Target Mitigated Event Likelihood (TMEL) of 1 x 10^{-5} per year for multiple fatalities along with an ignition probability of 0.1 and taking credit for pressure safety valve (PSV) on the column (PFD_{avg} of 1 x 10^{-2}), this would result in a SIL 1 with Risk Reduction Factor (RRF) 10 (target PFD_{avg} of 1 x 10^{-1}). However, if the initiating event frequency for pump trip is changed to 3.33 x 10^{-1} per year (once in 3 years), this translates to SIL 1 with RRF of 33 (target PFD_{avg} of 3 x 10^{-2}). This example demonstrates the dependence of initiating event frequency selection on the target PFD_{avg}.

2.2 Consequence Severity and Conditional modifiers

The TMEL assigned to the SIF is directly associated with the degree or the severity of the consequence assumed (e.g. single fatality: 1 x 10^{-4} per year, multiple fatalities: 1 x 10^{-5} per year). This in turn has a direct impact on the SIL assigned to the SIF as well as its target PFD_{avg}. Therefore, careful consideration must be given to the selection of the consequence severity during a LOPA review. For instance, the impact area or hazard zone due to a loss of containment event can be determined using consequence modelling [7], the result from which can be used to assign the severity (e.g. single fatality vs multiple fatalities). Additionally, this can be used to determine the relevant conditional modifiers such as occupancy factor within the impact zone. This approach is more reliable than making coarse assumptions which may introduce uncertainties in the analysis. Such refinement can also be applied to other modifiers such as ignition probability where using a more specific value representative of the hazardous fluid and/or the facility features may be appropriate as compared to use of generic values.

If we revisit the example of typically distillation column presented in Section 2.1; considering an initiating event frequency of 3.33 x 10^{-1} per year and assuming a TMEL of 1 x 10^{-4} per year (single fatality), an ignition probability of 0.1 and taking credit for PSV on the column (PFD_{avg}

of 1 x 10^{-2}), this would result in a RRF of 3 (target PFD_{avg} of 3 x 10^{-1}), which is below SIL 1 requirement. However, if a higher severity (TMEL: 1 x 10^{-5} per year corresponding to multiple fatalities) is selected based on consequence modelling results and an ignition probability of 0.3 is selected based on refinement, the assigned SIL will increase significantly to SIL 2 with RRF 100 (target PFD_{avg} of 1 x 10^{-2}). Although this example shows an increase in the assigned SIL, the application of detailed assessment for determining severity could also show a reduction.

2.3 Independent Protection Layers

Independent Protection Layers (IPLs) are devices, systems or actions that prevent the realization of an undesired consequence (e.g. the consequence the SIF is safeguarding against). Each IPL has an associated PFD_{avg} which is assigned and used in the LOPA methodology for SIL Determination. For example, process deviation alarm with sufficient time for operator intervention has a PFD_{avg} of 1 x 10^{-1}, an independent BPCS control has a PFD_{avg} of 1 x 10^{-1} while a PSV has a PFD_{avg} of 1 x 10^{-2}.

2.3.1 Effect of Common Cause Failures

One of the fundamental assumptions in the LOPA approach is that these IPLs are completely independent and do not share any common cause failures (CCF). In reality, multiple protection layers may be impacted or their performance may be affected due to similar environmental conditions or deviations from desired process parameters. Corrosion or deposition leading to simultaneous degradation of a pressure safety valve and a shutdown valve is one such example. Other examples of CCF include scenarios where multiple PSVs are required to meet total relief requirements, or where an alarm and control, although independent safeguards, both reside on the BPCS. Additionally, dual check valves for preventing backflow could also be impacted by CCF. Although such scenarios may take a period of time to develop and the likelihood may be low, such CCF can have a direct impact on the PFD_{avg}. For example, consider a case where the safety system comprises a SIF and an IPL and the target PFD_{avg} for this system is 1 x 10^{-3}. The PFD_{avg} of the IPL is 5 x 10^{-2}. If the SIF and IPL fail independently only, then the target PFD_{avg} required of the SIF is (1 x 10^{-3}) / (5 x 10^{-2}) = 2 x 10^{-2}. However on inclusion of CCF, even if a low probability of CCF is assumed (1% of the PFD_{avg} of the IPL), the target PFD_{avg} for the SIF then becomes 1 x 10^{-2}.

The authors do acknowledge that for the purpose of time-effective LOPA review, it may not be practical to dwell deeper into CCF for IPLs as it may require an in-depth analysis (for which sufficient details may also not be available at the time of LOPA review). However, it is important to be aware of the inherent uncertainties introduced in the analysis by not accounting for the CCF associated with IPLs.

2.3.2 Effect of Multiple Protection Layers

There is another inherent uncertainty associated with IPLs apart from CCF. The PFD_{avg} for any protection layer (whether mechanical protection, alarm with operator response, BPCS or SIF) is not static but increases over time until a required test or maintenance activity is conducted to check for faults [4]. This relationship between PFD and time is shown graphically in Figure 2. In LOPA, the PFD_{avg} values of the IPLs are multiplied together to estimate the PFD_{avg} of the overall safety system [4]. This calculation, while simple to perform, may tend to underestimate the overall PFD_{avg}. In a more quantitative analysis of reliability, the time-dependent PFDs of each protection layer are multiplied together to yield a result similar to Figure 3, which shows the PFD of the safety system. The PFD_{avg} for the system can then be calculated.

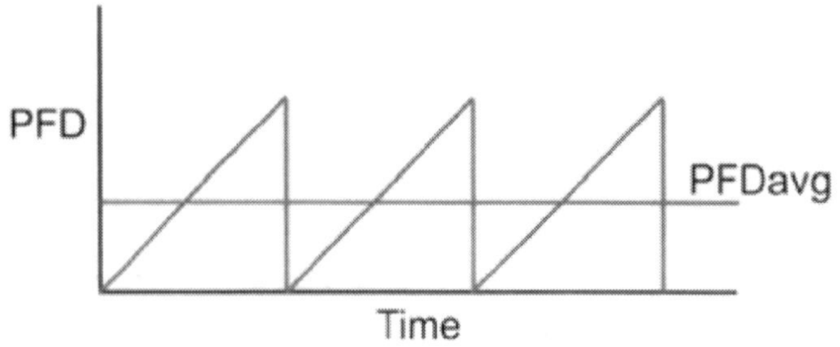

Figure 2. PFD as a function of time for a single protection layer. Periodic testing / maintenance was assumed to detect all faults in this example.

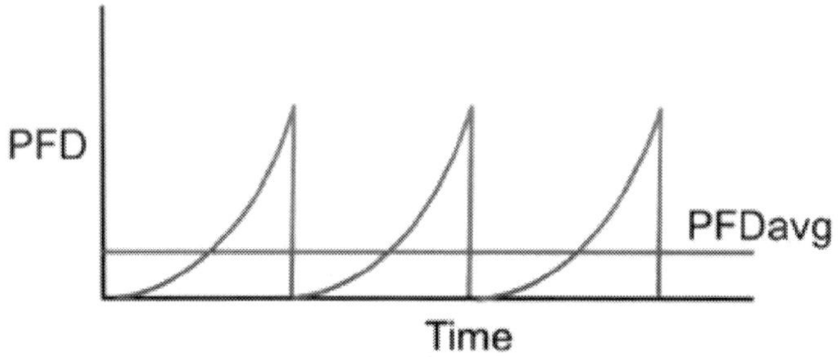

Figure 3. PFD as a function of time for a safety system with two protection layers (one SIF and one other IPL). Proof testing was assumed to detect all faults in this example.

To illustrate the difference in the results provided by LOPA and a more rigorous analysis, the same example from Section 2.3.1 is used. LOPA estimates the PFD_{avg} of the SIF to be 2×10^{-2} (excluding CCF), considering one IPL with a PFD_{avg} of 5×10^{-2} per year and a target overall PFD for the system of 1×10^{-3} (see calculation in Section 2.3.1). Using the PDS method of reliability analysis [8] gives a SIF PFD_{avg} of 1.5×10^{-2}, i.e. LOPA overestimates the target PFD_{avg} by 33%. As the number of protection layers increases, this error becomes more significant, as shown in Table 1. For two or more IPLs with a SIF, the error in the target PFD_{avg} for the SIF is at least 100% using the PDS method; i.e. the LOPA value for target PFD_{avg} is double that given by the PDS method. This gross overestimation in the target PFD_{avg} for the SIF suggests that the LOPA method may not be sufficiently accurate where two or more IPLs are considered. In such cases a more rigorous analysis may be conducted outside of the LOPA session (refer Section 3). Such rigorous approach, if adopted, may also help reduce the error for cases where only one IPL is applicable.

Table 1: Effect of the number of IPLs on the error in PFD_{avg} for a safety system as calculated by LOPA

Number of IPLs	Error in target PFD_{avg} for SIF
0	0%
1	33%
2	100%
3	220%
n	$\frac{2^{n+1}}{n+2} - 1$

2.4 *Assessing the cumulative uncertainty for multiple parameters in LOPA*

The above sections have discussed how the uncertainty of each individual parameter affects the target PFD_{avg} calculation. In practice, however, the cumulative effect of the uncertainties for multiple parameters also needs to be considered. As a simple illustration and keeping in line with the discussion above, the same example of the distillation column high high pressure SIF is used. As shown in the table below, the changes in various parameters results in a significant difference in the target PFD_{avg} requirements. One additional parameter presented in the table below is the use of PFD_{avg} of 3 x 10^{-2} for IPLs considering multiple PSVs are required to meet relief requirements. As expected, the required RRF is much more stringent when the uncertainties are modelled. However, as highlighted in earlier sections, depending on the SIF under review, in some cases, the requirements may also be less stringent upon further refinement.

Table 2: Effect of cumulative uncertainty of multiple parameters in LOPA

Description*	Initiating Event Frequency	IPL	Ign. Prob.	Total Mitigated Event Frequency	TMEL	Target PFD_{avg}	RRF
Case 1	1.00E-01	1.00E-02	1.00E-01	1.00E-04	1.00E-04	1.00E+00	1
Case 2	3.33E-01	1.00E-02	3.00E-01	1.00E-03	1.00E-04	1.00E-01	10
Case 3	1.00E-01	1.00E-02	1.00E-01	1.00E-04	1.00E-05	1.00E-01	10
Case 4	3.33E-01	1.00E-02	3.00E-01	1.00E-03	1.00E-05	1.00E-02	100
Case 5	1.00E-01	3.00E-02	1.00E-01	3.00E-04	1.00E-04	3.33E-01	3
Case 6	3.33E-01	3.00E-02	3.00E-01	9.00E-04	1.00E-04	1.11E-01	9

3 Approaches for handling uncertainty in LOPA

Based on the various modes of introduction of uncertainties in the LOPA analysis as discussed in the preceding sections, the question that arises is whether the direct use of target PFD_{avg} from LOPA as input to SIF design is optimal or is it desirable to investigate possible ways to refine the target PFD_{avg} from LOPA to ensure the SIF is not under-designed. Some ways to refine the target PFD_{avg} are discussed in the following paragraphs.

One proposed approach for dealing with uncertainty without compromising the simplicity that makes LOPA an attractive method is to conduct a more rigorous risk assessment outside of the LOPA session for high consequence / high risk scenarios to confirm that the SIF reliability target PFD_{avg} obtained from LOPA is sufficient. Such an assessment may include a quantitative reliability analysis such as a Fault Tree Analysis or Markov Model to obtain a more accurate estimate of the required target PFD_{avg}. These methods can also take into account common cause failures and the time dependency of PFD. In fact the most recent version of IEC 61511 [1] requires that a quantitative methodology be used for cases where the required risk reduction is greater than 10,000 (target PFD_{avg} from LOPA is equal to or lower than 1×10^{-4} per year). By using LOPA as a screening tool, this strategy can ensure that high consequence / high risk scenarios receive the more rigorous assessment they deserve, while for lower risk scenarios the more time-efficient LOPA can be considered sufficient.

Alternatively, a number of different cases can be considered for each LOPA scenario, such as in Table 2. Each case considers different values for the various parameters used in the LOPA calculation, resulting in a range of RRFs, from the more conservative to the more liberal. Taking Table 2 as an example, the results range from an RRF of 1 to 100. A final RRF at some point in between these two extreme cases may be selected for the SIF and carried forward into the SIF design. In this way, a level of uncertainty can be accounted for without being overly conservative.

Another approach is to introduce a "Safety Factor" to the reliability target set during LOPA. In this approach, it is recommended that the target Risk Reduction Factor obtained from LOPA should be increased by a safety factor to give a more stringent target for the SIF design to meet. For instance, if the target based on results from LOPA is RRF 100, the SIF design can conservatively consider a target PFD_{avg} associated with RRF 300. Similarly, if the target is RRF 500, then SIF design may consider RRF 800. However, when the RRF is at or close to borderline values, the SIF design may consider the next interval, e.g. for RRF 800 and above, the SIF design may consider RRF 1300 or alternately the RRF used can be justified by conducting a more rigorous quantitative analysis as mentioned above. Some engineering judgement should be applied when selecting this safety factor depending on the level of conservatism adopted in the initial LOPA assessment. As can be inferred from Table 2 above, when uncertainties are accounted for, the target PFD_{avg} can actually vary by one order of magnitude. Applying a safety factor to the results will therefore account for the uncertainties.

4 What PFD_{avg} to assume when Risk Graph Method is adopted for SIL Determination?

Risk Graph Method [1] is the alternate method commonly used for SIL Determination. Unlike the semi-quantitative LOPA approach, which specifies a numerical reliability target (PFD_{avg}) for the SIF in addition to the assigned SIL, the Risk Graph approach relies on the qualitative definitions for the parameters involved to deduce the assigned SIL without setting an actual numerical target (PFD_{avg}). Furthermore, this target PFD_{avg} value from Risk Graph is provided as a range. Hence, where Risk Graph method is used, it is not an uncommon practice to use the upper bound of the PFD_{avg} associated with the assigned SIL. For example, the LOPA method may provide an RRF of 40 (PFD_{avg} of 2.5×10^{-2} per year) for a SIL 1 function, while the same SIL 1 function following the Risk Graph method may adopt the upper bound RRF of 10 (PFD_{avg} of 1×10^{-1} per year).

Consider the same example of a distillation column as discussed in Section 2. If the column high high pressure SIF is assessed using the Risk Graph approach assuming that a trip of Reflux Pump could be expected once a year, taking credit for the pressure safety valve on the column as an IPL (credit of 100) and taking no credit for occupancy and avoidance; the resulting analysis show that Risk Graph method also determines a SIL 2 which is similar to the results from LOPA are accounting for uncertainties. The key difference being the overall PFD_{avg} of 3.33×10^{-3} per year (RRF 300) obtained from the LOPA which will be more stringent than the PFD_{avg} of 1×10^{-2} per year (RRF 100) in case the upper bound value of SIL 2 is used.

The above example shows that using the output from the Risk Graph directly may lead to under-prediction of the target PFD_{avg} and thereby leading to under design of the SIF. The concept of "Safety Factor" as discussed in Section 3 may therefore also be applied where Risk Graph is used. Based on this, it is recommended that the SIF design can use the mid-point of the upper bound and lower bound values based on PFD_{avg} range applicable to the SIF corresponding to the assigned SIL. The safety factor may also be selected on a case by case basis depending on the severity of the associated consequences (for example, using a value above the mid-point of upper and lower bound may be justified for the most severe consequences).

5 Uncertainties in SIL reliability analysis

The uncertainties in terms of obtaining an appropriate target PFD_{avg} for a SIF using LOPA methodology are discussed in Section 2. It is however, worth noting that potential for introduction of uncertainties into the analysis is not limited to this step alone. Moving on in the safety life cycle, the next step after LOPA involves a reliability analysis to ensure that the target PFD_{avg} is met for each SIF. This activity is commonly termed as "SIL Verification" in the industry, but strictly with reference to IEC 61511, it is considered within the domain of "SIS Design and Engineering". The target PFD_{avg} that is considered for the reliability analysis should be the same as the target PFD_{avg} that was calculated in LOPA (a safety factor may be applied as described in Section 3). Where Risk Graph methodology is applied, the target PFD_{avg} can be a mid-point value in the assigned SIL band as suggested in Section 4.

This reliability analysis also involves a number of input parameters, some of which may introduce uncertainties into the analysis. It is important to understand and evaluate the nature of these uncertainties associated with input parameters such as failure rate data for SIF components, beta factor for common causes and proof test coverage factors in SIF reliability analysis as discussed in the subsequent sub-sections. In this section, PFD_{avg} results that were calculated using exSILentia software are presented to illustrate the effects of the input parameters on the PFD_{avg}. The numerical PFD_{avg} results are only applicable for the specific systems and assumptions used in the calculations.

5.1 Failure Rate Data

Failure rate data for the individual SIF components is the basis of the reliability analysis. It is therefore of utmost importance that the data used is based on sufficient sample size and is representative of the conditions in which the SIF is applied. The quality of the data used as input parameters to reliability models has a large effect on the accuracy of the results.

Ideally, an appropriate method for determining the failure rate is to collect field data from a large, randomized sample of units. The operational time until failure should be recorded for each unit. Such data mining however; is a long term project, requiring the co-operation of multiple operators, ideally from a wide range of different facilities and operating environments. For this reason, such studies are rare and normally "generic data", i.e. data applicable for a broad type of component (such as pressure transmitter or shut-down valve) is used rather than data for a specific model or a specific environment. The OREDA database [5] is one example, which provides such generic data. Alternately, some vendors may commission a field study to collect failure data specific to their components or build their own failure data based on reported failures. However, given the smaller sample size as well as the gaps in ensuring that each failure is reported appropriately by the operating facilities, there will be uncertainties in the reported data and these values may tend to be on the optimistic side. For example, while the failure rate for a ball valve from a generic data source is 9.8×10^{-7} per hour [9], several vendors of ball valves report failure rates of less than 1.0×10^{-7} per hour. Another issue is that such data may only be sourced from a limited number of facilities operating in a specific environment and hence, this data may not be applied to a range of facility types or operating environments.

One approach to iron out these uncertainties is to conduct a Failure Mode and Effect Analysis which is a workshop activity bringing together a group of experienced personnel (operations, maintenance, instrumentation, design, etc.) for brainstorming the possible failure modes associated with each component and its associated failure rate using operational experience and field data. The quality of such an exercise is highly dependent on the experience of the participants and their engineering judgement. Hence, even though conducting such an exercise will be fruitful, some margin of error is still expected.

5.2 Beta Factor

Typically, the overall reliability of a SIF is estimated based on both independent and dependent failures. For functions with redundant components, multiple components may fail simultaneously due to a common failure mode (e.g. stress, maintenance errors, etc.). These dependent failures are the percentage of the failures attributed to the common causes and are denoted as Beta Factor in the reliability analysis.

Taking note of the uncertainties in failure data for components as highlighted in earlier sections, it is worth highlighting that the chances of availability of detailed field data for beta factors covering a large sample size is even lower, thereby leading to a greater degree of uncertainty. IEC 61508 [2] advocates a method for estimation of the beta factor using a questionnaire and a scoring system. Based on the scoring system, typically the beta factor value ranges from 1 - 10%. This is a somewhat subjective approach. In reality, beta factors higher than 10% have also been reported in the industry [10]. This may however represent an extreme operating environment pertaining to a particular facility. Therefore, some engineering judgement is required.

It is to be noted that beta factor has a direct impact on the PFD_{avg} achieved for systems with redundant components. Since common cause failures are often the dominating contributor to PFD_{avg}, an increase in beta factor leads to a significant increase in PFD_{avg}. This relationship can be seen in the PFD_{avg} results given in Figure 4, for SIFs with various configurations (1oo2, 2oo3, etc.). Hence, selection of a low beta factor may lead to underestimating the PFD_{avg} and potentially also under design of a SIF. For configurations without redundancy (e.g. 2oo2), the beta factor has little impact on the PFD_{avg}.

In the author's experience, some references and guidelines from operating companies provide recommended beta factor values. However, taking note of the findings presented in this section, it should be ensured that such recommended values have a sound basis and are as accurate as possible. Over-estimating beta factor value may lead to over-design of the SIF which in turn also has an associated cost impact. Conversely, under-estimating the beta factor may lead to SIF being under-designed.

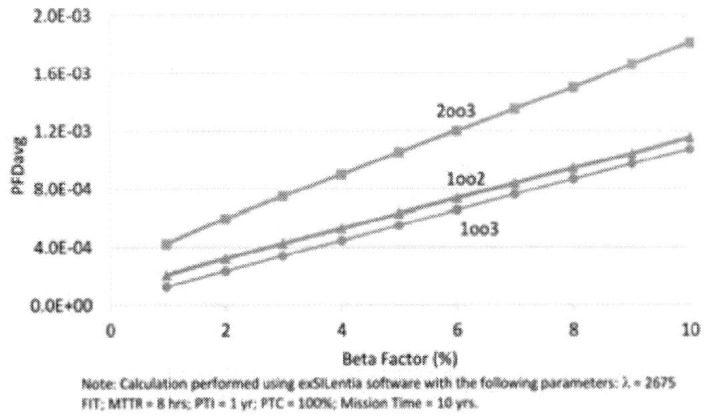

Figure 4. Impact of Beta Factor on PFD_{avg} calculation

5.3 Proof Test Interval

A proof test is performed periodically to detect failures and restore the system / component to an "as good as new" condition. The actual proof test interval may be governed by factors associated with SIF design and operating philosophy specific to a facility. However, the selected interval forms the basis of SIF design since the reliability of the SIF can be compromised with regard to the desired target PFD_{avg} in case the proof test interval used for SIF design is not followed. The uncertainty in this regard lies in whether the recommended proof test interval is rigidly followed or is only conducted on as-need basis in the actual operational phase. Operators may tend to postpone a shutdown for testing purposes if it may result in a loss of production, thereby compromising the SIF reliability. The emphasis to rigidly adopt the recommended proof test intervals therefore rests with the Operating Company and the kinds of safety culture they advocate or follow.

5.4 Proof Test Coverage

One of the important parameters associated with proof tests is the coverage factor which represents the fraction of the potential failures that will be identified during a full proof test. A perfect proof test (i.e. coverage factor of 100%) is unlikely. As far as the authors are aware, there is also no field data available to determine the effectiveness of proof tests. Vendors may recommend a coverage factor based on a detailed FMEA exercise for the associated components. Such an FMEA can identify which failure modes will be successfully captured by a proof test, and thus estimate the coverage factor. Proof test coverage factors can sometimes be lower for components with good diagnostic coverage because the remaining failure modes that are undetected by diagnostics are more difficult for a proof test to detect. In the author's experience, typical coverage factors range from 80 - 99% for sensors and logic solvers and 70 - 80% for final elements (e.g. shutdown valves). While this is currently the best approach for determining the coverage factor, the uncertainty in the result is usually not estimated.

The proof test coverage factor has a significant impact on the calculated PFD_{avg} of a SIF element, as can be seen in Figure 5. For example, for an element with a 1oo1 configuration, decreasing the coverage factor from 90% to 80% resulted in a 46% increase in the PFD_{avg}. Given this effect, caution should be exercised when selecting coverage factors for use in SIF design since they will impact the risk reduction that can be achieved. The actual test performance should ideally be verified to match the assumptions (this becomes especially critical when higher factors e.g. 90% or above are used).

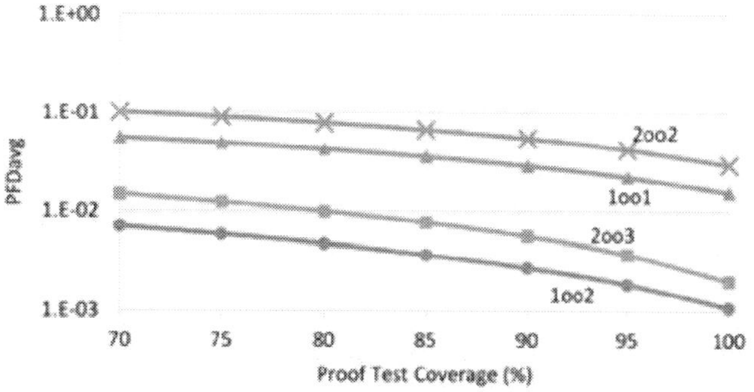

Note: Calculation performed using exSILentia software with the following parameters: λ = 1810 FIT; MTTR = 8 hrs; PTI = 2 yrs; Mission Time = 20 yrs; beta factor = 5%

Figure 5. Effect of Proof Test Coverage Factor on PFD_{avg} for various voting configurations

An alternative method for considering failures that are not detected by a proof test is to include a "probability of test independent failures", or P_{TIF}. This approach, which is advocated in the

PDS method [8], differs from the proof test coverage approach in that P_{TIF} is a constant contributor to the PFD_{avg}, whereas the effect of proof test coverage on the PFD_{avg} increases over time. Even when using P_{TIF}, however, there remains uncertainty in the value used for this parameter, with implications on the accuracy of the calculated PFD_{avg}.

5.5 SIF Operational Life

Operational life of a SIF refers to the life time of the various components of the SIF, sometimes also denoted as mission time. This also serves as a key input in SIF design. In the author's experience, typically a mission time in the range of 20 to 30 years is assumed. Lower mission times such as 5 or 10 years may be impractical in terms of actual implementation on site since this implies that the corresponding components will be replaced or completely overhauled after a 5 or 10 year period, which also has a significant cost impact. This assumption on operational life of the SIF has a direct impact on the risk reduction achieved by a SIF, provided that all other parameters are kept constant. Figures 6 and 7 show that a higher mission time is likely to result in a higher PFD_{avg} (or in other terms, a lower risk reduction). Figure 7 shows that the relationship between mission time and PFD_{avg} is almost linear, with a steeper gradient at a lower proof test coverage. This shows that the mission time has a greater impact on PFD_{avg} when the proof test coverage is lower. If the proof test coverage is taken as 100%, the mission time has no effect on PFD_{avg}. However, as discussed in Section 5.4, 100% proof test coverage is not likely to be realistic.

While operators may typically consider the cost of replacing equipment when deciding upon mission time, consideration should also be given to the effect that mission time will have on the reliability of SIFs. A shorter mission time can help SIFs to meet the target PFD_{avg} and may even allow a lower level of redundancy to be used for the SIF components. This is particularly relevant for SIL 3 rated SIFs, which may require a high level of redundancy as well as high reliability of components, resulting in a high capital cost. Reducing the mission time may help optimize the capital cost of the SIF. A cost benefit analysis may be conducted to justify optimization of the mission time. However, some engineering judgement must be exercised in order to keep the mission time within practical limits. In this regard, 15 years may be considered as a practical lower limit for mission time but this may vary on a case by case basis also depending on the owner's operating philosophy. In such cases, it is important to be aware that in case the capital replacement is not performed after 15 years, this will potentially leave the system exposed to higher probability of failures.

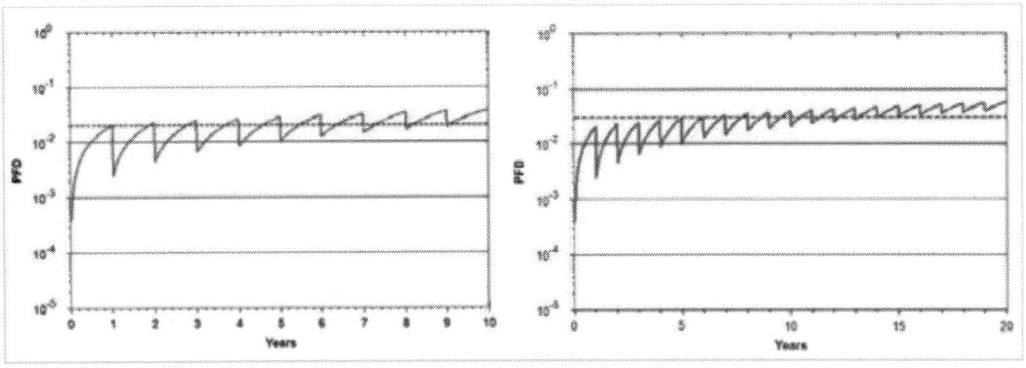

Figure 6. PFD_{avg} vs SIF Operational Life / Mission Time (Years)

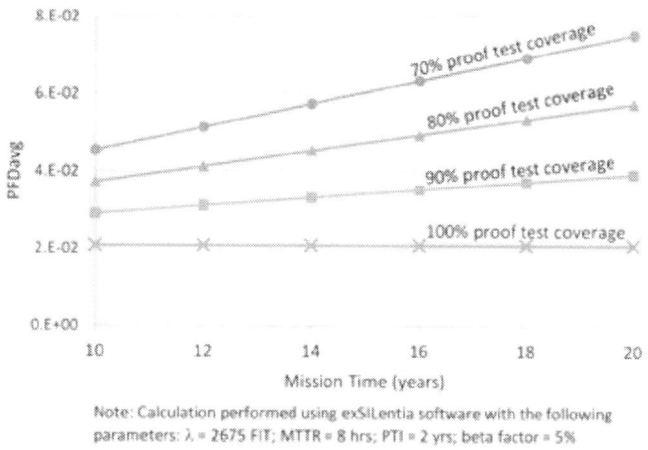

Figure 7. Effect of Mission Time on PFD_{avg} for a SIF with 1oo1 configuration

5.6 Assessing the cumulative uncertainty for multiple parameters in SIF Reliability Analysis

The above sections discuss how the uncertainty of each individual parameter affects the PFD_{avg} calculation. In practice, however, the cumulative effect of the uncertainties for multiple parameters also needs to be considered. As a simple illustration, Table 3 presents PFD_{avg} values calculated for a high pressure SIF with 1oo2 component configuration, using three different combinations of beta factor and proof test coverage, representing a best case, base case and worst case scenario. The beta factor was varied between 1% and 10%, while the proof test coverage was varied between 70% and 90%, which are ranges that can be reasonably expected for these parameters. The PFD_{avg} for the worst case scenario was more than double that of the base case and more than five times that of the best case. In these scenarios, beta factor was found to be the main contributor to the variation in PFD_{avg}, as the uncertainty in the beta factor may be +100% / -60%, whereas the proof test coverage may typically have an uncertainty of only ±13%.

Table 3: PFD_{avg} for a high pressure SIF in three scenarios

Scenario	Beta Factor	Proof Test Coverage	PFD_{avg}	Variation from base case
Best Case	1%	90%	2.93E-04	-70%
Base Case	5%	80%	9.88E-04	0
Worst Case	10%	70%	2.10E-03	+113%

Note: Calculation performed using exSILentia software with the following parameters: λ = 600 FIT; MTTR = 8 hrs; PTI =3 yrs; Mission Time = 15 yrs.

6 Conclusion

In this paper, the authors have highlighted multiple modes through which uncertainties are introduced into the LOPA review and subsequently in the reliability analysis for SIF design. Through working examples, this paper demonstrated that without accounting for such uncertainties, the SIF can easily be under-designed or over-engineered. The emphasis on accounting for such uncertainties is also highlighted in the most recent edition of IEC 61511. This paper investigated further into the nature of uncertainties associated with various input parameters associated with LOPA and SIF reliability analysis including approaches using which such uncertainties can be accounted appropriately.

7 Acknowledgement

The authors wish to acknowledge the support received from IRESC including some of the information included in this paper, as well as the contribution from Ms. Nicole Yeung.

8 References

[1] The International Electrotechnical Commission (IEC), Functional Safety – Safety Instrumented Systems for the Process Industry Sector, IEC-61511, Second Edition

[2] The International Electrotechnical Commission (IEC), Functional Safety of Electrical/ Electronic/ Programmable Electronic Safety Related Systems, IEC-61508, First Edition

[3] Guidelines for Enabling Conditions and Conditional Modifiers in Layer of Protection Analysis, CCPS

[4] Layer of Protection Analysis, Simplified Process Risk Assessment, CCPS

[5] Offshore and Onshore Reliability Data (OREDA), Volume 1 – Topside Equipment, 6th Edition, 2015

[6] Akshat Khirwal, Venkatesh S., Consideration of Cumulative Frequency of Demand Scenarios in SIL/LOPA Review, Hazards Asia Pacific, Kuala Lumpur, Malaysia, 2015

[7] Akshat Khirwal, Venkatesh S., Inherently Safer Design & Robust Protection Layers for High Consequence Hazards, Mary Kay O'Connor Process Safety Center, 19th Annual International Symposium, College Station, Texas, USA, 2016

[8] Reliability Prediction Method for Safety Instrumented Systems, PDS Method Handbook, SINTEF, 2010 edition

[9] Safety Equipment Reliability Handbook, Exida, Third Edition, 2007

[10] S. Hauge et al., Common cause failures in safety-instrumented systems: Using field experience from the petroleum industry. Reliab. Eng. Sys. Saf. 151 (2016), 34-45.

479264 Fundamental Research at Purdue Process Safety Research Center

Abstract:

Fundamental Research at Purdue Process Safety Research Center

Significant industrial incidents continue to occur such as fires, explosions and gas releases resulting in serious injuries, loss of life, and damage to the environment and facilities. Such incidents are not solely germane to the oil & gas industry, but occur in a wide variety of industries including agriculture, pharmaceuticals, and manufacturing. Two years ago the 'Purdue Process Safety and Assurance Center' (P2SAC) was established within the School of Chemical Engineering to conduct fundamental research to prevent such incidents. The presentation will address the four research projects the Center is currently pursuing with funding from its initial Sponsors: BP, Chevron, ExxonMobil, Honeywell, Eli Lilly and Shell.

The initial projects include:

- **optimal placement of gas detectors to prevent fires and explosions**: the research uses CFD-based dispersion simulations and rigorous mathematics to perform optimal placement over hundreds of thousands of leak scenarios and potential detector locations.

- **design & synthesis of catalysts to prevent undesired hazardous byproducts**: zeolite catalysts are widely used in the refining industry and certain structures can promote undesired byproducts leading to potential over-pressures and explosions. Zeolite synthesis and characterization methods are being developed.

- **real time process optimization to reduce risk**: Novel continuous manufacturing processes can reduce the risk related to powder handling. Continuous manufacturing, as with pharmaceuticals, requires efficient process monitoring and control.

- **application of inherently safer separation technology** to significantly reduce processing volume, solvent use, and extent of unit operations. Advanced mixed solvent extraction & simulated moving bed techniques are used to recover polycarbonates and flame retardants from plastic waste.

Indirect benefits from the Center are the depth of the core Chemical Process Safety course required of all undergraduates, as well as undergraduate process safety research initiatives in such as areas as dusts, reactive chemicals and nano-materials.

Current focus is on growing the Center and encouraging participation by a wide variety of industries such as pharmaceuticals, agriculture and manufacturing, in addition to oil & gas, and chemicals.

10-13-16

479277 Bringing New Life to Process Safety

Abstract:

<>Abstract:

Whilst risk cannot be eliminated, it can be better managed. The world of Process Safety Management continues to evolve as organizations find better ways of managing risks. In the simplest form, managing the risk is to reducing inherent process risk to the tolerable level.

First step in managing risks is to identify source of Hazards and quantifying risk associated to Hazards. Traditionally organizations heavily rely on expert judgment and brainstorming methodologies like HAZOP for finding operational and design risks. In addition, quality of such analysis are directly related to the precision of assumptions and information from operation and maintenance of the equipment under control.

In this session we will explore how organizations are using new techniques such as dynamic risk visualization, cumulative risk techniques, operational risk assessments and cyber Hazop's to advance the boundaries of process safety management.

This paper will explore how existing systems and data sources, risk visualization and intelligence techniques are being used to bring static process safety data to life. Flat "1 dimensional" information such as lists, forms and databases are now being transformed into "2 dimensional" dynamic information such as plot plans, barrier models, risk matrix, bow ties etc. that provide greater context of relationships between information.

This information will resonate with engineers and technician who rather working with meaningful information of operation and maintenance systems instead of some abstract numbers and concepts that need extensive expertise in the reliability engineering.

We also will look at how advances in Operational Risk Assessments and Cumulative Risk allow operators to determine in real time if the overall cumulative risk is tolerable before actions are taken, devices are removed from service, maintenance activities are deferred etc. In other words, leading indicators and metrics will be in disposal of operators to proactively facilitate decision making for managing risk. This could avoid incidents that developing in the operation due to degradation of control and safety systems or deviations from initial operation and maintenance procedures and may lead to severe incidents.

And finally, since new technologies and architectures for instrumentation, control and safety systems introduced new risks of cyber intrusions and attacks that need to be considered in the risk management and the world of security standards (ISA 62443-3-2 Security Risk Assessment and System Design) and safety standards (ISA-TR84.00.09-2013 Security countermeasures related to Safety Instrumented Systems and Associated IACS) advanced in the last few years, we will look at how Cyber Process Hazard Analysis (PHA) techniques are being applied to provide the same systematic approach to security as traditionally applied to safety.

"Good process safety management does not happen by chance and require constant active engagement", in this paper we suggested novel approaches that can revive and improve the process safety.

479367 Developing an Effective System for Action Item Resolution

Abstract:

PSM action items are are regularly identified through many different means in a typical chemical processing facility. Knowing what to do with these action items and how best to proceed is no easy task. Not all action items are created equally, or even with the same goal in mind. An action item identified in a PHA may present a direct hazard to an employee, while an action item generated from a pressure relief systems study may present a documentation concern, a deviation from current practices, a direct hazard to an employee, or any combination thereof. Many facilities have a system in place for shepherding action items to completion, but these systems typically assume all action items are created equally. In this paper, we will demonstrate the importance of having a system that handles each action item in the most effective way; from evaluating the nature of the concern, considering the risks associated with the action item, developing the action plan, to completing the action plan. We will also cover whose input should be considered at each stage in the process to help ensure the best solutions possible are developed.

13TH GLOBAL CONGRESS ON PROCESS SAFETY

Where Does Management of Change (MOC) Begin?

Juliana Schmitz
Praxair, Inc.
Av. Pastor Martin Luther King Jr., 126, bloco 10/ala A,
Corporate Shopping Nova América, Rio de Janeiro, RJ, Brazil
juliana_schmitz@praxair.com

Patti Jones
Praxair, Inc.
175 E Park Dr, Tonawanda, NY, 14150, USA
patti_jones@praxair.com

Gabriel Drummond
Praxair, Inc.
Av. Pastor Martin Luther King Jr., 126, bloco 10/ala A,
Corporate Shopping Nova América, Rio de Janeiro, RJ, Brazil
gabriel_drummond@praxair.com

Prepared for Presentation at
American Institute of Chemical Engineers
2017 Spring Meeting and 13th Global Congress on Process Safety
San Antonio, TX
March 26 – 29, 2017

Where Does Management of Change (MOC) Begin?

Juliana Schmitz
Praxair, Inc.
Av. Pastor Martin Luther King Jr., 126, bloco 10/ala A, Corporate Shopping Nova América, Rio de Janeiro, RJ, Brazil
juliana_schmitz@praxair.com

Patti Jones
Praxair, Inc.
175 E Park Dr, Tonawanda, NY, 14150, USA
patti_jones@praxair.com

Gabriel Drummond
Praxair, Inc.
Av. Pastor Martin Luther King Jr., 126, bloco 10/ala A, Corporate Shopping Nova América, Rio de Janeiro, RJ, Brazil
gabriel_drummond@praxair.com

Keywords: MOC, engineering, process safety

Abstract

Management of Change is often affiliated with an operating facility. However, several projects have the common challenge of being developed within the constraints of an aggressive schedule, considering the time from project kick-off until unit start-up. In order to comply with the deadlines, early releases of engineering design documentation become subject to changes. During project development, these changes may come as scope changes, supplier updates, etc. that may affect more than one engineering discipline and any on-going construction/assembly activities. Another change source may be non-conformances identified during the construction/assembly phase. In this case, considering project milestones and costs, an option taken into account is to leave the system or portion the way it is instead of re-working to match the original design. For all of the given cases, a robust Engineering Management of Change (EMOC) process must be in place throughout the project life cycle to guarantee modifications are properly identified, communicated, and evaluated by technical experts for possible effects on process and design safety prior to implementation, or at a minimum, prior to startup of the system/subsystem. In addition, an EMOC process gives the opportunity to analyze the reasons for the modifications in order to avoid increased costs and design errors on future projects, leading to improved quality control. This paper presents some previous accidents related to management of changes, and addresses EMOC implementation, work process and elements of a successful EMOC program.

1 Introduction

Preliminary design activities begin early in the process to define scope, cost and product delivery requirements. Changes are a natural part of these efforts.

As the project progresses through the engineering detail design phase, a process hazard analysis (PHA) is conducted. At this moment, a multidisciplinary team evaluates if the process is safe and if any modifications are required. The project team is informed of the revised documentation and proceeds with each skill activity.

Subsequent changes may occur as, for example: updated supplier documentation, the team that will effectively operate the plant in the future may raise a new need for the project, some may be identified during detail design requiring changes and change requests during the construction phase. Any modification after the PHA may not be properly communicated to the adequate professionals and analyzed, resulting in a possible unsafe condition.

This paper focuses on project execution phases until plant startup and turnover to operations where the Management of Change (MOC) work process begins. It also explores the challenges identified when developing an Engineering Management of Change (EMOC) work process and shares an approach taken within Praxair, Inc.

2 Why is the Engineering Management of Change Important?

Process Safety is a relatively new engineering area, being developed from the early 60's and mostly based on catastrophic incidents, which negatively affected the industry. Kerin [5] confirms this point of view and lists a series of landmark events, their key learnings and influences in process safety. These events had different causes, mostly related to poor management of various safety areas, such as hazard identification, training, maintenance, management of change (MOC) and others.

The numerous and notable MOC related accidents that happened in the industry over the last century showed the world the need of a well-developed and maintained MOC procedure.

2.1 Titanic, 1912

The sinking of the Titanic, in 1912, is a well-known catastrophic incident. "The design of the ship included 44 water tight compartments giving rise to the perception that the ship was unsinkable and in the event of an accident would be its own lifeboat." [2] An accelerated construction schedule was the catalyst in making changes that proved to be fatal. A robust management of change process and a cultivated safety culture may have prevented the incident.

2.2 Flixborough, 1974

Venart [3] describes the Flixborough accident, which occurred on June 1st, 1974, causing 28 fatalities and several property damages near the Nypro plant in North Lincolnshire, UK. In this plant, there was a process area responsible to produce caprolactam, an intermediary for the Nylon production. Caprolactam production was conducted through the oxidation of the cyclohexane, a highly flammable chemical. There were six reactors in series where this oxidation reaction was performed.

Two months before the explosion, a crack was observed in the 5th reactor of the chain. With the objective of reducing the impact in production, the responsible plant personnel decided to connect directly the 4th and 6th reactors while the 5th was removed for maintenance. To proceed with the change, a "dog-leg" pipe was used to connect the reactors 4 and 6, while the original installation considered a straight pipe between these reactors. This is a modification that clearly changed the characteristics of fluid flow between reactors 4 and 6. The dog-leg line support was a temporary installation, made of scaffolding bars. These changes were never evaluated in detail and during start-up. On June 1st, the bellows connected to the reactors flanges failed due to an overstress and tons of cyclohexane was released into the atmosphere. With the vapor dispersion, an ignition source was found in the plant, and a vapor cloud explosion occurred resulting in fatalities, as well as significant property damage. [4]

2.3 CONDEA Vista Company, 1998

U.S. Chemical Safety and Hazard Investigation Board [5] promotes a discussion about the reaction vessel explosion in CONDEA Vista Company, on October 13th, 1998. In this unit, linear alkyl benzene was produced by mixing aluminum chloride and hydrocarbons. Around 3 months before the incident the process was modified, involving a mixing of powdered aluminum with hydrogen chloride to form the aluminum chloride. After this change was implemented, the bottom nozzle of the reactor got plugged due to the formation of a sludge residue, which the operators were not able to reach and remove during maintenance. Considering this scenario, the chemist of the site and the process engineer decided to use water to dissolve the sludge and to absorb the heat generated due to an exothermic reaction between the water and the sludge. This process was started and once it was apparently under control, the process engineer and chemist left the plant.

The shift supervisor decided to make a new change and inject steam at the nozzle. The process supervisor who was not in the plant agreed with the new procedure. At the shift turnover, the day shift supervisor talked with the night shift supervisor about the steam injection. The night shift supervisor took responsibility to inject steam and oriented an operator to execute this procedure.

After a few minutes of continuous steam injection, the reactor exploded, leaving three workers injured and property damage estimated in $13 million USD. Among other shortcomings related to the catalyst change that resulted in sludge formation, U.S. Chemical Safety and Hazard Investigation Board **[5]** affirms that this accident highlights the value of having an authorization or approval step as part of an MOC system for abnormal situations. If such a procedure had been in place, a technical manager would have reviewed the proposed procedure involving steam and may have detected its deficiencies.

2.4 Prevention

Among others, these accidents demonstrate the importance of Management of Change in process safety. This was emphasized when MOC was included as one of the main pillars of the Process Safety Management (PSM) standard, issued by the U.S. Occupational Safety and Health

Administration and of the Risk Management Plan of the Environmental Protection Agency (EPA).

However, not only changes performed after the plant startup will result in safety accidents. Engineering Management of Change (EMOC) is also a key area to prevent accidents and should be addressed in a proper manner by industrial companies.

One example of a poor EMOC related event can be an updated drawing sent by an equipment supplier during the project execution phase indicating a decrease of a vessel Maximum Allowable Working Pressure (MAWP). If this modification is not communicated to the proper professionals, the pressure relief device set point designed for the equipment may not be adjusted as required, which can lead to lack of vessel overpressure protection and possible vessel rupture.

Another example can be a change of a pipe routing during project phase, without revision of the flexibility study. This could result in pipe over stress, leading to a line rupture and energy release.

Several other examples of poor EMOC related accidents could be discussed showing proper management of changes during engineering is indispensable for the safety of the plant, workers and environment. Figure 1 indicates some sources of modifications that could impact the safe startup/operation of the system.

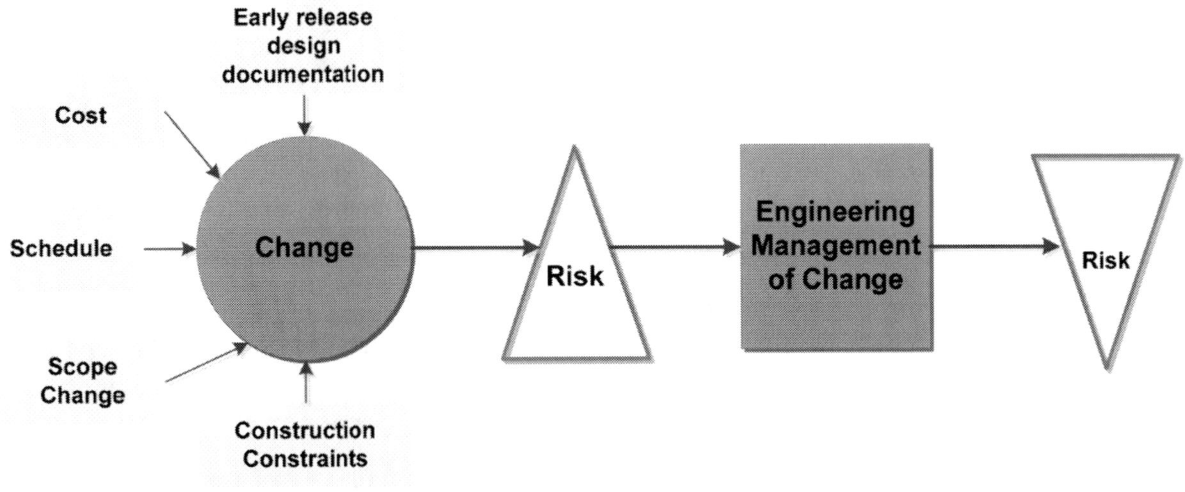

Figure 1 Integrated management of change system

Unfortunately, EMOC use is not as widespread as MOC, which still gives space for industrial accidents to happen. The main purpose of defining a work process for managing changes, even during the project development phases, is to guarantee all changes are adequately communicated, reviewed, new hazards are identified and analyzed and a mitigation plan is defined to ensure that a safe installation and operation is delivered to the customer.

3 EMOC Implementation

A change is known as a deviation from the original design. Modifications may come as scope changes, supplier updates, or other sources, and they may affect more than one engineering discipline and any on-going construction/assembly activities.

"Changes in construction projects are very common and likely to occur from different sources, by various causes, at any stage of a project, and may have considerable negative impacts on item such as costs and schedule delays." [6] If there is no formalized change management process, impacts not only on cost and schedule may be observed, but also process safety events, as mentioned on the previous section of this paper.

According to CCPS [7], the level of detail covered by the management of change system should be based on

- The hazards or risks of the process
- The existing process safety culture at the location

The company should develop a system for the control of modifications implemented on the project under design, construction, commissioning or startup phases. This work process should include approval of the change by a qualified professional or team, depending on the complexity of the hazards associated.

Based on the hazards involved in the specific process, the type of changes that need evaluation by an engineering expert from the appropriate discipline, as well as changes that do not, shall be defined when designing the specific EMOC work process for a certain company. For example; changes that can affect safety systems, critical operating parameters, shutdown set points, relief valve settings, materials of construction, etc. should undergo EMOC approval. Minor corrections to documentation or clarifications may not.

The management of change work process may not be the same in every industry and this does not mean one is better than another. The culture of the company and the location may require different applications of the MOC activities. [7]

3.1 Critical Phase

Changes during construction inherently introduce unexpected risk, especially under schedule and cost pressure. Some changes that can affect process safety may not be obvious to construction personnel. A simple yet important example is relocating a support on the discharge of a relief valve due to interference. Another may be substituting flange bolt material. Regardless of the origin of the change, evaluation and approval should take place prior to starting up the system/subsystem. Identification of all sources which can generate an engineering change must be considered as a natural part of the overall work process.

4 Proposed EMOC Work Process

Typically, during the engineering phase, several design safety meetings are conducted, such as P&ID Reviews, Layout Reviews, PHAs and FSSs. These reviews shall be conducted with the participation of all engineering disciplines related to the plant design. After these project landmark meetings, several changes may occur and their evaluation is a critical part of risk management throughout the project life cycle. At this moment, after the main safety reviews are performed, the EMOC work process is expected to begin.

As shown on Figure 2, all changes related to the project design are identified and communicated by the professional who initiated the modification, to a single point of contact within the project team.

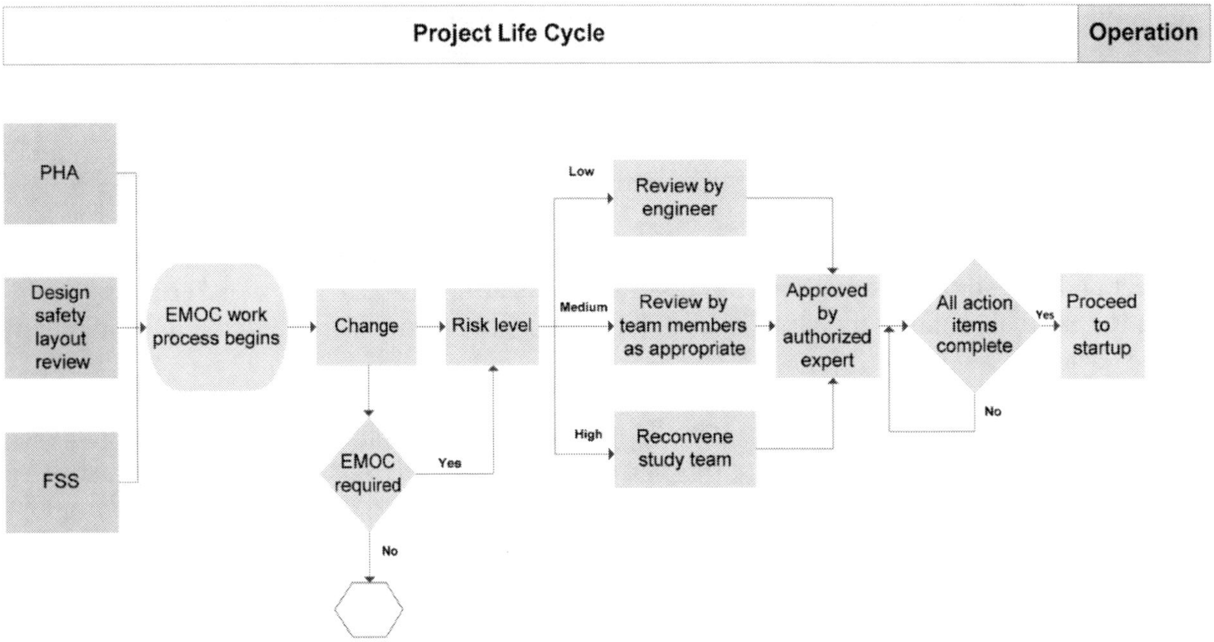

Figure 2 High-level EMOC Workflow Diagram

In cases where a modification is only a correction, clarification or replacement in a kind, there is no need to go through the full EMOC work process. The change can be recorded in a simple spreadsheet containing the type of change, a brief explanation and a mitigation plan to make sure nothing has been missed.

The modification should then be classified as low, medium or high risk per the EMOC owner and the level of review is defined based on the associated risk. The risk ranking criteria shall be described on the company's EMOC program; e.g.:

- Low: addition of a local pressure indicator on the design for maintenance purposes only.
- Medium: manual isolation valve removal, while valve not indicated at PHA as part of a hazard scenario.

- High: customer addition of a tie in point near the pipeline battery limit, which could lead to misdirected or reverse flow, with potential for process safety impact.

After reviewing the change, an approval should be required by a Subject Matter Expert (SME) before proceeding to change implementation and mitigation plan. The plant will be allowed to start-up only after all EMOC items were reviewed and approved.

It will be responsibility of the company implementing EMOC to determine which kind of changes will require a complete review, the risk categories that will be available to be assigned, both qualitatively and/or quantitatively and the required level of review according to these risks. In the example presented in Figure 2, a low risk results in a review by an engineer, a medium risk results in a review by the team members and a high-risk change will require the engineering team involved in the HazOp, Design Safety Layout review or the FSS evaluations to reconvene.

Several different approaches are possible, as long as they are simple, easy to understand by those responsible for implementation and provide a robust EMOC work process.

5 Elements of a Successful Engineering Management of Change Program

Successful EMOC programs define the key elements applicable to their industry. Some will be common such as commitment from leadership, roles and responsibilities, etc. Included below are some elements to take into consideration.

- Commitment from Leadership;
- Roles and responsibilities as they apply to specific skill sets (i.e. process engineering, controls system engineering, manufacturing, construction management, etc.);
- Clear definition of changes that must undergo engineering evaluation, including examples;
- Definition of changes that are not considered EMOC related;
- Risk level associated with the change;
- Approvers limited to subject matter experts;
- A streamlined work process focused on function over form, aligned with the various phases of the project life cycle;
- Identification of change sources;
- In process verification at predetermined milestones;
- Robust training that includes internal and external examples accompanied by a test with feasible scenarios;
- Associated PSI updated accordingly;

- Auditable;
- Evergreen
  - Periodic calibration of the work process;
  - Input from those responsible to implement the process;
  - Captures internal data to enhance understanding of what constitutes a change that requires EMOC review.
- Lessons learned captured and implemented on future projects;
- Leverage key learnings from external events in where MOC was identified as the cause or a contributing factor.

6 Conclusion

An aggressive schedule may lead to early release of design documentation that is still subject to change, which could result in the execution of PHA studies earlier than the optimal period. Although the early study may provide an opportunity to identify key risks earlier in the process, a significant number of action items and changes will probably occur, which may affect the integrity of the PHA.

A well-defined and implemented EMOC program is an essential part of risk management. The methodology applied allows systematic control of project changes which helps to ensure process, personnel and environmental safety. An additional benefit is facilitating a proactive means to analyze the underlying cause for unplanned changes. These lessons learned can assist on future projects to manage cost and improved quality control.

The main purpose of the management of change is to avoid loss of primary containment (LOPC), especially of hazardous chemicals, which could lead to catastrophic consequence. Once the EMOC process is successfully implemented, the reduction of quality issues will become a natural part of the process, resulting in a safer and higher quality facility/assembly/operation. An effective Engineering Management of Change work process is critical to the safe commissioning and startup of a chemical plant.

7 References

[1] T. Kerin. The evolution of process safety standards and legislation following landmark events – What have we Learnt?, Process Safety Progress 35 (2016), 165-170.

[2] CCPS. *Incidents that Define Process Safety*, Center for Chemical Process Safety, American Institute of Chemical Engineers, New Jersey, 2008.

[3] J.E.S Venart, Flixborough: The explosion and its aftermath. Process Safety and Environmental Protection 82 (B2): 105 – 127 (2004).

[4] C. Sadee, D.E. Samuels and T.P. O'Brien, The Characteristics of the explosion of cyclohexane at the Nypro (UK) Flixborough plant on June 1974, Journal of Occupational Accidents, 1 (1976/1977) 203 – 205.

[5] U.S. Chemical Safety and Hazard Investigation Board, Management of Change (No. 2001-04-SB) (2001).

[6] Q. Hao et all, Change Management in Construction Process, International Conference on Information Technology in Construction, Santiago, Chile, 2008

[7] CCPS. *Guidelines for Management of Change*, Center for Chemical Process Safety, American Institute of Chemical Engineers, New Jersey, 2008.

479400 Effectively Utilizing Calarp/RMP/PSM Contractors to Assist in Optimizing Process Safety Culture, Compliance with Standards, and Workforce Involvement

Abstract:

Since the inception of Occupational Safety and Health Administration (OSHA) Process Safety Management (PSM) in 1992, facilities have frequently contracted process safety consultants to satisfy regulatory needs, as well as assist in complying with industry best practices.

The primary objective of this paper is to provide tangible suggestions for effectively utilizing process safety contractors for assisting in optimizing process safety culture, compliance with regulatory requirements and industry best practices, as well as tips to improve workforce involvement and coordination. The core of this paper will detail various management styles of the prevention program and an evaluation of influences on process safety culture based on varying degrees of involvement from the facility and usage of contractors. This paper aims to provide an in-depth analysis of the pros and cons of contractor involvement and its efficacy relative to the scale of the facility and intricacy of corporate structure based on anecdotal evidence and industry experience.

Key topics include:

- Overview of different types of process safety culture in relation to prevention program management and contractor usage,
- The Challenge - Striking an optimal balance between contractor use and facility involvement whilst maintaining economic viability,
- The Challenge for Multi-Facility Corporations - Ensuring changes in corporate standards are successfully communicated to all children facilities ranging from documentation updates and organizational changes,
- Suggestions and Improvements - Implementation of effective prevention program management systems to ensure workforce involvement and accountability, as well as practical allocation of contractor resources to accomplish critical tasks whilst balancing facility resource commitment.

A Streamlined Approach for Full Compliance with SIF Implementation Standards

William G. Bridges, President
PROCESS IMPROVEMENT INSTITUTE, INC. (PII)
1321 Waterside Lane, Knoxville, TN 37922
Phone: (865) 675-3458
Fax: (865) 622-6800
e-mail: wbridges@piii.com

Art Dowell III, Principal Engineer
PROCESS IMPROVEMENT INSTITUTE, INC. (PII)
2437 Bay Area Blvd PMB 260
Houston, TX 77058
Phone: 713-865-6135
e-mail: adowell@piii.com

Prepared for Presentation at
13th Global Congress on Process Safety
San Antonio, TX
March 27-29, 2017

UNPUBLISHED

13TH GLOBAL CONGRESS ON PROCESS SAFETY

A Streamlined Approach for Full Compliance with SIL Implementation Standards

William G. Bridges, President
PROCESS IMPROVEMENT INSTITUTE, INC. (PII)
e-mail: wbridges@piii.com

Art Dowell III, Principal Engineer
PROCESS IMPROVEMENT INSTITUTE, INC. (PII)
e-mail: adowell@piii.com

Keywords: Safety Instrumented System, SIS, Safety Integrity Level, SIL, LOPA, Layer of Protection Analysis, Independent Protection Layers, IPLs, human factors

Abstract

Many companies put FAR too much redundant effort into determining what SIL (safety integrity level) is needed and then verifying the SIF (safety instrumented function) design will give the SIL targeted. This paper shows how to apply the qualitative definition of independent protection layers (IPLs) within the setting of a process hazard analysis (PHA) to get most of the gain from Layer of Protection Analysis (*LOPA)* without doing a LOPA (without using numerical values). We show how we use a PHA team to identify when a SIF is needed and to select the proper target SIL. This portion of the SIL evaluation and the identification and labeling of the IPLs during the PHA/HAZOP does not take any longer than a normal PHA/HAZOP, once the right habits are established. Note that this approach eliminates the need for a separate SIL Evaluation Study to identify the SIFs and select the target SIL. Then, this paper describes how to perform the SIL Verification and Safety Requirements Specification (SRS) remotely, again without the need for a redundant team meeting. This approach has been used at many sites and for thousands of SIFs.

Background

Typically, hazardous consequences, their causes, and safeguards are identified in a PHA/HAZOP (process hazard analysis / hazard and operability analysis)[1]. The PHA may determine if additional protection layers are required or if additional analysis is required.[2]

In many organizations, a separate LOPA (layer of protection analysis)[3] meeting is held to determine which HAZOP safeguards are independent protection layers (IPLs) and to determine if additional IPLs are needed to meet the risk tolerance criteria. If SIFs are needed, the required risk reduction (PFD, probability of failure on demand) is determined in LOPA, which then sets the SIL.

Some organizations may even have a separate SIL evaluation study, which is certainly redundant to a well-run PHA and is wasteful if the PHA is properly staffed and managed and/or if any subsequent LOPA are performed well.

Fundamentals of SIL Assessment

The lifecycle requirements for safety instrumented functions are specified in industry standards IEC 61511 (ANSI/ISA 84.00.01-2004, Part 1)[4,5]. The required steps to specify the SIS (safety instrumented system) are:

1. Safety Instrumented Function (SIF) identification.
2. Determining the Safety Integrity Level (SIL) for each SIF.
3. Designing the SIF to meet the required SIL.
   SIL Verification Calculation (actually, this calculation is iterative with step 3, but the end calculation is a deliverable that proves 3 is correct). The SIL standards allow the following methods to use to determine the required SIL (ANSI/ISA 84.00.01-2004, Part 3, Section 3.8)[4]:
   - Fault Tree Analysis (FTA), Full Quantitative Risk Analysis (QRA)
   - LOPA – simplified quantitative analysis
   - Qualitatively within PHA (HAZOP, etc.)

     "A qualitative method may be used as a first pass to determine the required SIL of all SIFs. Those which are assigned a SIL 3 or 4 by this method should then be considered in greater detail using a quantitative method to gain a more rigorous understanding of their required safety integrity." – *ANSI/ISA 84.00.01-2004, Part 3, Section 3.8*[4]

Table 1 compares risk analysis approaches for Steps 1 and 2.

FTA and QRA are very detailed and labor-intensive and would be required for complicated systems, particularly where there is not complete independence between safeguards.

LOPA is a simplified analysis tool that requires fewer resources, but is heavily dependent on the requirement that all the safeguards be completely independent.

However, for many processes, there are a number of scenarios (cause leading to a consequence) where the PHA team can readily determine sufficient independent protection layers are already in place.

Table 1: Comparison of Risk Analysis Approaches

Approach	Assessment Methods	Risk Judgment Method	Risk Judgment Method	Estimated Range of the Results
Qualitative Only	HAZOP, FEMA	Expert Voting, focusing on site data	Capable >95% of time	Plus or minus 1/2 order of magnitude
Simplified Quantitative	LOPA, Risk Graph	Multiplication of statistical averages of general failure rate data; with broad assumptions on management systems	Needed on about 5% of the scenarios	Plus or minus 1 order of magnitude
Full Quantitative	FTA, ETA, QRA, HRA		Needed for less than 0.01% of the scenarios	Plus or minus 1 order of magnitude

Figure 1 contrasts the typical approach to SIL assessments with the optimized streamlined approach.

Typical SIL Determination versus Streamlined Approach

In the typical approach,

- **Step 1**: The PHA/HAZOP is performed using a multidisciplinary team of engineers, operators, instrumentation staff, maintenance staff, and perhaps vendors. Then in a separate SIL study, a multidisciplinary team of engineers, instrumentation staff, maintenance staff, and perhaps vendors brainstorm the hazard scenarios (again) and decide which are candidates for SIFs. A semi quantitative approach, such as RiskGraph, or layer of protection analysis, determines if an SIF is required.
- **Step 2:** The RiskGraph dictates the required SIL, or the LOPA compares the mitigated risk to the risk tolerance criteria and determines the SIL from the order-of-magnitude risk reduction needed.

In contrast, in the streamlined approach,

- **Step 1:** The PHA/HAZOP is performed using a multidisciplinary team of engineers, operators, instrumentation staff, maintenance staff, and perhaps vendors (to this point the same as the typical approach). Within the PHA/HAZOP, the team estimates the risk of each scenario using qualitative judgment (based on experience, knowledge, and memory of site-specific data). The PHA/HAZOP team then qualitatively determines if the best risk reduction method is an SIF.
- **Step 2:** The PHA/HAZOP team qualitatively determines the SIL needed (typically by judging the level of redundancy needed based on the value they place on each IPL and the new or modified SIF needed)

Figure 1: Comparison of the Typical and the Optimized Streamlined Approach from PII for SIL Assessments (Steps 1 and 2)

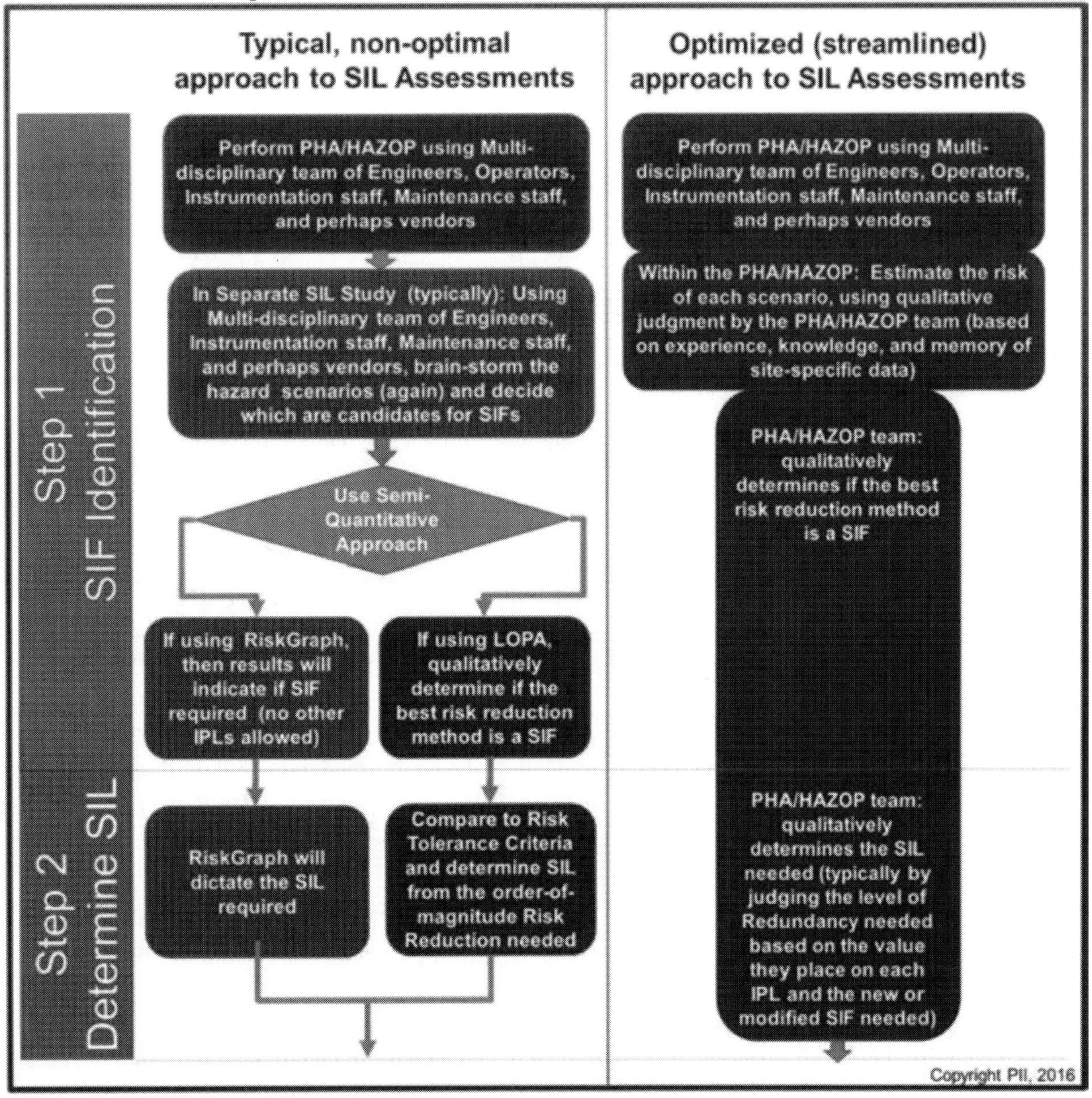

Figure 2 also contrasts the typical approach to SIF design and verification with the streamlined approach. Steps 3 and 4 are iterative.

In the typical approach,

- **Steps 3 and 4**:
    - Decide on basic architecture of SIF, from experience with other SIFs.
    - Decide on vendors of components
    - Provide full design of SIFs from experience

- Perform trial SIL verification calculations using software such as exSILentia™ using actual component data
- If the target SIL is achieved, issue the SIL verification calculation report and the safety requirement specification (SRS)
- If the target SIL is not achieved, revise the design and recalculate.

Figure 2: Comparison of the Typical and the Optimized Streamlined Approach from PII for SIL Assessments (Steps 3 and 4)

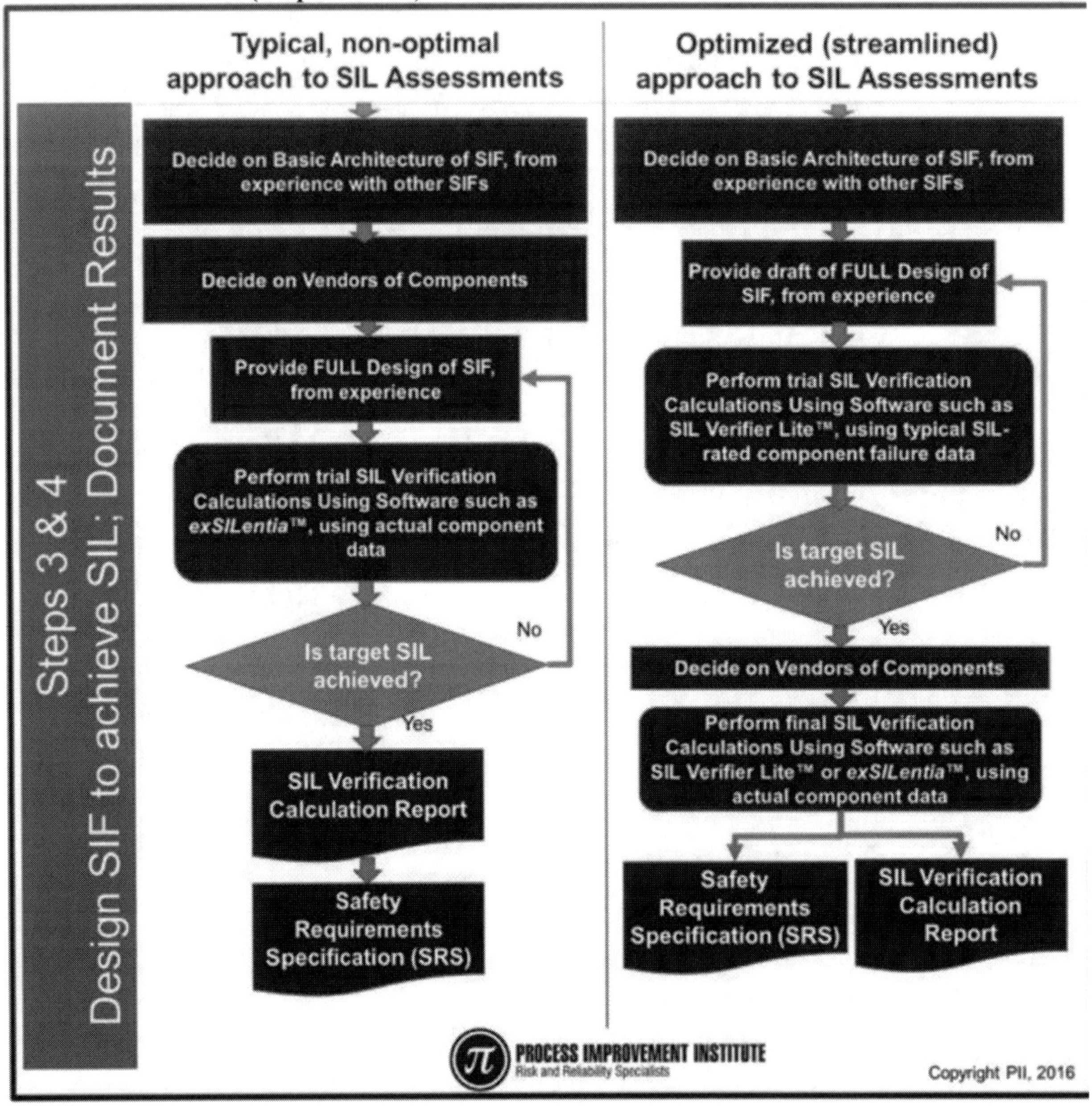

In the streamlined approach,

- **Steps 3 and 4**:
  - Decide on basic architecture of SIF, from experience with other SIFs.

- Provide draft of full design of SIF, from experience.
- Perform trial SIL verification calculations using software such as SIL Verifier Lite™ using typical SIL-rated component failure data.
- If the target SIL is not achieved, revise the design and recalculate.
- If the target SIL is achieved,
  - Issue the SIL verification calculation report and the safety requirement specification (SRS).
  - Decide on vendors of components.
  - Perform final SIL Verification Calculations Using Software such as *SIL Verifier Lite*™ (from PII) or *exSILentia*™ (from exida) using actual component data.
  - Issue the SIL verification calculation report and the safety requirement specification (SRS).

The advantage of the streamlined approach is determination of the need for an SIF and of its SIL quicker and with fewer resources than the typical approach.

Requirements for the Qualitative Streamlined Approach for SIL Determination

For the qualitative streamlined approach to be used effectively, there are important requirements for the PHA/HAZOP team.

- PHA/HAZOP team leader is knowledgeable and proficient in LOPA.
- PHA/HAZOP team members have the following information in memory:
  - Consequence severity.
  - Initiating cause frequency.
  - Independent protection layer (IPL capability).
  - Risk tolerance criteria.

An experienced PHA team, who is knowledgeable in the process technology, can effectively and efficiently analyze a scenario and mentally perform a layer of protection analysis. The team determines the consequence severity and the initiating cause frequency. A highly competent team understands which safeguards are truly independent protection layers and can judge the risk reduction provided by each. Then the team can determine if additional IPLs are required to meet the risk tolerance criteria. For some scenarios, an SIF may be required and the team can determine the required probability of failure on demand for the SIF and can thus determine the required SIL.

A best practice is that the PHA team recommends the functional description and the SIL for the proposed SIF. The functional description does not specify the details of the SIF design, rather it specifies what process variables should be considered and what action should be taken to detect the upset and to prevent the consequence. For example, the functional description may say "monitor the level of a vessel and stop the flow into the vessel to prevent overflow".

If the PHA team is uncertain, or if the scenario is complex, then the PHA can recommend a detailed LOPA be done for the scenario.

These conclusions are based on more than 8000 PHA/HAZOPs, 3000 LOPAs, and about 100 QRAs and HRAs. We have observed PHA participants becoming knowledgeable about LOPA and IPL requirements. The participants then began to apply the principles of LOPA spontaneously in the PHA. With good leadership, the PHA team could then easily move to the full streamlined approach.

The experience and competency of the PHA leader and the PHA team is extremely important. If the PHA team is not comfortable with the principles of LOPA, and especially the principles of IPLs, the streamlined approach is not appropriate. Likewise, the PHA team must have the required information and understanding in its memory. Otherwise, the qualitative approach can degenerate into the "arm wrestling" arguments that PHA teams used qualitatively to determine SIL requirements before the introduction of LOPA.

Suggestions for the Streamlined Approach for SIF Design

Likewise, there is a streamlined approach for the SIF design. Organizations have developed a recipe book of typical SIF designs for SIL 1, SIL 2, and SIL 3. Depending on the required SIL, the designer can choose a design from the recipe book (as described in Figure 2, "from experience"). The recipe book provides a starting point for the detailed SIF design. Figure 3 shows a small sample of SIF designs that could be included in a recipe book.

Figure 3: Typical SIF Architectures for Information

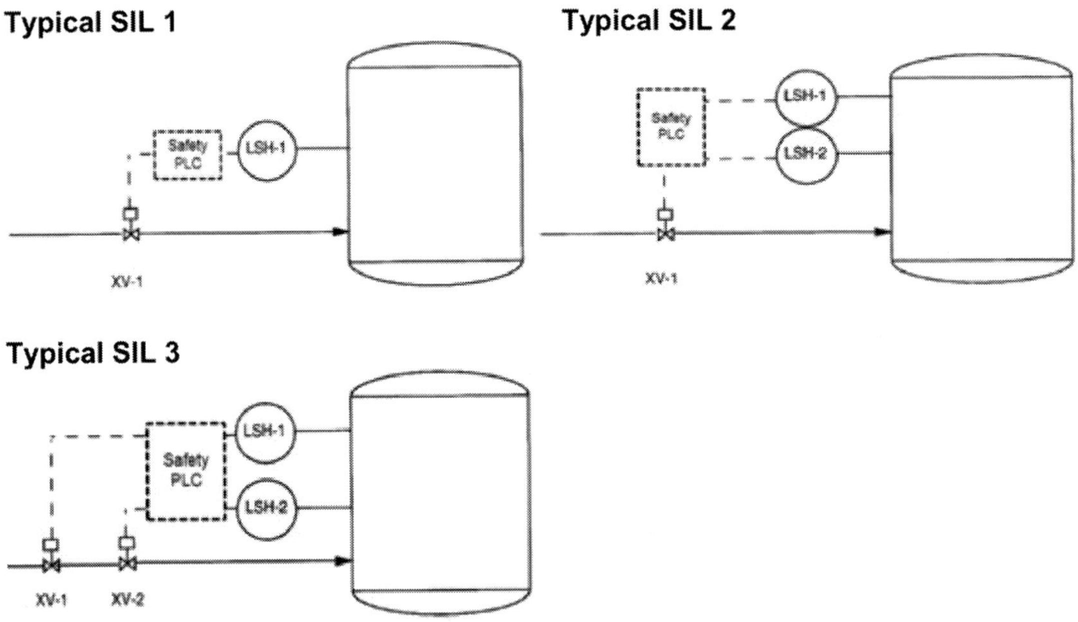

Case Study 1

An example of the streamlined approach is summarized in Table 2 for a facility with storage spheres (in China; the PHA and other analyses were performed in 2014). The table shows two deviations from the PHA. IPLs are identified and existing SIFs are assigned an SIL, based on the architecture. (Note that the achieved SIL for each of the SIFs (existing or recommended) must be confirmed by a SIL verification calculation that includes the failure rate of the components, the voting architecture, and the proof test interval.)

For the high-level deviation, the high-level SIF was assigned SIL 1. The overflow through the pressure equalization line to other spheres was determined to be an IPL.

For the low level deviation, the low level indication and low level alarm were determined not to be an IPL because they are part of the initiating cause. The PHA team concluded that feeding from two spheres at all times is an IPL for this facility because of the unlikelihood for both spheres to have low level at the same time. The PHA team also thought that an additional SIF level alarm with operator response to switch tanks within 60 minutes could be a possible IPL, if the action of the operator is quick enough. The team wrote a recommendation to ensure that the operator response could be an IPL.

In this case study, the PHA team was able to determine in the PHA meeting which safeguards were IPLs. They also determined the nominal SIL for existing SIFs (to be confirmed in a later SIL verification). And the team determined that operator response could be an IPL if there was a documented procedure and an annual drill. For these two deviations, the PHA team did the work that would be done in a later LOPA meeting in the typical approach.

Next, the SIL Verification was performed for the SIL 1 SIF (and all other SIFs in the unit). This SIL Verification was accomplished using exSILentia™ software and the analysis and documentation took less than 2 hours.

Thus, the entire requirements for a SIL identification, SIL target level determination, and SIL Verification were accomplished very quickly compared to typical approaches.

Table 2: Storage Sphere PHA

No.: 2 XXXX storage spheres xxx-T-XX A/B/C/D/E/F/G/H/I/J/K/L (1 of 12)					
#	**Dev.**	**Causes**	**Consequences**	**Safeguards**	**Recommendations**
2.1	High level	Too much flow to one sphere from XX Plant (through their pump; about 40 bar MDH)	High pressure (see 2.5)	High level SIF with level sensors voted 2oo2, to close inlet valve - **SIL 1** Overflow thru pressure equalization line to other spheres (through normally open [NO] valve) - **IPL**	
2.2	Low level	Failing to switch from the sphere with low level in time (based on level indication)	Low/no flow - Liquid from spheres through high pressure product pumps to the vaporizer (see 4.2) Low/no flow - Unqualified liquid from spheres back to Plant (see 6.2)	Level indication and low level alarm, inspected each year, per government regulation **(not IPL; part of the cause)** 9 other spheres with possibly enough level to switch to Feeding from two spheres at all times, so unlikely for BOTH spheres to have low level at the same time - **IPL** Two level indication from SIS level transmitter, with low level alarm, with more than 60 min available to switch tanks (SIF driven alarm and response) - **possible IPL, if action of the operator is quick enough**	Rec 4. Make sure the Human IPL of response to low level in all spheres and tanks is described in a trouble-shooting guide (like an SOP) and practiced once per year per unit operator. This will make this response a valid IPL.

Case Study 2

In 2014 and 2015, an identical approach to that of Case Study 1 (the streamlined approach of this paper) was performed for a large gas producer in South America. In that work, the PHAs for more than 500 nodes plus all procedures of all operating modes was completed in about 10 weeks of PHA meetings. The PHA team identified more than 800 SIFs, along with their target SILs, using the approach in Steps 1 and 2.

Next, staff within PII but separate from the PHA team performed about 60 LOPA to clarify certain issues, some related to SIFs. But less than 5% of the SIFs were established with LOPA; 95% were established with the PHA alone.

Finally, staff separate from the PHA performed SIL Verification of the 800+ SIFs, using exSILentia™. The verification was for a combination of SIL 1 and SIL 2 SIFs; a couple of SIL 3 SIFs were also analyzed and debated. The relatively few LOPAs took about 2-3 hours each (due to the complexity of the 5% of scenarios that went to LOPA) and each SIL Verification took about 2-3 hours to perform and document.

Caution – the issue of human factors for SIL 2 and SIL 3 SIFs[6]

There is an important aspect of important aspect of SIFs and the possible achieved SIL that is not addressed in this paper. Relying on a high integrity SIF, without accounting for human error during interventions, is also a waste of resources. You may believe that you have reduced the risk by a factor of 100 for a SIL 2 SIF, or a factor of 1000 for a SIL 3 SIF, when in fact the human factors during commissioning and proof testing may degrade the overall risk reduction factor to about 10. If the human factor is accounted for in the SIF design phase and the SIL verification phase, then the human error can also be prevented to allow the higher reduction in risk that is targeted.

Conclusions

Based on experience in a multitude of PHAs, LOPAs, QRAs, and HRAs, we have seen that experienced, competent PHA teams with an experienced, competent leader can effectively apply the principles of LOPA to many scenarios during the PHA meeting. The team can determine the approximate SIL of existing SIFs, the team can determine which safeguards are truly IPLs, and the team can recommend additional IPLs, including SIFs with specific SILs. If the scenario is too complex, or the PHA team is uncertain, the team can recommend a detailed LOPA. The streamlined qualitative makes better use of resources and time than the typical approach with separate meetings for PHA and LOPA.

It is critical that the PHA leader is experienced and competent in LOPA. It is critical that the PHA team has in its memory the data for consequence severity, initiating cause frequency, IPL requirements, and the risk tolerance criteria.

Acronyms Used

AIChE– American Institute of Chemical Engineers
CCPS – Center for Chemical Process Safety (an AIChE technology alliance)
HAZOP – Hazard and Operability Analysis
HRA – Human Reliability Analysis
IEC – International Electrotechnical Commission
IEF – Initiating Event Frequency
ETA – Event Tree Analysis
FMEA – Failure Mode Effect Analysis
FTA – Fault Tree Analysis
HRA – Human Reliability Analysis
IPL - Independent Protection Layer
LOPA – Layer of Protection Analysis
PHA – Process Hazard Analysis
PII – Process Improvement Institute, Inc.
QRA – Quantitative Risk Analysis
SIF – Safety Instrumented Function
SIL – Safety Integrity Level
SIS – Safety Instrumented System
SRS – Safety Requirements Specification

References

1. *Guidelines for Hazard Evaluation Procedures*, 2008, CCPS/AIChE.
2. Bridges, W., & Dowell, A., "Identify SIF and Specify Necessary SIL, and other IPLs, as part of PHA/HAZOP," *12th Global Congress on Process Safety*, AIChE, 2015.
3. *Layer of Protection Analysis: Simplified Process Risk Assessment,* 2001, CCPS/AIChE.
4. *Functional Safety: Safety Instrumented Systems for the Process Industry Sector - Part 3: Guidance for the Determination of the Required Safety Integrity Levels*; ANSI/ISA 84.00.01 Part 3, 2004.
5. IEC 61511, *Functional Safety: Safety Instrumented Systems for the Process Industry Sector - Part 1: Framework, Definitions, System, Hardware and Software Requirements*, International Electrotechnical Commission
6. Bridges, W., and Thomas, H., "Accounting for Human Error Probability in SIL Verification Calculations," *8th Global Congress on Process Safety,* Houston, AIChE, April 2012.

479509 The Art of the Audit: Being Audit Ready, Everyday

Abstract:

Have you ever made an appointment for an annual physical and suddenly join a gym? We all know we should be exercising every day, but it becomes more important before going in for that checkup. In reality, not exercising has consequences to your health and not your doctors, so why do we do this? To impress them, instead of doing it to take care of ourselves? This phenomena of doing things you do not normally do in preparation for an event is unfortunately how too many of us treat audits. We hear an auditor is coming and end up running around trying to organize paperwork, convince the operators to clean up the plant, and everyone in management becomes much more attentive. Does this behavior really set the proper tone to develop a safety culture? Auditors are there to check up on our compliance of rules that benefit only us to follow, particularly in the field of safety.

A successful operation will be 'Audit Ready' every day. 'Audit ready' means that operations are in compliance with rules, personnel are safe and exhibit safe behaviors, and all of the team members understand why they perform the functions of their job. Compliance tools, personnel awareness and routine self-audits are three key topics in becoming ready for an audit every day.

Personnel awareness is among the most vital areas to ensuring safe and effective operations, as well as successful audits. Audits involve people from across almost all departments, and as such mirror the requirements of multifunctional teams in everyday operations.

Compliance tools are important to track work that is being completed, and create convenient locations to store data which proves compliance. This can include tracking the progress of MOCs through weekly notifications, using databases to track when procedures need to be reviewed, and more. Visibility of what needs to be accomplished is important.

Routine self-audits help us to find gaps in compliance, and reduce complacency. These can include a simple spot check of one area of compliance, or be more comprehensive, and there are pros and cons to both. A spot check keeps everyone on their toes and takes a short enough time such that they can be done more frequently.

Ensuring sustainable records and checking everyday habits will go a long way to creating a safe operating environment, and with that a workforce that is audit ready, every day.

479546 Economic Justification for Better Maintenance in a Safety System

Abstract:

It is not difficult to think of numerous ways in which human factors might impact functional safety. For example, imperfect repair, improper calibration, faulty installation, etc. can all contribute to decreases in functional safety performance. However, if actions are to be taken to mitigate these impacts then it is essential to have some quantitative measure of how this impacts the PFDavg, so that the actions' effectiveness can be assessed. We will show how a simple measure system called the Site Safety Index™ (SSI) is used to adjust safety metrics, including the PFDavg and the MTTFS, on a site specific basis. The lifecycle economic impact of these metrics is then used to show how improved operations and maintenance can be justified. Economic justification for better maintenance can be easily verified through this approach being further mindful of the safety culture.

479547 How Important Is Realistic Failure Rates for Safety Instrumented Functions?

Abstract:

A given Safety Instrumented Function design must be verified to meet the required SIL level by calculating probability of failure on demand. However, this does require realistic failure rate data for each device. Fortunately, many device manufacturers are getting their products certified per IEC 61508 and many of the certificates now have failure rate data. Other manufacturers supply analysis reports with failure rate data. But how good is this data supplied by the manufacturers? All such data must be checked to make certain it is applicable for the process industries as some data may have been derived based on machine safety or automotive applications.

This paper will show some examples of manufacturer provided failure rate data and a methodology created to validate device type failure rate data for the process industries.

479917 Enabling Sustainable Relief System Management Programs through Independent Reviews

Abstract:

Overpressure protection and flare systems usually represent the last line of protection for personnel and the environment in industrial facilities operating under high temperatures and pressures. During the design phase of either Greenfield or Brownfield projects, assessing the basis for major relief systems and making clear recommendations to reduce the risks associated with each system has helped companies realize substantial safety and economic benefits. Independent and safety focused reviews, with an emphasis on value added solutions has enabled organizations to set the right foundation for a sustainable relief system management program. This, in turn, has helped enable the facilities in meeting the ever-changing regulatory needs in a timely manner.

This paper will give a typical breakdown of the workflow process for each of the entities involved in the project - the client, the contractor, and the independent reviewer. Four distinct work activities are typically identified for such a review - development of preliminary overpressure protection, engineering design of relief systems, validation of the design, and validation of as-built documentation. After each stage, the output from the independent analysis is compared with the contractor's work and updates are performed before continuing to the next stage. This paper will also give an overview of these stages and discuss the various steps and value added for each stage of the project.

Jack Rabbit II: Experimental Large-Scale Chlorine Release Trials

Mark Whitmire[1], Shannon Fox[2], Thomas Spicer[3], Leo Stockham[4], Steven Hanna[5], Joseph Chang[6], Michael D. Sohn[7], Thomas Mazzola[8], Ronald Meris[9], Damon Nicholson[10], Andy Byrnes[11]

[1]Noblis, Falls Church, VA, USA; [2]DHS S&T Chemical Security and Analysis Center, Aberdeen Proving Ground, MD, USA; [3]University of Arkansas, Fayetteville, AR, USA; [4] Leidos, Albuquerque, NM, USA; [5]Hanna Consultants, Kennebunkport, ME, USA; [6]ANSER, Falls Church VA, USA; [7]Lawrence Berkeley National Laboratory, Berkeley CA, USA; [8] Engility, Lorton, VA, USA; [9]DTRA, Fort Belvoir, VA, USA; [10]Dugway Proving Ground, UT, USA; [11]Utah Valley University, Provo, UT USA

INTRODUCTION

Chlorine is a potent toxic inhalation hazard (TIH) chemical, and each year millions of tons of it are transported in bulk via road, water, and rail. Consequently, an accidental or deliberate chlorine release near a populated area may cause a considerable number of fatalities, and would create other health, safety, and environmental problems. Hazard prediction models are inconsistent with the evidence and observations from chlorine disasters, and the understanding of such releases is otherwise insufficient to guide and inform an effective emergency response. So beginning in 2010, DHS and its Health, Safety and Environment (HSE) partners and stakeholders planned the Jack Rabbit (JR) Program. The program was led by the DHS Chemical Security Analysis Center (CSAC) located at Aberdeen Proving Grounds-Edgewood Area MD.

OVERVIEW

The Jack Rabbit (JR) program is comprised of a series of projects that includes JRI (2010), JRII Phase 1 (2015), and JRII Phase 2 (2016). The JRI (Fox and Storwold, 2011) project was completed to develop, test, and evaluate a mechanism for the controlled, rapid release of liquefied, pressurized gases (specifically, ammonia and chlorine), to observe the transport and dispersion of a disseminated vapor and aerosol cloud, and to investigate chemical deposition and reactions with water and soil. The project also evaluated instruments, test methods, and strategies that may be used to study more industrial scale tests. Using either chlorine or ammonia, two 'pilot' tests and eight 'record' tests of either 1 or 2 tons were completed. For chlorine, among other findings, the results demonstrated that downwind transport and turbulent mixing are initially resisted by a dense persistent gas and aerosol cloud, that source phenomena were non-linear with increasing release volumes, and that the reactivity with soil containing water and organic matter is an important removal mechanism.

JACK RABBIT II

JRII leveraged and extended the work of JRI with controlled chlorine field experiments on a scale corresponding to railcars and tank-trucks; that is, 5 to 20 tons each. The goals were to collect data on the release source, measure chemical reactions with the environment, observe and evaluate exposure effects on equipment and infrastructure, and assess urban impacts using a Mock Urban Environment (MUE) or an unobstructed dissemination and downwind dispersion environment, both located on the Urban Test Grid (UTG) at Dugway Proving Grounds UT. Data was collected in real time via ground based and unmanned aerial vehicle systems, video cameras, and spectroscopic instruments, in situ point detectors, path integrated detectors, and concentration and dosage determination instruments. The data was thoroughly and qualitatively reviewed and shared with all project participants in prescribed and common formats. The generated data and findings will improve the understanding of and operational hazard prediction modelling for TIH releases, and lead to more effective emergency response and training, national preparedness, and mitigation strategies.

URBAN TEST GRID (UTG)

The UTG (Nicholson, Hedrick, and Lian, N; 2017) was the designation for the JRII chlorine release site surveyed and built specifically for the JRII tests; it was at Dugway Proving Grounds west of Granite Mountain on Goodyear Road. It included the playa and adjacent support areas, concrete pad, gravel pad, and surveyed grid extending 11 km downrange and aligned roughly south to north.

The UTG included a circular, 0.5 ft. deep, 82 ft. diameter, rebar reinforced concrete disseminator vessel pad with 3 embedded temperature sensor stations, and a 1 inch steel lip to contain chlorine liquid. The center of the circular pad was the center of the UTG site, Northing 4445633.945, Easting 288109.182 (Latitude, 40.134347; Longitude, -113.487076). The concrete pad rested on a 24 inch deep, 400 ft. × 600 ft. gravel pad level with the concrete pad and symmetric about it.

Figure 1. Chlorine release from the disseminator on the concrete pad, with CONEX container array on the gravel pad of MUE, looking upwind (south) from atop the 2 x 3 CONEX container stack. The dense cool cloud remains low as it moves downwind and laterally around and through the Mock Urban Environment made from CONEX containers.

Winds predominate from approximately 165° from June to September, so the UTG was oriented about a centerline from that direction. The UTG included the playa within a 90° angle bisected by the UTG centerline with vertex at the disseminator vessel location, and extending to the 11 km arc. The "Near Field" was the UTG region from the disseminator up to and including the 2 km arc and inside the 90°

angle; the "Extended Field" was the UTG region past the 2 km arc to the 11 km arc and inside the 90° angle.

The Mock Urban Environment (MUE) was constructed on the gravel pad and surrounding the concrete disseminator pad using a staggered row array of 86 CONEX containers (CONtainer EXpress shipping containers, either 18 or 40 ft. long) to observe and study the transport and dispersion of dense gas around and above urban structures. The array included a two wide by three high (2 x 3) stack of six 20 ft.-CONEX containers located directly downwind of the source, and two 2-axle trailers nearby with insulation, dry wall, and trailer skirting for infiltration studies. Aside from the CONEX containers and trailers, up to seven vehicles were included in the array. Three fire trucks, one ambulance, and three cars were located on the north edge of the gravel pad during Trials 1-3. During Trial 4, four vehicles were placed on the playa past the 250 m arc; during trial 5, three vehicles were moved between the rows of CONEX containers but adjacent to the concrete disseminator pad, and three vehicles remained on the playa past the 250 m arc.

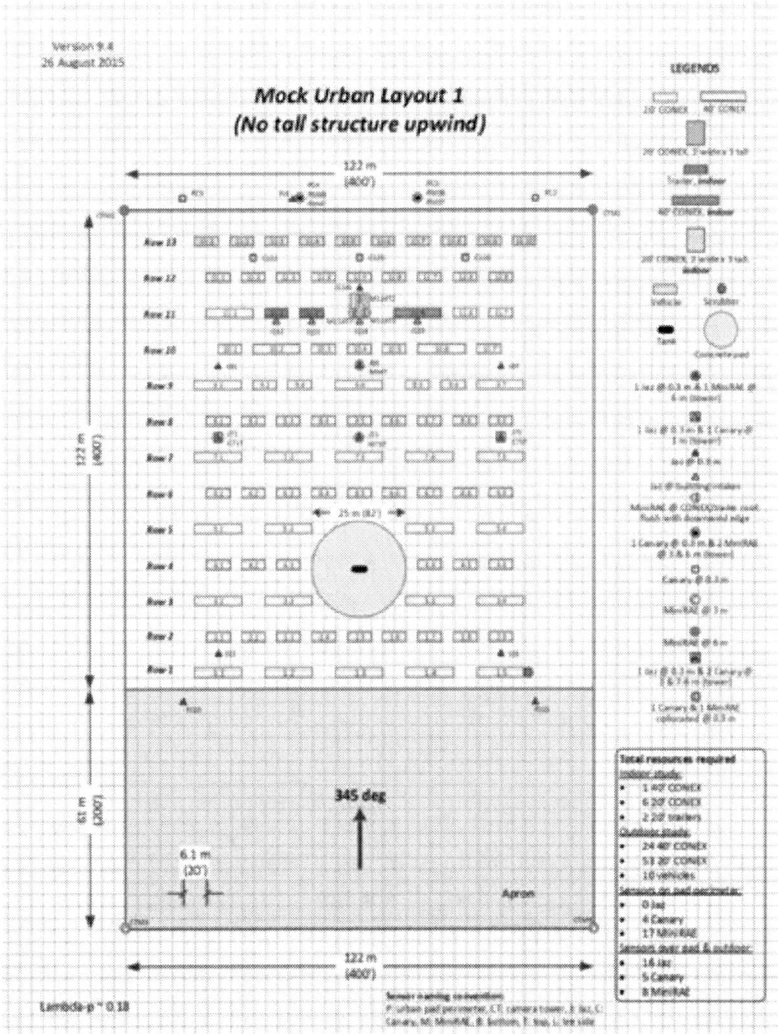

Figure 2 Mock Urban Environment (MUE) schematic showing gravel pad around circular pad for Chlorine disseminator; CONEX containers and instrument locations are also shown.

Instrument locations for the UTG "Near Field" and "Extended Field" were surveyed, recorded, and reported. Point sensors included about 170 MiniRAE 3000s equipped with volatile organic compound

(VOC) photoionization detectors; about 60 ToxiRAE Pro units; a Gasmet™ DX-4000 (FTIR) gas analyzer; 28 Ultraviolet (UV) Canary units for UTG monitoring and indoor studies; and 16 Ocean Optics Jaz instruments for UTG monitoring and indoor studies. Standoff Detectors included two Cerex Monitoring Solutions UV Sentry chlorine-specific integrated path detector, three Ultraviolet (UV) Light Detection and Ranging (LiDAR) systems for monitoring the moving chlorine gas cloud, and elastic backscatter LiDAR systems for monitoring the moving aerosol cloud. A GRIMM Aerosol portable laser aerosol spectrometer and dust monitor was also used for particle sizing model using laser scattering techniques. Other equipment included remotely operated high-definition (HD), high-speed (HS), and FLIR® IR Camera SC660 micro-bolometer cameras positioned throughout the test grid to visualize and record the chlorine release and resulting plume.

Meteorology equipment included 34 Portable Weather Information Display Systems (PWIDS) collecting data at 2 m; three 32 m towers each equipped with sonic anemometers collecting data at heights between 2 and 32 m; and 31 3D ultrasonic anemometer to gather wind and turbulence statistics, and fluxes of heat and momentum. Energy Balance Station (EBS) provided in situ measurements of the surface vertical fluxes of sensible heat, latent heat, net radiation, and soil surface heat flux. The MiniSODAR™ high-frequency Doppler acoustic sounding system designed to measure the atmospheric vertical wind profile from 15 to 200 m. Numerical gridded forecast meteorology using embedded observations and real-time meteorological data from the test grid were used for HPAC modelling leading up to each trial and to determine "GO/NO GO" test status.

DOWNWIND DISPERSION AND TRANSPORT ENVIRONMENT

For tests in 2016, and other than two vehicles, a CONEX container, and a modular office trailer used for follow-up infiltration experiments, the UTG was cleared of the simulated buildings and obstructions used in 2015. The project goals were to monitor and collect the cloud concentration data from the initial phase of a very dense, two-phase cloud near the source to dispersion of the vapor cloud farther downwind and up to 11 km away using point detectors and stand-off spectroscopic instruments.

DISSEMINATION

The disseminator vessel for use as the chlorine release source during the tests was located at the center of the concrete pad with its major axis perpendicular to the UTG centerline. The 2000 gal. tank was mounted in a 3-dimensional dynamic mass measurement load cell system. Other vessel mounted instrumentation included a redundant system for determining the remaining mass of chlorine in the vessel, differential pressure sensors on the vertical axis at various depths in and below the vapor space, guided wave radar liquid level measurement, and instrumentation for absolute pressure and direct contact temperature profiles inside the vessel. At 80-90% operating capacity, the vessel could contain from 8.5 to 10 tons chlorine.

Each release was accomplished using a blind flange with exploding bolts rated to safely hold the pressure of a vessel loaded with liquid chlorine at ambient temperature. To blow the flange and release the chlorine, the bolts were remotely detonated; when the bolts broke, the blind flange was blown off with the aid of the chlorine weight and head pressure, and chlorine flowed through the vessel's circular opening. A quick release valve was not used to simulate the release because of the likelihood of internal flashing flow in the piping and valves; that is, the mass release rate would be less than what would be observed from a breach of containment. In 2015, every release was directly down (180°) through a six inch hole. In 2016, each release was either directly down (Test 6); at 135° and downwind (Test 7), or directly up (Test 8). After Tests 7 and 8 any liquid chlorine remaining in the vessel was discharged via a remotely operated valve in the bottom of the tank. Before inert gas purging and chlorine refill operations, the ejected blind flange was replaced using standard bolts. The last release was from a tanker truck that had been removed from service, and containing up 19 t chlorine. The tanker truck's skin was breached with a remotely activated explosive shape charge.

Figure 3. Test Release 6; 10 Tons chlorine on August 31, 2016. A moment after the release the liquid chlorine may be seen splashing onto the concrete pad. As the chlorine aerosolizes and vaporizes it cools the surroundings and condenses water vapor.

OBSERVATIONS AND FINDINGS

From the JRII trials the program observed that:

- Upon breaching the tank, liquid chlorine exits the tank and rapidly flashes to a two phase aerosol and vapor cloud with a liquid layer deposited on the pad, ultimately all vaporizing and transporting downwind.
- The escaping chlorine was found to auto-refrigerate to its -34 degree Celsius boiling point and lower due to further evaporative cooling outside the tank.
- The force of the escaping chlorine jet generated strong initial turbulence and momentum to drive the expanding cloud over nearby Conex structures.
- Evacuation of the tank during 5-ton releases took approximately 35 seconds.
- Evacuation of the tank during 9-ton releases took approximately 50 seconds.
- Structures were observed to channel the escaping and expanding cloud laterally.
- Significant upwind movement of the chlorine clouds was observed due to the momentum of the escaping gas, expansion of the gas, and gravity-slumping.
- Visible clouds were observed to persist in UTG for approximately 20-30 minutes depending upon release amount and winds.
- Long visible clouds observed to stretch from UTG to approximately 2 km as downwind dispersion progressed.

- Data sets from Jack Rabbit II have been quality-assured, finalized, and recently released, enabling in-depth studies to begin on cloud concentrations, movement, and behavior to be quantified.

EMERGENCY RESPONSE EXPERIMENTS

For each test, researchers from Utah Valley University placed two "Witness Boards" on surveyed arcs adjacent to the disseminator in the urban environment to observe the effect of liquid chlorine and high concentrations of chlorine gas (approaching 25%) on common building components, urban surfaces, and fire equipment. Go-Pro® and HD color surveillance cameras captured the exposure on video. The selection of the test materials emphasized high organic content (creosote treated telephone pole and railroad tie, road asphalt, shingles, rubber, and bare wood), but copper, aluminum, galvanized steel, and carbon steel coupons were included. The copper and carbon steel surfaces began to be discolored 1-2 hours after the exposure, and the chlorine sometimes dissolved the petroleum based materials.

INDOOR INFILTRATION STUDIES

The Defense Threat Reduction Agency (DTRA) is evaluating various hazardous airborne contaminant applications for transition to DOD CBRN consequence models. Their current indoor application models presume that any contaminant is inert and neutrally buoyant, thus the models may not be useful for predicting the indoor effects of a dense, reactive gas like chlorine. Lawrence Berkeley National Laboratories (LBNL) proposed a series of experiments to DTRA and the JR program to assess the applicability of the models for predicting the indoor concentration of chlorine in the event of an outdoor chlorine release.

LBNL successfully completed a series of indoor infiltration experiments in 2015 in support of the DTRA effort. The LBNL experiments were completed to understand and characterize the dominant infiltration pathways for reactive dense gases into buildings, the uptake of such gases into structures, and the re-emission of them from the structures after the peak of the plume had passed. Three single-story CONEX structures, a single three-story CONEX structure, and 2 modular office trailers placed on the gravel pad were used to represent indoor office, school, or home environments. Follow-on studies were completed in 2016 using a single-story CONEX structure at the same, but unobstructed site. The CONEX structures were configured for controlled experiments of airflow and transport in single- or multi-story structures, and modified to include insulation and temperature-control. Chlorine penetration into and through the structure, and transfer of it across the building envelope was monitored via controlled artificial "leaks" and exhaust fans. The model offices were furnished with realistic insulation, surface materials, windows, and HVAC system characteristics, thus the simulated modular offices helped to relate the data from the controlled CONEX structures to real buildings. The data may be used for calibration, verification and/or validation of the indoor model, and to support DOD concepts of operation for sheltering in the event of an outdoor CBRN release.

PROGRAM STATUS

The program is entering a new phase after completing field testing. In 2017, the program will be processing and reporting data from 2015 and 2016, and transitioning it for permanent data storage, and to stakeholders. CSAC will also be presenting a JR overview, narrative, and plans in presentations to interested groups. The program will be proposing, designing, managing, and performing small and large scale experiments during chamber testing, and reducing and analyzing data to demonstrate or elucidate the reaction of chlorine with various substrates to suggest large-scale toxic inhalation hazard (TIH) chemical release mitigation strategies, inform emergency response strategies that may reduce the impact to affected populations and infrastructure, calculate the value of modeling parameters, and otherwise support stakeholders interests.

ACKNOWLEDGEMENTS
This research has been supported by the US Department of Homeland Security (DHS) and the US Defence Threat Reduction Agency (DTRA). Additional support (funding, personnel and instruments) is coming from numerous partners, with the list growing with time. We appreciate guidance from many scientists who are members of the planning groups and/or have provided comments on the scientific approach.

REFERENCES

1. Nicholson, D., Hedrick, A., and Lian, N., Final Data Report for Jack Rabbit (JRII), 2017

2. Fox, S. and D. Storwold, 2011: Project Jack Rabbit: Field Tests. Chemical Security Analysis Center, Science and Technology Directorate, U.S. Department of Homeland Security. CSAC 11-006, 162 pp.

479394 Functional Safety - Pitfalls and Good Practices

Abstract:

The implementation of effective functional safety is valuable in ensuring that safety critical control and protective systems are appropriately specified, designed, installed and maintained. Through the International Electro-technical Commission (IEC) 61508 and 61511 suite of functional safety standards, the concepts of the Safety Lifecycle and Safety Integrity Levels (SIL) are now well known. However, DNV GL's experience from numerous projects across different industry sectors suggests that there remain challenges to be overcome.

DNV GL has conducted a number of Functional Safety Assessments (FSAs) on various electrical/ electronic and programmable electronic safety systems in the process sector. The FSAs are a mandatory element of IEC 61511 which provide an independent judgment on the integrity of the safety system and should be performed by competent independent Functional Safety Practitioners. At least one FSA must be performed on a Safety Instrumented System and should be completed prior to the introduction of hazards. Through DNV GL's experiences in functional safety assessments this paper will discuss;

- How to avoid common mistakes in the management and implementation of functional safety;
- Examples of good practice in the application of functional safety;
- How good practice can help optimize test regimes, increase plant availability and reduce maintenance workload whilst ensuring the system continues to provide the required risk reduction.

General Method for Uncertainty Evaluation of Safety Integrity Level (SIL) Calculations – Part 2 Analytical Methods

Raymond "Randy" Freeman
S&PP Consulting

For Presentation at the American Institute of Chemical Engineers
2017 Spring Meeting in San Antonio, Texas -- March 26-30, 2017

Abstract

The IEC 61511 standard requires a verification calculation that a proposed design for a safety instrumented function (SIF) achieves the desired safety integrity level (SIL). The evaluation of the safety integrity level of a new or existing safety instrumented system requires detailed calculations based on the failure rates of the device and the planned maintenance/testing cycle for the system. The failure rates of the devices are often taken from standard failure rate tabulations of equipment. The maintenance and testing plans are developed based on plant experience. All of the data used in the SIL calculations are uncertain. This paper develops a general method for uncertainty analysis of the SIL calculations. The general method is based on the application of probability theory - variance contribution analysis (VCA) – to the equations presented in ISA TR 84.00.02-2115. An example is worked to demonstrate the methodology.

Background

The calculation of the probability of failure on demand (PFD) is a common engineering task when designing an interlock or safety system that is to be in compliance with IEC 61511 [Ref. 1]. The calculation of the PFD is often done using approximate equations defined in the ISA TR84.00.02 technical report [Ref. 2]. The simplified equation method in the ISA report is commonly used and is based on the use of reliability block diagrams where the field sensors, Safety Instrumented System (SIS) logic solver and final control elements are considered independent of each other in the sense of not sharing common devices or systems. The PFD is then calculated using the failure rates of the devices, planned test intervals, vendor supplied estimates on diagnostic coverage of the devices and an allowance for the potential for common cause failures. Almost all of these parameters are uncertain. The failure rate data is often taken from generic data sources which show wide ranges in the observed values.

Because of the uncertainty in the parameters, the design engineer makes allowances in the design by the use of safety factors or rules of thumb to improve the chances that the final interlock installation will work as intended. Since each engineer has a different set of safety factors and rules of thumb, two designs may differ significantly in the way a hazard is controlled.

A more formal method for handling the underlying uncertainty in the calculation of the PFD of an interlock is needed. Previously, Freeman and Summers [Ref. 3] published an uncertainty analysis of the PFD of an interlock. Two different methods were used in this analysis:

- Monte Carlo Simulation
- Variance Contribution Analysis (VCA)

Monte Carlo Simulation requires that the engineer build a model of the interlock using specialized computer software. The use of VCA requires that the sensitivity of the interlock model be determined either by numerical methods or by direct analytical calculations. The Freeman paper demonstrates that VCA can be used for the uncertainty analysis. However, the paper does not present a complete analysis method that can be applied to any system defined in the ISA technical report TR84.00.02. The goal of this paper is to develop a general set of analytical equations that will allow VCA to be used in uncertainty analysis of any interlock developed per the IEC 61511 standard.

Review of Interlock Design

The IEC standard looks at an interlock as a series of three major components (see Figure 1). First is the sensor set which sends an indication of an abnormal event to a logic solver. The logic solver determines is the signal from the sensor meets the conditions required to activate (trip) the interlock. If the interlock is to be activated, a signal is sent to the final control elements to take action (stop flow, close valve, etc) to prevent a process safety event from occurring.

Based on the devices selected the design engineer then evaluates whether the proposed interlock design meets the design criterion for probability of failure on demand. Typically, the design engineer is given the target PFD as a statement such as "provide a SIL-2 interlock to stop flow on high level in a vessel." A SIL-2 interlock requires a PFD of at most 0.01 or a risk reduction factor of at least 100. The needed risk reduction is often determined in a Layer of Protection Analysis (LOPA) [Ref. 4] and represents the managements decision on how risk is to be managed in a system. The resulting SIL-2 interlock design must have a high likelihood of achieving the target risk reduction as the LOPA team may be relying on it to be part of an overall risk reduction plan to prevent a process safety incident.

The ISA technical report (TR 84.00.02) [Ref. 2] provides shortcut methods for the evaluation of the PFD of an interlock design. Tables 1 and 2 are taken from the ISA technical report [Ref. 2]. Typical parameters need to compute the PFD include:

- DC is the diagnostic coverage;
- DI is the diagnostic interval;
- TI is the proof test interval,
- λ^D is the dangerous failure rate;
- MTTR is the mean time to restore the system to operation
- β is the common cause failure parameter that is always is between 0 and 1

In practice most if not all of these parameters are uncertain. Using parameter values obtained from plant records, vendor data or generic data bases the design engineer proceeds to develop the model for the interlock.

What is needed is a means to rapidly evaluate the impact of parameter uncertainty on the interlock PFD. The remainder of this paper applies the methods of variance contribution analysis (VCA) to determine the mean (expected value) and the standard deviation of the interlock PFD for a given design.

Table 1. Simplified PFD_{avg} formulas for Non-Repairable System without considering CCF, Diagnostics or MTTR.			
Configuration	**Function**	**PFD based on "Average before" failure rate**	**Appendix A Equation Number**
1oo1	F_1 =	$\left[\lambda^D \times \frac{TI}{2}\right]$	A-1
1oo2	F_2 =	$\left[(\lambda^D)^2 \times \frac{TI^2}{4}\right]$	A- 2
1oo3	F_3 =	$\left[(\lambda^D)^3 \times \frac{TI^3}{8}\right]$	A-3
2oo2	F_4 =	$\left[\lambda^D \times TI\right]$	A-4
2oo3	F5 =	$\frac{3}{4} \times \left[(\lambda^D)^2 \times TI^2\right]$	A-5
3oo3	F6 =	$3 \times \left[\lambda^D \times \frac{TI}{2}\right]$	A-6

Where TI is the proof test interval, λ^D is the dangerous failure rate.

Table 2. Simplified PFD_{avg} Formulas for Repairable System considering CCF, Diagnostics and MTTR			
Configuration	**Function**	**PFD based on "Average before" failure rate**	**Appendix A Equation Number**
1oo1	F_7 =	$\left[\frac{(1-DC)\times\lambda^D\times TI}{2}+\frac{DC\times\lambda^D\times DI}{2}+\lambda^D\times MTTR\right]$	A-24
1oo2	F_8 =	$\left[\frac{(1-DC)\times(1-\beta)\times\lambda^D\times TI}{2}+\frac{DC\times(1-\beta)\times\lambda^D\times DI}{2}+(1-\beta)\times\lambda^D\times MTTR\right]^2+\left[\frac{(1-DC)\times\beta\times\lambda^D\times TI}{2}+\frac{DC\times\beta\times\lambda^D\times DI}{2}+\beta\times\lambda^D\times MTTR\right]$	A- 25
1oo3	F_9 =	$\left[\frac{(1-DC)\times(1-\beta)\times\lambda^D\times TI}{2}+\frac{DC\times(1-\beta)\times\lambda^D\times DI}{2}+(1-\beta)\times\lambda^D\times MTTR\right]^3+\left[\frac{(1-DC)\times\beta\times\lambda^D\times TI}{2}+\frac{DC\times\beta\times\lambda^D\times DI}{2}+\beta\times\lambda^D\times MTTR\right]$	A-26
2oo2	F_{10} =	$2\times\left[\frac{(1-DC)\times\lambda^D\times TI}{2}+\frac{DC\times\lambda^D\times DI}{2}+\lambda^D\times MTTR\right]$	A-27

Table 2. Simplified PFD_{avg} Formulas for Repairable System considering CCF, Diagnostics and MTTR			
Configuration	**Function**	**PFD based on "Average before" failure rate**	**Appendix A Equation Number**
2oo3	F_{11} =	$3\times\left[\frac{(1-DC)\times(1-\beta)\times\lambda^{D}\times TI}{2}+\frac{DC\times(1-\beta)\times\lambda^{D}\times DI}{2}+(1-\beta)\times\lambda^{D}\times MTTR\right]^{2}+\left[\frac{(1-DC)\times\beta\times\lambda^{D}\times TI}{2}+\frac{DC\times\beta\times\lambda^{D}\times DI}{2}+\beta\times\lambda^{D}\times MTTR\right]$	A-28
3oo3	F_{12} =	$3\times\left[\frac{(1-DC)\times\lambda^{D}\times TI}{2}+\frac{DC\times\lambda^{D}\times DI}{2}+\lambda^{D}\times MTTR\right]$	A-29

Review of Variance Contribution Analysis (VCA) Methodology

The mean and variance of a function of random variables can be approximated using the method described by Haugen [Ref. 5] and applied by Freeman [Ref. 3, 6, 7]. Define an arbitrary function of a set of random variables, x_i, as:

Let

$$Y = F(x_i) \qquad \text{(Eq 1)}$$

The mean of Y can be estimated using the following approximation:

$$E(Y) = F[\, E(x_i)] \qquad \text{(Eq 2)}$$

Where:
$E(Y)$ = expected value of random variable Y = mean of Y
$E(x_i)$ = expected value of random variable x_i = mean of x_i

The variance of Y can likewise be estimated as:

$$V(Y) = \sum_{i=1}^{n} \left[\frac{\partial Y}{\partial x_i}\right]^2 V(x_i) \qquad \text{(Eq 3)}$$

Where:

$V(Y)$ = variance of random variable Y as defined above in Equation 1

$V(x_i)$ = variance of random variable x_i as defined above in Equation 1

Note that the variance is simply the square of the standard deviation. Using the variance will simplify the mathematics that is described below. The contribution of each independent random variable to the overall variance in the function is:

$$V(Y \text{ from } x_i) = \left[\frac{\partial Y}{\partial x_i}\right]^2 V(x_i) \qquad \text{(Eq 4)}$$

The relative contribution of each term to the overall variance V(Y) is a measure of the importance in the uncertainty in the particular random variable, x_i. In effect, this is a sensitivity analysis combined with a uncertainty evaluation. The variance contribution combines the sensitivity in the answer to changes in the uncertain random variable, x_i, with a measure of the uncertainty in the random variable, x_i. The overall variance in Y is found by summing the sensitivity weighted variances from each random variable.

Recommended Interlock PFD Uncertainty Analysis Method

Previously, Freeman and Summers [Ref. 3] suggested a framework for the inclusion of uncertainty analysis in the calculations completed for a new interlock per the ISA TR [Ref. 2]. The following uncertainty analysis method has been expanded to incorporate critical decisions that the process management must make in the design process.

1. Complete the interlock design using the methods outlined in IEC 61511 [Ref. 1].

2. Review with the process system management and determine if the proposed interlock should be considered repairable or non-repairable. A simple flow chart for this decision making is presented in Figure 2. The basic question is: "can the interlock be repaired safely while the process operates." This is a management question and management should be the one that decides the answer to this important question.

3. Create interlock performance equation as the mathematical model for the combination of sensor, logic solver and final control elements using the methods outlined in ISA technical report (TR84.00.02) [Ref. 2]. Equations needed to evaluate the uncertainty of the models specified in the ISS report are developed in Appendix A of this paper. For non-repairable systems, Table 3 can be used. For more complex systems such as non-repairable redundant systems with the potential for common cause failures, use the repairable equations of Table 4 and set DI, DC, and MTTR all equal to zero. For systems where common cause failures (CCF), Diagnostics and repair are to be considered, use the recommendations of Table 4. There are design situation where the equations developed in Appendix A will not exactly apply. Specifically, designs where the system contains redundancy subject to common cause failures (CCF) that should be considered non-repairable during operation. The failure models and the equations needed to evaluate the uncertainty in these systems are developed in Appendix B to this paper.

Table 3. Roadmap for Mean and Variance of Non-Repairable System Without Considering CCF, Diagnostics and MTTR			
Configuration	**Function**	**Mean PFD Using Appendix A Equation**	**PFD Variance Using Appendix A Equation**
1oo1	F_1 =	A-9	A-18
1oo2	F_2 =	A- 10	A- 19
1oo3	F_3 =	A-11	A-20
2oo2	F_4 =	A-12	A-21
2oo3	F_5 =	A-13	A-22
3oo3	F_6 =	A-14	A-23

Table 4. Roadmap for Mean and Variance of Repairable System Considering CCF, Diagnostics and MTTR			
Configuration	**Function**	**Mean PFD Using Appendix A Equation**	**PFD Variance Using Appendix A Equation**
1oo1	F_7 =	A-30	A-44
1oo2	F_8 =	A- 31	A- 86
1oo3	F_9 =	A-32	A-94
2oo2	F_{10} =	A-33	A-48
2oo3	F_{11} =	A-34	A-102
3oo3	F_{12} =	A-35	A-50

Note that in all of the calculations indicated by the equations referenced in Table 4, the expected value of any the random variables is used to complete the calculations.

4. Define the uncertainty in the parameters and variables of the interlock model specified in step 2. The uncertainty can be given as the upper and lower range of the possible values (uniform probability distribution), as the upper, lower and recommended values (triangular distribution), or as a mean and standard deviation (normal distribution). See the example above for guidance in the evaluation of safety instrumented system interlocks.

5. Compute the expected value of each variable in the interlock performance equation. The equations for the mean and variance for the uniform, triangular and normal probability distributions are presented for various probability distributions in Vose [Ref. 8]. The example interlock calculations used a triangular distribution to represent the uncertainty in the parameters.

6. Compute the expected value or mean of the interlock PFD using the mean value of each of the variables in Step 5. See Tables 4 and 5 for recommended equations.

7. Compute the sensitivity of the result from the interlock performance equation by use of the partial derivative of the basic interlock performance equation with respect to each of the variables as presented in Appendix A.

8. Compute the variance of the interlock performance equation PFD by use of the variance contribution using equations from Tables 3 or 4. This entails multiplying the variance of each of the uncertain variables in the basic interlock performance equation by the square of its sensitivity (obtained in step 7), as evaluated at the variable mean. Sum the resulting terms to obtain the overall variance of the PFD in the interlock performance equation. See Tables 4 and 5 for recommended equations.

9. Determine the level of risk that the owner/operator wishes to take that the final interlock will not work. Note that this is a management decision not an instrument engineer decision! In this paper, the 95% level of risk reduction has been used:
   - 5% chance of failure or a 95% chance of the interlock achieving the desired risk reduction

10. Assuming that the interlock owner operator wishes to take a low risk (5%) of the interlock failing to achieve its design target PFD, compute the 95% upper confidence limit on the computed PFD by use of the standard normal factor, Z, [Ref. 9] as:

$$Z = \left[\frac{xi - E(x)}{\sigma}\right] \qquad \text{(Eq 5)}$$

Where:

σ = standard deviation of the PFD of the interlock of interest from the interlock performance equation obtained from step 7. Note that the variance of a random variable is the square of the standard deviation of the random variable.

E(x) = the expected value of the PFD of the interlock of interest from the interlock performance equation obtained from step 5

For the 95% upper limit, Z = 1.645. Rearranging Eq. 5 allows for the direct calculation of the corresponding value of the 95% upper confidence limit on the PFD as:

$$X95\% = 1.645\ \sigma + E(x) \qquad \text{(Eq 6)}$$

Where:

X95% = the upper 95% limit on the computed PFD of the interlock of interest from the interlock performance equation.

Compare the 95% upper confidence limit on the PFD of the interlock of concern with that established as the desired PFD for risk reduction. If the 95% confidence of the RRF is greater than the desired RRF, the design is complete. If not, revise the design or change inspection test intervals to achieve the desired RRF. If it is not possible to achieve the desired target RRF economically, revisit the LOPA study accordingly to incorporate better information obtained in the uncertainty analysis. Improve the integrity of the LOPA IPLs or identify additional IPLs to drive the risk to a tolerable level. Continue this process until the computed RRFs are greater than the desired RRFs for risk reduction and risk management.

Example Interlock Problem

Previously, Freeman and Summers [Ref. 3] published an example in the use of VCA to determine the likelihood that the interlock would provide the risk reduction desired by the management and the LOPA team. The process system is shown in Figure 3. The process uses a compressor to increase the pressure of a process stream prior to additional processing. The gas being compressed is toxic and flammable. Of concern is a slug of liquid being sent to the compressor. If a liquid slug is sent to the compressor, significant damage to the compressor is expected with probable seal damage and a subsequent release of flammable and toxic material into the work area. A large fire and/or explosion could result if the release were to be ignited. If the release is not ignited, the nearby workers could be exposed to the toxic gas resulting in death or injury. A Layer of Protection Analysis (LOPA) review of this system has been completed. Among several recommendations, the LOPA team recommended the installation of a SIL-2 high level interlock in the Compressor Knock Out Drum to stop the compressor prior to liquids entering the system.

Figure 1 shows a simplified diagram for the interlock. Three level sensors are provided in the Compressor Knock Out Drum. The SIS logic solver will use 2oo3 voting to detect high level in the Compressor Knock Out Drum. The SIS logic solver will also monitor the difference in level signal from each of the three level sensors and will activate an alarm if the deviation is excessive. Two independent methods of stopping the compressor are provided. The SIS will directly signal the motor controller on the compressor to stop. In addition, the SIS logic solver will also signal two additional relays to open the power supply to the compressor motor. These two different shutdown will both be activated in the event of high level in the Compressor Knock Out Drum. Either one of the two shutdowns is capable of stopping the compressor by turning off the electric power supply to the motor.

Models

The first step in the calculation of the "goodness" of an interlock is to establish the model to be used in the calculations. Note that the sensors are 2oo3 voting and the final control elements are each 1oo1. Appendix A presents the equations for various models that can be used for describing this system.

The overall probability of failure on demand (PFD) of the interlock is given as:

PFD = PFDs + PFDsis + PFDfce (Eq 7)

Where:

PFD = Probability of failure on demand of the interlock as a whole

PFDs = Probability of failure on demand of the sensors (voting as 2oo3)

PFDsis = Probability of failure on demand of the SIS logic solver

PFDfce = Probability of failure on demand of the final control elements.

Since there are two final control elements arranged in series, the PFDfce = sum of the PFDs of the final control elements (Relay PFD and MCC PFD).

PFDce = PFDr + PFDmcc (Eq 8)

Where:

PFDr = Probability of failure on demand of the two relays voting as 1oo2 to shutdown gas compressor.

PFDmcc = Probability of failure on demand of the MCC to shutdown the gas compressor

For this example, the following selections are made to model the performance of the interlock.

Model for Sensors (2oo3)

The sensors are considered repairable as one sensor can be replaced while the other two function and provide continuous functionality of the interlock. There are three sensors that will be used in a 2oo3 voting system. From Appendix A, using Equation A-7, the model for the sensors becomes:

$$PFDsavg = 3\times\left[\frac{(1-DCs)\times(1-\beta s)\times\lambda^{Ds}\times TIs}{2}+\frac{DCs\times(1-\beta s)\times\lambda^{Ds}\times DIs}{2}+(1-\beta s)\times\lambda^{Ds}\times MTTRs\right]^2+$$

$$\left[\frac{(1-DCs)\times\beta s\times\lambda^{Ds}\times TIs}{2}+\frac{DCs\times\beta s\times\lambda^{D}\times DIs}{2}+\beta s\times\lambda^{Ds}\times MTTRs\right] \qquad \text{(Eq 9)}$$

Where:

PFD_{savg} is the average probability of failure on demand of the sensors

DC_s is the diagnostic coverage for sensor failure

DI_s is the diagnostic interval for the sensors

$MTTR_s$ is the mean time to restore the sensors to functionality given a sensor failure

TI_s is the test interval for the sensors

β_s is the common cause failure parameter

λ^{Ds} is the failure rate to a dangerous condition for the sensors

MTTR is the mean time to restore the system from the time that failure occurs

Model for SIS Logic Solver

Use a fixed probability of failure on demand as specified by the vendor. Unless detailed design information is provided by the SIS logic solver vendor, this will be the normal default condition for most studies. For this problem a typical PFD of 1.30×10^{-4} was selected to represent the SIS Logic Solver.

Model for Final Control Elements

There are two separate paths to shutdown the gas compressor. First is by the SIS logic solver commanding the MCC to shutdown power to the the gas compressor motor. The second is for the SIS logic solver to issue a shutdown command to two interposing relays (R1 and R2) which will cause the power to the gas compressor motor to stop. Two different models are needed to the final control elements. These two subsystems are considered non-repairable in this analysis,

Model for Interposing Relays (1oo2)

There are two interposing relays (R1 and R2 in the interlock). From Appendix A, use Equation A-25 setting parameters DC=DI=MTTR=0 for the non-repairable interposing relays, the model becomes:

$$PFDravg = \left[\frac{(1-\beta r)\times\lambda^{Dr}\times TIr}{2}+\right]^2+\left[\frac{\beta r\times\lambda^{Dr}\times TIr}{2}+\right] \quad \text{(Eq 10)}$$

Where:

PFD_{ravg} is the average probability of failure on demand of the relays voting 1oo2 to shutoff the gas compressor.

TI_r is the test interval for the relays

β_r is the common cause failure parameter

λ^{Dr} is the failure rate to a dangerous condition for the relays

Model for MCC (1oo1)

From Appendix A and using equation A-1 for the non-repairable MCC, the model becomes:

$$PFDmccavg = \frac{\lambda^{Dmcc} * TImcc}{2} \quad \text{(Eq 11)}$$

Where:

PFD_{mccavg} is the average probability of failure on demand of the MCC to shutoff the gas compressor.

TI_{mcc} is the test interval for the MCC

$\lambda^{Dmcc}$is the failure rate to a dangerous condition for the MCC

Data

The calculation of the PFD of the interlock requires a set of data to be used to represent the system. Tables 5, 6, and 7 are taken from the Freeman-Summers paper [Ref. 3] and presents the data used to represent the interlock system. Note that these data were originally taken from generic data sources and do not represent any particular device or system.

Table 5. Uncertainty Data for Level Sensors Used in Example Interlock

(All variable probability distributions assumed to be triangularly distributed)

Variable	Min	Mode	Max	Mean	Variance
λ^{Ds}- Fail Dangerous Rate	2.84×10^{-3} Yr^{-1}	5×10^{-3} Yr^{-1}	8.5×10^{-3} Yr^{-1}	5.45×10^{-3} Yr^{-1}	1.36×10^{-6} Yr^{-1}
DC_s – Diagnostic Coverage *	0.8	0.9	0.99	0.897	1.51×10^{-3}
DI_s – Diagnostic Interval **	5.71×10^{-5}	5.71×10^{-5}	5.71×10^{-5}	5.71×10^{-5}	0
TI_s – Test Interval	4 Yr	5 Yr	6 Yr	5 Yr	5.56×10^{-2} Yr^2
β_s- Common Cause Failure Fraction *	0	0.02	0.1	0.04	4.67×10^{-4}
$MTTR_s$ - Mean Time to Restore ***	1.37×10^{-3} Yr	8.22×10^{-3} Yr	1.92×10^{-2} Yr	9.59×10^{-3} Yr	1.34×10^{-5} Yr^2

* Unit less

** 0.5 hours, assumed to be deterministic

***Min of 12 hr, Mode of 72 hours, Max of 168 hrs

Table 6. Uncertainty Data for Relays Used in Example Interlock (All variable probability distributions assumed to be triangularly distributed)					
Variable	**Min**	**Mode**	**Max**	**Mean**	**Variance**
λ^{Dr}- Fail Dangerous Rate	$8.76x10^{-9}$ Yr^{-1}	$2.00x10^{-3}$ Yr^{-1}	$4.73x10^{-2}$ Yr^{-1}	$1.64x10^{-2}$ Yr^{-1}	$1.19x10^{-4}$ Yr^{-2}
DC_r – Diagnostic Coverage *	NA	NA	NA	NA	NA
DI_r – Diagnostic Interval*	NA	NA	NA	NA	NA
TI_r – Test Interval	1 Yr	1 Yr	2 Yr	1.33 Yr	0.056 Yr^2
β_r- Common Cause Failure Fraction	0	0.02	0.1	0.04	$4.67x10^{-4}$
$MTTR_r$ - Mean Time to Restore*	NA	NA	NA	NA	NA

* Not used in relay model

Table 7. Uncertainty Data for MCC Used in Example Interlock (All variable probability distributions assumed to be triangularly distributed)					
Variable	**Min**	**Mode**	**Max**	**Mean**	**Variance**
λ^{Dr}- Fail Dangerous Rate	$1.74x10^{-4}$ Yr^{-1}	$1.31x10^{-3}$ Yr^{-1}	$3.00x10^{-2}$ Yr^{-1}	$1.05x10^{-2}$ Yr^{-1}	$4.76x10^{-5}$ Yr^{-2}
DC_r – Diagnostic Coverage*	NA	NA	NA	NA	NA
DI_r – Diagnostic Interval*	NA	NA	NA	NA	NA
TI_r – Test Interval	1 Yr	1 Yr	2 Yr	1.33 Yr	$5.6x10^{-2}$ Yr^2
β_r- Common Cause Failure Fraction*	NA	NA	NA	NA	NA
$MTTR_r$ - Mean Time to Restore*	NA	NA	NA	NA	NA

* Not used in MCC model

Uncertainty Analysis Results

Using the VCA method to estimate the PFD of the interlock design yields the results shown in Table 8.

Table 8. Results of VCA on Example Interlock Design

Subsystem	Mean PFD	Variance of PFD
Sensors	6.41E-5	1.7657E-9
Logic Solver	1.34E-4	0
Relays	5.46E-4	2.0333E-7
MCC	6.98E-3	4.52E-5
Totals	**7.73E-3**	**4.54E-5**

Note: Calculations are carried to an adequate number of places to allow for the resulting PFD to be determined. Rounding to significant number of figures is done at the end.

The mean PFD of 0.00773 implies a mean risk reduction factor of:

RRF = 1/PFD = 1/0.00773 = 129 (Eq 11)

The variance of 4.54E-5 implies a standard deviation in the PFD of

Std Dev = $(\text{Variance})^{1/2}$ = 0.006737 (Eq 12)

The corresponding 95% level of confidence in the PFD:

X95% = 1.645 std Dev + Mean PFD (Eq 13)

X95% = 1.645 (0.006737) + 0.00773= 0.01881 (Eq 14)

The RRF for 95% certain PFD is

RRF95% = 1/0.01881 => 53 (Eq 15)

This is essentially the same result previously reported by Freeman and Summers [Ref. 3] using either Monte Carlo Simulation or numerical approximation methods for the VCA sensitivities. The calculated PFD at the 95% level of 0.01881 indicates that there is only a 5% chance that the interlock will provide an RRF worse that 53. Since this level is only SIL-1 capable and not
SIL-2 capable as desired by management, a revised design will be needed to achieve the desired risk reduction or a different protective measure will be needed to control risk.

Acknowledgement

This paper is an expanded version of a paper entitled ''General Method for Uncertainty Evaluation of Safety Integrity Level (SIL) Calculations'' presented at the Mary Kay O'Connor 2016 Process Safety Symposium, October 26, 2016, College Station, TX. The author wishes to thank those individuals who gave comments and suggestions for improving this article.

References

1. International Society of Automation, Functional Safety: Safety Instrumented Systems for the Process Sector—Parts 1, 2, and 3, ANSI/ISA 84.00.01, 2004 [USA implementation of IEC 61511].
2. International Society of Automation, *Safety Integrity Level (SIL) Verification of Safety Instrumented Functions*, ISA TR84.00.02, Research Triangle Park, NC. (2015).
3. Freeman, R. And Summers, A, "Evaluation of Uncertainty in Safety Integrity Level Calculations," *Process Safety Progress*, Vol 35, No 4, pp 341-348, December 2016. Published Wiley Online Library (wileyonlinelibrary.com). DOI 10.1002/prs.11805, 23 January 2016
4. Center for Chemical Process Safety, *Layer of Protection Analysis – Simplified Process Risk Assessment*, American Institute of Chemical Engineers, New York, 2001
5. Edward B. Haugen, *Probabilistic Approaches to Design*, John Wiley, New York, 1968
6. R.A. Freeman, "Quantifying LOPA Uncertainty," *Process Safety Progress*, Vol 31, No. 3, pp 240-247, 2012
7. R. A. Freeman, "Simplified Uncertainty Analysis of Layer of Protection Analysis Results," *Process Safety Progress*, Vol. 32, No 4, pp 351-360, December 2013, Published online in Wiley Online Library (wileyonlinelibrary.com) DOI 10.1002/ prs11585.pdf, 3 June 2013
8. David Vose, *Risk Analysis – A Quantitative Guide*, 3rd Edition, John Wiley, 2008
9. Paul L. Meyer, *Introductory Probability and Statistical Applications*, John Wiley, New York, 1972

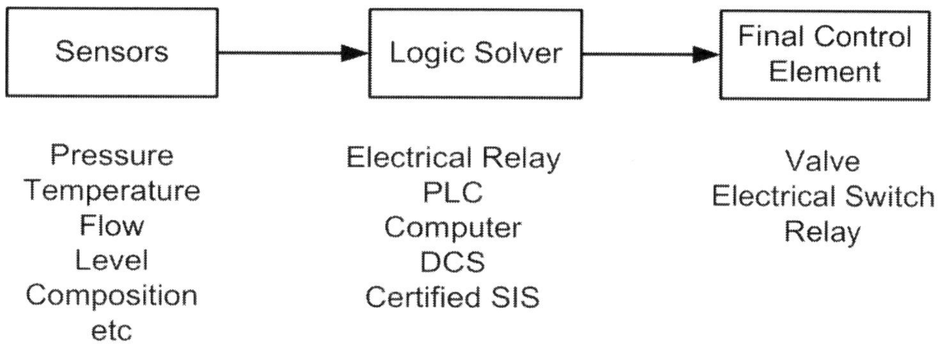

Figure 1. Typical Components of an SIS Interlock

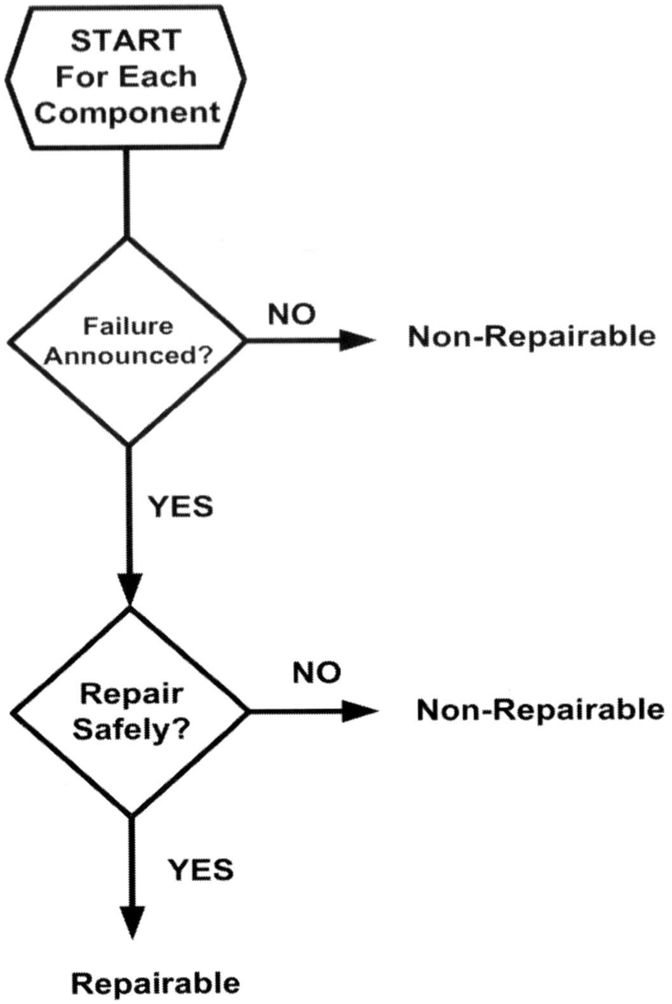

Figure 2. Decision Tree for Determination of Repairable or Non- Repairable

Figure 3. P&ID Sketch for Example Interlock

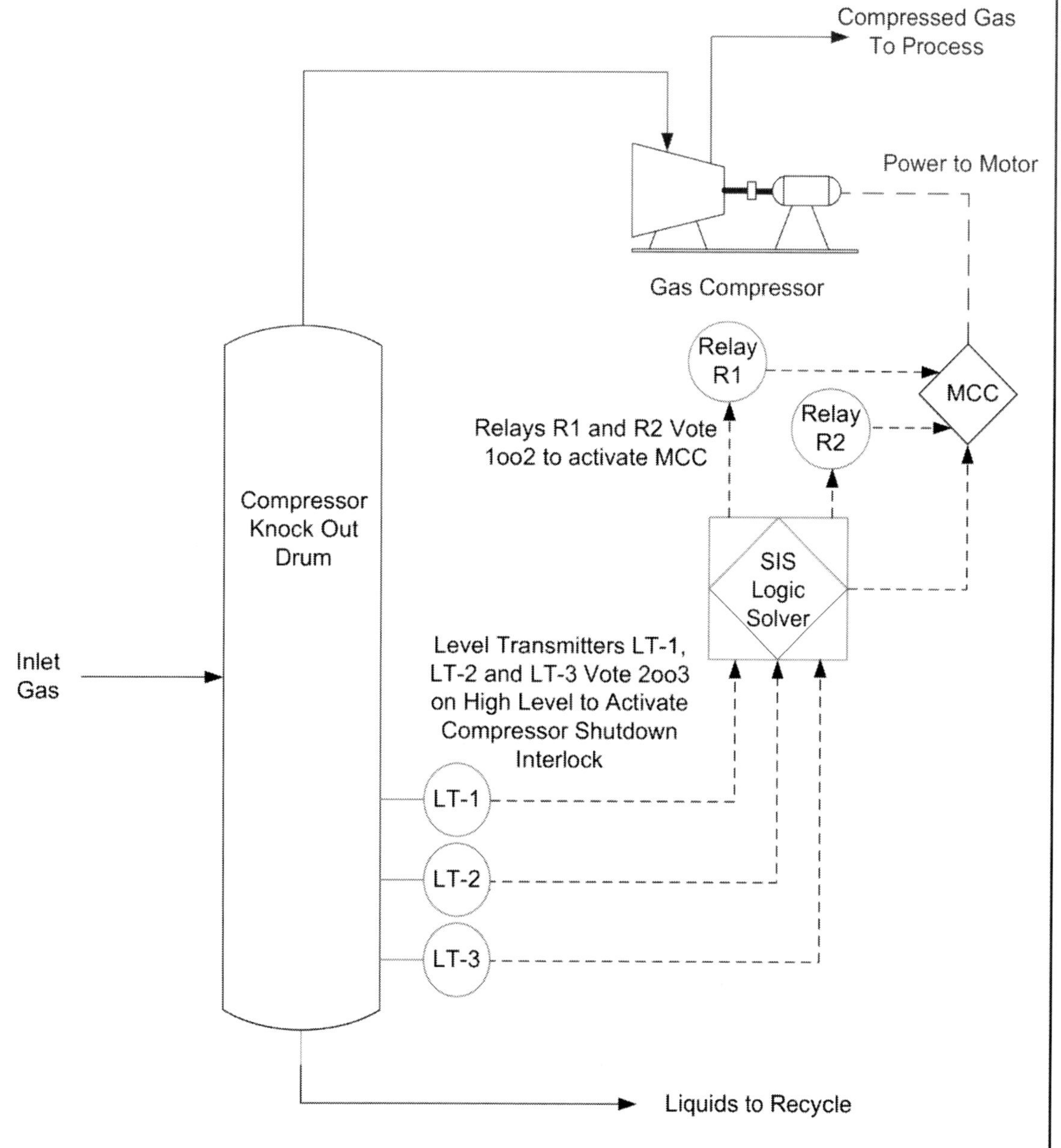

APPENDIX A -- DEVELOPMENT OF UNCERTAINTY EQUATIONS

This appendix develops the equations needed to evaluate the uncertainty in the SIL PFD as computed using the methods outlined in the ISA Technical Report [Ref. 2]. This appendix is divided into two sections:

- Non-Repairable Systems
- Repairable Systems

The uncertainty equations for other configurations not covered in the ISA report are developed in Appendix B.

NON-REPAIRABLE SYSTEMS

For systems that are considered non-repairable, the simplified equations of Table A.1 (taken from the ISA Technical Report [Ref. 2]) are used for the analysis.

Table A.1 Simplified PFD_{avg}formulas for Non-Repairable System without considering CCF, Diagnostics or MTTR.

Configuration	Function	PFD based on "Average before" failure rate	Equation Number
1oo1	F_1 =	$\left[\lambda^D \times \frac{TI}{2}\right]$	A-1
1oo2	F_2 =	$\left[(\lambda^D)^2 \times \frac{TI^2}{4}\right]$	A- 2
1oo3	F_3 =	$\left[(\lambda^D)^3 \times \frac{TI^3}{8}\right]$	A-3
2oo2	F_4 =	$\left[\lambda^D \times TI\right]$	A-4
2oo3	F5 =	$\frac{3}{4} \times \left[(\lambda^D)^2 \times TI^2\right]$	A-5
3oo3	F6 =	$3 \times \left[\lambda^D \times \frac{TI}{2}\right]$	A-6

Where TI is the proof test interval, λ^D is the dangerous failure rate.

The expected value of the probability of failure on demand is found by calculating the PFD using the expected value (mean) of the random variables.

Let y a function of some random variables, x_i, as:

$$y = F(x_i) \quad \text{(Eq A-7)}$$

The mean of y can be estimated using the following approximation:

$$E(y) = F[E(x_i)] \quad \text{(Eq A-8)}$$

Where:

$E(y)$ = expected value of random variable y = mean of y

$E(x_i)$ = expected value of random variable x_i = mean of x_i

For the configurations defined in Table A.1, the corresponding expected PFD are presented in Table A.2.

Table A.2 Expected Value of PFD_{avg} formulas for Non-Repairable System without considering CCF, Diagnostics or MTTR.

Configuration	Function	Expected PFD based on "Average before" failure rate	Equation Number
1oo1	$E(F_1)$ =	$\left[E(\lambda^D)\times\frac{E(\text{TI})}{2}\right]$	A-9
1oo2	$E(F_2)$ =	$\left[(E(\lambda^D))^2\times\frac{E(\text{TI})^2}{4}\right]$	A- 10
1oo3	$E(F_3)$ =	$\left[(E(\lambda^D))^3\times\frac{E(\text{TI})^3}{8}\right]$	A-11
2oo2	$E(F_4)$ =	$\left[E(\lambda^D)\times E(\text{TI})\right]$	A-12
2oo3	$E(F_5)$ =	$\frac{3}{4}\times\left[(E(\lambda^D))^2\times E(\text{TI})^2\right]$	A-13
3oo3	$E(F_6)$ =	$3\times\left[E(\lambda^D)\times\frac{E(\text{TI})}{2}\right]$	A-14

Where E(x) is the expected value of random variable x.

The variance of the PFD is found by taking the weighted sum of the variance of the random variables. The weighting function is the sensitivity of the PFD with respect to random variable. In mathematical terms the variance of a function may be found in terms of the variance of the random variables as:

Once again let

$$y = F(x_i) \tag{Eq A-15}$$

The variance of Y can likewise be estimated as:

$$V(y) = \sum_{i=1}^{n} \left[\frac{\partial y}{\partial x_i}\right]^2 V(x_i) \tag{Eq A-16}$$

Where:

V(y) = variance of random variable y as defined above

$V(x_i)$ = variance of random variable x_i as defined above

The sensitivity of y with respect to a random variable x_i is:

$$\text{Sensitivity of y with respect to } x_i = \left[\frac{\partial y}{\partial x_i}\right] \tag{Eq A-17}$$

For the configurations defined in Table A.1, the corresponding variance of the PFD are presented in Table A.3.

Table A.3 Variance of PFD_{avg} formulas for Non-Repairable System without considering CCF, Diagnostics or MTTR.			
Configuration	**Function**	**Variance in Function F_i based on "Average before" failure rate**	**Equation Number**
1oo1	$V(F_1) =$	$\left[\frac{E(TI)}{2}\right]^2 V(\lambda^D) + \left[\frac{E(\lambda^D)}{2}\right]^2 V(TI)$	A-18
1oo2	$V(F_2) =$	$\left[E(\lambda^D)^2 \frac{E(TI)^2}{2}\right]^2 V(\lambda^D) +$ $\left[E(\lambda^D)^2 \frac{E(TI)}{2}\right]^2 V(TI)$	A- 19
1oo3	$V(F_3) =$	$\left[3 \times E(\lambda^D)^2 \frac{E(TI)^3}{2}\right]^2 V(\lambda^D) +$ $\left[\frac{3}{8} E(\lambda^D)^3 E(TI)^2\right]^2 V(TI)$	A-20
2oo2	$V(F_4) =$	$E(TI)^2 V(\lambda^D) + E(\lambda^D)^2 \times V(TI)$	A-21
2oo3	$V(F_5) =$	$\left[\frac{3}{2} \times E(\lambda^D) \times E(TI)^2\right]^2 V(\lambda^D) +$ $\left[\frac{3}{2} \times E(\lambda^D)^2 \times E(TI)\right]^2 V(TI)$	A-22
3oo3	$V(F_6) =$	$\left[\frac{3}{2} \times E(TI)\right]^2 V(\lambda^D) +$ $\left[\frac{3}{2} \times E(\lambda^D)\right]^2 V(TI)$	A-23

Note that for the calculation of the variance functions, V(Fi), the random variables are evaluated at the expected value (mean).

REPAIRABLE SYSTEMS

For systems that are considered repairable, the simplified equations of Table A.4 are used for the analysis. The expected value of the probability of failure on demand (PFD) is found by substituting the expected value of each of the random variables into the corresponding equation. The expected value of the PFD for each of the system configurations is presented in Table A.5

Table A.4 Simplified PFD_{avg} Formulas for Repairable System considering CCF, Diagnostics and MTTR

Configuration	Function	PFD based on "Average before" failure rate	Equation Number
1oo1	$F_7 =$	$\left[\frac{(1-DC)\times\lambda^D\times TI}{2}+\frac{DC\times\lambda^D\times DI}{2}+\lambda^D\times MTTR\right]$	A-24
1oo2	$F_8 =$	$\left[\frac{(1-DC)\times(1-\beta)\times\lambda^D\times TI}{2}+\frac{DC\times(1-\beta)\times\lambda^D\times DI}{2}+(1-\beta)\times\lambda^D\times MTTR\right]^2+\left[\frac{(1-DC)\times\beta\times\lambda^D\times TI}{2}+\frac{DC\times\beta\times\lambda^D\times DI}{2}+\beta\times\lambda^D\times MTTR\right]$	A- 25
1oo3	$F_9 =$	$\left[\frac{(1-DC)\times(1-\beta)\times\lambda^D\times TI}{2}+\frac{DC\times(1-\beta)\times\lambda^D\times DI}{2}+(1-\beta)\times\lambda^D\times MTTR\right]^3+\left[\frac{(1-DC)\times\beta\times\lambda^D\times TI}{2}+\frac{DC\times\beta\times\lambda^D\times DI}{2}+\beta\times\lambda^D\times MTTR\right]$	A-26
2oo2	$F_{10} =$	$2\times\left[\frac{(1-DC)\times\lambda^D\times TI}{2}+\frac{DC\times\lambda^D\times DI}{2}+\lambda^D\times MTTR\right]$	A-27

Table A.4 Simplified PFD_{avg} Formulas for Repairable System considering CCF, Diagnostics and MTTR			
Configuration	**Function**	**PFD based on "Average before" failure rate**	**Equation Number**
2oo3	F_{11} =	$3\times\left[\frac{(1-DC)\times(1-\beta)\times\lambda^D\times TI}{2}+\frac{DC\times(1-\beta)\times\lambda^D\times DI}{2}+(1-\beta)\times\lambda^D\times MTTR\right]^2+\left[\frac{(1-DC)\times\beta\times\lambda^D\times TI}{2}+\frac{DC\times\beta\times\lambda^D\times DI}{2}+\beta\times\lambda^D\times MTTR\right]$	A-28
3oo3	F_{12} =	$3\times\left[\frac{(1-DC)\times\lambda^D\times TI}{2}+\frac{DC\times\lambda^D\times DI}{2}+\lambda^D\times MTTR\right]$	A-29

Where:

DC is the diagnostic coverage;

DI is the diagnostic interval;

TI is the proof test interval,

λ^D is the dangerous failure rate;

MTTR is the mean time to restore the system to operation

β is the common cause failure parameter that is always is between 0 and 1

Table A.5 Expected PFD Formulas for Repairable System considering CCF, Diagnostics and MTTR			
Configuration	**Function**	**PFD based on "Average before" failure rate**	**Equation Number**
1oo1	E(F$_7$) =	$\left[\frac{(1-E(DC))\times E(\lambda^D)\times E(TI))}{2}+\frac{E(DC)\times E(\lambda^D)\times E(DI)}{2} + E(\lambda^D)\times E(MTTR)\right]$	A-30
1oo2	E(F$_8$) =	$\left[\frac{(1-E(DC))\times(1-E(\beta))\times E(\lambda^D)\times E(TI)}{2}+\frac{E(DC)\times(1-E(\beta))\times E(\lambda^D)\times E(DI)}{2}+(1-E(\beta))\times E(\lambda^D)\times E(MTTR)\right]^2 + \left[\frac{(1-E(DC))\times E(\beta)\times E(\lambda^D)\times E(TI)}{2}+\frac{E(DC)\times E(\beta)\times E(\lambda^D)\times E(DI}{2})+E(\beta)\times E(\lambda^D)\times E(MTTR)\right]$	A- 31
1oo3	E(F$_9$) =	$\left[\frac{(1-E(DC))\times(1-E(\beta))\times E(\lambda^D)\times E(TI)}{2}+\frac{E(DC)\times(1-E(\beta))\times E(\lambda^D)\times E(DI)}{2}+(1-E(\beta))\times E(\lambda^D)\times E(MTTR)\right]^3 + \left[\frac{(1-E(DC))\times E(\beta)\times E(\lambda^D)\times E(TI)}{2}+\frac{E(DC)\times E(\beta)\times E(\lambda^D)\times E(DI)}{2}+E(\beta)\times E(\lambda^D)\times E(MTTR)\right]$	A-32

Table A.5 Expected PFD Formulas for Repairable System considering CCF, Diagnostics and MTTR			
Configuration	**Function**	**PFD based on "Average before" failure rate**	**Equation Number**
2oo2	$E(F_{10}) =$	$2\times\left[\frac{(1-E(DC))\times E(\lambda^D)\times E(TI))}{2}+\frac{E(DC)\times E(\lambda^D)\times E(DI)}{2}+E(\lambda^D)\times E(MTTR)\right]$	A-33
2oo3	$E(F_{11}) =$	$3\times\left[\frac{(1-E(DC))\times(1-E(\beta))\times E(\lambda^D)\times E(TI)}{2}+\frac{E(DC)\times(1-E(\beta))\times E(\lambda^D)\times E(DI)}{2}+(1-E(\beta))\times E(\lambda^D)\times E(MTTR)\right]^2+\left[\frac{(1-E(DC))\times E(\beta)\times E(\lambda^D)\times E(TI)}{2}+\frac{E(DC)\times E(\beta)\times E(\lambda^D)\times E(DI)}{2}+E(\beta)\times E(\lambda^D)\times E(MTTR)\right]$	A-34
3oo3	$E(F_{12}) =$	$3\times\left[\frac{(1-E(DC))\times E(\lambda^D)\times E(TI))}{2}+\frac{E(DC)\times E(\lambda^D)\times E(DI)}{2}+E(\lambda^D)\times E(MTTR)\right]$	A-35

The variance of the PFD of the repairable systems is found in the same manner as that in the non-repairable cases. Once again let:

$$y = F(x_i) \quad \text{(Eq A-36)}$$

The variance of Y can likewise be estimated as:

$$V(y) = \sum_{i=1}^{n} \left[\frac{\partial y}{\partial x_i}\right]^2 V(x_i) \quad \text{(Eq A-37)}$$

Where:

V(y) = variance of random variable y as defined above

$V(x_i)$ = variance of random variable x_i as defined above

The sensitivity of y with respect to a random variable x_i is:

In the case of the repairable systems, there are several more potentially uncertain or random variables:

- DC is the diagnostic coverage;
- DI is the diagnostic interval;
- TI is the proof test interval,
- λ^D is the dangerous failure rate;
- MTTR is the mean time to restore the system to operation
- β is the common cause failure parameter that is always is between 0 and 1

We must evaluate the partial derivative of the PFD function with respect to each and then combine them using equation A-37. The partial derivatives of function F7 with respect to each random variable are found as follows

$$\frac{\partial F_7}{\partial DC} = \frac{-\lambda^D \times TI}{2} + \frac{\lambda^D \times DI}{2} \quad \text{(Eq A-38)}$$

$$\frac{\partial F_7}{\partial \lambda^D} = \frac{(1-DC) \times TI}{2} + \frac{DC \times DI}{2} + MTTR \quad \text{(Eq A-39)}$$

$$\frac{\partial F_7}{\partial TI} = \frac{(1-DC) \times \lambda^D}{2} \quad \text{(Eq A-40)}$$

$$\frac{\partial F_7}{\partial DI} = \frac{DC \times \lambda^D}{2} \quad \text{(Eq A-41)}$$

$$\frac{\partial F_7}{\partial MTTR} = \lambda^D \qquad \text{(Eq A-42)}$$

We can now compute the variance of function F7 as:

$$V\left(F_7\right) = \sum_{i=1}^{n} \left[\frac{\partial F_7}{\partial x_i}\right]^2 V(x_i) \qquad \text{(Eq A-43)}$$

$$V\left(F_7\right) = \left[\frac{-\lambda^D \times TI}{2} + \frac{\lambda^D \times DI}{2}\right]^2 \text{V(DC)} + \left[\frac{(1-DC)\times TI}{2} + \frac{DC \times DI}{2} + MTTR\right]^2 \text{V(}\lambda^D\text{)+}$$

$$\left[\frac{(1-DC)\times \lambda^D}{2}\right]^2 \text{V(TI)} + \left[\frac{DC \times \lambda^D}{2}\right]^2 \text{(DI)} + \left[\lambda^D\right]^2 \text{V(MTTR)} \qquad \text{(Eq A-44)}$$

Equation 44 represents the variance of the probability of failure on demand of the 1001 configuration when the system is repairable. We note the following relationships between the 1001, 2oo2 and 3oo3 configurations:

$F_{10} = 2\ F_7$ (Eq A-45)

$F_{12} = 3\ F_7$ (Eq A-46)

We can directly determine the variance of functions F_{10} and F_{12} from the properties of the variance operator, V(X).

$V(F_{10}) = V(2\ F_7)$ (Eq A-47)

$V(F_{10}) = 4\ V(F_7)$ (Eq A-48)

$V(F_{12}) = V(3\ F_7)$ (Eq A-49)

$V(F_{12}) = 9\ V(F_7)$ (Eq A-50)

The determination of the variance of functions F_8, F_{10}, and F_{11} is done in a similar manner. To simplify the presentation of the derivation of variance of functions F_8, F_{10}, and F_{11}, we introduce two functions H and Q as:

$$H = \left[\frac{(1-DC)\times(1-\beta)\times\lambda^D \times TI}{2} + \frac{DC\times(1-\beta)\times\lambda^D\times DI}{2} + (1-\beta)\times\lambda^D\times MTTR\right] \qquad \text{(Eq A-51)}$$

$$Q = \left[\frac{(1-DC)\times\beta\times\lambda^D\times TI}{2} + \frac{DC\times\beta\times\lambda^D\times DI}{2} + \beta\times\lambda^D\times MTTR\right] \qquad \text{(Eq A-52)}$$

We note the following relationships:

$F_8 = H^2 + Q$ (Eq A-53)

$F_9 = H^3 + Q$ (Eq A-54)

$F_{11} = 3\ H^2 + Q$ (Eq A-55)

The variance of F_8, F_9 and F_{11} are found as:

$V(F_8\ ) = V(H^2 + Q)$ (Eq A-56)

$V(F_9\ ) = V(H^3 + Q)$ (Eq A-57)

$V(F_{11}) = V(3\ H^2 + Q)$ (Eq A-58)

We will need the sensitivity of F_8, F_9 and F_{11} to each of the potential random variables.

$$\frac{\partial F_8}{\partial x} = \frac{\partial(H^2 + Q)}{\partial x} = \frac{\partial(H^2)}{\partial x} + \frac{\partial(Q)}{\partial x} \quad \text{(Eq A-59)}$$

Likewise

$$\frac{\partial F_9}{\partial x} = \frac{\partial(H^3 + Q)}{\partial x} = \frac{\partial(H^3)}{\partial x} + \frac{\partial(Q)}{\partial x} \quad \text{(Eq A-60)}$$

$$\frac{\partial F_{11}}{\partial x} = \frac{\partial 3(H^2 + Q)}{\partial x} = 3\frac{\partial(H^2)}{\partial x} + 3\frac{\partial(Q)}{\partial x} \quad \text{(Eq A-61)}$$

Using the chain rule for partial derivatives, we now write:

$$\frac{\partial F_8}{\partial x} = 2H\frac{\partial(H)}{\partial x} + \frac{\partial(Q)}{\partial x} \quad \text{(Eq A-62)}$$

$$\frac{\partial F_9}{\partial x} = 3H^2\ \frac{\partial(H)}{\partial x} + \frac{\partial(Q)}{\partial x} \quad \text{(Eq A-63)}$$

$$\frac{\partial F_{11}}{\partial x} = 6H\frac{\partial(H)}{\partial x} + \frac{\partial(Q)}{\partial x} \quad \text{(Eq A-64)}$$

We now evaluate the derivatives of function H with respect to each potential random variables:

- DC is the diagnostic coverage;
- DI is the diagnostic interval;

- TI is the proof test interval,
- λ^D is the dangerous failure rate;
- MTTR is the mean time to restore the system to operation
- β is the common cause failure parameter that is always is between 0 and 1

$$\frac{\partial(H)}{\partial DC} = \frac{-(1-\beta)\times\lambda^D\times TI}{2} + \frac{(1-\beta)\times\lambda^D\times DI}{2} \quad \text{(Eq A-65)}$$

$$\frac{\partial(H)}{\partial TI} = \frac{(1-DC)\times(1-\beta)\times\lambda^D}{2} \quad \text{(Eq A-66)}$$

$$\frac{\partial(H)}{\partial \lambda^D} = \begin{bmatrix} \dfrac{(1-DC)\times(1-\beta)\times TI}{2} + \\ \dfrac{DC\times(1-\beta)\times DI}{2} + (1-\beta)\times MTTR \end{bmatrix} \quad \text{(Eq A-67)}$$

$$\frac{\partial(H)}{\partial MTTR} = (1-\beta)\times\lambda^D \quad \text{(Eq A-68)}$$

$$\frac{\partial(H)}{\partial \beta} = \begin{bmatrix} \dfrac{(1-DC)\times(-1)\times\lambda^D\times TI}{2} + \\ \dfrac{DC\times(-1)\times\lambda^D\times DI}{2} + (-1)\times\lambda^D\times MTTR \end{bmatrix} \quad \text{(Eq A-69)}$$

$$\frac{\partial(H)}{\partial DI} = \frac{DC\times(1-\beta)\times\lambda^D}{2} \quad \text{(Eq A-70)}$$

In a similar manner we evaluate the partial derivatives of function Q with respect to each potential random variables.

$$Q = \left[\frac{(1-DC) \times \beta \times \lambda^D \times TI}{2} + \frac{DC \times \beta \times \lambda^D \times DI}{2} + \beta \times \lambda^D \times MTTR \right] \quad \text{(Eq A-71)}$$

- DC is the diagnostic coverage;
- DI is the diagnostic interval;
- TI is the proof test interval,
- λ^D is the dangerous failure rate;
- MTTR is the mean time to restore the system to operation
- β is the common cause failure parameter that is always is between 0 and 1

$$\frac{\partial(Q)}{\partial DC} = \frac{(-1) \times \beta \times \lambda^D \times TI}{2} + \frac{\beta \times \lambda^D \times DI}{2} \quad \text{(Eq A-72)}$$

$$\frac{\partial(Q)}{\partial TI} = \frac{(1-DC) \times \beta \times \lambda^D}{2} \quad \text{(Eq A-73)}$$

$$\frac{\partial(Q)}{\partial \lambda^D} = \left[\frac{(1-DC) \times \beta \times TI}{2} + \frac{DC \times \beta \times DI}{2} + \beta \times MTTR \right] \quad \text{(Eq A-74)}$$

$$\frac{\partial(Q)}{\partial MTTR} = \beta \times \lambda^D \times MTTR \quad \text{(Eq A-75)}$$

$$\frac{\partial(Q)}{\partial \beta} = \left[\frac{(1-DC) \times \lambda^D \times TI}{2} + \frac{DC \times \lambda^D \times DI}{2} + \lambda^D \times MTTR \right] \quad \text{(Eq A-76)}$$

$$\frac{\partial(Q)}{\partial DI} = \frac{DC \times \beta \times \lambda^D}{2} \quad \text{(Eq A-77)}$$

We can now determine the sensitivity of the functions F8, F9 , F11 to the random variables of interest. Restating equation A-62 for sensitivity of F8 with respect to the random variables.

$$\frac{\partial F_8}{\partial x} = 2H \frac{\partial(H)}{\partial x} + \frac{\partial(Q)}{\partial x} \quad \text{(Eq A-78)}$$

For F_8 we find

$$\frac{\partial F_8}{\partial x} = 2\mathrm{H}\frac{\partial(\mathrm{H})}{\partial x} + \frac{\partial(\mathrm{Q})}{\partial x} \qquad \text{(Eq A-80)}$$

$$\frac{\partial F_8}{\partial DC} = 2\left[\begin{array}{l}\frac{(1-DC)\times(1-\beta)\times\lambda^D\times TI}{2}+\\ \frac{DC\times(1-\beta)\times\lambda^D\times DI}{2}+(1-\beta)\times\lambda^D\times MTTR\end{array}\right]\times$$
$$\left[\frac{-(1-\beta)\times\lambda^D\times TI}{2}+\frac{(1-\beta)\times\lambda^D\times DI}{2}\right]+$$
$$\frac{(-1)\times\beta\times\lambda^D\times TI}{2}+\frac{\beta\times\lambda^D\times DI}{2} \qquad \text{(Eq A-79)}$$

$$\frac{\partial F_8}{\partial TI} = 2\left[\begin{array}{l}\frac{(1-DC)\times(1-\beta)\times\lambda^D\times TI}{2}+\\ \frac{DC\times(1-\beta)\times\lambda^D\times DI}{2}+(1-\beta)\times\lambda^D\times MTTR\end{array}\right]\times\left[\frac{(1-DC)\times(1-\beta)\times\lambda^D}{2}\right]+$$
$$\frac{(1-DC)\times\beta\times\lambda^D}{2} \qquad \text{(Eq A-81)}$$

$$\frac{\partial F_8}{\partial MTTR} = 2\left[\begin{array}{l}\frac{(1-DC)\times(1-\beta)\times\lambda^D\times TI}{2}+\\ \frac{DC\times(1-\beta)\times\lambda^D\times DI}{2}+(1-\beta)\times\lambda^D\times MTTR\end{array}\right]\times$$
$$\left[(1-\beta)\times\lambda^D\right]+\beta\times\lambda^D\times MTTR \qquad \text{(Eq A-82)}$$

$$\frac{\partial F_8}{\partial \beta} = 2\left[\begin{array}{l}\frac{(1-DC)\times(1-\beta)\times\lambda^D\times TI}{2}+\\ \frac{DC\times(1-\beta)\times\lambda^D\times DI}{2}+(1-\beta)\times\lambda^D\times MTTR\end{array}\right]\times$$

$$\left[\left[\begin{array}{l}\frac{(1-DC)\times(-1)\times\lambda^D\times TI}{2}+\\ \frac{DC\times(-1)\times\lambda^D\times DI}{2}+(-1)\times\lambda^D\times MTTR\end{array}\right]\right]+$$

$$\left[\frac{(1-DC)\times\lambda^D\times TI}{2}+\frac{DC\times\lambda^D\times DI}{2}+\lambda^D\times MTTR\right] \qquad \text{(Eq A-83)}$$

$$\frac{\partial F_8}{\partial DI} = 2\left[\begin{array}{l}\frac{(1-DC)\times(1-\beta)\times\lambda^D\times TI}{2}+\\ \frac{DC\times(1-\beta)\times\lambda^D\times DI}{2}+(1-\beta)\times\lambda^D\times MTTR\end{array}\right]\times\left[\frac{DC\times(1-\beta)\times\lambda^D}{2}\right]+$$

$$\frac{DC\times\beta\times\lambda^D}{2} \qquad \text{(Eq A-84)}$$

$$\left[\frac{\partial F_8}{\partial \lambda^D}\right] = 2\left[\begin{array}{l}\frac{(1-DC)\times(1-\beta)\times\lambda^D\times TI}{2}+\\ \frac{DC\times(1-\beta)\times\lambda^D\times DI}{2}+(1-\beta)\times\lambda^D\times MTTR\end{array}\right]\times\left[\begin{array}{l}\frac{(1-DC)\times(1-\beta)\times TI}{2}+\\ \frac{DC\times(1-\beta)\times DI}{2}+(1-\beta)\times MTTR\end{array}\right]+$$

$$\left[\frac{(1-DC)\times\beta\times TI}{2}+\frac{DC\times\beta\times DI}{2}+\beta\times MTTR\right] \qquad \text{(Eq A-85)}$$

Using the sensitivities for F8 calculated above, the variance in F8 due to uncertain or random variables is found as:

$$V(F_8) = \left[\frac{\partial F_8}{\partial DC}\right]^2 V(DC) + \left[\frac{\partial F_8}{\partial TI}\right]^2 V(TI) + \left[\frac{\partial F_8}{\partial \lambda^D}\right]^2 V(\lambda^D) +$$

$$\left[\frac{\partial F_8}{\partial MTTR}\right]^2 V(MTTR) + \left[\frac{\partial F_8}{\partial \beta}\right]^2 V(\beta) + \left[\frac{\partial F_8}{\partial DI}\right]^2 V(DI) \qquad \text{(Eq A-86)}$$

Using the equation A-63, we now determine the variance of F_9.

$$\frac{\partial F_9}{\partial x} = 3\text{H}^2 \frac{\partial(\text{H})}{\partial x} + \frac{\partial(\text{Q})}{\partial x} \qquad \text{(Eq A-87)}$$

$$\frac{\partial F_9}{\partial DC} = 3\left[\begin{array}{l} \dfrac{(1-DC)\times(1-\beta)\times\lambda^D\times TI}{2} + \\ \dfrac{DC\times(1-\beta)\times\lambda^D\times DI}{2} + (1-\beta)\times\lambda^D\times MTTR \end{array}\right]^2 \times$$

$$\left[\frac{-(1-\beta)\times\lambda^D\times TI}{2} + \frac{(1-\beta)\times\lambda^D\times DI}{2}\right] +$$

$$\frac{(-1)\times\beta\times\lambda^D\times TI}{2} + \frac{\beta\times\lambda^D\times DI}{2} \qquad \text{(Eq A-88)}$$

$$\frac{\partial F_9}{\partial TI} = 3\left[\frac{(1-DC)\times(1-\beta)\times\lambda^D\times TI}{2} + \frac{DC\times(1-\beta)\times\lambda^D\times DI}{2} + (1-\beta)\times\lambda^D\times MTTR\right]^2 \times \left[\frac{(1-DC)\times(1-\beta)\times\lambda^D}{2}\right] +$$

$$\frac{(1-DC)\times\beta\times\lambda^D}{2} \qquad \text{(Eq A-89)}$$

$$\frac{\partial F_9}{\partial MTTR} = 3\left[\frac{(1-DC)\times(1-\beta)\times\lambda^D\times TI}{2} + \frac{DC\times(1-\beta)\times\lambda^D\times DI}{2} + (1-\beta)\times\lambda^D\times MTTR\right]^2 \times \left[(1-\beta)\times\lambda^D\right] +$$

$$\beta\times\lambda^D\times MTTR \qquad \text{(Eq A-90)}$$

$$\frac{\partial F_9}{\partial \beta} = 3\left[\frac{(1-DC)\times(1-\beta)\times\lambda^D\times TI}{2} + \frac{DC\times(1-\beta)\times\lambda^D\times DI}{2} + (1-\beta)\times\lambda^D\times MTTR\right]^2 \times$$

$$\left[\left[\frac{(1-DC)\times(-1)\times\lambda^D\times TI}{2} + \frac{DC\times(-1)\times\lambda^D\times DI}{2} + (-1)\times\lambda^D\times MTTR\right]\right] +$$

$$\left[\frac{(1-DC)\times\lambda^D\times TI}{2} + \frac{DC\times\lambda^D\times DI}{2} + \lambda^D\times MTTR\right] \qquad \text{(Eq A-91)}$$

$$\frac{\partial F_9}{\partial DI} = 3\left[\begin{array}{l}\frac{(1-DC)\times(1-\beta)\times\lambda^D\times TI}{2}+ \\ \frac{DC\times(1-\beta)\times\lambda^D\times DI}{2}+(1-\beta)\times\lambda^D\times MTTR\end{array}\right]^2 \times\left[\frac{DC\times(1-\beta)\times\lambda^D}{2}\right]+$$

$$\frac{DC\times\beta\times\lambda^D}{2} \qquad \text{(Eq A-92)}$$

$$\left[\frac{\partial F_9}{\partial\lambda^D}\right] = 3\times\left[\begin{array}{l}\frac{(1-DC)\times(1-\beta)\times\lambda^D\times TI}{2}+ \\ \frac{DC\times(1-\beta)\times\lambda^D\times DI}{2}+(1-\beta)\times\lambda^D\times MTTR\end{array}\right]^2\times$$

$$\left[\begin{array}{l}\frac{(1-DC)\times(1-\beta)\times TI}{2}+ \\ \frac{DC\times(1-\beta)\times DI}{2}+(1-\beta)\times MTTR\end{array}\right]+$$

$$\left[\frac{(1-DC)\times\beta\times TI}{2}+\frac{DC\times\beta\times DI}{2}+\beta\times MTTR\right] \qquad \text{(Eq A-93)}$$

Using the sensitivities for F_9 calculated above, the variance in F_9 due to uncertain or random variables is found as:

$$V(F_9) = \left[\frac{\partial F_9}{\partial DC}\right]^2 V(DC) + \left[\frac{\partial F_9}{\partial TI}\right]^2 V(TI) + \left[\frac{\partial F_9}{\partial\lambda^D}\right]^2 V(\lambda^D) +$$

$$\left[\frac{\partial F_9}{\partial MTTR}\right]^2 V(MTTR) + \left[\frac{\partial F_9}{\partial\beta}\right]^2 V(\beta) + \left[\frac{\partial F_9}{\partial DI}\right]^2 V(DI) \qquad \text{(Eq A-94)}$$

Finally, in a similar manner the variance in F_{11} is computed as:

$$\frac{\partial F_{11}}{\partial x} = 6\text{H}\frac{\partial(\text{H})}{\partial x} + \frac{\partial(\text{Q})}{\partial x} \quad \text{(Eq A-95)}$$

$$\frac{\partial F_{11}}{\partial DC} = 6\left[\begin{array}{l}\dfrac{(1-DC)\times(1-\beta)\times\lambda^{D}\times TI}{2} + \\ \dfrac{DC\times(1-\beta)\times\lambda^{D}\times DI}{2} + (1-\beta)\times\lambda^{D}\times MTTR\end{array}\right] \times$$

$$\left[\frac{-(1-\beta)\times\lambda^{D}\times TI}{2} + \frac{(1-\beta)\times\lambda^{D}\times DI}{2}\right] +$$

$$\frac{(-1)\times\beta\times\lambda^{D}\times TI}{2} + \frac{\beta\times\lambda^{D}\times DI}{2} \quad \text{(Eq A-96)}$$

$$\frac{\partial F_{11}}{\partial TI} = 6\left[\begin{array}{l}\dfrac{(1-DC)\times(1-\beta)\times\lambda^{D}\times TI}{2} + \\ \dfrac{DC\times(1-\beta)\times\lambda^{D}\times DI}{2} + (1-\beta)\times\lambda^{D}\times MTTR\end{array}\right] \times \left[\frac{(1-DC)\times(1-\beta)\times\lambda^{D}}{2}\right] +$$

$$\frac{(1-DC)\times\beta\times\lambda^{D}}{2} \quad \text{(Eq A-97)}$$

$$\frac{\partial F_{11}}{\partial MTTR} = 6\left[\begin{array}{l}\dfrac{(1-DC)\times(1-\beta)\times\lambda^{D}\times TI}{2} + \\ \dfrac{DC\times(1-\beta)\times\lambda^{D}\times DI}{2} + (1-\beta)\times\lambda^{D}\times MTTR\end{array}\right] \times \left[(1-\beta)\times\lambda^{D}\right] +$$

$$\beta\times\lambda^{D}\times MTTR \quad \text{(Eq A-98)}$$

$$\frac{\partial F_{11}}{\partial \beta}=6\left[\begin{array}{l}\frac{(1-DC)\times(1-\beta)\times\lambda^D\times TI}{2}+\\ \frac{DC\times(1-\beta)\times\lambda^D\times DI}{2}+(1-\beta)\times\lambda^D\times MTTR\end{array}\right]\times$$

$$\left[\left[\begin{array}{l}\frac{(1-DC)\times(-1)\times\lambda^D\times TI}{2}+\\ \frac{DC\times(-1)\times\lambda^D\times DI}{2}+(-1)\times\lambda^D\times MTTR\end{array}\right]\right]+$$

$$\left[\frac{(1-DC)\times\lambda^D\times TI}{2}+\frac{DC\times\lambda^D\times DI}{2}+\lambda^D\times MTTR\right] \quad \text{(Eq A-99)}$$

$$\frac{\partial F_{11}}{\partial DI}=6\left[\begin{array}{l}\frac{(1-DC)\times(1-\beta)\times\lambda^D\times TI}{2}+\\ \frac{DC\times(1-\beta)\times\lambda^D\times DI}{2}+(1-\beta)\times\lambda^D\times MTTR\end{array}\right]\times\left[\frac{DC\times(1-\beta)\times\lambda^D}{2}\right]+$$

$$\frac{DC\times\beta\times\lambda^D}{2} \quad \text{(Eq A-100)}$$

$$\left[\frac{\partial F_{11}}{\partial \lambda^D}\right]=6\left[\begin{array}{l}\frac{(1-DC)\times(1-\beta)\times\lambda^D\times TI}{2}+\\ \frac{DC\times(1-\beta)\times\lambda^D\times DI}{2}+(1-\beta)\times\lambda^D\times MTTR\end{array}\right]\times$$

$$\left[\begin{array}{l}\frac{(1-DC)\times(1-\beta)\times TI}{2}+\\ \frac{DC\times(1-\beta)\times DI}{2}+(1-\beta)\times MTTR\end{array}\right]+$$

$$\left[\frac{(1-DC)\times\beta\times TI}{2}+\frac{DC\times\beta\times DI}{2}+\beta\times MTTR\right] \quad \text{(Eq A-101)}$$

Using the sensitivities for F_{11} calculated above, the variance in F_{11} due to uncertain or random variables is found as:

$$V(F_{11}) = \left[\frac{\partial F_{11}}{\partial DC}\right]^2 V(DC) + \left[\frac{\partial F_{11}}{\partial TI}\right]^2 V(TI) + \left[\frac{\partial F_{11}}{\partial \lambda^D}\right]^2 V(\lambda^D) +$$

$$\left[\frac{\partial F_{11}}{\partial MTTR}\right]^2 V(MTTR) + \left[\frac{\partial F_{11}}{\partial \beta}\right]^2 V(\beta) + \left[\frac{\partial F_{11}}{\partial DI}\right]^2 V(DI) \qquad \text{(Eq A-102)}$$

PROCEDURE FOR DETERMINATION OF VARIANCE FOR REPAIRABLE 1oo2, 1oo3 AND 2oo3 SYSTEMS

Based on the above derivations we may now write the formulas for the variance of various configurations of repairable equipment.

Step 1. Determine the configuration of equipment to be studied.

Step 2. Evaluate Function H using the expected value of all random or uncertain variables, equation A-51

Step 3. Evaluate Function Q using the expected value of all random or uncertain variables, equation A-52

Step 4. Evaluate the sensitivity of function H with respect to the random variables using equations A-65 thru A-70.

Step 5. Evaluate the sensitivity of function Q with respect to the random variables using equations A-72 thru A-77.

Step 6. Evaluate the variance as:

a. For 1oo2 systems use equation A-86 for $V(F_8)$

b. For 1oo3 systems use equation A-94 for $V(F_9)$

c. For 2oo3 systems use equation A-102 for $V(F_{11})$

The end result of the above calculations is the determination of the mean and variance of the configuration used in the sensor, logic solver or final control element systems. Table 4 of the main body of this paper presents a cross reference roadmap for the determination of the mean and variance of various of the probability of failure on demand (PFD) of various hardware configurations. The mean and variance for the particular configuration is then returned to the overall calculations of the PFD of the interlock.

APPENDIX B -- UNCERTAINTY EQUATIONS FOR CONFIGURATIONS NOT COVERED IN ISA TR 84.00.02

The ISA Technical report [Ref. 2] does not define the calculation equations for the PFDavg for non-repairable systems where duplicated devices create the potential for common cause failures. This appendix derives the equations needed to evaluate the uncertainty for non-repairable systems with the potential for common cause failures.

To derive the needed functions for the non-repairable systems with common cause failure (CCF) potential, we start with the equations for the repairable system with common cause failure allowed (Table A.4). First we note that a system composed of only one set of components (1oo1) without duplication has no potential for a common cause failure. In addition, a review of the equations of Table 2 finds that systems 1oo1, 2oo2 and 3oo3 do not contain the common cause failure (CCF) parameter β. Therefore, systems 1oo1, 2oo2, and 3oo3 can be eliminated from consideration as common cause failures cannot lead to failure of the specified interlocks.

For the other situations [1oo2, 1oo3 and 2oo3] we note that no repair will be allowed during the mission of the non-repairable system. This observation implies that the diagnostic coverage (DC), the diagnostic interval (DI) and the mean time for restoration (MTTR) are all zero (0). We will set DC = DI = MTTR = 0

Substitution of the zero value for DC, DI and MTTR into the equations of Table A.4 yields the non-repairable system PFD equations where CCF is allowed. Table B.1 presents the results of these algebraic substitutions.

Table B.1 Simplified PFD_{avg} formulas for Non-Repairable System Allowing for CCF

Configuration	Function	PFD based on "Average before" failure rate	Equation Number
1oo2	F_{13} =	$\left[\dfrac{(1-\beta)\times\lambda^D\times TI}{2}\right]^2+\left[\dfrac{\beta\times\lambda^D\times TI}{2}\right]$	B-1
1oo3	F_{14} =	$\left[\dfrac{(1-\beta)\times\lambda^D\times TI}{2}\right]^3+\left[\dfrac{\beta\times\lambda^D\times TI}{2}\right]$	B-2
2oo3	F_{15} =	$3\times\left[\dfrac{(1-\beta)\times\lambda^D\times TI}{2}\right]^2+\left[\dfrac{\beta\times\lambda^D\times TI}{2}\right]$	B-3

Where TI is the proof test interval, λ^D is the dangerous failure rate.

The expected value of the probability of failure on demand is found by calculating the PFD using the expected value (mean) of the random variables.

Let y a function of some random variables, x_i, as:

$$y = F(x_i) \quad \text{(Eq B-4)}$$

The mean of y can be estimated using the following approximation:

$$E(y) = F[E(x_i)] \quad \text{(Eq B-5)}$$

Where:

$E(y)$ = expected value of random variable y = mean of y

$E(x_i)$ = expected value of random variable x_i = mean of x_i

For the configurations defined in Table A.1, the corresponding expected PFD are presented in Table B.2.

Table B.2 Expected Value of PFD_{avg} formulas for Non-Repairable System Allowing CCF			
Configuration	**Function**	**Expected PFD based on "Average before" failure rate**	**Equation Number**
1oo2	$E(F_{13})$ =	$\left[\frac{(1-E(\beta))\times E(\lambda^D)\times E(TI)}{2}+\right]^2 + \left[\frac{E(\beta)\times E(\lambda^D)\times E(TI)}{2}\right]$	B-6
1oo3	$E(F_{14})$ =	$\left[\frac{(1-E(\beta))\times E(\lambda^D)\times E(TI)}{2}+\right]^3 + \left[\frac{E(\beta)\times E(\lambda^D)\times E(TI)}{2}\right]$	B-7
2oo3	$E(F_{15})$ =	$3\times\left[\frac{(1-E(\beta))\times E(\lambda^D)\times E(TI)}{2}+\right]^2 + \left[\frac{E(\beta)\times E(\lambda^D)\times E(TI)}{2}\right]$	B-8

Where E(x) is the expected value of random variable x.

The variance of the PFD is found by taking the weighted sum of the variance of the random variables. The weighting function is the sensitivity of the PFD with respect to random variable. In mathematical terms the variance of a function may be found in terms of the variance of the random variables as:

Once again let

$$y = F(x_i) \qquad \text{(Eq B-9)}$$

The variance of Y can likewise be estimated as:

$$V(y) = \sum_{i=1}^{n} \left[\frac{\partial y}{\partial x_i}\right]^2 V(x_i) \qquad \text{(Eq B-10)}$$

Where:

V(y) = variance of random variable y as defined above

$V(x_i)$ = variance of random variable x_i as defined above

The sensitivity of y with respect to a random variable x_i is:

$$\text{Sensitivity of y with respect to } x_i = \left[\frac{\partial y}{\partial x_i}\right] \qquad \text{(Eq B-11)}$$

The variance of the PFD for the repairable cases 1oo2, 1oo3 and 2oo3 are presented in equations of Appendix A as:

- 1oo2, the variance is specified by equation A-86
- 1oo3, the variance is specified by equation A-94
- 2oo3, the variance is specified by equation A-102

These three equations are the basis upon which the equations for the non-repairable system with common cause failure (CCF) are developed.

Variance of PFD for Non-Repairable 1oo2 with CCF

To derive the variance for F_{13} we start with the general equation for the variance as:

$$V(F_{13}) = \left[\frac{\partial F_{13}}{\partial DC}\right]^2 V(DC) + \left[\frac{\partial F_{13}}{\partial TI}\right]^2 V(TI) + \left[\frac{\partial F_{13}}{\partial \lambda^D}\right]^2 V(\lambda^D) +$$

$$\left[\frac{\partial F_{13}}{\partial MTTR}\right]^2 V(MTTR) + \left[\frac{\partial F_{13}}{\partial \beta}\right]^2 V(\beta) + \left[\frac{\partial F_{13}}{\partial DI}\right]^2 V(DI) \qquad \text{(Eq B-12)}$$

The variance of a constant is zero. Thus, setting DC=DI=MTTR =0 results in the variance of these variables as becoming zero.

$$V(DC) = V(DI) = V(MTTR) = 0 \qquad \text{(Eq B-13)}$$

Thus, the variance for the 1oo2 case becomes:

$$V(F_{13}) = \left[\frac{\partial F_{13}}{\partial TI}\right]^2 V(TI) + \left[\frac{\partial F_{13}}{\partial \lambda^D}\right]^2 V(\lambda^D) + \left[\frac{\partial F_{13}}{\partial \beta}\right]^2 V(\beta) \qquad \text{(Eq B-14)}$$

The evaluation of the sensitive's of the PFD may now be taken from appendix A equations as:

$\left[\frac{\partial F_{13}}{\partial TI}\right]$ from equation A-81

$\left[\frac{\partial F_{13}}{\partial \lambda^D}\right]$ from equation A-85

$\left[\frac{\partial F_{13}}{\partial \beta}\right]$ from equation A-83

Substituting DC=DI=MTTR=0 into these sensitivity equations yields:

$$\frac{\partial F_{13}}{\partial TI} = 2\left[\frac{(1-\beta)\times\lambda^D\times TI}{2} + \right]\times\left[\frac{(1-\beta)\times\lambda^D}{2}\right] + \frac{\beta\times\lambda^D}{2} \qquad \text{(Eq B-15)}$$

$$\frac{\partial F_{13}}{\partial \lambda^D} = 2\left[\frac{(1-\beta)\times\lambda^D\times TI}{2} + \right]\times\left[\frac{(1-\beta)\times TI}{2}\right] + \left[\frac{\beta\times TI}{2}\right] \qquad \text{(Eq B-16)}$$

$$\frac{\partial F_{13}}{\partial \beta} = 2\left[\frac{(1-\beta)\times\lambda^D\times TI}{2}\right]\times\left[\frac{(-1)\times\lambda^D\times TI}{2}\right] + \left[\frac{\lambda^D\times TI}{2}\right] \qquad \text{(Eq B-17)}$$

Variance of PFD for Non-Repairable 1oo3 with CCF

To derive the variance for F_{14} we start with the general equation for the variance as:

$$V(F_{14}) = \left[\frac{\partial F_{14}}{\partial DC}\right]^2 V(DC) + \left[\frac{\partial F_{14}}{\partial TI}\right]^2 V(TI) + \left[\frac{\partial F_{14}}{\partial \lambda^D}\right]^2 V(\lambda^D) +$$

$$\left[\frac{\partial F_{14}}{\partial MTTR}\right]^2 V(MTTR) + \left[\frac{\partial F_{14}}{\partial \beta}\right]^2 V(\beta) + \left[\frac{\partial F_{14}}{\partial DI}\right]^2 V(DI) \qquad \text{(Eq B-18)}$$

The variance of a constant is zero. Thus, setting DC=DI=MTTR =0 results in the variance of these variables as becoming zero.

$$V(DC) = V(DI) = V(MTTR) = 0 \qquad \text{(Eq B-19)}$$

Thus, the variance for the 1oo3 case becomes:

$$V(F_{14}) = \left[\frac{\partial F_{14}}{\partial TI}\right]^2 V(TI) + \left[\frac{\partial F_{14}}{\partial \lambda^D}\right]^2 V(\lambda^D) + \left[\frac{\partial F_{14}}{\partial \beta}\right]^2 V(\beta) \qquad \text{(Eq B-20)}$$

The evaluation of the sensitive's of the PFD may now be taken from appendix A equations as:

$\left[\frac{\partial F_{14}}{\partial TI}\right]$ from equation A-89

$\left[\frac{\partial F_{14}}{\partial \lambda^D}\right]$ from equation A-93

$\left[\frac{\partial F_{14}}{\partial \beta}\right]$ from equation A-91

Substituting DC=DI=MTTR=0 into these sensitivity equations yields:

$$\frac{\partial F_{14}}{\partial TI} = 3\left[\frac{(1-\beta)\times\lambda^D\times TI}{2} + \right]^2 \times \left[\frac{(1-\beta)\times\lambda^D}{2}\right] + \frac{\beta\times\lambda^D}{2} \qquad \text{(Eq B-21)}$$

$$\frac{\partial F_{14}}{\partial \lambda^D} = 3\left[\frac{(1-\beta)\times\lambda^D\times TI}{2}\right]^2 \times \left[\frac{(1-\beta)\times TI}{2}\right] + \left[\frac{\beta\times TI}{2}\right] \qquad \text{(Eq B-22)}$$

$$\frac{\partial F_{14}}{\partial \beta} = 3\left[\frac{(1-\beta)\times \lambda^{D} \times TI}{2}\right]^{2} \times \left[\frac{(-1)\times \lambda^{D} \times TI}{2}\right] + \left[\frac{\lambda^{D} \times TI}{2}\right] \qquad \text{(Eq B-23)}$$

Variance of PFD for Non-Repairable 2oo3 with CCF

To derive the variance for F_{15} we start with the general equation for the variance as:

$$V(F_{15}) = \left[\frac{\partial F_{15}}{\partial DC}\right]^{2} V(DC) + \left[\frac{\partial F_{15}}{\partial TI}\right]^{2} V(TI) + \left[\frac{\partial F_{15}}{\partial \lambda^{D}}\right]^{2} V(\lambda^{D}) +$$

$$\left[\frac{\partial F_{15}}{\partial MTTR}\right]^{2} V(MTTR) + \left[\frac{\partial F_{15}}{\partial \beta}\right]^{2} V(\beta) + \left[\frac{\partial F_{15}}{\partial DI}\right]^{2} V(DI) \qquad \text{(Eq B-24)}$$

The variance of a constant is zero. Thus, setting DC=DI=MTTR =0 results in the variance of these variables as becoming zero.

$$V(DC) = V(DI) = V(MTTR) = 0 \qquad \text{(Eq B-25)}$$

Thus, the variance for the 2oo3 case becomes:

$$V(F_{15}) = \left[\frac{\partial F_{15}}{\partial TI}\right]^{2} V(TI) + \left[\frac{\partial F_{15}}{\partial \lambda^{D}}\right]^{2} V(\lambda^{D}) + \left[\frac{\partial F_{15}}{\partial \beta}\right]^{2} V(\beta) \qquad \text{(Eq B-26)}$$

The evaluation of the sensitive's of the PFD may now be taken from appendix A equations as:

$\left[\frac{\partial F_{15}}{\partial TI}\right]$ from equation A-97

$\left[\frac{\partial F_{15}}{\partial \lambda^{D}}\right]$ from equation A-101

$\left[\frac{\partial F_{15}}{\partial \beta}\right]$ from equation A-99

Substituting DC=DI=MTTR=0 into these sensitivity equations yields:

$$\frac{\partial F_{15}}{\partial TI} = 6\left[\frac{(1-\beta)\times\lambda^{D}\times TI}{2}+\right]\times\left[\frac{(1-\beta)\times\lambda^{D}}{2}\right]+\frac{\beta\times\lambda^{D}}{2} \quad \text{(Eq B-27)}$$

$$\frac{\partial F_{15}}{\partial \lambda^{D}} = 6\left[\frac{(1-\beta)\times\lambda^{D}\times TI}{2}\right]\times\left[\frac{(1-\beta)\times TI}{2}\right]+\left[\frac{\beta\times TI}{2}\right] \quad \text{(Eq B-28)}$$

$$\frac{\partial F_{15}}{\partial \beta} = 6\left[\frac{(1-\beta)\times\lambda^{D}\times TI}{2}\right]\times\left[\frac{(-1)\times\lambda^{D}\times TI}{2}\right]+\left[\frac{\lambda^{D}\times TI}{2}\right] \quad \text{(Eq B-29)}$$

The evaluation of the functions for the variance are always done using the expected value of the variable. For example when the variable TI appears in an equation, the numerical evaluation of the function would be using E(TI) where E(TI) is the expected value of the random variable TI.

Comparison of Batch versus Continuous Process in Pharmaceutical Industry Based on Safety Consideration

Shiqi Chen
Mary Kay O'Connor Process Safety Center
Artie McFerrin Department of Chemical Engineering
Texas A&M University
College Station, Texas 77843-3122, USA Email address
shiqi_chen@tamu.edu

Yogesh Koirala, M. Sam Mannan
Mary Kay O'Connor Process Safety Center
Artie McFerrin Department of Chemical Engineering
Texas A&M University
College Station, Texas 77843-3122, USA Email address
ykoirala@tamu.edu, mannan@tamu.edu

Prepared for Presentation at
American Institute of Chemical Engineers
2017 Spring Meeting and 13th Global Congress on Process Safety
San Antonio, TX
March 26 – 29, 2017

Comparison of Batch versus Continuous Process in Pharmaceutical Industry Based on Safety Consideration

Shiqi Chen
Mary Kay O'Connor Process Safety Center
Artie McFerrin Department of Chemical Engineering
Texas A&M University
College Station, Texas 77843-3122, USA Email address
shiqi_chen@tamu.edu

Yogesh Koirala, M. Sam Mannan
Mary Kay O'Connor Process Safety Center
Artie McFerrin Department of Chemical Engineering

Keywords: Continuous Process, Hazard Identification, Pharmaceutical Manufacturing

Abstract

As opposed to the petro-chemical and bulk chemical industry, where continuous processes are widely applied, the pharmaceutical industry still primarily relies on traditional batch process due to the complexity of product, multi-step operation and low-volume production. Considering these conditions, the versatile batch process is more appropriate. Nowadays, driven by the contradiction between increasing demand for drugs and inefficient batch production mode, there is a trend in pharmaceutical industry, that is the transformation from traditional batch process to novel continuous process. Related projects and research that is aimed at analyzing this transition are conducted in worldwide, and the scale of these studies ranges from lab-scale reactions to overall arrangement of the factories.

Although continuous pharmaceutical process is thriving, the safety issue in this field is not promoted at the same time. Since the continuous pharmaceutical process is a novel technology, little information can be provided to evaluate its safety level. Moreover, process conditions are usually intensified for continuous process comparing to batch process. It also may bring potential risks and make continuous manufacturing inappropriate for some of pharmaceutical productions.

This research provides a comprehensive comparison for batch versus continuous pharmaceutical process by application of Dow's Fire and Explosion Index. In addition to this conventional safety evaluation, influences from production efficiency and specialties in pharmaceutical production are integrated into the comparison. Production of 2-methyl benzimidazole and peracetic acid via batch and continuous processes are conducted in this research. Via these integrative and systematic studies, safety levels of batch and continuous processes is analyzed and compared. The ways in which process conditions, production efficiency, and other requirements influence safety level for different production modes are illustrated.

1 Introduction

Continuous process is widely applied in petro-chemical and bulk chemical industry for its high production rate, automated operation and saving in cost[1]. In contrast, the pharmaceutical industry still primarily relies on traditional batch process. This aspect stems from intrinsic limitations in active pharmaceutical ingredients (APIs) production. Complexity of medical substances means multistep synthesis. An average of 8.1 steps is needed for one API synthesis[2]. Another limitation is production volume. Demand for drugs are not comparable with most of the bulk chemicals. These limitations make batch process dominate pharmaceutical industry for a long time.

However, due to the gradual expansion of the pharmaceutical market, batch process is behind the time because of its low production rate caused by low yield and inefficient operation mode. Driven by this contradiction between increasing demand for drugs and inefficient production mode, many pharmaceutical manufacturers are trying to shift from a traditional batch process to a novel continuous process. Experimental and industrial trials that are aimed at this transitioning have been done and exhibit the practicability of this new production mode. Several companies such as Lonza, Phoenix Chemical, and Corning make significant progress in this area[3-5]. Novel continuous process exhibits great potential for the pharmaceutical industry.

"Which process should be applied?" and "Is there enough benefit to move forward?" are the questions that are raised when an innovative process is introduced. To answer these questions, comprehensive comparison between batch process and continuous process is required.

This comparison should be specific in the early stages. For example, for a given substance, based on its intrinsic properties and demand in market, reasonable comparison is required to make a selection between batch and continuous process. After composing a conclusive summary that is based on number of different specific cases, the result of these comparisons can be modified and applied to generate a practical process screening methodology.

For the questions mentioned above, economic benefits, safety and environmental impacts should all be considered to get a reliable answer. In the batch versus continuous process comparisons, related economic analysis and environmental analysis have been done by many researchers[6-8]. However, there remains gaps in research and understanding of safety considerations in comparison of batch versus continuous process. Therefore, the need arises for related comparison within safety consideration and systematic methodology for safety evaluation.

In the following sessions, the development in continuous pharmaceutical process, the previous studies in process screening, selecting and comparing, and concept of inherent safety assessment are discussed. The statement of research objective, the description of methodology and results of case studies are illuminated as well.

2 Literature Survey

To decide whether to head to novel continuous process or not, evaluations of batch and continuous process are in demand for pharmaceutical companies.

In preliminary stage of the evaluation, fundamental information about reaction kinetics, heat transmission and mass transfer is collected and analyzed. The objective of this investigation is looking into the technical feasibility of chosen reaction. After these basic assessments for intrinsic properties of reaction, far-reaching analyze in economic, environmental and safety aspect can be achieved. In this more comprehensive comparison based on overall picture, more factors such as expenditure, wastewater treatment and loss control measure should be covered. Among these factors, which factor should be selected as criterion, how these factors interact, and what's the correlation between these factors and comparison results are questions require to be answered.

To answer these questions, some researchers put efforts in individual comparisons of manufacturing for specific active pharmaceutical ingredient (API). Most of these studies focus on economic advantage of continuous process. Spencer D. Schaber et al.[12] estimated the capital, operating and the overall cost for production of a given API from synthesis to tablet formulation by integrated continuous and batch pharmaceutical manufacturing respectively. Based upon the equal annual yield, the result shows that the capital expenditures for continuous production could be markedly lower than those of batch mode. Differences in operating and overall cost are not such significant but potential saving by material handling and labor in continuous process is explicitly illustrated in the study. Hessel, V. et al. compared continuous vs. batch pharmaceutical manufacturing for production of ibuprofen and artemisinin on the standpoint of cost and environmental impact. Capital expenditure and operating expenditure savings are up to 57.0% and 51.6% respectively for ibuprofen, while capital expenditure and operating expenditure savings are up to 19.6% and 29.3% respectively for artemisinin[6, 8]. These study shows the potential economic benefits in application of continuous pharmaceutical processes.

In addition to process comparison, some researchers developed general criterions or created systematic methodologies for process selection. Dominique M. Roberge et al.[13] classified reactions used in the fine and pharmaceutical industry into 4 categories according to kinetic properties. Based on this classification, the very fast reactions (reactions time less than 1s) and the fast reactions (reactions occurring from 1s to10min) would benefit the most from continuous process. These two types of reactions occupy 44% of all reactions investigated. Ryan L. Harman and Jonathan P. McMullen[14] developed a decision roadmap of flow process for use in lab-scale. Considering objectives of the study, the intrinsic properties of the chemistry and requirements for mixing, heat transferring, related questions are asked at the various nodes in this flow map to get the final choice. Gary S. Calabrese and Sergio Pissavini[15] established an conceptual methodology to evaluate applicability of continuous reactor in the form of reaction properties test. They measured applicability of flow reactor via series of questions. These questions covered issues in reaction properties and some requirements on process. Three "zones" that tagged with different level of applicability are defined. Different questions are included for each "zone". For any given reaction, the more answers with "yes" in the "zone", the higher possibility of this reaction to be at the level of applicability that the "zone" stands for.

3 Problem Statement

Although advantages of continuous process in pharmaceutical manufacturing are illustrated by many studies, it is still too early to claim that continuous process is the right direction to go forward. Several gaps in the literature survey are required to be focused on.

Firstly, most of process selection methodologies are confined to analyzing separate properties of reaction taken in process, rather than evaluation in a systematic way. For example, small inventory is usually identified as one of significant advantage of continuous process, however, it is possible that to achieve similar yield in the corresponding batch process, the reaction in continuous process requires more intensified conditions like higher temperature, higher pressure, and higher concentration of hazardous chemicals. Selection and comparison by independent and individual parameters without integration is not enough.

Secondly, existing comparisons between batch and continuous processes mainly focus on economic aspect, while lacking combination of safety issue and other important aspects from real production. Although the driving force for the application of this novel process is the economic interests, safety issues directly determine its feasibility. A choice between batch and continuous process needs to be made based on business performance within acceptable safety levels. Those comparisons that merely consider the profits are not enough.

Thirdly, the existing studies are limited to fixed production system. Potential problems in a different scale are not considered. In the pharmaceutical industry, gradually increasing production scales is required which is different from the bulk chemical industry. Hence, safety of process conditions needs to be guaranteed for all scale production. Comparison only based on fixed scale is not comprehensive.

Therefore, the primary objective of this study is to provide a comprehensive and systematic comparison between batch and continuous process used in pharmaceutical industry. Then on the basis of this analysis, this study will evaluate how main factors considered in Dow F&EI impact the safety level for these two different processes. Finally, this study will provide support to further improve existing methodology for process screening and selection for pharmaceutical industry.

It is worth mentioning that there are other risks such as chemical exposure for pharmaceutical manufacturing, however, this study mainly focuses on potential fire, explosion and reactivity incident. In addition, comparison in this study focuses on intrinsic properties of the reaction and process which are the major hazards of the reaction system. Therefore, other factors like loss control measures and emergency equipment are supposed to be effective for both batch and continuous systems.

4 Methodology

To measure the realistic fire, explosion and reactivity potential of batch and continuous process, Dow's Fire and Explosion Index (F&EI) methodology is applied as hazard index in this study, since it is the most widely used hazard index that recognized by the chemical industry[18].

From the first edition of Dow's Fire & Explosion Index Hazard Classification Guide in 1964, this index has been continuously improving though the past 50 years. The last version is the seventh edition that published in 1994. The purposes of this index include forecasting quantitative damage of potential explosion, reactivity incidents and fire; determining hazardous equipment which is possible to give rise to incidents in the system; conveying this F&EI risk potential to related managements and organizations. Since the aim of this study is to identify major hazard that affect selection, application of Dow F&EI will be mainly focus on its function of hazard identification.

In the Dow's Fire & Explosion Index Hazard Classification Guide, economic analysis is confined to financial loss after an incident. Correlation between safety and economic gain from production is not built. Actually, this kind of connection could be established via yield of reactions and geometry and inventory of units, since these factors take important role in both safety assessment and process gain. In the current study, this correlation between safety and profit is constructed on the foundation of Dow's Fire & Explosion Index.

5 Case Study I: Benzimidazole Derivative Synthesis

5.1 overview

Benzimidazole derivative is an important type of intermediates in pharmaceutical industry. It is widely exhibited in a variety of pharmaceutical substance, such as proton pump inhibitors, antimicrobials, antivirals, anticancer, anti-inflammatory, hormone modulators as well as depressants, antidiabetics and so on[22]. Famous drug with benzimidazole nucleus includes Omeprazole and Albendazole. Omeprazole is used to treat gastroesophageal reflux disease and peptic ulcer disease[23], and Albendazole is an effective drug in treatment of various parasitic worm infestations[24].

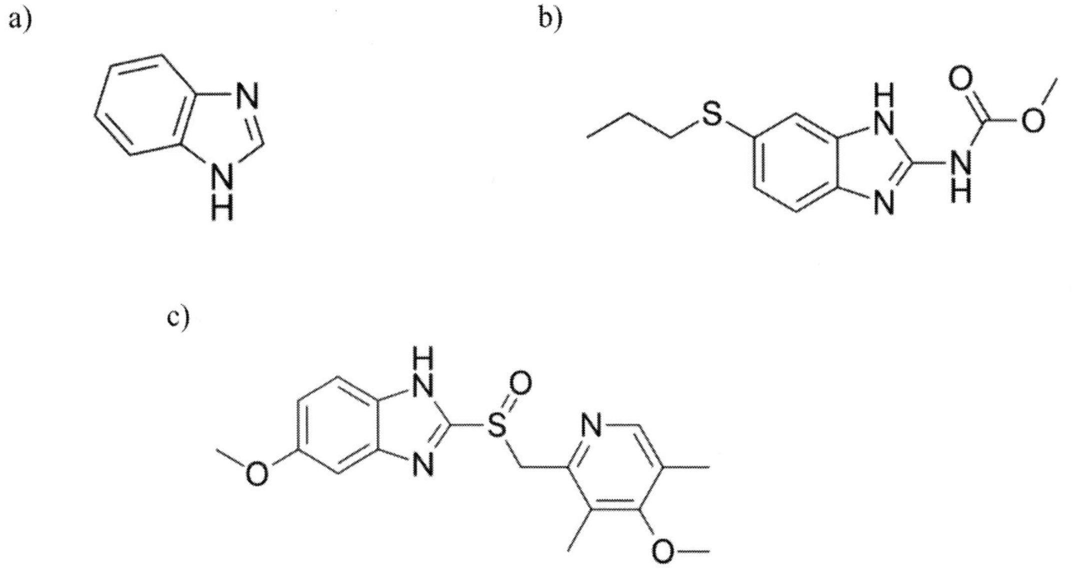

Figure 1. a) Molecular Structure of Benzimidazole, b) Molecular Structure of Albendazole, c) Molecular Structure of Omeprazole

2-methyl benzimidazole is a pharmaceutical-potential substance that belongs to this benzomidazole ring system. The condensation of o-phenylenediamines with carboxylic acids is one of the most common synthesis route for 2-methyl benzimidazole[25]. Conditions and production comparisons of o-phenylenediamines condensation by batch vs. continuous processes are showed in Table 1. These fundamental parameters are used for safety evaluation by Dow's F&EI in the following study.

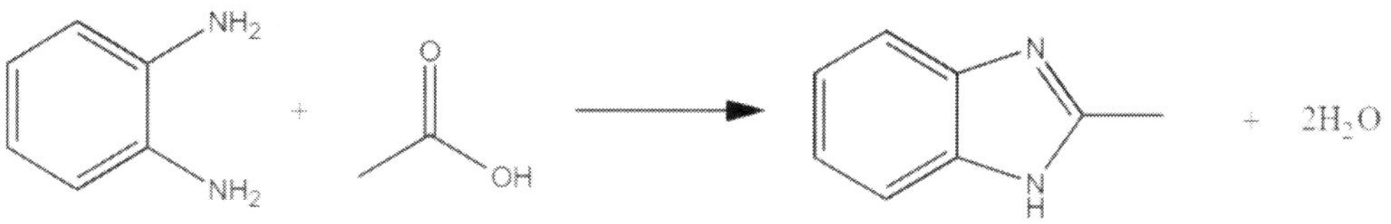

Figure 2. Synthesis of 2-methyl Benzimidazole by Condensation of o-phenylenediamine

Table 1. Production Conditions of Batch and Continuous Processes for Synthesis of 2-methyl benzimidazole

	Batch Process	Continuous Process
Temperature (℃)	200	270
Pressure (psi)	145	1885
Reaction Time (min)	5	0.5
Yield (%)	98	94
Substrate Concentration (M)	5	1
Reaction Volume (ml)	1000	4

In order to concentrate on study in intrinsic differences between two processes, and simplify the hazard identification in F&EI, several assumptions are adopted:

1. The reaction mass is perfect mixed in reactors;
2. Adjunct cooling systems in both processes are in perfect performances;
3. Emergency equipment is same and in perfect performances in both processes;
4. Scalabilities of both processes are good enough that reaction conditions remain same in all scales.

5.2. Safety Comparison Result

The comparison is conducted based on material properties and process features. Related explanation is stated below.

5.2.1. comparison on material

In this study, impacts from intrinsic properties of reactants are same for both processes. Material factors of all substances are listed in Table 2. According to Dow's Fire and Explosion Index Hazard Classification Guide, impact from the most hazardous material dominates the material factor of the whole process, therefore only impact from acetic acid is considered for both processes. In addition, although temperatures and composition are different in two processes and may lead to different chemical properties, these features are neglected in this part and contained in process comparison.

Table 2. Material Factor for Batch and Continuous Reactor

	Batch Reactor	Continuous Reactor
o-phenylenediamine	10	10
2-methylbenzimidazole	1	1
Acetic Acid	14	14

5.2.2. comparison on process

First of all, it's worth mentioning that albeit lots of distinctions exist between batch and continuous processes, comparisons in this study are confined to synthesis in reactor on account of its core role in the whole process: operation mode of reactor – continuously or intermittently – determines whether the process is continuous or not from the start. Therefore, the reactor is selected as the pertinent process unit without conventional unit screening procedure.

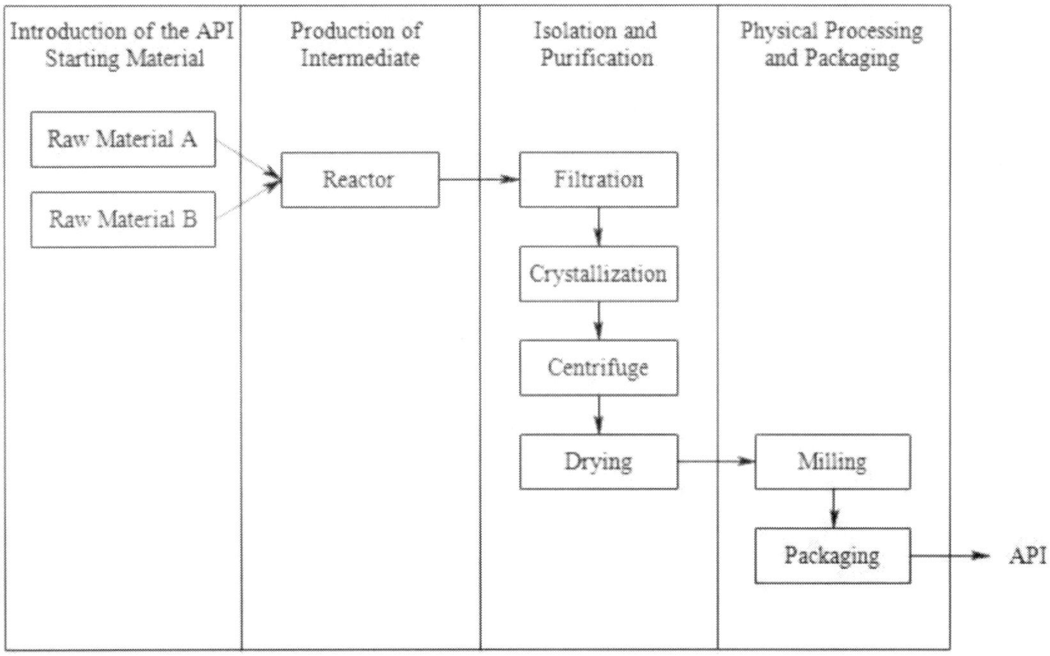

Figure 3. Procedure for Pharmaceutical Production

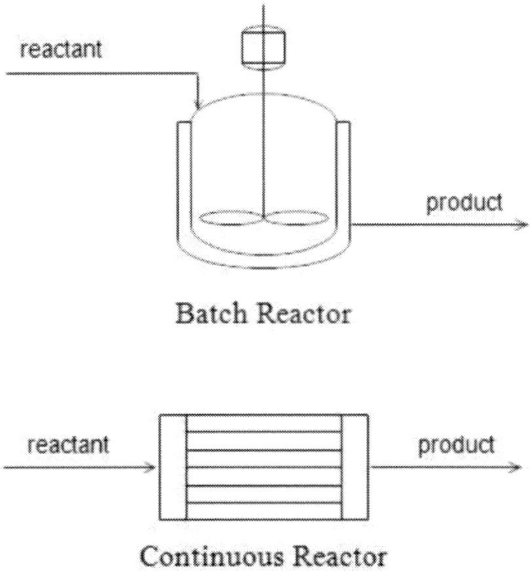

Figure 4. Differences in Structure between Batch and Continuous Reactors

On the foundation of safety evaluation by Dow's Fire & Explosion Index, quantitative judges between batch and continuous process are illustrated through comparison in sensitivity and productivity within safety consideration.

This sensitivity analysis focus on correlation between safety level and inventory of reactor. Inventory of unit plays an important role in the safety. In addition, this sensitivity analysis can also provide supportive information for scaling up procedure.

Inventory of unit plays an important part in the safety via quantity of flammable or unstable material. This potential is considered in both the General Process Hazards (F1) and the Special Process Hazards (F2) according to Dow's F&EI Hazard Classification Guide. In general, more the flammable or unstable material, higher the risk of fire and explosion within the process. For different unit (process unit or storage unit) and different phase (liquid, gas or solid), different correlation formula between inventory and penalty is given in the Special Process Hazards (F2) part of the Dow's F&EI Guide. Moreover, other issues such as whether these materials are processed indoor or outdoor, what's the temperature to handle these materials influences the safety level as well. These questions are included in the General Process Hazards (F1) part.

In this sensitivity analysis, F&EI that combines both the general and the special process hazards is expressed as a dependent variable that changes with the variation of total inventory of reactor. Other operation conditions remain the same for different inventory. Correlations between the F&EI and material inventory for batch and continuous process are shown in the Figure 5.

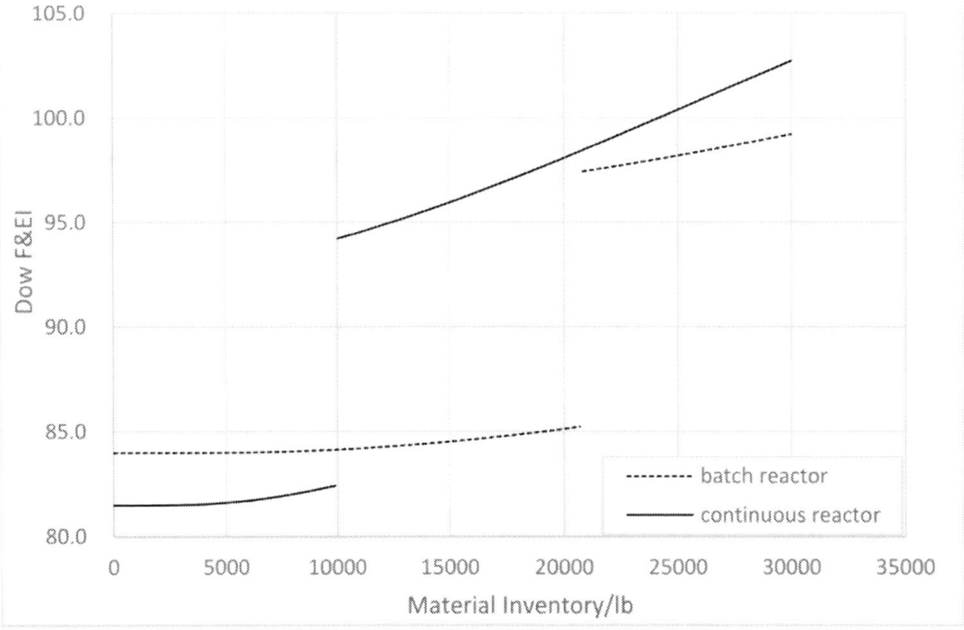

Figure 5. Analysis of Sensitivity to Inventory for Batch and Continuous Reactors Based on 2-methyl Benzimidazole Synthesis

for 0 lb < Inventory < 10000 lb:

F&EI of batch reactor is slightly greater than continuous reactor. This difference mainly comes from difference in usage of stirrer and reaction pressure. Due to requirement for homogeneous mixture, stirrer is necessary in batch reactor, while such a stirrer is not needed (or not allowed) in a continuous tubular reactor. Thus, stirrer, as a hazardous rotating equipment, adds extra penalty to batch process. As for difference in pressure, higher penalty is given to continuous process for its intensified condition. From the result, influence from rotating equipment is greater than that from pressure in this case.

Another observation is that, the magnitude of the difference in the F&EI between two reactors decreases with the growing inventory. This trend stems from the increasing impact from quantity of hazardous material. When inventory approaches zero, difference in F&EI is mainly caused by stirrer and distinct pressure aforementioned. As inventory increases, influence from unstable material extends. This influence depends on composition of mixture, or more specifically, concentration of acetic acid that defines the Material Factor (MF). Based on information of reactions, molar concentrations of acetic acid in batch and continuous reactor are 8.31M and 15.66M respectively. Thus, for similar total inventory, penalty from hazardous material in continuous reactor is greater than batch reactor.

when 10000 lb < Inventory < 20833 lb:

F&EI of continuous reactor is signally higher than batch reactor, and this difference grows along with the increasing inventory. In this range, part of influences come from rotating equipment, pressure and increasing inventory that mentioned above. More importantly, quantity of flammable and unstable liquids handled above boiling point exceeds threshold of 10,000lb for continuous reactor, therefore give a vertical rise to F&EI for continuous reactor, as shown in the Figure 5. This threshold is illustrated clearly in Dow's F&EI Hazard Classification Guide: when quantity of flammable and unstable liquids above its boiling point is less than 10,000 lb, a penalty of 0.60 is determined. A penalty of 0.90 is given when this liquids quantity above its boiling point is greater than 10,000 lb. In this case study, operation temperature of continuous reactor (270℃) is higher than boiling point of both o-phenylenediamine (257℃) and acetic acid (118℃), while operation temperature of batch reactor (200℃) merely higher than boiling point of acetic acid (118℃). Therefore, for inventory in the range of 10,000lb to 20833lb, total mass in continuous reactor is defined as unsafe, while merely acetic acid mass is defined as unsafe in the batch reactor. This divergence leads to different penalty in this range of inventory.

when 20833 lb < Inventory < 30000 lb:

F&EI of batch reactor drastically increases. However, F&EI of continuous reactor is still slightly greater than batch reactor, and the magnitude of this difference increases with the growing inventory.

This vertical rise in the F&EI of batch reactor is also caused by critical quantity of material above the boiling point. When total quantity in batch reactor goes beyond 20,833lb, quantity of flammable and unstable substance - acetic acid - also goes beyond threshold of 10,000 lb. Thus, penalty from material above boiling point is similar for both of processes.

The higher F&EI of continuous reactor originated from quantity of flammable and unstable material in this range of inventory. Since higher concentration of acetic acid is applied in continuous reactor, more proportion of substance is identified as unsafe for continuous reactor, thus higher penalty is given. This explanation is similar to that in range of low inventory (< 10,000 lb). However, unlike appearance in the low inventory range, influence from these flammable and unstable materials exceeds impact from pressure and stirrer. Therefore, continuous reactor becomes more dangerous for this range of total inventory.

It's worth mentioning that higher inventory is not considered here. Upper limit of inventory (30,000 lb) is set based on the general highest inventory of reactor used in pharmaceutical manufacturing.

Degree of hazard for batch versus continuous reactor used in this case study is shown in Table 3. In the range of inventory investigated, batch reactor is identified as moderate hazardous for a broader range of inventory, while continuous reactor is identified as intermediate hazardous for a broader range of inventory. Even so, It's worth mentioning that the F&EI values for both reactors are in a relative safe range (F&EI > 128 indicates heavy even severe hazard).

Table 3. Degree of Hazard for Batch and Continuous Reactors

Material Inventory X	Batch Reactor	Continuous Reactor
X < 15,152 lb	Moderate	Moderate
15,152 lb < X < 20,833 lb	Moderate	Intermediate
20,833 lb < X < 30,000 lb	Intermediate	Intermediate

In summary, F&EI for both batch and continuous reactors increase as material inventory increases. Moreover, F&EI of continuous reactor is more sensitive to change of inventory comparing to batch reactor. This higher sensitivity caused by comprehensive influence from higher concentration of unstable and flammable material, higher operation temperature and higher pressure it applied.

Evaluation in this part mainly focuses on safety of batch versus continuous process base on productivity considerations. Emphasis on productivity scale of this study primary comes from flexibility required for pharmaceutical production. Pharmaceutical industry is highly supply driven. Requirement on production rate is changed according to demand of market all the time. Therefore, better understanding on correlation between safety and productivity provides practicable support for research in process selection.

To approach this target, safety level is evaluated for the process at the same productivity capacity. In this part, F&EI is regarded as indicator for safety evaluation as above. Annual production is set as indicator for productivity. Related formula is shown below.

General calculations for productivity capacity are:

$$\text{Annual Production} = \text{Product Yield} \times \text{Annual Operation Time}$$

$$\text{Product Yield} = \text{Mass Flow of Reagent} \times \text{Conversion}$$

Mass flow is supposed to be proportional to reactor volume as set in the laboratory trial.

According to definition of fine chemical, annual production of fine chemical is limited to 1,000 tons per year[26]. Therefore, 1,000 tons per year is set as the upper limit in this evaluation.

As for product yield and annual operation time, the former is provided in Table 1, and expression for the latter is formed referring to the normal operation time[27].

Annual operation time for batch reactor is:

$$\text{Annual Operaton Time} = 300\ \text{day/yr} \times 24\ \text{hr/day}$$

Annual operation time for continuous reactor is:

$$\text{Annual Operation Time} = 8000\ \text{hr/yr}$$

Normally, in pharmaceutical industry, multi products are produced in one production line. Therefore, the same product cannot occupy the producing line for the whole year. In this study, 2-methyl benzimidazole is assumed to be produced in 2 months of one year.

In addition, circulating time in batch reactor need to be considered. According to the literature[11], the ratio of reaction time to total operation time for this reaction is shown below. Excepted reaction, other operations include filling, cleaning and cooling and other additional work.

$$\frac{\text{Reaction Time}}{\text{Total Operation Time}} = 0.185$$

Based on these formulas and assumptions, correlation between annual productions and F&EI is shown in the Figure 6.

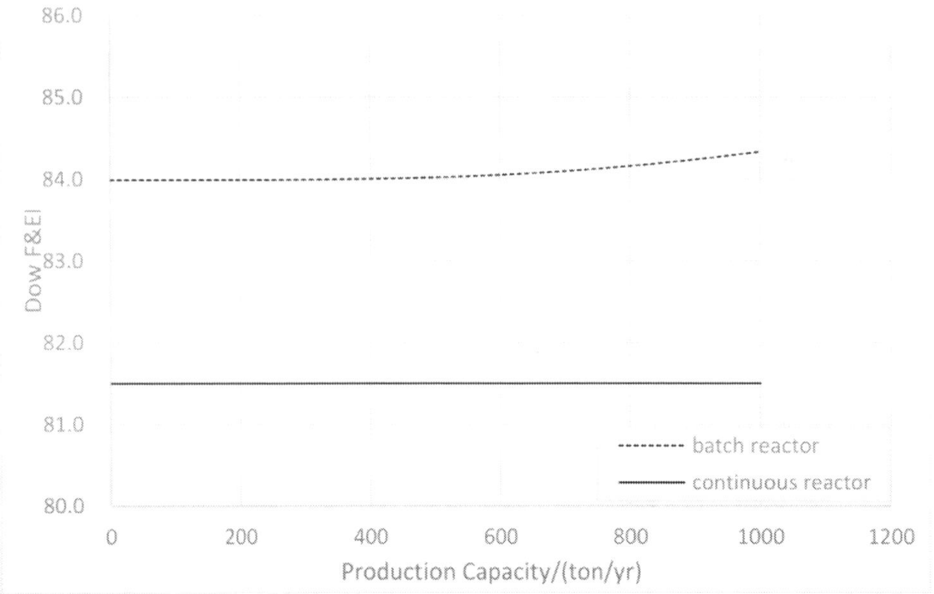

Figure 6. Analysis of Safety Based on Production for Batch and Continuous Reactors Based on 2-methyl Benzimidazole Synthesis

According to the Figure 6, F&EI value is constant in the range of practicable annual production (0 ~ 1000 tons per year). Plus, the F&EI for both reactors corresponds to a moderate level of hazard (61 < F&EI < 96).

This low F&EI resulted from relatively low demand for production. According to the sensitivity analysis, F&EI for both batch and continuous reactor increase as material inventory increases. Both processes could be intermediate hazardous once its inventory is higher than a specific value. However, these values are unapproachable in practical pharmaceutical manufacturing. Since demand for fine chemical, always less than 1,000 tons per year, confines scale of production.

Another observation is that, the F&EI of continuous reactor is always lower than that of batch reactor for same production. Plus, slight rise appears for batch reactor when production capacity exceeds 500 tons per year. These appearances predicate consistent superiority in safety for continuous reactor than batch reactor. Explanation is stated below:

Firstly, as stated in the sensitivity analysis, settled elements, such as operation pressure, temperature and use of rotating equipment, take major account for the difference at small-scale production. Advantages of continuous process in these aspects determine its superiority in small-scale pharmaceutical production.

Secondly, continuous reactor possesses higher space-time yield compared to batch reactor. This advantage stems from the intrinsic property of continuous operation mode - the low residence time and high flow throughput. That is to say, it costs fewer time and lower inventory for continuous reactor to produce same amount of product than batch reactor. These benefits in production efficiency influence safety level indirectly but significantly: Smaller inventory means fewer unstable materials in process, and shorter operation time means less chance to fail. This point is not considered in separate safety evaluation, but can be clearly illustrated by correlation between production capacity and F&EI. In this case, when demand on production growing, demand on inventory grows faster for batch reactor than continuous reactor. This faster rise in inventory lead to faster rise in amount of unstable material in process so as to the faster rise in F&EI value. It explains why slight rise appears for batch reactor when production capacity exceeds 500 tons per year.

In fact, the comparable batch and continuous reactor are rarely at the same level of volume. Referring to literature, for commercial production of fine chemical, the volumes of continuous reactor are usually in a range of 0.1 to 3 liter, while the volume of batch reactor is usually around the range of 2 to 10 cubic meters[6]. For this case, based on assumption of same annual production, ratio of batch reactor inventory and continuous reactor inventory is as much as one hundred.

6 Case Study II: Oxidation with Hydrogen Peroxide

6.1. overview

Hydrogen peroxide is one kind of widely used oxidation agent in pharmaceutical production. Typical reaction involving hydrogen peroxide includes Baeyer–Villiger oxidations for synthesis of lactone[9], N-oxidation of alkyl pyridine for synthesis of alkyl pyridine N-oxidant[28] and benzylic hydro-peroxide rearrangement for synthesis of Brivanib alaninate[29].

This common oxidation agent is also used in production of peracetic acid. Peracetic acid is produced via oxidation of acetic acid by hydrogen peroxide with sulfuric acid as the catalyst. Conditions for peracetic acid production by batch vs. continuous process are showed in Table 4[30].

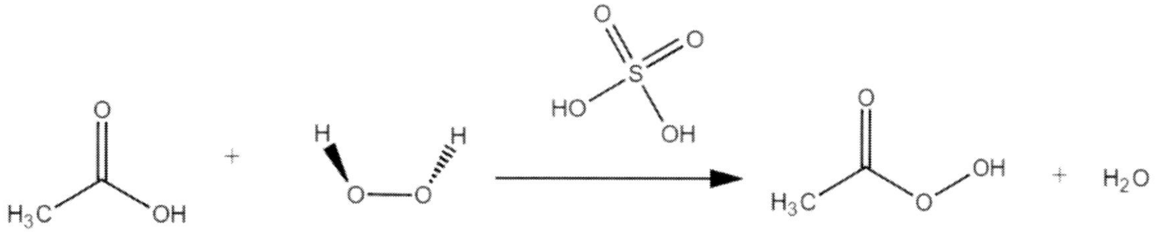

Figure 7. Synthesis of Peracetic Acid by Oxidation of Acetic Acid

Table 4. Production Conditions of Batch and Continuous Process for Synthesis of Peracetic Acid

	Batch Process	Continuous Process
Temperature (℃)	60	80
Pressure (psi)	14.5	14.5
Reaction Time (min)	30	15
Yield (kg/h)	170	20
Acetic Acid Concentration (M)	7.3	7.3
Reaction Volume (L)	4000	10

Assumption for this case study is listed below:

1. The reaction mass is perfect mixed in reactors.
2. Since the temperature for decomposition of hydrogen peroxide in peracetic acid synthesis reaction is higher than 120 ℃ [31], which is larger than operation temperature for both processes in the case study, decomposition of hydrogen peroxide is assumed to be inexistent.
3. Sulfuric acid is dissociated completely to H^+ and HSO_4^-, in addition, concentration of H^+ is assumed to be constant of 0.67 mol/L during the reactions for both processes.
4. Influence from sulfuric acid is not considered in safety evaluation.
5. Emergency equipment and cooling system are perfectly performed for both of processes.
6. Scalabilities of both processes are good enough that reaction conditions remain same in all scales.

6.2. safety comparison result

The comparison is conducted based on material property and process feature. Related explanations are stated in the following paragraph.

6.2.1. comparison on material

In this case, impact from material within the reactor is same for both of processes. Material factors for reactants involved are listed in Table 5. According to Dow's Fire and Explosion Index Hazard Classification Guide, influence from the most hazardous material dominates the material factor. Although material factor values are same for hydrogen peroxide and peracetic acid, since concentration of hydrogen peroxide is higher than peracetic acid, hydrogen peroxide is chosen to be the dominant material. Therefore, only impact from hydrogen peroxide is considered for both processes. Change on substances' property from temperatures are neglected in this part and reconsidered in process comparison.

Table 5. Material Factor for Batch and Continuous Reactor

	Batch Reactor	Continuous Reactor
Acetic Acid	14	14
Hydrogen Peroxide	24	24
Peracetic Acid	24	24

6.2.2. comparison on process

The major difference between these two production processes come from distinctions in reaction conditions and reactor constructions. These differences are shown in Table 4. Based on safety evaluation by Dow's Fire & Explosion Index, quantitative judges between batch and continuous process are illustrated through comparison in sensitivity and productivity within safety consideration.

Procedure for sensitivity comparison is the same as the first case study. Correlations between the F&EI and material inventory for batch and continuous process are shown in the Figure 8.

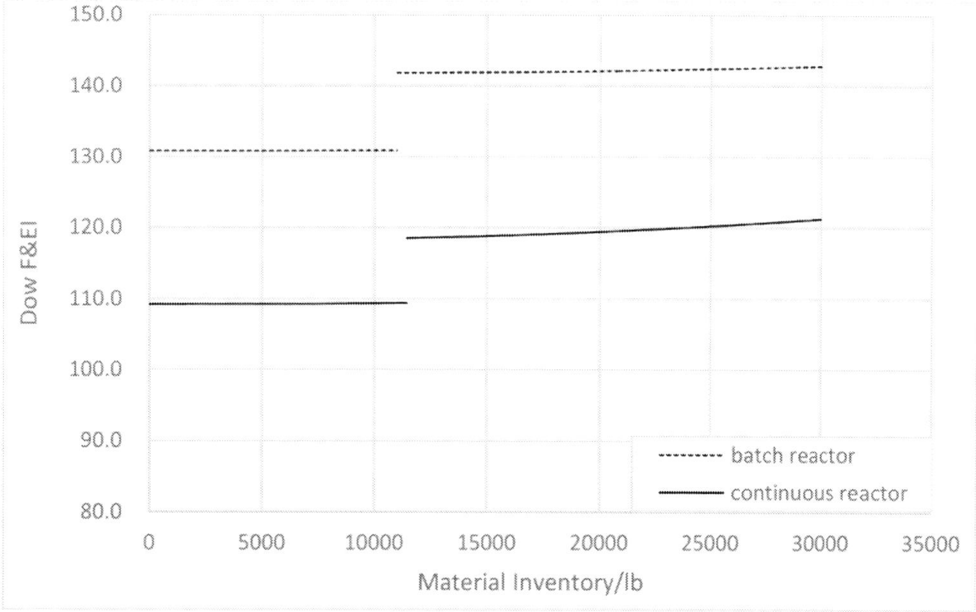

Figure 8. Analysis of Sensitivity to Inventory for Batch and Continuous Reactors Based on Peracetic Acid Synthesis

for 0 lb < Inventory < 11000 lb:

F&EI of batch reactor is larger than continuous reactor. This difference is almost caused by usage of stirrer in batch reactor. Application of such a rotating equipment increases penalty for batch process.

when 11000 lb < Inventory < 11430 lb:

F&EI of batch reactor is still higher than continuous reactor, but this difference grows drastically due to the sharp transition in F&EI of batch reactor. This transitions are caused by quantity of flammable liquids that handled above flashing point exceeds threshold of 10,000lb. This threshold is illustrated clearly in Dow's F&EI Hazard Classification Guide: when quantity of flammable liquids above their flashing point is less than 10,000 lb, a penalty of 0.30 is determined. A penalty of 0.45 will be given when this quantity of flammable liquids above their flashing point is larger than 10,000 lb. For peracetic acid production, operation temperature for batch reactor (60℃) and continuous reactor (80℃) is higher than flashing point of both acetic acid (39℃) and peracetic acid(40.5℃). Therefore, the value of inventory for higher penalty is determined by compositions of peracetic acid and acetic acid in reactors. Since concentrations of these two materials is slightly higher for batch comparing to continuous reactor, batch reactor approaches this transition in F&EI at a smaller total inventory.

when 11430 lb < Inventory < 30000 lb:

F&EI of continuous reactor exceed the transition caused by quantity of unsafe material above flashing point. However, it is still lower than F&EI of batch reactor.

Compared to the first case study, value of slope for both processes is similar low. It results in the consistent superiority of continuous reactor in safety level compared to batch reactor. This similarity in slopes comes from similarity in operation conditions. Differences only appear in operation temperature and hydrogen peroxide concentration. These differences are not such significant that lead to huge divergence between slopes. Thus, difference in F&EI value between two reactors does not change along with increasing inventory.

Degree of hazard of this case study is exhibited in Table 6. Although marked transitions appears for both reactor, safety levels for two reactors do not change based on scaling up of reactor size. The safety levels are majorly decided by settled parameters, such as operation pressure and usage of stirrer. These parameters do not change with inventory as variable. Hence, inventory does not play a significant role in judgement on degree of hazard.

Table 6. Degree of Hazard for Batch and Continuous Reactors

Reactor Type	Batch Reactor	Continuous Reactor
Degree of Hazard	Heavy	Intermediate

In sum, F&EI for both batch and continuous reactor increase as material inventory increases due to the increases of quantity above flashing point. However, since operation conditions are relatively mild and similar for these two reactors, both reactor do not exhibit high sensitivity when inventory is growing gradually.

For evaluation based upon productivity, assumption for practical production are set as following:

1. Confined by the conversion of this reaction route and volume of reactor, upper limit for annual production is settled to be 580 tons per year.
2. Same as the first case study, referring to normal operation time for this two operation mode[27], annual operation time for batch reactor is 8760 hour per day, and annual operation time for continuous reactor is 8000 hour per day.
3. Peracetic acid is assumed to be produced for 2 months in one year.
4. According to the literature[30, 32], the ratio of reaction time to total operation time for this reaction is 0.143

Based upon these formulas and assumptions, correlation between annual productions and F&EI is shown in the Figure 9.

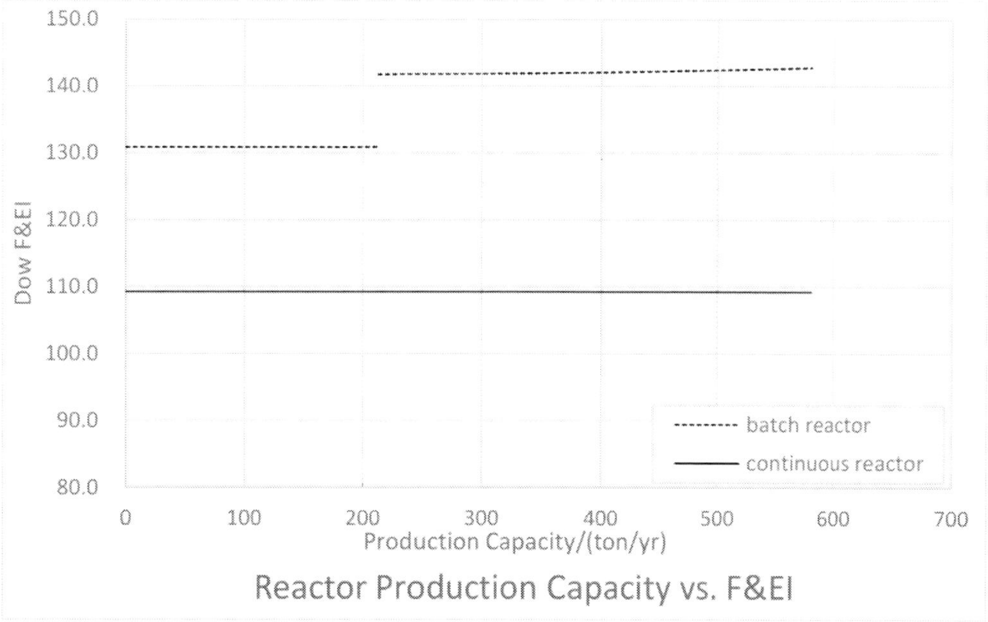

Figure 9. Analysis of Safety Based on Production for Batch and Continuous Reactors Based on Peracetic Acid Synthesis

According to the Figure 9, F&EI is stabilized at 109 in the range of practicable annual production for continuous reactor. For batch reactor, the F&EI value rise drastically when the annual production is higher than 220 tons per year.

For the batch production mode, relatively low production efficiency leads to substantial increase in reactor volume when demand on production grows. This increase in reactor volume is so significant that exceed the threshold for large quantity of flammable and unstable material above flashing point. Therefore, the F&EI value jump from 131 to 143 when the annual production is higher than 220 tons per year.

For continuous production mode, high production efficiency causes only a few change in volume hence only few change in F&EI in as well. This difference in F&EI between batch and continuous reactor means that, to produce the same amount of peracetic acid in the given period, the volume of continuous reactor is much smaller than batch reactor.

7 Result Summary

Figure 10 represents the comparisons of two case studies. The X-axis represents two different case studies, and the Y-axis represents the average Dow's F&EI value which can be used to define the degree of hazard.

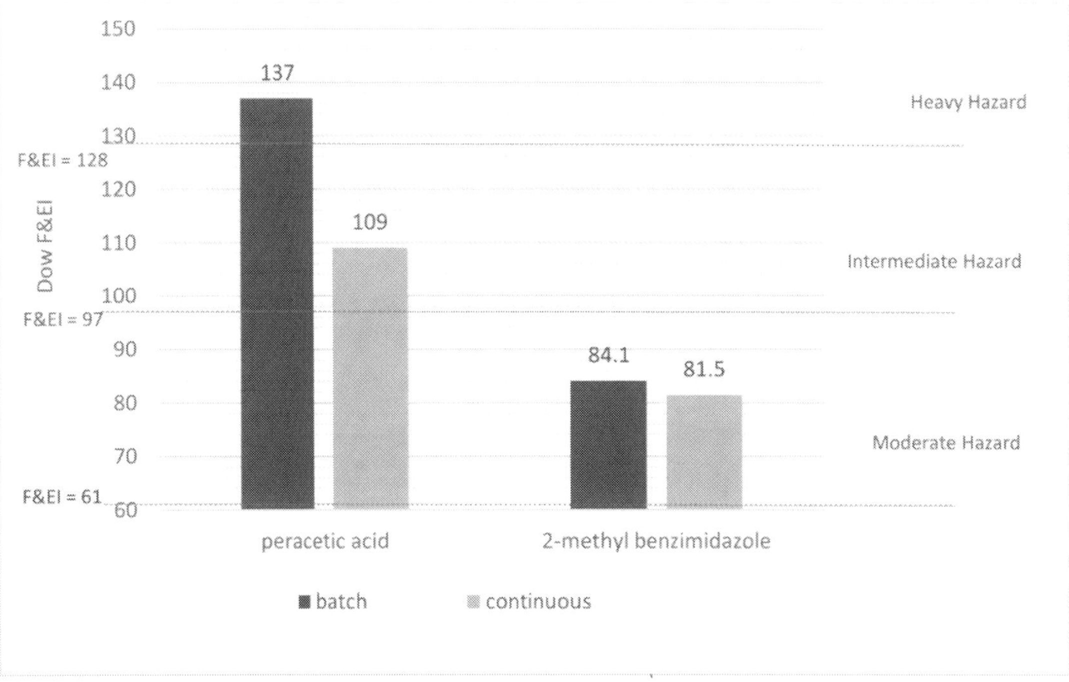

Figure 10. Comparison of Safety Level by Batch and Continuous Productions for Two Case Studies

Regardless of the reactor type, the safety level is predetermined by properties of the synthesis reaction. Although distinctions in F&EI appear when using different reactors for both cases, peracetic acid production is generally more hazardous than 2-methyl benzimidazole production, no matter which reactor it uses. This higher level of danger comes from the intrinsic properties, which consist of the flammability and reactivity of the chemicals involved in the synthesis of peracetic acid.

Taking batch and continuous reactors into consideration, distinction in F&EI comes from the differences in operation conditions and production efficiency. These varying factors such as temperature, pressure, and yield do not influence F&EI as individual factors. Instead, they are correlated and impact the F&EI as an ensemble. Difference in these ensemble decides the variance in F&EI value between reactors, then determines to which extent the selection of reactor can make a difference. For this study, the variance in F&EI value between batch and continuous reactor is greater for peracetic acid production (variance in F&EI = 28) than 2-methyl benzimidazole production (variance in F&EI = 2.6). Therefore, selecting a reactor is more important for peracetic acid comparing to 2-methyl benzimidazole production.

The importance of choosing a reactor is different for the two cases, the F&EI value for a continuous reactor is lower than a batch reactor for both of these cases. This consistency exhibits the superiority of a continuous reactor in safety.

8 Conclusion

A comprehensive comparison between the batch and continuous process used in the pharmaceutical industry based upon Dow's F&EI is made in this study. Two typical pharmaceutical productions, the production of 2-methyl benzimidazole and the production of peracetic acid are analyzed. Safety levels for batch and continuous reactors are compared based on both inventory and annual production being used as variables. The conclusions on the foundation of these two case studies are stated below:

1. The comparison of batch versus continuous processes in the pharmaceutical industry is complex. It's determined by reaction properties, operation conditions, structures of equipment, production efficiency, scale of production, and many other factors. To get a conclusive comparison between these two production modes, more systematic studies based on the integrated batch and continuous process need to be conducted for safety consideration.

2. Just focusing on the reactor in which part of API synthesis occurs, the continuous reactor is more sensitive to inventory in general. The magnitude of this sensitivity is determined by operation conditions such as operation temperature, operation pressure and concentration of hazardous material. This sensitivity to inventory might lead to problems for continuous production in a relatively large scale.

3. However, when considering practical pharmaceutical production, which is always in a small scale (no larger than 1000 tons per year), hazardous large-volume continuous reactor is not necessary. In addition, the space time yield of continuous production mode is usually higher than the batch production mode, application of a continuous reactor is relatively safer when compared to batch reactors.

9 References

1. Plumb, K., Continuous processing in the pharmaceutical industry: changing the mind set. *Chemical Engineering Research and Design* **2005,** *83* (6), 730-738.

2. Laird, T., What Type of Reactions Do Process Chemists Use on Scale? *Organic Process Research & Development* **2006,** *10* (5), 851-852.

3. Braune, S.; Poechlauer, P.; Reintjens, R.; Steinhofer, S.; Winter, M.; Lobet, O.; Guidat, R.; Woehl, P.; Guermeur, C., Selective nitration in a microreactor for pharmaceutical production under cGMP conditions. *Chimica oggi* **2009,** *27* (1), 26-29.

4. Proctor, L. D.; Warr, A. J., Development of a Continuous Process for the Industrial Generation of Diazomethane1. *Organic process research & development* **2002,** *6* (6), 884-892.

5. Roberge, D. M.; Gottsponer, M.; Eyholzer, M.; Kockmann, N., Industrial design, scale-up, and use of microreactors. *Chemistry Today* **2009,** *27* (4), 8-11.

6. Hessel, V.; Vural Gürsel, I.; Wang, Q.; Noel, T.; Lang, J., Potential analysis of smart flow processing and micro process technology for fastening process development: use of chemistry and process design as intensification fields. *Chemical Engineering & Technology* **2012,** *35* (7), 1184-1204.

7. Schaber, S. D.; Gerogiorgis, D. I.; Ramachandran, R.; Evans, J. M.; Barton, P. I.; Trout, B. L., Economic analysis of integrated continuous and batch pharmaceutical manufacturing: a case study. *Industrial & Engineering Chemistry Research* **2011,** *50* (17), 10083-10092.

8. Jolliffe, H. G.; Gerogiorgis, D. I., Plantwide design and economic evaluation of two continuous pharmaceutical manufacturing (CPM) cases: ibuprofen and artemisinin. *Computers & Chemical Engineering* **2016,** *91*, 269-288.

9. Gutmann, B.; Cantillo, D.; Kappe, C. O., Continuous‐Flow Technology—A Tool for the Safe Manufacturing of Active Pharmaceutical Ingredients. *Angewandte Chemie International Edition* **2015,** *54* (23), 6688-6728.

10. Wakami, H.; Yoshida, J.-i., Grignard exchange reaction using a microflow system: from bench to pilot plant. *Organic process research & development* **2005,** *9* (6), 787-791.

11. Damm, M.; Glasnov, T. N.; Kappe, C. O., Translating high-temperature microwave chemistry to scalable continuous flow processes. *Organic Process Research & Development* **2009,** *14* (1), 215-224.

12. Schaber, S. D.; Gerogiorgis, D. I.; Ramachandran, R.; Evans, J. M. B.; Barton, P. I.; Trout, B. L., Economic Analysis of Integrated Continuous and Batch Pharmaceutical Manufacturing: A Case Study. *Ind Eng Chem Res* **2011,** *50* (17), 10083-10092.

13. Roberge, D. M.; Ducry, L.; Bieler, N.; Cretton, P.; Zimmermann, B., Microreactor technology: A revolution for the fine chemical and pharmaceutical industries? *Chem Eng Technol* **2005,** *28* (3), 318-323.

14. Hartman, R. L.; McMullen, J. P.; Jensen, K. F., Deciding Whether To Go with the Flow: Evaluating the Merits of Flow Reactors for Synthesis. *Angew Chem Int Edit* **2011,** *50* (33), 7502-7519.

15. Calabrese, G. S.; Pissavini, S., From Batch to Continuous Flow Processing in Chemicals Manufacturing. *Aiche J* **2011,** *57* (4), 828-834.

16. Kletz, T. A., What you don't have, can't leak. *Chemistry and Industry* **1978,** *6*, 287-292.

17. Kletz, T. A., *Plant design for safety: a user-friendly approach.* Taylor & Francis: 1991.

18. Fire, D., Explosion Index Hazard Classification Guide. *AICHE New York* **1994**.

19. Lawrence, D. Quantifying inherent safety of chemical process routes. © Duncan Lawrence, 1996.

20. Heikkilä, A.-M., *Inherent safety in process plant design: an index-based approach.* VTT Technical Research Centre of Finland: 1999.

21. Gentile, M.; Rogers, W.; Mannan, M., Development of a fuzzy logic-based inherent safety index. *Process Safety and Environmental Protection* **2003,** *81* (6), 444-456.

22. Bansal, Y.; Silakari, O., The therapeutic journey of benzimidazoles: a review. *Bioorganic & medicinal chemistry* **2012,** *20* (21), 6208-6236.

23. omeprazole. https://www.drugs.com/monograph/omeprazole.html.

24. albendazole. https://www.drugs.com/monograph/albendazole.html.

25. Patil, A.; Ganguly, S.; Surana, S., A systematic review of benzimidazole derivatives as an antiulcer agent. *Rasayan J Chem* **2008,** *1* (3), 447-460.

26. Patt, R.; Kordsachia, O.; Süttinger, R.; Ohtani, Y.; Hoesch, J.; Ehrler, P.; Eichinger, R.; Holik, H.; Hamm, U.; Rohmann, M., Ullmann's encyclopedia of industrial chemistry. Wiley: 2002.

27. Annual Production. http://www.eng-tips.com/viewthread.cfm?qid=389883.

28. Pineda-Solano, A.; Saenz-Noval, L.; Nayak, S.; Waldram, S.; Papadaki, M.; Mannan, M. S., Inherently safer reactors: Improved efficiency of 3-picoline N-oxidation in the temperature range 110–125° C. *Process Safety and Environmental Protection* **2012,** *90* (5), 404-410.

29. LaPorte, T. L.; Spangler, L.; Hamedi, M.; Lobben, P.; Chan, S. H.; Muslehiddinoglu, J.; Wang, S. S., Development of a Continuous Plug Flow Process for Preparation of a Key Intermediate for Brivanib Alaninate. *Organic Process Research & Development* **2014,** *18* (11), 1492-1502.

30. Ebrahimi, F.; Kolehmainen, E.; Turunen, I., Safety advantages of on-site microprocesses. *Organic Process Research & Development* **2009,** *13* (5), 965-969.

31. Kadla, J. F.; Chang, H., The reactions of peroxides with lignin and lignin model compounds. ACS Publications: 2001.

32. Ebrahimi, F.; Kolehmainen, E.; Oinas, P.; Hietapelto, V.; Turunen, I., Production of unstable percarboxylic acids in a microstructured reactor. *Chemical engineering journal* **2011,** *167* (2), 713-717.

478238 Risk Based Decision Making Methods for Evaluating Complex Technologies, a Comparative Analysis on Pipelines and Railroad for Crude Oil Transport

Abstract:

Decision making for long-term capital intensive projects involve large number of uncertainties. Traditional risk assessments are based on probabilities which are estimated from limited data sets and provide limited flexibility for complex technologies. Further, the environment and specific conditions under which the data is obtained may be unknown and yield to erroneous extrapolations. The method proposed in this work utilizes a novel method for estimation of the risk based on the net present value of technologies and provides the basis for a decision support tool based on the expected value theorem. This is demonstrated using the case of the safety and economic crisis in onshore crude oil transport in the continental United States. The exponential growth of crude oil production in the continental United States has led to crisis in the mid-stream industry with existing pipeline infrastructure under pressure to service the increased production. This fact is reflected in the 4000% increase in the carloads in a span of five years from 2009 to 2013. This has been compounded by the volatile fluctuation in crude oil price from $100 to $30 over the past few years. This has put enormous constraints on capital decision making. Since railroad infrastructure albeit aged, already exists in many regions, loading and unloading crude oil is relatively easy and their connection to the refining centers along the Gulf of Mexico apart from sites in the east and west coast make it an option to many producers. In this paper a novel decision making model is proposed using the concept of 'Net present value' of a technology to evaluate the option between railroad and as suitable mid-stream alternatives for the transportation of crude oil. The model is built on using a prescriptive decision making technique for the quantification of factors affecting the decision, incorporating also the utility of wealth of the decision maker. Monte-Carlo simulation is further used to obtain results and trends based on variation of initial parameters. This analysis can provide a basis for making decisions given the complex environment of crude oil transport operations and can be extended to a variety of geographies given that they satisfy the requirements of the model. The method can also be further extended to other technology evaluation given appropriate factors can be identified to build the model.

479492 Key Modeling Parameters and Process Safety Information for Effective Atmospheric Dispersion Modeling

Abstract:

Dispersion modeling is an important part of Quantitative risk assessment (QRA) documentation. Dispersion modeling can be performed using commercially available software. Variety of modeling parameters and Process Safety Information (PSI) data are needed for atmospheric dispersion modeling. Some of the parameters include weather, facility geographical location, flammability and toxicity data of the service fluid, radiation intensity, software tool, and model. Some PSI data needs include process and instrumentation diagrams (PIDs), Process Flow Diagrams (PFDs), relief system evaluation documentation, Heat and Material Balance (H&MB), and equipment data sheets. This paper describes the key parameters and their selection for effective dispersion modeling.

479373 Implementation of Hazard Mitigation Solutions: Consequence Vs. Risk-Based Building Specifications

Abstract:

Implementation of Hazard Mitigation Solutions: Consequence vs. Risk-Based Building Specifications

Travis Holland, Joshua Bruce-Black

Baker Engineering and Risk Consultants

3330 Oakwell Court, Suite 100

San Antonio, TX 78218

tholland@bakerrisk.com, jbblack@bakerrisk.com

Abstract

There are various strategies to mitigate occupant vulnerability and risk after completion of a consequence based facility siting study (FSS) or quantitative risk assessment (QRA). One potentially cost-effective mitigation strategy is to focus on reducing the occupant vulnerability and risk to personnel by upgrading existing buildings or designing new buildings. A building specification is commonly used to relay a small set of building design parameters based on the results of a consequence based FSS to engineers or contractors. However, with the industry shift towards risk based studies (QRAs), there has been a growing perspective that the information required to develop an appropriate building specification is more complicated given the increased number of variables involved in a QRA. Using the development of a new control room specification as an example, this paper will compare and contrast the steps and benefits/drawbacks involved in developing an appropriate building design basis employing a consequence and risk based approach. This paper will highlight how establishing tolerance criteria, emergency response strategy, personal protective equipment (PPE), and building occupancy early in the project definition stage promotes effective streamlined communication in the specification and reduces the differences between consequence and risk-based building specifications.

Generalization of Modeling Along-Wind Dispersion

Jessica M. Morris
Tom Spicer, Ph.D., P.E.
Ralph E. Martin Department of Chemical Engineering
University of Arkansas, Fayetteville, AR
Jmm065@uark.edu

Prepared for Presentation at
American Institute of Chemical Engineers
2017 Spring Meeting and 13th Global Congress on Process Safety
San Antonio, TX
March 26 – 29, 2017

Generalization of Modeling Along-Wind Dispersion

Jessica M. Morris
Tom Spicer, Ph.D., P.E.
Ralph E. Martin Department of Chemical Engineering
University of Arkansas, Fayetteville, AR
Jmm065@uark.edu

Keywords: along-wind, dispersion, finite duration, time-varying, atmosphere, releases, modeling

Abstract

The method to model along-wind dispersion is an important aspect of currently used commercial atmospheric dispersion models when considering time-limited or time-varying releases of toxic or flammable materials. Past wind tunnel experimental data has shown the validity of this approach. Inclusion of the effects of along-wind dispersion has proven to be critically important to modeling available field-scale data with current models for such time-varying releases. Many investigators have considered modeling along-wind dispersion from a theoretical point of view as well as basic parameterization of available field test data. This paper reviews past work and proposes a non-dimensionalized parameterization which shows improved agreement with available field-scale data.

1 Introduction

Continued world-wide industrial development has brought the wide spread use of industrial chemicals, many of which are hazardous. Whether released accidentally (or intentionally), air borne hazardous chemicals effect large populations quickly as evidenced by several industrial chemical accidents (e.g., Bhopal[1], Flixborough[2], Seveso[3]). Understanding and modeling hazardous chemical releases is vital to protecting the public. Many researchers have worked to assess these hazards using physical and computational models. Predicting the potential impact of hazardous gas releases in the atmosphere is imperative to properly determine where transport, storage, and use of such chemicals should be located. In the case of an actual loss of containment, understanding where air borne chemicals can travel is important to determine the proper response to an emergency.

While models exist to address these questions, there are opportunities to improve modeling methods particularly for (realistic) finite duration and time varying hazardous gas releases, which depend strongly on parameterizing the effects of along-wind dispersion (the elongation of the cloud in the along-wind direction). Modeling methods for along-wind dispersion are frequently based on established theoretical approaches which provide guidance for analyzing currently available field scale data. This paper considers experimental data and established theoretical methods to consider how the effects of along-wind dispersion can be improved.

2 Theoretical Description

2.1 Theoretical Models

Along-wind dispersion can be calculated under many different conditions and assumptions. There have been many established theories accounting for different conditions and assumptions. Scientists choose the along-wind dispersion coefficient as either a function of time or distance. Majority of scientists choose the along-wind dispersion coefficient as a function of the friction velocity. Some of the most prominent theories about along-wind dispersion are discussed below.

2.2 Batchelor

Batchelor (1964)[4] determined the Lagrangian Similarity Hypothesis. This hypothesis relates Eulerian and Lagrangian quantities when deriving theoretical solutions to problems. Applying known properties of fluid flow to the atmosphere, Batchelor was able to establish a generalized equations under neutral stability conditions, with no restraints on maximum height, for center line concentration at ground level and wind speed dependence with height. Batchelor derived many equations that are the foundation for modern theoretical equations commonly utilized. Batchelor's relationship between turbulence in a constant stress region on a boundary layer is a function of the friction velocity and time. This relationship and the Lagrangian Similarity Hypothesis serve as a template for along-wind dispersion equations to come.

Release height can be interpreted differently on its effect on dispersion. Restating Batchelor's similarity theory, "the statistical properties of the velocity of a marked fluid particle at time, t, after release at the ground depend only on friction velocity, u_* and t." This theory is then extended to the case of an elevated release: "The statistical properties of the velocity of a marked particle at time, t, after release at height, h, are the same as those of a particle released at the ground (z=0) at the instance $-t_1$, provided $t \gg t_1$ where t_1 is expected to be of the order of magnitude of the time-scale of the turbulence at height h, that is, of order h/u_*". Under Batchelor's similarity theory, after a sufficient amount of time, the height of release can be ignored.

2.3 Chatwin

Chatwin (1968)[5] was the first mathematician to theoretically derive a full equation with numerical values for the along-wind dispersion coefficient. He also derived the shear stress component of the along-wind dispersion equation. Chatwin used Batchelor's similarity theory along with the assumptions of the ability of turbulence to be described as an eddy diffusivity to solve for along-wind dispersion, σ_x. He makes further assumptions of neutral stability, ground release, and an unbounded boundary layer. The generalized equations for both overall along-wind dispersion, σ_x , and wind shear, σ_{xs} , by Chatwin are as followed

$$\sigma_x = 0.803 \frac{u_* t}{\kappa} \quad (1)$$

$$\sigma_{xs} = 0.596 \frac{u_* t}{\kappa} \quad (2)$$

where u_* is the friction velocity and κ is the van Karman's constant, approximately equal to 0.41.

2.4 *Van Ulden*

Van Ulden (1992)[6] derived an equation under slightly different dispersion definitions. Van Ulden made the distinction between absolute and relative dispersion. Wheatley and Chatwin were able to authenticate their equations with the assumption of a hypothetical atmosphere: where no large horizontal eddies exist. Taking into consideration this distinction of absolute and relative dispersion, van Ulden derived a new equation for σ_x, the total variance of the concentration distribution in the wind direction

$$\sigma_x^{\ 2} = \sigma_c^{\ 2} + \sigma_d^{\ 2} \quad (3)$$

where σ_c is the deterministic part (or centroid variance), and σ_d is the random part (or diffusive variance). The total longitudinal variance, σ_x, is dependent on three separate quantities: centroid variance, diffusive variance due to shear effects, and direct dispersion. These three variables were able to be solved for shear, centroid, and diffusive effects respectively as a function of friction velocity and time

$$\sigma_s = 0.5 \frac{u_* t}{\kappa} \quad (4)$$

$$\sigma_c \simeq 0.54 \frac{u_* t}{\kappa} \quad (5)$$

$$\sigma_d = 0.67 \frac{u_* t}{\kappa}. \quad (6)$$

Under the assumption that $\sigma_c \simeq \sigma_s = 0.5 \frac{u_* t}{\kappa}$, these variables can be substituted into equation (3) to get the total longitudinal variance. An exact equation for σ_x is obtained:

$$\sigma_x = 0.84 \frac{u_* t}{\kappa} \quad (7)$$

Van Ulden not only derived an along-wind dispersion equation under neutral conditions from a ground release in an unbounded boundary layer, but also for an along-wind dispersion equation for very stable conditions. Under very stable conditions, the wind speed profile and eddy diffusivities change. When applying the new constraints for wind speed and eddy diffusivities in a different stability class, the total variance of concentration distribution in the wind direction is shown to be:

$$\sigma_x = \{(\frac{3}{4} - \frac{16}{9\pi}) \frac{5 u_*^{\ 3} t^3}{kL}\}^{1/2} \simeq 0.43 (\frac{5 u_*^{\ 3} t^3}{kL})^{1/2}. \quad (8)$$

where L is the Monin-Obukov length.

2.5 Summary

The relationship between advection cloud speed, time, and distance to determine the along-wind diffusion coefficient is complex. Changing conditions such as stability class, surface roughness, wind speed, function of distance or time, release height, and other parameters all have an effect on determining the along-wind diffusion coefficient. With the varying conditions, not all the equations are able to be integrated together into an overarching equation meeting all conditions. Different assumptions or criteria result in diverse equations for the same diffusion coefficient. An equation that takes into account all of these parameters is desired.

3 Correlation of Field Data by Hanna and Franzese

The 12 field and wind tunnel tests mentioned prior were analyzed by Hanna and Franzese to determine a suitable along-wind dispersion coefficient that resulted in an equation comparable to Chatwin's theoretical equation[5]**Error! Bookmark not defined.**. Results of the data and the predicted equation are shown below in Figure 1.

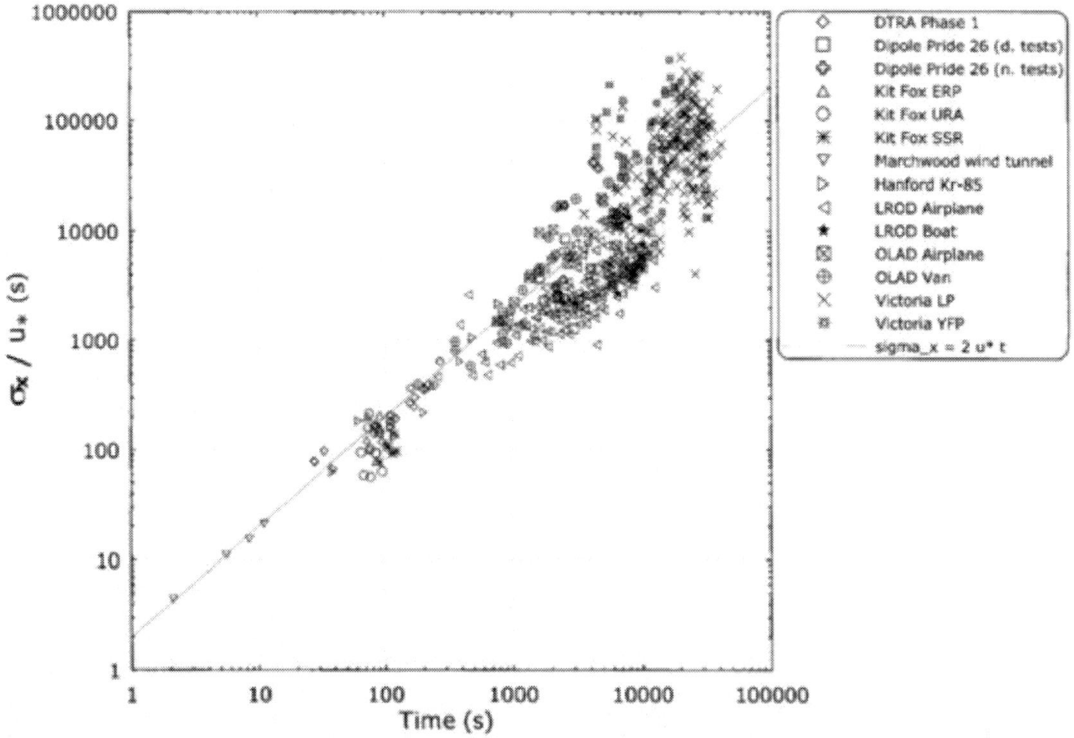

Figure 1. Hanna and Franzese results of along-wind dispersion coefficient divided by friction velocity and plotted vs time. Line represents $\sigma_x = 2u_* t$.[8]

4 Reconsideration of Field Data Using Dimensionless Groups

Figure 1 shows a correlation of the along-wind dispersion coefficient with the friction velocity as a function of time. While this is linear, their data can be up to a factor of 10 away from the predicted equation. There is also a noticeable trend of data beneath the predicted line when time is less than 10,000 seconds and above the predicted line when time is greater than 10,000 seconds. This is believed to be an effect of the near field vs. the far field. This equation for the along-wind dispersion coefficient does not take into effect the near field and far field which results in a very large portion data over predicted from 1,000-10,000 seconds and a large portion of data under predicted from 10,000 seconds and above. An improved equation to improve the correlation accounting for the error and trends is desired.

A non-dimensionalized equation was created to better predict the along-wind dispersion coefficient. The parameters of the non-dimensionalized equation were based off theory and assumptions regarding trends seen with atmospheric dispersion and along-wind dispersion. The non-dimensionalized equation allows for the along-wind dispersion coefficient to be predicted based on wind speed, distance, and stability class for each test in addition to the previous parameters of time and friction velocity.

The non-dimensionalized equation takes into account multiple theories and trends. Batchelor's Lagrangian Similarity Theory predicts σ_x to be a function of friction velocity, u_* , and time, t, while ignoring the release height[4]. Van Ulden predicts separate equations for both neutral and stable conditions[6]. Surface roughness has been shown to have an effect on wind speed[7]. The overall non-dimensionalized equation for along-wind dispersion is shown below

$$\frac{\sigma_x}{x} = \left(\frac{ut}{x}\right)^p \left(\frac{u_*}{u}\right)^q C_1 \qquad (9)$$

where x is the distance in meters and p, q, and C_1 are all dimensionless constants. The $\left(\frac{ut}{x}\right)$ term accounts for the effect of wind speed, u, time, t, and distance, x. The $\left(\frac{u_*}{u}\right)$ term is to take into effect the wind speed, u, and friction velocity, u_*, which ultimately represents the stability class.

Data was accumulated and analyzed by Stephen Hanna and Pasqual Franzese in 1999 from 11 field tests and 1 wind tunnel test[8]. The accumulation of these 12 tests produced 446 data sets over varying conditions. Other tests were accumulated and stored in the Modeler's Data Archive[9] for future reference. The following tests were analyzed in hopes of a better fit for along-wind dispersion in atmospheric modeling. Of the 446 data sets, only six data points were not used in the analysis due to lack of recorded friction velocity and/or wind speed. Reference heights for recording the wind speed from each of the data sets vary. A constant reference height of 10m was implemented across all data sets for a more standardized data analysis. Data sets used in this analysis are described below.

4.1 DTRA Phase 1

The DTRA Phase 1 tests were carried out from the 9th -26th of September in 1996 at the Dugway Proving Ground on Pad 11 producing 23 ensemble averages[10]. Propylene was instantaneous released from air cannons at ground level. Each air cannon produced 0.36kg of propylene in a period of 0.2-0.4 seconds. DigiPIDs and Ultra-Violet Ion Collectors were both used at distances of 200 m, 300 m, 400 m, 800 m, and 1200 m along a fixed sampling line 1.5m above the ground. Increasing distances resulted in an increased number of air cannons to result in an increased mass of propylene for the increased distance. 1 air cannon was used for distances of 200 m and 300 m, 2 air cannons for 400 m, 4 air cannons for 800 m, and 8 air cannons for 1200 m. Wind speed is referenced at a height of 2m which ranged from 2-7 m/s. The Great Salt Lake Desert produced a surface roughness of less than 1mm.

4.2 Dipole Pride

Dipole Pride tests were grouped into day and night tests from the 4th -20th of November in 1996 at the Department of Energy Nevada Test Site – Yucca flat. SF_6 was released from 2 vertically mounted, pressurized cylinders on the back of a 5-ton flatbed truck. These point sources released 8-21.6 kg of mass in a duration of 1-3 seconds[11]. 90 whole air samplers, with 15 minute time averaging, and 6 continuous TGA-4000 SF_6 analyzers were placed in 3 lines at distances of 1 km, 10 km and 20 km[12]. Wind speeds were referenced at a height of 10 m resulting in wind speeds of 3-5 m/s during the day and 2-5 m/s during the night. Surface roughness at the Yucca flat was around 32 mm. The 23 total tests produced 10 adequate day releases and 5 adequate night releases.

4.3 Kit Fox (ERP, URA, SSR)

All Kit Fox releases took place at the National Spill Test Facility at the Nevada Test Site. The Equivalent Roughness Pattern (ERP) and Uniform Roughness Array (URA) tests took place from the 24th -31st of August in 1995 whereas the Smooth Surface Rough (SSR) tests took place the 11th -15th of September in 1995.[13] The tests were similar in all conditions other than surface roughness. The Kit Fox ERP tests had a surface roughness of 0.2 m, the Kit Fox URA tests had a surface roughness of 0.02 m, and the Kit Fox SSR tests had a surface roughness of 2 mm[14]. Carbon Dioxide was released from a ground level area source with dimensions of 1.5 m x 1.5 m x 1 m at a rate of 1 m/s resulting in 1.5 kg/s – 4 kg/s being released[15]. Concentrations were measured at distances of 25 m, 50 m, 100 m, and 225 m. 84 high and low infrared concentration sensors, solid state infrared CO_2 sensors, and sample bag Nova sensors, were all used to detect concentrations of CO_2 with an averaging time of around 20 seconds[16]. Wind speeds were referenced at a height of 2 m which resulted in wind speeds of 2-3 m/s for the ERP and SSR test and 3-4 m/s for the URA tests. All tests took place right before sunset which resulted in lower wind speeds with neutral to stable stability classes. The ERP and SSR tests resulted in 6 sufficient trials each and the URA tests resulted in 9 sufficient trials to be used in this analysis.

4.4 Hanford Kr-85

The Hanford Fr-85[17] tests took place from 18 August to 8 November 1967 at the Battelle Pacific Northwest Laboratory. A vial containing Krypton 85 at ambient pressure was crushed by a guillotine-like device at ground level, instantaneously releasing a 1 m puff of the radioactive gas. Concentration was detected by 64 halogen-quenched Geiger-Muller tubes at a distance of 200 m and an elevation of 1.5 m above the ground. 24 halogen-quenched Geiger-Muller tubes were positioned on 6 towers at a distance of 800 m at elevations of 0.8 m, 4.6 m, 10.7 m and 21.3 m. The time averaging associated with the Geiger-Muller tubes was 4.8 seconds. 8 trials were performed, but only 6 releases remained within the boundary of the array of sensors. The wind speed was taken at a reference height of 1.5 m which resulted in wind speeds of 1-8 m/s. The roughness was around 30 mm due to the sagebrush and steppe grass.

4.5 LROD

The Long Range Overwater Dispersion (LROD)[18] tests took place from 13th -26th of July in 1993 at the Pacific Missile Range Facility in Kauai, Hawaii. A crosswind line source of pressurized, liquid SF_6 was released from a high pressure hose from the back of a plane at an elevation of 90 m at a rate of 9.6 g/meter and a speed of 129 m/s over water. A total duration of the release of SF_6 was around 13 minutes for the 100 km distance. Concentration was measured at ground level from boats along with planes at various elevations using electron capture detectors (ECD) and TGA-4000 whole air samplers. The averaging time was around 30-120 seconds for the sequential bag samples. Noise associated with the continuous analyzer resulted in limited use. Concentration was detected from the plane by flying parallel to the dissemination line and 45 km south of the sampling line (5 km, 15 km, 30 km, 60 km, and 100 km). The plane, equipped with a continuous SF_6 analyzer, would fly to the dissemination line then back to 25 m. It would slowly ascend from 25 m to 2500 m for various heights. Testing over water resulted in a surface roughness of 0.02-0.86 mm. A reference height of 10 m was used for wind speed which resulted in speeds of 3-11 m/s. The LROD tests produced 11 ensemble averages.

4.6 OLAD

The Over Land Along-wind Dispersion (OLAD)[19] tests were similar to the LROD tests, except over land. These tests took place from the 8th -25th of September in 1997 at the Dugway Proving Ground, West Desert Test Center. SF_6 was released from both plane and truck line sources. The plane released 100 kg of SF_6 at an elevation of 100 m from a N_2 pressurized tank at a rate of 33.3 kg/min for a distance of 20 km. The truck released 15 kg of SF_6 at an elevation of 3 m from a mounted tank at a rate of 1.5 kg/min for a distance of 10 km. Concentrations were recorded with TGA-4000 whole air samplers, with time averaging of 15 minutes, and 2 continuous electron capture detector (ECD) analyzers at the end of each line at distances of 2 km, 5 km, and 10 km for the truck releases and 10 km, 15 km, and 20 km for the plane releases. Wind speed was measured at a height of 2 m which resulted in speeds of 0.5 – 7 m/s. The surface roughness was measured around 30 mm. These tests resulted in 3 ensemble averages for the aircraft release and 8 surface ensemble averages for the truck release.

4.7 Marchwood wind-tunnel

The Marchwood[20] wind tunnel tests are the only tests that took place in a wind tunnel. A mixture of propane and helium, resulting in neutral buoyancy, was released from a 15 mm ground level point source. The mixture was released for a duration of 0.5 – 12 seconds. 150-300 releases were summarized into 4 ensemble averages. A single FID measured the concentration with a 5 ms sampling interval at locations of 1 m, 3 m, 5 m, and 7 m. Wind speeds ranged from 1-4 m/s at neutral stability class. Surface roughness was measured at 0.3 mm for the wind tunnel.

4.8 Oceanside

The Oceanside[21] tests took place at the shoreline of Oceanside and Del Mar in Southern California. 20 μm particles of green and yellow fluorescent pigment (FP) were released by plane, boat, and land with standardized Dugway Proving Ground equipment. For the plane releases, the DPG model D-1 released 1500 g/min of FP for a duration of 10 minutes at an elevations of 61 m, 122 m, and 152 m over a 40 km distance. For the boat releases, a DPG Mark IV was used to release 110 g/min for a duration of 2 hours at sea level over 40 km. For instantaneous land releases, the DPG Mark IX was used at three different inland locations to release 10-50 g of FP instantaneously at ground level. Concentration of FP were recorded from rotorod samplers, sequential samplers, vertical samplers, 100 m square grids, dense sampling lines, and metronics drum impact samplers. Time averaging ranged from 5 minutes at coastal regions to 15 minutes at inland conditions. Surface roughness ranged from 30-50 m due to the coastal shoreline. The wind speed reference height varied from 8-120 m which resulted in wind speeds ranging from 1-5 m/s.

4.9 Fort Wayne

Fort Wayne[22] releases took place in the Summer and Fall of 1963 in Fort Wayne, Indiana. An aerosol of fluorescent pigment (FP) was released via 2 aircrafts in the form of an elevated, 91-244 m, line source. Dry green and yellow FP were released from the planes via FP dispensers supplied by the Dugway Proving Ground. The amount of release was measured by the difference in weight, before and after. The release was upwind of the 180,000 population urban setting with a river running through the city and agriculture surrounding. Rotorods, Gelman throw-away plastic filter holders, and Gelman paper tape samplers were used for concentration measurements of the FP. The tape provided an averaging time of 5 minutes where the filter holders had an averaging time of 30 minutes. Wind speed was measured at the release height and resulted in speeds ranging from 5-16 m/s.

4.10 Short Range Diffusion Experiment Series

The Short Range Diffusion Experiment Series[23] took place in 1979 in Japan. A 200 m x 200 m vacant lot covered in turf and surrounded by trees was used for the releases. NO_x was released instantaneously from a balloon at an elevation of 1.5 m. Chemiluminescence NO_x meters were used to detect concentration with a sampling time of <2 minutes. Sample lines were chosen at distances of 60 m and 100 m. Wind speeds were referenced at heights of 0.5 m, 3 m, and 9 m which resulted in speeds of 1-2 m/s.

4.11 Results and Discussion

A multivariable regression analysis, also known as multiple regression, is an extension of linear regression[24]. A regression analysis determines the relationships among variables.

With a multivariable regression analysis, the relationships between multiple independent variables can be determined. A regression analysis is a useful tool in understanding the changes of one variable with respect to others. The form of a typical multi-variable regression analysis is shown below

$$Y = a + b_1x_1 + b_2x_2 + \cdots + b_nx_n \qquad (10)$$

A multi-variable regression analysis can be applied to the non-dimensionalized equation. The non-dimensionalized equation is rearranged with the natural log of each variable to perform the multi-variable regression analysis. This equation is shown below

$$ln\left(\frac{\sigma_x}{x}\right) = p\,ln\left(\frac{ut}{x}\right) + q\,ln\left(\frac{u_*}{u}\right) + C_2 \qquad (11)$$

where C_2= ln(C_1). A multivariable regression analysis with known data is used with this equation to determine the constants for p, q, and C_1. An analysis is done for the overall data and categorized into stability classes using macros in Excel[25].

Multiple parameters were analyzed for optimal results using the multivariable regression analysis. Data was quantified and analyzed based on individual tests, surface roughness, friction velocity, Monin-Obukhov length, distance, time, near field, and far field.

Data from all 12 experimental data sets referenced above were analyzed with a multi-variable regression analysis to determine appropriate exponents and a constant for an optimum non-dimensionalized equation. While many parameters were originally analyzed for the optimum non-dimensionalized equation, the quantification of near field and far field as a function of time was chosen as the most prominent parameter. The optimum non-dimensionalized equation was parameterized by near field with time less than 10,000 seconds and far field with time greater than 10,000 seconds. These equations are shown below in Table 1. The ratio, α, of the along-wind dispersion coefficient, σ_x, for the data vs σ_x for the predicted non-dimensionalized equation is used for comparison of the difference in equations. A non-parametric confidence interval was determined for the non-Gaussian distribution of α.[26] Hanna and Franzese's along-wind dispersion coefficient was analyzed in the same manner in Table 1.

Table 1. Comparison of along-wind dispersion equations with 90% confidence intervals

Model	Equation	90% Confidence Interval of $\alpha = \frac{\sigma_x\ (predicted)}{\sigma_x\ (data)}$	Range of Confidence Interval
Hanna and Franzese	$\sigma_x = 2u_* t$	$1.63 < \alpha < 1.88$	26%
Near Field Non-Dimensionalized	$\frac{\sigma_x}{x} = 1.72\left(\frac{ut}{x}\right)^{1.18}\left(\frac{u_*}{u}\right)^{1.10}$	$1.16 < \alpha < 1.31$	16%
Far Field Non-Dimensionalized	$\frac{\sigma_x}{x} = 0.05\left(\frac{ut}{x}\right)^{0.67}\left(\frac{u_*}{u}\right)^{-0.27}$	$0.85 < \alpha < 1.05$	20%

From Table 1, the range of the 90% confidence intervals for each of the equations is shown. The non-dimensionalized equations have a tighter correlation of the predicted values. The 90% confidence intervals shows that Hanna's predicted along-wind dispersion equation has a trend over predict the σ_x values. An interesting result from the multi-variable regression analysis is the determination of the coefficient for the $\left(\frac{u_*}{u}\right)$ term. The -0.27 exponential value essentially confirms that the stability class has no effect on the along-wind dispersion coefficient after a specific amount of time. A discontinuity is understood between the near field and far field non-dimensionalized equation and is currently being evaluated.

The results of the non-dimensionalized predicted σ_x values were divided by the friction velocity and plotted against the original data to determine their relationship. This relationship is shown below in Figure 2.

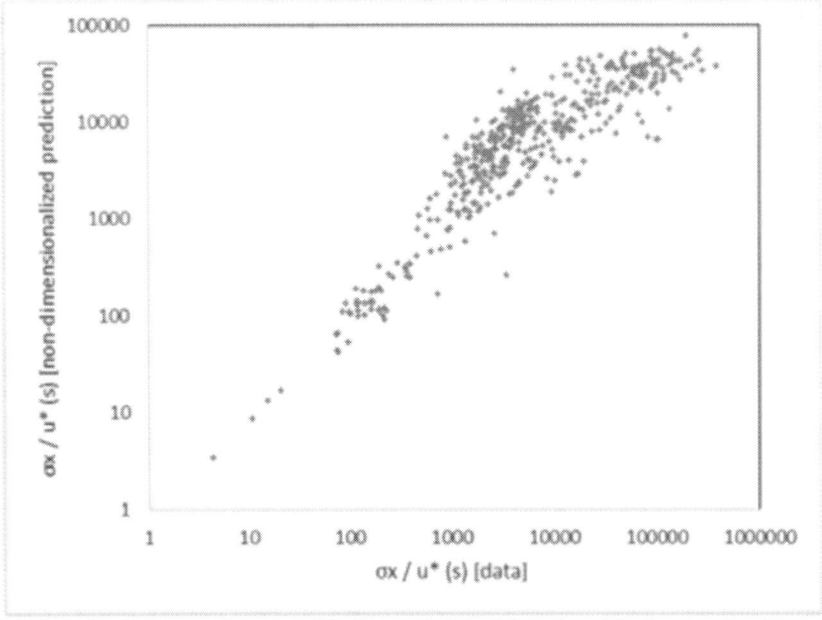

Figure 2. Along-wind dispersion coefficient divided by the friction velocity of the non-dimensionalized modeled values vs the original data.

As shown in Figure 2, the correlation between the data and non-dimensionalized predicted values shows a 45^{0} line. This correlates a good fit for the predicted σ_x values when compared to the original σ_x values.

A visual representation of the non-dimensionalized predicted σ_x values, Hanna and Franzese's σ_x values, and the original σ_x values as a function of time are all shown below in Figure 3.

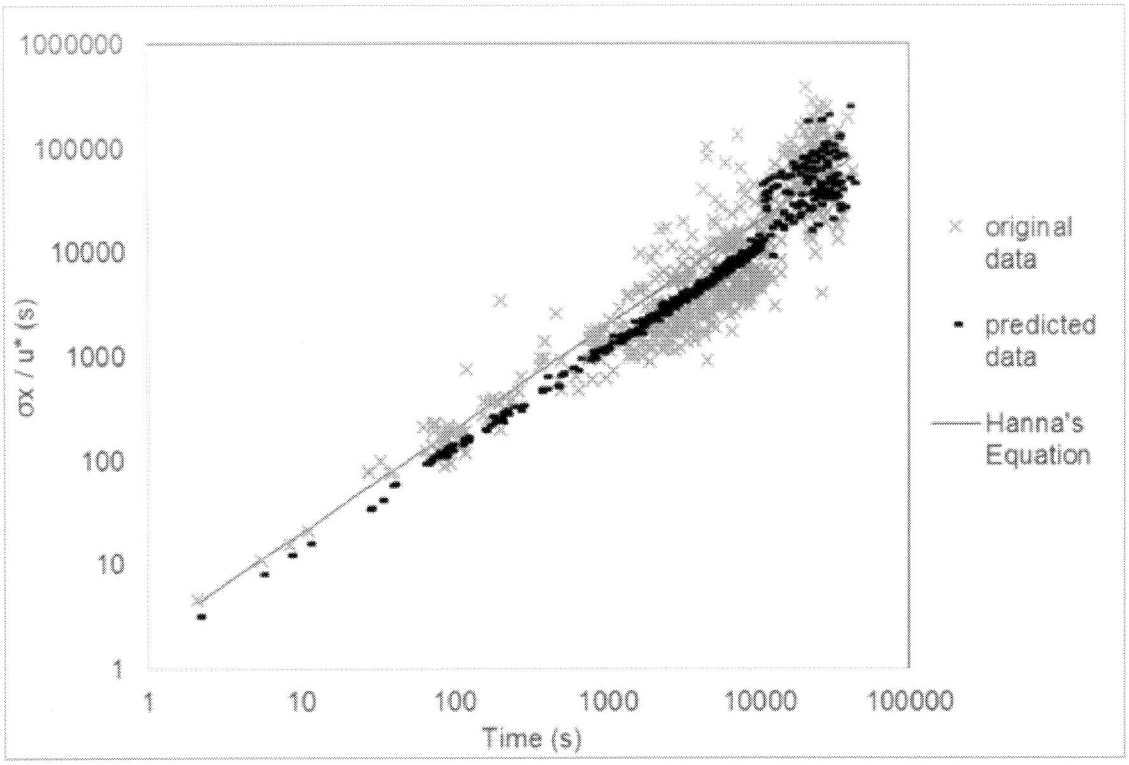

Figure 3. Time vs. the along-wind dispersion coefficient divided by the friction velocity as a function of the original data, non-dimensionalized predicted data, and Hanna's predicted data.

From this figure, the predicted values from both Hanna and Franzese's predicted data and the non-dimensionalized predicted data are shown in comparison with the original data. The non-dimensionalized equation takes into account additional parameters, other than time, which results in the spread of σ_x values as a function of time and additional parameters in comparison to the single line by the $\sigma_x = 2u_*t$ along-wind dispersion equation. The non-dimensionalized equation also takes into effect the near field and far field. This is clearly seen at times greater than 10,000 seconds where the σ_x values are spread over a wider range. Overall, the non-dimensionalized equation for along-wind dispersion proves better than the previous $\sigma_x = 2u_*t$ correlation.

5 Summary

Along-wind dispersion is critical when modelling toxic, flammable, or hazardous time-limited or finite-duration releases into the atmosphere. While many models exist from a theoretical standpoint or parameterization of experimental data, there is no overarching correlation to predict the along-wind dispersion coefficient. This paper took into effect established theoretical equations validated with experimental data to determine a better equation to model the effects of along-wind dispersion. A successful non-dimensionalized equation parameterized by the near field and far field was determined and validated against available field-scale data which showed improvement in comparison to previous equations. The non-dimensionalized equation for along-wind dispersion solved for in this paper will improve the existing atmospheric modeling with along-wind dispersion.

6 Acknowledgments

We gratefully thank our colleagues from DNV GL, Henk Witlox and Maria Fernandez, who provided insight and expertise that greatly assisted the research. We also thank Steven R. Hanna and Pasquale Franzese for sharing their accumulated data for further analysis.

7 References

[1] Boybeyi, Zafer, Sethu Raman, and Paolo Zannetti. "Numerical investigation of possible role of local meteorology in Bhopal gas accident." *Atmospheric Environment* 29.4 (1995): 479-496.

[2] Høiset, S., et al. "Flixborough revisited—an explosion simulation approach." *Journal of hazardous materials* 77.1 (2000): 1-9.

[3] De Marchi, Bruna, Silvio Funtowicz, and Jerome Ravetz. "Seveso: A paradoxical classic disaster." *The long road to recovery: Community responses to industrial disaster* (1996): 86-120.

[4] Batchelor, G. K. "Diffusion from sources in a turbulent boundary layer." *Arch. Mech. Stosow* 3.6 (1964): 661-670.

[5] Chatwin, P. C. "The dispersion of a puff of passive contaminant in the constant stress region." Quarterly Journal of the Royal Meteorological Society 94.401 (1968): 350-360.

[6] Van Ulden, A. P. "A surface-layer similarity model for the dispersion of a skewed passive puff near the ground." *Atmospheric Environment. Part A. General Topics* 26.4 (1992): 681-692.

[7] DNV GL. DNV GL Theory Manual for Unified Dispersion Model. London UK 2015.

[8] Hanna, Steven R., and Pasquale Franzese. "Alongwind dispersion—a simple similarity formula compared with observations at 11 field sites and in one wind tunnel." *Journal of Applied Meteorology* 39.10 (2000): 1700-1714.

[9] Chang, Joseph C., and Steven R. Hanna. "Modelers' Data Archive – A Collection of Atmospheric Transport and Dispersion Data Sets." (2010).

[10] Biltoft, C., Phase I of DSWA transport and dispersion model validation study. DPG-FR-97-058, (1997): 156.

[11] Biltoft, C., Dipole Pride 26: Phase II of DSWA transport and dispersion model validation study. DPG-FR-98-001, (1998): 130.
[12] Watson, T. B, R. E. Keislar, B. Reese, D. H. George, and C. A. Biltoft, The DSWA Dipole Pride 26 field experiment. NOAA Tech. Memo. ERL ARL-225, (1998): 99.
[13] King, S. Bruce. The Kit Fox Field Demonstration Project and Data Set. *International Conference and Workshop on Modeling the Consequences of Accidental Releases of Hazardous Materials.* (1999).
[14] Spicer, T. and Havens, J. Description and Analysis of Atmospheric Dispersion Tests Conducted by EPA. *International Conference and Workshop on Modeling the Consequences of Accidental Releases of Hazardous Materials.* (1999).
[15] Sheesley, D. et al.. Chemical Dispersion Modeling Project, 1995 Kit Fox Field Demonstration DOE Nevada Hazardous Spill Center. Volume I – Experiment Description and Data Processing. (1998).
[16] Hanna, S. R., and K. W. Steinberg, Studies of dense gas dispersion from short-duration transient releases over rough surfaces during stable conditions. *Air Pollution Modeling and Its Applications XI,* Plenum Press, (1996): 481–490.
[17] Nickola, P. W., J. D. Ludwick, and J. V. Ramsdell Jr., An inert gas tracer system for monitoring the real-time history of a diffusing plume or puff. *J. Appl. Meteor.,*9, (1970): 621–626.
[18] Bowers, J. F., G. E. Start, R. G. Carter, T. B. Watson, K. L. Clawson, and T. L. Crawford, Experimental design and results for the Long-Range Overwater Diffusion (LROD) Experiment. DPG/JCP-94/012, (1994): 531.
[19] Biltoft, C., S. D. Turley, T. B. Watson, G. H. Crescenti, and R. G. Carter, Over-Land Atmospheric Dispersion (OLAD) test summary and analysis. WDTC Doc. WDTC/JCP-99/048, (1999): 38.
[20] Robins, A. G., and J. E. Fackrell, An experimental study of the dispersion of short-duration emissions in a turbulent boundary layer. *Air Pollution VI,* C. A. Brebbia, C. F. Ratto, and H. Powers, Eds., WIT Press, (1998): 697–707.
[21] Smith, and B. L. Niemann, Shoreline diffusion program, Oceanside, CA. Vols. I and II. MRI-64-FR-860, (1969): 230 and 456.
[22] Hilst, G. R., and N. E. Bowne, A study of the diffusion of aerosols released from aerial line sources upwind of an urban complex. Vols. I and II. Rep. DA-42-007-AMC-38®, (1966): 231 and 583.
[23] Sato, J., F. Kimura, and T. Yoshikawa, The longitudinal spread of puff in the short range diffusion experiment. *Pap. Meteor. Geophys.,*32, (1981): 155–162.
[24] Draper, Norman Richard, Harry Smith, and Elizabeth Pownell. *Applied regression analysis.* Vol. 3. New York: Wiley, (1966).
[25] Davis, Richard A. "Practical Numerical Methods for Chemical Engineers: Using Excel with VBA". Lexington, KY, (2015).
[26] Conver, W. J. "Practical Nonparametric Statistics". Wiley, New York, (1971).

478500 Review and Analyze the Effects of Parameters for H_2s Gas Mixture Dispersion Models: Degadis and CFD Models

Abstract:

Due to the increasing demands of energy, natural gas explorers and developers are interested in exploring sour gas fields, which have a high concentration of hydrogen sulfide (H_2S). Because H_2S is highly corrosive and could reduce the heating values of natural gas, H_2S needs to be removed from natural gas before transfer. Then the removed H_2S gas will be converted to sulfide as the products in the Claus process. The H_2S mixture (*e.g.*, H_2S and CO_2) may cause corrosion to pipelines and thus, could lead to H_2S gas mixture release.

H_2S mixture behaves as a heavy gas, which travels down to the ground causing harm to the local population and the environment. In this paper, the DEnse GAs DISpersion (DEGADIS) model and the Computational Fluid Dynamics (CFD) model (*e.g.*, ANSYS Fluent) are implemented for the H_2S mixture simulation, and assumptions in the models are also studied. The effects of topography and meteorological parameters such as atmospheric temperature, wind speed, and humidity are investigated. Finally, the simulation results from the DEGADIS model and the CFD model for flat terrain with constant wind velocity are compared.

The simulation results show that the DEGADIS model, which is suitable for flat terrain and constant wind conditions, can be applied to simple terrain geography and stable atmospheric conditions. Whereas the CFD model, which considers unsteady wind speed and turbulence near the obstacles, can be utilized for complex topography. The series of simulation results can provide guidance for the risk assessment of H_2S mixture releases under different topographies and meteorological conditions.

478998 Considerations When Justifying the Purchase of an Operator Training Simulator

Abstract:

Highly skilled plant operators ensure higher levels of plant safety, reliability and profitability. An Operator Training Simulator accelerates both the time to competence and raises the competency level of an operator. The ARC Advisory Group recently surveyed companies who have invested in Operator Training Simulators (OTS). Two critical results from the survey are that over sixty five percent of those companies report fewer abnormal situations, fewer trips, and incidence avoidance attributable to the use of OTS. Over sixty percent of these companies report improved safety and reduced risk and lower liability attributable to the use of the OTS.

Many leading companies have embraced the use of Operator Training Simulators across their Enterprise however many facilities and companies have yet to embrace the idea. The most common reason given for not investing in such a solution is that the cost of an OTS cannot be justified. The nuclear industry mandates that a nuclear power plant must have an OTS whereas the American Petroleum Institute only recommends the use of Training Simulators. The Process industries overall will benefit by the greater adoption of training simulators and this paper discusses the different ways that can be used to justify the investment in an OTS.

479500 Consequence Analysis of Gas Explosion in Underground Common Utility Tunnels

Abstract:

The purpose of this study is to figure out the effects of blast overpressure in gas piping leakage in underground common utility tunnels through the explosion simulation tool FLACS (Flame Acceleration Simulator) and find and apply methods that can reduce the risk.

Common utility tunnels are facilities to jointly bury pipes in underground spaces. Placing gas piping in common utility tunnels has the advantages of reducing the risk of corrosion and preventing the physical damage due to civil engineering works and excavating work on the ground. However, when an explosion has occurred in a common utility tunnel, the consequences could be larger in that it is a closed space. Nevertheless, studies on underground common utility tunnels are currently limited to fire accidents. Therefore, risks due to explosions should be quantitatively studied.

Therefore, in this study, the effects of explosions were analyzed assuming cases where gases leaked from gas piping in underground common utility tunnels exploded. In addition, blast overpressures were evaluated for diverse scenarios regarding ignition sources, leakage locations, and discharge rates and the results were statistically derived. Blast overpressures high enough to cause dangerous situations of the underground common utility tunnels were observed. Methods that can reduce the risk, that is, methods that can reduce the frequency of explosions, and methods for designs of internal structures of common utility tunnels that can mitigate the consequences will be presented.

479158 Air Cooler Explosion Analysis Using CFD

Abstract:

Air coolers or fin-fans are located in most process plants. They cool process fluids using forced air flow over radiators containing the fluid. Air coolers initially appear to offer many features that would result in a significant explosion source should a flammable cloud reach the cooler. The radiator coils are closely spaced pipes which will accelerate flames. The metal box represents confinement and the turbulent air flow produced by the fans will accelerate flames similar to how congestion produces turbulence ahead of a flame front. Also, the flow of air can draw rich flammables near the ground up into the air cooler.

This study examines the severity of an air cooler explosion using computational fluid dynamics (CFD). The commercial software FLACS is used to examine the parameters mentioned above (congestion, confinement, and turbulence) and their effect on explosion severity. Comparisons between FLACS and the Multi-Energy Method are made and recommendations provided for assessing air coolers in non-CFD facility siting studies.

Revisions to the QMEFS Vapor Cloud Explosion Model

Jeffrey D. Marx and Benjamin R. Ishii
Quest Consultants Inc.®
908 26th Avenue N.W.
Norman, OK 73069
jdm@questconsult.com / bri@questconsult.com

Prepared for Presentation at
American Institute of Chemical Engineers
2017 Spring Meeting and 13th Global Congress on Process Safety
San Antonio, TX
March 26 – 29, 2017

Revisions to the QMEFS Vapor Cloud Explosion Model

Jeffrey D. Marx and Benjamin R. Ishii
Quest Consultants Inc.®
908 26th Avenue N.W.
Norman, OK 73069
jdm@questconsult.com / bri@questconsult.com

Keywords: vapor cloud explosion (VCE), Baker-Strehlow-Tang (BST) model, estimating flame speeds

Abstract

The Quest model for estimating flame speeds (QMEFS) was introduced in 2007, has since been published, and has been used by Quest and other companies to provide a detailed method for vapor cloud explosion (VCE) predictions. QMEFS provides a set of parameters that predict a flame speed, which then is used with the Baker-Strehlow-Tang (BST) blast curves to predict overpressure and impulse. The original intent of the QMEFS model was to provide better characterization of confined or congested areas, as well as fuel reactivities, which fit between the low, medium, and high categories offered by the BST model, so that the large differences in predictions between categories could be eliminated. In this way it provided a more refined system for predicting the consequences of a VCE, while remaining anchored within the flame speeds provided by the BST matrix.

In the years since QMEFS' publication, Baker has added to the BST flame speed matrix. In addition, use of the model has highlighted some of the areas where the model could be improved. This paper shows several ways in which the model has been enhanced to encompass a wider range of conditions than are covered by the BST model.

1 Introduction

Any release of a flammable fluid in a petrochemical facility has the potential to generate a flammable vapor cloud that, if ignited, could produce a vapor cloud explosion (VCE). Most VCEs that generate damaging levels of overpressure originate in what is called a potential explosion site (PES), where confinement and congestion contribute to flame acceleration and the generation of overpressure. When flammable vapors enter the PES and are ignited, the possibility of human injury/death, asset damage, or event escalation becomes a concern. The concern for human injury or death is most often addressed in the form of a building siting study. Because people are somewhat less likely to be injured or killed when outside, as compared to when inside a building, the siting study focuses on the potential impacts to buildings within and around petrochemical facilities. It then becomes the task of process safety professionals to estimate the potential for VCE events, and their resulting overpressure impacts on buildings.

The Baker-Strehlow-Tang (BST) model and the TNO Multi-energy (ME) model use blast curves to estimate the overpressure impacts generated by an exploding vapor cloud. Both approaches require the user to estimate the strength of the explosion as a function of the reactivity of the flammable material and the degree of confinement or congestion present in the cloud. This information is then used to determine which strength curve (in the case of the TNO model) or flame speed curve (in the case of the BST model) is used to calculate the overpressure of the explosion as a function of distance from the center of the explosion. It is not clear, however, how best to choose the correct curve when modeling an explosion.

There have been several papers published regarding blast curve selection for the ME model. These methods are generally presented as a qualitative ruleset which acts as an index into which of the 10 curves should be used. The BST model provides a simple set of guidelines that result in a selection of a flame speed, which corresponds to an overpressure vs. distance relationship. Three parameters are used to determine the flame speed: reactivity of the flammable gas, the degree of confinement of the flammable cloud, and the degree of obstruction due to obstacles within the flammable cloud. These parameters are represented by a set of what are effectively low, medium, and high choices for each parameter.

While the BST model provides reasonable guidance on choosing the explosion parameters, there are still areas that could be improved. One problem is that the BST flame speed matrix includes large jumps in flame speed between categories of confinement, reactivity, or congestion. This was first addressed by Baker in a 1998 paper [1] where a new confinement category, 2½-D, was added for those situations that were more confined than the 3-D case, but less so than the 2-D case. The flame speeds used for the 2½-D confinement were simply the arithmetic average of the flame speeds for the 2-D and 3-D cases for a given reactivity and congestion class. Even with this extension, large discontinuities remained among the categories for flame speed according to the prescribed methodology. These discontinuities drive the need for a new method to determine flame speed based upon quantifiable properties of the flammable gas and its surroundings that varies smoothly across the range of conditions that are found in actual process plants.

The creation of the QMEFS model [2] improved upon the basic BST scheme by combining it with the GAMES correlations [3] and providing a numerical resolution to the BST matrix. Integration of the GAMES model allows the strength of the explosion to be defined by numeric input parameters and avoids the low/medium/high categorizations.

2 Parameters Affecting Flame Speed

With the incorporation of the GAMES correlations, the resulting explosion flame speed is characterized by five parameters:

- Laminar burning velocity (LBV) is used as the characteristic property for fuel reactivity

- Volume blockage ratio (VBR) and average obstacle diameter define the level of congestion
- Confinement is characterized as the number of confining planes, or walls, that are available to confine the expanding gases.
- Flame run-up distance is used to characterize the flame acceleration path

The last part of explosion characterization is the explosive energy of the fuel/air mixture, which is defined by the gas concentration, the combustion energy of the fuel, and the volume occupied by the fuel/air mixture.

2.1 Reactivity

Fuel reactivity is a measure of the propensity of the flame front to accelerate and create overpressures or potentially undergo a deflagration-to-detonation transition (DDT). The BST model originally classified reactivity as high, medium, or low based on the fuel in the fuel-air mixture. The original rule set [1] defined propylene oxide, ethylene oxide, and hydrogen (or materials with a 0.8 m/s LBV or larger) as high reactivity; methane and carbon monoxide were classified as low reactivity, and everything else as medium reactivity. According to the current BST model, materials having a laminar flame speed greater than 0.75 m/s are considered high reactivity (to accommodate ethylene being high reactivity) while those having a laminar flame speed below 0.4 m/s are considered low reactivity [1], although some BST publications list the low boundary as 0.45 m/s. In 2016, the BST categories were expanded to include a very-low category for fuels like ammonia, which has an LBV in the range of 0.07 m/s to 0.15 m/s [4].

In the experimental studies that form the basis of the BST model, ammonia, methane, propane, and ethylene were used to define the very low, low, medium, and high reactivity categories. These categories produce confusion due to how they create boundaries between the BST categories. For example, the LBV of ethylene, depending on the cited reference, ranges between 0.64 and 0.83 m/s. With the current definition of 0.75 m/s as the high reactivity boundary, this makes ethylene a "borderline" fuel – one that may either be defined as medium or high reactivity. However, ethylene is explicitly categorized as a high reactivity material, as their experimental work used ethylene as the high reactivity material [5]. This then begs the question: what about ethylene oxide (LBV 1.0 m/s) or acetylene (LBV 1.6 m/s)? These two materials are clearly more reactive than ethylene, but are still categorized as high reactivity. The BST model effectively says that an ethylene explosion will be as severe as an acetylene explosion, when all other parameters are held constant.

2.2 Congestion

In the original BST model, low obstacle density was defined as having an area blockage ratio (ABR) of less than 10%, while high obstacle density was defined as an ABR of 40% or greater, and everything in between is considered medium [6]. Later BST guidelines have used the volume blockage ratio (VBR) as the parameter for classifying obstruction [5]. In practice both blockage ratios will affect how fast a flame accelerates, but, for a uniform obstacle field, the two are related simply by the pitch-to-diameter ratio of the obstacles. If the obstacles included in an ABR are not repeated quasi-uniformly throughout the obstacle field, then their effect would be more accurately portrayed by VBR.

The classifications of low, medium, and high are often too coarse for proper assignment of congestion to a particular PES. In the experimental programs that helped to define the BST model, VBRs of 1.5%, 4.3%, and 5.7% were used to represent the low, medium, and high categories. As with other VCE parameters, the differences between these categories leaves significant room for variation between the categories. In addition, no definition of 0% VBR is provided, nor is any relationship to VBRs higher than 5.7%.

2.3 Confinement

The effect of the confinement of the flammable cloud is taken into account by determining the number of dimensions in which the burning gas may expand. 3-D expansion allows the burning cloud to expand freely in all directions and results in the slowest flame acceleration and lowest overpressures. 2-D expansion, such as a flame between two flat plates, generates higher overpressures because the combustion gases have fewer directions in which to expand resulting is a higher flame acceleration.

Confinement is included in the BST model by identifying the number of dimensions that are available to the products of combustion for expansion. Expansion into free space (except for the plane of the ground) is considered 3-D expansion, expansion between two parallel planes is considered 2-D expansion, and expansion in a pipe is considered 1-D expansion. The 1-D case was removed in the more recent publications discussing the BST model [5]. To handle the case of a frangible or partially-confining plane, such as a very closely spaced pipe rack, the BST model added a 2.5-D classification which simply averaged the 2-D and 3-D flame speed results [5].

While the classification is fairly straightforward, especially after the 2.5-D category was added, some situations are still difficult to define. These include partial walls, pieces of equipment that may form more confinement than congestion, or surfaces that are porous such as grated decks.

2.4 Other Factors

Several researchers [5, 7, 8] have acknowledged that the overall dimensions of the flammable vapor cloud before ignition directly affect the final flame speed and consequent overpressures of the vapor cloud explosion. Since a flame will accelerate until it reaches a maximum sustainable value (or undergoes a deflagration to detonation transition, a DDT), the maximum dimension of a flammable cloud is expected to be an important variable in the creation of overpressure. As a point of reference, the length available for flame acceleration in the more recent BST tests was 15 meters [5].

In addition, research [7, 8] suggests that the scale of obstacles within the congested area also affects the maximum flame speed that may be achieved. As a flame front moves around and past obstacles, the turbulence that is generated can increase the flame surface area and accelerate combustion. The characteristic obstacle diameter for the BST tests was 2 inches (0.0508 meters); other test series (including the MERGE/EMERGE tests, varied the average obstacle diameter).

2.5 Parameters Comparison

The BST parameters discussed above are listed in Table 1, along with the corresponding numerical values that are equivalent to those applied in the QMEFS model. When the numerical values in Table 1 are applied in the QMEFS model, the results will match the predictions of the BST model.

Table 1. BST Flame Speed Parameters

BST Parameter	BST Category	Model Parameter	Corresponding Parameter Value
Reactivity	Very-Low (e.g., Ammonia)	S_l	0.08 m/s
	Low (e.g., Methane)		0.37 m/s
	Medium (e.g., Propane)		0.43 m/s
	High (e.g., Ethylene)		0.75 m/s
Obstacle Density	Low	*VBR*	0.015
	Medium		0.043
	High		0.057
Expansion	3-D	*nPlanes*	1
	2.5-D		1.5
	2-D		2

3 Model for Estimating Flame Speed

The QMEFS model was created to overcome some of the limitations described above. By incorporation of the GAMES correlations and the overpressure/impulse curves from the BST method, a model that provides the capability for more detailed descriptions of explosion scenarios has been created. Thus, QMEFS incorporates the modeling methodologies and experimental data sets that have been used in both the BST model and in the GAMES project.

3.1 Fundamental Correlations

The fundamental correlation is based on correlations developed from the GAMES project [3]. These correlations related overpressure to parameters based on the MERGE and EMERGE experiments [7,8]. These correlations can be combined with acoustic theory to relate flame speed to explosion parameters [2] to produce the equations below.

For no confining surfaces other than the ground (3D expansion):

$$2.4 * \frac{M_f^2}{1+M_f} * p_0 = 0.84 * \left(\frac{VBR*L_f}{D}\right)^{2.75} * S_l^{2.7} * D^{0.7} \tag{1}$$

For confinement between two flat planes (2D expansion):

$$2.4 * \frac{M_f^2}{1+M_f} * p_0 = 3.38 * \left(\frac{VBR*L_f}{D}\right)^{2.25} * S_l^{2.7} * D^{0.7} \tag{2}$$

where:

VBR = volume blockage ratio
L_f = maximum distance a flame may propagate in the obstructed region (i.e., the run-up distance), m
D = average obstacle diameter, m
S_l = laminar burning velocity (LBV) of the flammable gas, m/s
p_0 = the ambient pressure, bara
M_f = the flame speed relative to a fixed observer, expressed as a Mach number

To correct the flame speed to account for confinement in spaces other than one or two planes, a curve fit of the ratio of burned cloud to unburned cloud radii versus the number of confining planes was made. A curve fit was generated with an expansion ratio (α) equal to 7, a typical expansion ratio for common hydrocarbon fuels in air. This is also the value assumed in the derivation of portions of the BST model [9]. The equation resulting from this curve fit is:

$$NPF = \frac{1}{0.5035 - 0.0757 \cdot nPlanes} - 1.338 \tag{3}$$

where:

NPF = flame speed correction factor
$nPlanes$ = number of confining planes

For true 3-D spherical expansion, $nPlanes = 0$. For a typical explosion at grade level, $nPlanes = 1$. For 2-D expansion, such as a flame propagating between a floor and a ceiling, $nPlanes = 2$. For 1-D expansion, such as a flame front in a pipe, $nPlanes = 4$. Planes that are not completely solid or rigid may be accounted for using a fraction of a plane.

The flame speed correction factor, *NPF*, does not fully account for increases in flame speed as the number of confining planes increases. This is accomplished by applying the following factor to the flame speed when *nPlanes* is equal to 2:

$$\beta = \frac{M_{f,2D}}{M_{f,3D} \cdot NPF(nPlanes = 2)} \tag{4}$$

This can be implemented for nPlanes =1 to nPlanes = 4:

nPlanes = 1

$$M_f = M_{f\,3D} * NPF \tag{5}$$

1 > *nPlanes* < 2

$$M_f = M_{f\,3D} * NPF * \beta \tag{6}$$

nPlanes >= 2

$$M_f = M_{f\,3D} * NPF * [1 + (\beta - 1) * (nPlanes - 1)] \tag{7}$$

3.2 Flame Speed Limits

The equations above are implemented in the QMEFS model to define the flame speed (M_f) and then coupled with the latest BST flame speed matrix [4]. One limitation to the GAMES correlations is that they can predict unrealistic flame speeds. This is a direct result of the correlations fitting the data within the scope of the MERGE and EMERGE tests, and not being designed to extend beyond that scope. By coupling with the BST model, this problem is partially resolved because the BST matrix provides a set of upper limits to flame speeds (and by extension, the explosion strength).

Upon examination, it can be seen the GAMES correlations will predict higher and higher flame speed values as parameters such as L_f or S_l, increase. In reality, for a given obstacle configuration, geometry, and flammable gas mixture, there is a limit to the flame speed that can be achieved. This flame speed may be either subsonic or the cloud may undergo a deflagration-to-detonation transition (DDT) in which the flame front becomes a detonation propagating at roughly the Chapman-Jouguet (C-J) detonation velocity for the flammable mixture. For most common flammable hydrocarbons in air, this speed is roughly 1800 m/s or $M_f = 5.2$. For purposes of this model, any flame that accelerates to a calculated velocity greater than $M_f = 3.0$ is considered to have undergone DDT [10, 11] and the flame speed is set to the C-J velocity, $M_f = 5.2$.

The flame speeds suggested by the latest BST model are applied as the upper limit flame speeds to be used in this model, as they have been "scaled up" to account for industrial scale flammable clouds [5] and account for DDTs.

The result of these correlations is a model that estimates the flame speed based on VBR, D, S_l, L_f and the number of confining planes. The resulting flame speed is compared to the published BST model, which provides a matrix of flame speeds as a function of reactivity, obstacle density, and flame expansion. The BST matrix values are used as maximum flame speed values for the set of parameters given in Table 1. Values between the BST matrix elements are calculated by linear interpolation (as was done for Baker's 2½-D flame expansion category). This methodology provides an extended, systematic approach for estimating flame speeds resulting from the combustion of a flammable cloud in an obstructed and/or confined region. Predicted flame speeds are used with the existing BST blast curves to produce estimates of overpressure at a distance from the explosion source.

3.3 Expansion of the BST Matrix

The range of the QMEFS model is limited by the upper and lower bounds of the BST matrix elements for congestion, confinement, and fuel reactivity. Through application of this model to various petrochemical projects, the need to expand the matrix became obvious. As described above, the model is built to provide predictions for the *nPlanes* (confinement) parameter in the range of 1.0 to 4.0 (the model is not validated nor trusted above this value). Values of average obstacle diameter and flame run-up distance are built into the GAMES correlations, so naturally are accounted for within reasonable limits. This leaves the congestion (VBR) and fuel reactivity parameters as places to potentially expand the model.

The intent of model expansion is to widen the range of congestion and reactivity inputs so that it removes questions about its use in the "below low" and "above high" regions. With proper expansion of the model described above, the predicted flame speed following ignition of a flammable cloud may then be used with the BST blast curves to estimate the overpressure impacts in the area surrounding a vapor cloud explosion.

To accomplish this expansion, additional categories were added to the BST matrix for congestion and reactivity. This follows Baker's own expansion into the **very-low reactivity** category when ammonia explosion experimental results were published [4].

For a **very-high reactivity** category, hydrogen was assumed as the characteristic fuel, and represents a true upper bound for LBV. Expansion to the very-high category, for parts of the matrix with high flame speeds is simple because DDTs are expected in these conditions already. For other areas, published data was required for the basis of the expansion.

In the case of the **very-low congestion** category, open air (3-D flame expansion) explosion tests were performed by Harris and Wickens [12] as well as Sato and Iwabuchi [13] for a variety of materials. The **very-high congestion** category, several of the BFETS series of tests were used [14] due to their use of a VBR in the 10% range.

The expanded matrix applied in the new QMEFS model is presented in Table 2. The latest BST flame speed matrix values are outlined by bold lines in Table 2. Values outside these areas have been added as part of the QMEFS expansion.

While a 1-D confinement is not expected to be applied in large-scale industrial scenarios, the *nPlanes* = 5 (1-D) confinement category is included in this model to provide an interpolation point beyond *nPlanes* = 2 (2-D flame expansion).

Table 2. Expanded Flame Speed Matrix

Confinement	Reactivity	Congestion				
		Very-Low	Low	Medium	High	Very-High
1D	Very-High	DDT	DDT	DDT	DDT	DDT
	High	DDT	DDT	DDT	DDT	DDT
	Medium	1.029	1.029	1.765	2.265	DDT
	Low	0.5	0.5 *	1.029	2 *	DDT
	Very-Low	0.023	0.023	0.124	0.168	0.303
2D	Very-High	DDT	DDT	DDT	DDT	DDT
	High	0.588	0.588	DDT	DDT	DDT
	Medium	0.47	0.47	0.66	1.6	DDT
	Low	0.079	0.079	0.47	0.66	1.244
	Very-Low	0.008	0.008	0.049	0.069	0.130
3D	Very-High	0.5	DDT	DDT	DDT	DDT
	High	0.06	0.36	DDT	DDT	DDT
	Medium	0.03	0.11	0.44	0.5	1.084
	Low	0.025	0.026	0.23	0.34	0.678
	Very-Low	0.003	0.003	0.024	0.036	0.073

* Values adjusted from BST publication for model consistency

4 Application

It is helpful to test the QMEFS model outside of the published BST matrix, in the areas where the QMEFS matrix has been expanded. Large scale hydrogen experiments were performed in a hemispherical test rig with no congestion. The test volume was 300 m^3 with a base of 10.4 meters diameter [13] and center ignition. The tests were modeled with the latest BST model and the expanded BST model. The model parameters listed in Table 3 were applied to these calculations. Figure 1 presents the overpressure in relation to distance from center of the fuel-air mixture, for the test results and both BST and QMEFS predictions. The BST model does not have a category for a hydrogen vapor cloud with no congestion, thus the scenario falls into the high fuel reactivity and low congestion portion of the matrix. The result is an underprediction of the overpressure. The BST high fuel reactivity works well for fuel such as ethylene (LBV = 0.76 m/s), but will often underpredict the impacts of a hydrogen (LBV = 3 m/s) explosion. The expanded matrix allows a higher fuel reactivity category and a lower congestion category, and consequently a better prediction of the explosion impacts. As seen in Figure 1, the QMEFS result is a slight overprediction of the overpressure.

Table 3. Modeling Parameters for an Unconfined Hydrogen Explosion

Parameter	BST	Expanded QMEFS
Volume [m^3]	300	300
nPlanes	1	1
Reactivity	High	Very-High
VBR	N/A	0
Congestion	low	N/A
Run-up distance [m]	N/A	5.2

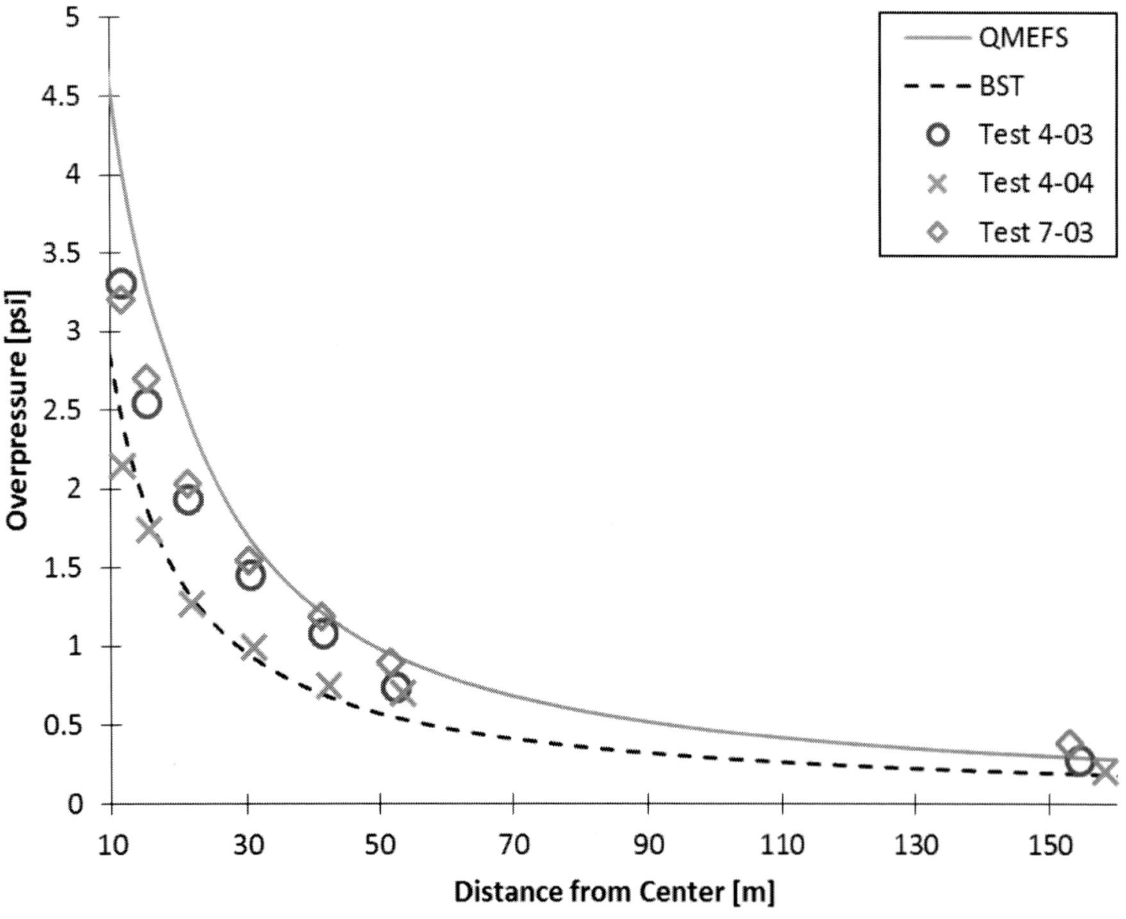

Figure 1. Overpressure vs. Distance for Open-Air Hydrogen Tests

As discussed in the paper, the distance that is available for the flame to accelerate, i.e. the run-up distance, can have a significant effect on the flame speed attained in a flammable gas cloud. The flame speeds presented in the original Baker-Strehlow (BS) model [6] were generated in a test rig whose largest dimension was less than 6 feet (1.8 meters) [5]. The flame speeds presented in the 2005 BST paper are based on tests where the largest dimension was 48 feet (14.6 meters) with the published flame speed results "scaled up" to account for the maximum size of a typical industrial plant [5]. One test of the QMEFS model is to determine the flame speed versus run-up distance for the test configurations used in the newest BST model when specific parameters are varied.

Figure 2 shows the results for the 3-D flame expansion (*nPlanes* = 1), medium reactivity (S_l = 0.43 m/s) case. The curves plotted in Figure 2 show six VBR values – the high (5.7%), medium (4.3%), and low (1.5%) congestion values used in the BST experiments, one intermediate value to show how the model behaves as the VBR is adjusted, a fifth value (1%) below the BST low category to demonstrate the behavior in the "less-than-low" congestion range, and the sixth value (10%) above the BST high category to demonstrate the behavior in the "higher-than-high" congestion range.

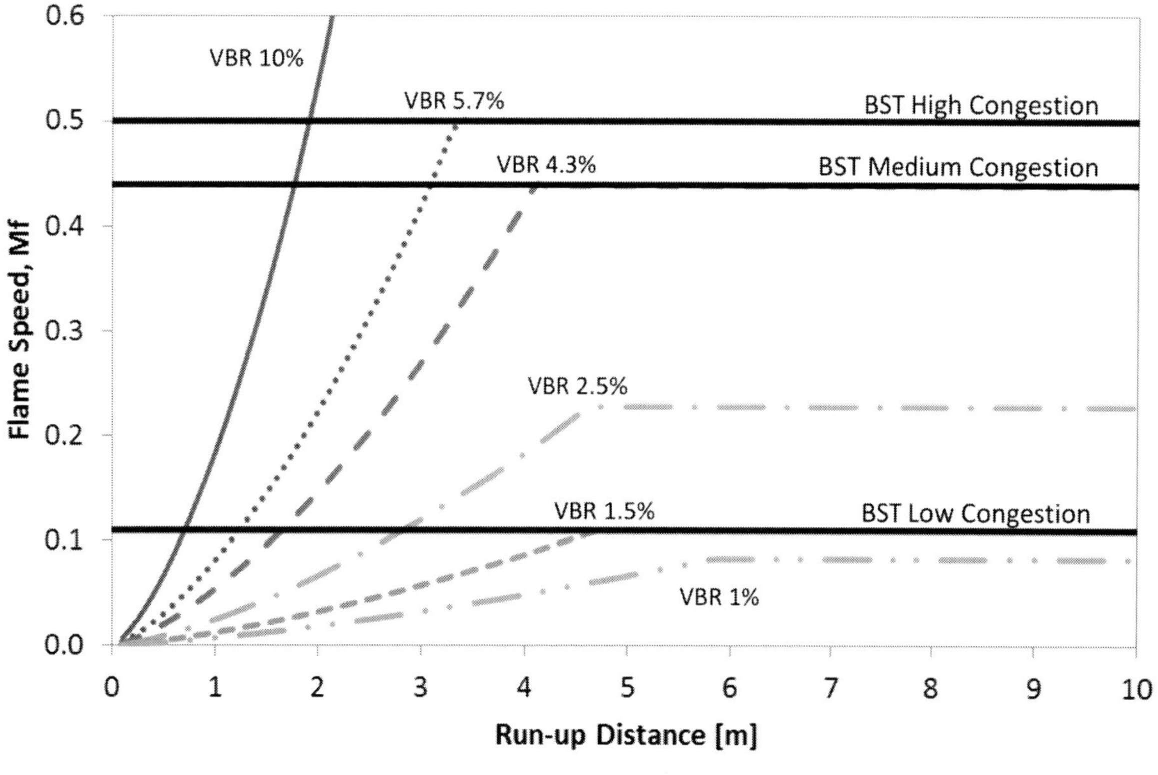

Figure 2. Flame Speed vs. Run-Up Distance; 3-D, Propane, D = 2"

Figure 3 demonstrates how flame speed varies with *nPlanes* and LBV. The basis of the figure is an explosion involving a medium VBR (0.043) and a 4-inch average obstacle diameter. The *nPlanes* parameter is varied between 1.0 (3-D flame expansion) and 4.0 (the maximum value recommend for use in the QMEFS model). The curves plotted in Figure 3 show five materials with various LBV values – ammonia (0.08 m/s), methane (0.37 m/s), propane (0.43 m/s), ethylene (0.76

m/s) and ethylene oxide (0.91 m/s). In the BST model ethylene oxide and ethylene are grouped together into the high category, but these two fuels have different burning properties as can be seen with their LBV values. Thus, by accounting for the LBV of the fuel, the new model provides a finer resolution for flame speed. In addition, as seen in Figure 3, the QMEFS model provides more resolution for the flame expansion parameter through the use of the *nPlanes* variable.

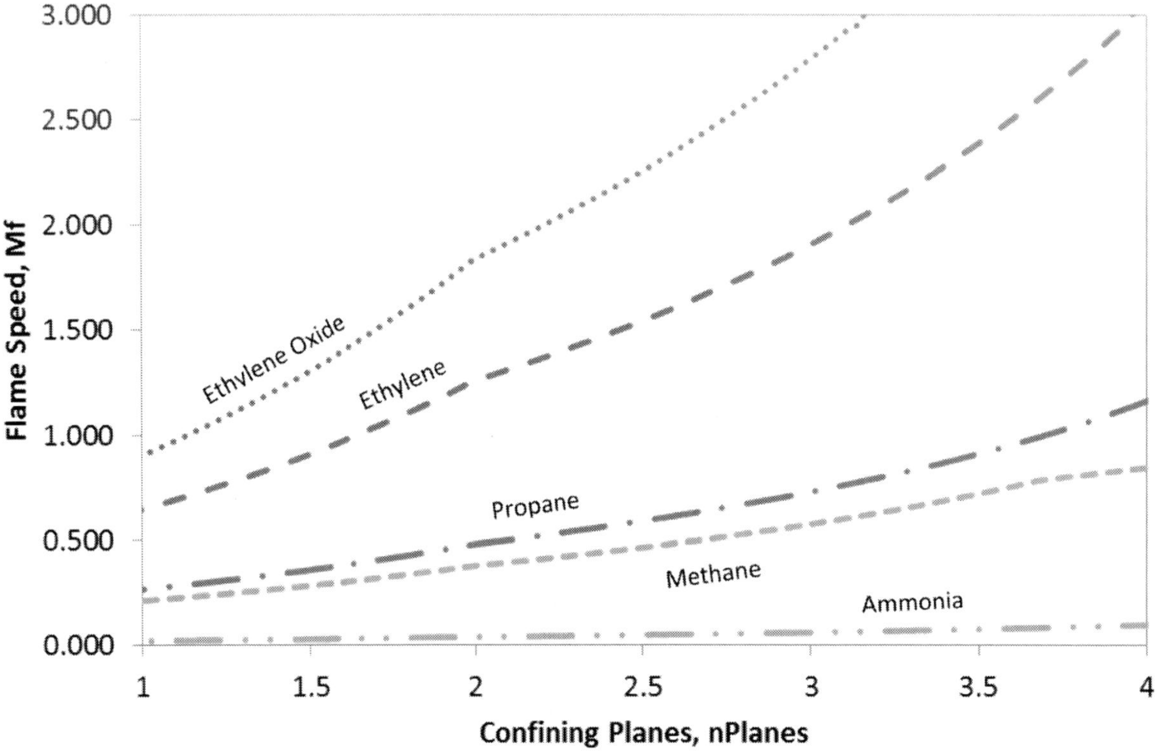

Figure 3. Flame Speed vs. Confining Planes; Medium VBR, D = 4"

5 Conclusions

The model described in this paper provides an improved method to describe the characteristics of a vapor cloud explosion over a wide range of conditions. It combines elements of the work conducted by Baker to develop the BST model, as well as the European MERGE/EMERGE projects; thus incorporating data from the two largest, modern vapor cloud explosion test projects.

Because the BST categories are resolved into numeric ranges, a higher resolution solution space and a clearer set of measureable inputs are achieved. With the ability to model the reactivity of a flammable gas cloud based on laminar flame speed instead of a low, medium, or high classification, the model is able to more accurately describe a wide range of flammable gases and mixtures. Expansion of the constraining flame speed categories to very-low and very-high then provides an even wider range of possibilities.

With the QMEFS matrix expanded to congestion (VBR) categories of very-low and very-high, the model provides a wider range of predictive capability for explosion scenarios. These expansions and enhancements are still confined by the flame speeds in the BST system, and are applied within the set of curves for overpressure and impulse. Thus, with the QMEFS model presented here, process safety experts have a more versatile tool to predict VCE impacts in petrochemical plants.

6 References

[1] Baker, Q.A., C.M. Doolittle, G.A., Fitzgerald, and M.J.Tang, "Recent Developments in the Baker-Strehlow VCE Analysis Methodology," *Process Safety Progress*, v.17, No.4 (1998).

[2] Melton. T.A. and Marx J.D., "Estimating Flame Speeds for Use with BST Blast Curves," Process Safety Progress, v. 28, March, (2009).

[3] Mercx, W.P.M., A.C. van den Berg, and D. van Leeuwen, A*pplication of correlations to quantify the source strength of vapour cloud explosions in realistic situations*. Final report for the project: 'GAMES', TNO report PML 1998-C53, Rijswijk, The Netherlands (1998)

[4] Thomas, K.J., Malik. D.R., Downes, A.M., Rodges, S., "Ammonia Vapor Cloud Explosion Testing and Implications for Facility Siting Studies, Consequence Analyses and Risk Assessments," *50th Loss Prevention Symposium*, AIChE, April, (2016).

[5] Pierorazio, A.J., J.K.Thomas, Q.A.Baker, and D.E.Ketchum, "An Update to the Baker-Strehlow-Tang Vapor Cloud Explosion Prediction Methodology Flame Speed Table," *Process Safety Progress*, v.24, no.1 (2005).

[6] Baker, Q.A., M.J. Tang, E. Scheier, and G.J. Silva, "Vapor Cloud Explosion Analysis," *28th Loss Prevention Symposium*, AIChE, April, (1994).

[7] Mercx, W.P.M., editor, *Extended Modelling and Experimental Research into Gas Explosions*, Final summary report of the project EMERGE, CEC contract EV5VCT930274, European Commission, Directorate General XII, Brussels, Belgium (1997).

[8] Mercx, W.P.M., A.C. van den Berg, C.J. Hayhurst, N.J. Robertson, and K.C. Moran (2000), *Developments in Vapour Cloud Explosion Blast Modeling*, Journal of Hazardous Materials 71 (2000) 301-319.

[9] Tang, M.J. and Q.A.Baker, "A New Set of Blast Curves from Vapor Cloud Explosion," *Process Safety Progress*, v.18, no.4 (1999).

[10] Oran, E.S. and V.M. Gamezo, "Origins of the deflagration-to-detonation transition in gas-phase combustion," *Combustion and Flame*, 148 (2007).

[11] Vasil'ev, A.A., "Estimation of Critical Conditions for the Detonation-to-Deflagration Transition," *Combustion, Explosion, and Shock Waves*, v.42, no. 2 (2006).

[12] Harris, R.J. Wickens, M.J., "Understanding Vapor Cloud Explosions – An Experimental Study," The Institute of Gas Engineers 55th Autumn Meeting, (1989).

[13] Sato, Y., Iwabuchi, H., Groethe, M., Merilo, E., Chiba, S., "Experiments on Hydrogen Deflagration," Journal of Power Sources v.159 (2006).

[14] Health & Safety Executive, "Explosions in Full Scale Offshore Module Geometries Main Report," Offshore Technology Report – OTO 1999 043. Prepared by BG Technology, April, (2000).

479444 Do We Need a Quantitative Risk Assessment (QRA) If My Consequence Analysis Facility Siting Presents a Problem?

Abstract:

There are three steps that a regular facility siting study (FSS) can go through. The first step is the basic separation table approach descrived in API 752. The second step is the consequence analysis facility siting and the third and final one is the quantitative risk assessment. It turns out, however, that the vast majority of the consequence analysis FSS studies quickly demonstrate that the impact to buildings can be more significant that what have been defined as acceptable. For such cases, the analysis normally turns into the QRA, exceedance curves, building structural assessment, etc. The idea of this paper is to cover a intermediate step in early stages to use computational fluid dynamics (CFD) technique in the early stage of the analysis without compromising the time schedule and the budget, this would include dispersion, radiation and explosion modeling. The more accurate results obtained with CFD modelling lead to cloud sizes which are mainly less conservative than traditional methods. The objective of this paper is to go a setp beyond a previous analysis for the dispersion using CFD, but rather make a full evaluation of a FSS using CFD.

Proposed Method for Investigating the Reactivity of Chlorine with Environmental Materials under Relevant and Controlled Conditions

Audrey Feuvrier, PhD.
Ralph E. Martin Department of Chemical Engineering
University of Arkansas, Fayetteville, AR
audrey.feuvrier@uark.edu

Tom Spicer, Ph.D., P.E.
Ralph E. Martin Department of Chemical Engineering
University of Arkansas, Fayetteville, AR
tos@uark.edu

Prepared for Poster Presentation at
American Institute of Chemical Engineers
2017 Spring Meeting and 13th Global Congress on Process Safety
San Antonio, TX
March 26 – 29, 2017

Proposed Method for Investigating the Reactivity of Chlorine with Environmental Materials under Relevant and Controlled Conditions

Audrey Feuvrier, PhD.
Ralph E. Martin Department of Chemical Engineering
University of Arkansas, Fayetteville, AR
audrey.feuvrier@uark.edu

Tom Spicer, Ph.D., P.E.
Ralph E. Martin Department of Chemical Engineering
University of Arkansas, Fayetteville, AR

Keywords: Chlorine reactivity, chlorine release, dry deposition, Jack Rabbit II

Abstract

Many common toxic industrial chemicals (TICs) such as chlorine and ammonia are very reactive with commonly encountered materials in the environment. The amount of TIC material reacted in the environment is an important factor in the impact of a release. Typically characterized with a deposition rate in atmospheric dispersion models, the reaction rate and total deposition quantities have predominately been studied in field scale experiments. Laboratory experiments have also been undertaken, but past investigations have inadequately quantified the available turbulence which can importantly impact surface reaction rates. The state of the art description of surface deposition represents the process as a series of resistances including the effect of atmospheric turbulence and surface reaction(s) depending on the surface material (substrate). Sponsored by the Chemical Security Analysis Center (CSAC) of the U.S. Department of Homeland Security and Transport Canada, an experimental program is underway to test potential substrates under known inlet gas concentrations with levels of turbulence that are comparable to those of the atmosphere. The objective is to develop a test methodology for subjecting a substrate to a reactive gas flow of known TIH concentration with turbulence levels consistent with those in the atmosphere. With data from repeat experiments, deposition of a reactive TIH chemical can be quantified with the proposed experimental protocol.

1 Introduction

Estimating the quantity of a Toxic Industrial Chemical (TIC) removed from the environment through chemical processes is critical to proper consequence assessment of accidental (or intentional) releases, particularly in urban areas [1]. TIC removal mechanisms play an important role in the rational assessment of risk, as these phenomena may significantly and quickly reduce the TIC concentration near a release [2,3], particularly for a reactive chemical such as chlorine. Typically characterized as dry deposition, reactions of chlorine with the environment was not included in the computational models used to predict the consequences of large releases of chlorine in the past [2], but new efforts are underway to characterize chlorine reactivity and incorporate such effects in consequence and risk assessments.

While species like sulfur dioxide and iodine have been the subject of extensive experimental studies [3], studies about deposition involving chlorine are scarce and show a great deal of variability. The high reactivity of chlorine as well as its toxicity make it difficult to study in a laboratory environment, and consequently, many of the studies are conducted as field studies [4]–[6], which restricts test conditions to a limited set of atmospheric conditions. In one set of laboratory experiments, Hill [7] studied the effect of deposition of several TICs on an alfalfa canopy in a chamber where humidity, light, and wind speed were controlled. He showed that chlorine deposition is rapid on alfalfa and that chamber air velocity has a significant effect on the chlorine uptake, but the level of turbulence in the chamber is unknown. Other experimental studies gave promising results such as [8]–[10], but these studies involved no other movement of the flow than the one produced by the mixing device or the circuit used to monitor the concentration level. The state of the art description of surface deposition, [11], represents the process as a series of resistances including the effect of atmospheric turbulence and surface reaction(s) depending on the surface material (substrate).

In this context, an experimental program, sponsored by the Chemical Security Analysis Center (CSAC) of the U.S. Department of Homeland Security and Transport Canada is underway to test potential substrates in a controlled environment. The project includes the design and characterization of a mixing chamber as well as development of a protocol for the measurement of chlorine removal rate for different samples of soil and vegetation. Concentration, velocity, and turbulence measurements will be made in the working section to confirm that the flow is uniform and consistent. It is also desired that the flow in the working section of the apparatus is consistent with the atmospheric flow with target turbulence intensities on the order of 10%. Working section flow speed will be variable, with a maximum in the section of 8 m/s, and expected velocities between 0.5 to 3 m/s. Passive flow conditioning will be used to eliminate non-uniformities in the flow. Recognizing that chlorine gas concentrations as high as 10% or 100,000 ppm may be used, apparatus materials of construction will be carefully chosen to resist damage and also to not remove significant amounts of chlorine during a test. Finally, to limit the hazards involved with the use of chlorine at high concentration level, the apparatus will need to fit under a 6 ft long (about 1.8 m) fume hood.

In the following, the proposed designs for the apparatus and its main elements are introduced. The choice of the mixer, fan, and materials will be discussed. In the second part of the article, the future protocol of measurement will be presented. Focus will be put on the means of measurement used and the characterization of the chamber. The steps to determine chlorine removal rates for several samples types under different environmental conditions will be considered.

2 Proposed design

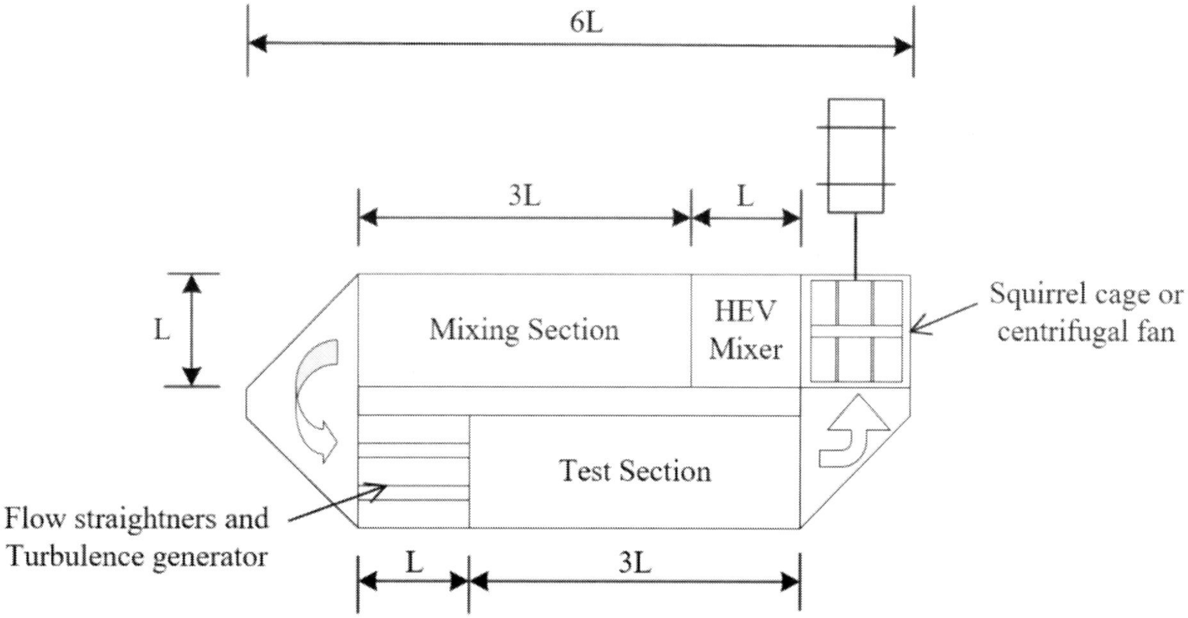

Figure 1. Preliminary apparatus design

Figure 2 shows the preliminary design of the apparatus following the criteria laid down previously using a simple closed return design. The test or working section is placed under the mixing section to facilitate access to the test section through a lab hood sash. The characteristic length, L, chosen to fit in a standard 6 ft hood, will be between 0.25 and 0.3 m, giving a total volume for the apparatus of between 0.16 and 0.28 m^3. A standard lecture bottle contains 454 g of chlorine, which would be sufficient to provide chlorine for 8 to 14 tests depending on the length L based on a maximum concentration of 100,000 ppm. The use of lecture bottles limits the quantity of chlorine required in the lab during the experiments and effectively eliminates the hazard associated with the storage of large quantities of chlorine on site. These volume values are approximate because it may be convenient to use a circular conduit for mixing and a square cross-section for the working section. A circular conduit reduces the required volume of the mixing section (with some minor reduction in power consumption), and a working test section with flat surfaces facilitates optical measurements during characterization and sample placement in the testing phase.

2.1 Mixer and flow conditioner

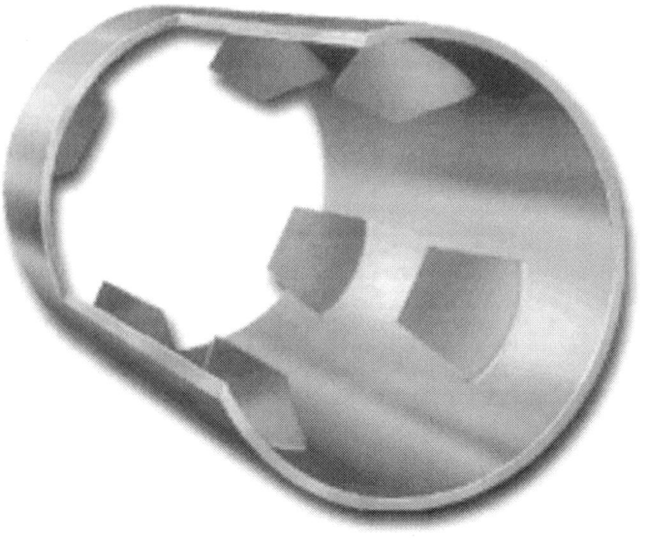

Figure 2. Typical High Efficiency Vortex (HEV) mixer [12].

Figure 3 shows a typical High Efficiency Vortex (HEV) mixer which is placed downstream of the fan. This type of mixer was selected because it provides efficient mixing with turbulent flow with minimal pressure drop. For a Kenics™ HEV mixer, Figure 4 shows the relative coefficient of variation, COV_R , defined as the quotient between the outlet and inlet coefficient of variation, as a function of the Reynolds number (defined as $U_\infty L/\upsilon$ where U_∞ represent the characteristic velocity in the tunnel, L is the characteristic length and υ is the kinematic viscosity of the gas). The coefficient of variation for the mixing of two fluid represents the quotient of the standard deviation of the concentration and the mean concentration. Therefore, the lower the COV_R, the better the mixture quality. The gray zone on Figure 4 represents the range of Reynolds number of this study. Based on Myers et al [13], the pressure drop for our (custom made) apparatus was estimated to be between 1.7 and 17.3 Pa, making it negligible. The HEV mixing process rely on the production of hairpin vortices downstream of the blades of the mixer combined with a pair of counter rotating longitudinal vortices [14]. A mixing section downstream of the device and equal to three times the length of the mixer is recommended to provide for sufficient time for the vortices to complete the mixing process and consequently obtain the predicted COV_R. Additional turbulence generated between the mixing section and the test section will only improve the COV_R before the flow enters the test section..

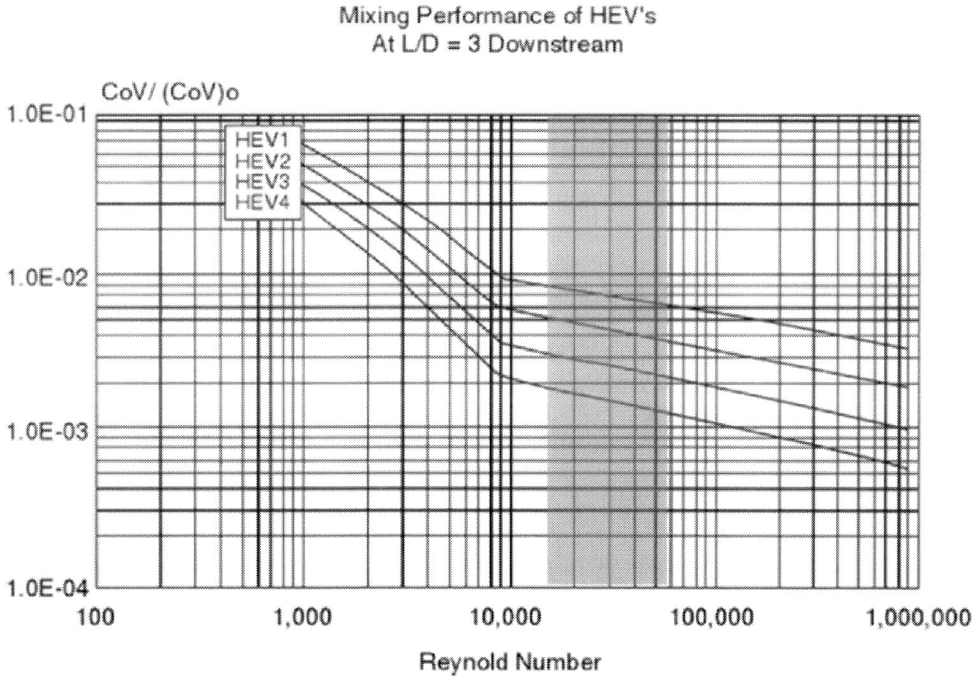

Figure 3. Relative coefficient of variation (COV_R) vs. Reynolds Number and number of tabs row for the Kenics HEV, [12].

The corners currently consist of two 45° bends with a short duct between them, and the 45° bends are preferred to 90° bends to prevent the accumulation of gases in the dead space. To help maintain good flow distribution in the channel, sets of vanes will be placed in each bend. Operating at low wind speed allows us to consider simple circular cross sectional shapes for the vanes [15].

After the two first corners, the flow conditioner section is planned to eliminate large eddies that might be present in the flow ensuring uniformity of the flow and simultaneously raising the turbulence intensity up to 10%. To achieve this, we plan to use a combination of screens or perforated plates. Samples will be placed in the test section through a removable side with provision for a tray in the floor of the test section to accommodate soil and/or plant roots. After experiments, ventilation of the apparatus will be achieved through two hatches on each bend between the mixing and flow conditioner sections.

2.2 The fan

On Figure 2, the motor of the fan is represented outside of the apparatus because chlorine would be expected to react and corrode any of its exposed surfaces. The motor shaft will be supported by Nylon or Teflon bearings located in the side of the apparatus to isolate the motor from chlorine exposure. A centrifugal (squirrel cage) fan is being considered to allow placement of the fan above the apparatus. The fan blades and shaft exposed to chlorine will be treated to limit damage due to chlorine exposure as discussed below.

2.3 Building materials

During the first phase of the experimentation, the apparatus will only use a mixture of air and an inert gases will be used for optic velocity and turbulence measurements and concentration measurements (including a propylene tracer). During this preliminary phase, the structure of the test section will be made with plywood, but two of its walls will be made with polycarbonate to allow visualization and optical measurements of the velocity in the section.

In the second phase of the experimentation involving chlorine, all elements in contact with the test mixture must be made with a material or coated with a material capable of resisting high chlorine concentrations. In addition to extending the workable lifetime of the apparatus, reducing the potential for chlorine to react with materials in the test chamber is also desired so that the reduction in chlorine gas concentration can be attributed to reaction with the sample tested as opposed to the chamber itself. In the case of the fan blades and shaft, a commercial coating is planned (such as Cerakote or titanium). Titanium is known to be highly reactive with dry chlorine, however 1% of water may be enough for passivation after mechanical damage to titanium under normal conditions [16]. For the mixing section, a metal or plastic (PVC) duct will be used since no optical access is required. The structure of the corners and the test section will be made with plywood. In the second phase, polycarbonate may be replaced by plywood. The flow straighteners will likely be made of a suitable metal. All surfaces except the fan blades and shaft are planned to be coated with Viton™, a coating material widely used in applications with chlorine in the chemical industry. We are currently evaluating whether the Viton™ coating under consideration will adhere sufficiently to plywood, and if it does not, then the entire apparatus may need to be fabricated from sheet metal with all internal surfaces painted with Viton™.

3 Protocol for future experimentations

3.1 Experimental resources and characterization of the mixing chamber

Several instruments will be used during the characterization of the chamber and concentration measurement. During both phases of the experimental measurements, the temperature, pressure and humidity of the ambient air will be monitored.

For the characterization of the chamber, the first step will be verifying that the flow entering the test section is uniform and fully mixed through visualization. The working section will be equipped with LED lights and a ROSCO 1500 fogger machine will provide the fog. The fog will be introduce downstream of the fan, at the same point planned for injection of the gases. Video recordings of the fog in the working section will be made with several velocities. After visualization experiments, velocity measurements will be made with a 2D-LDV (Laser Doppler Velocimetry) system using an Argon Ion 300 mW air-cooled laser. This single point technique is non-intrusive and allows simultaneous measurement of two components of the velocity with a good spatial and temporal resolution. Finally, the concentration measurements will be made using an Aurora Scientific mini-PID (photo ionization detectors) The PID measurement method relies on photoionization of a volatile organic compound (in this case a propylene tracer). Small access points will have to be made in the test section for the PID. In the characterization phase, a simulant gas mixture (using sulfur hexafluoride) with be used to reproduce the molecular weight of chlorine. Sulfur hexafluoride is non-toxic, non-flammable, non-reactive and can be used in the PID system.

In conclusion, the characterization of the mixing chamber will assess the quality of the mixing and the flow in the working section, the steps are listed below :

1. Visualization of the flow going in the test section.
2. Velocity profiles taken in the test section will ensure the uniformity of the flow and that the turbulence intensity level is respected.
3. Concentration profiles will provide information on the uniformity of the concentration of the simulant gases in the chamber.

3.2 Concentration measurements and evaluation of the reactivity of chlorine with different samples

After the characterization of the working section, the apparatus will be modified to undergo tests with high concentration of chlorine. The changes have been described in section 2.3. Furthermore, the tunnel will have to allow the access of the JAZ sensors in order to measure the concentration of chlorine. As with characterization of the chamber, the evaluation of the reactivity of chlorine with different samples will request several steps listed hereunder.

1. Characterization of the response of the apparatus to chlorine : Tests will be done without samples, so that the chlorine concentration decay due to the apparatus itself can be measured. A correction factor will be calculated and implemented in the following concentration measurements. Multiple tests will be done, in order to ensure of the repeatability of the measurements.
2. Removal rate estimation : Samples will be placed in the tray, allowing the removal rate of these samples to be estimated through concentration measurements.
3. Analysis of the results : The measurement of the concentration decay in the chamber and the study of the samples put in contact with chlorine will provide valid information in order to calculate the removal rate of chlorine for these different samples. Analytical techniques for determining the amount of chlorine uptake by the test material will depend on the material.

4 Conclusion

In this article, the design of an innovative apparatus or mixing chamber and its main elements have been presented in details. One of the main difficulties encountered was to scale the device to fit under a six feet long fume hood to limit the hazards involved with the use of chlorine at high concentration level in the laboratory. One other obstacle was to find materials and coating suitable and able to resist chlorine in the long run. The future means and procedures of characterization and measurement have been introduced as well. Once build and characterized, this apparatus will provide new data sets for chlorine deposition on vegetal samples and small objects in conditions similar to the atmospheric environment. In the end, this study aim at the improvement of the understanding and computational modeling of toxic inhalation hazards and the consequences to affected population and infrastructure.

5 Aknowledgment

This project is supported by the Chemical Security Analysis Center (CSAC) of the U.S. Department of Homeland Security and Transport Canada. The authors wishes to thank Katherine Elmore for her assistance with this project.

6 References

[1] M. B. Dillon, "The role of deposition in limiting the hazard extent of dense-gas plumes," *J. Hazard. Mater.*, vol. 164, no. 2–3, pp. 1293–1303, 2009.

[2] S. Hanna, S. Dharmavaram, I. Zhang, J., Sykes, H. Witlox, S. Khajehnajafi, and K. Koslan, "Comparison of Six Widely-Used Dense Gas Dispersion Models for Three Actual Railcar Accident," in *Process Safety Progress*, 2008, vol. 27, no. 3, p. 248–259.

[3] S. P. Arya, *Air pollution meteorology and dispersion*, 6th ed. 1999.

[4] J. P. Dimbour, A. Dandrieux, and G. Dusserre, "Reduction of chlorine concentrations by using a greenbelt," *J. Loss Prev. Process Ind.*, vol. 15, no. 5, pp. 329–334, 2002.

[5] G. Dusserre and J. Ollivier, "Small scale field experiments of chlorine dispersion," *J. Loss Prev. Process Ind.*, vol. 15, pp. 5–10, 2002.

[6] J. D. Hearn, R. Weber, R. Nichols, M. V. Henley, and S. Fox, "Deposition of Cl_2 on soils during outdoor releases," *J. Hazard. Mater.*, vol. 252, pp. 107–114, 2013.

[7] A. C. Hill, "Vegetation: A Sink for Atmospheric Pollutants," *J. Air Pollut. Control Assoc.*, vol. 21, no. 6, pp. 341–346, Jun. 1971.

[8] J. Hearn, J. Eichler, C. Hare, and M. Henley, "Cl_2 deposition on soil matrices," *J. Hazard. Mater.*, vol. 237–238, pp. 307–314, 2012.

[9] J. Hearn, J. Eichler, C. Hare, and M. Henley, "Effect of soil moisture on chlorine deposition," *J. Hazard. Mater.*, vol. 267, pp. 81–87, 2014.

[10] D. Brown, "The removal of toxic gases from the air by water, plants and soils.," 2006.

[11] M. L. Wesely and B. B. Hicks, "A review of the current status of knowledge on dry deposition," *Atmos. Environ.*, vol. 34, pp. 2261–2282, 2000.

[12] J. R. Couper, W. R. Penney, and J. R. Fair, *Chemical process equipment revised 2E: selection and design.* Gulf Professional Publishing, 2009.

[13] K. J. Myers, A. Bakker, and D. Ryan, "Avoid agitation by selecting static mixers," *Chem. Eng. Prog.*, vol. 93, no. 6, pp. 28–38, 1997.

[14] W. J. Gretta and C. R. Smith, "The Flow Structure and Statistics of a Passive Mixing Tab," *J. Fluids Eng.*, vol. 115, no. 2, p. 255, 1993.

[15] R. H. Perry, D. W. Green, and J. O. Maloney, *Perry's chemical engineers' handbook*. McGraw-Hill, 2008.

[16] Timet corporation, "Corrosion resistance of titanium," 1997.

Fire Proofing Cost Reduction Measures for LNG Plant

Raminaidu Girada
Senior Process Safety Engineer, Bechtel Oil, Gas & Chemicals, Inc.
3000 Post Oak Boulevard, Houston, TX 77056
rgirada@bechtel.com

Jegan Babu Arumugakan Thuraiswamy
Senior Process Safety Engineer, Bechtel Oil, Gas & Chemicals, Inc.
3000 Post Oak Boulevard, Houston, TX 77056
jbarumug@bechtel.com

Nino Nicotra
Process Safety Engineering Chief, Bechtel Oil, Gas & Chemicals, Inc.
3000 Post Oak Boulevard, Houston, TX 77056
anicotra@bechtel.com

Prepared for Presentation at
American Institute of Chemical Engineers
2017 Spring Meeting and 13th Global Congress on Process Safety
San Antonio, TX
March 26 – 29, 2017

Fire Proofing Cost Reduction Measures for LNG Plant

Raminaidu Girada
Senior Process Safety Engineer, Bechtel Oil, Gas & Chemicals, Inc.
3000 Post Oak Boulevard, Houston, TX 77056
rgirada@bechtel.com

Jegan Babu Arumugakan Thuraiswamy
Bechtel Oil, Gas & Chemicals, Inc.
3000 Post Oak Boulevard, Houston, TX 77056
jbarumug@bechtel.com

Nino Nicotra
Process Safety Engineering Chief, Bechtel Oil, Gas & Chemicals, Inc.
3000 Post Oak Boulevard, Houston, TX 77056
anicotra@bechtel.com

Keywords: Consequence Modeling, Liquefied Natural Gas (LNG) & Liquefied Petroleum Gas (LPG) and Safe Design

Abstract

The current industry practice in designing fireproofing requirements for LNG facilities follows the traditional approach of "Liquid Pool Fire". The common industry practice for fire proofing design criteria defines a fire potential equipment (FPE) as the equipment that handles flammable liquid above a specific threshold volume. The FPE generates fire scenario envelope whose dimension ranges are recommended by API 2218. Equipment and structural supports that falls inside the fire envelope are considered for fire proofing.

LNG plants are handling products with very low boiling point, and any leak to atmosphere results in an instant flashing into vapor phase or aerosols. The results from our recent works on pool fire consequence modelling has proved that most of the currently considered fire potential equipment seldom create pool fire. This consequence modeling results clearly demonstrates that the current fire proofing design approach is overdesign, that is not effective. The apparent saving in man-hours for not modeling the liquid release will result in a considerable amount of engineering efforts and material cost being spent on the design and installation of fire proofing for equipment and structure with potentially no real benefits to the inherent safety of the process plant. Considering the current economic trend, a change in approach using consequence modelling will effectively result in considerable cost savings in major LNG projects.

1 Introduction

Currently in LNG projects, fireproofing is considered for equipment supports, pipe rack support, structural supports and other auxiliary critical hydrocarbon carrying pipe supports within a fire proofing zone to improve its structural integrity during a fire. The principal value of fire proofing is realized during the early stages of a fire by providing additional time to fire fighters to control or extinguish a fire thus by delaying additional loss of containment which may escalate the incident. The fireproofing requirement is determined by identifying the fire potential equipment and the associated fireproofing zones. Fire potential equipment is defined as an equipment holding flammable liquid above a specific threshold volume, which will have the potential to create a liquid pool and subsequent pool fire if ignited. Per API 2218, the fireproofing zone is defined for every fire potential equipment, that extends 20 ft. to 40 ft. (6 m to 12 m) horizontally and 20 ft. to 40 ft. (6 m to 12 m) vertically from the source of liquid fuel. As a typical approach most projects consider fireproofing zones that extend 30 ft. (9m) horizontally and 30 ft. (9m) vertically from the source.

The above mentioned qualitative approach does not reflect the realistic possibility of pool fire scenario. Therefore, a quantitative analysis may be performed to analyze whether a pool fire is realistic for equipment handling multiple refrigerants and LNG. This analysis has been performed on some process equipment handling multiple refrigerants and LNG using Canary software by Quest Consultants.

2 General Technical Guidance

2.1 Scope of Fire Proofing

Fire scenario envelopes for an LNG plant are based on pool fire scenarios. Jet fires are typically not considered as credible for defining fire scenario envelops for fireproofing as described below:

- Jet fires are handled by Prevention, Intervention and Protection. Prevention is by designing the facility in compliance with general fire prevention activities such as those included in NFPA 59A and in the broad application of Process Safety Management(PSM). Once a pressurized jet fire occurs, the intervention and protection can be achieved by a high flow rate firewater stream directed on the point of impingement. This has been proved as an effective mitigation measure in preventing escalation with Jet fires. This will be achieved by strategically locating the fire water monitors around high risk process area, where jet fire is foreseen as a credible scenario.

- Emergency isolation and depressurization system will be provided to lower the pressure of specific isolated sections of the plant that may be involved in a localized jet fire scenario and to de-inventory of gas to flare.

- Efforts on similar LNG projects to extend the fire proofing systems for jet fire scenarios led to enormous impractical level of fireproofing requirement. This could also introduce other hazards such as corrosion.

- Jet fire represents a significant risk or concern for personnel life, working on offshore installations whereas property protection is the basis of pool fire for on-shore LNG facility or petroleum refinery or petrochemical facilities.

2.2 Pool Fire Formations

A pool fire may result via a number of scenarios and it begins typically with the release of flammable material from process equipment. If the material is liquid, stored at a temperature below its normal boiling point, the liquid will collect in a pool. The geometry of the pool is dictated by the surroundings (i.e., diking), but an unconstrained pool in an open, flat area is possible, particularly if the liquid quantity spilled is inadequate to completely fill the diked area. If the liquid is stored under pressure above its normal boiling point, then a fraction of the liquid will flash into vapor, with un-flashed liquid remaining to form a pool in the vicinity of the release.

The pool of flammable liquid may be ignited creating a pool fire. If the release of flammable material from a pressurized process equipment continues, then a jet fire is also likely. If the ignition source occurs at the very beginning of the release, then inadequate time is available for the liquid to form a pool and only a jet fire will result.

Pressurized liquefied gases such as LNG, methane, ethylene and propane will in general disperse as mists or vapor, but may possibly form a small pool if the source operates at low pressure conditions. An exception has to be made for butanes, pentanes and heavier hydrocarbons, which always create pool.

Refer to the following **Figure 1** for pictorial view of the pool fire formation.

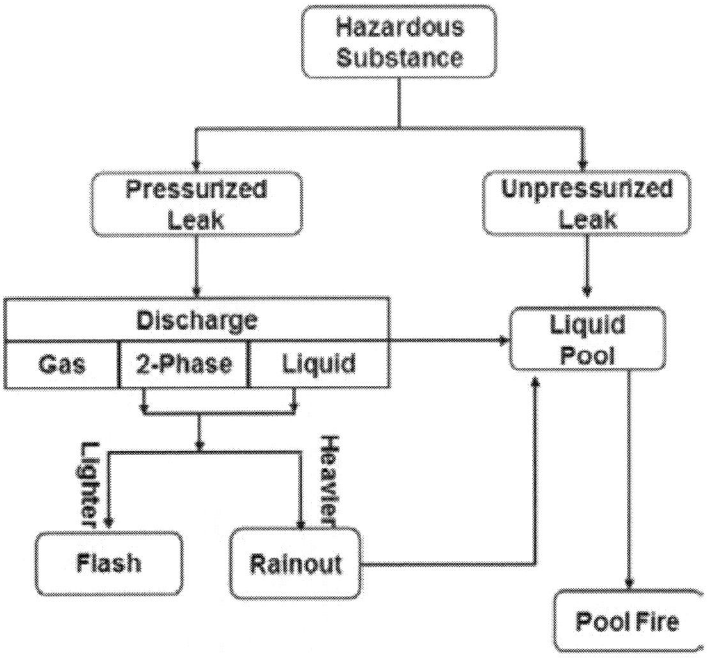

Figure 1: Pool Fire Formation

2.3 Determination of Pool Fire scenario

This section provides an overview of the methodology for determining pool fire existence.

In case of accidental release of flammable materials such as Pentanes and heavier hydrocarbons create liquid pool irrespective of their operating conditions. However, the lighter hydrocarbons such as butane and propane may create liquid pool only under specific operating conditions. The potential for creating liquid pool is based on certain parameters as described below:

- **Butanes**

  Butane releases potentially create liquid pool except for high pressure releases and slight sub cooling of release. Further consequence analysis is required to check if a liquid pool is created from the release.

- **Propane and Ethylene**

  Ethylene and propane liquids may or may not pool, depending on conditions. It has been found that a pool formation can be correlated on several parameters: (1) vessel operating pressure, (2) size of rupture hole for self-leakage vessel, (3) the amount of subcooling, which is also important but the assumption is the liquid will be slightly subcooled, and (4) confinement of *fire hazard area*. Together, these variables establish the size of the pool.

  Higher pressure release are more likely to vaporize and mist, and low pressure releases are more likely to pool. Based on a slightly subcooled liquid at pressure, the analysis is divided into three regimes (1) no liquid pools, (2) highly likely to form a significant pool fire, and (3) possible pool fire. Depending on the size of the release, slope, temperature and composition, a consequence model may be necessary to determine if a pool fire is warranted.

- **LNG**

  Similar to propane and ethylene liquids, LNG may or may not pool depending on several parameters: (1) vessel operating pressure, (2) size of rupture hole, (3) the amount of subcooling, and (4) confinement.

Therefore, a quantitative approach was performed in the following section.

3 Pool Fire Case Study

In this case study, the following light flammable fluids are modelled: methane, propane, ethylene and LNG. Based on the current fire proofing design approach a total of 28 equipment of the LNG plant that handles the above fluids are qualified as Fire Potential Equipment. All those leakage scenarios are modelled using Canary modelling tool, based on its operating condition.

3.1 Leak Scenarios

To provide a spill rate using CANARY, one has to postulate a leak scenario.

Typical hole size choices range from small (3-10mm) to full bore ruptures (diameter of pipe). Hole size selection can define the general level of risk (larger, high consequence releases are significantly less frequent than small, low consequence release).

CCPS book of "Guideline for Fire Protection in Chemical, Petrochemical, and Hydrocarbon Processing Facilities" defined hole sizes as small (3-10 mm), medium (10-50 mm), large (50-150 mm), and full-bore ruptures. Discussion around these hole sizes also points out that the large and rupture releases are not always the worst-case events due to release duration and dosage effects that should be discounted for.

The Canadian Standard Association (CSA) standards Z276 specifies design spills which define the potential extent of hazard zones that are not allowed to impact the public. Z276 defines, among other (smaller) release scenarios, that the design spill should be "The maximum flow of LNG that could be spilled, from the largest branch connection, 60 mm (2-inch) or smaller..."

Based on the above references, 2-inch is considered as an appropriate hole size for consequence modeling. Therefore, the pool fire modeling has been performed for 2-inch hole size. However, sensitivity analysis has been done considering 6-inch hole to check if larger hole size can result in pool formation. The modeling results for both 2-inch and 6-inch are presented below.

The release elevation was considered to be constant at 1 m height (the lowest height is deemed to result in higher consequences). The release directions in all scenarios were assumed to be horizontal. The consequence modeling input data shown below were used for the calculations. All scenarios were modeled using CANARY software developed by Quest. The software is widely used in industrial consequence modelling studies.

3.2 Atmospheric Conditions

The following more stringent atmospheric conditions that favors/sustains the pool scenario are considered:

Wind Speed: 2 m/s

Stability Class: F

Ambient temperature: 80 °F

3.3 Liquid Methane Release at High Operating Pressure

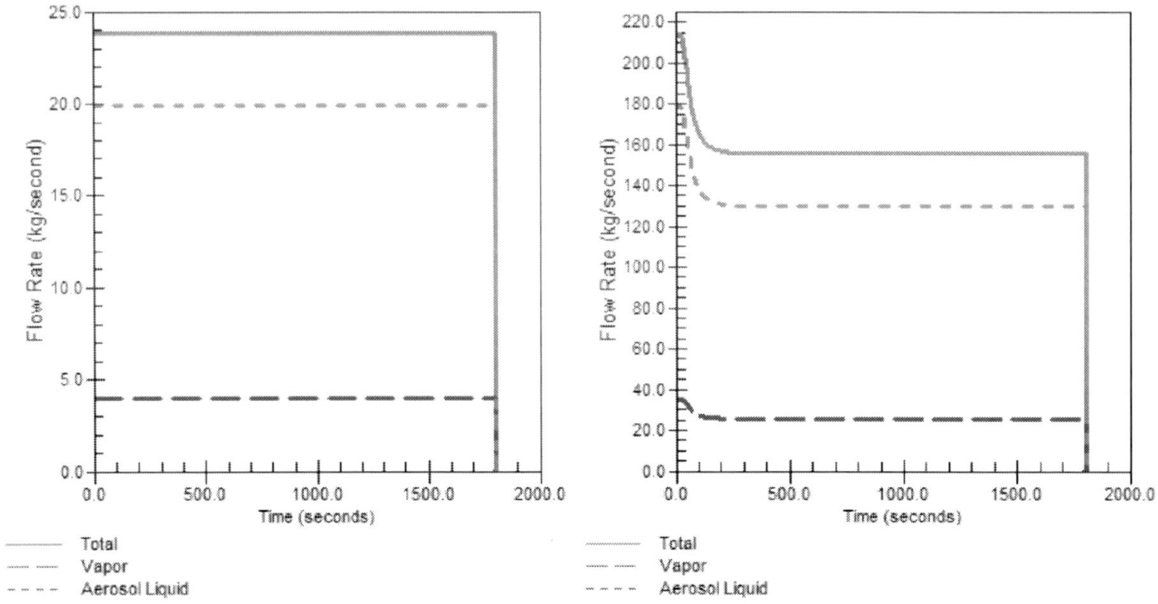

Figure 2: Liquid Methane Release – 2-inch (left figure) & 6-inch (right figure)

Consequence modeling for a liquid methane release was performed at specific process conditions for 2-inch and 6-inch hole size. As shown in the above figures, most of the release is flashing into aerosol liquid and remaining release is in vapor phase. No liquid pool is generated from the liquid methane release.

3.4 Liquid Propane Release

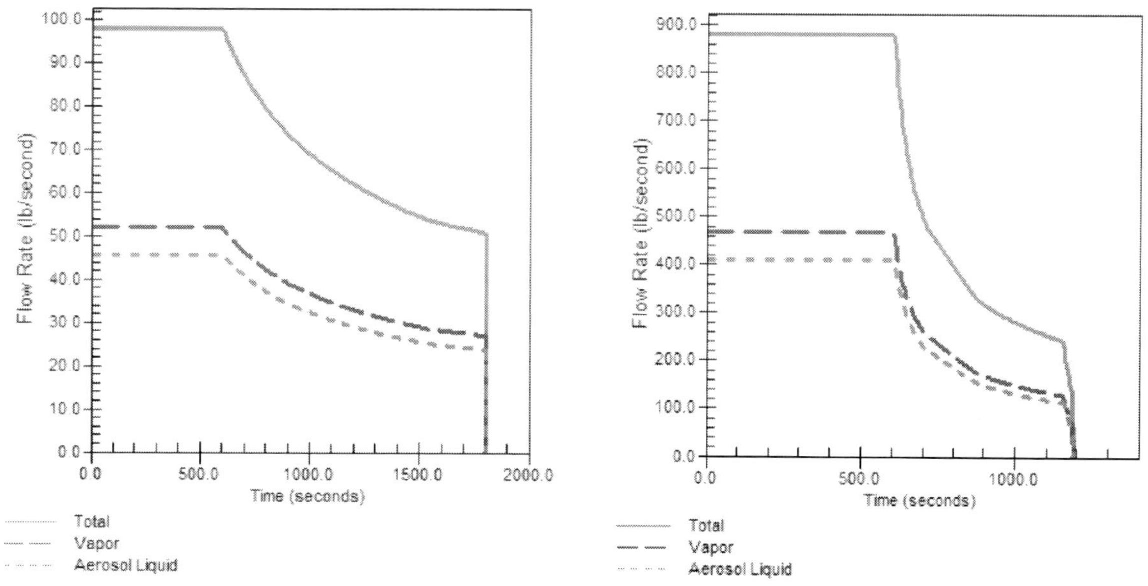

Figure 3: Propane Release – 2-inch (left figure) & 6-inch (right figure)

Consequence modeling for a propane release was performed at specific process conditions for 2-inch and 6-inch hole size. As shown in the above figures, the release is flashing into aerosol liquid and vapor phase. No liquid pool is generated from the propane release.

3.5 Liquid Ethylene Release

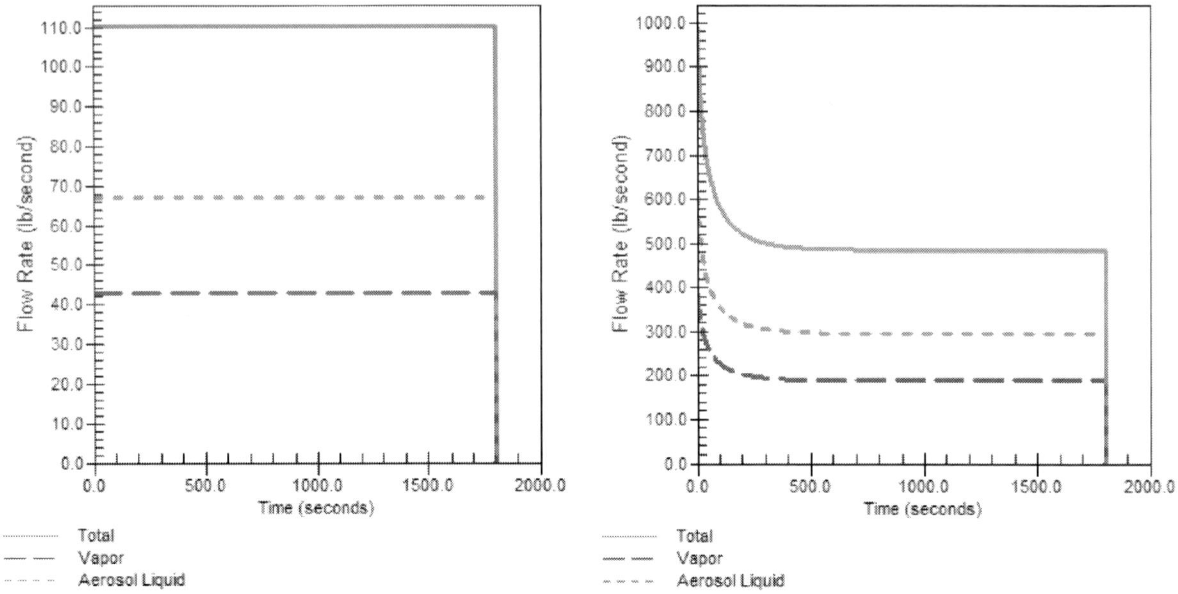

Figure 4: Ethylene Release – 2-inch (left figure) & 6-inch (right figure)

Consequence modeling for ethylene release was performed at specific operating process conditions for 2-inch and 6-inch hole size. As shown in the above figures, the release is flashing into aerosol liquid and vapor phase. No liquid pool is generated from the refrigerant release i.e. Ethylene release.

3.6 *LNG Release at Low Operating Pressure condition*

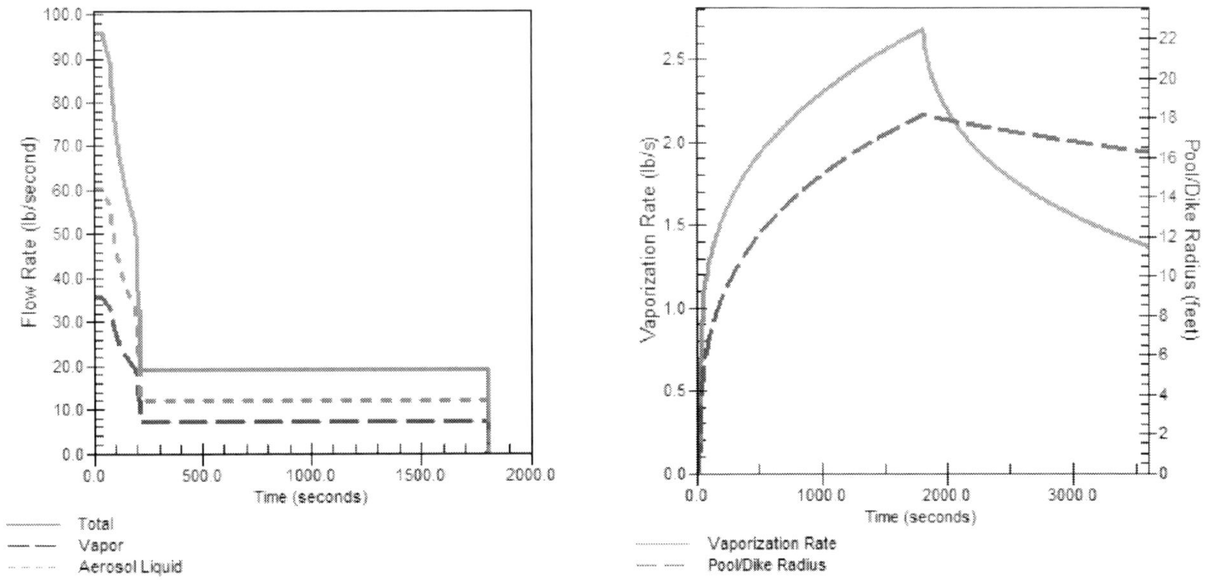

Figure 5: Liquefied Natural Gas Release – 2-inch (left figure) & 6-inch (right figure)

Consequence modeling for a liquefied methane was performed at the process conditions of an equipment for 2-inch. and 6-inch hole size. As shown in the above left side figure, the release for 2-inch hole size is flashing into aerosol liquid and vapor phase, without creating any pool. However, as shown in the right side figure, a release from 6in. hole size from the same equipment is creating a liquid pool of 16 ft. radius.

4 Conclusions

Consequence modeling results showed that out of the total twenty-eight (28) equipment of LNG liquefaction plant that were normally considered as Fire Potential Equipment (FPE) based on specific threshold volume, thirteen (13) equipment that handles methane, propane, ethylene at some specific operating conditions (e.g. low or medium or high pressure) did not create any liquid pool. Instead they flash along forming a large fraction of aerosol liquids, which never creates liquid pool.

The pool fire modeling was performed for a 2-inch release and a sensitivity analysis was carried out for 6-inch release to check the possibility of liquid pool formation for larger size releases for all the 28 equipment. The 6-inch release did not make a major difference in pool fire modeling results except for one equipment which contain LNG at low operating pressure condition. As stated earlier, thirteen (13) out of twenty-eight (28) equipment had not produced a liquid pool irrespective of the release size.

One (1) out of the remaining fifteen (15) equipment, that handles LNG at lower pressure condition did not generate liquid pool at 2-inch release, however it has generated liquid pool for a larger size (6-inch) release. Therefore, it is recommended to perform the consequence modeling for larger hole size to check the liquid pool formation.

The consequence modeling proved that 46% of the total equipment containing liquid such as methane or propane or ethylene or LNG had not produced a liquid pool. Therefore, considering the current economic trend, consequence modeling based approach for determining fireproofing requirements in conjunction with API 2218 requirement can be used in major LNG projects to justify considerable cost savings.

5 References

[1] J.D. Marx and A. Nicotra, "Is a Two-Inch Hole Adequate for Siting Study?", 12th Global Congress on Process Safety, American Institute of Chemical Engineers, Houston, Texas, 2016.

[2] API. "Fireproofing Practices in Petroleum and Petrochemical Processing Plants", American Petroleum Institute, 2013.

[3] CCPS (2003), "Guideline for Fire Protection in Chemical, Petrochemical, and Hydrocarbon Processing Facilities", Center for Chemical Process Safety of the American Institute of Chemical Engineers, 3 Park Avenue, New York, 10016-5991. (ISBN 0-8169-0898-2).

[4] CAN/CSA Z276, "Liquefied natural gas (LNG) - Production, Storage, and Handling", National Standard of Canada, 2015.

Simplistic Yet Insightful Way to Present Operational Risks in Order to Facilitate Risk Discussions

Don Byrne
DuPont Sustainable Solutions
CRP 730, 974 Centre Road, Wilmington, Delaware
Donald.F.Byrne@dupont.com

Alfonsius Ariawan
DuPont Sustainable Solutions
1919 Minnesota Court, Mississauga, Ontario
Alfonsius.B.Ariawan@dupont.com

Prepared for Presentation at
American Institute of Chemical Engineers
2017 Spring Meeting and 13th Global Congress on Process Safety
San Antonio, TX
March 26 – 29, 2017

Simplistic Yet Insightful Way to Present Operational Risks in Order to Facilitate Risk Discussions

Don Byrne
DuPont Sustainable Solutions
CRP 730, 974 Centre Road, Wilmington, Delaware
Donald.F.Byrne@dupont.com

Alfonsius Ariawan
DuPont Sustainable Solutions
1919 Minnesota Court, Mississauga, Ontario
Alfonsius.B.Ariawan@dupont.com

Keywords: Risk Awareness, Risk Matrix, Risk Prioritization

Abstract

Risk matrices can be used as an effective engagement tool to help personnel internalize the concept of risk and risk prioritization. We have used risk matrices outside of the technical application of Process Hazard Analysis in simple exercises to broaden risk awareness at client organizations. In our experience, consensus on potential consequences can usually be achieved relatively easily. However, defining a meaningful likelihood or frequency of the risk can be quite challenging. It requires many assumptions and can be misleading. Using incident data, for example, often results in the underestimation of frequency as actual incidents may not have occurred yet. Using usually grossly under-reported near-miss data can also result in the same. Our experience suggests that such disagreement can be utilized, with frequency represented as a range in the risk matrix to provide insightful information and rich discussion on the potential for incidents.

1. Introduction

The importance of risk matrix has not been well defined and/or understood by many organizations. A risk matrix is a useful tool for organizations to characterize their risk profile and help prioritize risk management efforts. It has a paradoxical intent. On one hand, it helps organizations increase their level of confidence that risks are under control. On the other hand, it is intended to create a sense of vulnerability to prevent complacency based on past experience. Ultimately, each organization has to understand that the purpose of risk profiling using a risk matrix is to be able to differentially manage risks rather than treating all risks equally.

There is no one risk matrix that fits all organizations. For each organization, a single matrix should be defined with scales that are consistent with the nature of the risks and the organization's tolerance to them. By plotting the risks onto the matrix, we can quickly identify risk items that require immediate attention. The convention is to show each risk as a single point in the matrix. While this is the generally accepted practice, however, it invites us to subconsciously think of each point as being 'precise', even when we know that there could be significant subjectivity in the characterization of these risks.

In our work, we have found an alternate way to represent risks using a risk matrix that has helped organizations better appreciate their relative risks. It also provides additional information to help prioritize and reduce risks. In fact, our definitions of risk frequency have helped uncover several risk items that require immediate attention that may otherwise have been are categorized as being 'low' in priority.

Due to the proprietary nature of this approach, we are unfortunately unable to share key Figures to support the content of this paper. Please contact the author directly for further discussion.

2. Hazards and Risks

We define hazards as any source of potential damage, harm or adverse health under certain work conditions. Loosely speaking, we can categorize hazards into three general categories: process-based, equipment-based, and activity-based. Process-based hazards include hazards that are due to materials and processing conditions in a certain operating environment. Equipment-based hazards refer to hazards associated with the failure of equipment or infrastructure. Lastly, activity-based hazards are hazards arising from tasks that people perform.

Similarly, we have process-based, equipment-based and activity-based risks. Level of risks is determined by considering the likelihood, or frequency of a potential incident resulting from the hazard against the severity of its worst (reasonable) potential consequence.

Typically, risk assessments are conducted by a small group of personnel, usually by way of brainstorming or discussing the consequence and likelihood of various hazardous events, such as in Process Hazard Analysis sessions. As such, the level of experience within the risk assessment team is a critical determinant of the output quality. Our experience suggests that it is relatively easy to gain consensus on the severity of a potential event, assuming there is an unambiguous definition of each level of consequence category. However, the discussion regarding event likelihood/frequency can be quite subjective, with disagreement typically arising between historical consideration of actual events and their potential likelihood, especially when controls are included in the discussion (i.e. mitigated risks). Often, we resort to tools such as Layers of Protection Analysis or Fault Tree Analysis to help with the assessment of event likelihood. However, for lower hazard operations, such tools may be too complicated or seen as unnecessary.

3. Alternate Representation of Risks Using Risk Matrix

In our recent work with clients, we have represented risks in a slightly modified manner using the risk matrix. Instead of a single point item, risk is plotted as a vertical line in the frequency vs. severity matrix. The worst reasonable potential severity of incident is defined, and the vertical line is placed in that region of the x-axis. Note that the x-axis is not continuous. It is comprised of several regions or "buckets" (Minor, Moderate, Serious, Critical, etc.). All risks within this region have the same worst reasonable potential severity.

The length of the line spans over a range bounded by the Opportunity Frequency on the high end and the 'Oops!' Frequency on the low end.

The Opportunity Frequency is straightforward and easily calculated. It is determined by how often the associated activity occurs and the number of hazard exposures in the activity. The hazard has one or more opportunities to cause harm every time the activity occurs. It can be astonishing how many times a year an activity that could result in a fatality can occur at a site in a year. We have seen Opportunity Frequencies that exceed 1,000,000 times per year. An incredibly good performance rate is required to assure no incidents!

In the context of continuous operations, estimating the number of times an 'activity' is performed is not meaningful. In this case, defining Opportunity Frequency is irrelevant. Representing a continuous operation with a "Very High" Opportunity Frequency is an adequate approximation. For this, we modify the vertical axis of the risk matrix to include a 'Continuous' frequency threshold.

The highly technically named "Oops!" frequency is more subjective and requires more explanation. We started out calling this the Near Miss frequency, recognizing that it is only a matter of probability that the Near Miss could be an actual incident. The thought was this would be a closer approximation of true risk than actual incidents. But this quickly raised some resistance. Most know that the vast majority of Near Misses are not reported, and using Near Miss data would under-state the risk.

We settled on 'Oops!' Frequency to reflect any near misses, close calls, and other "Aw Shucks" moments that could have easily been an incident. This definition requires that several of the operators and mechanics that do the work are involved in the estimating this frequency. While there is often a wide range of opinions on the 'Oops!' Frequency initially, we have rarely seen a situation where the group could not come to a consensus (+/- 50 %). A glance at a risk matrix should satisfy most people that +/- 50 % is accurate enough, since the y-axis is a logarithmic scale.

In organizations with good safety performance, the number of prior incidents involving the specific hazard can be low or zero. When using incident data to represent the risk frequency, organizations may mistakenly dismiss the risk as being 'unlikely' or non-existent. The people doing the work know better. They know that they "come close" to an incident more frequently than other people may realize. The 'Oops!' Frequency expresses that.

4. Interpretation

The representation of frequency as lines as opposed to single points enables a rich discussion within the risk assessment team and others in the operations group. These lines between the 'Oops!' Frequency and the Opportunity Frequency represent "the opportunities for more incidents to occur"! For example, if Operational Discipline declines due to fatigue, frustration, or other circumstances, the real-time "Oops!" frequency can climb right up the line. This is a scary thought, if only administrative controls are in place.

Why some lines are longer than others can also make for interesting discussions. It might seem that sufficient controls are in place if lines are longer. As discussed above, for risk with high Opportunity Frequency, a longer line does not assure adequate performance. The 'Oops!' Frequency provides the best measure of the need to reduce risk in this case. However, lowering Opportunity Frequency by process or work changes will also likely lower the 'Oops!' Frequency. Therefore, both frequencies can be used to attack risk reduction!

A short line suggests that inadequate controls are in place. In the extreme case, the Opportunity Frequency and the 'Oops!' frequency fall on top of one another. This suggests that each time an activity is conducted or a process is run, some kind of a close-call happens. One can see how this provides additional information by which risk can be prioritized and reduced. Representing risks in this manner highlights those cases where controls may not be adequate and require more immediate attention.

In a recent application at a client site, a hazard associated with the loading of a chemical was characterized as having a potential consequence of multiple fatalities. The site viewed the risk as "Low", however, as there had been no incidents in the 8 years that the activity was conducted. The loading activity took place weekly, translating to an opportunity frequency of 52x/year. When we discussed the "Oops!" frequency with the site, however, they reported that a release of chemical cloud each time the loading occurred – i.e. 52x/year! The overlap of Opportunity Frequency and 'Oops!' Frequency drew immediate attention. The 'Oops!' frequency is clearly unacceptable for a potential consequence of multiple fatalities. The site immediately elevated this risk to be addressed by additional controls.

The number of hazards and their associated risks are large for any significant operation. Understanding how to prioritize risk reduction efforts, and ensuring that the highest risk is addressed first is important. Figure 1 shows an example of the resulting risk dashboard developed to assist one of our clients with risk prioritization. It provides an un-conventional likelihood-severity risk characterization, presenting both the Opportunity Frequency and the 'Oops!' Frequency vs. Potential Consequence in a format which allows each area of the site to "see" the highest risks – and select them for Risk Reduction work.

5. Summary

In this paper, we describe an alternate way to present risks using a risk matrix. Rather than showing a risk as a single point in a risk matrix, we propose to represent risk by way of a line, spanning the two extreme frequencies defined as the "Oops! frequency" at the low end and the "Opportunity frequency" at the high end, plotted in the appropriate severity column. To be meaningful, the "Oops! frequency" needs to include near-misses, close calls, unusual or unexpected situations and actual incidents. Both reported and unreported cases need to be considered (hence, the importance of having operators participating in the analysis). Presenting risks in this manner provides a better representation of the risks for risk prioritization, and additional insights on how to approach risk reduction work.

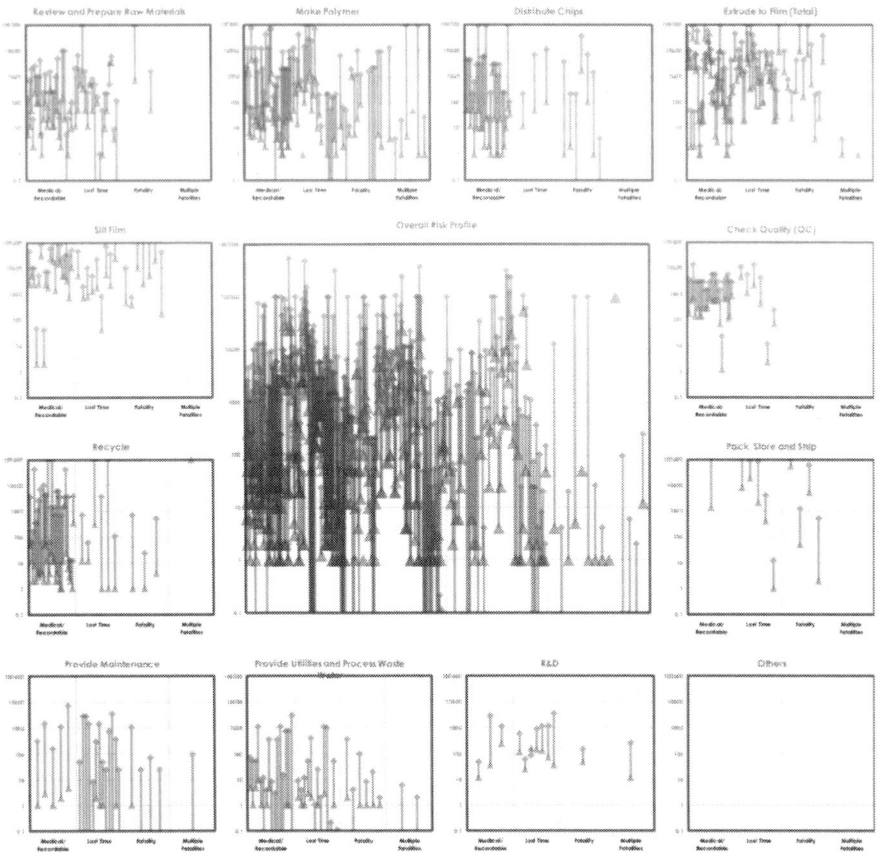

Figure 1. Risk dashboard allowing for prioritization of risks using an un-conventional likelihood-severity risk characterization, presenting both the Opportunity Frequency and the 'Oops!' Frequency vs. Potential Consequence

Uncertain About Uncertainty

John Joseph Crump; P.E., J.D., Member-AIChE
CPI — Consulting & Professional Instruction, LLC

3772 Jardin Street
Houston, TX 77005
USA

Calle Republica de Chile 21
Colonia Cinco de Diciembre
Puerto Vallarta, Jalisco 48350
Mexico

JohnCrump@GMAVT.Net

Prepared for Presentation at
American Institute of Chemical Engineers
2017 Spring Meeting and 13th Global Congress on Process Safety
San Antonio Texas
March 27-29, 2017

UNPUBLISHED

Uncertain About Uncertainty

John Joseph Crump; P.E., J.D., Member-AIChE
CPI — Consulting & Professional Instruction, LLC

3772 Jardin Street
Houston, TX 77005
USA

Calle Republica de Chile 21
Colonia Cinco de Diciembre
Puerto Vallarta, Jalisco 48350
Mexico

JohnCrump@GMAVT.Net

Keywords: Process Safety, Understand Hazard & Risk, Hazard Identification and Risk Analysis, Define probability; Subjective probability

Abstract

It is well-known that considerable uncertainty characterizes Process Safety (PS) with respect both to frequency of undesired conditions and events and to the consequences thereof. In the extreme (and in all sorts of human endeavors), the uncertainty that surrounds our sense of "likelihood" may give rise to the Precautionary Principle or may even cloak a Black Swan. This short paper describes three general approaches to defining probability and focuses on the ambiguity that characterizes qualitative or subjective probability. Particularly with Risk-Based Process Safety (RBPS), PS project teams very commonly use subjective probability, because it is simple and often because more definitive data are scarce — but what does it mean when a team of 7 members, or maybe just 2, decides that the chance a condition or event may occur is "rare," or "likely," or "a toss-up?" In fact, this problem is so common that we hardly consider it in the multitude of decisions that we make every day. This paper explores ambiguities inherent in such terms and presents results from group tests that demonstrate that, absent some disciplined definition, individual players may, and often do, assign substantially different values in making those judgments. These results show that it is very important to use clear, agreed and consistent definitions for the subjective probabilities that will apply in a project — whether a PHA, a MOC, a PSSR, or any other project element where uncertainty lurks. Process Safety teams must think carefully about how to define uncertainties as they characterize facilities and conditions that may enable or cause incidents, particularly given that the incidents of primary concern may be extremely unlikely events but are absolutely catastrophic.

1. Introduction

Mark Lijek: *What are our chances?*
Mendez: *Chances are good.*
Mark Lijek: *"Good?" Well, what's the …uh…. What's the number value of "good?" 30% chance of being publicly executed?* [1]

Making decisions is like speaking prose — people do it all the time, knowingly or unknowingly. [2]

We all do it. Every day. Our minds and guts are constantly calibrating and judging likelihood and risk — ranging from minute to gigantic, from insignificant to catastrophic. Along the way, heuristics and cognitive biases serve as guideposts, short cuts and sometimes blinders that get us to what we hope are predominately good analyses and decisions. For Process Safety and its various codifications, [3] that human process and our resulting judgments play vital roles in Process Hazards Analyses (PHAs), in Employee Participation, in Human Factors and in many other elements. Those heuristics and cognitive biases help us to move efficiently through myriad decisions that we face every day, but sometimes (as Tversky and Kahneman showed in their papers) they lead us astray and into error, and *those* errors are *our* errors — human errors. Therefore, probability estimates play a vital role in Process Safety, but they also can confound and conflate uncertainties that inevitably arise. In addition, Process Safety itself was born out of conditions that compound such problems of human judgment — catastrophic events that were (or at least seemed) extremely improbable.

This paper examines the role of subjective probability in that process, starting with a brief exploration of risk — of the elements and nature of uncertainties therein — progressing to define different sources of probability information and then comparing objective vs. subjective probability and errors estimating extreme chances and finally considering the quantitative values people assign to the qualitative terms they typically use to describe uncertainty. It ends by offering some thoughts about how to control those human tendencies that can lead to errors of estimation and judgment

2. What is "Risk?"

Most commonly, "Risk" is calculated according to Bernoulli's mathematical expectation formulation — by "which the value, or utility, of an uncertain prospect is obtained by

---

[1] Dialog from the award-winning film Argo (at 1:05:50). Directed by Ben Affleck; written by Chris Terrio; released in the USA October 12, 2012. Quotations from Argo are courtesy of Warner Bros. Entertainment, Inc., with sincere thanks.

[2] Kahneman, Daniel, and Amos Tversky. *Choices, Values, and Frames*. American Psychologist (April 1984, Vol. 34, No. 4) at p. 341.

[3] Process Safety codifications include Risk-Based Process Safety ("RBPS") from CCPS, OSHA's Process Safety Management ("PSM") rule in the USA (29 CFR §1910.119), and other codes and regulations of a number of countries across the globe.

adding the utilities of the possible outcomes, each weighted by its probability." [4] That equation for risk is —

$$\textbf{Risk} = f(\textbf{Utility, Probability}) = \Sigma(\textbf{Consequence}_n \times \textbf{Probability}_n) \quad \textbf{(1)}$$

The classic example of Bernoullian expectation is a game of chance, where the stakes and the odds are well-known and the expected outcome can be calculated. Such games are often used to illustrate risk. Equation (1) is common, straightforward and simple (and serves well as a pedagogical tool), but is far from adequate for real applications, as in Process Safety. Everyone knows from daily personal experience that there is more to risk. As Tversky and Kahneman cautioned in their first paper, "[t]he heart of the gambler's fallacy [5] is a misconception of the fairness of the laws of chance." ... "People have erroneous intuitions about the laws of chance." [5]

More broadly than any gambler's intuition, in recent decades numerous psychologists have produced abundant research tying risk to psychology, in addition to and beyond Bernoulli's formulation. Therefore, a proposition offered here is that a more complete statement of "Risk," particularly when used as "Risk Analysis" or "Risk Management" must require a function of at least 3 variables —

$$\textbf{Risk} = g(\textbf{Consequence, Probability, Psychology}) \quad \textbf{(2)}$$

where — **Risk** (as well as analysis and control thereof) is some function of:
Consequence (or utility, result, outcome, ...) and
Probability (or uncertainty, frequency, chance, likelihood, ...) and
Psychology (or tolerance, attitude, acceptance, perceptions, ...).

One reaction to Tversky and Kahneman's skepticism is to recognize that all three variables must be considered in any practical and real valuation of risk. Therefore, Figure 1 depicts "Risk" as a three-legged stool. Analysis and management must honor that structure: Remove any one of those legs, and the stool will fall. Additionally, practical applications of risk assessment and control (particularly for Process Safety) are further conflated with human imposition of laws, rules, regulations, organizational policies ... and

Figure 1. Risk as a 3-Legged Stool

4 Kahneman and Tversky. Op Cit. (April 1984) at pp. 344-345.

5 Tversky, Amos, and Daniel Kahneman. *Belief in the Law of Small Numbers*, Psychological Bulletin (1971, Vol. 75, No. 2) at pp. 105-106. Tversky and Kahneman define the gambler's fallacy: The gambler "act[s] as if every segment of [some] random sequence [in a game of chance] must reflect the true proportion: if the sequence has strayed from the population proportion, a corrective bias in the other direction is expected. [Take flipping a coin as an example game of chance.] Even the fairest of coins [has no] memory and moral sense [, and] cannot be as fair as the gambler expects it to be."

enforcement ... and with comparisons of expert opinion vs. lay opinion, individual action vs. group dynamics, and perhaps yet other factors.[6]

Expert opinion often targets relatively narrow technical issues. However, even experts are subject to natural biases and psychological errors in judging probability. [7] Thus, it is foreseeable that other stakeholders (e.g., other employees who may not have expert knowledge about specific operations, regulators, the public, and certainly the media) very commonly draw — knowingly or unknowingly — on psychological sources (perceptions, feelings, emotions and so forth) in judging risks. Furthermore, the first foundation block of the CCPS Risk-Based Process Safety (RBPS) system directs practitioners to interact with and to understand all sorts of others (with elements around culture, competency, workforce involvement and access, and stakeholder outreach).

Thus, psychology is essential to any practical definition of "Risk," and it is worthwhile to consider the diversity of factors that researchers have found can influence risk awareness, sensitivity, analyses and management. A review of literature on psychology of risk would be beyond the scope of this paper, but psychologists have published a great deal of literature on the topic. Over decades, Kahneman and Tversky, Fischhoff, Lichtenstein, Slovic, Thaler, and many other psychologists have explored a huge array of cognitive biases. Wikipedia (a simple example and one resource for initial learning about psychological factors in risk) has a short introductory article that lists dozens of such biases — habits of thinking and human cognitive tendencies that can efficiently lead to reasonable solutions but that also can lead to systematic deviations from intended rationality and good judgment. [8]

An alternative of particular interest in the context of this study and of Equation (2), is the 2-dimensional Cartesian plotting system Paul Slovic developed to map interactions of 17 characteristics that influence common perceptions of "Risk." [9] Figure 2 is a simplified adaptation of that published map (but listing 19 characteristics). In related graphics, [9] Slovic plots 91 hazardous activities, products and industries on that map and displays related differences in demand for regulation. Slovic has used that graphic over at least 20 years with changes from time to time to the numbers of hazards and mapped locations as risk factors have evolved. Elsewhere, Vincent Covello produced a similar but separate list of 19 factors that affect public perceptions of risk (Table 1). [10] Thus, the psychological dimension of "Risk" is structured out of many factors that must evolve over time for any individual or group in response to many cognitive, motivational, emotional, environmental, and other contextual factors and sub-factors.

---

6 Also, people increasingly worry today about intelligent, motivated activities that constitute deliberate forms of human hazard (i.e., terrorism or simple vandalism).

7 Tversky and Kahneman. Op. Cit. (1971) *passim*. This is a key point of their paper.

8 Wikipedia. *List of Cognitive Biases* (Updated 25 December 2016). Retrieved 26 December 2016 from https://en.wikipedia.org/wiki/List_of_cognitive_biases.

9 Slovic, Paul, Baruch Fischhoff, and Sarah Lichtenstein. *Why Study Risk Perception?* Risk Analysis (June 1982) pp. 83-93.

10 Howard Margolis has questioned whether Covello's list is "far too large." He concluded that it is "not logically impossible [but nevertheless is] highly improbable" for all 19 of Covello's factors to be statistically significant. *See* Margolis, Dealing with Risk (Univ. of Chicago Press, 1996) at p. 105 and footnote 6 thereon.

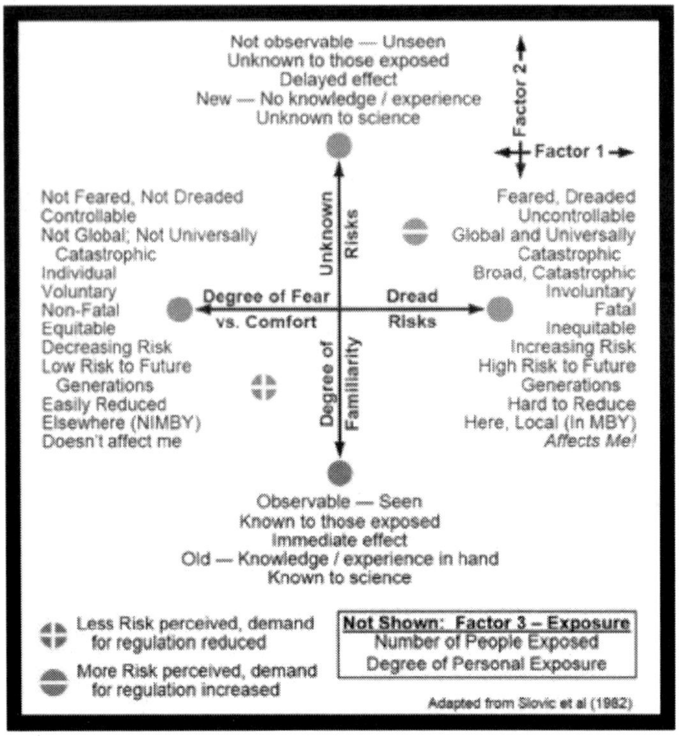

Figure 2. Structure for 2 of 3 Dimensions (or Factors) for Mapping Hazards (Activities, Products and Industries) to 18 Characteristics of Risk Perception.
Slovic used this structure to map hazards and attitudes regarding risk and regulation. [9]

Compared with Subjects who participate in studies of risk psychology and who are commonly drawn from ranges of accessible population groups, most actors in Process Safety have significant expertise around risks that arise in processing facilities, and their analyses tend to be more objectively technical. However, Slovic warns:

> [D]ifficulties in understanding probabilistic processes, biased media coverage, misleading personal experiences, and anxieties generated by life's gambles cause uncertainty to be denied, risks to be misjudged (sometimes overestimated and sometimes underestimated), and judgments of fact to be held in unwarranted confidence. Unfortunately, experts' judgments appear to be prone to many of the same biases as those of laypersons .... [Furthermore,] initial views are resistant to change [even with new evidence] because they influence [how] subsequent information is interpreted. [9]

Elsewhere, Slovic characterizes "risk analysis [as] a political enterprise as well as a scientific one" that inevitably drags "issues of values, process, power, and trust into the picture" [11] — whether we like it or not. Therefore, it is important to be aware that psychological factors inevitably and naturally affect analysts and decision-makers, as

---

[11] Slovic, Paul, and Elke Weber. *Perception of Risk Posed by Extreme Events.* Paper discussed at the conference "Risk Management Strategies in an Uncertain World," April 2002. Since risk management proceeds from risk assessment, this quote must apply to both.

Factors Important in Risk Perception and Evaluation		
19 Factors	Conditions that tend to increase public concerns	Conditions that tend to reduce public concerns
1. Catastrophic potential	Fatalities and injuries grouped in time & space	Fatalities scattered and random
2. Familiarity	Unfamiliar	Familiar
3. Understanding	Mechanisms or process not understood	Mechanisms or process understood
4. Uncertainty	Risks scientifically unknown or uncertain	Risks known to science
5. Controllability (personal)	Uncontrollable	Controllable
6. Voluntariness of exposure	Involuntary	Voluntary
7. Effects on children	Children specifically at risk	Children not specifically at risk
8. Effects manifestation	Delayed effects	Immediate effects
9. Effects on future generations	Risk to future generations	No risk to future generations
10. Victim identity	Identifiable victims	Statistical victims
11. Dread	Effects dreaded	Effects not dreaded
12. Trust in institutions	Lack of trust in responsible institutions	Trust in responsible institutions
13. Media attention	Much media attention	Little media attention
14. Accident history	Major and sometimes minor accidents	No major or minor accidents
15. Equity	Inequitable distribution of risks and benefits	Equitable distribution of risks and benefits
16. Benefits	Unclear benefits	Clear benefits
17. Reversibility	Effects irreversible	Effects reversible
18. Personal stake	Individual personally at risk	Individual not personally at risk
19. Origin	Caused by human actions or failures	Caused by acts of nature or God

Table 1. Vincent Covello's List of 19 Factors Characterizing Risk [12]

[12] This list is from Vincent Covello as quoted by Howard Margolis, Dealing with Risk (Univ. of Chicago Press, 1996) at p. 28. It originated in *Risk Comparisons and Risk Communication* in Kasperson and Stallen, Communicating Risks to the Public (Kluwer, 1991) at p. 112, and it is drawn mainly from work of Slovic, Fischhoff and others. *See also* a nearly-identical list in "Appendix C" of Risk Communication, Risk Statistics, and Risk Comparisons: A Manual for Plant Managers (1988), by Covello et al. for the Chemical Manufacturers' Association (CMA, now the ACC or American Chemistry Council). The CMA version does not include factor 18 (personal stake).

well as the lay public — regardless of their intentions, consciously or otherwise. In addition, technical experts often hold widely different and deeply rooted opinions and perceptions of (and tolerance for) risk as compared with other stakeholders — particularly inspectors, other government officials, the media, and the public.

Most often in Process Safety, the project boundaries exclude overtly political issues and terrorism, [13] and the same will be true for this paper. Moreover, a full examination of the various psychological elements, heuristics and biases that drive human response to uncertainty would be far beyond the goals of this paper. Therefore, the sections that follow will review the interplay of probability and psychology within the risk function described in Equation (2). In doing so, the goal is to show that — although human intuition, patterns or modes of thinking and psychological issues of concern come to all of us quite naturally as elements of human acuity and efficiency — even the most savvy professionals inevitably fall prey at times to the imprecision, inaccuracy, ambiguity and mental error that the same factors impose on estimates and valuations that are essential to decision-making for businesses and process facilities in general and for Process Safety in particular.

3. Where Do Probability Values Come From?

Returning for simplicity for the moment to the conventional Bernoullian expectation formula [Equation (1), which is obviously a subset of the full range of functions that could fit within Equation (2)], it may appear at first glance that "uncertainty" in risk is loaded in "probability," the first variable. One might be tempted to regard "consequences" deterministically. Indeed, objective information is available in some cases (e.g., especially games of chance) to determine the probability of some event and to know its outcome. However, in the practical world both questions of whether a triggering event will occur and questions of what consequence(s) may ensue are clouded with uncertainty. Particularly in Process Safety — where very improbable but catastrophic events are of greatest concern — limits of resources, knowledge and experience typically force analysts to estimate the likelihood of an event, and it often is equally difficult to forecast which consequence(s) will result in a given incident and how severe the outcome may be. Altogether, the combined uncertainties inspire the familiar "bow-tie" analysis.

The range of possibilities is illustrated in Figure 3, where data are adequate to analyze risk only in the upper-left corner. Then, taking that upper-left corner as the origin and moving from top to bottom along the ordinate itself, the likely consequences are clear (or at least imagined), but, regardless of the initial level of uncertainty as to likelihood in the upper-left corner, it becomes increasingly uncertain from top to bottom whether some event of concern may trigger any outcomes. In the lower-left corner, an analyst has little or no information to characterize probability but can readily imagine some adverse consequence(s). The Precautionary Principle lives there.

---

[13] *For a contrary example, see*: Nolan, Dennis P. Safety and Security Review for the Process Industries (William Andrew, 2008).

In contrast, starting again at the upper-left corner and moving horizontally along the abscissa to the upper-right corner, the probability that some event of concern may occur is fixed, but it becomes increasingly uncertain what (if any) consequence(s) may arise from that event. In both of those cases (i.e., in both lower-left and upper-right corners), ignorance (as to whether some triggering event might happen and / or about what outcome(s) might result) suggests a cautionary approach for risk decisions.

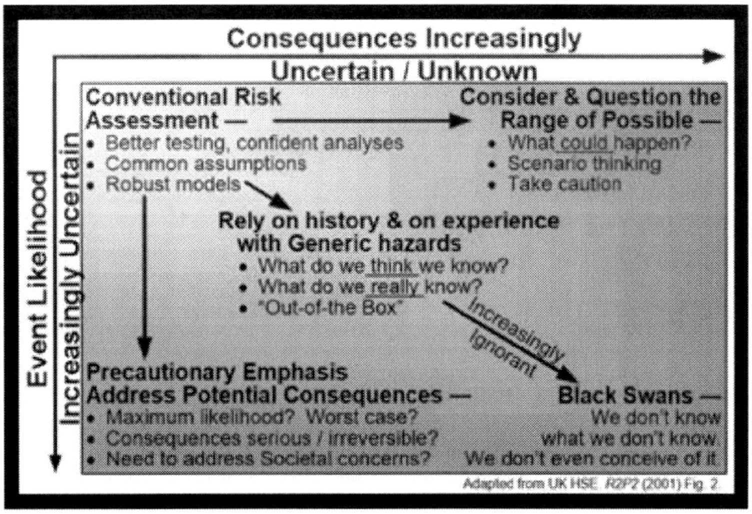

Figure 3. Event and Consequence Uncertainties Inherent in Risk and Decisions

Of course Process Safety is no game of chance, and very often facilities operate in the center of Figure 3. There, one has some notion how likely an event may be to trigger an incident and what consequence(s) might result, but the data are imperfect and analysts urge each other to think "outside-of-the-box" — applying history, experience, and imagination to guide judgment of risk. In such cases, however, people somehow evaluate — either analytically or most often by qualitative estimates, scenarios or other subjective / intuitive valuations — both the chance of that some set of triggering conditions may arise and the likelihood of specific consequences. When, on the other hand, ingenuity and experience fail, companies find themselves in the lower-right corner of Figure 3, at the mercy of dreaded Black Swans. [14]

[14] *See* Taleb, Nassim Nicholas. The Black Swan: The Impact of the Highly Improbable (Random House, 2007). *See also* a series of four articles by John Murphy and James Conner about Black Swans in the context of Process Safety in Process Safety Progress (2011, 2012, 2014, and 2016).

Taleb defines a "Black Swan" as an event that:

1. Is an Unexpected Rarity, an Outlier to our experiences — Nothing in the past suggests that the event will happen, or at least nothing suggests that convincingly. Prior to occurrence, it is considered as extraordinarily unlikely, if it is even conceived.
2. Has Extreme Impact (Major Consequences) — Taleb opines that practically anything of real significance in this world, anything that changes life (anything in science, art, politics, romance, ...), results from a rare Black Swan event, at least for those seriously affected.
3. Is Subsequently Explained — Even though Black Swans are unexpected and unpredicted, victims and / or beneficiaries explain it all after the fact, suggesting (falsely) that someone really foresaw and understood the situation all along. In his book, Taleb refers to such explanations as "narrative fallacies" or "stories." "Scenarios" could be a synonym.

Therefore, it is important to value and to understand probabilities, both for triggering events and for consequences. But what *are* the chances? Where do they come from? How "good" are they? How does one find them … or make them up?

3.1 Types of Probability

Well, there are three generic sources or types of probability: Classical or Theoretical Probability, Empirical Probability or Relative Frequency, and Subjective Probability. These three generic sources are illustrated in Table 2.

Classical or Theoretical Probability that Event "A" will occur [Pr(A)] provides quantitative, mathematical results from abstract or scientific analyses expressed as:

$$\textit{Classical or Theoretical } \Pr(A) = \boxed{\text{Calculate the } \frac{\#\text{ Ways "A" Can Occur}}{\#\text{ Ways Experiment for "A" Can Proceed (or \# Pathways "A" Can Follow)}}} \quad (3)$$

[15]

Empirical Probability or Relative Frequency also provides quantitative results, but these come from experimental data analyzed as:

$$\textit{Empirical } \Pr(A) \textit{ or Relative Frequency} = \boxed{\text{Record the } \frac{\#\text{ Times "A" Occurred}}{\#\text{ Times Tried or Ran Experiment for "A"}}} \quad (4)$$

[16]

Both Classical and Empirical Probability yield quantitative results that are "objective," to the degree one trusts the analysis or the experiments, but Subjective Probability is different. It arises as a personal qualitative and / or semi-quantitative expression of what one believes or thinks will happen: How likely is it that Event A will occur, and how likely is it that particular consequences will result? Obviously, this is a judgment call —

$$\textit{Subjective } \Pr(A) = \boxed{\text{How do you "}\textit{feel}\text{" about Event A?}} \quad (5)$$

Analysts' feelings raise the psychological issues described in the previous section, and often they translate such qualitative values into numerical estimates for use in decision

---

[15] Here is a common, simple example: What is the probability of rolling "5" with two dice? Answer: Four combinations of two dice will result in "5," and each roll of two dice will result in one of 36 combinations of the dice. The probability of rolling "5" is 4/36 = 11.1111%. The probability distribution function yields a triangular distribution, with tail probabilities of 2.77778% for rolling either "2" or "12" and a peak probability (the mean, the median and the mode) at 16.6667% for rolling"7."

[16] Here is a common, simple example: What is the probability of rolling "5" with two dice? Answer: Take two dice and roll them many times (say 10,000 times, for example), recording the results of each roll. Count the number of times "5" resulted. That value is the relative frequency per the total number of rolls (10,000 or whatever). Divide the relative frequency (total number of "5s") by the total number of rolls (10,000 or whatever) to get the probability of rolling "5." If the total number of rolls is large enough, the result should be stable, and the probability of rolling "5" should be close to 11.1%.

100% or One (1)	Certainty — No question that Event "A" will occur.
50% or One-Half (½)	It is exactly as likely as not that Event "A" will occur.
0% or Zero (0 or ∅)	No chance at all that Event "A" will occur.

¿¿ What do we mean by "the *Probability of Event A*" [Pr(*A*)] ??

Classical Probability (Theoretical / Math)	**Empirical (Practical) Probability (Relative Frequency)**	**Subjective Probability**
Examples: ▶ By inspection, chance of flipping Heads or Tails with a "fair" coin is ½ — IS IT?? [17] ▶ By inspection, the probability of rolling "3" with a 6-sided die is 1/6 … assuming a "fair" die [*i.e.*, equally likely (¿?) to roll each side] ▶ Pascal's 17th century binomial distribution solution to the "Problem of Points" as to how two equally-skilled players should split winnings from a game of chance if forced to terminate prematurely. Pascal's solution was for the card game *balla*, but it applies much more broadly. Along with that, he invented the Theory of Probability.	**Examples:** ▶ Inspectors found 4 out of 1,000 plastic cups defective ⇒ Chance of finding a defective cup ~ 4/1000 = 0.4% ▶ Flip a coin or roll two dice many times. Count and plot the results to learn the probability distribution. (Note — "Many times" means "continue until the collected data produce a reasonably smooth or stable estimate of the Probability Distribution Function (PDF)." In view of the fact that coins and dice have no memory from trial to trial, this can easily be >1,000 times.)	**Examples:** ▶ Percent chance of success for some first-time venture — *e.g.*: Contain an oil spill, Introduce new product, Develop a project in a foreign country. ▶ Fuzzier: Who will win the World Cup? ▶ Most Fuzzy: Will you marry your high school sweetheart? ▶ Fuzziest: Will that company succeed … or fail? ▶ Also Fuzziest: Who will win the election? ⇒ Expectation or Level of Belief — **This is the most common meaning of "Pr(A)."**

Table 2. What Do We Mean by "*Probability*?" — Three generic sources of values

[17] *See* Klarreich, Erica. *Toss Out the Toss-Up: Bias in heads-or-tails*. Science: Science News Online (2/28/2004; Vol. 165, No. 9).

making. In accord with the opening quotation from Kahneman and Tversky, everyone does so routinely, *knowingly or unknowingly* — regardless of intent, regardless of profession or skill and regardless of whether a "correct" or "true" solution exists (e.g., as in estimating real, calculable probabilities).[18]

This is one thing when people are playing games. Then they can follow their hopes and dreams, to the degree that wallets allow. In contrast, the goal is to make judgments for Process Safety and other business purposes as objective, accurate, consistent and repeatable as reasonably possible.

4. Weighing Decisions — Objective vs. Subjective Probability

Unfortunately, objectivity and accuracy often elude people (even trained scientists) as they value subjective probability and then apply it as a weighting factor in decision-making. In the paper already cited, Kahneman and Tversky challenge the commonly assumed Bernoullian expectation rule.[19] For that expectation rule to hold, the resulting prospect value should be a linear function of the probability. However, intuitively the function is non-linear, and the characteristic response plot of quantified subjective probabilities (or decision weights as Kahneman and Tversky called them) as a function of objective probability confirms that suspicion, as supported "by a wide range of experiments and field studies".[20]

Figure 4[19, 20] illustrates a number of issues about the difference between objective quantified probabilities and comparable subjective probabilities. People often over-estimate low probability values (i.e., Low Pr $\Rightarrow$ $Pr_{Subj} > Pr_{Obj}$) and under-estimate high probability values (i.e., High Pr $\Rightarrow$ $Pr_{Subj} < Pr_{Obj}$), but commonly $Pr_{Subj} < Pr_{Obj}$ for more than half of the range (i.e., $Pr_{Subj} = Pr_{Obj}$ at $Pr < ½$). More important, the slopes of both curves (Pr_{Subj} and Pr_{Obj}) are always positive, but the slope of Pr_{Subj} is less than or equal to 1 except at or near the endpoints. Therefore, subjective probability is always less sensitive than objective probability, except in the extremes of the curves, where the subjective probability function manifests a dichotomous or bi-modal behavior, crossing what Kahneman and Tversky called a "category-boundary effect [where a] change from impossibility to possibility [at $Pr_{Obj} = 0$] or from possibility to certainty [at $Pr_{Obj} = 1$] has a bigger impact than a comparable change [anywhere else along] the scale."[19]

The implication of Figure 4 is that our choices of subjective risk-weighting factors (i.e., Pr_{Subj}) tend to encourage risk-seeking behavior at low values of Pr_{Obj} and risk averse behavior at higher values of Pr_{Obj} when weighing gains and the opposite when weighing losses — even though actual probability (Pr_{Obj}) is near to certainty (0 and 1) in those cases. Kahneman and Tversky warned that "people are often risk seeking in dealing

---

18 *See* various papers by Tversky and Kahneman, including, but not limited to: Tversky and Kahneman Op. Cit. (1971) and Kahneman and Tversky, *Subjective Probability: A Judgment of Representativeness* (Cognitive Psychology, 1972) pp. 430-454.

19 Kahneman and Tversky. *Choices, Values, and Frames.* American Psychologist (April 1984, Vol. 34, No. 4) at pp. 344-345.

20 Howard Margolis. Dealing with Risk (1996). p. 81.

with improbable gains [as in lotteries] and risk averse in dealing with unlikely losses [as with inflated insurance premiums for a low-probability risk]." [19] However, at both extreme ends of the Pr_{Subj} curve, people make bi-modal choices: Some decision-makers apply their subjective weighting factors (however large or small) to near-certain situations, but others reach points where they decide that some probability is so small as to be negligible or so large as to be a sure thing and set Pr_{Subj} to 0 or 1 accordingly.

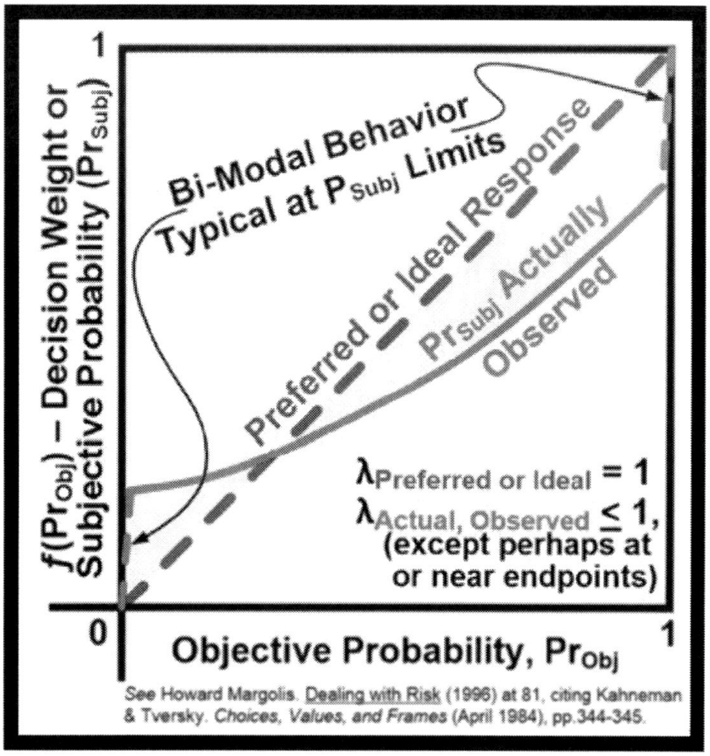

Figure 4. Compare Objective (or "True") Probabilities *vs.* Subjective Probability or Weighting Factor Values

For business decisions of all sorts, it is substantially preferable to apply weighting values that are as close as possible to the objective, true probabilities and to provide separately for contingencies, safety margins and other conservative allowances. Then the decision-maker can be confident that the basis is clear and not unduly risky or risk averse. Moreover, Process Safety is most concerned about events that are highly unlikely but that have utterly catastrophic consequences. Therefore, it is important for analysts to use probability values that are as realistic and accurate as reasonably possible. Otherwise, those risks may be handled in design, construction or operation somewhat like a lottery.

5. How Objective is Subjective Probability? — Two Examples

This section reviews a pair of examples drawn from the literature to support the claim that Pr_{Subj} is nonlinear as illustrated in Figure 4. The first one comes from Lichtenstein

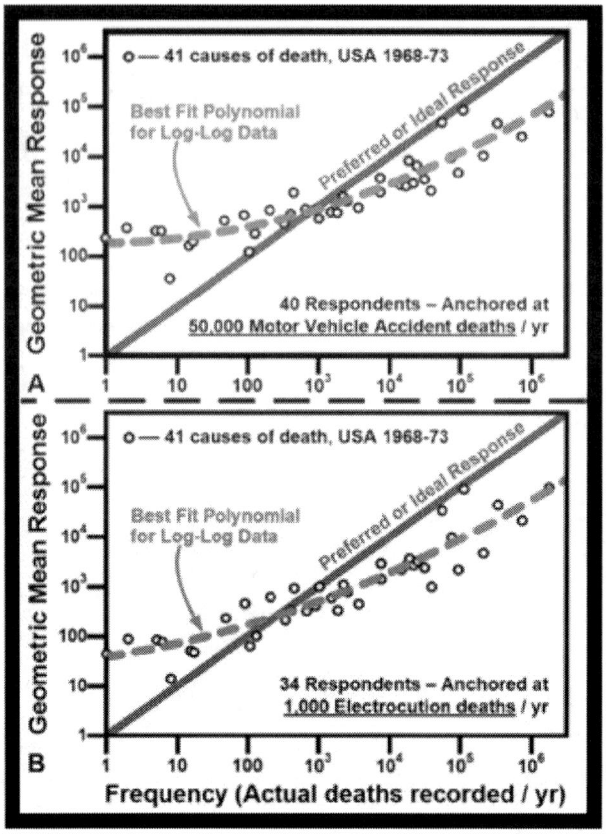

Figure 5. Survey data replotted from data published by Lichtenstein, et al (1978).

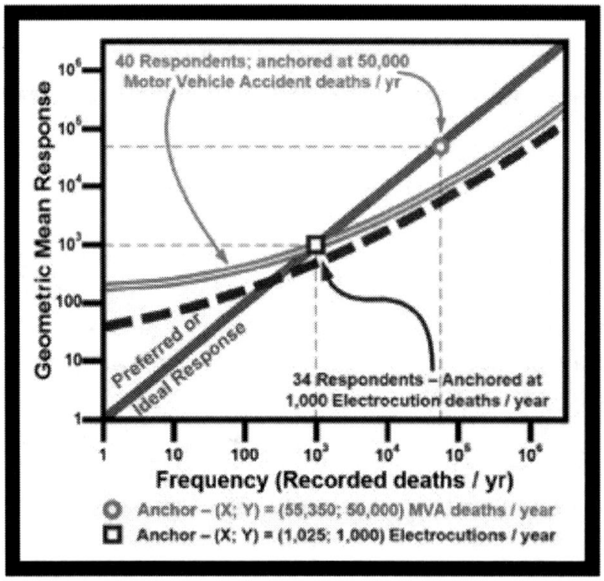

Figure 6. Evidence of Various Natural Heuristics and Biases (e.g., Anchoring, Adjustment and Availability) Found in Lichtenstein's Study of Estimation of Causes of Deaths. Survey data replotted from data in Lichtenstein, et al (1978).

et al, who published results of tests shown in Figures 5 and 6, which were redrawn from data published in that paper. [21] In autumn of 1974, Lichtenstein asked two small groups of Subjects (all university students) to estimate the frequency of death in the USA from 41 causes, a number of which happened to be similar to hazards of concern for Process Safety — e.g., Fireworks, Fire & Flame, Electrocution, Motor Vehicle Accidents, Accidental Falls and other.

Lichtenstein subsequently compared the Respondents' estimates against recorded frequency data from the National Center for Health Statistics, 1968-1973. She plotted her results as log-log frequency data rather than as linear probabilities (as Kahneman and Tversky originally postulated), but that does not matter, because the fundamental points still hold: Despite the uncertainty surrounding small samples [22] (Lichtenstein's two groups were 40 and 34 respondents — and, of course, Process Safety teams are almost always much smaller than that.), the results show the same sort of deviation from a correct or ideal response as what Kahneman and Tversky postulated. Subjects over-valued low frequency causes of death (judging them to be more likely that they are, in fact) and under-valued high frequency causes of death (Figure 5).

The plotted results also show the influence of heuristics or short-cuts that Subjects systematically and commonly bring to such exercises in judgment. One group (40 Respondents) was told that 50,000 Motor Vehicle Accident deaths occur annually (The actual number was 55,350.), and, separately, the other group (34 Respondents) was told that 1,000 Electrocution deaths occur annually (The actual number was 1,025.). Estimating differences attributed to Tversky and Kahneman's "Anchoring" and "Adjustment" heuristics are illustrated in Figure 6. Also, in line with the "Availability" heuristic, Subjects gave better estimates (in some cases very good judgments) for certain causes of death about which they could be expected to have some common knowledge, experience and memory (e.g., Firearms, Motor Vehicle Accidents, and Electrocution) as compared with estimates for other causes about which they had heard but with which they likely had little or no personal experience (e.g., Fireworks, Accidental Falls, Botulism, Smallpox, and various other diseases).

In sum, Lichtenstein's study is a real case that illustrates concepts in Figure 4, as postulated in various papers by Kahneman and Tversky. Their data take the problem

[21] Lichtenstein, Sarah, Paul Slovic, Baruch Fischhoff, Mark Layman and Barbara Coombs. *Judged Frequency of Lethal Events*. J Exp Psychol: Human Learning and Memory (November 1978, Vol. 4, No. 6) pp. 551-578.

[22] Starting in their first paper with the comments previously quoted about the gambler's fallacy, Tversky and Kahneman were always skeptical of research conclusions based on small samples. They found that "[e]ven statisticians were not good intuitive statisticians." "[L]ike us, [they] greatly exaggerated the likelihood that the original result of an experiment would be successfully replicated even with a small sample." That "strong bias [in favor of] small samples [is] part of a larger story: we are prone to exaggerate the consistency and coherence of what we see." "Jumping to conclusions is a safer sport in the world of our imagination than it is in reality." Kahneman, Daniel. Thinking, Fast and Slow (Farrar, Straus and Giroux, Kindle Edition, 2011) pp. 5, 6, 114 and 118.

very close to the low-frequency end point, with death due to smallpox [23]. However, they did not demonstrate the bi-modal behavior described for the end points of Figure 4.

The second example (Figure 7) does demonstrate that bi-modal behavior for a low-probability scenario. This example originated in insurance markets where that subjective behavior has been demonstrated as postulated, and this case is said to be based again both on experiments and on field studies. [24] When potential customers were offered insurance against a middle-probability (40%) loss of $1,000 (which means the Bernoullian expected value, or "EV," of possible loss was $400 = 40% × $1,000), insurance customers generally were willing to pay premiums that one might consider "reasonable" (i.e., roughly normally distributed around a mean of $400). However, when offered insurance for a much higher ($40,000) potential loss but with a relatively small likelihood (1%, much nearer the lower probability end point, but still quite high by Process Safety standards), the customers split into roughly three groups: Some (but substantially fewer) were willing to pay premiums at about $400 (= 1% × $40,000, still the same EV). However, many consumers either were significantly risk averse and paid substantially higher premiums or dropped the insurance altogether (risk-seeking, apparently deciding, consciously or not, to assign 0% chance to the potential loss).

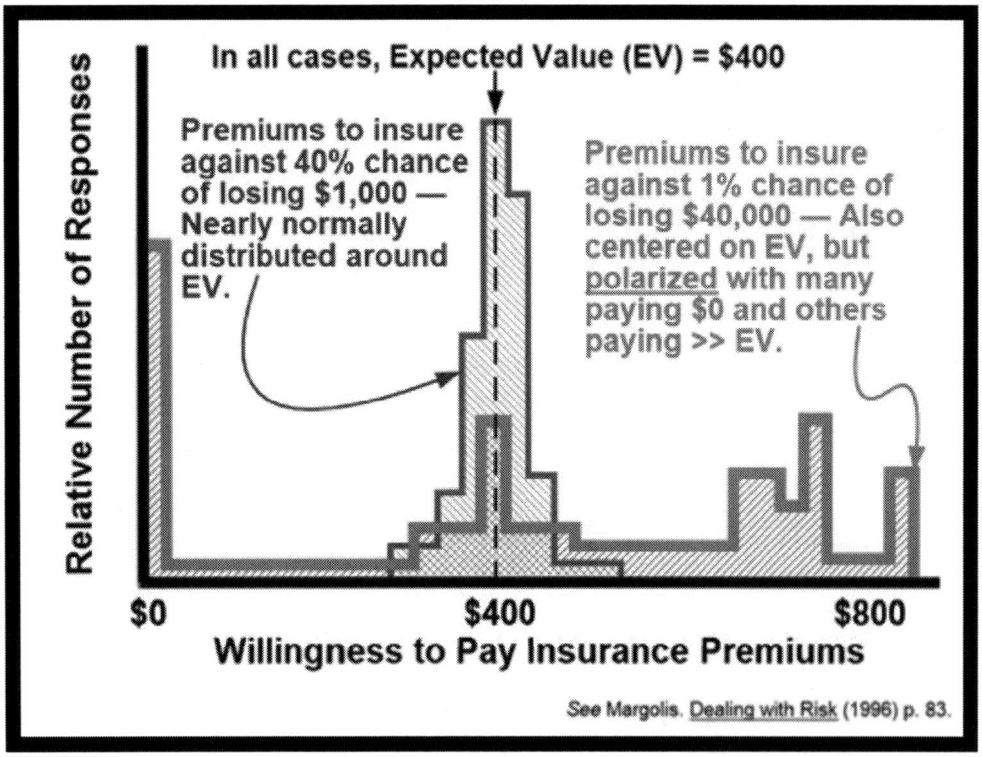

Figure 7. Bi-Modal Subjective Response to Insurance Options

[23] Lichtenstein reported the actual recorded frequency as zero, and there have been no deaths due to smallpox in the USA to date since 1949, according to Wikipedia. However, MVA-anchored Subjects estimated the frequency at 88, and Electrocution-anchored Subjects estimated the frequency at 37.

[24] Howard Margolis. Dealing with Risk (1996). p. 83.

6. Qualitative Descriptions for Subjective Probabilities

The two studies reviewed in Section 5 show that Respondents' judgments regarding subjective probability, a key weighting factor for decisions, can be significantly different from the true probabilities, due to systematic judgment errors that people make naturally. However, people — analysts and decision-makers in processing industries and in everyday life — like subjective probability; subjective probability is good. This is particularly true in Process Safety. Figure 8 illustrates a key reason. Intuitively speaking, qualitative risk assessments typically are easier to complete and cost less. As risk assessment demands move from qualitative studies to semi-quantitative valuations to rigorous quantitative alternatives, the amount of effort, the intellectual stress and demand, the analytical risk (i.e., the risk of technical mistakes — e.g., due to limited data, to judgment errors, or to calculation errors) and the cost all climb. Simultaneously, decision makers and other stakeholders (managers, regulators, investors, the media and the public) find it more and more difficult to understand details and to interpret results of analyses that are more and more sophisticated.

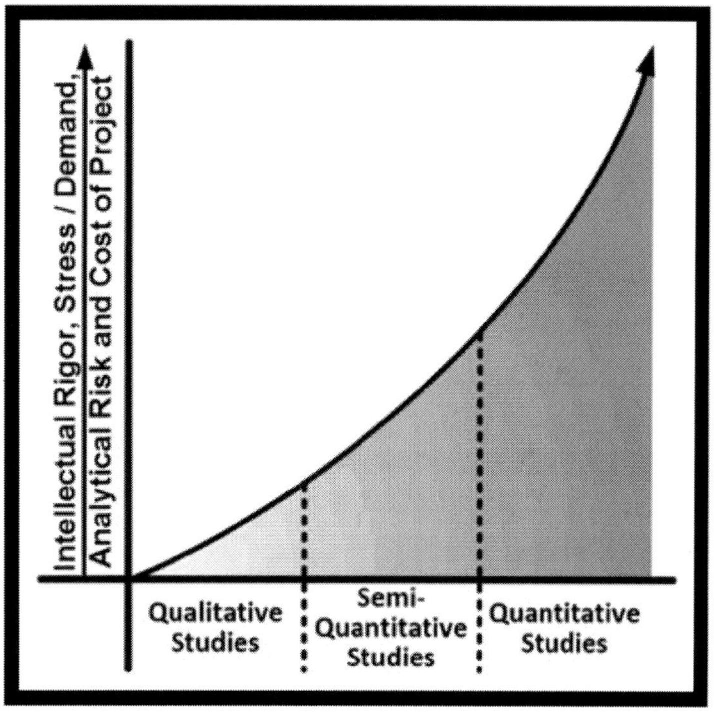

Figure 8. Qualitative Studies Are Easier and Less Risky, with Exceptions

For those sorts of very practical reasons, qualitative analyses (with subjective probabilities) most often are favored. But what does that mean? Imagine a PHA team, tasked with analyzing some process node, trying to identify hazards and risks that components might fail. There may be a "good chance" for any given piece of equipment to fail with respect to a given operating parameter. For another piece of equipment, failure with respect to some parameter may be "a toss up" or "possible" or "improbable" or .... What do those terms mean?

Consider Figure 9, an ordered list of qualitative terms that a team might use to identify subjective probabilities (or decision weighting factors as Kahneman and Tversky put it). There is no claim that the list is comprehensive. At a glance, a reader might disagree with some ordering or may search for missing terms that he / she thinks should be included. Maybe one asks, "Where is '*maybe*'?" Those are valid issues — but, for the moment, ask instead: "What do those terms mean? What quantitative value(s) attach to each of them?" Before reading on or looking at Figures 10 and 11, readers might take a moment to think about that list critically and to give their own meanings or values to some of its terms. Then, the subsequent three sub-sections of this paper will review values that people have assigned to some of these qualitative descriptors. The results will show that imprecision and ambiguity are inherent in such qualitative terminology.

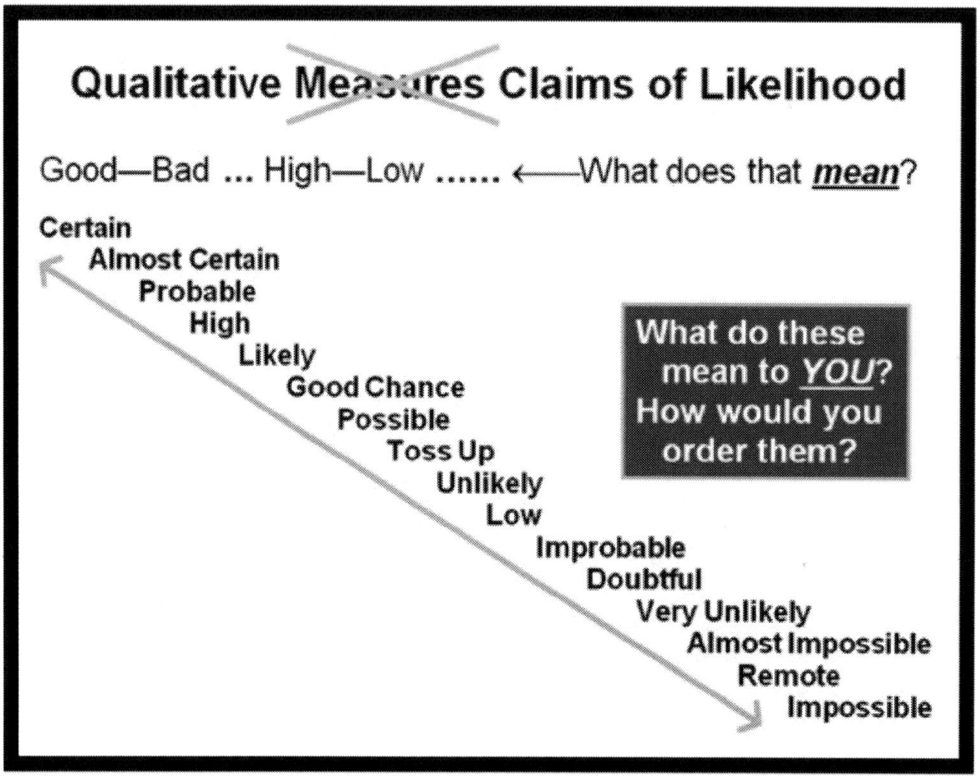

Figure 9. Qualitative Terms Indicating Probability — What Do These Mean?

6.1 What Do Those Terms Mean? — Wallsten's Findings

This is not a new question, and various psychologists and others have considered it. One in particular is Thomas S. Wallsten at the University of Maryland Department of Psychology. A simple graphical summary of his findings is presented in Figure 10. [25]

[25] Figure 10 is redrawn from Morgan, M. Granger. *Use (and Abuse) of Expert Elicitation in Support of Decision Making for Public Policy* (PNAS, Vol. 111, No. 20, 05/20/2014) p. 7177, which used (redrew) Figure 4 in Wallsten, T.S., et al. *Measuring the Vague Meanings of Probability Terms.* J Exp Psychol Gen (December 1986, Vol. 155, No. 4). pp. 348–365.

To a certain degree, the reported ranges of response should seem logical, but ranges associated with the various terms overlap broadly. Note particularly that the range for "Possible" covers almost 100% of the probability domain. How should those terms be interpreted and distinguished when one of them embraces almost all of the theoretical range of probability or when multiple terms overlap for part, but not all, of that range?

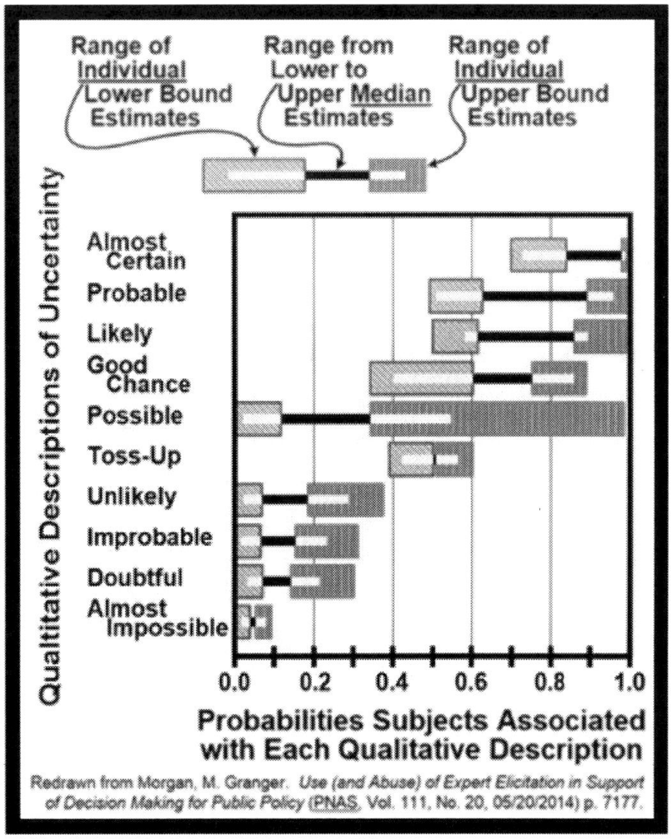

Redrawn from Morgan, M. Granger. *Use (and Abuse) of Expert Elicitation in Support of Decision Making for Public Policy* (PNAS, Vol. 111, No. 20, 05/20/2014) p. 7177.

Figure 10. Range of Numerical Values That 20 Respondents Attached to Qualitative Descriptions of Probability, Absent of Any Specific Context

Wallsten's sample came from 20 graduate students, selected from social science and business programs on the grounds that "they represent a population of people who think seriously about communicating degrees of uncertainty and who generally do so with non-numerical phrases." [26] Therefore, those respondents had more awareness and skill around the issue at hand than one might expect from a random sample of the general public. Wallsten collected his data over three 1½ hour sessions spaced over a week's time — a practice session and two data-collection sessions — and he took pains to assure his respondents followed certain procedures so as to obtain data that passed statistical tests of validity, which they did pass with flying colors.

[26] Wallsten, T.S., et al. *Measuring the Vague Meanings of Probability Terms.* J Exp Psychol Gen (December 1986, Vol. 155, No. 4). p. 352.

6.2 What Do Those Terms Mean? — Boot Camp Comparison

To provide the statistical precision Wallsten's study required takes time and resources that often are not available. However, similar results were obtained quickly and at no cost from students at a CCPS RBPS Boot Camp for comparison with Wallsten's results. This was a simple survey of 24 engineers, operators and other specialists employed at a processing plant in the USA and enrolled in an in-house Boot Camp in July, 2016. No attempt was made to duplicate Wallsten's multiple sessions and other precautions for statistical validity. Instead, Respondents just wrote down what values they would assign to the same list of ten qualitative descriptions of uncertainty.

The Boot Camp survey results are summarized in Figure 11 as ranges of Respondent choices, and the pattern of collected data is quite similar to Wallsten's results. The data were accepted at simple face value, but still they raise important issues relative to valuing qualitative terms for use of subjective probabilities (or weighting factors) in Process Safety projects.

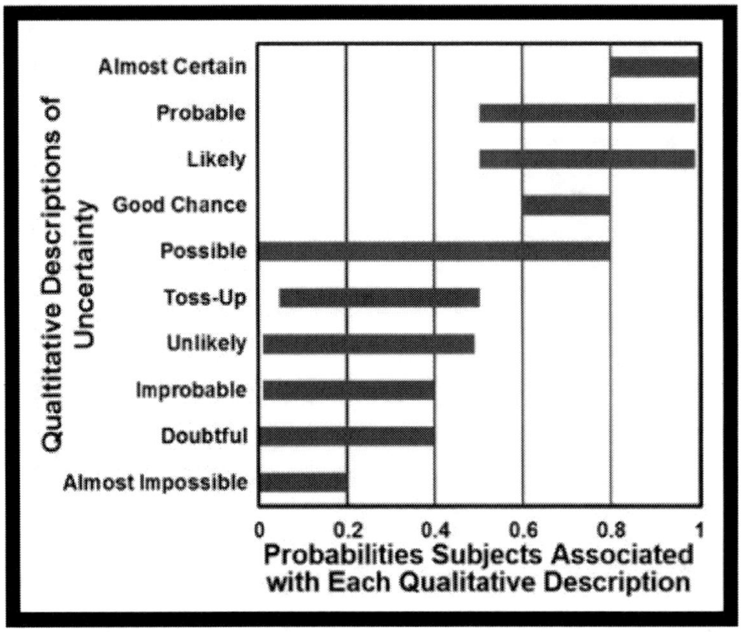

Figure 11. Results of a 2016 Test to Compare with Figure 10.
Bars represent test result ranges with 24 students in a 2016 CCPS RBPS Boot Camp. Subjects were asked what numerical values they would attach to the same 10 qualitative likelihood terms that Wallsten used in Figure 10, with no other context.

Although Respondents were asked to specify choices of minimum, most likely and maximum numerical values for each term, some only gave one value per term; others gave a range or the three values, as requested. Some people responded with linear values (e.g., round percentages of chance); others responded with logarithmic values (10^{-3}, 10^{-5}, and so on). Nevertheless, the pattern of collected data was rather similar to Wallsten's results. Naturally, many terms are logically arrayed from Pr_{Subj} = 0 to 1, but again Respondents assigned numerical values that overlapped each other hugely from term to term, just as they had for Wallsten's study. Again, the most extreme range was for "Possible," which spanned almost the entire domain of probability, a range similar to but a bit narrower than Wallsten's. Then, some results were quite different from Wallsten's. Values assigned for "Toss Up" ranged from 0.05 to 0.5, which is substantially lower than the range assigned by Wallsten's respondents (about 0.39 to 0.6 in Figure 10) and quite different from what a reader might expect.

Overall, the Boot Camp message is quite clear: Even when survey Subjects are well-educated and accustomed to dealing with qualitative descriptions of probability, interpretation shows that subjective probabilities are vague and ambiguous. It is chancy both to quantify terms one-by-one and to compare terms, one to another.

6.3 What Do Those Terms Mean? — An Answer from "Argo"

For one final reflection of vague qualitative probability descriptions, consider the valuation Mark Lijek gave to a "good chance" in the quote from Argo that introduced this paper — *"What's the number value of 'good?' 30% chance of being publicly executed?"* First, this illustrates a case of framing information and decisions. "Good" connotes something desirable, such as a successful escape for Lijek and the other diplomats who hid for weeks in the Canadian ambassador's residence, but Lijek's dialog conflates a "good chance" with public execution, which does not sound so "good." The 30% chance of execution in the public square leaves 70% chance for all other potential outcomes, including both successful escape and simple, unexplained permanent disappearance. Does that feel "good?" Second, that dialog was actually included in the film, but at least one version of the screenplay is different, with Lijek asking, *"What's the number value of 'good?' 30 percent chance of success? 80 percent?"* [27] That 30%-80% range for a "good chance" is significantly broader than that from the Boot Camp Respondents (60%-80% in Figure 11) and is somewhat similar to, but nevertheless different from, results of Wallsten's respondents (about 0.34 to 0.89 in Figure 10).

Wallsten would be among the first to accept the ambiguity both of qualitative descriptions in general and individual differences among them: "The meanings of non-numerical probability phrases, even to an individual, are almost assuredly not fixed over contexts. Thus, the precise numerical aspects of the [data shown in Figures 10 and 11] should not be taken too seriously." [28] However, for a similarly broad range of "good chance" to emerge from a major motion picture simply underscores the common ambiguity of efforts to capture subjective probabilities with qualitative terms.

In sum, qualitative descriptions of likelihood are imprecise, vague and ambiguous. Comparisons show that individuals judge probability terms quite differently, and that includes people whose training, experience and work teach them the hazards of

---

[27] Dialog from the award-winning film Argo (Scene 183, at p. 70). Directed by Ben Affleck; written by Chris Terrio; released in the USA October 12, 2012. Quotations from Argo are courtesy of Warner Bros. Entertainment, Inc., with sincere thanks.

[28] Wallsten, T.S., et al. Ibid. p. 362.

ambiguity and the importance of probability, precision and risk. [29] Nevertheless, everyone uses these qualitative values to weigh decisions — everyone, every day.

Most people, including technical experts, generally prefer to communicate uncertainties with non-numerical terms (as in Figure 9) rather than with numerical probabilities. [30] Often this reflects limits in their confidence in the precision, clarity and consistency of the information they are conveying, as well as other contextual factors, such as —

- Framing;
- Source, credibility, and type of data (i.e., anecdotal / experiential, model output, direct / indirect observations, theoretical analysis and so forth);
- Volume of information (scarce or abundant);
- Any concern that the information may be instable;
- Competing explanations for some phenomenon;
- Degree of consensus within a team, community, or technical specialty; and
- Other state of knowledge or quality of data factors.

Also, it is frequently suggested that most people feel they better understand words than numbers. This is very important. "We pay more attention to the content of messages than to information about their reliability, and as a result end up with a view of the world around us that is simpler and more coherent than the data justify. Jumping to conclusions is a safer sport in the world of our imagination than it is in reality." … "[H]umans constantly fool [them]selves by constructing flimsy accounts of the past and believing they are true." [31]

7. Some Concluding Thoughts

The literature cited here and the results of the Boot Camp survey together illustrate what people intuitively know: Subjective estimates of probabilities and qualitative descriptions of probability are substantially subject to common, natural individual biases and are ambiguous, imprecise and vague.

The discussion has shown two sets of problems that arise from subjective probabilities (or decision weighting factors). The first set of problems is illustrated in Figure 4 —

- People (including skilled analysts) naturally tend to over-weight low probabilities.
    - If potential gains are at risk, people may assign a higher subjective probability than is justified for a condition, event or consequence and then gamble with risk-seeking behavior as the probability approaches zero; or

[29] In various tests, Kahneman and Tversky found that even trained statisticians make simple errors of intuition in valuing probabilities and in making other judgments.

[30] Wallsten T.S., et al. Ibid. p. 348.

[31] Kahneman, Daniel. Thinking, Fast and Slow (Farrar, Straus and Giroux, Kindle Edition, 2011) pp. 118 and 199. The last sentence quoted paraphrases a comment by Taleb in The Black Swan (2007).

- If risks of interest concern potential losses (Process Safety), people might be more cautious than is necessary or reasonable (although "better safe than sorry" is a recommended attitude for Process Safety).

- Also, people naturally tend to under-weight high probabilities.
    - If potential gains are at risk, people may assign a lower subjective probability than is justified for a condition, event or consequence and then pursue an excessively cautious policy, leaving value on the table due to risk-averse behavior as the probability approaches certainty of 1.0.
    - If risks of interest concern potential losses (Process Safety), people may be inadequately cautious, even as probabilities approach 100% and even though they might not really intend to gamble.

Then, in the alternative illustrated by Figures 4 and 7 and of particular concern in the context of Process Safety, it may be that — because probabilities are extremely small (maybe 10^{-4} or 10^{-5}) and because actual experience with catastrophic events is so rare — some analysts or project team members might decide to ignore some hazard and to treat the probability of an event as zero. Examples might include —

- Poor driving or work practices;
- Inadequate SPCC designs;
- Rushed procedures;
- Failure to recognize and to report a near miss;
- Risky short-cuts and normalization of risk; and
- Any practice that places economic, other business or personal concerns at or above the importance of Safety, Health & Environmental concerns.

Those are situations when someone might hear some cliché: "It will never happen here;" or "That's the way it's gonna be!" Such comments are always red flags.

A second set of problems arises from a host of factors that one might summarize as limitations to our data and resources, combined with natural tendencies and ordinary communication. As a result, it is very common to use qualitative descriptions for subjective probabilities in analyzing hazards and risks — but, whether recognized or not, those descriptions are typically inadequate, vague, imprecise and ambiguous. Further, they are not necessarily reproducible, even within the vague and ambiguous limits of qualitative analyses, as shown by comparison of results from Wallsten (Figure 10) vs. the Boot Camp (Figure 11).

These findings apply across the spectrum of business and to Risk-Based Process Safety in particular. However, analysis of uncertainty is a recent addition to the intellectual toolbox. Pascal completed the first quantitative risk analysis less than 300 years ago, after humans had built natural deterministic foundations for thought ... and potential error ... over millennia. Now, everywhere, stakeholders of all sorts consider and weigh their decisions and continue to use subjective, qualitative analyses routinely. Thus, we are stuck with imperfect subjective probabilities. What are we to do?

Solutions must start with awareness and build from that into policies, training, culture and communication with all stakeholders. Paraphrasing Tversky and Kahneman in their first paper: Even if biases and imperfections of intuition, of subjective thinking and of qualitative language cannot be unlearned and inevitably will affect communications and limit clarity and bias decisions, Process Safety stakeholders "can learn to recognize [that those elements exist and are hazardous] and to take the necessary precautions."[32]

Organizations as a whole and team leaders in particular should arm the workforce and other stakeholders to take those precautions — with orientation, training and preparation as assignments begin so that individuals and teams are alert to this human error hazard. Ultimately, it would be best to build appreciation of limitations of subjective probability and of qualitative terminology into organizational culture. Goals could include developing an appreciation that we live and work in a probabilistic, risk-based world — not a deterministic one, whether we like it or not. Personnel should be sensitized to the dangers of natural decision-weighting biases and to limitations of ambiguous, vague terminology and of deterministic thinking. It will be beneficial for the workforce as a whole to appreciate issues around all types of probability, but especially those that arise from subjective probability, decision weighting factors, and qualitative terminology.

Process Safety training should include specific examination of what "Probability" and "Risk" mean, of the psychological character of "Risk," and of the multi-dimensionality of uncertainty (as to likelihood, as to outcomes, as to psychological elements, and as to any other issues). Workforce and other team players need good, defensible subjective probability definitions that are distinct down to exceptionally small values (10^{-6} or less). Organizations should settle agreed definitions of qualitative terminology and should communicate those definitions to employees and other stakeholders as a matter of policy.

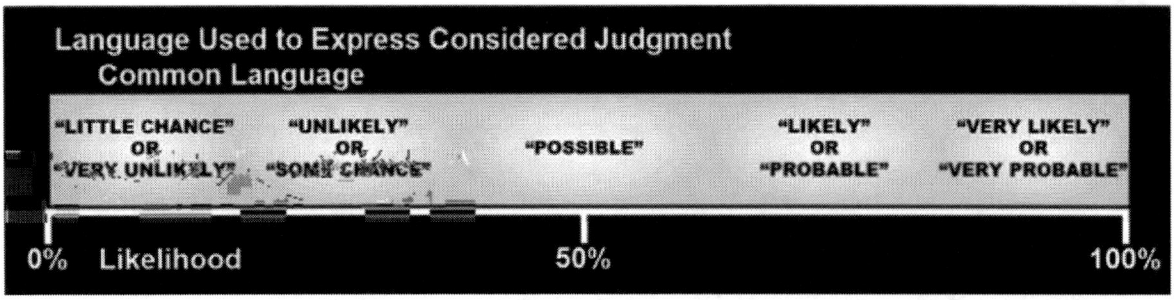

Figure 12. Mapping of Qualitative Probability Terms into Quantitative Probabilities. Used by the U.S. National Assessment Synthesis Team in two reports.[33]

---

[32] Tversky and Kahneman. Op Cit. (1971). p. 110. "Even if the bias [of natural but misleading statistical intuitions] cannot be unlearned, students can learn to recognize its existence and take the necessary precautions."

[33] Morgan, M. Granger, et al. CCSP 5.2. Best Practice Approaches for Characterizing, Communicating, and Incorporating Scientific Uncertainty in Decision Making. *A Report by the Climate Change Science Program and the Subcommittee on Global Change Research* (National Oceanic and Atmospheric Administration, Washington, 2009). pp. 27, 28, 72.

Organizations may find it useful to create visual tools to help promote common definitions for subjective probability estimates. Examples of such efforts can be found in the literature. As a case in point, Figures 12 and 13 and Table 3 are examples of attempts USA climate-change researchers made to establish an organizational Framework for Associating Common Language with Subjective Probability.

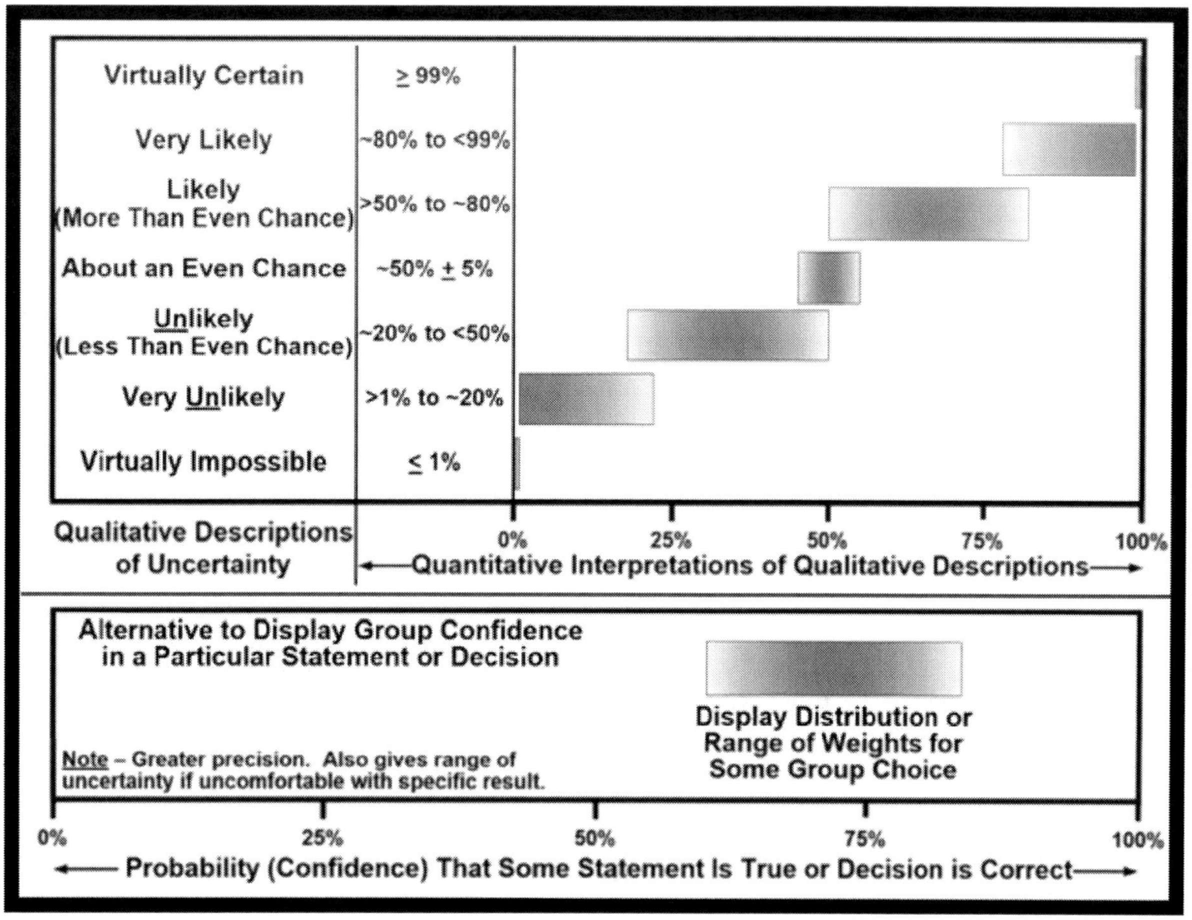

Figure 13. One Alternative for Associating Common Language with Subjective Probability. Redrawn from recommended Framework in Figures 9.1 and 9.2, p. 72, of Morgan, et al (2009). [32]

Those figures and tables may be useful as pedagogical demonstrations, but they are not adequate for Process Safety, because they do not extend to the very low probabilities that arise in the Process Safety context, where the likelihood of potential catastrophes may be on the order of 10^{-3} to 10^{-6}. A Risk Matrix, which many organizations already use, can be a better tool. [34] A matrix can be designed to accommodate as low a probability as desired; several sets of synonymous terms can be

[34] But Thomas et al describe flaws inherent in risk matrices and recommend against using them, or at least that they be used with caution. *See* Thomas, Philip, Reidar B. Bratvold and J. Eric Bickel. *The Risk of Using Risk Matrices* (SPE 166269-MS, 2013).

used, to help people understand distinctions among values (e.g., 10^{-3} vs. 10^{-6}) and to demonstrate number ranges assigned to various terms ("unusual" or "rare" or whatever); and the tool itself combines considerations of uncertainties both as to events and as to consequences.

Table 3-A. Mapping of Probability Words into Quantitative Subjective Probability Judgments, Used by WG I and II of the Intergovernmental Panel on Climate Change Third Assessment (IPCC 2001).

Word	Probability Range
Virtually certain	>0.99
Very likely	0.9-0.99
Likely	0.66-0.9
Medium likelihood	0.33-0.66
Unlikely	0.1-0.33
Very unlikely	0.01-0.1
Exceptionally unlikely	<0.01

Note: In Morgan, M. Granger, et al. Op Cit. (2009), this is Table 2.1 on p. 27.

Table 3-B. Mapping of Probability Words into Quantitative Subjective Judgments of Confidence as Used in the U.N. Intergovernmental Panel on Climate Change Fourth Assessment (IPCC, 2005, 2007).

Word	Probability Range
Very High Confidence	At least 9 chances out of 10
High Confidence	About 8 chances out of 10
Medium Confidence	About 5 chances out of 10
Low Confidence	About 2 chances out of 10
Very Low Confidence	Less than 1 chance out of 10

Note: In Morgan, M. Granger, et al. Op Cit. (2009), this is Table 2.2 on p. 28.

Table 3 (A and B). Mapping of Probability Words into Quantitative Subjective Probability Ranges as Used in U.N. IPCC Reports [35]

Guidance notes for Lead Authors of the IPCC Fourth Assessment Report on Addressing Uncertainties (IPCC, 2005) include both of these tables.

As an alternative, one example of a risk matrix is attached at the end of this paper. Many examples of such matrices are available in the literature, but this one is included

[35] Morgan, M. Granger, et al. Op Cit. (2009). pp. 27, 28.

here purely as a pedagogical example to show how several different but synonymous descriptions of probability might be combined to help users understand the very low probabilities of concern in Process Safety and the qualitative terms that might be used to define them. Also, different categories of outcome can be presented to help users grapple with uncertainty as to severity and with combinations of consequences that can result (i.e., combinations of losses for safety and health, environmental protection, product or service quality, business reputation, economic loss or other issues — But caution: Safety, health and environment issues are always dominant concerns. Never infer any direct or indirect translation among the consequence categories.).

In the context of the discussion of uncertainty here, one shortfall of a risk matrix is that it does not explicitly include each possible outcome for each consequence category in Bernoulli's expected value [Equation (1)]. It lists a spectrum of possible outcomes for each given category, but does not weight and sum each one into the calculated result for that category. Therefore, a risk matrix underplays the role of uncertainty in attempts to foresee the impact of an event. Thomas et al [34] list other risk matrix flaws and even suggest that it may be better to do nothing. However, it can be helpful for analysts and managers dispassionately considering what might happen in a risky world.

In conclusion, awareness and understanding of human error hazards that lurk within subjective probability and qualitative assessment terminology should help Process Safety practitioners and others to cope better with the uncertainties that surround and characterize efforts to weigh options and make decisions, as one must in everyday life and particularly in Process Safety. It is an important, value-added step forward to add risk-based processes to toolkits for Process Safety and other business needs, but it is also important to recognize the boundaries that heuristics, cognitive biases and common qualitative terminology inevitably and naturally impose on dealing subjectively with uncertainty. However vague or ambiguous the practical language of uncertainty may be, we all follow risk-based practices — consciously or unconsciously — every time we make a decision. It is better to do so conscious of limits and to strive as best possible to use meaningful language and at the same time to make accurate and consistent judgments. As always, we need to maintain sharp vigilance and a continuing sense of vulnerability and to remember that Black Swans will occur no matter what the numbers say, no matter how our gut intuitions may beguile us.

All our knowledge, sense, and sight lie in deepest darkness shrouded .... [36]

8. Acknowledgements

Sincere thanks to John Murphy and John Herber for their suggestion of and encouragement for writing this paper. Also, sincere thanks to John Murphy, John Herber and Dr. Rand B. Krumland for their insightful reviews and helpful comments. The quotations from Argo are courtesy of Warner Bros. Entertainment, Inc. Sincere thanks to Warner Bros. Entertainment for that courtesy.

---

[36] Clausnitzer, Tobias, et al. Hymn 440 (a 17th century hymn), Episcopal Hymnal 1982.

9. References

[1] Argo. Motion picture directed by Ben Affleck; written by Chris Terrio; released in the USA 12 October 2012. Two screenplays available — One retrieved 11 Jan 2017 from http://screenplayexplorer.com/wp-content/scripts/argo.pdf, and the other retrieved from http://www.springfieldspringfield.co.uk/movie_script.php?movie=argo on 16 Jan 2017. The quotations referenced here from Argo are courtesy of Warner Bros. Entertainment, Inc., with sincere thanks.

[2] Covello, Vincent. The List of 19 Factors Characterizing Risk given here as Table 1, is from Margolis, Howard, Dealing with Risk (Univ. of Chicago Press, 1996) at p. 28. Margolis quotes Covello in *Risk Comparisons and Risk Communication* in Kasperson and Stallen, Communicating Risks to the Public (Kluwer, 1991) at p. 112. The list is drawn mainly from work of Slovic, Fischhoff and others. *See also* a nearly-identical list in "Appendix C" of Risk Communication, Risk Statistics, and Risk Comparisons: A Manual for Plant Managers (1988), by Covello et al. for the Chemical Manufacturers' Association (CMA, now the ACC or American Chemistry Council). The CMA version does not include factor 18 (personal stake) and was retrieved 15 February 2017 from http://www.psandman.com/articles/cma-appc.htm.

[3] Klarreich, Erica. *Toss Out the Toss-Up: Bias in heads-or-tails.* Science: Science News Online (2/28/2004; Vol. 165, No. 9). Retrieved 4/4/2016 from https://www.sciencenews.org/article/toss-out-toss-bias-heads-or-tails?mode=magazine&context=443.

[4] Health & Safety Executive (the UK HSE). Reducing Risks, Protecting People: HSE's Decision-Making Process ("R2P2") (Her Majesty's Stationery Office, 2001). Retrieved 24 February 2017 from http://www.hse.gov.uk/risk/theory/r2p2.pdf.

[5] Kahneman, Daniel. Thinking, Fast and Slow (Farrar, Straus and Giroux, Kindle Edition, 2011).

[6] Kahneman, Daniel, and Amos Tversky. *Choices, Values, and Frames.* American Psychologist (Vol. 34, No. 4, April 1984). pp. 344-345. Retrieved 25 January 2017 from http://web.missouri.edu/~segerti/capstone/choicesvalues.pdf.

[7] Lichtenstein, Sarah, Paul Slovic, Baruch Fischhoff, Mark Layman and Barbara Coombs. *Judged Frequency of Lethal Events.* J Exp Psychol: Human Learning and Memory (November 1978, Vol. 4, No. 6) pp. 551-578. Retrieved 8 February 2017 from https://www.researchgate.net/publication/232557428_Judged_Frequency_of_Lethal_Events

[8] Margolis, Howard. Dealing with Risk (1996). p. 81-85 and *passim.*

[9] Morgan, M. Granger, et al. CCSP 5.2. Best Practice Approaches for Characterizing, Communicating, and Incorporating Scientific Uncertainty in Decision Making. *A Report by the Climate Change Science Program and the Subcommittee on Global Change Research* (National Oceanic and Atmospheric Administration, Washington, 2009). Retrieved 3 February 2017 from https://downloads.globalchange.gov/sap/sap5-2/sap5-2-final-report-all.pdf.

[10] Morgan, M. Granger. *Use (and Abuse) of Expert Elicitation in Support of Decision Making for Public Policy* (PNAS, Vol. 111, No. 20, 20 May 2014). p. 7177. Retrieved 31 January 2017 from http://www.pnas.org/content/111/20/7176.full.pdf.

[11] Murphy, John F. *The Black Swan: LOPA and Inherent Safety Cannot Prevent All Rare and Catastrophic Incidents* (Process Safety Progress, 2011) pp. 1-2.

[12] Murphy, John F. and Jim Conner. *Beware of the Black Swan: The Limitations of Risk Analysis for Predicting the Extreme Impact of Rare Process Safety Incidents* (Process Safety Progress, December 2012, Vol.31, No.4) pp. 330-333.

[13] Murphy, John F. and Jim Conner. *Black Swans, White Swans, and 50 Shades of Grey: Remembering the Lessons Learned from Catastrophic Process Safety Incidents* (Process Safety Progress, 2014) pp. 1-5.

[14] Murphy, John F. *Surviving the Black Swan, Strategies for Process Safety Specialists, and Companies to Survive Unpredicted Catastrophic Events* (Process Safety Progress, March 2016, Vol.35, No.1) pp. 13-17.

[15] Nolan, Dennis P. Safety and Security Review for the Process Industries (William Andrew, 2008).

[16] Slovic, Paul, and Elke Weber. *Perception of Risk Posed by Extreme Events.* Paper discussed at the conference "Risk Management Strategies in an Uncertain World," April 2002. Retrieved 2/15/2017 from https://www.ldeo.columbia.edu/chrr/documents/meetings/roundtable/white_papers/slovic_wp.pdf.

[17] Slovic, Paul, Baruch Fischhoff, and Sarah Lichtenstein. *Why Study Risk Perception?* Risk Analysis (June 1982) pp. 83-93. Retrieved from http://sds.hss.cmu.edu/risk/articles/WhyStudyRiskPercep.pdf.

[18] Taleb, Nassim Nicholas. The Black Swan: The Impact of the Highly Improbable (Random House, 2007).

[19] Thomas, Philip, Reidar B. Bratvold and J. Eric Bickel. *The Risk of Using Risk Matrices* (SPE 166269-MS, 2013). Retrieved 8/6/2014 from http://www.spe.org/jpt/pdf/tech_papers/14-08/SPE166269.pdf.

[20] Tversky, Amos, and Daniel Kahneman. *Belief in the Law of Small Numbers,* Psychological Bulletin (1971, Vol. 75, No. 2). pp. 105-110. Retrieved 15 February 2017 from http://citeseerx.ist.psu.edu/viewdoc/download;jsessionid=2066528F5C42D74A39F502C618704297?doi=10.1.1.399.1696&rep=rep1&type=pdf.

[21] Wallsten, Thomas S., David V. Budescu, Amnon Rapoport, Rami Zwick, Barbara Forsyth. *Measuring the Vague Meanings of Probability Terms.* J Exp Psychol Gen (December 1986, Vol. 155, No. 4). pp. 348–365. Retrieved 31 January 2017 from http://psycnet.apa.org/journals/xge/115/4/348.html?uid=1987-09386-001.

[22] Wikipedia. *List of Cognitive Biases* (Updated 25 December 2016). Retrieved 26 December 2016 from https://en.wikipedia.org/wiki/List_of_cognitive_biases.

Example Combined, Multi-Function Risk Matrix and Memory Jogger

Frequency or Likelihood or Probability — Alternative Expressions of Likelihood (Increasing Likelihood ↑)

Alternative Expressions of Likelihood				Likelihood	1	2	3	4	5
1 > Probability > 0.1	Daily to Once in 1 yr to Once in 10 yrs	Continuous to often to occasional occurrence	Unpredictable, but this will happen	5	Fix It ALARP III	Intolerable IV	Intolerable IV	Intolerable IV	Intolerable IV
0.1 > Probability > 0.001	Once in 10 yrs to Once in 1,000 yrs	Occasional to probable to improbable occurrence	It can happen to this unit or company, but unusual	4	Mitigate Impact ALARP II	Fix It ALARP III	Fix It ALARP III	Intolerable IV	Intolerable IV
0.001 > Probability > 0.0001	Once in 1,000 yrs to Once in 10,000 yrs	Improbable to rare occurrence	Occurs within this industry, so we can conceive of it	3	Least Risky I	Mitigate Impact ALARP II	Mitigate Impact ALARP II	Fix It ALARP III	Intolerable IV
0.0001 > Probability > 0.000001	Once in 10,000 yrs to Once in a million yrs	Rare to remote to incredible occurrence	Seemingly unlikely or novel threat, except for common facilities	2	Least Risky I	Least Risky I	Mitigate Impact ALARP II	Fix It ALARP III	Intolerable IV
0.000001 > Probability	Less than Once in a million yrs	Negligible	Disastrously fateful or unforeseeably freak event	1	Least Risky I	Least Risky I	Mitigate Impact ALARP II	Mitigate Impact ALARP II	Fix It ALARP III

—— Increasingly Severe Consequences ——→

Consequences for Various Sources of Risk and Value (All Consequences relate back to Workplacce Errors & Poor Practices.)

Source	1	2	3	4	5
Safety and Health of Public and / or Employees →	Near miss, but no injury or health issues (public or employee)	First aid cases or Slight health problems with no lost time	Minor injury(ies) or Health problem(s) — May have some lost time	Serious injury(ies) or Major health problem(s) with lost time	≥ 1 Fatality or Total disability(ies) and Severe health problem(s)
Environmental Protection or Damage →	Near Miss (Negigible, Internal, Confined to unit, No off-site pollution)	Minor Incident (Public complaints about Odor, Drainage, Noise, etc)	Significant local–regionall ncident (Spill or release, Exceedence, Recording error, etc)	Major and / or Repeated Events with wide notice (Major spill or release, Violation, etc)	Massive Damage with international notice (Major spill–release, CERCLA Site, Disaster, etc)
Quality of Products and / or Services →	Some Product off-spec or Service fails standards	Customer complaints, verbal or written	Significant complaints and threats to business relationships	Business lost as important or major customers cancel orders	Lose substantial market share; Impacts deals, leases and contracts
Reputation →	Embarassing, but Slight affect	Noteworthy impact on local or community reputation	Noteworthy impact on state, provincial or regional reputation	Regional or national reputation damaged and some business lost	International reputation & business loss; Also legal, regulatory impacts
Economic Loss → (Includes — Natural resource damages; Public, private or company property damages; Inventory losses, production or service interruption; Fines, penalties and restitution; Impacts of regulatory changes, etc.)	Slight damage–loss of < $10,000	Noticable damage–loss of $10,000 to $100,000	Significant damage–loss of $100M to $1MM	Major damage–loss of $1MM to $10MM	Disastrous damage–loss of > $10MM
			Regulation & reputation losses limit future		

←—— Consequence Categories ——→

Frequency or Likelihood Descriptions and Categories

Description (Return Period)	Events / Year Value	Events / Year Log_{10}	Prob Class
Daily ("Continuous")	365	2.6 (or > 0)	
Weekly ("Continuous")	52	1.7 (or > 0)	
Monthly ("Continuous")	12	1.1 (or > 0)	
Quarterly ("Continuous")	4	0.6 (or > 0)	5
Annually	1.0E+00	0	
3 years	3.3E-01	–0.5	
10 years	1.0E-01	–1	
30 years	3.3E-02	–1.5	
100 years	1.0E-02	–2	4
300 years	3.3E-03	–2.5	
1,000 years	1.0E-03	–3	
3,000 years	3.3E-04	–3.5	3
10,000 years	1.0E-04	–4	
30,000 years	3.3E-05	–4.5	
100,000 years	1.0E-05	–5	2
300,000 years	3.3E-06	–5.5	
1,000,000 years	1.0E-06	–6	
> 1,000,000 years	<1.0E-06	< –6	1

Decision Authority for Risk Categories

Color	Risk Category, Risk Rank and Description		Decision Authority
IV	IV	Severe / High	Senior Management
III	III	Serious	Plant / Unit Management
II	III	Medium	Program / Section Manager
I	I	Least Risky	Others As Delegated

Recommendations for Applying This Tool

1. **Obey all Laws, Rules and Regulations.**
   - **Honor legal, policy and ethical obligations.**
   - **Mitigating Quality, Reputation & Economic losses is at local management discretion.**
2. **Never infer any direct or indirect translation among various Consequence categories.**
3. **Use this tool to help both to understand and to communicate risks and to guide decision processes for risky situations.**
4. **Control your risks. Use tools & processes objectively, deliberately and unemotionally.**
5. **Know the limits of your risk authority —**
   - **Get authorization or concurrence when you reach risk limits.**

Caution — Only a pedagogical tool. Not intended for actual project use.

A Look into Cumulative Risk Assessment: Past, Present and Future

Syeda Zohra Halim
Mary Kay O'Connor Process Safety Center
Artie McFerrin Department of Chemical Engineering
Texas A&M University
College Station, Texas 77843-3122, USA
zohrahalim@tamu.edu

Yogesh Koirala and M. Sam Mannan
Mary Kay O'Connor Process Safety Center
Artie McFerrin Department of Chemical Engineering
Texas A&M University
College Station, Texas 77843-3122, USA
koirala@tamu.edu and mannan@tamu.edu

Prepared for Presentation at
American Institute of Chemical Engineers
2017Spring Meeting and 13thGlobal Congress on Process Safety
San Antonio, TX
March 26 – 29, 2017

A Look into Cumulative Risk Assessment: Past, Present and Future

Syeda Zohra Halim
Mary Kay O'Connor Process Safety Center
Artie McFerrin Department of Chemical Engineering
Texas A&M University
College Station, Texas 77843-3122, USA
zohrahalim@tamu.edu

Yogesh Koirala and M. Sam Mannan
Mary Kay O'Connor Process Safety Center
Artie McFerrin Department of Chemical Engineering
Texas A&M University, College Station

Keywords: Cumulative Risk Assessment, Cause Effect, Systems Approach, Complex Systems

Abstract

The effort towards carrying out quantitative risk analysis in today's sizeable facilities has mostly been focused on considering technical safety systems with little emphasis given to human errors. The consequences from the cumulative effect of a number of deviations that arise from equipment, procedure or a person not doing an assigned work properly can become catastrophic. In order to assess the total risk, we need to consider technical, operational, human and organizational factors together. This paper provides a review of past attempts and progresses made at bringing all of these factors together into one risk assessment framework based on leading articles published in various journals. A comparative study has been carried out in an attempt to distinguish between the approaches based on whether they are qualitative, semi- quantitative or quantitative and if they address the complexity and size of the systems, temporal aspects and dependencies of components and uncertainties of parameter estimation. The paper aims at providing a direction for future research towards better assessment of cumulative risk in today's chemical plants.

1 Introduction

It is well accepted that a good understanding of risk is essential for day-to-day safer operations of a chemical process facility. Safer operations with minimal risk ensure continued production, profitability and survival in a competitive market without unforeseen mishaps [1]. The risk arising from a single deviation of equipment, person or operation from a defined safe state of a system is usually well understood and maintained. However, large numbers of deviations can exist throughout a plant and it is also essential to understand their interaction and their holistic effect on the overall risk of the facility. This 'cumulative risk' may result from a variety of origins such as reduced equipment reliability due to maintenance backlog, aging assets, supply delays, shortage of staff, changes in management or regulations *etc.* [2] Although such deviations may not impose an immediate threat to the system, they can exist without being addressed and culminate to a risk beyond the acceptable range. The disasters at Deepwater Horizon [3] and Texas City Refinery [4] teach how unaddressed deviations ultimately led to a catastrophic incident that caused loss of life and large financial losses. Thus, it is essential to have an understanding of the holistic effect of risk by conducting a cumulative risk assessment.

Cumulative risk assessment can be a challenging task because deviations can arise from all quarters of the facility and can be technical, operational, human or organizational in nature. Environmental or external factors can add to the challenge as well. Quantification of risk is essential to obtain a precise understanding of such risk: they provide a magnitude of the risk showing us whether it is within an acceptable range and what causes led to such high risk values. This further enables work prioritization through pinpoint determination of where, when and what mitigative measures are required. Traditionally, risk assessment methods had focused only on the technical aspects of a system mainly because of the ease of a quantitative representation of their failure [1]. However, over the last few decades, with realization sprouting mainly from numerous catastrophic incidents, there has been a growing trend to incorporate human and organizational factors (HOF) into the traditional risk assessment methods. Even so, a large gap exists in merging the more abstract HOF with the more quantitative technical factors of a system [1]. A study reported by Skogdalen *et al.* showed that of the 15 offshore installations considered, the majority conducted QRAs with "HOF Explained" and considered HOF only to some extent and not separately [5]. None of them fell in the category "HOF explained, models adjusted and included in the overall risk management" where HOF were given as much importance as the technical factors and formed an integral part of the overall risk management system. The biggest challenge in getting a holistic look into risk is consideration of all factors together: technical, operational, human and organizational.

Complexity of current day facilities adds to the challenge of assessing cumulative risk. Deviations arising from the large number of equipments in operation, standby, under repair, *etc.* and subgroups such as operators, maintenance workers or contractors working under decentralized management systems will have dependencies and linear and non linear interactions between them. This will make cumulative risk assessment even more complicated. Also, a dynamic system's response to an initial perturbation evolves over time as the system components interact with each other and for this, the temporal aspect of the plant also needs to be considered. Data used for risk assessment will have to be precise; although generic data is often used. This adds uncertainty to the assessment and hence, imposes another challenge towards accurate understanding of the risk. Cumulative risk assessment would thus require merging all deviations and understanding of how their changes and interactions would affect the safety level of the system.

Quantitative risk assessment requires hazard identification, estimation of frequency, consequence analysis followed by estimation of the risk from these [6]. Current paper focuses only on selected methods of risk assessment that gives an understanding of categorization of approaches to risk assessment: the cause-effect approach and systems approach. Emphasis is given on estimation (whether qualitative, semi quantitative or quantitative) of the probability of failure of all types of deviations while considering component dependencies, dynamicity and data uncertainty. This paper is not intended to provide an exhaustive review of all risk assessment methodologies, nor is it to provide an in-depth analysis of any particular methodology; it focuses on putting forth key approaches of representative methodologies in a way that will provide a review of the path of development of risk assessment. Section 2 provides an outline of how the analysis has been conducted and the basis of selection of methods for comparison. Section 3 describes briefly the various selected methods and Section 4 provides a comparison in a tabular form with discussion about the road map to the future. Section 5 provides a conclusion of the paper and future work.

2. Comparison Framework and Selection of Methods

The methods that are commonly applied in the industry for risk assessment purposes are based on causal chain and event modeling methods in which a system is decomposed for analysis and the components (machine, human, and organization) undergo linear interaction and follow a sequence of failure that leads to an incident. Various risk assessment methods have been developed based on this approach [8]. The other approach that has been gaining momentum recently is the systems approach which analyzes the outcome of a system as emergent where the whole is larger than the sum of the individual parts. Non-linear interaction and self-organization with evolving dynamics are well handled by this approach [7].

Current work provides a review of how risk assessment is conducted under these two broad categories and the methodologies discussed provides a general idea of how these approaches give a look at risk, bringing forth their pros and cons in doing so. As mentioned, the study is focused on proving an overview of approaches which will enable development of a quantitative cumulative risk assessment model that:

- Merges deviations that are technical in nature with those that are human or organizational
- Handles complex systems with non-linear interaction
- Considers dependency of components
- Realizes the dynamic nature of the system
- Provides information about uncertainty of parameter estimation

If used proactively, it is also possible to use incident investigation methods discussed here for risk assessment. Selection of methods discussed in this paper is based on generality of the methods so that they can be applied for a holistic look into risk. Only well known and accepted models which have recurred in literature over time are mentioned.

3. Risk Assessment Methods

3.1 Causal chain and event modeling approaches

3.1.1 Event Cause Trees:

Fault trees, event trees, and bow-ties have long been used in the industry for risk assessment purposes. These cause and effect models describe linear propagation of failure through the system. Although the conventional ones can provide simple, quantitative and graphical representation of the failure states of a system whereby the effect of multiple deviations can be observed, they find limited application for dynamic systems with dependent or common cause failures. Many approaches to modification of these methods have been published in literature [8, 9] in an attempt to provide a more realistic modeling. Recent literature shows an increased trend in the application of tools such as Bayesian Network [1, 10-14] and Petri Nets [15-17] to model fault trees, event trees and bowties to increase their capacity to handle dependencies. These also shows prospect in real time updating of information by connecting various databases to the Bayesian network [12].

3.1.2 Hybrid Causal Logic:

An extended approach of event cause trees mentioned before has led to the development of the hybrid causal logic (HCL), which provides a framework for attaching Bayesian belief network (BBN) to feed information to connected fault trees and event trees and vice versa [18]. Fault trees allow tackling of technical factors while the connection of BBN allows for incorporation of risk influencing factors (RIF), which are introduced by the human and organizational aspects of the system. The method is simplified by utilization of expert judgment to express causal relationship between the RIFs. This allows development of a model that gives a graphical representation of dependencies. Although allowing for detailed causal relationship modeling, methods such as the HCL are resource intensive and the dependence on opinions of experts can bring variation and uncertainty in the risk evaluation. A similar approach was used in developing the Risk_OMT project which brought in human activities and their contribution to the risk [19, 20]. Both approaches are modifications of the Barrier and Operational Risk Analysis (BORA) model [21].

3.1.3 I-Risk:

The integrated risk (I-Risk) analysis methodology contains three main components: the technical model, the management model, and their interface [22]. The technical model is developed based on Master Logic Diagrams to identify immediate causes of failure followed by the development of event tree and/or a fault tree in order to quantitatively analyze the identified causes of failure. The management model which covers functions such as operations and maintenance is quantified based on audits. At the interface, guidance is provided to assign weightage on how the management model will fit with the technical model.

Although I-Risk qualitatively as well as quantitatively enables the managerial aspects to be incorporated with the technical ones, the need for having an audit team to be able to identify management tasks and quantify them is a large challenge. Consequently, frequent updating of the risk assessment may not be possible.

There are numerous other methods that have been developed based on cause-effect approaches with slight changes made depending on application [23, 24]. But all in all, they more or less follow an approach similar to the ones mentioned above. An overall look into the limitations of cause and effect models shows their inability to capture differences in risk when there is a variation in the success and failure events or variations in event timing whereby scenarios can change [8]. Variations in parameters that are not explicitly modeled are not considered in the risk, which can have a profound effect whenever the system is changed/ modified. Consequently, the dynamic aspect of the plant cannot be considered. These approaches generally provide a modular look at risk, falling short of capturing the factors at systems level that influence the risk. However, these approaches do help provide a simplified look at a complex problem and thus finds larger application in the real world [18].

3.2 Systems Approaches

3.2.1 Accimap:

Rasmussen and Svedung were the first to draw a safety control structure within a socio-technical system which provided a graphical representation of the causal flow of events showing failures to result from contributions across six nested levels of decision-making (Government, Regulators, Company, Management, Staff and Work) [25, 26]. They pointed out that risk management is a control task and proper co-ordination and bidirectional flow of information is needed between all levels of the socio-technical system. Instead of analyzing risk only horizontally at the bottom levels, this procedure enabled a look at the higher levels in the system to identify their contribution and draw logical causal diagram to show their influence on the risk. This revealed the remote causes coming from higher levels in the hierarchy and disseminated complete attention given to frontline operation. In that context, the whole look into risk was obtained and areas where improvement was needed were identified. The Generic Accimap, created by combination of several Accimaps, provided an understanding of the generalized safety level and the decision bodies in the system [27].

Accimap has been used to analyze different kinds of incidents in several domains and provides a systematic approach to identify gaps in a system, be it technical or HOF related. However, it requires sufficient data and analyst skill, the map itself can become complicated for a simple system and it does not provide any direct information about what measures need to be taken or what barriers need to be placed to prevent and/or mitigate the risk [28].

3.2.2 System Theoretic Accident Modeling and Processes (STAMP)

With modification of Rasmussen's approaches, System Theoretic Accident Modeling and Processes (STAMP) is based on three basic components: safety constraints, hierarchical control levels and process models with control loops [29-31]. Leveson emphasizes that complex systems are dynamic and incidents result not from failure of individual components but because of violation or inadequate control of safety constraints. STAMP views the system as having hierarchical levels of control and constraints indicating that each level in the system contributes to the system's safety or to incidents. Loops are generated when each level enforces safety constraints to a lower level while feedback regarding proper/improper implementation of these constraints are fed bottom up to ensure proper placement of the enforced constraints. Thus feedback and control operations replace the traditional chain of events model. Inadequate safety controls can arise from deficiency or absence of constraints, improper or inadequate execution of control actions at lower levels, and inadequate or missing feedback communication within the various levels of the system. Based on STAMP, the System Theoretic Process Analysis (STPA) enables identification of a system's weaknesses via a list of guided questions which then enables determination of component flaws in the control loop could cause such a flaw from which appropriate preventive measures can be identified [32]. The method can be used at any stage of the process life cycle to identify and repair system degradation.

STAMP models dynamic systems which may migrate to a state of increased risk over time due to external disturbances, component failure or dysfunctional interaction of components. It especially focuses on human and organizational subsystems. Although it has been applied in a variety of domains, quantification has not yet been possible. Identification of all loops in the system is enormous task if at all possible.

3.2.3 Functional Resonance Analysis Method (FRAM)

While STAMP is based on a specific causal model (a feedback control system), Functional Resonance Analysis Method (FRAM) is based on the specific theory of functional resonance [33]. FRAM provides a socio-technical approach with principles of resilience engineering to look at the safe state of a system when it is subjected to dynamic operational conditions. Hollnagel explains four principles on which FRAM stands: 1. Outcomes of a process can be different (failure or success) but the underlying processes are not: the reason why some operations fail are the same reason why they succeed; 2. Human and organizations make approximate adjustments to meet existing conditions and their performances can vary which may lead to failure or success; 3. Combinations of variations can have nonlinear effects and the system's outcomes are emergent and not resultant; and 4. Consequences of such variations can be exemplified via tight couplings or functional resonance which cannot be understood simply through cause and effect analysis [34, 35]. FRAM decomposes a system's functions into modules characterized by six parameters: inputs, resource, controls, time, precondition, and output.

FRAM provides a step-by-step guide by identifying system functions, assessing and evaluating the potential variability of the functions, identifying the possible way the variability of one function can affect other functions and spread through the system (functional resonance) and finally identify countermeasures that will enable the system to maintain a safe state. Because it focuses on the likelihood of functional variability instead of probability of component failure, FRAM has not yet been quantified although recently it has been used with Bayesian network as discussed later [36].

4. Discussions

Table 1 shows a comparison of the various models according to the two approaches discussed in section 3. The very nature of cause-effect models limits projection of non-linear relationships that arise from feedback loops generated from complex interaction of components within the system and this prevents prediction of unexpected emergent phenomena. At the same time complex systems, being open to the environment (or higher levels), are subject to change and transforms itself to adapt to the changes with time (self organization), which makes such systems dynamic [7]. Organizations are complex systems where numerous non-linear cause-effect relationships make it difficult to model, let alone quantify. As shown for the cause-effect models, quantification of organizational factors are difficult, if not impossible, with heavy reliance on scoring methods which brings with them large uncertainty and ambiguity. At the same time, attempting to look at technical, operational, human or organizational aspects separately and then merging them later will not incorporate their emergent effect, leading to a deficit in the risk assessment model. Also, when systems evolve or change, the previous cause-effect relationships change, the data used become obsolete and so, the system will no longer be correctly represented.

Table 1. Various approaches are compared based on their prospect for use in cumulative risk assessment

Method	Approach	Quantifiable	Dynamic	Dependency	Data uncertainty understood/ Precision
ET/ FT/ BT	Cause and Effect	Yes	No	No	Yes
HCL/ Risk_OMT	Cause and Effect	Yes (Hybrid)	No	Partially	Yes
I-Risk	Cause and Effect	Yes (Hybrid)	No	Partially	Yes
Accimap	Systems	No	Yes	Yes	No
STAMP/ STPA	Systems	No	Yes	Yes	No
FRAM	Systems	No	Yes	Yes	No

On the other hand, attempting a systems approach brings about different kinds of limitations. These models require extensive resources and demand deep understanding of the system. It is difficult to identify all links within the large hierarchical system, both horizontally as well as vertically. Quantification from these models is the present challenge of risk assessment.

The cause and effect models have found wide application in the industry while the systems approach remains more theoretical. Several tools developed for industrial application for real time risk assessment have their basis on the cause and effect approach and focus on the status of barriers in the operation of a facility [2, 36]. Data gathered from various databases are brought together based on expert judgment and provides a semi-quantitative look at cumulative risk that can be used for work prioritization. However, such tools, being structured upon expert opinion, cannot capturing the evolving or changing nature of the organization or system and are not capable of capturing unfamiliar sequences or unexpected events that are unforeseen by experts. They fail to predict risk for tightly coupled systems with complex interactions such as that in chemical facilities.

Recent research seeks to uncover the value of using different approaches together so that the benefit of systems approach can be used to improve causal models. One such method uses FRAM to provide an in depth operational perspective to human factor in a Bayesian network [37]. In the study, FRAM looked into the process functionally and was able to modify the chain of probable causal events (initially shown by fault tree and Bayesian network) by focusing on what the human errors could be and what influenced those errors, thus enabling identification of required modification of the system. Thus, while Bayesian network provided a chain of probable causal events, FRAM was used to provide a stronger holistic incident explanation for a better understanding that allowed further strengthening of the network to achieve a more reliable risk assessment.

When used synergistically, a better look at risk is possible and here lies the future of cumulative risk assessment. Bayesian Network and Petri Nets with their strong modeling formalism show promising aspects to aid in this endeavor. Bayesian Network attains probabilistic results through inference of connected nodes. In Petri Nets, for large systems, probabilistic values can be obtained by statistical analysis of the behavior of process via simulation. Both tools achieve their common goal through different approaches which gives them their inherent pros and cons [17]. In order to carry out cumulative risk assessment, both methods have provisions for modeling the interaction of various deviations. While Bayesian network allows consideration of dependencies and updating of probabilities, petri nets allows modeling of cyclic processes and modeling of sequential operation. Future research work may focus on developing an integrated model for cumulative risk assessment using such tools.

5. Conclusions

In order to obtain a complete look at risk, it is essential to have a holistic look at all factors, including technical, operational, human and organizational, that contribute to it. Thus cumulative risk assessment requires realization of the complexity, dynamicity and interrelation of all the deviations arising from these factors of the system. Approaches of looking at risk have been broadly categorized as being cause-effect or systems.

Current study reviewed event cause trees and models such as the hybrid causal logic and I-risk which are extended versions of such models, to provide a generalized idea about how these are being modified and applied for risk assessment purposes. The general approach is to develop event trees for technical factors and use a scoring method to merge human and organizational factors with it. Although these provide a deterministic approach towards risk, they have limitations in being able to handle the dynamic aspect of the facility as well as failing to capture all the inter dependence of the components.

The systems approach which is based on the emergence theory that the total is larger than the sum of the individual parts of risk, have failed at providing a simple quantitative model. Accimap provides a causal structure that included hierarchy of management system; STAMP focuses on control loops and constraints to identify how things fail while FRAM helps determine how functional variation in a coupled system can lead to failure or success. However, these approaches are difficult to use due to the substantial amount of work that is needed to identify all links within the system. Even then, the future lies in developing processes that will enable to synergistically use both these approaches so that the cause-effect model is adapted by the systems approach in order to obtain a cumulative look at risk.

6. References

[1] P. Weber, G. Medina-Oliva, C. Simon, and B. Lung, "Overview on Bayesian networks applications for dependability, risk analysis and maintenance areas." *Eng. Appl. Artif. Intel.* 25.4 (2012): 671-682

[2] M. Neill, "Improving Risk-based Decision Making by Connecting PSM Systems to Day-to-Day Plant Operations", *Global Congress on Process Safety, AIChE Spring Meeting*, (2015).

[3] CSB. "Investigation Report: Refinery Explosion and Fire", U.S. Chemical Safety and Hazard Investigation Board, Report No. 2005-04-I-TX, (2007).

[4] CSB. "Investigation Report Executive Summary: Drilling Rig Explosion and Fire at Macondo Well", U.S. Chemical Safety and Hazard Investigation Board, Report No. 2010-10-I-OS, (2016).

[5] J.E. Skogdalen, and J.E. Vinnem. "Quantitative risk analysis offshore—human and organizational factors." *Reliab. Eng. Syst. Safe.* 96.4 (2011): 468-479.

[6] Khan, Faisal, Samith Rathnayaka, and Salim Ahmed. "Methods and models in process safety and risk management: past, present and future." *Process Safety and Environmental Protection* 98 (2015): 116-147.

[7] J.C.L. Coze, "Are organisations too complex to be integrated in technical risk assessment and current safety auditing?." *Safety Sci.* 43.8 (2005): 613-638.

[8] N. Siu, "Risk assessment for dynamic systems: an overview." *Reliab. Eng. Syst. Safe.* 43.1 (1994): 43-73.

[9] J.C. Dugan, S.J. Bavuso, and M.A. Boyd, "Dynamic fault-tree models for fault-tolerant computer systems." *IEEE T Reliab.* 41.3 (1992): 363-377.

[10] A. Bobbio, L. Portinale, M. Minichino, and E. Ciancamerla, "Improving the analysis of dependable systems by mapping fault trees into Bayesian networks." *Reliab. Eng. Syst. Safe.* 71.3 (2001): 249-260G.

[11] Bearfield, and W. Marsh, "Generalising event trees using Bayesian networks with a case study of train derailment." *International Conference on Computer Safety, Reliability, and Security*. Springer Berlin Heidelberg, (2005).

[12] N. Khakzad, F. Khan, and P. Amyotte, "Dynamic safety analysis of process systems by mapping bow-tie into Bayesian network." *Process Saf. Environ.* 91.1 (2013): 46-53.

[13] N. Khakzad, F. Khan, and P. Amyotte, P., "Risk-based design of process systems using discrete-time Bayesian networks." *Reliab. Eng. Syst. Safe.* 109 (2013): 5-17.

[14] S.M.H. Hosseini, and M. Takahashi, "Combining static/dynamic fault trees and event trees using Bayesian networks." *International Conference on Computer Safety, Reliability, and Security*. Springer Berlin Heidelberg, (2007).

[15] O. Nývlt, and M. Rausand, "Dependencies in event trees analyzed by Petri nets." *Reliab. Eng. Syst. Safe.* 104 (2012): 45-57.

[16] O. Nývlt, L. Ferkl, and S. Haugen, "Stochastic Coloured Petri Nets as a modelling language for complex Event Trees." *Nutritional Care of the Patient with Gastrointestinal Disease* (2015): 201.

[17] S.Z. Halim, H. Pasman, Y. Koirala, M.S. Mannan, "Probabilistic Methods of Quantitative Risk Analysis: A Case Study with Bayesian Network and Petri Nets Approaches", *Mary Kay O'Connor International Symposium* (2016).

[18] W. Røed, A. Mosleh, J.E. Vinnem, and T. Aven "On the use of the hybrid causal logic method in offshore risk analysis." *Reliab. Eng. Syst. Safe.* 94.2 (2009): 445-455.

[19] J.E. Vinnem, R. Bye, B.A. Gran, T. Kongsvik, O.M. Nyheim, E.H. Okstad, J. Seljelid, and J. Vatn, "Risk modelling of maintenance work on major process equipment on offshore petroleum installations." *J Loss Prevent. Proc.* 25.2 (2012): 274-292.

[20] B.A. Gran, R. Bye, O.M, Nyheim, E.H. Okstad, J. Seljelid, S. Sklet, J. Vatn, and J.E. Vinnem, "Evaluation of the Risk OMT model for maintenance work on major offshore process equipment." *J Loss Prevent. Proc.* 25.3 (2012): 582-593.

[21] J.E. Vinnem, J. Seljelid, S. Haugen, S. Sklet, and T. Aven. "Generalized methodology for operational risk analysis of offshore installations." *P I Mech. Eng.O-J Ris.* 223.1 (2009): 87-97.

[22] I.A. Papazoglou, L.J. Bellamy, A.R. Hale, O.N. Aneziris, B.J.M. Ale, J.G. Post, and J.I.H. Oh. "I-Risk: development of an integrated technical and management risk methodology for chemical installations." *J Loss Prevent Proc.* 16.6 (2003): 575-591.

[23] H.J. Pasman. "Risk Analysis and Control for Industrial Processes-Gas, Oil and Chemicals: A System Perspective for Assessing and Avoiding Low-probability, High-consequence Events". Butterworth-Heinemann, (2015).

[24] K. Øien. "A framework for the establishment of organizational risk indicators." *Reliab. Eng. Syst. Safe.* 74.2 (2001): 147-167.

[25] J. Rasmussen, "Risk management in a dynamic society: a modelling problem." *Safety Sci.* 27.2 (1997): 183-213.

[26] I. Svedung, and J. Rasmussen, "Proactive risk management in a dynamic society." *Karlstad: Swedish Rescue Services Agency* (2000).

[27] I. Svedung, and J. Rasmussen, "Graphic representation of accident scenarios: mapping system structure and the causation of accidents." *Safety Sci.* 40.5 (2002): 397-417.

[28] P.M. Salmon, C. Miranda, and J.T. Margaret, "Systems-based accident analysis methods: A comparison of Accimap, HFACS, and STAMP." *Safety Sci.* 50.4 (2012): 1158-1170.

[29] N. Leveson, "A systems approach to risk management through leading safety indicators." *Reliab. Eng. Syst. Safe.* 136 (2015): 17-34.

[30] N. Leveson, *Engineering a safer world: Systems thinking applied to safety.* MIT press, (2011).

[31] H. Altabbakh, M.A. Alkazimi, S, Murray, and K. Grantham, "STAMP–Holistic system safety approach or just another risk model?" *J Loss Prevent Proc.* 32 (2014): 109-119.

[32] J. Thomas, and D. Suo. "STPA-based Method to Identify and Control Feature Interactions in Large Complex Systems." *Procedia Eng.* 128 (2015): 12-14.

[33] E. Hollnagel, and J. Speziali. "Study on Developments in Accident Investigation Methods: A Survey of the "State-of-the-Art". SKI Report 2008:50 (Swedish Nuclear Power Inspectorate), (2008).

[34] E. Hollnagel, "FRAM, the functional resonance analysis method: modelling complex socio-technical systems". Ashgate Publishing, Ltd., (2012).

[35] E. Hollnagel, "An Application of the Functional Resonance Analysis Method (FRAM) to Risk Assessment of Organisational Change". Report No. 2013:09, Swedish Radiation Safety Authority, (2012).

[36] A. Blacklaw, A. Ward, and K. Cassidy, "The Cumulative Risk Assessment Barrier Model", SPE 146255, (2011)

[37] D. Smith, B. Veitch, F. Khan, and R. Taylor, "Understanding industrial safety: Comparing Fault tree, Bayesian network, and FRAM approaches." *J Loss Prevent Proc.* 45 (2017): 88-101.

Sustaining Process Safety during Difficult Economic Times

John W. Herber
Process Hazard Management Services, LLC
11 Moonlight Bay
Stillwater, MN 55082
john@phms.us.com

Prepared for Presentation at
American Institute of Chemical Engineers
2017 Spring Meeting
13th Global Congress on Process Safety
San Antonio, Texas
March 27-29, 2017

UNPUBLISHED

Managing Process Safety during Difficult Economic Times

John W. Herber
Process Hazard Management Services, LLC
11 Moonlight Bay
Stillwater, MN 55082
john@phms.us.com

Keywords: Economic Downturn, Process Safety, Business Case for Process Safety, Process Safety Warning Signs, Cost Control, Cost Reduction

Abstract

A number of economic factors have collided recently: the step drop in the price of oil, uncertain US and global economies, and mixed signals from economic and political leaders of many countries. All of these conditions have placed companies, especially those affected by oil prices under difficult economic stress. Companies are taking radical actions to, reduce capital and operational spending and in general, reduce cost.

These economic pressures can be onerous on all areas of a business, but especially on the production floor-level operations. Cost-cutting dicta come from the executive levels, demanding a significant reduction in operating expenses or capital expenditures even when production levels are stable or increasing. While all areas of a company feel the impacts, key process safety impacts are felt on the shop floor; by the exact people, process safety is meant to protect, the operators, maintenance personnel and their supervisors. During these times of extreme economic pressure, it is imperative that process safety systems are not allowed to lose effectiveness.

This paper is intended to serve as a reminder of some impacts of cost-cutting and to present tools to combat the loss of process safety effectiveness.

1. Introduction

Economic challenges come from many situations: general economic downturns such as the recession of 2005-8, sector challenges such as the drop in crude oil price in mid-2014, or pressures from poor company performance. No matter what the causes, economic challenges are usually managed with swift edicts to cut operating and capital budgets, reduce non-essential activities, and reduce headcount. These edicts are often pronounced from a high level and echo through the halls of the company. They seldom provide much guidance or detail, but it is clear that they must be achieved – or else. Rarely are potential impacts on risk recognized as part of these edicts, and even more rarely do the edicts verbalize the need to maintain safe and sustainable operations.

These bold proclamations affect all areas and levels of the company – production level employees through management as well as suppliers, customers, and contractors. The purpose of this paper is to explore the impacts of economic downturns, to examine the negative impacts on process safety implementation, and to provide tools to maintain support of those activities during these challenging periods.

Human factors research shows that people will do what they perceive is the intent of the message. That means that a quickly-crafted note to operations to cut costs may result in a number of differing approaches, some of which will be wrong or even dangerous. People set and achieve goals based on the potential rewards or punishments. Past performance can and will impact the outcomes of today's request to cut costs. When cost reductions are announced, human nature drives us to revisit what was done the last time such an announcement was made and to replicate that. In the past, there may have been rewards for groups who achieved the cost cutting goals (carrot) and negative impacts on those who didn't comply with the terms of the announcement (stick). The carrot and stick model has many potential outcomes, not all of which are positive. To achieve the desired outcome, leaders need to craft their communications with care and include some imperatives to ensure that key systems, especially process safety systems, do not degrade and expose the operation to more risk than is acceptable.

To be fair, there are also some positive outcomes from an economic downturn or challenge. One cost-cutting measure often employed is to reduce inventories. Less hazardous material on-site generally equates to lower risk, unless that material is critical to the safe operation of the process (e.g., caustic for scrubbers or foam for fire protection systems). Another potential risk reduction may come from removing older, less efficient equipment from service. Temporary mothballing equipment requires a solid review of the shutdown procedure and how the system will be monitored in the idled state. This should be covered by an MOC review, and revisited regularly until the system is back in service or permanently removed.

> Case study: During a risk management survey, a recommendation was made to reduce the inventory of highly flammable gas by reducing the number of large, spherical tanks. The site's response was that the inventory was necessary for business continuity purposes. Ironically, a year later, one of the spheres did not pass inspection and required replacement at a significant cost. Suddenly, that portion of the inventory was expendable, and the tank was not replaced.

2. Impacts on Employees

As mentioned above, cost-cutting measures impact employees at all levels. From the control room to the board room, all are under pressure to meet the challenges presented. This section will list some of those impacts at several levels. A perception that permeates headcount reductions is "a body is a body" meaning that it is the number of people, not their skills or qualifications, that must be managed. This is dangerous; remaining senior employees may no longer be trained or certified to handle tasks outside their current duties, even if they had performed them in the past. Secondly, if contractor activity is also limited, maintenance employees may be asked to take on new inspection or repair activities for which they are not properly trained or certified. Simply "reshuffling the deck" of employees creates new challenges and probably increases risk.

A. Front-line employees

For the purpose of this paper, operators, maintenance technicians, warehouse people, those closest to the processes, will be considered front-line employees. Headcount reductions are often the first area considered during challenging times, and hourly employees can be the first affected. There are two ways employees leave: retirement and lay-offs or furloughs. Those who choose to retire take with them with years of experience and knowledge of processes and incidents. Their experience and judgement have stabilized the operation for some time. They may have been there when the plant or unit was built. Many times, the true loss of knowledge is not exposed until a process is challenged by some duress, and no one in the operation knows how to handle it. At the other end of the demographics, lower-seniority employees are usually on the early wave of lay-offs. While they may not have years of experience, their duties take them closest to the process and, therefore, they serve as a major safeguard. These employees may lack the financial assets to survive long-term unemployment. They may use their mobility to seek employment elsewhere and will not be returning if/when a recall occurs.

B. Technical employees

Technical employees are vulnerable to the same retirement/layoff activities as front-line employees. Retirement offers will drain experience out of the organization, and the resulting instability may cause newer employees to seek other opportunities. With their departure goes technical expertise and process experience that was not formally captured in technical notebooks. It can be more difficult to replace technical employees than front-line employees due to recruiting efforts and relocation needs. Replacing their lost skill, unique process knowledge, and experiences can present additional risks and expenses.

C. Management

From the worker level, it may seem that management is not greatly impacted by economic downturns. That is not true. Management receives a cost-cutting challenge and is duty-bound to implement it. They need to sort out what they can do and still maintain a safe operation. Developing a plan to enact effective and safe cost-cutting measures would best be done by a cross-functional team, where all the impacts could be analyzed prior to implementation. Unfortunately, this seldom occurs. The urgency forces immediate action, and broad impacts on work groups are developed in secrecy to avoid labor relations issues. So management does the best they can with the information provided to them in the timeline allowed, but safety may not appear on their radar.

D. Addressing loss of talent

Loss of talent during economically challenging times is a given; however, there are tools to reduce the risk posed by the talent drain. Succession planning is a tool used in good times to develop and promote those who are demonstrating skill in their areas of expertise. This same succession plan should provide the basis for a review of potential or actual skill deficiencies caused by layoffs or retirements. A recent article in Chemical Engineering Progress, "Look to the Future with Succession Planning[4]," outlines a six-step procedure to develop an effective succession plan.

3. Impacts on Process Safety Systems

A. Urgency

Economic pressures can generate many issues and affect many areas of an operation, but one that permeates most areas is urgency. The perception, or perhaps misperception, is that doing a task faster saves money. However, if there isn't time to do it right the first time, how will there be time to do it over? People often confuse action and progress. A colleague refers to this as "rearranging deck chairs on the Titanic." It looks like something is getting done, but in reality, it is nothing of value. Urgency often stems from management's imperative to reduce costs without any qualifying counterbalances, such as properly managing risk to the operation.

> Case study: A PHA action item required the design of a portable foam system to reduce the risk of ignition after a flammable spill. The team found a foam nozzle and, with little planning or foresight, quickly identified a water source and made connections during a brief downtime. Testing the nozzle, they found the available pressure was not sufficient for the nozzle to work properly. More research during the design phase would have avoided this error and eliminated the cost of redesign and additional downtime for reinstallation. However, the real cost may have been additional time without a mitigative safeguard.

B. Modified risk parameters

Since the target of process safety systems is prevention of infrequent, but catastrophic events, it is easy to change a safety parameter (e.g., an inspection frequency), since any impacts will be felt in the future. This may be fine for a unit that operates at lower throughput or is shutdown completely. However, has the overall risk tolerance of the organization changed? That risk tolerance was developed for and approved by some level of executive management to protect all phases of the operation. By "tweaking" the interpretation, more risk is being assumed, but without any knowledge by those who are ultimately responsible for managing it.

An example of increased risk from modified risk parameters would be a PHA team that lowers the risk of a recommendation because of its cost. The risk of a scenario doesn't change just because economics become difficult. The unseen risk factors of a change may not be reviewed for another five years or during the next revalidation. Perhaps it may not be addressed then if the revalidation team is unaware of the altered interpretation of the risk matrix.

C. Management of Changes (MOC)

On the surface, one would think that MOC activity would decline along with lower capital spending. However, in many cases, the rush to cut costs (urgency) drives many activities aimed at improvement that should, but may not, receive MOC reviews. Companies may be tempted is to make a series of changes where each builds on the previous change. Taken separately, each is minor, but if the risk of overall change is not reviewed, the process can take baby steps right off the cliff.

Another area of change occurs when many operational issues are rapidly altered following a decision to shut down a process. The existing shutdown procedure may be adequate for a short-term shutdown (weeks), but not adequate for one of longer duration (months or longer).

Often, past cost-cutting measures are dusted off and tried again. It can be tempting not to perform an MOC review since the change was reviewed when initially tried. However, this idea is fraught with error. Many other changes may have occurred in the interim that introduced new deviations that weren't considered "back when."

D. MOC of Personnel (MOCP)

This may be a part of a company's MOC process or may be handled by the Human Resources department (HR). Either way, the reviews to personnel, staffing levels, and reassignments must identify the impacts on the safety of the process. Personnel changes are typically made quickly to reduce stress and anxiety on all levels of the workforce. While well-intentioned, this places the review of the impacts and implementation of new safeguards after the fact. During the interim, the process is at higher risk.

E. Monitoring process safety performance gaps

Most people have plenty of work to do and may not have sufficient time to also monitor potential process safety gaps. However, if the economic pressures reduce the vigilance for process safety, the risk of an incident increases rapidly. Existing metrics are one way to measure and report on impacts of the economic challenges. It may be inaccurate to blame every uptick in metrics on these pressures, but if there is a clear cause and effect, it deserves increased visibility. Another monitoring mechanism uses workers at all levels to monitor and report any weaknesses. While this should already be in effect before any economic challenges, it becomes even more important afterward. Use regular safety meetings to ask questions about issues that are of concern. Some ideas for these questions can be found in several CCPS books such as the "Process Safety Culture" chapter of Guidelines for Risk Based Process Safety[1] and "Appendix A" in, Recognizing Catastrophic Incident Warning Signs in the Process Industries[2].

F. Qualified outside resources (inspectors, PHA facilitators, auditors,)

Another personnel-related impact is the limited use of outside contractors. Three areas of potential impact are Asset Integrity (AI) inspections, PHA facilitation, and Audit team composition. AI inspections are an expensive but essential part of maintaining the reliability of the process equipment. Some employees may be trained to perform certain inspection activities, but other inspections require industry certifications (e.g., API piping and vessel inspection). Economics may drive rebidding inspection contracts, but price should not be the sole factor when awarding these activities.

> Case Study
> A plant that outsourced their piping thickness inspections to a contractor who would provide a thickness report – a list of piping locations and the thickness at each one. During an audit, the numbers were reviewed and several piping locations had unacceptable wall thickness. The plant was aghast that the contractor hadn't told them they had thin pipe. However, by accepting the lower bid, they unknowingly accepted responsibility for analyzing the readings and identifying problem areas.

Contractors who directly impact process safety activities include outside PHA facilitators, incident investigators, and audit team members. While not required by most regulations, these functions provide the outside perspective needed to avoid degradation of process safety systems and to ensure that neither urgency nor "group-think"[3] adversely impact the final product. An outside PHA facilitator should guide the team to examine certain deviations, especially in the light of declining safeguards, and to consistently apply the risk matrix to prioritize action items and recommendations.

Similarly, the outside person who validates the incident investigation needs to ensure that the urgency to complete the investigation did not compromise its effectiveness. For instance, were the methodologies used commensurate with the complexity of the incident? Did the team actually uncover management system failures? Did they just stop at contributing causes?

Lastly, the outside auditor's role is to provide external validation of the audit process and outcomes. If the audit team consists entirely of company employees, it might easily overlook a process safety system weakness, since all think the current status is acceptable. In reality, the outsider may take a broader view of several audit gaps and point out a major overriding system failure. For example, take the situation where PID quality is weak, design calculations are missing, and PSI updates are not completed after MOCs. This could indicate that the group responsible for PSI updates doesn't understand their role in process safety. This results in a significantly different audit finding than several findings concerning poor quality PSI.

4. Selling Process Safety Initiatives in Good Times and Bad

Selling process safety improvements and closing performance gaps are not affected by economic difficulties; however, the work needed to successfully sell them can be much higher when resources are scarce. Five attributes "sell" process safety: proactivity, credibility, analysis, communication, and follow-up. Before announcing the impacts of economic constraints, leaders should identify the existing performance gap and the desired end state. To avoid reversal of the announced change, management should carefully clarify of the goals of the changes and communicate acceptable paths to those goals.

A. Proactivity

Once a process safety performance gap or potential challenge is uncovered, action must be initiated without denigrating the decision or those who made it. This means researching on the change, an honest evaluation, not a slanted witch hunt. First, read or review the announcement that was made; what did it really say? This will reveal one of the following things: Either the change is not as negative as originally thought and its impact on risk is relatively minor, or there are major impacts on risk to the workers and the company that deserve attention. In the first situation, by understanding the change, it can be addressed promptly with the affected employees and the real impacts on the operational risk explained. In the second option, by understanding the announced directive, a better business case for required countermeasures can be developed. If after review, there is insufficient direction or understanding, clarification should be requested.

Being proactive takes courage because taking a contrary position may have negative impacts. Decide how best to handle this; once the journey begins, maintain the diligence to sustain the effort. Human nature lessens the urgency as the initiating event slips further into the past. Connect with other people, preferably from other areas and/or functions, to help steer the effort and provide balance to any further analysis. Where corporate experts exist, use their expertise and garner their support.

B. Credibility

There may be only one chance to make an impact on this issue; use it well. That means that all claims and concerns must be grounded in facts and numbers. Opinions are great, but they are personal and can be emotionally based. Making the business case for issues with long-term implications like process safety takes more hard data. This is a major mission of the steering team mention in the section above. They provide insights and a balanced view of the problem and its impacts. Don't try to "cherry-pick" the data by selecting only numbers that support your perspective. Analyze all sides of the issue and admit any positive aspects to the change.

What data is needed? That depends on the proposed change, but typically, cost, timing and impact are required. If the issue has become emotional, it is difficult to separate emotion from fact, but that separation is necessary to maintain credibility. Overly optimistic or pessimistic claims quickly erode the logic needed to build the business case. "You only get one chance to make a first impression", so make the best effort to assemble the data in the time allotted. All budgets and timelines need to be created and reviewed by the steering team. This may be a good time to use data ranges versus hard numbers. Also, know what you don't know. There are things that are difficult to predict or quantify. Typically, it is best to use established tools such as the company's risk tolerance or risk matrix. Find reasonable scenarios and do the math. There are also things that are unknowable, such as impacts on company reputation.

C. Analysis

Managers and especially executives are not searching for problems; they expect or demand solutions. It is difficult to perceive an executive's perspective, so it is best to offer options. Options must include the zero option – maintaining the existing methods. It must be considered since it usually is the lowest cost, most expedient, and politically acceptable. While this option may seem unacceptable, the true consequences of inaction may not be understood by management. By having these consequences articulated, management may recognize some critical gaps that had not been previously noted. From this baseline, all other options can be presented. How many options should be considered? There is no magic answer to this. However, three would seem to be the minimum: the zero option, the optimum or best case, and a compromise position somewhere in between. Presenting too many options can be overwhelming; do not expect management to sort through ten options. If the steering team is caught up in "analysis paralysis," it's time to move from developing more options to culling them to a smaller number for presentation to management.

Each option needs an honest, accurate analysis that can be placed side-by-side for comparison and decision-making. Cost and timing are obvious components of this; however, because process safety manages risk, risk impacts need to be addressed during the evaluation. Risk is a function of frequency and consequences. Both terms of the risk equation need to be considered. Extending the frequency of an asset integrity inspection has a greater impact on frequency, but the consequences of the potential loss of containment may not change. Again, referring to the company's accepted risk tolerance is imperative.

Seldom does an idea go from the operating floor to executive approval in one step. The proposal will be reviewed by a series of approvers along the way. At each review, the message will be refined and hopefully perfected. And at each level, new sponsors are added to the group thus building a coalition of support. There may be additional advocates or people whose support is key to successful approval. While building the alliance may delay the apparent timing of approval, the extra time spent may make the difference between approval and rejection.

D. Communication for a positive outcome

The intent of this paper is not to provide a treatise on making convincing presentations; there are many resources for that. However, there are several recommendations for improving the odds of success.

1. Know the audience
2. Present succinctly
3. Admit gaps & ask for support
4. Present the plan and accept the outcome

First – know the audience. How? Most of us are not on a first-name basis with the senior leaders of the company. However, this is a benefit of using the coalition. As the idea is developed, it is vetted by those further up the company organization chart. At each review, you gain insight on what will make the "sales pitch" better and how the ultimate approver thinks. Having an 80-page technical report on the issue may be necessary for an accurate analysis, but will be of little value to senior leaders. They need the executive summary – a page or two, preferably a bulleted list of the issues, options, and recommendations. Leaders with a technical background may have different concerns from one with purely business experience.

The presenter must be clearly state and understand the topic of the presentation. Some leaders are open to positions contrary to theirs; others are not. This frames how the discussion should be handled. The intent is not to openly disagree with them, but to present risk perspectives and alternatives while achieving the goals of the challenge or change. Also, be open to a compromise position; allowing an alternative approach to be piloted and evaluated later is a success.

Second, be succinct. Know the allotted timeline for the presentation and hit the mark. Allow time for questions and interruptions. Provide examples the approver will be familiar with -- past incidents or technology implemented elsewhere in the company. Practice the presentation so you have a solid grip on the content and timing.

Third, admit preparation gaps; approvers have an innate ability to ask a question that you or the steering team didn't consider. This is where the time spent analyzing options bears fruit. You will have the answers to reduced budget or extended timing. However, this is not the time to ad lib. If the question is outside of the options that were considered, admit it. Then ask for permission to review the new request and return a timely, well-developed response. Better to get a "maybe" today with the option for a "yes" tomorrow than a flat "NO". The purpose of the presentation is to ask for the leaders' support. At some levels, that means a formal motion for board or committee approval. In others, it may require approval of capital or other resources. Be sure that the final outcome is stated and understood, and thank the group for their time and consideration.

Lastly, accept the decision of the approver and agree to support it. This is easy if it goes as planned. It is much more difficult to support a negative outcome since the team has invested a lot of time and emotional capital. While the team may dislike the outcome, recognize that, in the decision, the leader/approver has accepted the responsibility for the new risk level. Since senior leaders approve the risk matrix or risk tolerance criteria, they also approve deviations from it. That is why including the risk impacts in the presentation is imperative.

E. Follow-Up

Managing and reporting on the effort is an aspect that is easily overlooked when developing the overall plan. Who will manage the budget and timeline? Who will communicate progress or problems to management and how? What metrics will ensure a successful outcome? All of these issues must be rolled into a workable reporting system. Do not fall into the trap of "no news is good news." Regular, proactive, accurate communication has several benefits. It holds the team accountable for their responsibilities. It allows problems to be addressed when they are smaller and more correctable. It builds the credibility of the team and raises the confidence of the approvers that the effort and risk are being managed to their desired level.

If the approved option is considered temporary, the project ends when the temporary change is made permanent, or the temporary condition/safeguards are returned to the original condition. For example, an asset inspection frequency is changed in the procedures or returned to the earlier frequency. Improper documentation of these decisions may result in incidents, or at best, audit findings. If the approved approach was to pilot a concept for later evaluation, state that the pilot is complete, and recommend how to adjust and expand it to correct the entire situation.

5. Case Study

An internal PSM audit found that many of the PIDs for PSM-covered processes were deficient (inaccurate, multiple mark-ups, inconsistent format, etc.) The audit action item assigned to the site engineering manager was to update all deficient PIDs. The manager had several options: ignore it and hope that it became someone else's problem; assign it to a junior engineer meanwhile hoping it got done; or take responsibility for the project. Fortunately, the last option was chosen and a plan was developed. That plan required several steps: determining the extent of the gaps, searching for resource options to close the gaps, estimating and funding the resources needed, making field reviews, updating, and final acceptance by the users.

Gap assessment

Of the several aspects of this activity, the first is to find the PIDs for all PSM-covered processes listed by the audit team. This meant all versions: those in the field, on engineering desks, in maintenance and any electronic versions that may have been developed. This activity was delegated to the process engineers through their respective managers. It took several months, but the real size of the gaps was quantified and, of course, was much larger than initially suspected. Knowing that number provided a complete scope for updates.

Resource availability – design skill

The engineering manager had options for resourcing the updates: two internal engineering groups or an outside contract design firm. Taking the scope to managers of internal design groups resulted in both internal groups admitting that they would have to outsource the work anyway. That left the site engineering manager to find a suitable contractor to perform the updates, from field walk-downs to verification of the existing process configuration to final revisions filed properly.

Resource availability – financial

Fortunately, the engineering manager had been through a similar exercise in the past and had a reasonable estimate for the cost. However, the company procedures required a formal authorization for expenditures, which meant approval by the Vice President of Engineering. That VP had just issued an edict that all funding of outside engineering was cancelled until further notice, an apparent brick wall.

Building support

The engineering manager took steps to secure support across all groups before proceeding with the plan. There was no doubt that the site manager supported this project; his plant was not in compliance with company policy and several federal regulations. The engineering manager's boss in corporate engineering was also supportive; Engineering is responsible to provide all technical support for PSM. The Corporate Safety Department led the audit, so they were aware of the gaps and the potential impacts of poor PSI. All that was left was to bring a formal request to the VP of Engineering to fund the update project. In the light of the veto of outside resources, the proposal had to be properly developed and presented.

The formal plan had to address four issues:

The problem: the site is out of compliance with PSM due to poor management of PIDs.
The options: remain out of compliance (not acceptable), use internal resources (none available) or outsource to a contract design firm (contrary to VP's decision).
The cost: estimated based on past updates
The timing: This provided some relief. The company's PSM policy differentiated processes covered by regulations from those covered by company risk management practices. Priority could be placed on the PSM-covered processes first, and those covered by company risk next. However, to close the audit finding, all PIDs had to be updated.

The meeting with the engineering VP went well; he had very few questions. He knew his PSM responsibilities were to enable compliance for all plants, so the plan was approved with a minor revisions.

Follow-Up

The audit closure plan required quarterly updates on progress. By making these updates as formal reports, all concerned parties knew the project's progress. While it took over a year to complete all the PID updates, both the schedule and the budget were achieved.

A second activity was initiated by the plant. To ensure that this effort was never needed again, the MOC and PSSR (Pre-Startup Safety Review) procedures were revised. Creating red-lined drawings was added to the PSSR checklist, and PID updates had to be completed and filed before the MOC was considered closed.

6. Summary

Economic pressures are part of competitive business and economic cycles. They are difficult to predict, and they challenge the status quo. However, to manage risk effectively, process safety practitioners need to have a continuous improvement mentality and have viable options prepared in any economic climate.

A. Broad, sweeping edicts on cost-cutting will continue to be made and the impacts on process safety must be always be considered. Preferably, analysis precedes such edicts, but it definitely must follow them so that risk does not exceed acceptable levels.

B. Any plan to maintain process safety effectiveness needs to be proactive and credible, include options, and be effectively presented to the approver(s).

C. Selling process safety in any economic environment is challenging but is even more difficult when resources are constrained. Making a solid, logical argument for continuing or improving process safety is imperative. Use a broad spectrum of resources and supporters to develop an accurate proposal and an effective presentation. The decision of the approver(s), whether favorable or not, must be supported.

D. Once a plan is approved, communicate the progress, whether good or bad, to those who need to know and who can provide support to maintain acceptable progress.

E. Changes made to process safety systems anytime must be captured by an MOC process. This allows the proposed changes to be evaluated and properly approved for the future.

7. Acknowledgements

The author would like to thank Kathleen Clemmer for her detailed review of the English and grammar.

8. References

[1] *CCPS Guidelines for Risk Based Process Safety,* Wiley, 2007

[2] *CCPS Recognizing Catastrophic Incident Warning Signs in the Process Industries,* Wiley, 2012

[3] Group think: a pattern of thought characterized by self-deception, forced manufacture of consent, and conformity to group values and ethics. Merriam-Webster on-line Dictionary, M-W.com

[4] McDermott & Marshall, "Look to the Future with Succession Planning," *Chemical Engineering Progress,* December 2016

13TH GLOBAL CONGRESS ON PROCESS SAFETY

And Now For Something Completely Different

Michael P. Broadribb
Baker Engineering and Risk Consultants, Inc.
3330 Oakwell Court, San Antonio, Texas 78218-3024
mbroadribb@bakerrisk.com

Prepared for Presentation at
American Institute of Chemical Engineers
2017 Spring Meeting and 13th Global Congress on Process Safety
San Antonio, TX
March 26 – 29, 2017

And Now For Something Completely Different

Michael P. Broadribb
Baker Engineering and Risk Consultants, Inc.
3330 Oakwell Court, San Antonio, Texas 78218-3024
mbroadribb@bakerrisk.com

Keywords: process safety management, risk based process safety

Abstract

Process safety management systems are generally credited with reducing major accident risk in the process industries. While those systems have been around for many years—the EU introduced the Seveso Directive in 1982, and US OSHA promulgated the PSM regulation ten years later— many companies today are challenged by inadequate and/or stagnant process safety management system performance. Several companies have experienced major incidents, particularly when they are subject to prescriptive regulations such as OSHA PSM.

Reliance on compliance with prescriptive regulations is no guarantee of preventing major incidents. No regulator anywhere has ever thought of everything that can cause a process safety incident and then written a regulation to safeguard against that occurrence. Good process safety requires that hazards are identified, the risks of these hazards are understood, and those risks are managed by ***"doing the right thing"***. This frequently requires additional activities that go beyond regulatory compliance.

In 2007, CCPS introduced Risk Based Process Safety (RBPS), which builds on lessons learned since the early 1980s, and applies management system principles of Plan, Do, Check, Act. Being risk-based, RBPS is not prescriptive. It recognizes that all hazards and risks are NOT equal, and focuses more resources on the higher hazards and risks that have been identified. This enables finite company resources to be optimally deployed to efficiently improve process safety performance. The twenty elements of RBPS go beyond OSHA PSM to include other elements, such as risk analysis and performance metrics.

Some companies have embraced RBPS and moved away from compliance-driven process safety management systems. However, RBPS is primarily focused on process plant operations. While some elements of RBPS can be applied to large and small projects, RBPS and its guidance doesn't comprehensively address projects or certain other aspects of process safety.

The paper highlights 'some completely different' process safety elements for which no RBPS guidance is currently available, and discusses the application of a more comprehensive RBPS approach to fill these gaps.

1 Introduction

Process safety practices[1] have been around since at least the industrial revolution in the 1800's. For example, the UK Government set up a Select Committee in 1817 to "consider means of preventing mischief of explosion on steamboats to the danger or destruction of His Majesty's Subjects" [1]. This was followed by the Factory Act in 1833 (and multiple subsequent amendments) that appointed Factory Inspectors, whose responsibilities were to check features of what we now know as 'process safety', including asset integrity, and 'occupational safety'. Despite these early initiatives, process safety incidents continued to occur, and two Boiler Explosions Acts were promulgated in 1882 and 1890 to report and investigate explosions, and to require installation of overpressure protection.

Meanwhile a similar picture emerged in the USA. Between 1816 and 1848 there were 3270 fatalities due to boiler explosions involving steamboats. The first federal safety regulation requiring semi-annual visual inspection of steamboat boilers was enacted in 1838. A second Boiler Safety statute was passed in 1852 that required annual pressure testing (at 1.5 x design rating), quality of boiler plates, provision of safety valves, licensing of engineers and pilots, and the provision of fire hoses and lifeboats. This latter regulation had the first elements of a process safety management system, i.e. asset integrity, quality management, protective devices, competency, and emergency response. Nevertheless, incidents continued to occur, such as a boiler explosion on the grossly overloaded Mississippi steamboat "Sultana" that killed 1700 to 1800 passengers in 1865.

Since these early rules, a multitude of laws and regulations addressing aspects of process safety have been enacted, some of the most notable being:

- The Alkali, etc. Works Regulation Act, UK, 1906
- Explosives Transportation Act, USA, 1908 (amended 1909, 1921)
- The Chemical Works Regulations, UK, 1922
- The Petroleum (Consolidation) Act, UK, 1928
- Tank Vessels Act, USA, 1936
- Nuclear Installations Act, UK, 1959
- The 'Dangerous Substances' Directive 67/548/EEC, EC, 1967
- The Occupational Safety and Health Act, USA, 1970
- The Health and Safety at Work Act, UK, 1974
- The 'Explosive Atmospheres' Directive 76/117/EEC, EC, 1976

As a result of a continuing litany of major disasters (Flixborough, Seveso, Bhopal, Mexico City,

[1] The terminology 'process safety' was not widely recognized until the late twentieth century, but early safety regulations covered aspects of what we now know as occupational safety and process safety.

Piper Alpha, Philips Pasadena and others) in the 1970's and 80's, process safety practices have matured into regulated management systems, culminating in the EU Seveso Directive [2] in 1982 (with updates in 1996 and 2012), the US OSHA PSM regulation [3] in 1992, and the US EPA RMP regulation [4] in 1999 (amendment pending). Other countries have also introduced regulations or guidelines that address aspects of process safety, e.g. Australia, Canada, and India. Process safety regulations are widely credited with reducing major accident risk in the chemical, oil, and gas industries. However, major incidents still occur, and many companies today are challenged by inadequate and/or stagnant process safety management system performance.

2 Compliance with Regulations

Government agencies in some countries have tried to force the application of process safety through regulation. Several companies have experienced major incidents, even when in broad compliance with prescriptive regulations, such as OSHA PSM. In fact, dependence on complying with regulations is no guarantee of preventing or mitigating a major incident. Regulations are generally written in response to one or more major incidents, such as the steamboat boiler explosions mentioned earlier. Such prescription can never anticipate every hazard/risk, and the means of reducing risks, but sets minimum requirements. It is impossible for a regulator to imagine ***everything*** that can cause a future process safety incident, and write regulations to safeguard against those events. In effect, regulators will always be playing catch up.

Some companies are still committed to a philosophy of reliance on regulatory compliance, and do not understand that a change of culture is necessary to prevent all major incidents. Effective process safety performance requires that:

- All hazards be identified,
- The risks associated with these hazards be understood, and
- These risks be managed by ***"doing the right thing"***.

In order to properly manage risk, companies should recognize that it frequently requires additional activities that go beyond regulatory compliance. Only by embracing a *'beyond compliance'* culture can companies expect to prevent major incidents.

3 Risk Based Process Safety

In 2007, the Center for Chemical Process Safety (CCPS) introduced guidance on Risk Based Process Safety (RBPS) [5], which built on lessons learned from process safety management systems since the early 1980's. RBPS applies the management system principles of the Shewhart cycle (a.k.a. Deming cycle), which is a 4-step plan–do–check–act (PDCA) model for continuous improvement.

RBPS is not prescriptive, and recognizes that all hazards and risks are **not** created equal. The elements can be implemented at varying levels of rigor to optimize process safety management performance, efficiency, and effectiveness. Therefore, RBPS focuses more resources on the higher hazards and risks that have been identified, enabling finite company resources to be optimally deployed to efficiently reduce risk and improve process safety performance. A company should evaluate each element and may choose to implement some aspects, based on consideration of their risks and existing management system. Hence, local circumstance may result in different RBPS activities based on perceived needs, resource requirements, and existing safety culture.

Some companies have embraced RBPS and moved away from compliance-driven process safety management systems. The twenty elements of RBPS go beyond a process safety management system, such as OSHA PSM, to widen its scope and include other elements, e.g. risk analysis, performance indicators/metrics, and management review. The new elements (additional to OSHA PSM) help eliminate gaps and inconsistencies that have contributed to process safety management system failures.

4 Something Completely Different

If one accepts that a compliance culture (i) is not the answer to preventing incidents and (ii) the risks need to be properly managed, why would one believe that the 20 elements of RBPS would be sufficient? Why not 21, 25, or more elements? Certainly the existing 20 elements represent a significant improvement over some earlier regulations, and may be adequate for some process operations. However, proper management 'by doing the right thing' often involves other activities that are not included in RBPS guidance or thoroughly covered by the existing 20 elements. Some of these activities are discussed below.

4.1 Engineering Design

The existing RBPS elements are primarily focused on process plant operations, and do not address the design and engineering of the plants in any significant detail. For example, the *Process Knowledge Management* element covers types of process safety information to be managed, updated and retained, but not its generation. CCPS has published separate guidance on engineering design [6].

4.1.1 Inherently Safer Design

The *Hazard Identification & Risk Analysis* element mentions inherent safety issues without further explanation. Other elements cover engineering codes and standards (*Compliance with Standards*), and hazard and technical reviews of plant changes (*Management of Change*). However, these elements do not adequately address process safety during the earliest stage of design when process safety issues can be addressed at the lowest cost and with the greatest effect.

The principles of inherently safer design (ISD) are an important consideration in optimizing risk reduction during process plant design [7]. The traditional principles of ISD are:

- Minimize: Reduce inventory of hazardous material
- Substitute: Replace one material with another of lower hazard
- Moderate: Reduce operating severity
- Simplify: Eliminate problems by design rather than adding complexity

Inherently safer systems feature a lower overall residual risk that must be managed by the Operator implementing the elements of process safety, i.e. a reduced reliance on process safety management. As a result, inherently safer systems are less likely to incur serious process incidents.

4.1.2 Risk Management

The primary focus of the existing RBPS element, *Hazard Identification & Risk Analysis*, is on evaluating hazards and risks [5]. It does briefly address risk-based decision-making, but the emphasis is on risk ranking, an 'as low as reasonably practicable' (ALARP) approach, and comparison against risk tolerability criteria. Other CCPS publications also provide partial guidance on decision-making [7, 8].

When making risk management decisions on design options for new projects and process modifications subject to MOC, the initial approach should consider (i) ISD principles (see 4.1.1 above), and (ii) hierarchies for risk reduction and reliability/availability. The philosophy for risk reduction should be based on the following hierarchy:

HAZARD ELIMINATION is better than
PREVENTION is better than
CONTROL is better than
MITIGATION is better than
EMERGENCY RESPONSE

This risk reduction philosophy adopts some of the principles of ISD. For example, 'hazard elimination' corresponds to 'substitute', although the hazard is not completely eliminated but replaced by a lower hazard. Also 'control' corresponds to 'minimize' in that the consequence (or likelihood) is less, e.g. a smaller fire of shorter duration.

The other risk reduction stages involve 'prevention' (hazard is prevented from being realized, e.g. corrosion resistant alloy instead of carbon steel), 'mitigation' (people are protected, e.g. blast resistant building), and 'emergency response' which, as the last resort, is self-evident.

The reliability/availability hierarchy [9] is:

PASSIVE measures are more reliable than
ACTIVE measures are more reliable than
ADMINISTRATIVE or PROCEDURAL measures

This hierarchy should be used when considering options for additional safety measures to address findings from hazard evaluation studies (e.g. HAZOP) that indicate inadequate safeguards. For example, 'passive' measures include dikes/curbs, drainage, blast walls, and fire-proofing, while 'active' measures include process trips, interlocks, and firewater systems. The least reliable measures involve human intervention, e.g. operating procedures and response to process alarms.

Use of the ALARP approach and risk tolerability criteria should only be necessary on a selective basis if decisions on ISD and these hierarchies involve significant cost and/or require quantitative risk analysis (QRA).

4.1.3 Capital Projects

While some elements of RBPS can be applied to large and small projects, RBPS and its guidance doesn't comprehensively address certain aspects of process safety for capital projects. The author is writing a future CCPS publication that will address this omission.

Management of large capital projects (and some smaller ones) in the process industries generally uses a 'stage gate' philosophy over the project lifecycle. Basically, the life cycle of the project is divided into a number of discrete stages (or phases). As part of the project governance process a virtual gate is placed between successive stages. When the project reaches a gate at the end of a particular stage, an impartial *gatekeeper* judges whether the project still meets the business needs, has adequately delivered the stage objectives, and should continue to the next stage.

To assist the gatekeeper, it is common for the project team to schedule a technical peer review and develop some form of decision support package at each stage gate. Process safety should feature prominently in each stage gate review to ensure that process safety (and EHS) risks are being adequately managed by the project.

The stage gate review is conducted by an independent multi-discipline team (including a process safety engineer) that is familiar with similar facility/process/technology. The team requires a detailed protocol and/or checklist, the content of which varies by the stage of the project lifecycle. At the conclusion of the review, the team will make recommendations for any improvements needed, and indicate to the gatekeeper, based on process safety, whether the project is ready to proceed to the next stage.

Thus, the addition of stage gate reviews, and their associated protocols, would make a beneficial supplement to RBPS as an assurance practice that process safety is being adequately prioritized and addressed by project management. When the future CCPS guideline book on capital projects is completed, the author will likely present another technical paper highlighting other process safety features that should be adopted for projects.

4.2 Human Factors

Human factors (HF) comprises ergonomics (man-machine/system interface), and human performance (human error, human reliability). RBPS does not specifically address the former, but the *Operating Procedures* and *Training & Performance Assurance* elements describe ways to improve human performance [5].

4.2.1 Ergonomics

Ergonomics is the study of people's efficiency in their working environment, i.e. how they interact with a machine or system of work. In the process industries workers can often be exposed to an environment of heat/cold, noise, chemical emissions, electrical and other energy sources, working at height, manual handling, and other challenges. One cannot design the human element, but the job, machinery, operations, and working environment should be designed so that they are compatible with human capacities and limitations. The application of ergonomic engineering at the design stage (or through modification) can reduce risk by considering a range of issues including, but not limited to:

- Layout of DCS displays and controls,
- Alarm flood/overload,
- Accessibility of valves, gauges, sample points, and other equipment for routine tasks by operators,
- Accessibility of process equipment for non-routine tasks by maintenance technicians,
- Labeling of piping and equipment, and warning signage.

Ergonomic design of these and other tasks can reduce human error, and improve human performance. Many companies use very simple HF checklists to meet regulatory requirements, but these are generally inadequate to properly manage risks. CCPS has published a concept book with a lengthy HF checklist that is certainly better [10]. The book also briefly introduces methods to improve human performance, but more detail is necessary for process safety practitioners to fully understand the methodologies. Several companies have developed more substantial internal guidance, and CCPS should consider a future *Guideline* project to gather and promote good practice in this field.

4.2.2 Abnormal Situation Management

A particular example of human performance is when an 'abnormal situation' occurs, i.e. one or more process upsets occur and the automated control system (typically a DCS) cannot cope. This requires an intervention from an operator(s) to augment the control system in order to avoid a production/quality impact or a potentially catastrophic outcome.

The RBPS *Operating Procedures* element includes the safe operating limits, consequences of deviation from safe limits, and the actions required to correct a deviation. However, with new technology and the increasing complexity of some integrated process plants, imagining ***all*** the possible causes of an upset is extremely difficult. If not immediately identified and correctly resolved, a process upset can develop and change over time in a dynamic process control situation complicating intervention requirements. Therefore, the 'actions to correct deviations' may or may not address the cause of the abnormal situation, and guide the operator(s) in executing compensatory or corrective actions in a timely and efficient manner.

An abnormal situation can vary between (i) operating parameters merely being off target to (ii) escalating to a complete loss of control to (iii) a potential catastrophic emergency. The speed with which this progression occurs means that often only the control board operator is available to respond, if he/she recognizes the problem in the first place. Technical and supervisory staff may have a greater understanding of the process technology, but may not always be present during shift hours. Therefore, the ability of the control board operator to trouble-shoot and correctly diagnose the problem is paramount.

Extensive training, especially with a DCS simulator, can undoubtedly improve the performance of the control board operator. However, these days there is a whole science around improving human performance in relation to abnormal situation management (ASM), including, but not limited to:

- Operator situation awareness,
- Real-time support to control board operator,
- Operator information and communication systems,
- Human-machine interface (HMI) design and ergonomic control rooms,
- Alarm management and operator response to multiple alarms,
- Early detection of process upsets.

While a joint industry project[2] already exists, the topic of ASM would make an appropriate future CCPS project to provide basic guidance. This would also supplement CCPS's recent publication on safe automation [11]. ASM would be a worthwhile addition to RBPS as part of Human Factors and/or Operating Procedures.

4.3 Case for Safety

In some jurisdictions, operators of onshore and offshore installations are required to prepare a Safety Case or Safety Report to describe the inherent major accident hazards, the studies undertaken to evaluate their risks, and the measures employed to manage the risks or mitigate their potential consequences [12, 13].

Even where there are no regulatory requirements, some companies have elected to prepare a voluntary 'Case for Safety' in order to provide the future Operations team with a single integrated user-friendly document that contains a summary of:

- Inherent major accident hazards,
- Design hazard management decisions,
- Diverse range of design safety information,
    - e.g. specifications and performance standards of all the design safety measures,
- Residual risk (ALARP),
- Means of managing the residual risk to ensure health and safety.
    - Process safety management requirements
    - Administrative/procedural measures that Operations should implement

Failure to understand and maintain these safety measures throughout the facility lifecycle would increase risks that might result in a major accident. Therefore, the Case for Safety has the following benefits: (i) improved understanding of hazards and risks; (ii) enhanced knowledge of technical and administrative risk reduction measures; and (iii) likely reduction in major incidents or their consequences.

---

[2] The ASM Consortium® is a joint industry project involving companies in the chemical and oil industries.

Finally, the Case for Safety document could be read in one day by a new manager or employee, who would then have a good understanding of the platform design, its residual risk, and how the process safety management system manages those risks. Thus, it would be an excellent training tool.

The Case for Safety is a worthwhile addition to the *Process Knowledge Management* element and/or a new RBPS element on Engineering Design/Capital Projects.

4.4 Process Upset Investigation

The RBPS *Incident Investigation* element requires the user to define the types of incidents to be investigated, and offers examples, e.g. activation of safeguards, such as safety critical systems [5]. Other industry guidance also refers to 'tier 3 near-miss' indicators, such as activation of relief valves and interlocks [14, 15].

Today, some companies do not investigate the activation of protective devices, some of which are often the final element in preventing a catastrophic incident, such as a pressure safety valve (PSV) or high integrity pressure protection system (HIPPS). Inspection, testing, and preventive maintenance (ITPM) can improve their reliability. However, even with frequent ITPM, none of these systems can achieve 100% reliability, and could fail to operate correctly on demand.

In the circumstances even if a safeguard was not activated, perhaps companies should also investigate process upsets (see Abnormal Situation Management above) to learn lessons regarding ASM and upset prevention.

4.5 Sharing and Learning Lessons

The RBPS *Incident Investigation* element requires the user to communicate investigation findings internally to all affected personnel (including contractors) whose job tasks are relevant to the findings. Companies are also encouraged to share findings externally with relevant facilities and industry groups.

Most companies with a transparent, blame free safety culture have systems to share findings and lessons from investigations. In fact, some companies share summaries of dozens of incidents. However, how many 'repeat incidents' occur that are similar to prior events? Repeat incidents are an indicator that (i) the original investigation did not identify *true* root causes (and hence develop meaningful recommendations) or (ii) the lessons were not learned. Often the latter is the case. Organizations need to do more than share lessons – they need to **learn** the lessons. Sharing is easy, learning is more difficult.

There are 3 distinct levels of learning:

- Ability to recall the knowledge (i.e. incident lesson),
- Ability to apply the knowledge in similar circumstances as the original incident,
- Ability to apply the knowledge in dissimilar circumstances.

Many organizations circulate incident summaries and discuss some of them in regular safety meetings. However, organizations still have difficulty sustainably learning the lessons from previous incidents. Process operators (and other personnel) periodically attend refresher training, but many companies do not 'refresh' the refresher training to include incident lessons. Even if refresher training includes new information on the process (e.g. changes subject to MOC), the attention often paid to this new knowledge, and verification of understanding, is minimal.

One means of capturing the most significant lessons for the future is to formally document them within the process safety management system. The lesson could be added to the management system for the appropriate element. For example, if a type of valve should not be used in a certain duty, it could be documented in a revised corporate engineering standard. Similarly, other lessons could be included in revisions to operating procedures, asset integrity practices, or other elements.

Nevertheless, more effective training to *really* learn from incidents is needed to help prevent incidents in the future. Further guidance on more effective training should be developed by CCPS. *Sharing Lessons Learned* would be a worthwhile addition as a new RBPS element to provide guidance on effective learning techniques for not only lessons from incidents, but also lessons from audits and other aspects of process plant operations.

5 Conclusions

Regulatory compliance will not guarantee good process safety performance and avoidance of major incidents. Hazards must be identified and their risks managed, which invariably means going beyond compliance. The twenty elements of RBPS include requirements that go beyond the regulations of some jurisdictions, and are an excellent source of guidance. Some organizations have embraced RBPS, but a few are starting to recognize that managing risks may involve process safety activities not adequately covered in the existing 20 elements.

The author has suggested a number of opportunities to add new elements or extend guidance on existing RBPS elements to provide an even more comprehensive source of guidance. In particular, new elements on Engineering Design (inherently safer design, risk management, capital projects), Human Factors (ergonomics, abnormal situation management), and Sharing Lessons Learned would be worthy additions.

It is hoped that this paper will raise awareness of the necessity to go beyond compliance, as a significant number of organizations are still committed to a compliance only mindset. It is also hoped that the paper will spur debate around what else is *something completely different* and necessary for good process safety? This paper is a start, but what has the author missed?

6 References

[1] Parliamentary Sessional Papers, *Report* (1817), VI, 223, United Kingdom, 1817.

[2] EU. *Council Directive 82/501/EEC of 24 June 1982 on the major-accident hazards of certain industrial activities*, European Commission, European Union, 1982.

(Subsequently superseded by *Council Directive 96/82/EC*, 1996; and *Directive 2012/18/EU of the European Parliament and of the Council*, 2012).

[3] OSHA. *Process Safety Management of Highly Hazardous Chemicals*, 29 CFR 1910.119, U.S. Dept. of Labor, Occupational Safety & Health Administration, Washington D.C.

[4] EPA. *Risk Management Plan (RMP) Rule*, 40 CFR Part 68, U.S. Environmental Protection Agency, Washington D.C.

[5] CCPS. *Guidelines for Risk Based Process Safety*, Center for Chemical Process Safety, American Institute of Chemical Engineers, New York, NY, 2007.

[6] CCPS. *Guidelines for Engineering Design for Process Safety,* Center for Chemical Process Safety, American Institute of Chemical Engineers, New York, NY, 2012.

[7] CCPS. *Guidelines for Hazard Evaluation Procedures*, 3rd edition, Center for Chemical Process Safety, American Institute of Chemical Engineers, New York, NY, 2008.

[8] CCPS. *Guidelines for Chemical Process Quantitative Risk* Analysis, 2nd edition, Center for Chemical Process Safety, American Institute of Chemical Engineers, New York, NY, 1999.

[9] CCPS. *Inherently Safer Chemical Processes: A Life Cycle Approach*, 2nd edition, Center for Chemical Process Safety, American Institute of Chemical Engineers, New York, NY, 2009.

[10] CCPS. *Human Factors Methods for Improving Performance in the Process Industries*, Center for Chemical Process Safety, American Institute of Chemical Engineers, New York, NY, 2007.

[11] CCPS. *Guidelines for Safe Automation of Chemical Processes*, 2nd Edition, Center for Chemical Process Safety, American Institute of Chemical Engineers, New York, NY, 2016.

[12] UK. *The Control of Major Accident Hazards Regulations 2015*, Statutory Instrument 2015 No. 483, United Kingdom, 2015.

[13] UK. *The Offshore Installations (Safety Case) Regulations 2005*, Statutory Instrument 2005 No. 3117, United Kingdom, 2005.

[14] CCPS. *Guidelines for Process Safety Metrics*, Center for Chemical Process Safety, American Institute of Chemical Engineers, New York, NY, 2010.

[15] API. *Process Safety Performance Indicators for the Refining and Petrochemical Industries*, API RP 754, 2nd edition, American Petroleum Institute, Washington, D.C., 2016.

479221 Supervisors and Mid-Managers at the Center of Safety

Abstract:

Supervisors and Mid-Managers at the Center of Safety

We believe no one has more power to improve safety than our supervisors and mid-managers. Our paper shows how we are reaching them with new communication supporting their efforts to lead safety on the front line.

We were inspired by research from Professor Sicco van As working with eight Dutch chemical plants and finding a correlation of -0.64 between informal supervisor communication and the likelihood of serious process safety accidents. Taking this insight to heart, we are building communication making it easier for supervisors and mid-managers to understand and talk about our safety policies and procedures.

When we looked at our existing safety communication, we found our content was mostly accurate and up to date. On the other hand, the communication of this content was something from the 1950s. Our policies and procedures looked like they were written on a typewriter. A typewriter is 1950s technology. None of the advances in communication that have occurred in the last 70 years are evident in our old safety communication.

We are building new safety communication using the latest research:

- reducing writing complexity by 50% (measured by Flesch-Kincaid grade levels)
- graphical layout (associated with increased comprehension between 100% and 600%)
- line length of 9 centimeters (correlated with highest reading accuracy)
- fonts specifically designed for reading online (Verdana & Georgia)
- color (increases time spent on page by 20%)
- clear and precise information to reduce the potential of human error
- pictures of incidents (increases compliance by 50%)

Our paper will highlight "before" and "after" examples of safety communication while showing the measured improvements in frontline comprehension.

13TH GLOBAL CONGRESS ON PROCESS SAFETY

Managing Risk through Integrity Operating Windows

Russ Davis
Mistras Group
4000 Underwood, LaPorte, TX 77571
russ.davis@mistrasgroup.com

Prepared for Presentation at
American Institute of Chemical Engineers
2017 Spring Meeting and 13th Global Congress on Process Safety
San Antonio, TX
March 26 – 29, 2017

Managing Risk through Integrity Operating Windows

Russ Davis
Mistras Group
4000 Underwood, LaPorte, TX 77571
russ.davis@mistrasgroup.com

Keywords: integrity operating window, criticality assessment, risk based inspection (RBI)

Abstract

A comprehensive asset integrity management (AIM) program is essential to reducing risk of operations to as low as practicable. Risk profiles should be developed for each equipment item that has a potential influence on process safety.

A comprehensive asset integrity program is developed based on an analysis of all potential damage mechanisms that could adversely affect the process equipment. Inspection and test methodologies are then selected for the detection of all potential damage. A programmatic approach to getting the most safe and reliable longevity from the process equipment is then developed. Process equipment life-cycles can be effectively managed.

The first step in the development of a holistic AIM program is understanding all potential damage mechanisms that could affect the process equipment. A program has to be implemented for the management of the program in order to effectively reduce risk or operation. Understanding the operating parameters involved in the analysis of damage mechanisms and program development is critical to the effectiveness of the AIM program.

An Integrity Operating Window (IOW) program is the analysis of the parameters necessary to control the process within the limits utilized for the development of the AIM program. By implementing an effective IOW program a company can control potential damage to process equipment and can recognize changes to the process that affect the values used in development of the AIM program. Deviations from AIM safe operating limits can be recognized and efforts can be implemented to return the process to within the integrity operating window.

This paper will demonstrate the value of an IOW program and discuss the implementation of a program.

Introduction

Historically asset integrity management (AIM) programs are developed based on a given set of operational parameters. Process conditions such as the various chemical components of process streams, temperature, pressure, and flow rates are analyzed by mechanical integrity (MI) subject matter experts (SME's). The potential damage mechanisms associated with the process conditions are identified, inspection and test methods and testing frequencies are established. Reliability and MI groups manage the AIMS program based on the analysis that was performed at a time when the program elements were established. When these operational parameters change the rate of equipment degradation may be changed as well. The asset integrity program can become ineffective.

To maintain the integrity of the process equipment in refining and petrochemical facilities several process safety management systems are required. Currently, most of these systems are oriented toward a rigorous mechanical integrity program to predict or prevent failures of pressure equipment. The key PSM support systems include, implementation of IOW's, effective training and management of the IOW programs and the integration of the management of change (MOC) program to identify those changes that affect process equipment. The IOW program is designed to monitor and control process variables that can impact the likelihood of failure.

Defining Integrity Operating Windows

Integrity operating windows (IOW's) are established limits for process variables (parameters) that can affect the integrity of the equipment if the process operations deviates for the established limits for a predetermined length of time (includes critical, standard, and informational IOW's).[1] The purposes of IOW's is to establish, implement and maintain a program for the identification of potential damage mechanism that may adversely affect the process and identify the safe operating limits with the express purposes of avoiding equipment degradation leading to a loss of containment. The ultimate goal of IOW's is to lower the risk of operating plant process equipment.

Integrity operating windows are typically defined by their criticality. API RP 584 divides IOW's into three categories:

1. IOW critical limit
2. IOW standard limit
3. IOW information limit

An IOW critical limit would be one that if exceeded could lead to rapid deterioration of process equipment. A critical limit would require immediate operator action to return the process variable back to a predetermined parameter in order to prevent potential equipment damage in a short time frame. An example of a critical limit could be a pH excursion in a process stream. An extremely low pH could quickly damage pressure equipment and could reasonable lead to loss of containment in a short time frame.

An IOW standard limit is an integrity parameter that if exceeded over a specific time frame could cause increased corrosion rates or eventually lead to cracking or other damage to the material of construction. Standard limits are typically very time based. The time required to adversely impact the equipment generally defines the response to a standard limit IOW. The consequence of damage to the equipment associated with the IOW will also define operator response or actions by the SME for that equipment item.

IOW information limit is generally used by the SME to predict the long term integrity of equipment or to analyze impact on the asset integrity program. Informational IOW's may be parameters that may or may not be affected by operations. These IOW's do not typically require operational responses but may be utilized to assess future repairs of turnaround/shutdown frequencies. Information IOW's can be used to evaluate the process assumptions used to establish risk based inspection (RBI) programs.

Critical limit IOW has also been referred to as a safe operating or safety critical limits by the different companies. Standard limit IOW's have been referred to as key operating limits or reliability limits. Information limit may also be referred to as corrosion control limits depending on the industry in which it is used.

Developing an IOW program

The first step to developing an IOW program is to evaluate all potential damage/corrosion mechanisms that may adversely impact the process equipment. A diverse team of professionals will evaluate the operating parameters of each process stream, i.e., temperature, pressure, stream constituents, abrasiveness, etc. By understanding the potential corrosivity of the process conditions and then evaluating the impact on equipment materials of construction a damage/corrosion map can be developed. The map will be further used to later evaluate what indicators are currently provided to detect exceedance of an IOW limit. This team of SME's can then define the limits for each damage mechanism

After all damage mechanisms have been identified the consequence of failure and the likelihood of failure need to be analyzed and understood. Consequence of failure data can be gathered from process hazard analysis data or by consequence modeling. The equipment failures are then risk ranked based on risk associated with equipment failures.

The third step to an IOW program is typically an evaluation of alarms, indications, and procedures necessary to recognize exceedance of the IOW limit.

Step four is to define the criticality of the operating limit and define the priority of the IOW limit. A determination should be made to categorize IOW's as:

- Critical
- Standard
- Informational

Step five is the documentation of each IOW and the development of proper responses to IOW's alarms and notifications. Procedures or work instructions should define the roles and responses necessary to return the process to a reliably safe state or to further assess the response to limit exceedance, depending on criticality.

Step six is the training of responsible personnel on IOW implementation and timely response to IOW's alarms, alerts, and information.

Step seven is integrating the IOW program into the rest of the plant operations, maintenance and reliability programs.

The last step to an IOW program is the revalidation of the IOW's. Revalidation consists of reviewing each IOW for effectiveness and avoidance of spurious alarms and notifications.

Damage/Corrosion Analysis

A comprehensive analysis must be performed by a team with knowledgeable in the process, operating parameters, and corrosion analysis and damage mechanisms. This team can assess the potential damage that can adversely affect the equipment based on the material of construction. This team will assess the operating parameters such and temperature, stream constituents, pressure, vibration, abrasiveness, etc. each process stream. Once the team has identified potential corrosive streams and the equipment material of construction the limits associated with a damage mechanism are identified. The team may have an output report such as the following:

System	Description	Color	API 570 Pipe Class	Corrosion Rate mpy	Damage Mechanism	Limits
100	(NH) Ammonia NH3 Vapor		1	.5	NH3 SCC	Carbon Steel in aqueous ammonia Non-PWHT
101	(NL) Ammonia Liquid		2	1	Ammonium Chloride Corrosion	Concentration of ammonium salts and temperature
102	(NM) Ammonia, CO2, H2O Vapor, Ammonia Recycle, HP		2	3	NH3 SCC	Carbon Steel in aqueous ammonia Non-PWHT
300	Tail Gas		2	3	CO2 Corrosion	<300°F Liquid phase where CO2 condenses from vapor phase
500	(NG) Natural Gas		2	.5	CO2 Corrosion	<300°F Liquid phase where CO2 condenses from vapor phase
501	(FG) Fuel Gas		2	.5	CO2 Corrosion	<300°F Liquid phase where CO2 condenses from vapor phase
502	Hydrogen Line		2	.5	CO2 Corrosion	<300°F Liquid phase where CO2 condenses from vapor phase

Following the identification of potential damage mechanisms associated with the various systems and limit of the damage mechanism this data can be depicted on plant PFD's.

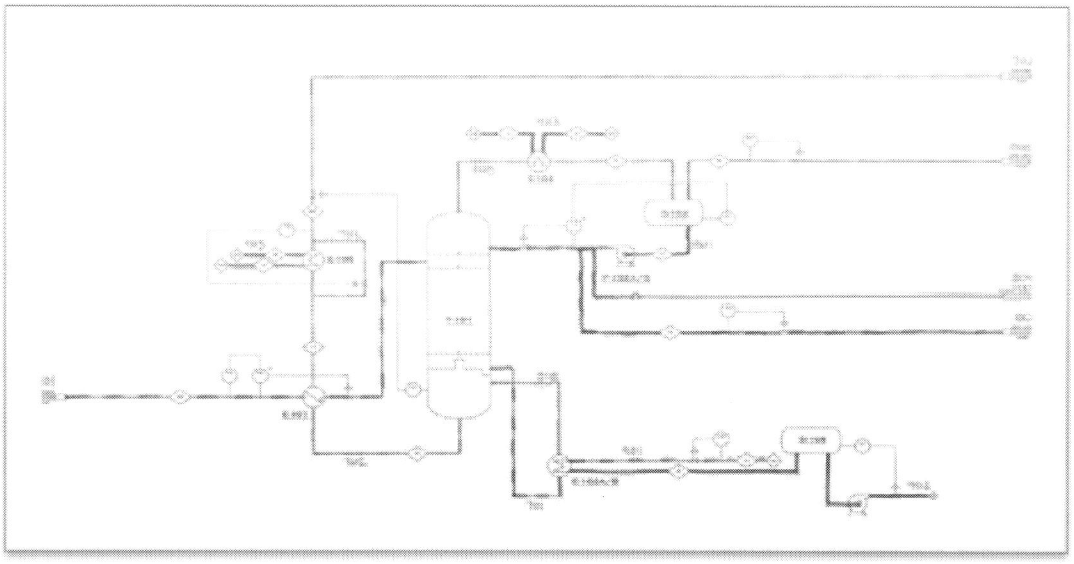

Risk Ranking

The risk ranking process consists of analysis of the consequence of failure and the probability of failure for the equipment within each system as noted above. Consequence of failure data may come from process hazard analysis (PHA) data or consequence modeling and analysis may be performed as an element of the risk assessment. Once a risk analysis is completed ranking can be performed and the risk rank associated with each system or subsystem can be documented.

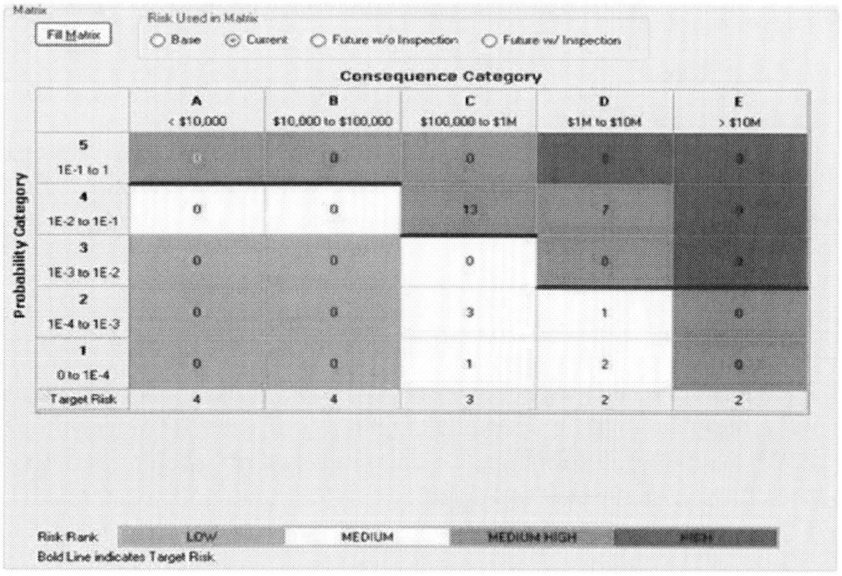

The risk ranking data will be used to categorize the IOW limits. High risk events will require a Critical IOW, medium risk may be require a Standard IOW and low risk events may be categorized as requiring Informational IOW only.

Evaluate

By understanding the limits associated with potential damage mechanisms that may affect the equipment an evaluation of what alarms and notifications will be required for operations, reliability and mechanical integrity groups can be performed. Process parameters such as:

- Temperature
- Pressure
- Flow
- Stream constituents
- Water content
- Chlorides
- Sulfur
- pH

can be fully assessed. By understanding what needs to be measured we can now look at how to measure these parameters.

Define the IOW

At this step a determination of the criticality of the IOW limit is performed and the priority of the IOW is determined.

- Critical IOW – alarm requiring operator or SME timely response to bring the process back within the integrity operating window
- Standard IOW – Typically includes an alert to operations personnel and to Reliability SME. Standard IOW's usually have a time frame associated with them. If an integrity limit is exceeded for a set time equipment will suffer damage.
- Informational IOW – Information conveyed from field gathered data to the reliability SME. This information may require changes to the RBI assumptions, frequency of inspection, or NDE methodology considerations

Documentation & Training

Just like every other aspect of a process safety program documentation is critical. The IOW program must include documentation of each IOW and proper response to the IOW alarm/alert/information. This information should be incorporated into the operating procedures as well as reliability program documentation.

Roles and responsibilities should be properly defined for operations, engineering, reliability and mechanical integrity groups. All of this information should be proceduralized and all responsible personnel should be trained on IOW implementation and timely response to IOW alarms/alerts/information.

IOW Integration

The IOW program must be integrated into the plant operations and maintenance as well as the reliability/mechanical integrity programs. The IOW program must be a fully integrated into engineering, and process engineering as well. Changes in feed stock, changes in temperatures and flow characteristics will impact the asset integrity program. Understanding the information returned from the IOW program is critical to how each facility operates from purchasing to final product.

Audit

The IOW program should be reviewed at least every 3 years. IOW's are a significant element of the overall asset integrity program and should be audited just like any other element of the PSM program. It is recommended that the initial IOW program be audited after the first year to ensure the objectives of the program have been met and to avoid spurious notifications that consume resources but do not add value to the program.

Conclusion

Getting the longest safe reliable life from each equipment item in the plant is critical to the profitability of a company. Process safety is an evergreen management program. Changes that affect the program design must be monitored and properly responded to in order to maintain risk as low as practicable. An integrity operating windows program is the necessary monitoring of the process and its potential impact on the integrity of process equipment. When process changes occur the asset integrity group may not always be informed. A slight temperature elevation can raise a corrosion rate significantly. Changes in sulfur content in a crude stream can lead to sulfidation damage in areas where the MI group had not anticipated. An integrated IOW program is essential to the mechanical integrity program.

References

[1] American Petroleum *Institute. Integrity Operation Windows*, RP 584, 1st ed. May 2014

[2] Center for Chemical Process Safety (CCPS), *Guidelines for Asset Integrity Management,* John Wiley & Sons, Inc., Hoboken, New Jersey, 2017

13TH GLOBAL CONGRESS ON PROCESS SAFETY

Human Reliability Analysis for Evaluation of Conduct of Operations and Training

Erin P. Collins
JENSEN HUGHES
333 E. 46th Street, #11L
New York, NY 10017
ecollins@jensenhughes.com

Bijan Najafi
JENSEN HUGHES
2105 S Bascom Ave, Suite 350
Campbell, CA 95008-3278
bnajafi@jensenhughes.com

Prepared for Presentation at
American Institute of Chemical Engineers
2017 Spring Meeting and 13th Global Congress on Process Safety
San Antonio, TX
March 26 – 29, 2017

Human Reliability Analysis for Evaluation of Conduct of Operations and Training

Erin P. Collins
JENSEN HUGHES
333 E. 46th Street, #11L
New York, NY 10017
ecollins@jensenhughes.com

Bijan Najafi
JENSEN HUGHES

Keywords: Human error, Human Reliability Analysis, Process Safety Culture

Abstract

Human reliability analysis (HRA) is defined as a structured approach used to identify potential human failure events (HFEs) and to systematically estimate the probability of those events using data, models, or expert judgment. HRA is most commonly performed as part of a larger quantitative risk assessment (QRA), but can also be performed as a part of incident investigation to evaluate the root causes for the incident and attempt to prevent their recurrence. There are several pieces to the risk assessment and risk management that allow facility operators to evaluate and measure what is actually occurring in their facilities, including the review of processes, procedures, and events to assess human reliability. This paper will discuss the process for evaluating human errors by obtaining process-specific information used for qualitative and quantitative HRA.

1 Introduction

Risk assessment and management can allow facility operators to evaluate and measure what is actually occurring in their facilities, including the review of processes, procedures, and events to assess human reliability.

Human reliability analysis (HRA) is defined as a structured approach used to identify potential human failure events (HFEs) and to systematically estimate the probability of those events using data, models, or expert judgment. The probabilities used to evaluate HFEs are known as human error probabilities or HEPs. HRA is most commonly performed as part of a larger quantitative risk assessment (QRA), but can also be performed as a part of incident investigation to evaluate the root causes for the incident and attempt to prevent their recurrence.

The purpose of this paper is to provide a brief overview of HRA techniques developed in the nuclear power industry over several decades and successful experience in applying these techniques to various industries, such as aerospace, chemical weapons destruction, HVAC system optimization, and nuclear fuel disposal facility design.

2 Principles of HRA

HRA provides a framework to evaluate, prioritize, quantify and improve the human interface with complex systems. HRA tools and techniques were pioneered by the commercial nuclear power industry, but are based on fundamental precepts of human reliability and data collection and analysis that are not technology specific. They have been applied to an increasing number of facilities across various different industries, such as aerospace [Ref. 1], chemical weapons destruction [Ref. 2], HVAC system optimization, nuclear fuel disposal facility design risk evaluation [Ref. 3], detection and suppression activities in response to fire events in the military [Ref. 4], and cargo loading of LPG tankers [Ref. 5]. This experience and repetition have allowed lessons learned to be identified and fed back into safe practices, the development of a safety culture, and the ability to meet standards [Ref. 6] and regulatory goals (which are quite stringent in the nuclear power industry).

This history and associated lessons learned indicates that wider application of HRA in the process industry could provide greater understanding of risks (both accident and economic), how to control them (by improving procedures, focusing maintenance and surveillance efforts on the actual errors), and how to develop a safety culture that focuses on human error prevention and mitigation of its potential consequences.

The steps of the HRA process discussed below are adapted from NUREG-1921, the EPRI/NRC-RES Fire Human Reliability Analysis Guidelines [Ref. 7].

The process begins with an identification and definition task to identify actions performed by various workers and to define the actions in terms of human failure events or HFEs. These actions may include both failure to perform a required task or undesired actions with deleterious impact on system performance and safety. For the process industry, this identification and definition step could be expected to be very closely tied with the subsequent step of qualitative analysis because the same information gathering that helps define the events also feeds into the development of an "operational story" or HFE narrative that assists in further understanding the events.

The analyst conducts this initial step by reviewing the relevant incident reports, procedures or training manuals in order to become familiar with the process and expected actions, as well as to identify preliminary assessments of the part of the process that could go wrong and why.

Site visits are made to allow the human reliability analyst the opportunity to see the site and equipment configurations first-hand and to conduct interviews with several knowledgeable workers and trainers with significant experience. The visit begins with an overview of the reason for the visit and the information that will be gathered. Interview forms are developed by the analyst prior to the visit to structure the interview process and later, to document the specific HRA-related input information gathered from the interviews. Then, walk-throughs of the facility are conducted to allow the analyst a clearer understanding of the worker interfaces and the location of components, as well as visual insights on the use of specific equipment and tools.

The information gathered through the document review and site visits provides input to the qualitative analysis.

A sound qualitative analysis allows the HRA to provide feedback to the facility operator on the factors contributing to both the success and failure of a worker action. This involves the evaluation of performance shaping factors (PSFs), which are situation-specific influences on human error probabilities (HEPs) and include:

- Cues and Indications that indicate a system problem
- Timing of the Task
- Procedures and Training
- Complexity of the Task
- Workload, Pressure, and Stress
- Human-Machine Interface
- Environmental Factors, such as heat or confined spaces
- Special Equipment, such as tools and protective gear

These PSFs are used to flesh out the HFE narrative and identify the key elements of the HFEs that require further understanding through the performance of subject matter expert interviews.

Finally, quantification or quantitative analysis is performed by applying the information gained from identification, definition and qualitative analysis of HFEs toward the estimation of HEPs. This provides site operations with the understanding of which human errors are most likely and how they can be prevented.

Quantification can be done at varying degrees of detail and rigor. The three levels of analysis performed in HRA are: Screening, Scoping and Detailed.

- Screening analysis involves the assignment of a preliminary and conservative expert judgment value as a means for "screening" out the risk-significant HFEs from the non-contributing ones.
- The Scoping approach, discussed in the NUREG-1921, was developed to provide additional rigor to a semi-screening process, but still results in conservative values for initial quantification.
- Detailed analysis is performed by evaluating the action steps, timing and PSFs in detail and assessing the probability of errors in the cognitive aspects, or diagnosis and decision-making process for the correct action to take, as well as the execution aspects, such as the equipment manipulation steps.

Several HRA quantification methods are available and it has been found to be most effective if the analyst evaluates the strengths and weaknesses of each method against the nature of the human error being evaluated to make the best match. Having a large HRA method toolkit to draw from allows the analyst the flexibility to assess the driving failure mechanism behind the error rather than simply filling out a checklist.

The following HRA quantification methods are most widely used in the nuclear power industry and are contained in the Electric Power Research Institute's (EPRI's) HRA Calculator [Ref. 8] software:

- Human cognitive reliability/operator reliability experiment (HCR/ORE) [Ref. 9], which focuses on the cognitive or decision making errors that are constrained by time;
- Cause-based decision tree method (CBDTM) [Ref. 9] that addresses cognitive errors as well but from the standpoint of the error-causing mechanisms, such as unclear instrumentation or lack of procedural direction;
- The technique for human error rate prediction (THERP) [Ref. 10], which has been used for many years for the evaluation of task performance failures (also called execution errors) that can occur either before or after an accident;
- Accident sequence evaluation program (ASEP) [Ref. 11] that provides a simplified version of THERP for human errors that contribute to but occur prior to the accident; and
- The SPAR-H method [Ref. 12], which only permits highly simplified assessments of cognitive and action errors (and has not been used in the latest generations of nuclear plant risk assessments since it does not allow the depth of analysis to support safety improvements).

For HRA applications other than nuclear power, such as chemical weapons destruction plants and nuclear fuel transport and storage, other quantification methods have been successfully applied and their results accepted during peer and regulatory agency reviews. These include the Cognitive Reliability Error Analysis Method (CREAM) [Ref. 13], and the Human Error Assessment and Reduction Technique (HEART) [Ref. 14]. As the name implies, CREAM focuses on the decision-making errors by providing HEPs for a standard set of task types, but also includes adjustment factors to evaluate facility and scenario-specific influences. HEART contains a list of generic task types and their HEP estimates that are modified using factors associated with 'error producing conditions'.

While the application of a method and generation of HEP values is important, the primary output of the HRA is the identification and prioritization of the human error issues of greatest risk significance and the development of risk insights. The benefit of conducting more detailed HRA is in gaining a better perspective on what influences human failure and therefore what can be done to improve human performance. Detailed analysis also allows for a better understanding and treatment of dependencies for complex systems with multiple human actions involving both errors of omission and errors of commission.

The decomposition of human error into cognitive vs. execution tasks and a detailed evaluation of these tasks has proven to be particularly important since the approaches to their improvement are different. For example, more instrumented processes attempt to provide information to the operator that allow him/her to diagnose the situation and take the appropriate action. In other cases, however, the process may not include such instrumentation or it may not be available due to failure, so diagnosis becomes a critical issue. Even when there is no instrumentation or early warning system provided for an action, the HRA can provide an understanding of the cues required for diagnosis and allow these to be included in training and procedures/manuals. These cues may even be sensory in nature, such as the sound, smell or sight of a process that is going awry. This does not mean that the HRA will recommend instrumenting everything. If correctly done, the HRA instead allows the most critical information used by the worker to be identified and suggests how it should be presented or emphasized to optimize performance.

The reliability of execution tasks is generally improved by providing a formal set of procedures and training that are continually reviewed and refined. However there may be instances where procedures and training are not yet formalized, such as when the facility is a new design or is in the pre-operational testing phase, or because of historical precedent or preference. In these instances, the worker action is evaluated in HRA as a so-called 'recovery' action, where the worker utilizes previous experience with tasks they consider to be similar and applies them to a new situation. These insights are important to evaluating the reliability of task performance and can even be subsequently captured in procedure modifications and training enhancements. The risk benefit assessment might even recommend conducting a design change to the human-machine interface to improve the execution task reliability.

In summary, HRA is

- Widely used
- Applied across a variety of industries, which demonstrates that it provides value
- Allows the user to quantify and prioritize the human influence on safety.
- Provides the means for improving process safety and reliability in the short term and a safety culture to reduce future errors over the longer term.

3 HRA Evaluation of Operations and Training

It is recognized that the HRA process discussed here would need to be evaluated against process facility design and operational characteristics. Some of these issues have been discussed in the AIChE CCPS Guidelines for Preventing Human Error in Process Safety [Ref. 15], but in the ten years since that guidance was issued, the Deepwater Horizon incident has occurred and much has been learned in various industries on the analysis and mitigation of human error using HRA.

HRA of commercial nuclear reactors has the benefit of a structured and well-reviewed set of procedures to use to identify the steps that will be taken by workers to mitigate an anomalous condition, as well as formalized training to form the basis for reliability of those actions. However, in the process industry, there may be significant training but no step-by-step procedures for the staff to follow while performing their tasks or perhaps the procedures are not followed as closely as they should be. In these cases, a review of historical event data or interviews with training staff and workers can identify the human interfaces with the systems and process.

Site interviews give the HRA an opportunity to become familiar with the conduct of operations, namely:

- The hierarchy of organizations, operators and reporting chain of command, not only during normal operations but during incident response,
- The normal staffing profile and shifts
- Personnel responsibilities and roles and overlaps between them
- Incident response roles, e.g., release or fire response teams and their associated procedures

Review of prior incidents may reveal:

- Breakdowns in communications
- Gaps in operations and maintenance responsibilities and lapses in process and procedure verifications and checks
- Less than optimal operations and maintenance interfaces with equipment

In addition, review of records of equipment failure can show patterns of recurring maintenance issues that could ultimately lead to major failures. Looking at the root causes of these failures can help identify whether they are in fact equipment related or are actually related to faulty maintenance practices.

Regarding training, conducting an HRA can help to identify:

- How well the key issues and emphases of training are being understood and implemented
- Whether actual practices are consistent with procedures and training materials
- How frequently training is being conducted on the key risk significant tasks and whether that frequency is sufficient
- Whether timed training sessions are being conducted for tasks that involve a time constraint for performance
- How well cognition (diagnosis and decision-making) is assisted by the training program/process in terms of providing the cues to look for, flow diagrams for analyzing anomalous conditions and response evaluation

- Plans that are in place to capture experience of retiring members of the workforce to carry corporate knowledge on to new hires.

4 HRA Interface with PHA and LOPA

Process Hazard Analysis methods such as What If? Analysis, Checklists, Failure Modes and Effects Analysis (FMEA) and Hazard and Operability Studies (HAZOP) are proven techniques for evaluating the key hazard and risk issues of a process and the ways that personnel intervene, for good or for ill.

As discussed in the CCPS Guidelines for Hazard Evaluation Procedures [Ref. 16]: "A What If/Checklist Analysis may consider issues concerning the potential for an operator to open a valve instead of closing it during an upset condition. Typical HAZOP analyses frequently include general operator errors as causes of process deviations."

In 1983, Swain and Guttman [Ref. 10] identified the following standard human error types related to worker actions that are still very much in use today in HRA:

- Error of Omission
  - Leaves out a task or a step in a task
- Error of Commission
  - Selection error
    - Selects wrong control
    - Mispositions control
    - Issues wrong command or information
  - Time error
    - Too early
    - Too late
    - Wrong sequence
  - Qualitative error
    - Too little
    - Too much

The Guidelines for Initiating Events (IEs) and Independent Protection Layers (IPLs) in Layer of Protection Analysis (LOPA) [Ref. 17] noted that these human error descriptions have the potential to "be useful to PHA teams to facilitate discussions of what types of errors can occur in a given process and what human error initiating events and scenarios can be considered in LOPA." Having someone with an HRA perspective on the PHA team is beneficial for identifying and defining worker caused hazards, as well as potential corrective actions and preventive measures.

In this way, the HRA component of the PHA can identify both the worker related IEs and IPLs, such as these primary categories identified in the LOPA IE IPL Guidelines:

- "Errors in following procedures, such as the startup of a unit, that initiate a LOPA scenario
- Errors in responding to a call for action that, if otherwise performed incorrectly, would have prevented the consequence from occurring."

Both these categories of worker error can then be assessed further as part of a LOPA or independent assessment using HRA by evaluating the likelihood of causing the IE and the effectiveness of the IPL.

It is important to address what the LOPA Guidelines [Ref. 17] refers to as the "IPL response time", which involves the cue to the worker, diagnosis of the situation and decision-making regarding the plan of action, as well as the proper action execution. HRA provides quantification tools and techniques for evaluating each of these facets of the response time.

The HRA evaluates the available information to determine what can be used to identify and assess crew actions. If a risk model already exists, the model structure and results can indicate opportunities for crew recovery of equipment to mitigate an accident sequence. In addition, interviews of personnel with first-hand knowledge of the process, systems, and equipment can be used to identify the steps they would actually perform to complete a task or address an anomalous condition.

In each instance where HRA is applied to a new industry, an evaluation must be made regarding the degree of applicability of the standard error types, PSFs and quantification methods. Since the error types and PSFs have an underpinning in human factors, psychology and cognitive science, they generally apply regardless of the industry. For instance, a recent paper [Ref. 5] has applied the CREAM cognitive error evaluation technique to the cargo loading process of LPG tankers. However, it is still worthwhile to evaluate the applicability of HRA tools and techniques to ensure that industry-specific issues are being appropriately addressed.

For example, some tasks are more driven by cognition and others by execution and still others by a combination of the two. For some tasks, timing is not an issue and there are no particular constraints. Other tasks however may have a definite timeframe for performance which impacts the reliability of the task, particularly when task complexity is great and the time window for performance is small.

Even in cases when timeframe is not an issue, it is still important to understand the risk-driving error mechanisms and the PSFs. This provides input to the assessment of the probability of failure on demand (PFD) values used in LOPA or the HEPs used in probabilistic risk assessment and represents the contribution of the worker to the calculation of scenario and process risk.

5 Conclusions and Recommendations

In summary, the key issues for assessing tools, techniques and guidance for HRA are:

- Review of existing procedures and training material

- Consideration of historical event data
- Performance of site visits and worker interviews
- Identification and characterization of human error types [how well do existing ones apply and are there new ones; determine timing aspects, etc. like definition]
- Evaluation and adjustment of PSFs based on relevance
- Evaluation of HRA model applicability (i.e., development of a process flow chart such as in NUREG-1921 [Ref. 7])
- Documentation of results and insights

6 References

[1] Collins, E.P. and Fragola, J.R., "Advances in Human Error Evaluation", 1995 Reliability, Availability, Maintainability Symposium (RAMS), Washington, DC, January 1995.

[2] Umatilla Chemical Agent Disposal Facility Quantitative Risk Assessment, Report No. SAIC-00/2641. Volume I, Science Applications International Corporation (SAIC), Abingdon, MD, 2002.

[3] Yucca Mountain Project Canister Receipt and Closure Facility Reliability and Event Sequence Categorization Analysis, 060-PSA-CR00-00200-000-00B, Bechtel SAIC Company, LLC (BSC), November 2008.

[4] Garvey, M. (Dept. of the Navy), Joglar, F. and Collins, E.P. (Hughes Associates, Inc.), "HRA for Detection and Suppression Activities in Response to Fire Events", 2014 Annual Reliability and Maintainability Symposium (RAMS), Colorado Springs, CO, January 2014.

[5] Akyuz, E. and Celik, M., Application of CREAM human reliability model to cargo loading process of LPG tankers, Journal of Loss Prevention in the Process Industries, Vol. 34, pgs. 39-48, Elsevier Ltd., 2015.

[6] ASME/ANS RA-Sa-2009, Addenda to ASME/ANS RA-S-2008, "Standard for Level 1/Large Early Release Frequency Probabilistic Risk Assessment for Nuclear Power Plant Applications", The American Society of Mechanical Engineers (ASME), New York, New York, 2009.

[7] Electric Power Research Institute and U.S. Nuclear Regulatory Commission, EPRI/NRC-RES Fire HRA Guidelines, EPRI 1023001/ NUREG-1921, July 2012.

[8] Electric Power Research Institute, The EPRI HRA Calculator® Software Manual, EPRI 3002004030, Palo Alto, CA, May 2014.

[9] EPRI TR-100259, An Approach to the Analysis of Operator Actions in Probabilistic Risk Assessment, Electric Power Research Institute (EPRI), Palo Alto, CA, June 1992.

[10] Swain, A. D. and Guttman, H., Handbook of Human Reliability Analysis with Emphasis on Nuclear Power Plant Applications, NUREG/CR-1278, Washington, DC: U.S. Nuclear Regulatory Commission, August 1983.

[11] Swain, A., Accident Sequence Evaluation Program Human Reliability Analysis Procedure, NUREG/CR-4772, U.S. Nuclear Regulatory Commission, Washington, DC, February 1987.

[12] The SPAR-H Human Reliability Analysis Method, NUREG/CR-6883, Idaho National Engineering Laboratory for U.S. Nuclear Regulatory Commission, Washington, DC, August 2005.

[13] Hollnagel, E., Cognitive Reliability and Error Analysis Method - CREAM, Elsevier, New York, NY, 1998.

[14] Williams, J., HEART - A Proposed Method for Assessing and Reducing Human Error, 9th Advances in Reliability Technology Symposium, National Centre of Systems Reliability, University of Bradford, Bradford, UK, 1986.

[15] Guidelines for Preventing Human Error in Process Safety, American Institute of Chemical Engineers (AIChE), Center for Chemical Process Safety (CCPS), August 2004.

[16] Guidelines for Hazard Evaluation Procedures, Second Edition with Worked Examples, AIChE, CCPS, 1992.

[17] Guidelines for Initiating Events and Independent Protection Layers in Layer of Protection Analysis, AIChE, CCPS, John Wiley & Sons, Inc., 2015.

Process Safety in Operations – What's Missing?

Mike Neill
Petrotechnics USA Inc.
10497 Town & Country Way Suite 700, Houston, Texas 77024
Mike.Neill@petrotechnics.com

Prepared for Presentation at
American Institute of Chemical Engineers
2017 Spring Meeting and 13th Global Congress on Process Safety
San Antonio, TX
March 26 – 29, 2017

Process Safety In Operations – What's Missing?

Mike Neill
Petrotechnics USA Inc.
10497 Town & Country Way Suite 700, Houston, Texas 77024
Mike.neill@petrotechnics.com

Keywords: Process Safety, Operational Excellence, Safety in Operations

Abstract

Many Operators of Oil, Gas and Petrochemical facilities are striving to engage their organizations in being more mindful of Process Safety risk and barrier health. It is certainly true that in order to make daily decisions on whether the plant is safe to operate or if a program of intervention work is safe to execute, the current status of these must be considered. However, much of the information and data on the health of critical barriers is often stored in "silos" within the organization and not readily accessible. Information on risk is even more elusive, based on scenarios that are difficult to relate to daily operations and on assumptions about safety critical equipment reliabilities which may or may not be correct at the time. The reality of daily plant life is more complicated than our simple assumptions with complex simultaneous activities and scenarios, multiple impaired barriers, deviations, management of change (MOC) and cumulative impact on risk - all of which must be considered to get close to evaluating the true level process safety risk.

What is needed is a more simplistic approach to risk which uses a fundamental barrier model for all areas of the plant and can indicate when multiple barrier impairments are elevating risk at any time based on the real-time status of safety critical equipment, defeated safety systems, ongoing activities, etc.

This paper will explore this type of approach which is being used in various areas of the Oil and Gas Industry. It will also look at the emerging software technology platforms which can support this type of model integrating disparate data silos, managing activities and deviations within our risk control systems. They offer new visualizations of risk and activity on the plant, one which can be accessible across the organization better supporting critical decision making. The paper will also examine the different approaches to risk and the dilemma of simplifying risk pictures so than Operations people who may not be "risk experts" can still be effective in their risk-based decisions.

In this paper, we will also address a survey which was conducted by Petrotechnics and AIChE's CCPS 2016 Steering Committee members exploring what is missing in Process Safety. The responses show that only 10% believe Process Safety is being proactively managed in their organization, 30% don't have effective risk management systems in place, and 75% believe safety would be improved with real-time access to process safety risk information.

1 Introduction

Frontline operations are dynamic. It's an inherently hazardous place to work and manage. It's where the plant produces, activities occur, and operational risks and human interactions meet head on. At the heart of this melding is process safety management. Getting it right, making process safety function at the front line is critical. So too is engagement. Process Safety is not just for Safety Professionals or Risk experts. It is for everyone down the line making decisions which have an impact on whether every shift is free of major incidents.

Research by Oliver Wyman suggests, despite significant progress, the oil industry remains a dangerous place. A sobering statistic, according to the U.S. Bureau of Labor, the oil and gas fatality rate for 2013 was 24.2 deaths per 100,000 workers. That is 7.6 times higher than the all-industry rate of 3.2 deaths per 100,000 workers.

In a recent Petrotechnics survey, CCPS members and industry professionals provided insight on process safety: the challenges faced and how to solve them. The result?

> **60% feel vulnerable to events like the Texas City and Deepwater Horizon disasters as regulations have changed little.**

These are perhaps an indication of why oil, gas, and petrochemical companies are tirelessly working to redefine their approach to process safety. The real challenge is arming the operating staff with the right information at the right time so everyone can make the most effective operational decisions.

By taking a step back to look at how the complex interactions of frontline operations and process safety are managed, there is an opportunity to drive continuous improvement and Operational Excellence by reducing risk and improving productivity.

McKinsey and Co. reports, profound technological advancements are disrupting old ways of working in oil & gas and are enabling step change improvements in safety and productivity.

Key to these improvements is utilizing the masses of data currently collected in our Operations but residing in silos within our organization, out of reach of those people who need it to make decisions.

Indeed, the promise of big data in process safety could provide an early warning system that looks at potential signals and trends in facility operations data to make major accident hazard risk exposure visible, prominent and available in real-time. With this information, everyone could proactively make better operational decisions to prevent major accidents.

Big data promises to help deliver what we all want; an improvement of industrial process safety and the achievement of process safety excellence that keeps people safe and the plant running efficiently.

2 The Role of Process Safety Management

Process Safety has, for the most part, become a universally accepted approach for minimizing the risk of Major Accidents in our plants and facilities. Using its principles, we design our plants and facilities with multiple barriers to manage hazards, prevent incidents, or mitigate consequences when something goes wrong.

Our models for design are relatively simple event-based scenarios with a rigorous risk-based approach to determine multiple barrier requirements. PHA, LOPA, and Bow Tie analyses are some of the many processes we use. HAZOP is yet another technique where we introduce some of the realities of operations into the design. Where changes on the plant design are necessary, we use processes such as Management of Change to ensure the original design principles and standards are retained with any modification. Other than for MOC, risk calculations are typically not revisited more frequently than a 5-year cycle. All this seems to be acceptable until we look at the simplicity of the models being used and compare it with reality on the plant.

In contrast with the simple, single event scenarios of our risk models, day-to-day operations tend to be much more complex. There are many things happening simultaneously - all of which could be initiating or escalating events. We will frequently have multiple impairments of barriers or deviations from the risk control systems which will reduce their effectiveness.

Intervention work ongoing on the live plant often introduces additional hazards such as containment compromises, ignition sources and deliberate impairment of detection and protection systems. Non-routine operations such as plant start up pose additional risks.

It is a well-documented fact that the majority of accidents occur during such times than during periods of steady operations. These factors, plus simultaneous work, environmental conditions and human factors, all combine to impair the performance of our process safety barriers. The result of this is an increase in risk.

Sometimes we forget risk is a simple mathematical construct of the probability that an unwanted consequence will occur. We design barriers to reduce the probability of an event occurring or to mitigate the magnitude of the consequence. The risk reduction they provide is based on the likelihood that the barriers function effectively.

For "hardware" barriers, i.e. equipment or instrumentation, manufacturers provide reliability data which assumes adherence to their recommended inspection, calibration, preventative maintenance schemes.

Anything which negatively impacts performance of the barrier, including the deferment of inspection, testing and preventative maintenance, negates or reduces the risk reduction credit taken during design.

What was considered an acceptable risk in design may no longer be the case in practice yet we do not have satisfactory ways of routinely checking this. To use the design process and repeat all the risk calculations on a daily basis taking into account all prevailing impairments would amount to a herculean task.

How, therefore, do we assure ourselves that daily operational risks are acceptable or not?

3 The Growing Consensus: There are Challenges in Process Safety Management

Evidence suggests Process Safety maybe struggling to have a major impact in our industry.

The Principles of Process Safety are not new and have been widely practiced in major companies since the 1970's and '80's and even before then.

In most operating areas, regulation mandates this type of approach; however, as data shows, despite this focus on Process Safety, major accidents have not been eliminated. They still grab the headlines despite it being over 40 years since Flixborough, 32 years since Bhopal, 28 years since Piper Alpha, 11 years since Texas City and six years since Deepwater Horizon.

The 100 largest hydrocarbon loss events since 1974, shows over $1/3^{rd}$ of these events occurred within the last ten years according to the Marsh report (Ref 1). In the report it cites that the

circumstances of these events are not unforeseen – rather they were predicted -but the safeguards or barriers designed to prevent or mitigate simply failed to perform.

Another study by Swiss Re (Ref 2), 45 major incidents indicated failures such as mechanical integrity, operating practices, and activity management. This and other evidence suggest that our plants are not necessarily getting any safer. Major accidents are occurring as a result of our barriers failing to perform. Not only are the barriers failing, but the processes we put in place to ensure reliable barrier performance are also failing.

It's difficult to find fault with Process Safety approaches in the oil and gas industry. In theory, they are sound and should result in reducing the risk of a major accident occurring on a plant.

There is also evidence in the fact that major corporations with international reputations for safety and well managed operations are still having major incidents despite, one assumes, adopting the latest thinking on process safety. This suggests something is missing in how Process Safety is implemented and particularly how it is used in day-to-day operations.

As part of trying to understand what this might be, the Center for Chemical Process Safety (CCPS) steering group members identified some of the greatest challenges in process safety management.

Fig 2: Survey of CCPS Steering Group Members

Not surprising, regulatory compliance was cited as the leading challenge. As CCPS members shared, 10% believe Process Safety Management (PSM) is proactively addressed, and more than 77% believe it's managed as an obligation. It's easy to conclude that as an industry, we are very compliance driven rather than seeking more effective solutions to reduce risk. But CCPS' members said, regulations have changed little over the years, and 60% feel vulnerable to dangers and flaws leading to events like the Texas City disaster and Deepwater Horizon.

Developing a safety culture and driving process safety to the frontline were the next highest survey responses. This result highlights the difficulty of getting all people engaged in Process Safety so they can be the active eyes and ears on the plant.

Making PSM more understandable and more relevant is not restricted to frontline workers. Leadership and Senior Management are often just as disconnected.

It should be said that "safety culture" is a convenient label against which we can lay a lot of blame for major accidents. Though many struggle to define it, we know it partly encompasses decision-making which, when poorly made, is frequently identified as a contributing cause in accident investigation. What maybe missed in some of these investigations is recognition of the poor quality of information and data against which these decisions are often made.

Regarding managing safety-critical maintenance, CCPS members reported conflicting priorities challenge companies' ability to achieved planned maintenance. In fact, it was reported that less than 80% of scheduled maintenance is carried out as planned. It's no wonder why companies feel vulnerable to dangers and flaws.

What is not reflected in the survey, but requires mention, is the role of ageing assets. Like ourselves, all assets suffer from age - both on the inside and the outside. The older a plant is, the more likely it will suffer from integrity issues and demand more attention to keep it operating.

Some remote areas such as our offshore industry have special problems since the structures on which they are built need as much maintenance as the pipes and vessels. They also have constraints which become acute in old age, loss of revenue to fund the higher levels of maintenance and the limited bed space and logistics to support maintenance campaigns.

What can be said is that managing risk in such situations is more acute in that there will likely be multiple risk sources due to corrosion or general ageing of equipment.

4 What's missing in Process Safety Management?

So, what's missing in Process Safety Management? Based on the survey among CCPS members and on anecdotal evidence from discussion with Plant Operators over the past few years, we have determined the following list:

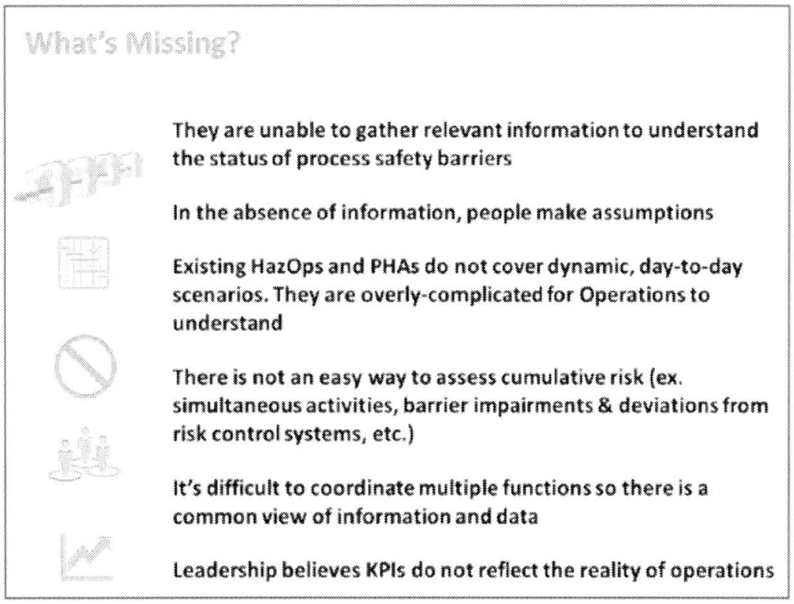

1) Many say they are unable to easily gather the relevant information to understand the current status of process safety barriers – the data is there but it is not easily accessible.
2) More than 30% surveyed admitted they don't have effective risk management systems in place.
3) In the absence of information, people make assumptions – at the start of each shift the Supervisor has to decide whether the plant is safe to operate and the work program for the day can be executed. Unless there is evidence to the contrary, they will assume that all barriers are in good working condition.
4) Existing HazOps or PHA's do not cover dynamic, day-to-day scenarios, (as was covered in section 2). Furthermore, they tend to be overly complicated for Operations to easily understand and relate to current conditions. A supervisor is typically faced with multiple activities planned for the day, many of them hazardous: Multiple impaired barriers, ongoing MOCs, defeats of safety systems, etc. They need to assimilate all this data and make judgements on whether the situation is safe for planned work to proceed or not.
5) There is no easy way to assess cumulative risk – i.e. simultaneous activities, barrier impairments & deviations from risk control systems.

6) It is difficult to co-ordinate multiple functions such as inspection; maintenance; reliability; asset integrity; Process Safety and Operations so there is a common view on risk and therefore a common view on priorities
7) Leadership complains the KPI dashboard shows everything being fine on the plant - but the reality is often very different. In many instances, our KPIs don't measure the outcome but the effort made to achieve the outcome. For example, 95% of inspections of our safety critical elements have been carried out. Sounds good, but what about the 5% remaining? Of the 95% inspections achieved, how many failed? How do I use this information to judge the level of risk on the plant?

To better understand some of these challenges, let's look at some of the multiple sources of risk whose aggregate impact can be summarized by the term "cumulative risk."

Day-to-day operations are characterized by a changing dynamic where there are multiple sources of potential risk which typically change daily, hourly and sometimes by the second. These multiple sources include the following:

- Barrier Impairments / rick control systems (RCS) deviations include damage and failure to physical barriers such as corrosion, mechanical failure, temporary repairs, procedure deviations and deferred inspections, testing or preventative maintenance.
- Operational activities such as equipment start up or shut down. Operator actions such as hydrocarbon sampling, filter change out, pig launching and receiving, etc. Excursions outside safe operating limits for the plant. Temporary defeats to safety systems. Process Isolations which do not meet company standards. Periods of instability on the processing plant.
- Working on the live plant introduces hazards such as ignition sources from hot work, breaking flanges, heavy lifting over live plant, excavation work, etc. These activities are

typically risk assessed and controls put in place (such as gas testing while conducting hot work) which mitigates but does not negate the risk.
- Simultaneous operations compound the risk when working in the same or proximal areas. Some of these activities may be managed separately from Plant Operations such as construction, project work, subsea operations, drilling or well service. So it is essential rules are defined for co-ordination but residual risks will always remain and more importantly they introduce work crews, possibly unfamiliar with the plant, into the hazard zone.
- Environmental conditions such as temperature extremes, wind, visibility, precipitation, electric storms all add risk potential and can compromise work programs
- Human factors such as competence, lack of experience, discipline, fatigue, distraction can compromise the quality of decision making at the frontline.

Faced with this array of potential risk sources, decision makers in the organization are faced with many questions.

Frontline Leadership:

- Is the plant safe to operate?
- Is the scheduled work activity for the day safe to execute?
- Am I set up for a safe day tomorrow?

Site Leadership and planners have to decide:

- Do my Frontline Leaders have all the information they need to safely and efficiently manage the day?
- Am I prioritizing activities which will reduce risk on the plant?
- How do I know things are safe on the plant?
- Are we executing too much work at the cost of increased risk?
- Do I need to intervene?

Executive Leadership must decide:

- What is happening to my corporate risk trending across all assets?
- How are all my risk control systems working?
- How are we balancing risk versus work efficiency?
- How do I know the information I am dealing with is real and current?
- Do I need to intervene?

These are fundamental questions to be answered but our information and data systems and our approaches to managing risk are not well set up to support such decisions. So, what's *really* missing? We think the answer to this can be summarized as follows:

- A simple model which allows us to evaluate risk from multiple sources
- A common "currency" of risk – more than just using the same risk matrix
- A simple real time visualization of barrier impairments and activities

- Connectivity of the model to real-time data being generated on safety critical equipment
- Easy tools to manage all activities and to be able to simply risk assess any arising situation
- More useful indicators or risk which are reflective of the real conditions on a plant
- The ability to see across an enterprise how risk is being managed to identify the overall corporate risk

5 An alternative way of managing risk across the organization

It's high time our industry takes a fresh view on risk management to address what's missing in process safety management.

Simple model for managing risk
Oil and gas operators (ref 3) are moving towards a fundamental barrier model, based on multiple layers of protection / mitigation. The idea is simple in that if you have impairments in several of the barriers (note these are not individual barriers but a collection of equipment, instrumentation or people driven processes which collectively fulfill the function of the barrier) your risk is increased because you are closer to a major accident occurring.

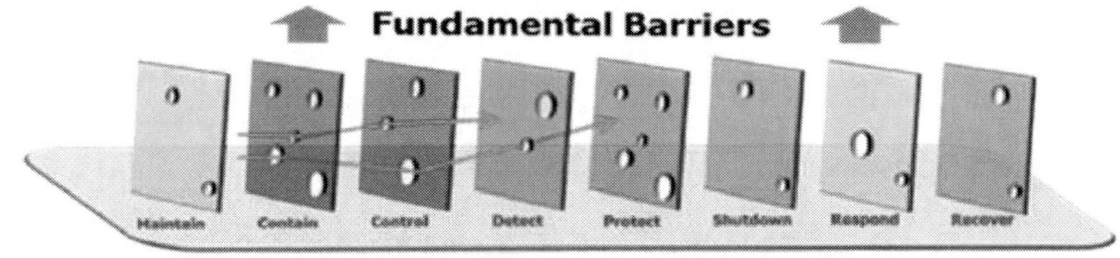

The diagram above shows this simple model using James Ransom's swiss cheese metaphor. The holes in the barriers represent impairments and when the holes line up an event can occur and escalate into a major accident. The degree of development of these risk pathways represents the level of risk.

A common currency of risk
The status of these barriers in each specific area of the plant, in the same time frame is related the risk of a specific major accident.

A simple example: If you were planning to perform hot work in a specific area at the same time you were function checking heat detectors you would have two barriers impaired: Ignition Control and Detection. If there was also a temporary repair on a low pressure hydrocarbon drain in the same area, that would be 3 barriers impaired so the risk of say a fire and explosion accident would be higher. If the fire pumps were also operating on reduced pressure then the Protection barrier would be another impairment increasing risk even further.

So you can see that this is a simple model where you can reflect the current status of safety critical equipment, ongoing activities or process operation happening in the one area and determine how many or to what degree these fundamental barriers are impaired.

Using an agreed mapping of equipment, different activity types, and different plant operations, etc. to specific barriers and using an agreed set of rules, an indication of the level of risk in any area of the plant can be determined simply and quickly. So, for example, if 3 barriers are impaired it is a medium level of risk, 4 barriers a high level of risk.

The advantage of this simplicity is that rules can be predetermined so the process can be somewhat automated. The other advantage is that it focuses people's attention on the real vulnerability – i.e. the barriers and the function they are supposed to perform.

Real time visualization

Technology can support real-time visualization and actively manage activities such as permitted work, energy isolations, temporary defeats, confined space work as well as operator activities. In addition, technology is capable of interrogating multiple data sources (such as production historians, inspection databases, incident tracking, etc. to collect real time status of equipment.

It can also interface with the maintenance management systems to execute (through permit control) work orders, to schedule activities and to view repair orders on safety critical equipment.

In other words, data on all components of risk can be collected and managed in near real time. These tools provide essential views into what is happening in terms of,

- **SPACE**, an operator can see the plant layout with icons indicating how plant activities and impairments align, changing the risk levels.
- **TIME**, an operator can see sequenced activities, deviations and scheduling conflicts, based on indicators of risk for current and future shifts
- **RISK**, for each shift, high or medium risk is predicted alongside specific barrier impairment. Operators can also enable quick testing to see the impact of deciding to repair specific impairments or to defer activities

Connect operational risk with real-time equipment status

There is a lot of data updated on a high frequency that can be queried in near real time. This starts to breakdown the functional silos in an organization allowing this information to be fed into the risk model to see the impact on barrier health.

The way this is done is by mapping the safety critical equipment register to fundamental barriers for a specific major accident hazard. When indications are received that they are not functioning correctly or that inspection and preventative maintenance is overdue, the model will show the relevant barrier impaired.

Manage all activities and assess risk

Again, technology has the ability to manage activities such as permitted work, energy isolations, JSA or task risk assessment, temporary defeats of safety systems, SIMOPS, as well as operator tasks, shift handovers, etc.

These activities are sources of risk in addition to deviations from risk control systems. There is, however, an element of choice in deciding the activities to be scheduled each shift as opposed to deviations which occur as the result of equipment failure and may not be remedied until the equipment is repaired. The repair of which will also be dependent on scheduling the appropriate activities when resources are available.

Sequencing activities has a direct impact on ongoing risk as well as the risk predicted for future shifts. Technology can help manage out elevated risk by optimal sequencing of activities.

Real risk indicators
Rather than relying on the traditional KPI's which only indirectly measure risk by measuring the performance of the risk control system, we can now use direct estimates of risk itself.

We are able to see specifically barriers which are impaired as a result of a deviation or an activity. We can also map these to the risk control systems so the function responsible can see directly the risk impact of any deviation.

Risk view across the enterprise
Through gathering data on risk and activity across multiple assets leadership can see for the first time real-time exposures to peak risks and be able to trend it against work activity efficiencies. In other words, we can see where assets maybe achieving higher rates of time on tools or plan execution but doing so by accepting higher levels of risk.

Also, through comparing performances, assets where consistently high risk exposures are experienced may be leading indicators or poor planning, high work load or resource constraints.

6 Conclusion

The fact that Process Safety in design seems well established but has still not eliminated or significantly reduced the frequency of major accidents across our industry, suggests that something is still missing.

In this paper we have proposed that this is in the area of Operations and managing the lifetime realities of the asset. Here the dynamic nature of Operations and the multiple contributing sources of major accident risk poses a real challenge to Operationalize Process Safety but considering the day-to-day decisions being made in this environment, it is imperative we see some improvements.

Building connections to disparate data trapped in functional silos combined with simpler models of risk provides opportunities to create new and better ways to visualize risk and support decisions being made throughout the organization. This is not based on conventional risk management approaches which are not possible in real time considering the complexity of our operations. Instead, a more pragmatic approach is proposed based on a fundamental barrier model. This simplification allows the data to be mapped in real time and to collect multiple sources of risk. Such solutions provide simple views of risk and activity that help support decision-making from the frontline to the executive office.

7 References.

[1] Marsh & McLennan Companies. "The 100 Largest Losses 1974-2015. Large property damage losses in the hydrocarbon industry, 24th edition", March 2016.

[2] "An analysis of common causes of losses in the Oil, Gas & Petrochemical Industries" Presented at IChemE Safety and Loss Prevention Special Interest Group, Ron Garvis Swiss Re Risk Engineering Services, July 7th 2015

[3] The Cumulative Risk Assessment Barrier Model, Alan Blacklaw, Al Ward, Kevin Cassity, BG Group, SPE, 146255, 2011

Using Process Safety Incident Warning Signs to Increase Hazard Understanding and Sense of Vulnerability

John W. Herber
CCPS Staff Consultant
john@phms.us.com

Jerry L. Jones
CCPS Staff Consultant
jjonespe@sbglobal.net

Brian D. Kelly
CCPS Staff Consultant
kellybd@telus.net

Prepared for Presentation at
American Institute of Chemical Engineers
2017 Spring Meeting
13th Global Congress on Process Safety
San Antonio Texas
March 27-29, 2017

UNPUBLISHED

Using Process Safety Incident Warning Signs to Increase Hazard Understanding and Sense of Vulnerability

John W. Herber
CCPS Staff Consultant
john@phms.us.com

Jerry L. Jones
CCPS Staff Consultant
jjonespe@sbglobal.net

Brian D. Kelly
CCPS Staff Consultant
kellybd@telus.net

Keywords: Process Safety, Hazard Awareness, Process Safety Incident Warning Signs, Vulnerability, Culture, Normalization of deviance

Abstract

Many factors across the Chemical Process Industry (CPI) are moving in directions that jeopardize long-term, stable, process operations. This places people, the environment and company assets at risk. Reduced oil prices are placing more cost pressure on the companies in that sector. Employee turnover at the operator, engineer and manager levels negatively impacts consistent operations and direction. A key difference between a longer term employee and one who has only been in their position for more than several years is awareness of process hazards. Unfortunately, we continue to learn our process safety experience the hard way, by having incidents. It is possible to shorten that learning curve by using past incidents, from inside the company or industry in general, as tools to expand hazard awareness. Other tools are the checklists from the "Recognizing Catastrophic Incident Warning Signs in the Process Industries" book [1]. This paper will present several methods for presenting incidents and a workshop format for using the warning signs checklists. The materials presented can be used for many situations: toolbox meetings, crew safety meetings, PHA review of past incidents or general process hazards awareness training. Many of the warning signs listed point to leadership gaps, which broadens the potential uses for these tools.

1. Introduction

The disciplined conduct of manufacturing operations in the chemical process industry (CPI) depends on people across all functions and at all levels to:

- Perform their job duties under an effective management system structure or framework that defines the business processes and practices to identify and analyze hazards, assess risks, and manage those risks to a tolerable level using multiple safeguards or controls. (Section 2)

- Develop an appropriate level of understanding of the process hazards and the multiple safeguards or layers of protection specified to control them for their areas of responsibility (Section 3),
- Maintain an appreciation of the organization's vulnerability to possible degradation or failure of the process safety risk management systems and existing process safeguards. (Section 4),
- Have the ability and commitment to help identify warning signs and correct substandard conditions, practices or actions that increase the vulnerability of the safeguards (Section 5).
- Correct substandard conditions noted during various assessments in a timely manner.

The intent of this paper is to describe several methods organizations can implement to improve the understanding of process hazards that includes knowledge of the existing safeguards or controls and the vulnerability of those controls to various types of failures.

In this discussion we focus on activities that can be implemented at the lower levels of the organization. While they need support of the facility's management team, they highly depend on engagement of the frontline supervisor, technical specialists (e.g. unit process engineers, reliability engineers, automation engineers, etc.), and the operators and maintenance personnel who are directly involved with the day-to-day operation and maintenance of the process.

2. Effective management systems require competency

Risk based process safety management systems are designed with the primary objective of preventing serious loss of containment/control events. This is accomplished through activities that decrease the probability of failures occurring, or through activities that mitigate the consequences of loss of control events when they occur. The management systems are vulnerable to degradation or failure because they depend on ongoing disciplined actions of many individuals across numerous functions.

Many different technical and administrative competencies are required by those across the workforce at any CPI manufacturing site to manage the hazards and risks using a business process to:

- identify and analyze process hazards and assess their risks (to people, the environment, property and production)
- determine what management systems and specific types of active and passive safeguards are needed to properly manage specific hazards and risks
- routinely check and verify that the preventative and mitigative systems / safeguards are operating well every day and that individuals are fulfilling their personal responsibilities to assure the reliable performance of those safeguards, and to
- maintain competencies necessary to understand process hazards, the controls that manage them and the vulnerabilities of those controls within the organizations risk tolerance.

Guidance on how to analyze and define, develop and maintain necessary process safety competencies across an organization is provided in the CCPS book "*Guidelines for Defining Process Safety Competency Requirements*" [2]

3. Developing appropriate understanding of process hazards and risks

A. General hazard awareness

A long-held tenet in both occupational safety and process safety is that a hazard that is not fully understood by key stakeholders across an organization may not be properly or effectively managed. Industry best practices to manage process hazards are anchored by a risk management business process such as described above. Although this approach is logical and in theory should be straight forward to implement, investigation findings from a long list of worldwide process incidents indicate that many organizations struggle with this business process. These struggles begin with ensuring appropriate process hazard awareness and understanding for all functions of the organization from design, through operation, and maintenance to management oversight.

In addition to understanding the magnitude of the hazards and risks, personnel across various functions must understand both the consequences of a top level event and the different types, purpose and limitations of the existing safeguards, and their roles and responsibilities to help ensure the effectiveness of such controls. To understand the entire risk profile requires information from various elements of the management system such as in Process Safety Information (PSI), Process Hazard Analyses (PHAs), Management of Change (MOC) reviews, operating and, maintenance procedures, training, as well as Emergency Preparedness/Response Plans, incident investigations and audits.

B. Factors impacting hazard awareness and understanding

There are many factors that influence hazard awareness and understanding. Some of these include: complexity of the process or system, how obvious the hazards are, resources available to search for and understand hazards. Similarly, management needs to consistently enforce their approaches for risk management and their expectations of individuals in different functions and at different levels within those function. Many of these factors are determined by the process safety culture. Changing a company's culture is similar to turning an ocean liner; it takes a lot of time and people working together toward a common, well-defined goal. However, raising the hazard awareness and understanding at the front line worker level is more akin to handling a small sailboat; it is easier to do in favorable conditions, but still possible even when strong headwinds or adverse conditions prevail.

Simply recognizing hazards will not get the expected reduction in incidents and risk that is desired. The hazards must be reported to those who can address the issues. Secondly, those who receive the reports of substandard conditions must take positive action. Not only do the reports need escalation to the appropriate level, they demand constant vigilance to ensure that corrective action is taken promptly. A third success factor is frequent updates to those who reported the condition so they understand that it was not lost. If any of these three factors is weak, it may result in a decline in attention to hazard reporting and cause damage to the process safety culture of the organization. This may be seen by a marked decline in hazard and near-miss reports.

C. Determining frequency of activities related to hazard reviews and discussion

How often to conduct hazard review activities cannot be rigidly quantified. It must be often enough to keep the level of hazard awareness high and people attentive to their surroundings. It must also allow time for problems to be identified, analyzed, and corrective actions implemented. Operations in which workers are required to patrol their assigned facilities and equipment on a regular basis (rounds) might have an advantage, but only if a rigorous process is followed. This normally requires considerable training and coaching. If major hazard awareness and review activities are conducted too frequently or with broadly varying themes and ideas, it can be confusing and create a form of complacency often referred to as "flavor of the day". The flavor of the day concept stems from focus on a given issue free short period of time, one day, and then moving on to some other issue the next. If important activities such as making inspections during rounds become mundane and are not acted upon, they become part of the background noise and their quality and effective seriously decline. Some hazard awareness activities, however, deserve regular mention. Safety discussions conducted such as the daily safety moment during toolbox talks are an example of frequent messages meant to highlight awareness of a specific issue for that day.

D. The important roles of experienced employees and those new to the organization

Experienced operators and maintenance specialists will often share their knowledge of past incidents and near misses to broaden hazard awareness and understanding and to map past efforts that have been made to reduce risk in the process. These "war stories" can carry more impact than hours of classroom training, because they are personal, and come from the heart. Also, to avoid losing key hazard and risk perspectives it is important to capture past experiences from senior employees while they are still in the workplace. Encouraging senior employees to document things that they have seen in the past can provide a library of incidents and experiences with far more breadth and detail than any company database.

Newer employees also have a role in the hazard identification and process. They may arrive at the site with little experience in chemical processing, or come from another CPI company with a significantly different risk paradigm. Newer employees may have experienced other chemical hazards and transfer their risk expectations to your operations. A newer employee may provide the benefit of objectivity and bring a fresh set of eyes to bear on hazards and risk. This objectivity needs to be nurtured and balanced between overreacting to smaller hazards, and not seeing larger or more technical issues. One way to calibrate a newer employee's risk understanding is to discuss the company's risk matrix. By using scenarios from the employees' past and current experience they can see how your company views these hazards and assesses the corresponding risk. The opinion of new employees needs to be valued through listening, understanding and explaining existing hazards and the safety systems that manage them.

E. Leveraging contractor experience

The contractors working in your operations have worked in other companies and other industries. They bring hazard and risk understanding from those past experiences the same way that new employees do. While all contractors may not share your exact risk tolerance, they deserve a safe workplace and have no intention of doing harm to employees - theirs or the host company's. If asked, contractors will often share issues or describe incidents they have seen in the past (but not cite the location). This provides an opportunity to expand the list of incident examples to use in safety meetings and training. It also initiates a positive discussion that lets the contractor know you are open to hearing about hazards found in their daily duties and that want to protect all people in that location.

4. Various techniques for maintaining a sense of vulnerability

There are a number of techniques can be used to establish and maintain a sense of vulnerability. The list below is not comprehensive, but provides insight into some techniques which are effective with frontline employees.

A. Company incident library

Most companies maintain a list of past incidents. The list may be in hardcopy or electronic. However, these corporate lists or even plant site lists, may not be all-inclusive. Past incidents, especially those that are minor may not have been formally captured, or well investigated. The details of these events may only reside in the memories of experienced employees and ways to capture these are listed in section 2 above. Despite its gaps, the existing incident list does provide material for developing case studies. These case studies occurred at locations or processes that current employees know and pass every day, and carry a greater impact on behaviors because of that. In some companies, there are seminal events, those that directly impacted the company and changed its safety culture. While these events may not be completely documented or well understood, often there are key lessons to be learned from them that must be kept alive well into the future. An example of this is an article describing an event that was researched, probable causes listed and lessons documented some 40 years later. [3]

In some cases, reviewing incident data from the past 2 to 5 years can highlight certain process safety system failures, hazards, or activities that are especially problematic. Using your data, provides greater impact. Discussions of past events often provide opportunities to gain additional insight from employees having direct experience with those incidents. It also provides an opportunity to clarify misunderstandings that may have developed over time you to poor documentation or communication.

Some companies maintain a book of the most serious events for the company or that site. These are shared with new employees, and with people who are placed in key leadership positions. Having such a book for review will impact those who read it, however if these events are explain by people with solid incident understanding or personal experience with them, that impact is significantly increased.

B. Hazard of the week

Frontline supervisors should be the primary source of safety advice and the most available safety resource for their employees. It is easy for a frontline supervisor to focus on a few hazards or safety issues such as proper PPE, following procedures, and hazards of the most prevalent raw materials. However, if there is no variety to that list of topics or issues, the message can become stale and employees soon tune it out. One way to combat this complacency is to broaden the repertoire of safety issues for topics available to the frontline supervisor. Building such a list of topics could be the collaborative effort of several frontline supervisors and technical resources. In a few brief meetings, they could quickly develop a long list of topics such as hazards or raw materials byproducts or even waste streams, how safety systems work, specific failure modes, or a past event. In the past, when the authors were in operational roles, these topics would have been recorded on index cards and randomly a card/topic selected to be the discussion point for that week. In the electronic age, this list could be recorded in a database form, sorted by keyword and based on circumstances or recent events, a topic selected for discussion that day.

The success of the hazard of the week technique, is especially dependent on the frontline supervisor's time on the work floor meeting with employees either one-on-one or in small groups. An article in Chemical Engineering Progress, "Learn by Wandering Around" [(4)] describes the benefits of time spent on the shop floor. Shop floor safety discussions provide another forum for workers to ask questions or suggest other safety issues or topics. Demands placed on frontline supervisors have not decreased with time, these discussions requires a concerted effort to be successful. It must be expected and reinforced by the supervisor's superiors.

C. The Process Safety Beacon

Using incident descriptions from various external sources is considered a best practice within many industry sectors. The "CCPS Process Safety Beacon"[(5)] is an easy mechanism to receive incident information from incidents at other companies on a regular Since 2001 CCPS has been generating the Beacon, a free, monthly process safety message that is shared globally. The formal title, "Process Safety Beacon- Messages for Manufacturing Personnel" defines its focus - process safety messages to frontline employees. Published in over 30 languages, its simple one page format makes it very suitable for posting on bulletin boards, as a worker handout, or a safety topic at a crew meeting. The format of the Beacon uses a process safety incident as a reminder of things that can happen because it already did elsewhere. The incident is presented in three sections:

- "What happened?"
- "Why did why did it happen?" or "Did you know"
  And
- "What can you do"

Since the Beacon is distributed electronically and can easily be forwarded to a broader audience across the company, sharing of the Beacon is not only allowed, it is encouraged. Discarding a Beacon from another industry or about a raw material not prevalent in your operations, as inapplicable would be an opportunity missed. The incidents described in the Beacon may involve several hazards and equipment or system failures. When reviewing these aspects of a Beacon, it may be necessary to look beyond the superficial, and deeper into the real message, then develop a list of materials, hazards, or system failures that are be similar to your operation. Again, this brings the hazard awareness message home to the frontline employee with additional impact. An article, "How can you use the *Beacon*" [(6)] lists additional ways to leverage this tool.

D. Bow tie diagrams

A bow tie diagram can capture a large amount of process safety information on a single diagram showing specific hazards, failure modes, potential major release events and their consequences. The Royal Dutch Shell and other companies have been implementing the bow tie methodology into their businesses and within the last 10 to 15 years. These diagrams are extremely effective visualization tools to help build process safety competencies and ensure that staff across many functions and at all levels are reminded of the potential top level events should key safeguards fail. Some organizations also incorporate use of the posted bow tie diagrams into safe work authorization discussions before any maintenance task begins involving safeguards to ensure that all involved understand which safeguards are going to disabled or shut down and the criticality of ensuring they are returned to service and will function as designed on demand.[(7)]

Numerous organizations have had their process area safety teams mine their process information data base (including the PHAs) and build simplified bow tie diagrams to post in their process control rooms which serve as a daily reminders of the safeguards that are between the hazards and a potential disaster. These diagrams also are frequently incorporated into new operator and maintenance training; they should also be a part of training all affected employees.

E. Use of incident warning signs to help recognize process safety vulnerabilities

"Prior to any major accident there are always warning signs which, had they been responded to, would have averted the incident." - Dr. Andrew Hopkins [(8)]

The CCPS concept book, "*Recognizing Catastrophic Incident Warning Signs in the Process Industries*" was published to help CPI companies recognize vulnerabilities within their process safety management practices and act on warning signs before a major process safety accident occurs An incident warning sign is defined in that book as "a subtle indicator of a problem that could lead to an incident." The book lists over 160 process safety incident warning signs that came from investigations of incidents in CPI. Had the vulnerabilities been recognized and addressed, many losses would have been avoided or of lesser magnitude. Looking for these incident warning signs is analogous to searching for the vulnerabilities in safeguards as depicted by the "holes" found in the slices of "Swiss cheese" in the barrier model presented by James Reason several decades ago.[(9)]
The warning signs fall into two broad categories:

- Observable (by sight, hearing, smell, touch) physical conditions associated with specific hazards or substandard conditions
- More subtle conditions associated in the organization management system issues which are more difficult to detect such as attitudes, behaviors and decisions or actions.

Most CPI organizations routinely conduct inspections to identify the physical warning signs but they don't always recognize the significance of the findings in relation to management system issues and "holes" in their safeguards or barriers. The more subtle or less obvious incident warning signs related to management system deficiencies are discussed in the next section with a focus on those that relate to culture and conduct of operations.

5. Proactive use of incident warning signs to detect and correct normalization of deviance or organizational drift

A. Normalization of Deviance (in this paper, the terms normalization of deviance and organizational drift are used synonymously)

Normalization of Deviance is the gradual erosion of standards of performance as a result of increased tolerance of nonconformance.[10] Management systems in manufacturing organizations address product quality, environmental performance, occupational safety, and process safety and are all vulnerable to normalization of deviance. Organizations drift from rigid disciplined adherence to established standards, procedures, rules and practices to a state where substandard conditions are condoned and become the norm. This phenomenon has been described in numerous investigation reports released publically over the last 30+ years. Normalization of deviance is typically a gradual process and every CPI manufacturing organization is not only susceptible to it, but very likely a victim of its effects somewhere in their operations. Having a situation where operating and maintenance staff more fully understand process hazards and the potential vulnerability of the safeguards to control those hazards should have a positive impact on individual behaviors and will undoubtedly reduce, but not totally eliminate normalization of deviance. Normalization of deviance is not limited to operations; it can be present in any function. Some examples are inconsistent support of process safety efforts by management or not documenting relief calculations in the PSI.

B. Using incident warning signs to detect and correct normalization of deviance

Normalization of deviance is detectable and can be addressed. In the CCPS *Guidelines for Risk Based Process Safety* [1] there is strong emphasis on the need to establish a safety culture that establishes and enforces high standards of performance across all elements of the management system including Procedures, Asset Integrity and Reliability, Management of Change, Conduct of Operations, etc. Conduct of Operations is the embodiment of an organization's values and principles in management systems that are developed, implemented, and maintained to:

- structure operational tasks in a manner consistent with the organization's risk tolerance,
- ensure that every task is performed deliberately and correctly,
- minimize variations in performance.

Its counterpart is Operational Discipline (OD), which is the performance of all tasks correctly every time. [12] Individuals demonstrate their commitment to process safety through OD; it is the execution of the COO system by individuals within the organization. A well-planned, rigorous, Conduct of Operations (COO) practice across a manufacturing site is considered to be a key deterrent to organizational drift. Actively looking for various incident warning signs is one approach to help ensure excellence in Conduct of Operations and the stated theme of the Catastrophic Incident Warning Signs book.

C. Leveraging the entire organization to identify incident warning signs and minimize organizational drift

To leverage the wide experience base and day-to-day direct involvement in site manufacturing activities, the entire workforce must be made aware of this phenomenon of normalization of deviance and be trained and empowered to look for it symptoms. Site leaders must also be open to receiving the "bad news" that warning signs have been identified and seek out advice on how to address the causes. The warning signs book provides a list of warning signs, an excellent tool to be used by employees across the organization from members of the site leadership team to process area safety team or committee members. They can examine the operations and culture for the manufacturing site or individual units and identify weaknesses that could lead to an event.

If several incident warning signs are uncovered in a given process safety element, it may signal a systemic problem with that element. Finding numerous warning signs across multiple categories may signal the need for significant improvements in the facility's process safety culture. The warning signs book can provide materials for a workshop where warning signs for a given process safety element are reviewed, identifying if that warning sign exists at the location, and how it is exemplified. A follow-on activity would be to identify common failings from several warning sign categories and to develop corrective action to improve process safety systems hazard awareness and process safety culture.

D. Workshops with CPI manufacturing companies on identifying incident warning signs

Workshops for identification of process safety incident warning signs have been facilitated by AIChE in focused process safety improvement training sessions for a variety of companies. Such workshops have also been incorporated into a broader scope AIChE process safety training course "Foundations of Risk Based Process Safety" also referred to as the "Process Safety Bootcamp." [13] The structure and content could be replicated in any company plant or unit. Appendix A of the Incident Warning Signs book lists all of the warning signs sorted by process safety element and presented in the book in the following groupings:

- Leadership and culture
- Competency and training
- Process safety knowledge mgt.
- Procedures for operation and maintenance
- Asset integrity and reliability
- Analyzing risk and managing change
- Learning from experience-- event reporting, investigations, metrics, and audits

A workshop could focus on one or more process safety elements and participants would do an in-depth review of the warning signs for the element or elements as members of breakout groups. A typical breakout session exercise used in the past involves each breakout group discussing examples of specific warning signs they are personally aware of within their current operations, or at "an unnamed site where they or a friend previously worked." Specific warning signs related to various management system elements are listed in Table 1 and groups may be assigned different warning signs to discuss.

Table 1: Incident warning signs list for use to detect normalization of deviance [14]

Leadership and Culture
• Management and workers either not aware of or not committed to safety standards and lack of enforcement of applicable standards and rules
• Poor housekeeping in production areas and support areas
• Accepting operation outside safe operating limits
• Tolerating varying shift operating practices
Procedures for Operation and Maintenance
• Failure to follow procedures tolerated
• Poor quality shift logs and/or inadequate shift turnover communication and rigor
• Chronic problems or inconsistencies with work permits
Asset Integrity and Reliability
• Bypassed alarms and safety systems—operation continues when safeguards known to be impaired
• High frequency of leaks
• Frequent temporary or substandard repairs
Analyzing Risk and Managing Change
• Instruments bypassed without adequate MOC
• Failure to recognize operational deviations and initiate MOC reviews
• Evidence that temporary changes have been made permanent without MOC
Learning from Experience
• Frequent releases or leaks, small fires, and other substandard conditions have become the "norm" and are accepted as tolerable events
• Incident investigations not always being done or superficial without root cause determination
• Symptoms or root causes being repeatedly found in incident investigations
• Repeat findings from previous audits

There are several workshop approaches; one is to use a recent process safety assessment or audit to identify process safety elements with the greatest number or most important deficiencies and review the incident warning signs associated with those elements. A second approach is reviewing the investigation of an incident or serious near-miss to find system deficiencies. Again, the incident could be reviewed against the list of warning signs to identify systems methods or attitudes that need to be changed. If an incident investigation team was enlightened enough to look for warning signs that existed prior to the event, they could compare these warning signs to the list in Appendix A to reverse engineer ideas for corrective actions.

E. Structure for a workshop using process safety incident warning signs

A successful workshop always starts with preparation. This means establishing the purpose of the workshop, how and when it will be conducted and listing the desired outcomes. It is also necessary to know how management will support the actions and recommendations from the workshop. Management support is essential to risk reduction; if you are unsure of the level of management support, you have two options: present the workshop concept to management and ask what they prefer to do when performance gaps or warning signs are identified, or abbreviate the workshop into an exercise for warning sign recognition and limit recommendations to the activities that can be implemented by those present during the workshop. While the second approach may seem to be a weak option, it prevents the participants from tackling issues that have little chance of being addressed immediately by those involved. If larger issues are identified, they can be captured for future review with management.

Determine what topics or elements are most important as part of the workshop planning activities, and focus the workshop on those. A workshop attempting to review all 160 warning signs from Appendix A will be overwhelming and frustrate the participants. However, resist the temptation to edit out warning signs from the review for a given element. Warning signs that are removed may be those most important to hazard awareness and risk reduction. Covering all process safety elements can be accomplished by conducting several workshops during a year, each focused on a particular group of elements.

The participants will arrive at the workshop with a broad range of expectations. A solid introduction is important to calibrate the teams on the intent, how the workshop will be conducted and how the conclusions of the workshop will be handled. Also, key to workshop success is explaining the process that will be employed to recognize warning signs and how management will address them.

The main part of the workshop should start with a review of the fundamentals for the process safety element(s) in question. The larger group may be divided into smaller working groups, who are then asked to review the list of warning signs pertinent to that process safety element. They should sort out those which they feel are most common to their operations. The effect or impact of this workshop can be expanded when groups are assigned different process safety elements and warning signs to review.

As the groups brainstorm their topic, they also develop their list of the warning signs they recognize in their areas. For each warning sign they need to list what they see, or hear that leads them to believe a performance gap exists. Typically the symptoms are, "I often see inconsistent use of procedures", "This checklist is not taken seriously", or "Operators tell me that problems listed on the rounds sheet, are never resolved." These statements provide substantial evidence that warning signs exist and a way to determine how to resolve them.

As each group reports their findings to the broader group, certain common themes may become apparent. These broader deficiencies or weaknesses be captured as a topic for further review or more in-depth study to develop corrective actions. One key attribute for sustaining hazard awareness activities is to communicate the outcomes of the activity and follow up with the participants so they understand the final resolution of their work.

6. Conclusions

A. Based on the incident history over the last 30 years, all have common attributes:

1. There were always management system deficiencies
2. There were warning signs that indicated weaknesses that lead to the event.
3. All were preventable.

B. The culture of the organization must drive for continuous improvement in hazard understanding and encourage active reporting of substandard conditions and timely resolution of them. These are responsibilities of everyone, no matter their position in the organization.

C. Maintaining a high sense of vulnerability is a key outcome of increased hazard understanding and will help workers at all levels be more sensitive and vigilant to substandard conditions that could indicate a weakness in controls and lead to an incident.

D. There are a number of activities that can be used to develop, increase, or maintain a high degree of hazard understanding and sense of vulnerability; among them:
1. Daily toolbox talks using the "hazard of the day" approach
2. Discussing past incidents that occurred in the organization or in similar processes elsewhere
3. Process Safety Beacon
4. Discussions based on review of bow tie diagrams
5. Workshops using the warning signs from the CCPS book.

E. Workshops using the warning signs approach should include many employee functions to ensure all perspectives are reviewed and discussed.

7. Acknowledgements

The authors would like to thank Eric Freiberger for his technical review and comments to improve this paper.

8. References

[1] CCPS *Recognizing Catastrophic Incident Warning Signs in the Process Industries* Wiley, 2012
[2] CCPS *Guidelines for Defining Process Safety Competency Requirements*, Wiley, 2015.
[3] Peter N. Lodal, "Distant Replay: What Can Reinvestigation of a 40-year-old Incident Tell You?" Chemical Engineering Progress September 2004
[4] Ross Topliff, "Learn by Walking Around" Chemical Engineering Progress, November 2011
[5] The Process Safety Beacon - Messages for Manufacturing Personnel https://www.aiche.org/ccps/resources/process-safety-beacon
[6] Hendershot, "How to use the *Beacon*", Chemical Engineering Progress, 2008
[7] B. K. Vaughen and K. Bloch, "Use the Bow Tie Diagram to Help Reduce Process Safety Risks", Chemical Engineering Progress, Dec. 2016
[8] Professor Andrew Hopkins, Australian National University
[9] James T. Reason, *Managing the risks of organizational accidents*. Aldershot, UK: Ashgate Publishing Limited, 1997
[10] AIChE CCPS Process Safety Glossary (https://www.aiche.org/ccps/resources/glossary/process-safety-glossary/normalization-deviance)
[11] CCPS *Guidelines for Risk Based Process Safety* Wiley, 2007
[12] CCPS *Conduct of Operations and Operational Discipline,* Wiley, 2011.
[13] https://www.aiche.org/academy/courses/ch910/foundations-process-safety
[14] Adapted from *Recognizing Catastrophic Incident Warning Signs in the Process Industries,* Wiley 2012

Other references not directly cited directly in this paper:

John W. Herber, "Quality Problems as Process Safety Warning Signs" Chemical Engineering Progress, June 2013
Deborah Luper," Create effective Process Safety Moments" Chemical Engineering Progress May 2010.

Systemic approach to improve learning from incidents

Juliana Guarguati Ariza
Mary Kay O'Connor Process Safety Center
Artie McFerrin Department of Chemical Engineering Texas A&M University
College Station, Texas 77843-3122, USA
Juliana_guarguati@tamu.edu

Noor Quddus, M. Sam Mannan
Mary Kay O'Connor Process Safety Center
Artie McFerrin Department of Chemical Engineering Texas A&M University
College Station, Texas 77843-3122, USA

Prepared for Presentation at
American Institute of Chemical Engineers
2017 Spring Meeting and 13th Global Congress on Process Safety
San Antonio, TX
March 26 – 29, 2017

Systemic approach to improve learning from incidents

Juliana Guarguati Ariza
Mary Kay O'Connor Process Safety Center
Artie McFerrin Department of Chemical
Engineering Texas A&M University
College Station, Texas 77843-3122, USA
Juliana_guarguati@tamu.edu

Noor Quddus, M. Sam Mannan
Mary Kay O'Connor Process Safety Center
Artie McFerrin Department of Chemical
Engineering Texas A&M University
College Station, Texas 77843-3122, USA

Keywords: Learning from incidents, organizational learning, learning from experience, safety knowledge, incident investigation process

Abstract

Learning from incidents has always been very challenging for industry and regulators due to the associated complexity in human and organizational behavior. In spite of the efforts and resources spent on incident investigation processes and organizational learning techniques, the outcome has not shown a significant improvement. Chemical process incidents such as Texas City Refinery explosion, T2 Laboratories explosion, and West Fertilizer explosion continue to occur. It is evident that as an industry we are failing to learn from past incidents. The question is if we are ready to learn, are we following the right learning approach? Companies should adopt a systematic learning approach where the collected knowledge is leveraged in order to enhance safety management systems and consequently prevent future incidents. The objective of this research is set to enhance our understanding on how a company and the industry as a whole learn from past incidents and define the key elements to improve the systematic learning. The limitations in the learning system are identified and a framework is developed for improving the internal learning process and transferring knowledge within the company. Finally, a proposed incident investigation process will be validated through a case study.

1. Introduction

During the last three decades, the industry has seen significant incidents that had been precursors of changes in regulations and standards in the United States and the rest of the world. However, there is a failure to successfully implement and enforce those changes and make them transcend over the years. Thus, incidents become part of the history and the industry is doomed to repeat the same mistakes. Without doubt, no incident has been as harmful as Bhopal disaster in 1984, due to the enormous impact in terms of fatalities, financial losses, environmental damage, and industry reputation. This incident made organizations realize the importance of learning from incidents not only inside the organization, but also learning from external incidents that may be applicable to their operations. While no other incident has been as devastating as Bhopal, there have been several disasters such as Texas City Refinery explosion, West Fertilizer explosion, and Tianjin explosion that have had significant impact on the industry and from which the same mistakes can be identified over and over again. The question is if organizations have the required information, are they following the right learning approach? Organizations should adopt a systemic learning approach where the collected safety knowledge is leveraged in order to enhance safety management systems and make the lessons learned part of their culture.

The study of how organizations learn from incidents and manage safety to prevent process safety incidents is not a new topic within the process safety research area. However, this field has gained more attention during the last sixteen years and more areas of study such as psychology, sociology, and engineering have been studying it. As a result, learning from incidents has become a fragmented field in which different approaches have been proposed for specific scenarios or situations such as a certain part of the learning process, the group of interest or the industry in which it is applied. Back in the 70s, the theory on organizational learning proposed by Argyris and Schön highlighted the importance of learning in order to detect and respond to undesirable events [1]. This theory is still accepted and has been widely recognized in this field despite current advances in organizational learning theories. Over the years, learning from incidents has been categorized into several areas of study such as lessons learned, incident investigation and analysis, learning from incident process, and conditions for learning [2]. With respect to the role of lessons learned into learning systems, several studies have suggested that significant improvement has been made in the development of lessons learned. However, the dissemination and implementation of those lessons are still challenging and inhibit the entire learning process [3-6]. Similarly, studies regarding incident investigation and analysis highlight the importance of identifying appropriate investigation techniques and suggested new investigation approaches to support this process [7-9]. Regarding learning from incidents processes, several frameworks have been proposed with the objective of increasing effectiveness in learning [8, 10-12]. Moreover, literature regarding condition for learning described relevant factors that influence the effectiveness of learning processes [13-19]. This paper aims to develop a comprehensive approach to integrate the main stages of the learning process, enhance incident investigation processes, and finally explain the approach with the help of a case study. This is done with the objective of providing a holistic and systemic view of the learning from incidents process and exploring the concept of knowledge management into safety learning systems. This paper incorporates a psychology and engineering viewpoint in order to provide an overall analysis of learning systems in the chemical and oil and gas industry.

2. Learning from incidents system

In a learning system, three entities can be identified as components that have the ability to learn: individual, organization, and industry, as depicted in Figure 1. The current paper focusesd on the learning based on previous incidents, the implementation of lessons learned, and good practices. Learning occurs from the smallest entity to the biggest one. Thus, once an incident occurs, the individuals involved in it are going to challenge their mental models and identify the potential causes of the incident and consequently learn from experience. However, due to the complexity of the human behavior, it can be expected that a small portion of the people who experience an incident, choose to ignore it.

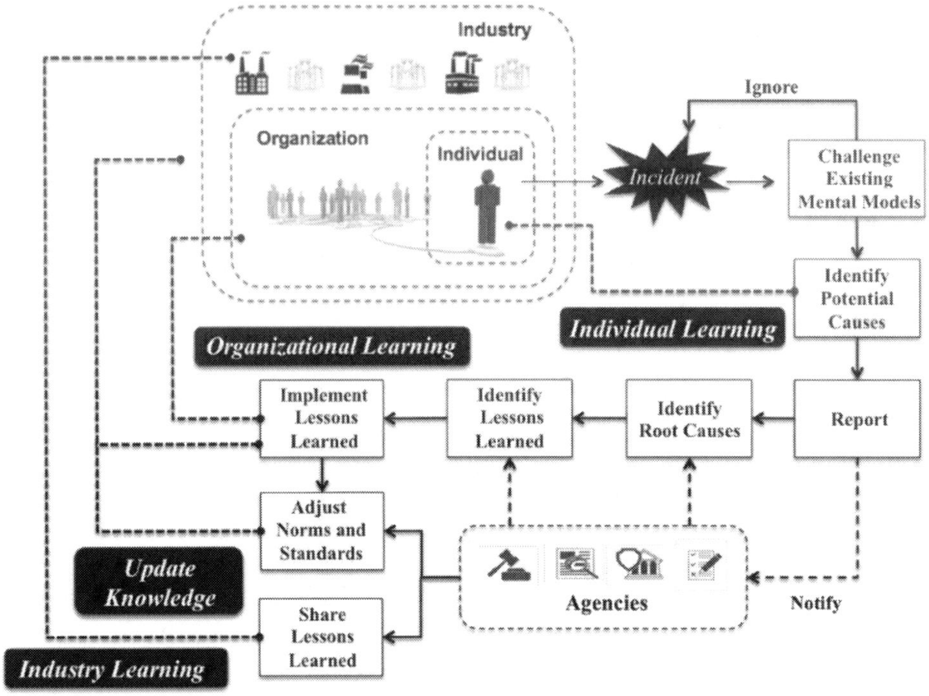

Figure 1. System of learning from incidents, adapted from [20]

Depending on the severity of the incident, reporting plays an important role in the learning system. In the case of near misses, the safety culture of the organization is challenged and the individuals involved in the near miss have to decide whether or not to report the event. On the contrary, in the case of a major incident, the process safety team is usually part of the immediate executive actions; therefore reporting is not required from the individuals that experienced the incidents to the safety team. Moreover, organizations need to notify the corresponding agencies and determine their role in the investigation process.

Once the incident is reported, the significance of the incident needs to be determined. The safety team evaluates the potential impact of the incident and based on the results, the safety team would determine whether or not the incident would be investigated in detail. This means to performs an incident investigation process, in which the root causes, lessons learned and recommendations are determined. Organizational learning occurs once individuals acquires and processes the new knowledge and collectively implements the lessons learned. The ultimate level of learning is the industry learning, in which the organizations share the lessons learned and transfer the existing knowledge across the industry.

2.1. Limitations on the learning system

Apart from explaining how the system works, the entities involved and their role in the learning process, the learning system gives a bigger picture of the main steps and sequence of the learning process. Based on figure 1, together with the available literature review in this field, it is evident that organizations still see knowledge as temporal factor that is inherent to the people involved in the organization, instead of being part of their management systems. Thus, organizations fail to transcend from individual learning to organizational learning. In addition, they fail to provide the required tools to enhance the individual learning process in order to ensure higher levels of retention and sense of ownership. Finally, the chemical and oil and gas industries fail to understand the need of putting all resources together in order to improve as a whole. In this context, several limitations that inhibit learning have been identified:

Identification of root causes: the incident investigation management systems set the basis to ensure learning from incidents, because it ensures that the fundamental causes of the incidents have been identified and managed properly. Organizations tend to identify direct causes as root causes of incidents. Thus, recommendations tend to solve superficial problems that do not prevent similar incidents in the future. Direct causes such as human error are usually identified as root cause, and investigation process ends placing the blame on operators instead of identifying failures in the management system that contributed to the execution of unsafe acts. This misinterpretation of concepts inhibits organizations from developing high quality incident reports and blocks the continuous improvement of management systems.

Promote reporting of near misses: Learning should be promoted from every valuable source of information. Acquiring data and information and analyzing it, gives the opportunity to learn and consequently prevent the occurrence of major incidents and avoid losses. In this context, reporting near misses give a significant amount of information to identify trends and high potential consequence incidents.

Ensure implementation of lessons learned: Organizations fail to support the implementation process and therefore, no actual learning is achieved. In the majority of the cases, learning is limited to the development of the report but actions are not executed.

Ensure sharing knowledge within the organization: The extracted learning from those investigations need to be spread out across the organization in order to ensure that people from different facilities, departments and even countries can get the same lessons and extract their knowledge to improve as an organization.

Use and understanding of databases and academic resources: Knowledge management has become a key element in organizational learning, in order to ensure that the data and information is stored and easily extracted when people need it. Additionally, external databases and academic resources can help organizations to extract new information and learn from external resources that handle similar processes. The main objective of those external resources is that organizations learn and improve as a whole, an industry, instead of confining learning to within an organization.

Validation of leading indicators: Process safety indicators provide evidence about the actual state of a particular process or activity. It helps organizations to get the right signals before an incident materializes and therefore, take appropriate actions to get the process state back to normal. Incident investigation should reevaluate the indicators of the process where the incident occurred and make sure operators can get the right information and understand how the process is operating.

3. Learning process

Based on an extensive review of the available literature and experience from subject matter experts on the field, a framework for the learning process for an organization has been developed. The framework suggests a standardized approach to leverage the safety knowledge within the organization. It includes seven steps to be followed in sequence in order to improve learning from incidents and experience. The steps included in the framework are intended to serve as a guideline for organizations that want to get a better understanding on how to obtain information and make it valuable. That is, transforming information into knowledge. In this sense, the learning process goal is to enhance, capture, refine, and implement safety knowledge so it can become part of the daily activities and culture of the organization.

In order to leverage the safety knowledge inside the organization, it is required to establish a corporate learning system, which should be accessible to all employees in the organization. The system should be easily understood and accessed when required. The corporate learning system has two main objectives. First, make new information and lessons learned part of the organization. This means, to make those lessons learned part of the existing resources in the organization such as procedures and guidelines. In this context, people can have access to lessons learned related with the information that they are looking for and at the time they want it. Secondly, the corporate learning system needs to support implementation and decision-making of new alternatives with the objective of making their facilities safer. In order to ensure that the information and knowledge that is being added will remain in the organization, it is important to document the whole process, make sure a proper management of change process is performed prior to any change, and link the reasons that support the alternatives. The last point is intended to ensure that people will understand the reasons of why the alternative was implemented at that time and the analysis behind it.

Sometimes people do not know how to take advantage of the available information inside the organization. They have access to a high volume of information but in many cases, they do not have the time to identify which information could be valuable for them. As a consequence, people lose relevant information that can be applied to their facilities. Therefore, transferring knowledge is a complex task, in which identifying the target and having a clear understanding of why it is important for them, becomes a crucial step to achieve learning.

The corporate learning system should be incorporated as part of the safety management system of the organization; the top management should support it, and all levels in the organization should be trained on it. The training would explain the benefits of it, the role of the corporate learning team and the people involved on it.

3.1. Corporate learning team

A corporate safety team should be in charge of managing the corporate learning system. The team has to be a group of subject matter experts with relevant knowledge in the processes, products and types of facilities inside the organization. The people in charge of the corporate learning system would serve as a source of knowledge for the different facilities and people in the organization. They would be the bridge between the information and knowledge that needs to be digested and implemented. Moreover, the team has the responsibility of filtering the information in order to ensure that the information that is being added is valid and reliable. It should be noted that the team is not intended to impose or enforce the implementation of lessons learned in the facilities; rather, the team is intended to guide facilities to get the right information and direct them in the right direction. The objective of the team is to support decision-making and provide the necessary resources to implement new alternatives. In this sense, the corporate learning team would be responsible to acquire new information, validate and analyze the applicability, disseminate the information to the intended targets, ensure the availability of the information, and support implementation as required.

3.2. Flow of information

Information and knowledge are important components in the management of any safety system. The system has to ensure the availability, accuracy and updated knowledge to support learning processes and decision-making. Inside the system two different types of information can be identified: internal and external information. Internal information refers to the information that is created inside the organization, such as incident investigation reports, good practices and performance indicators reports. Conversely, external information refers to the information that is obtained from agencies, databases, academic resources and other organizations. This information can be lessons learned, new standards, good practices, recommended practices, *etc.*

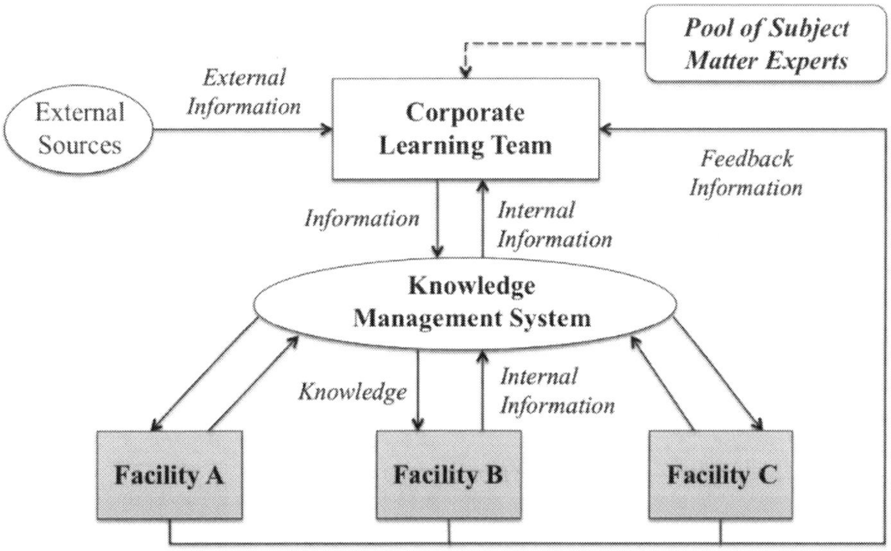

Figure 2. Flow of information in the corporate learning system

The corporate learning system has the challenge to effectively manage all channels and types of information at the right time and to the right target. Therefore, the system should be supported by a Document Management System (DMS), in which all the information would be uploaded easily and extracted as needed. In addition, the DMS needs to keep the information up to date and link the new information with the existing documents of the organization. The relevant information would then be incorporated into the DMS and linked to the existing documentations as required. Then, the team together with representatives of each facility, in which the knowledge is intend to be transferred, would work together to determine the best approach to implement that knowledge into the facility. After implementation, the facility would provide some feedback to the corporate learning team in order to ensure continuous improvement and verify the effectiveness of the process. This flow of information is represented in figure 2.

3.3. Learning process description

The learning process consists of seven steps: enhance internal information, acquire information, assess alternatives, define metrics, refine alternative, implement alternative, and monitor performance. The process describes how the corporate learning system would work and the key elements that need to be considered in each step. The first two steps refer to the acquisition of relevant data and information into the system and the last five steps refer to the transformation of that information into knowledge. A diagram of the process is presented in figure 3.

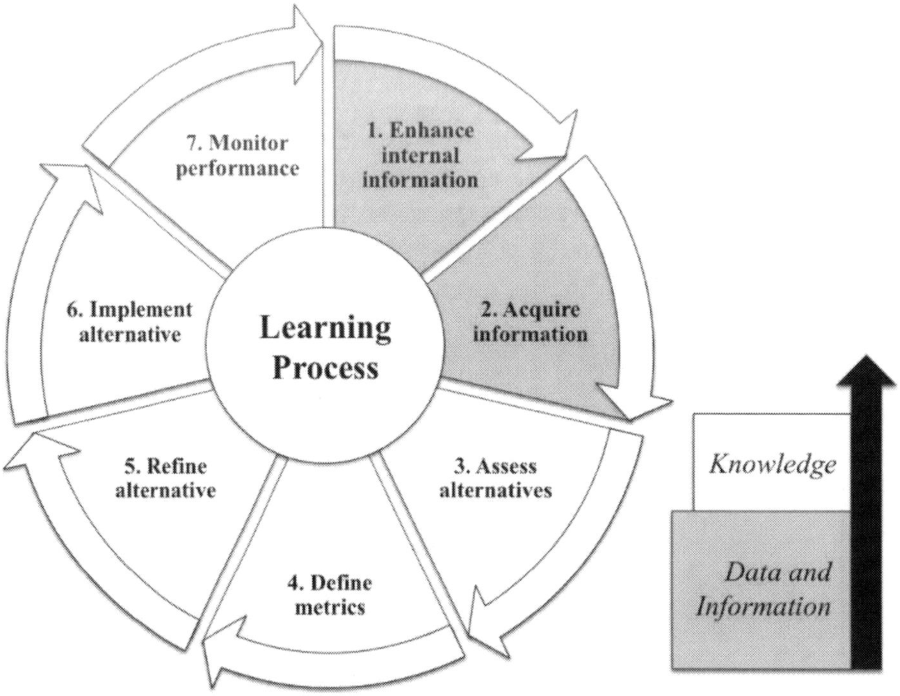

Figure 3. Steps of learning process

1. Enhance internal information: refers to the stage in which the incident investigation process is conducted and the final product of it can be improved through a more systemic analysis of the incident. Since incident investigations are one of the most powerful practices to learn from experience, organizations need to focus their attention on improving their incident investigation management programs to provide high-quality recommendations and ensure that organizational and cultural failures are identified [21, 22]. Incident investigation process refers to the sequence of stages addressed after an incident has occurred. Starting from the report of the incident, followed by the collection of evidence, incident investigation analysis, development of recommendations, communication of lessons learned, and follow-up of recommendations [23, 24]. Regardless of the incident investigation method used to investigate the incident, the main structure of the process remains constant. Thus, organizations can miss valuable information that should be analyzed during this process such as previous incidents, leading indicators of the process and how to incorporate this learning into their existing management systems. Figure 4 presents the proposed incident investigation process in which additional steps have been incorporated into the traditional process with the main goal of providing additional sources of information into the investigation. The grey boxes in the figure represent the steps that have been

included, while the white boxes refer to the original structure of the incident investigation processes.

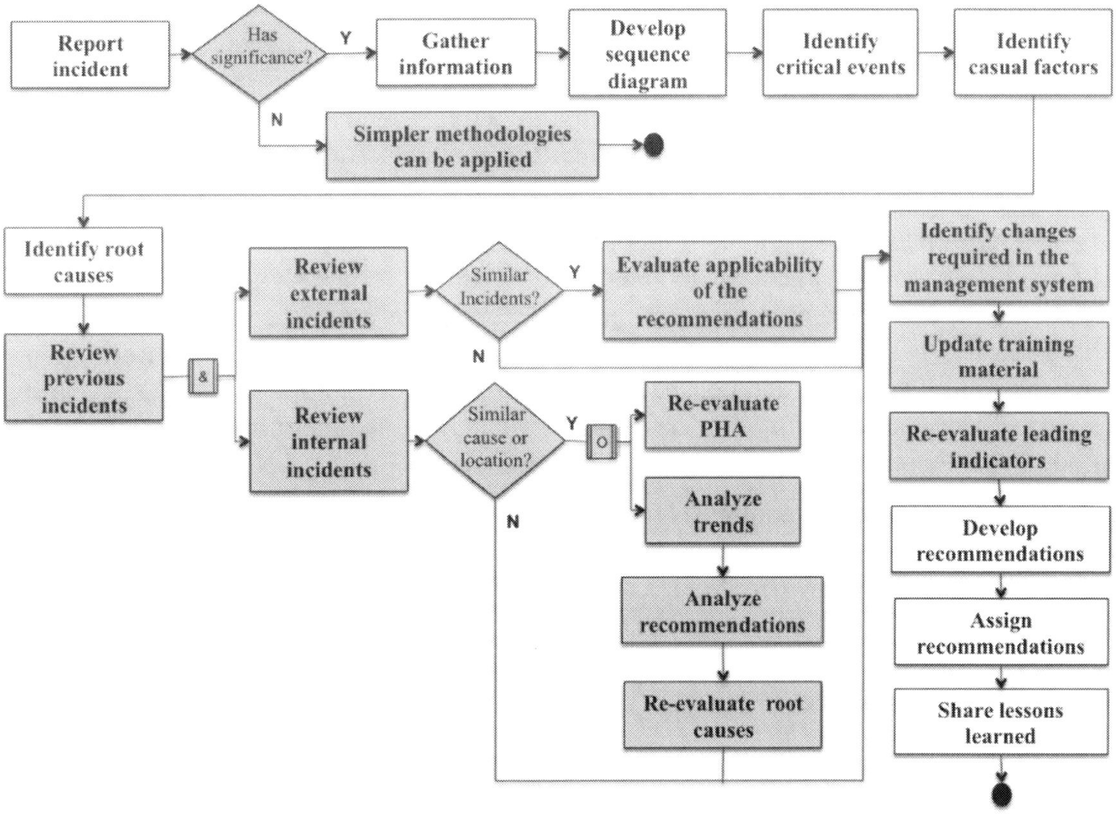

Figure 4. Incident investigation process

2. Acquire information: Once the organization can rely on the internal information that is created inside of it, the next step is to acquire additional information from different sources. In this sense, it is important to identify the potential sources from which relevant information can be extracted. As explained previously, there are two types of sources: internal and external. Internal information would be acquired from the different facilities within the organization. Likewise, internal information would be gained from all branch offices around the world that are part of the same organization. This information would be primarily lessons learned from incident investigations. However, information such as best practices and trend analysis reports can be considered as well. External information opens a wide window of opportunities from where new information can be acquired. The primary source of information would be available databases. Nowadays, there is still no single database available, which can provide full access to a large number of incidents in the chemical and oil and gas industry and, at the same time, have all the desirable features such as easily searchable, public, with multiple filters options, and with detailed analysis reports. Therefore, it is necessary to use a combination of multiple databases. Regulatory agencies such as the Environmental Protection Agency (EPA) have their own database, from which information regarding toxic releases can be found. The Chemical Safety Board [25] provides detailed analyses of process safety incidents. However, the amount of incidents investigated is limited. Some public databases such as eMARS developed by the Major Accident Hazards Bureau (MAHB) provides incidents reports involving highly hazardous substances. Likewise, private databases such as Process Safety Incident Database (PSID) created

by CCPS, provides incidents reports for organizations participating on it. Additional information can be acquired from external associations like the American Petroleum Institute (API) and the National Fire Protection Association (NFPA), which provides guidelines for standards and best practices. Finally, organizations also have the opportunity to established partnerships with academic entities in order to be involved and have access to trending investigations from first hand. This type of partnership can be established with universities/academic entities. It is evident the wide area from which safety information can be extracted. Thus, it is important that the corporate learning team set the basis and defines the scope and objectives of it. Moreover, those objectives have to be aligned with the organization's goals and business priorities.

3. Assess alternatives: Once the team has prioritized the data and information alternatives that could be valuable for the organization and fit the needs of it, they should identify the potential processes, facilities or business areas, where the alternative could be applied and would gain benefits from it. Those applicable facilities are the potential target, where the information could be translated into knowledge and consequently be implemented into their processes. During this process the corporate learning team has the responsibility of determining and classifying the potential targets and the channels of communication that are going to be used during the process.

4. Define metrics: This step requires asking the questions of what they want to achieve with this alternative? And how they are going to know that they are getting what they want? Defining performance indicators allow people to verify the system behavior and measure how much it has been changed with respect to the initial set point. Indicators should be determined prior to implementation in order to have a clear picture of what they want to achieve with it.

5. Refine alterative: Defining metrics and refining the alternative should be performed in parallel, because once the facility is fitting the alternative to their own processes, some modifications or alterations to the initial idea might be necessary. Refining the alternative refers to the process in which the facility has to define the limitations, assumptions, boundaries and required resources based on their own needs and capacity.

6. Implement the alternative: refers to the actual execution of the action plan that has been proposed by the facility. Each facility would be responsible for the execution and verification of it. However, the corporate learning team should verify periodically its current status. Likewise, the corporate learning team would support the whole process based on what has been agreed during the evaluation phase of the alternative.

7. Monitor performance: refers to continuous improvement in order to validate and refine the implemented alternatives. At this point, a detailed review should be performed to verify if the knowledge was successfully transferred, if the knowledge was appropriate, was analyzed in detail, accurate criteria was used, and the expected benefits were achieved.

4. Case Study

The proposed incident investigation process (Figure 4) presented as the first step of the learning process was validated through the development of a case study. For this study, an incident in the offshore industry was selected and analyzed within incorporated steps. The complete investigation was performed by the organization where the incident occurred. Therefore, the objective of this analysis was to identify improving opportunities of the final investigation report. The analysis comprises the analysis of two external incidents within the oil and gas industry and three incidents that occurred within the same organization.

4.1. Platform Fire and Explosion incident

A high-pressure fuel gas line ruptured on a platform releasing high quantities of high-pressure gas, which detonated within the next 5 seconds. The explosion and resulting fire killed 7 people and destroyed a significant part of the platform. The line was designed as a fuel gas separator bypass in order to provide operational flexibility. Thus, the line was used occasionally. Prior to the incident, the line had been inspected two years and eight years back in which a visual inspection, thickness measurement, and materials characterization were performed. For those inspections, the results were satisfactory and no reduction of the internal thickness was detected, with the exception of a weldolet that was changed due to a severe external corrosion. The investigation found that the line rupture was due to a severe localized reduction of the internal thickness of the line. The laboratory analysis determined that the reduction of the internal thickness was the result of a combination of a microbiologically influenced corrosion (MIC) and corrosion associated with high levels of hydrogen sulfide (H_2S).

4.2. Step by step analysis

The first step incorporated into the incident investigation process refers to the preliminary analysis in which the significance of the incident is determined. This means to categorize the incident based on the potential impact that the incident may have. This is done with the objective of determining the level of detail and complexity of the investigation. In this study, the significance of the incident was determined as very high due to the number of fatalities, injuries and property damaged. Thus, the incident was subject to further analysis during the investigation.

The following steps refer to the traditional incident investigation process in which the incident investigation team is in charge of the collection of evidence. Followed by the implementation of investigation methodologies to reconstruct the sequence of events, identify the critical and causal events and identify the root causes of the incident. For this case study, the results of these analyses were taken from the incident investigation report given by the organization where the incident occurred. Once the root causes of the incident have been identified, the proposed incident investigation process suggests a detailed review of previous incidents in order to evaluate the applicability of recommendations, analyze trends, and re-evaluate the identified root causes of the incident. Reviewing previous incidents take into consideration two different types of incidents: internal and external incidents.

4.2.1. Reviewing external incidents

In the external incident review, two incidents were considered into the case study. The first incident was BP North Slope Oil Spill in 2006 [26] and the second incident was Natural Gas pipeline Rupture and Fire in Carlsbad, New Mexico in 2000 [27]. These two incidents were selected based on the criteria of cause similarity compared with the incident being studied. Both incidents occurred in the oil and gas industry, occurred prior to the incident that is being analyzed, and were caused by internal corrosion in pipelines. The objective of reviewing external incidents is not only to analyze the applicability of their recommendations. It is also to analyze the current practices that others organizations are implementing, compare their own technology with respect to others organizations, and extend their existing knowledge regarding national and international standards and best practices that may be applicable to their facilities. Moreover, reviewing external incidents serves as tool for building trust and credibility with regard to the results that the incident investigation team is presenting. For this case, the analysis of external incidents into the investigation suggested improving opportunities to the final incident report developed by the organization and provided additional recommendations that can be implemented by the organization. Some of the obtained results are briefly summarized below.

- The final report of the incident developed by the organization presented a detailed analysis regarding the microbiologically influenced corrosion (MIC) and hydrogen sulfide corrosion(H_2S), supported by laboratory analysis and subject matter experts. However, the assimilation process from the operators, the supervisor and managers presented a challenge. Even though, the evidence was there, it was hard for them to believe "such unexpected event". Therefore, some representative statistics in the offshore industry were incorporated and analyzed in order to highlight the frequency of this type of incidents in the offshore industry and consequently failure to learn from them.

Additionally, the analysis recommended:

- The evaluation of a most suitable technology for internal corrosion monitoring in the fuel gas pipeline system, such as the implementation of smart pigging, ultrasonic testing, and/or gas monitoring systems. Likewise, the evaluation of the feasibility for implementation of state of the art technology such as molecular microbiological methods in order to monitor bacteria activity.
- Enforcing the management of change program for the implementation of any new technology or system in which the required training, responsibilities, procedures and channels of communication must be specified.
- Performing laboratory analysis of some of the portions of pipeline that have been replaced during inspections, in order to identify additional potential mechanisms of corrosion.
- Revising and updating the corrosion management program, policies and internal guidelines in order to ensure that are aligned and comply with all the applicable national and international standards. The investigation highlighted API standards. However, additional standards can be taken into consideration. Mentioned but not limit to: NACE SP0106-2006, NACE TM0106-2006, NACE SP0206-2006, ASME B31.8, ASME B31.8S, and ANSI GPTC Z380.1.
- Validating the corrosion management program in order to ensure that all types of corrosion threats are assessed and a risk assessment is performed. The program should incorporate the following threats: internal corrosion threat, external corrosion threat, safety/hazard threat, environmental threat and operability threat.
- Ensuring that safety management programs of the organization are aligned with the process hazard analysis and audit programs performed in the facility. Ensure appropriate channels of communication, training and qualified personnel in order to determine effective prevention and mitigation measures.

4.2.2. Reviewing internal incidents

For the internal incident review, three incidents were considered into the case study. All three incidents occurred on the same platform 3 months and one-year back prior to the incident that is being analyzed. While reviewing internal incidents, two approaches can be taken. First, analyzing if a significant number of incidents occurred at location of the current incident. The objective is to analyze the need of re-evaluating the process hazard analysis of the process. Secondly, analyzing similar incidents that had occurred in the same organization, analyze the trends and analyze the applicability of those recommendations to the incident that is being investigated. Finally, the incidents investigation team should go back to the identified root causes and re-evaluate them based on the identified findings. For this case, the analysis of internal incidents identified improving opportunities in the following areas:

- Although there is no record of a significant number of incidents in the same location where the incident occurred, the investigation revealed that the process hazard analysis performed in that process did not consider the different types of corrosion threats that can be recognized in fuel gas systems. Thus, it is required to re-evaluate the PHA and ensure that is executed by qualified personnel.
- The development of recommendations in previous incident investigation reports is limited to the resolution of direct causes such as physical or human causes (changing a valve, replacing a section or employee temporary suspension). Although, root causes analysis identified organizational failures as root causes of the incident, there is no evidence of action plans or recommendations for these types of causes.
- During the analysis of previous incidents, two out three of these incidents, emphasize a failure to properly analyze and take actions of the inspections results developed by third parties. The incident used for this case study exposes the same cause in which the supervisor in charge was not able to analyze inspections results due to lack of knowledge and training regarding this type of data. This is an illustration of how a lack of commitment with respect to the implementation of recommendations can lead to big catastrophes, such as the one that is being analyzed.
- The analysis identified lack of management involvement with respect to results of recommendations of minor incidents or near misses. As a result, the required resources and time for implementing recommendations and ensuring an organizational learning are limited.

4.2.3. Identifying changes required in the management system and updating the training material

Following the review of previous incidents, the next step in the proposed process is the identification of the changes needed in their organization's management systems, which are visible during the development of the incident investigation. This stage of the investigation aims to provide more insights into the identification of root causes and successfully uncover hidden faults of the system.

Through the analysis of the identified causes of the incident, review of previous internal and external incidents, and an analysis of the current state of their management system, relevant changes required in the management system were identified for this case study. First, the analysis revealed that changes in the current training needs matrix are necessary to ensure that personnel have the required knowledge, skills and experience to perform their responsibilities on the platform. This means to validate that the required training needs for a specific role is up to date and the identified gaps are filled by the organization. Similarly, the management of change program should be re-evaluated to ensure that changes in technologies or processes take into account the potential impact on training needs, roles responsibilities, and channels of communications among the organization. Moreover, the investigation suggested an overall change in the safety culture of the organization in which the level of engagement and awareness of the top management need to be re-evaluated, as well as the operational discipline and incident investigation program. Additional changes were recommended in the following areas: planning phase in the mechanical integrity program and qualification criteria for personnel in charge of the development of process hazard analysis.

Aligned with the analysis performed in this step, it is also important to evaluate which training material should be updated based on the technical gaps that were identified during the investigation. For instance, during the development of this case study, technical and organizational gaps such as internal corrosion monitoring program, identification of corrosion threats, and applicable corrosion standards and best practices we identified as current gaps in their training material. Similarly, the incident investigation training material should be validated in order to ensure that proper methodologies for root cause identification are implemented within the organization and the basis of an effective incident investigation program are covered in this training.

4.2.4. Closing up the process

The next step in the analysis is to re-evaluate the leading indicators of the process where the incident occurred with the objective of identifying if the process is showing the right signals and making sure that operators understand the data and information the system is giving. The case study identified that appropriate indicators were in place. However, performance indicators reports were usually presented without any analysis behind it, making difficult to understand the actual state of the process. Once the complete analysis has been performed, the next step is to develop and assign recommendations based on the identified findings. Finally, the recommendations can be implemented and lessons learned communicated.

This case study was developed with the objective of explaining the proposed incident investigation process and providing a clearer picture of how it can be implemented within the existing incident investigation programs. Moreover, it illustrates how investigations can be enhanced through a more detailed analysis in the identification of root causes and consequently achieves high-quality recommendations.

5. Discussion

This paper presents a systemic approach to improving learning from incidents, through the development of a corporate learning system, which provides guidelines of how corporate learning systems can be executed within the organization and how to support the implementation of safety knowledge inside of it. The framework could serve as a useful tool to disseminate and analyze safety knowledge across the organization. Likewise, it could be useful to support single and double-loop learning since it converges a detailed analysis of the underlying causes of incidents, provides the opportunity for analysis from different perspectives and the examination of potentially valuable information for the organization. Similarly, the proposed learning system gives some empirical and theoretical insights into the implementation of learning incidents into real organizations for the chemical and oil and gas industry. In this context, the framework aims to provide a more integrated perspective to this field by the incorporation of psychological and engineering inputs into the analysis. Moreover, this paper is intended to give a bigger picture of the learning from incidents process and overcome some of the limitations that the industry faces. Finally, the proposed incident investigation process was explained through the development of case study, which aims to validate the proposed steps and illustrate how it can be implemented into real incident investigations.

The resulting framework aims to overcome the identified limitations of the learning system. However, the scope of it is limited to individual and organizational learning. The authors of this paper believe that industry learning cannot be achieved if a well-established learning system is not in place in each organization. Thus, the study of this paper is restricted to enhance internal learning in an organizational level. The identified limitations have been addressed as follow:

- The implementation of lessons learned and sharing knowledge within the organization have been the main focus of this study. Even though, the available literature highlights the importance of sharing and implementing lessons learned [5, 28], less attention has been given on how to ensure that lessons learned are not implemented only in the place where the incident occurred and to ensure that the lessons are not only communicated and disseminated, but also analyzed and documented for all the facilities where it may be applicable. Similarly, sharing knowledge within the organization is encouraged in the corporate learning system by the flows of information in and out that are established from and to all facilities that are part of the organization.

- The use and understanding of academic resources and databases are encouraged in the second step of the proposed learning process: acquire information. The identification of external resources has been identified as the main stage of this process in order to get valuable safety information that may be applicable to the organization. Thus, the importance of using databases, external associations, and academic resources has been discussed.

- The identification of root causes is enhanced by the incorporation of additional steps into the traditional incident investigation process. These steps provide some guidelines to the incident investigation team in terms of identifying more insights into the analysis and more sources of information in order to have a bigger picture of the organizational, technical and cultural failures of the area where the incident occurred. Additionally, the proposed incident investigation process encouraged the validation of leading indicators of the process in order to ensure that operators and supervisors are getting the right signals to identify technical failures in the system.

6. Conclusion

A systemic framework for improving learning from incidents has been developed for the chemical and oil and gas industry based on the identified limitations of the learning system. It consists of seven steps that enable organizations to acquire relevant safety information, absorb and transform it into valuable safety knowledge that can be incorporated into the organizational management systems. The framework provides a holistic view of the learning from incident process and explores the concept of knowledge management into safety learning systems. Additionally, the limitations of the learning system and the main elements that organizations should take into account to improve learning have been discussed in order to enhance understanding of this field. Finally, the proposed incident investigation process was validated through the development of a case study in which an incident investigation report was enhanced by the identification of additional improving opportunities for the existing safety management system of the organization.

7. References

[1] Argyris, C. and D.A. Schön, *Organizational learning: a theory of action perspective*. Vol. 173. 1978: Addison-Wesley Reading, MA.

[2] Drupsteen, L., J. Groeneweg, and G.I. Zwetsloot, *Critical steps in learning from incidents: using learning potential in the process from reporting an incident to accident prevention*. International Journal of Occupational Safety and Ergonomics, 2013. **19**(1): p. 63-77.

[3] Kletz, T.A., *Lessons from disaster: How organizations have no memory and accidents recur*. 1993: IChemE.

[4] Duffield, S. and S.J. Whitty, *Developing a systemic lessons learned knowledge model for organisational learning through projects*. International Journal of Project Management, 2015. **33**(2): p. 311-324.

[5] Gordon, H.J. *Integrating learning into safety*. in *ASSE Professional Development Conference and Exposition*. 2006. American Society of Safety Engineers.

[6] Keegan, A. and J.R. Turner, *Quantity versus quality in project-based learning practices*. Management Learning, 2001. **32**(1): p. 77-98.

[7] Lindberg, A.-K., S.O. Hansson, and C. Rollenhagen, *Learning from accidents–what more do we need to know?* Safety Science, 2010. **48**(6): p. 714-721.

[8] Lukic, D., A. Littlejohn, and A. Margaryan, *A framework for learning from incidents in the workplace*. Safety Science, 2012. **50**(4): p. 950-957.

[9] Fahlbruch, B. and M. Schöbel, *SOL–Safety through organizational learning: A method for event analysis*. Safety Science, 2011. **49**(1): p. 27-31.

[10] Le Coze, J.C., *What have we learned about learning from accidents? Post-disasters reflections*. Safety Science, 2013. **51**(1): p. 441-453.

[11] Cooke, D.L. and T.R. Rohleder, *Learning from incidents: from normal accidents to high reliability*. System Dynamics Review, 2006. **22**(3): p. 213-239.

[12] Avnet, M.S., *A network-based approach to organizational culture and learning in system safety*. Procedia Computer Science, 2015. **44**: p. 588-598.

[13] Drupsteen, L. and F.W. Guldenmund, *What is learning? A review of the safety literature to define learning from incidents, accidents and disasters*. Journal of Contingencies and Crisis Management, 2014. **22**(2): p. 81-96.

[14] Pidgeon, N. and M. O'Leary, *Man-made disasters: why technology and organizations (sometimes) fail.* Safety Science, 2000. **34**(1): p. 15-30.

[15] Carroll, J.S. and B. Fahlbruch, *"The gift of failure: new approaches to analyzing and learning from events and near-misses." honoring the contributions of Bernhard Wilpert.* Safety Science, 2011. **49**(1): p. 1-4.

[16] Sanne, J.M., *Learning from adverse events in the nuclear power industry: Organizational learning, policy making and normalization.* Technology in Society, 2012. **34**(3): p. 239-250.

[17] Jacobsson, A., Å. Ek, and R. Akselsson, *Method for evaluating learning from incidents using the idea of "level of learning".* Journal of Loss Prevention in the Process Industries, 2011. **24**(4): p. 333-343.

[18] Hovden, J., F. Størseth, and R.K. Tinmannsvik, *Multilevel learning from accidents–Case studies in transport.* Safety Science, 2011. **49**(1): p. 98-105.

[19] Naot, Y.B.-H., R. Lipshitz, and M. Popper, *Discerning the quality of organizational learning.* Management Learning, 2004. **35**(4): p. 451-472.

[20] Pasman, H.J., *Risk Analysis and Control for Industrial Processes-Gas, Oil and Chemicals: A System Perspective for Assessing and Avoiding Low-probability, High-consequence Events.* 2015: Butterworth-Heinemann.

[21] CCPS, *Guidelines for investigating chemical process incidents.* Second edition ed. Vol. second edition. 2003: Center for Chemical Process Safety of the American Institute of Chemical Engineers.

[22] Energy Institute, L., *Guidance on investigating and analysing human and organisational factors aspects of incidents and accidents.* 2008: Energy Institute, London.

[23] Mannan, S., *Lees' Loss Prevention in the Process Industries, Hazard Identification, Assessment and Contro*, in *Lees' Loss Prevention in the Process Industries,*. 2005, Elsevier Inc.

[24] Sklet, S., *Methods for accident investigation.* 2002, Norwegian University of Science and Technology.

[25] CSB, U.S., *West fertilizer final investigation report* in *West fertilizer explosion and Ffre* 2016, U.S. Chemical Safety Board: U.S. Chemical Safety Board.

[26] Bailey, A., *BP: Learning from oil spill lessons*, in *Petroleum News.* 2006, Petroleum News.

[27] NTSB *Natural Gas Pipeline Rupture and Fire Near Carlsbad, New Mexico.* Pipeline Accident Report, 2003.

[28] Kletz, T.A., *Learning from experience.* Journal of Hazardous Materials, 2004. **115**(1): p. 1-8.

Leading and Lagging Indicators

Robert J. Weber, PE
President, PSRG

Sonny Sachdeva, PhD
Principal Risk Engineer / Training Director, PSRG

Process Safety and Reliability Group
800 W. Sam Houston Pkwy. South
Suite 107
Houston, Texas 77042-1908
Tel: 713-532-8800 800-250-8511 Fax: 713-532-8850
Email: psrghouston@psrg.com Web: www.psrg.com

13TH GLOBAL CONGRESS ON PROCESS SAFETY

Introduction

Process Safety and Reliability Group

"It should not be necessary for each generation to rediscover principles of process safety which the generation before discovered. We must learn from the experience of others rather than learn the hard way. We must pass on to the next generation a record of what we have learned."

~ Jesse C. Ducommun
Vice President, Manufacturing and a director of American Oil Company in 1961; Process Safety Pioneer

Content

- Indicators
  - Lagging
    - Characteristics
    - Evaluating
  - Leading
    - Characteristics
    - Evaluating
- Other Considerations
- Results of an Effective Leading indicators Program

What are Indicators?

Indicators are <u>measurements</u>:

- Achievement levels or amounts
  - 93%
- Presence or Absence of information or activity
  - Absent
- Quality
  - Meets XYZ standard

A comprehensive leading and lagging indicator program provides useful information for driving improvement and when acted upon, contributes to reducing risks of major hazards (by identifying underlying causes and taking action to prevent reoccurrences.

Examples of H&S Indicators

Process Safety and Reliability Group

Lagging and Leading Health and Safety Indicators

Lagging Indicators	Leading Indicators
Total lost work days	Safety/health meetings
Restricted work days	Supervisor training
Number of fatalities	Employee training
Injuries/illness rate	Number of inspections
Asset/property damage	No. of audits/surveys
Vehicle mishaps	No. of self-inspections
Near-miss incidents	Reward/recognition
Chemical releases	Employee turnover rate
WC trends and amount	Observations/accidents
Experience modification	Risk/hazard assessment

Why use Indicators?

Process Safety and Reliability Group

One way to improve the effectiveness of your safety process is to change the way it is measured and to act upon the information gained.

This is part of the standard Quality Loop.

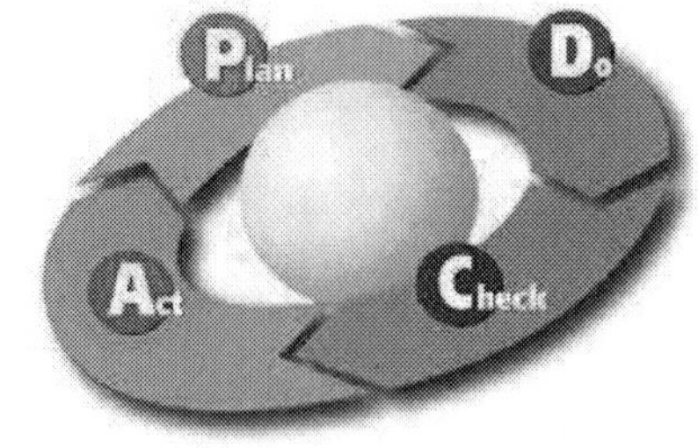

Checking leads to Acting leads to Learning!

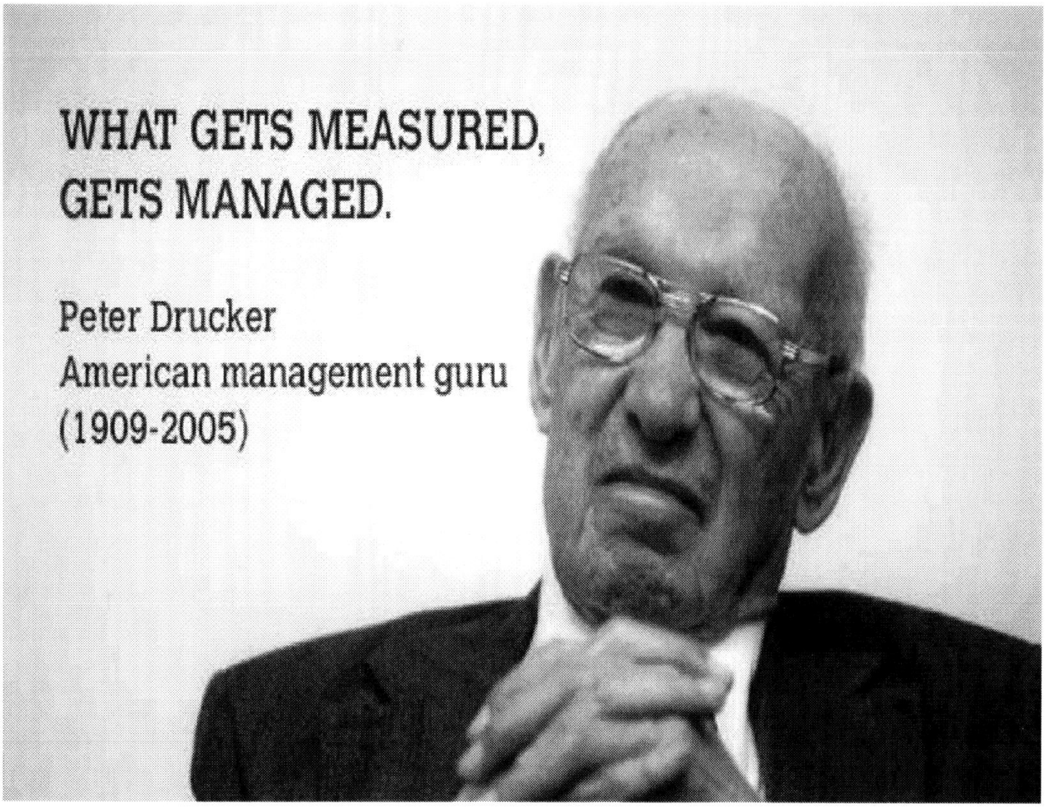

Process Safety and Reliability Group

Inputs and Outputs of Safety Management Programs

Inputs

(leading indicators)

Outcomes

(lagging indicators)

What is a Lagging Indicator?

Process Safety and Reliability Group

A Lagging Indicator is a Measurement that:

- Focuses on results at the end of a time period
- Normally characterizes historical performance

Characteristics of Lagging Indicators of Safety Performance

Process Safety and Reliability Group

- Measured at the end of a time period
- Easy to identify and capture
- Lag behind current system processes
- Are NOT normally predictive

Good for setting goals and objectives

Examples of Lagging Indicators of Safety Performance

Process Safety and Reliability Group

- Injury frequency and severity
- Lost workdays
- Restricted workdays
- Worker's compensation costs
- Spills
- Releases

Drawbacks of Lagging Indicators

Process Safety and Reliability Group

- Injuries / Illness / Property Damage has already occurred
- Easy to identify and measure; therefore, often over-relied upon by ignoring Leading Indicators
- Generally provide limited answers about relationships between causes and effects, so only broad accident preventive measures can be taken.

Low Incident Rate - Does Not Equal Low Risk

Inputs and Outputs of Safety Management Programs

Process Safety and Reliability Group

Inputs

(leading indicators)

Outcomes

(lagging indicators)

What is a Leading Indicator?

Process Safety and Reliability Group

A Leading Indicator is a Measurement that:

- Is believed to precede, and consequently reduce injury/incident rates.
- A measurement used for its ability to measure incremental progress, or quality, or to indicate the direction of future results (predictive).

Characteristics of Leading Indicators of Safety Performance

Process Safety and Reliability Group

- Difficult to identify and capture
- Can be developed from lagging indicators
  - Lagging – number of vehicle mishaps
  - Leading – number of times cell phone use while driving discussed in safety meetings
- May differ from current system processes
- Should be predictive
- Measure Behaviors

Good for identifying areas to on which to focus

Examples of Leading Indicators of Safety Performance

Process Safety and Reliability Group

- Numbers of near misses
- Numbers of team safety meetings
- Backlog on equipment inspections
- Numbers of safety suggestions
- Safety performance appraisals
- Inspections, action items overdue

Examples of Leading Indicators of Safety Performance

Process Safety and Reliability Group

- Inspections and Observations
- Employee Involvement in Program Implementation
- Employee turnover rate
- Risk/Hazard assessment
- Safety/Health meetings

Drawbacks of Leading Indicators

Process Safety and Reliability Group

- Difficult to identify and measure; therefore, ***often under-relied upon*** in favor of Lagging Indicators
- Reliance on Personnel Safety Leading Indicators may provide a false sense of security when not coupled with Process Safety and Culture Leading Indicators.

Low Prediction of Risk Does Not Equal Low Risk

Why use Leading Indicators?

Process Safety and Reliability Group

To Prevent Losses

- Leading indicators provide warning signs
  - Leading indicators are warning signs of failure in complex systems.
- Complex systems can be:
  - 1) processes,
  - 2) equipment executing processes,
  - 3) groups driving the processes.

All three can fail. All three can provide warning signals. Therefore, all three must be measured.

Leading Indicators

Process Safety and Reliability Group

An aggressive quality-controlled leading indicator program AT BEST:

- Provides an effective system to put the lessons learned from accidents, incidents and near misses into practice.

While it prevents incidents and lowers losses, it DOES NOT prevent all incidents.

Evaluating Leading Indicators

- Are the indicators being documented?
  - If not, are they documentable?
- What do the quality control checks on the indicators say?
  - If they indicate poor quality, identify why
- Reserve the right to not use the data if the data does not meet requirements.

GARBAGE IN EQUALS GARBAGE OUT (GIGO)

Evaluating Leading Indicators

PSRG has audited 100s of industrial plants around the world; of all sizes from boutique specialty chemical toll-plants to large-scale refineries.

Our audits indicate that a wide range (10-90%; avg. 70%) of small incidents that are investigated do NOT find contributing or root causes.

GARBAGE IN EQUALS GARBAGE OUT (GIGO)

Dynamics of Accident Causation

Process Safety and Reliability Group

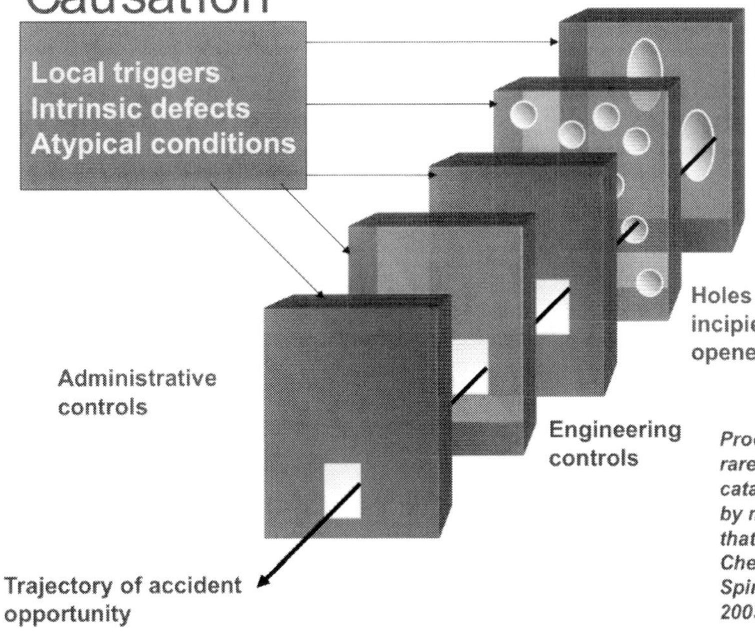

Process safety incidents are rarely caused by single catastrophic failure, but rather by multiple events or failures that coincide. Ref: Swiss Cheese (Static; Reason, 1990), Spinning Disk (Dynamic; Hart, 2003).

Incident Pyramid (API RP754)

Process Safety and Reliability Group

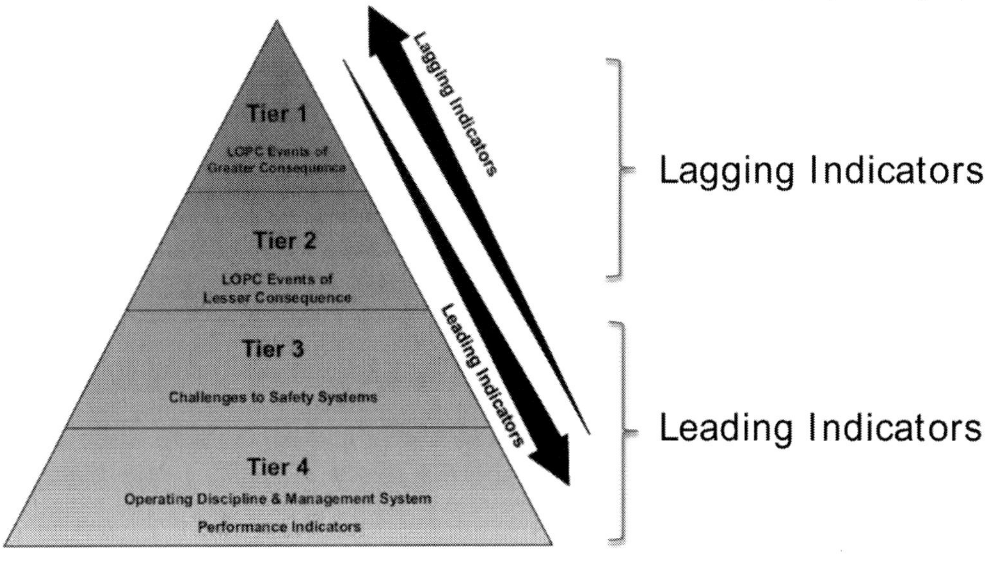

Originally based on concepts of Heinrich Accident Pyramid; introduced 1931 in insurance industry.

Statistical Analysis

- Statistics LIE when there is no data quality control – Garbage In Equals Garbage Out
- Trending is presentation, and needs quality control and analysis to be useful
- Statistical methods can show the relationship between latent (underlying) causes and the specific failures they create
  - For example, when an operational procedure change leads to subsequent injuries

Other Considerations

- Characteristics of Best In Class Safety Management Systems
- Characteristics of Best In Class Measurement of Safety Performance (i.e., use of indicators)

Safety Management Systems (SMS) – Best in Class Systems

Process Safety and Reliability Group

- Establish and apply a coherent set of measures of safety performance
- Link the set of safety performance measures coherently to the set of business performance measures

Safety Management Systems (SMS) – Best in Class Systems

Process Safety and Reliability Group

- Measuring indicators selected from the inputs to the safety management system requires a change in the way of thinking
- Companies who put greater effort into developing their SMS have better performance and this effort rubs off profitably onto other business systems

Measurement of Safety Performance – Best In Class

- Leaders in safety performance use metrics that effectively drive their continuous improvement efforts.
- Leading indicators (e.g. observations) are used to predict changes in safety performance
- Monitoring of the safety process versus program(s) implemented at all sites

Measurement of Safety Performance – Best In Class

- Performance targets are defined and communicated
- Review targets based on feedback from staff, metrics and benchmarking
- Safety performance targets should align with business targets

Measurement of Safety Performance – Best In Class

Process Safety and Reliability Group

- Discipline is consistent with well-developed guidelines
- Leading companies ensure in depth root cause analysis of accidents in formulating disciplinary actions
- Safety performance (operational) should be tied to the merit; this is consistent among leaders in business

Metrics – Employee Participation

Process Safety and Reliability Group

- Percentage of workers and managers trained on EP and their responsibilities
- Percentage of workers who have participated in key defined EP activities over the last 12 months. E.g.
  - Submitting a suggestion
  - Serving on a risk analysis team
  - Participating in an investigation
- Rate of submittal of worker suggestions
- Number of suggestions that have not been evaluated
- Number of accepted suggestions that have not been implemented
- Percentage of suggestions accepted

Metrics – PSI

Process Safety and Reliability Group

- Number or percent of blank records in PSI database
- Number of PHA recommendations that indicate of less than adequate PSI where information was available (but difficult to locate).
- Results of periodic surveys to determine if users of process knowledge believe it is accessible.
- If process safety information is web-based, the number or percent of dead links.
- Accuracy of process safety information during periodic reviews.
- Number of change requests initiated to "correct" process safety information.
- Engineering staff time spent on recreating process safety information.
- Average number of days required to have a drawing revised.
- Number of random checks of SDS files to determine if they are complete and current.

Metrics – PHA

Process Safety and Reliability Group

- Number of PHA overdue
- Number of audit findings related to PHA
- Percentage of recommendations for administrative controls, active engineered controls, passive and inherent safer alternatives – whether early life cycle reviews were effective
- Number of recommendations per study
- Number of unresolved recommendations
- Percentage of repeat recommendations
- Average time corrective actions require for completion.
- Percentage of recommendations rejected by management
- PHA action items overdue

37

Metrics – SOPs

- Number or percent of SOPs updated per year or staff hours spent in updating
- Number of incident/deficiency reports related to procedures that were unclear or not understood
- Number of procedures past due for review
- Mean time to correct or update procedures
- Number of audit findings related to procedures missing some element of required content.

38

Metrics – Training

- Percentage of incidents with training and performance as root causes.
- Number of SMEs providing training.
- Percentage change in training budget.
- Percentage of course deliveries audited.
- Number of workers of each type whose training is overdue.
- Percentage of workers who miss a scheduled training session.
- Percentage of training sessions that are offered on schedule.
- Time spent in training for individuals, shifts, departments and job functions.
- Percentage of workers who believe training is appropriate and effective.
- Average test scores for classes, individuals, shifts departments and job functions.
- Percentage of workers who require refresher training.

Metrics – Safe Work Practices

Process Safety and Reliability Group

- Percent compliance with training plan for activities related safe work practices.
- Progress towards making significant improvements to the existing system.
- Percent of permits completed correctly.
- Number of injuries/near-miss related to HW or other SWPs.
- Unsafe conditions or permit violations observed during routine audits (on a consistent basis).
- Number of causal factors identified by incident investigation teams related to failure of properly apply or follow the HW permit.
- Percent of scheduled job observations/audits performed.
- Average time between a request for a permit and when it is issued.
- Staff time spent writing, reviewing and approving safe work procedures.

Metrics – Contractors

Process Safety and Reliability Group

- Frequency of required waivers of qualification requirements for past safety performance or current safety program.
- Percentage of contracted firms that, based upon post-job evaluation, would be considered for future contracts.
- Percentage of required contractor training conducted on schedule.
- Percentage of contractor related incidents or near miss that were subjected to a root-cause analysis.
- Frequency of and percentage attendance for contractor safety meetings.
- Number of safety program improvement suggestions contributed by contractor personnel.
- Number of open contractor safety suggestions.

41

Metrics – MOC

Process Safety and Reliability Group

- Number of MOCs performed each month.
- Monthly average in the percent of work requests classified as change.
- Percentage of or variation in the number of changes processed on an emergency basis.
- Percentage of personnel involved in the MOC system who believe the system is effective.
- Average backlog of MOCs/active MOCs.
- Average amount of calendar time taken between MOC origination and authorization.
- Percentage of work orders/requests that were misclassified as R-I-K and were really changes.

42

Metrics - MOCs

Process Safety and Reliability Group

- Percentage of changes that were reviewed within MOC system but were reviewed incorrectly.
- Percentage of changes that were properly evaluated but did not have all authorization signatures.
- Percentage of MOCs reviewed that were not documented properly.
- Percentage of MOCs for which the drawings and procedures were not updated.
- Percentage of MOCs for which the workers were not informed or trained.
- Percentage of temporary MOCs for which temporary conditions were not corrected/restored to the original state at the deadline.

43

Metrics – M.I.

Process Safety and Reliability Group

- Safety critical equipment inspections completed on time
- Safety critical equipment deficiency management
  - Response to safety critical inspection findings (e.g., non-functional PRDs, SIS) to maintain continued safe operations.
- Number of inspectors/maintenance employees holding each type of required certification – skills gap
- Number of overdue ITPM tasks
- Number of emergency/unplanned work orders
- Work order backlog for the inspection group
- Total time charged to ITPM tasks each month
- Average time to address/correct deficiencies

44

Metrics – Incident Investigation

Process Safety and Reliability Group

- Number of repeat causes.
- Actual and potential losses from incidents.
- Number of qualified incident investigation leaders.
- Ratio of accident to near-miss reports by Shift, Department, or Unit.
- Number of incidents reported per unit time.
- Average time to initiate investigation.
- Average time to complete investigation reports.
- Average time to resolve recommendations.
- Number of times recommendation completion dates are revised.
- Number of lesson learned communications.
- Percentage of incident investigation reports where “root cause” identified.

ERP Metrics

- Number of meetings or other contacts with local emergency responders or the LEPC regarding emergency response plans.
- Number of meetings or contacts with community regarding how they will be notified of an emergency and what they should do they are notified.
- Number of trained ERT members of each shift.
- Number of unit that have up-to-date plans.
- Number of errors/omissions in ERP during drills and training.
- Percent of preventative maintenance work orders for ER equipment past due.
- Number of ERP past due to for periodic reviews.
- Percent failed tests or inspections on ER equipment.
- Percentage of ERT members who are up to date on ER training requirements.
- Percentage of drills that are conducted as scheduled.

Metrics – Compliance Audits

- Percentage of near miss and incident investigations identifying management system weakness that were not detected by prior audits.
- Percentage of audits completed according to schedule.
- Percentage of audits having few significant findings.
- Number of previous audits conducted by each audit team member.
- Average and maximum number of days overdue for open recommendations
- Number or percentage of unresolved audit recommendations
- Number of person-days required to complete an audit.
- Interval between completion of onsite work and completion of the audit report.
- Percentage of audit findings that are "repeat" findings.
- Percentage of recommendations that are rejected by the facility management.

Metrics – Fatigue Risk Management

Process Safety and Reliability Group

- Percentage of overtime
- Number of open shifts
- Number of extended shifts
- Number of consecutive shifts worked
- Number of exceptions

Process Safety and Reliability Group

Results of an Effective Program

CATERPILLAR

Using Proactive Leading Indicators to Create World-Class Safety

Caterpillar saved $450 million in 10 years by focusing on a strategic improvement project for safety

Caterpillar

- Traditional lagging safety metrics like OSHA recordable rates or lost-time case frequencies had been used for years to measure safety program success.
- While these metrics could help Caterpillar tell the score at the end of the game, they didn't help them understand the strengths and weaknesses of their safety efforts and cannot help managers predict future success.

Caterpillar

Process Safety and Reliability Group

- The critical elements of the Safety SIP included:
  - Enterprise-wide statement of safety culture.
  - Global process, tools and metrics.
  - Top-down leadership of and engagement with the process.
  - Clearly defined and linked roles and responsibilities.
  - Clearly defined accountability.
  - Consistent methods establishing targets and reporting performance.
  - Consistent criteria for prioritizing issues and aligning resources.
  - Recognition for positive behavior and performance.

Caterpillar

Process Safety and Reliability Group

What made the difference?

In order to create zero incident performance, the company used a process that included:

- Assessing the culture using one-on-one interviews and a safety perception survey.
- Building a plan by reporting and analyzing assessment results, training the safety steering team and agreeing on a strategic plan.
- Developing the process through safety culture excellence workshops, forming continuous improvement (CI) teams and coaching and leading CI teams.

Caterpillar

What made the difference?

- Implementing the process through pilot improvement projects and by implementing new processes.
- Checking the efficacy of the process by developing and utilizing an evaluation/feedback team. The process then went back to the assessment aspect should changes be necessary for continuous improvement.

Caterpillar

Caterpillar's Advice

- "Define, train, measure and recognize. Do it by involving safety leaders from all levels of the organization to pursue foundational safety processes, root cause incident resolution and accountabilities for every job."
- "... address safety the same way you address quality."

Confidential Chemical Co.

Leading indicator: Safety observations

Creating "eyes and ears" to prevent incidents "eyes and ears"

Confidential Chemical Co.

Prior to launching its Safety Observation System four years ago, the Company relied on a web-based reporting system for near misses. Only managers and supervisors had access to this system, resulting in a hierarchical, bureaucratic process.

They needed a more sophisticated system to handle near miss and hazard reporting.

Confidential Chemical Co.

Process Safety and Reliability Group

The new Safety Observation System was directly accessible to all employees and available in nearly twenty languages, providing Company with more detailed and timely information.

Employees reported not only "near misses" and incidents, but also any unsafe behaviors and conditions.

Confidential Chemical Co.

Process Safety and Reliability Group

Over 82,000 safety observations were reported in 2013, equaling roughly eight (8) observations per employee per year.

The business unit reduced the number of recordable injuries from 108 in the year 2010 to 54 in 2013.

During this same time period, the number of safety observations increased nearly 100%.

Confidential Chemical Co.

Company Advice:

"It's about staying ahead of the risks that cause the most injuries and creating vigilant observers of safety hazards at every level of the organization."

FLUOR®

Leading indicator: Leadership engagement

Management in action leads safety improvement

FLUOR

Process Safety and Reliability Group

Fluor established an effective suite of leading indicators that it assesses with its Corporate HSE Audit Tool. This tool includes measurements in the areas of program development and coordination; training, communication and HSE culture initiatives; field execution and management in action.

By measuring how well it performed in these key areas, the tool provided a leading indicator of future safety performance.

FLUOR

Process Safety and Reliability Group

The Corporate HSE Audit Tool was revamped in 2012 to put more weight on leading indicators.

The first six (6) months of audit scores revealed the greatest need for improvement in the management in action category.

FLUOR

These actions included:

- Managers' participation in field workers' orientation process and a demonstration of values by clearly defining management expectations in regard to safe work behavior, hazard and near miss reporting, and stop-work authority.
- Managers' simple, routine activities positively impacted the work on site and helped further promote and advance a positive HSE culture.

Conclusions

- Lagging indicators are easy to identify and measure, but are best used to define goals and objectives not the best predictive measure of future performance.
- Leading indicators are difficult to identify and measure, but can predict potential outcomes and are best used to indicate areas of focus
- Indicators should drive process safety performance and learnings.
- Indicators should be relatively easy to implement and easily understood by all stakeholders (e.g., workers and the public).

Thank you!

Process Safety and Reliability Group

800 W. Sam Houston Pkwy. South
Suite 107
Houston, Texas 77042-1908
Tel: 713-532-8800   800-250-8511 Fax: 713-532-8850
Email: psrghouston@psrg.com   Web: www.psrg.com

CELEBRATING
20 YEARS
1997 – 2017

Steve Arendt, P.E., Vice President - Global Oil, Gas, and Chemicals

40 years in process safety and risk assessment

Conducted 100s of PSM system design/implementations, audits, incident investigations, HAZID/HAZOP/LOPA/QRAs, best practice reviews, and safety culture evaluations

ABS Group manager for the BP-Baker Panel PSM reviews

80+ articles and books on PSM and risk management, including CCPS Guidelines on

- **Risk Based Process Safety**
- **Implementing PSM, 2nd edition**
- **Management of Change**
- **Guidelines for Hazard Evaluation Procedures, 2nd Edition**

CCPS Fellow and recipient of MKOPSC Merit Award

CCPS subcommittee member: 269-NOD; 275-PS Leadership; 276-SWP; 277-ISD

ABS Group

13TH GLOBAL CONGRESS ON PROCESS SAFETY

Discussion Outline

RBPS background
Essence of RBPS
Ways RBPS has been used
Misconceptions/misuse and possible future improvements
Closing remarks

ABS Group

13TH GLOBAL CONGRESS ON PROCESS SAFETY

RBPS Background

CCPS published its original 12 PSM elements in 1989 and followed it with 3 other management system books thru 1994

A lot of experience and lessons were learned from then thru 2005; CCPS wanted to update its PSM framework to be useful to industry as a thought and action leader for the next 20 years

RBPS came about for two reasons:

- **To be able to generate better results with fewer resources**
- **To provide a "target" and an approach for companies of all needs to implement, correct, and improve PSM systems**

ABS Group

13TH GLOBAL CONGRESS ON PROCESS SAFETY

RBPS Development Process

20 subcommittee members representing the process industry

16 peer reviewers

2+ year effort that:

- Surveyed PSM/ESH systems from around the world to identify good features
- Benchmarked with over 60 companies in three workshops to compile leading practices, improvement idea, and metrics

760-page guideline written by a team from ABS Consulting

ABS Group

13TH GLOBAL CONGRESS ON PROCESS SAFETY

RBPS Subcommittee and Co-Authors

Subcommittee

Jack McCavit	*CCPS Emeritus, Committee Chair*
Bob G. Perry	*Center for Chemical Process Safety*
Don Abrahamson	*Celanese Chemical*
Steve Arendt	*ABS Consulting*
Tim Blackford	*Chevron Energy Technology Company*
John Herber	*3M Company*
Dan Isaacson	*The Lubrizol Corporation*
Shakeel Kadri	*Air Products and Chemicals, Inc.*
Greg Keeports	*Rohm and Haas Company*
Jim Klein	*DuPont*
Pete Lodal	*Eastman Chemical Company*
Bill Marshall	*Eli Lilly and Company*
Darren Martin	*Shell Chemical Company*
Neil Maxson	*Bayer Material Science*
Lisa Morrison	*BP*
Karen Tancredi	*DuPont*
Tony Thompson	*Monsanto Company, Retired*
Scott Wallace	*Olin Corporation*
Roy Winkler	*INEOS Olefins and Polymers USA*
Gary York	*Rhodia, Inc., Retired*

Co-Authors

Steve Arendt
Bill Bradshaw, *Project Manager*
Walt Frank
Don Lorenzo
Lee Vanden Heuvel

ABS Group

13TH GLOBAL CONGRESS ON PROCESS SAFETY

Guidelines for Risk Based Process Safety

ABS Group

13TH GLOBAL CONGRESS ON PROCESS SAFETY

The Essence of Risk Based Process Safety

Management systems should be the simplest that they can be while still being fit-for-purpose

Consider the following issues when determining management system "rigor"

- **Perception of complexity, hazard, and risk**
- **Demand for the system results and the resources required to deliver them**
- **Current company/facility culture**

To design, correct, and improve process safety management activities

RBPS elements can be used throughout the company, plant or project life-cycle

ABS Group

13TH GLOBAL CONGRESS ON PROCESS SAFETY

RBPS System Element Resources

Element

- **Key Principles**
  - Essential Features
    - ➢ Possible Work Activities
      - ❖ Implementation Examples
        - Performance Improvement Practice
        - Efficiency Improvement Practice
    - ➢ List of metrics – LEADING and lagging
    - ➢ Management review items

ABS Group

13TH GLOBAL CONGRESS ON PROCESS SAFETY

9

RBPS Elements – Relationship to PSM

RBPS Element	New Element	Expanded Scope	Improved Practices
Process Safety Culture	✓		
Compliance to Standards	✓		
Process Safety Competency	✓		
Workforce Involvement		✓	✓
Stakeholder Outreach	✓		
Process Knowledge Management		✓	✓
Hazard Identification and Risk Analysis		✓	✓
Operating Procedures			✓
Safe Work Practices			✓
Asset Integrity and Reliability		✓	✓
Contractor Management			✓
Training and Performance			✓
Management of Change		✓	✓
Operational Readiness		✓	✓
Conduct of Operations	✓		
Emergency Management			✓
Incident Investigation			✓
Measurement and Metrics	✓		
Auditing			✓
Management Review and Continuous Improvement	✓		

ABS Group

13TH GLOBAL CONGRESS ON PROCESS SAFETY

10

Applications of RBPS during First 10 Years

Ways that RBPS has been used

Most frequently adopted new elements/practices

Most under-used elements and practices

Uncertainties from companies about how/whether certain PS aspects are addressed

Ideas on how to improve RBPS framework and approach

ABS Group

13TH GLOBAL CONGRESS ON PROCESS SAFETY

Ways that RBPS Has Been Used

Improve/optimize HSE/PS Management Systems
- Design/implement new system
- **Add new elements/upgrade existing elements with new element activities**
- **Fix broken system – metrics, work activities and implementation options**
- Improve existing system – improvement and MR ideas

As primary PSM framework

As PS competency training program framework

As best practice assessment/audit protocol

As organizational culture assessment basis

Basis for RCA structure

Built detailed key process workflow diagrams

Translated into other languages – Portuguese, Spanish, Chinese

Other applications? - Benchmarking Risk Based Process Safety Egg (RBPS Temple + 2 Universes) Against "The Art of War"

Most Frequent

Occasional Use

Rare

ABS Group

13TH GLOBAL CONGRESS ON PROCESS SAFETY

Misconceptions/Uncertainties

Use of "max" implementation options as basis for audit protocol

How is leadership commitment addressed

How is Human Factors addressed

Use of "all/many" element metrics

ABS Group

13TH GLOBAL CONGRESS ON PROCESS SAFETY

Underused Elements/Aspects

PKM – for knowledge

Stakeholder Outreach

CONOPs

Efficiency improvement ideas

Develop a key process/workflow for each HSE/PSM element using RBPS work activities as design toolbox

ABS Group

13TH GLOBAL CONGRESS ON PROCESS SAFETY

Improvement Ideas

Update/streamline leading indicator metrics

Performance improvement ideas

Efficiency improvement ideas

More explicit use for design organizations and capital project management

ABS Group

13TH GLOBAL CONGRESS ON PROCESS SAFETY

CCPS Guidelines for *Implementing PSM 2nd Ed*

Laundry

- Commit to Process Safety
- Understand Hazards & Risk
- Manage Risk
- Learn from Experience

RBPS Design Criteria

- Hazard/complexity/risk
- Resources
- Initial culture

Elements

- Key principles
- Essential features
- Work activities
- Implementation options
- Improvement ideas
- Metrics and MR

Gas Station

- Safety information
- Hazard analysis/JSA
- Procedures
- Training
- Emerge
- Investigation
- Audits

Low Hazard/Simple

- Ensure management cares/resources
- Ensure employees care
- Demonstrate stakeholder commitment
- Know what you operate
- Reduce or eliminate hazards
- risk
- Unde idual risk
- Know how to operate
- Know how to maintain
- Control changes
- Prepare/respond/manage incidents
- Monitor/act on internal learning
- Monitor/act on external sources

High Hazard/Complex

- Culture
- Standards
- Competency
- Involvement
- Outreach
- Knowledge
- Hazards/Risk
- Procedures
- SW Practices
- Integrity
- actors
- ormance
- MOC
- Readiness
- Discipli
- Emerg
- Audit
- Management Review

ABS Group

13TH GLOBAL CONGRESS ON PROCESS SAFETY

VISION 20/20 | CCPS

Industry Tenets

- Committed culture
- Vibrant management systems
- Disciplined adherence to standards
- Intentional competency development
- Enhanced application and sharing of lessons learned

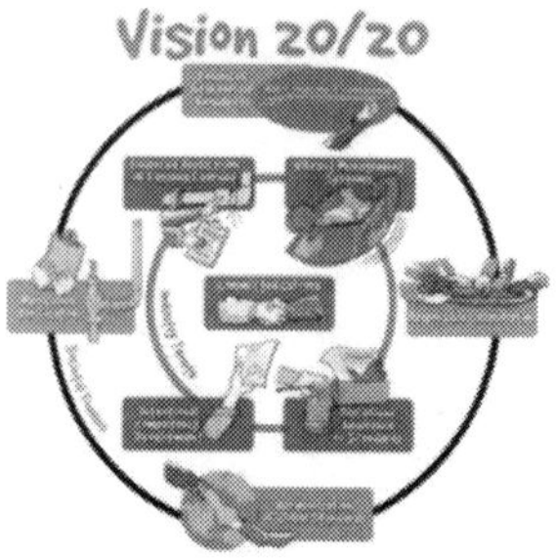

Societal Themes

- Enhanced Stakeholder Knowledge
- Responsible Collaboration
- Harmonization of Standards
- Meticulous Verification

Assessment Tool

Vision 20/20, developed by the Center for Chemical Process Safety (CCPS), looks into the not-too-distant future to describe how great process safety is delivered when it is collectively and fervently supported by industry, regulators, academia, and the community worldwide; driven by the five industry tenets; and enhanced by the four global societal themes.

ABS Group

13TH GLOBAL CONGRESS ON PROCESS SAFETY

VISION 20/20 | CCPS

Implementing the Industry Tenets…

Prepare	Assess	Plan	Perform	*Achieve*	Sustain
Present V20/20 to PSM Colleagues and Management	Complete the V20/20 Assessment Tool	Reinforce and Use Strong Elements as Building Blocks	Implement Action Plans	Complete Action Plans	Verify Management System Improvements
Make V20/20 a Regular Topic at PSM-Related Meetings	Identify Weak and Strong Sub-Elements (<2.5 or >3.5 Respectively)	Identify the Specific Improvements Needed	Monitor Status of Action Plan Implementation	Re-Assess V20/20 Implementation Status with the Assessment Tool	Develop Action Plans for Weak Sub-Elements and Individual Items
"Sprinkle" V20/20 into PSM Conversations	Report Results; Management Commits to Improve	Research Options to Improve (Reference Industry Documents)	Evaluate Effectiveness of Actions	Report & Celebrate Improvements	Implement Action Plans and Monitor Performance
Use V20/20 Logo on Internal Communications	Communicate Results Within Organization	Develop Specific Action Plans to Address Weak Areas	Capture & Communicate Learnings	Identify New Weak Sub-Elements and Weak Individual Items (<2)	Continual Improvement… Continue the Journey!

Today → *2020*

ABS Group

13TH GLOBAL CONGRESS ON PROCESS SAFETY

RBPS Trivia

Dedication	1
Bosses	2
RBPS Criteria	3
Pillars	4
Authors	5
Contributors	7
Figures	8
Reviewers	16
Elements	20
Subcommittee	20
Tables	26
Total People	48
Acronyms	65
Key Practices	78
Terms	99
References	117
Management Review Items	207
Index Terms	219
Improvement Ideas	306
Metrics	319
Essential Features	319
Work Activities	622
Pages	760
Implementation Options	2,042
Paragraphs	9,017
Lines	29,604
Words	254,233
Characters	1,450,964
Spaces	1,703,022
MSWord File Bytes	5,253,632
Priceless	RBPS source

ABS Group

13TH GLOBAL CONGRESS ON PROCESS SAFETY

Dedication

Guidelines for Risk Based Process Safety
is dedicated to the memory of

[illegible]
[illegible]

Les was among the first employees of the Center for Chemical Process Safety and was a major contributor to its success. All who knew and worked with Les remember him for his knowledge, his courage, and for his love of life. Although he is gone, his smile is still with us.

ABS Group

13TH GLOBAL CONGRESS ON PROCESS SAFETY

Closing Comments

Use RBPS to help "Manage generational change" in Process Safety

Rediscover use of the RBPS toolbox

Recommend CCPS implement online RBPS use/needs survey for near future to capture needs and improvement ideas for RBPS 2.0 or ...whatever...

ABS Group

13TH GLOBAL CONGRESS ON PROCESS SAFETY

ABS Group

22

Are your safeguards fast enough to be effective?

Eloise Roche
SIS-TECH Solutions, LP
Houston, TX
eroche@sis-tech.com

Angela Summers
SIS-TECH Solutions, LP
Houston, TX
asummers@sis-tech.com

Prepared for Presentation at
American Institute of Chemical Engineers
2017 Spring Meeting and 13th Global Congress on Process Safety
San Antonio, TX
March 26 – 29, 2017

Are your safeguards fast enough to be effective?

Eloise Roche
SIS-TECH Solutions, LP
Houston, TX
eroche@sis-tech.coms

Angela Summers
SIS-TECH Solutions, LP

Keywords: Safety Management System, Safety Interlocks, Safety Alarms, Safety Instrumentation Systems

Abstract

A core attribute of all safeguards is that they must be functional against the loss event in question and one of the essential elements of functionality is speed. How to determine the response time available for a given safeguard to complete its action is a common project question. Several instrumentation and controls design pitfalls should be avoided when specifying the response time. A variety of engineering practices, from basic to advanced, may be used to justify the selected value. Among these are:

- Expert judgment
- Extrapolation
- Mass and energy balance calculations
- First principles process modeling

Key technical issues include hardware response lags, process lags, application program delays, measurement error, and the design and safety margins. Careful attention to this specification is needed from the instrumentation and controls design team, for if an instrumented safeguard performs too slowly compared to the hazardous event, it provides no protection at all.

1 Introduction

Functional safety is achieved through implementation of effective process control systems that limit the likelihood of initiating a hazardous event and safeguards that manage the process safety risk remaining after inherently safer design principles have been applied to the process design. Functional safety projects can implement a wide variety of safeguards, including safety controls, alarms, and interlocks (SCAI) [1]. Successful execution of SCAI projects requires significant resource investment. A SCAI project typically involves hundreds of engineering hours to brainstorm hazard scenarios, identify functional safety gaps, and determine the best means to reduce risk. The safeguards may include the use of independent protection layers (IPL) such as a Safety Instrumented Function (SIF) executed within a Safety Instrumented System (SIS). While not all safeguards will meet the criteria of an IPL, an engineering task common to all instrumented safeguards is the specification of the safeguard setpoint and corresponding response time.

The term process safety time has been used in many publications to describe the time interval both for the escalation of the process deviation to the hazardous event and for the safeguard response to prevent the hazardous event. This paper uses terms recently adopted in industry practices:

> Process Safety Time (PST) [2]: Time period between a failure occurring in the process or the basic process control system (with the potential to give rise to a hazardous event) and the occurrence of the hazardous event if the SIF is not performed.
>
> IPL Response Time (IRT) [3, 4, 5]: The time available for the IPL to act given the specified setpoint.

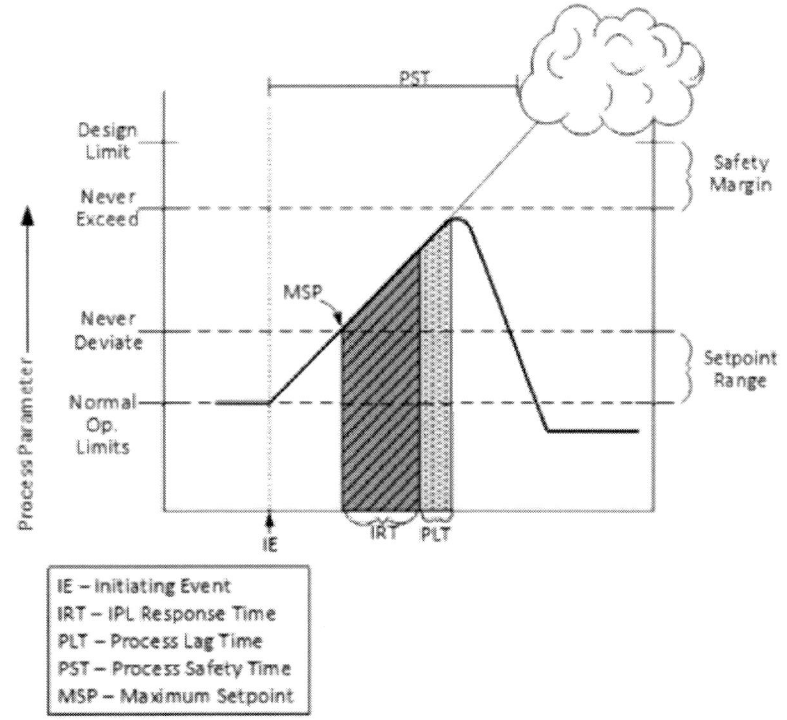

Figure 1: IPL Response Time Terminology (adapted from [3])

The most important distinction is that PST is a property of the process only. In contrast, the IRT (historically referred to as the allocated process safety time) is a property of both the process and the safeguard setpoint. The safeguard must detect the process deviation and complete its action within the IRT to prevent the hazardous event. The determination of the IRT is intrinsically related to the process and its response to the safeguard action, as this determination must account for the process escalation prior to action being taken and any process lag that exists due to retained mass and energy within the process (e.g. cooling of a vessel). Figure 1 depicts the difference between PST and IRT.

For efficient project execution, it is important that the setpoint and IRT specifications are made during front-end-loading. The response time of the final safeguard design should be estimated and compared to the IRT specification during detailed design reviews. Additionally, the safeguard response time measured during pre-startup safety tests is validated against the specified IRT when evaluating whether the functionality of the installed system is acceptable.

2 Understanding Process Response to Deviation

To determine the IRT, the engineering team first understands how the process reacts to the process deviation caused by the initiating source of the hazardous event. The PST may be only seconds, which limits the type of safeguards that can be effective. On the other hand, the process may take days to transition from the initiating source to the ultimate consequence, which could allow for a sequence of safeguards. A variety of engineering practices from basic to advanced may be used to better understand how the process responds to the process deviation and to the safeguard action.

2.1 Expert Judgment

Experience is a valuable teacher. Expert judgment involves specifying an IRT based on experience, whether personal or through industry incident reports, and technical analysis. Industry experience may be codified in consensus standards for a particular type of safeguard (e.g., a SIS) or process application (e.g., burner management standards). Expert judgment within a facility may also come from senior operators who have personal experience from historical events.

Expert judgment can be a very useful starting point for a project and may be the only method available for the preliminary IRT specification in early project phases. Expert judgment based on prior observation is often the IRT basis for very short PST events, such as the compressor overspeed reaction shown by process response curve A in Figure 2. However, relying on expert judgment alone may leave the operating facility with a relatively weak justification if a proof test or an incident calls the specified IRT in question. Consider a situation where a SIS has failed a proof test due to the failure mode - too slow of response. When the failure is investigated, the justification for the specified IRT is likely to be requested. Expert judgment is only as good as the individual's expertise.

2.2 Extrapolation

Extrapolation relies upon factored estimates based on equipment differences, such as vessel volume or pipe cross-sectional area, to estimate the IRT for one facility based on a known event in a very similar facility. Such scaling can be used to develop a preliminary IRT specification quickly and has a more clearly documented technical basis than expert judgment alone.

However, not all process responses scale simply. Events that involve multiple phases, energy transfer, or reactive chemistry often do not scale linearly. Furthermore, even minor changes in process equipment layout between the facility that experienced a previous event and the one that is under evaluation can result in potentially substantial differences in response to the same process deviation.

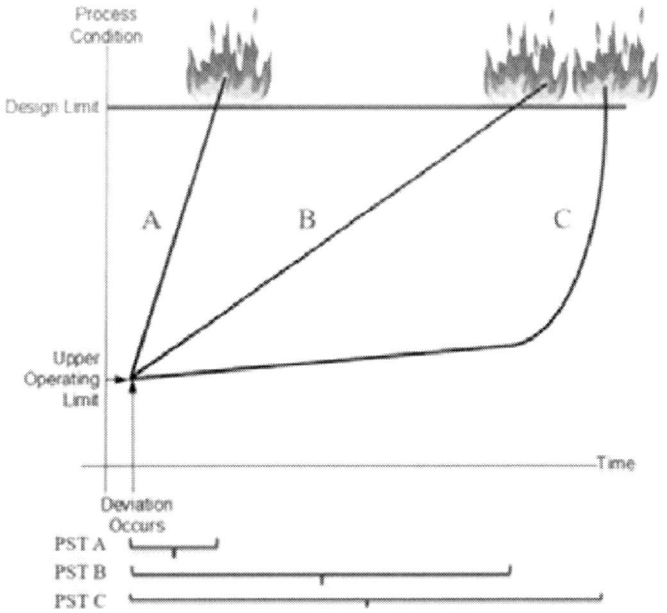

Figure 2: PST is based on process response to deviation

2.3 *Mass and Energy Balance Calculation*

The mass and energy balance tools typically used by process engineers during process equipment design can be helpful in the specification of IRT. In particular, iterative runs of a mass and energy balance might be used to predict the process response for relatively straightforward process deviations. For example, the time it takes to overfill a vertical vessel (which might have a response curve like the one depicted by process response curve B in Figure 2) or to pull in a dangerous level of oxygen into a tank after loss of a nitrogen pad can often be estimated using a simple spreadsheet. More complicated design simulators may be useful when the analysis must involve multiple pieces of equipment. For complex process deviations, however, the number of iterative runs of a static mass and energy balance can become burdensome. Due to the nature of the tools, this method is also unlikely to reveal short-term dynamic responses that occur under transitory operating conditions.

2.4 *First Principles Process Modeling*

First principles process modeling, sometimes referred to as dynamic simulation, models the response of multiple process parameters to one or more process perturbations as a function of time. In first principles process modeling, the actual process equipment dimensions, including piping configurations, and process chemistry are incorporated into the model. This increased degree of complexity can be used to model complex reactive or multi-phase process reactions and is more likely to reveal short duration dynamic responses that exceed the steady-state predicted values. Often these responses are non-linear and may have one or more points of inflection. For example, process response curve C in Figure 2 might reflect the complex temperature response resulting from the initiation of an auto-catalyzing degradation reaction. The process equipment and instrumentation information needed to build a useful model likely limits the use of this technique to the latter portion of detailed design. Verifying the process simulation against data collected during startup and normal operation helps increase confidence in the model accuracy.

3 Establishing Safety Margins

The design limit is a property of the equipment being challenged by the process deviation. Examples of design limits are the maximum allowable working pressure (MAWP) of a pressure vessel [6] or the critical high level [7] in a vessel. However, it is not an inherently safer practice to operate equipment all the way to the design limit. A safety margin, as shown in Figure 1, is usually established by the facility owner and provided to the project team during front-end loading. The minimum safety margin between the design limit and the never exceed limit is often expressed as a percentage in engineering units. Facilities with older or well-used equipment may establish a larger safety margin to account for uncertainty in the equipment's mechanical integrity. The never exceed limit is the maximum (or minimum) value the process parameter can achieve and the process still be considered safe for continued operation. The safeguard action must be successfully completed and successfully affect the process variable before the never exceed limit is reached. Designs which encroach upon this margin should be subject to review by process safety personnel to ensure the results can still satisfy the company risk criteria.

4 Understanding Process Lag

The same method(s) used to determine the process response to the deviation can usually be used to determine whether or not there will be a process lag after the proposed safeguard action is complete. In many cases, the process variable of concern will not stop moving in the dangerous direction the instant the final element of the safeguard (e.g., a valve or a motor) goes to the safe state. The level in a tank may continue to rise due to liquid rundown from material in an elevated pipe between the isolation valve and the tank inlet. The temperature and pressure in a distillation column may continue to rise due to the thermal energy stored in the metal of a reboiler, even though the reboiler steam has been shut off.

As shown in Figure 1, the existence of a process lag will require the safeguard action to be completed earlier in the process response curve to ensure the process variable does not violate the never exceed limit established by the design limit and the safety margin. Some scenarios are addressed by multiple independent safeguards that are designed to operate in a sequence. For these scenarios, the setpoints and IRTs should be specified so that the process response to one safeguard is complete before the next safeguard in the sequence is activated.

5 Estimating Safeguard Lag

Once the process response curve and process lag are understood, the time available for safeguard action after the process reaches a given setpoint can be determined. This is essential information for specifying the setpoint and IRT. The next step in the process is to determine the least conservative acceptable setpoint. For this task, the estimated response time of the safeguard must be determined. The estimation work can be divided into three tasks:

- Estimating the safeguard typical response time based on components
- Adding delays from the application program(s)
- Adjusting the setpoint for the additional lag caused by measurement error

5.1 Safeguard Device Response Time

The safeguard response time is determined by adding up the individual response times for the equipment needed to execute the required function, along with any time delays introduced within the application program (e.g., filters or delay timers). Detailed instrument and application program design information is often not available at the time of the preliminary hazards and risk analysis (H&RA), so the initial estimates are often made using conservative device response values and assuming no application program delays, such as those depicted in Figure 3. Once device data sheets and detailed application program design documents are available, the initial estimates can be updated.

Device Type	Typical Device Detection/Action Lag
Most analyzer installations (including delays from sampling system)	Tens of minutes
Typical temperature installations	2 to 10 minutes
Most other sensors	5 seconds or less
Large valves and variable frequency drive motors	Tens of seconds up to multiple minutes
Small valves and other motors	5 seconds or less
Programmable electronic logic solvers	3 seconds or less
Human response	10 to 40 minutes PLUS

Figure 3: Typical instrumentation response time [4]

Human response as a safeguard (e.g., an operator response to an alarm) introduces a large and potentially complex response time contribution. There are delays involved in recognizing the presence of an alarm, mentally relating that alarm to the response that must be taken, and overcoming any cognitive biases to taking the safe response action. To have a 90% probability of correct action in an alarm response for which the action is simple and involves no troubleshooting, at least 10 minutes of time for cognitive delays should be added to the estimated safeguard response time. If the response procedure does involve troubleshooting, the cognitive time should be increased to 40 minutes [8]. If the response action requires travelling out into the process facility, the time to get to the destination and take the final action should also be estimated and added to the estimated safeguard response time. The industry consensus values for operator response lags assume that performance shaping factors like training, work/rest schedule, and leadership culture regarding process safety activities are nominal. Deficiencies in any of the operations performance shaping factors can result in an increased human response lag.

5.2 Application Program Delays

In addition to the relatively fixed delays inherent to the physical devices that make up the safeguard, delay can also be introduced through the application program. Delays may be introduced on the safeguard activation, through enabling logic, and through filtering of the signal. Signal filtering for safety applications can be particularly concerning, as a first order filter introduces a delay equal to 2.3 times the time constant just to receive 90% of a signal change. Unless the IRT is particularly generous, it would be inherently safer to address erratic sensor signal problems through correction of the underlying electric system issue or through addition of an independent installation of redundant instrumentation with an NooN voting architecture rather than through the introduction of signal filtering.

The use of intelligent devices and wireless technologies in safeguards create an additional location to inspect for adjustable software delays. Modern sensor transmitters and final element actuators may include customizable filters or even contain activation logic for the safeguard loop, eliminating the need for a separate logic solver. Transmitted signals may also be slowed down through configurable I/O network buffers. As these delays are configurable, they will usually not be included in the base response time documented in the instrumentation datasheet. Ultimately, the design team must identify all sources of software delays and ensure they are accounted for when estimating the response time achieved by a given design.

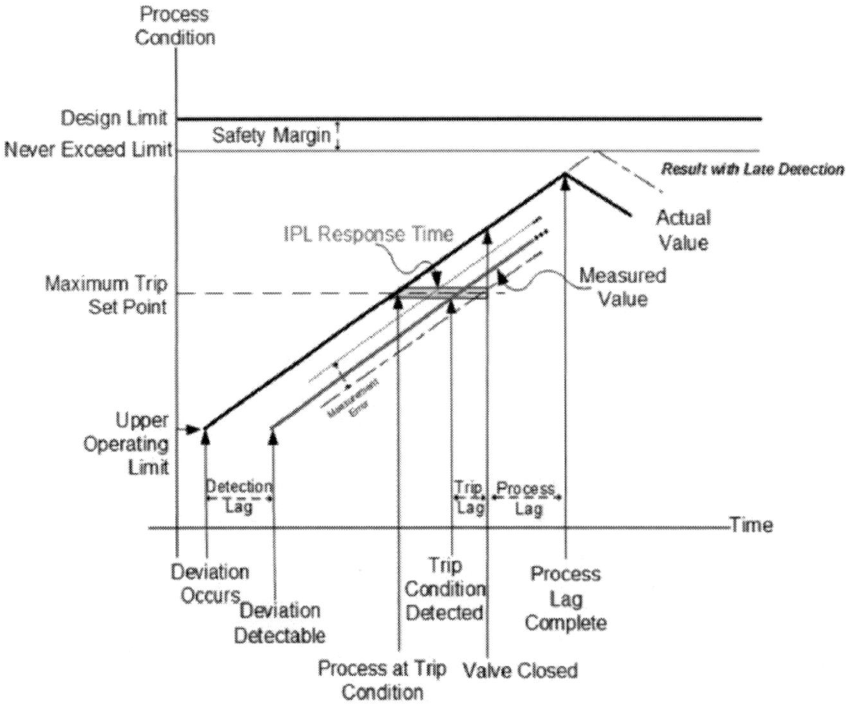

Figure 4: Adjusting setpoint for measurement error (adapted from [4])

5.3 Impact of Measurement Error on Setpoint Selection

Even for a new sensor that is correctly calibrated and configured, error is expected in the measurement. This normal amount of error is related to the specified accuracy of the device. In determining the final setpoint specification, it should be assumed that the sensor will be reading at the specified accuracy limit away from the setpoint. This would result in late safeguard actuation and the potential occurrence of the dangerous consequence. The setpoint should be adjusted to account for this error. As shown in Figure 4, this adjustment ensures that the process does not violate the never exceed limit even when the measurement is at the less safe extreme of its accuracy at the time of the demand.

6 Design Margins and Making the Final Setpoint/IRT Specification

Completion of the above tasks determines the least conservative setpoint value. That is, the highest setpoint value the safeguard could have and still be effective. Moving the setpoint close to the never exceed limit maximizes the operating margin and allows the maximum time for the operator to recover the process and prevent safeguard action, but allows the shortest time possible for the safeguard to act. However, it is rarely a good engineering practice to design a safeguard up to the least conservative (i.e., least safe) limitation.

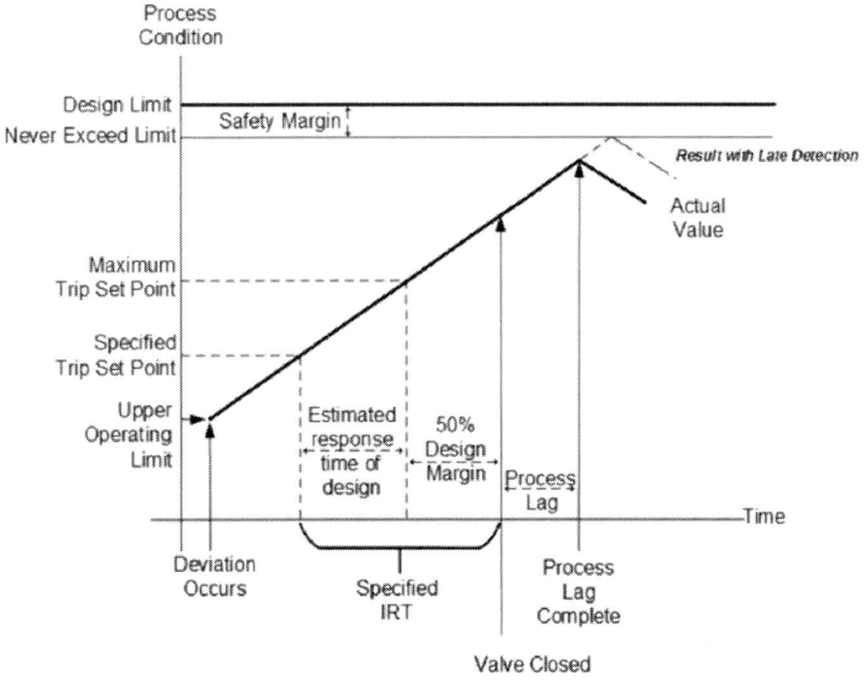

Figure 5: Estimated response time and design margin

Alternatively, the setpoint could be set very close to the normal operating limit, allowing for more response time. The generally accepted margin is to design the safeguard to work in half the time allowed in the specification (depicted in Figure 5). Using such a safety margin helps increase the likelihood that the safeguard is effective even when it does not operate quite as fast as expected once installed in the operating environment. However, having a safeguard setpoint too close to the operating limit can put a facility at risk of more frequent spurious trip events. Spurious trips will cause lost productivity and might come with their own inherent hazards, such as potential equipment damage or industrial hygiene hazards related to the clearing of equipment in preparation for startup.

Usually the optimal setpoint is somewhere between these two extremes. Balancing these two antagonistic design criteria and specifying the setpoint, as shown in Figure 6, often involves a team negotiation that includes members of operations, process safety, and engineering. Where the final selection does not allow a 50% design margin, the justification for the smaller design margin should be clearly documented. Based on the process response to the deviation, the final setpoint specification will determine the IRT specification.

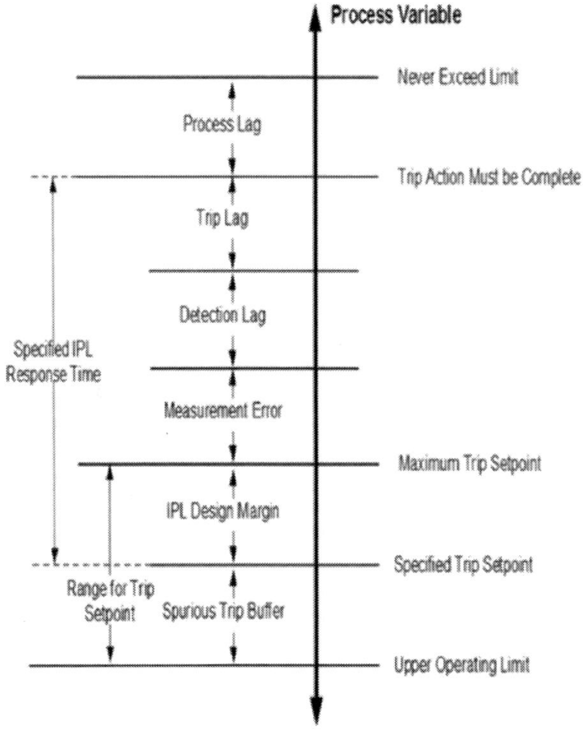

Figure 6: Balancing design margin and spurious trip buffer (adapted from [5] Annex Q)

7 Incorporation of IRT into Project Timeline

Due to the varying amount of engineering information required, the different techniques for determining the process response to deviation and specifying the response time of the designed safeguard have an optimum stage for their use during project front-end-loading and detailed design (shown in Figure 7). Just as an iterative process is the most effective for performing the H&RA activities, an iterative approach to the specification of setpoint and IRT can enable the project team to make more informed decisions during front-end-loading. Once engineering details for equipment, piping, application program and instrumentation are available, the design team can take advantage of more advanced techniques to verify and strengthen the documented technical justification for the specified values.

Incorporating this iterative approach into the project schedule is something that should be done during project scoping with the involvement of project management and facility leadership. Project milestone meetings, such as the conceptual design review and detailed design review, should confirm that the H&RA activities and safeguard specification details are at an appropriate level of development. Sufficient time and resources should be scheduled to ensure there will be clearly documented justification for the specified values, as seemingly minor changes in process conditions or equipment dimensions can result in a substantially different process response to deviation. This documentation will be necessary to support the management of change of future proposals for safeguard modification.

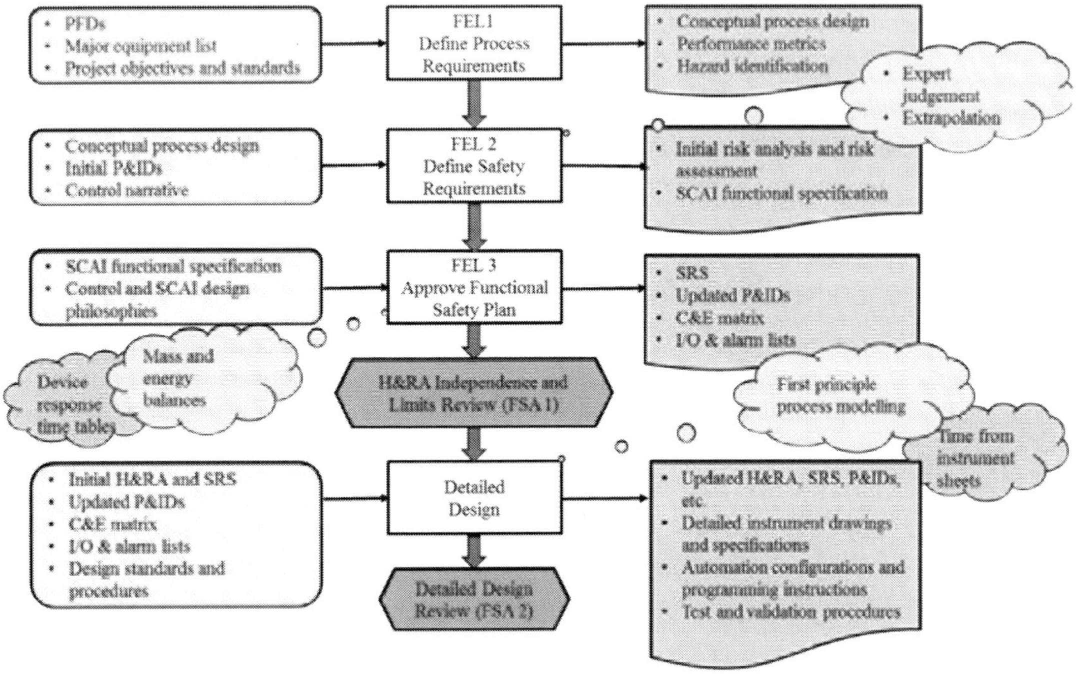

Figure 7: Incorporating response time techniques into project front-end-loading (FEL) and design

8 Conclusion

Understanding the process safety time for the process response to a hazardous event is useful when designing multiple safeguards against the same hazard, but the response time specified for a safeguard (the IRT) is a function of the specified setpoint and the process response curve. Multiple engineering techniques exist to analyze the process safety time for a given deviation event and any process lag that might exist after the safeguard action is complete. For best efficiency, these should be applied in an iterative process over the project front-end-loading and detailed design phases. The least conservative setpoint at which the safeguard will be effective is then determined, based on device lags, application programming delays, and adjustments for the expected amount of measurement error. At this point, the competing targets of the desired design margin and spurious trip buffer are used to select the final setpoint specification. Based on the process response curve and process lag, the specification of the setpoint established the IRT specification. The justification for these specifications, including the detailed process and equipment information used to derive them, should be clearly documented to enable future change proposals to be managed effectively.

9 References (examples given to show format).

[1] ANSI/ISA. 2012. Identification and Mechanical Integrity of Safety Controls, Alarms and Interlocks in the Process Industry, ANSI/ISA-84.91.01-2012. Research Triangle Park: ISA.

[2] IEC. 2016. Functional safety: Safety instrumented systems for the process industry sector - Part 1-3. IEC 61511. Geneva: IEC.

[3] CCPS. 2014. Guidelines for Initiating Events and Independent Protection Layers in Layers of Protection Analysis. New York: AIChE.

[4] CCPS. 2016. Guidelines for Safe Automation of Chemical Processes. New York, NY: Center for Chemical Process Safety.

[5] ISA. 2015. Guidelines for the Implementation of ANSI/ISA 84.00.01- Part 1, TR84.00.04-2015. Research Triangle Park: ISA.

[6] ASME. 2015. Boiler and Pressure Vessel Code. New York: ASME.

[7] API. 2012. Overfill Protection for Storage Tanks in Petroleum Facilities, 3rd Edition, Standard 2350. Washington, D.C.: API.

[8] CCPS. 2007. Guidelines for Safe and Reliable Instrumented Protective Systems. New York: AIChE.

Temporary Management of Change: A Problem or a Necessary Part of Managing Process Safety

Jody Olsen
JE Olsen Consulting LLC
Anchorage, AK
jodyeolsen@gmail.com

Prepared for Presentation at
American Institute of Chemical Engineers
2017 Spring Meeting and 13th Global Congress on Process Safety
San Antonio, TX
March 26 – 29, 2017

Temporary Management of Change: A Problem or a Necessary Part of Managing Process Safety

Jody Olsen
JE Olsen Consulting LLC
Anchorage, AK
jodyeolsen@gmail.com

Keywords: Management of Change, MoC, temporary, risk, OSHA

Abstract

Temporary Management of Change (MoC) projects seem to be the boogeyman of Process Safety Management (PSM). These temporary changes are frequently needed, but feared by operating company management, compliance groups, and lawyers. Some concerns regarding temporary MoCs are valid while other concerns may be based on false assumptions or incorrect interpretations of regulation. Most alarming, many of the concerns regarding temporary MoCs may lead to worse practices and the potential to introduce more risk into the business. This paper explores the reasons for this aversion to temporary MoCs and proposes a paradigm shift. Instead of considering permanence as the norm, consider that everything is actually temporary. The key is designing, operating, and maintaining for the appropriate window of time. The change will only be approved if the risk falls within accepted company protocol for the required duration of the change. The author will define five categories of temporary MoCs for the purpose of clarifying that one-size-fits-all duration limits and protocols are not appropriate. The author will address three myths regarding practices for managing temporary MoCs that may lead to unintended negative consequences. And finally, the author will propose preferred practices that avoid these unintended outcomes leading to more efficient work processes and safer operations. The need for temporary MoCs and the risks of mismanagement of those temporary MoCs must be well understood by both regulatory agencies and operating companies. Policies and procedures must drive practices that assure safe operation for any time horizon given real world impacts and constraints.

1 Introduction

Since the inception of process safety management (PSM) regulation [1], management of change (MoC) has been recognized as one of the most crucial elements of the regulation. Management of temporary changes is often highlighted as a critical and potentially more problematic type of change to be managed.

The incident at Flixborough, 1974 [2][3][4], is frequently cited as an early example of not only inappropriate management of change but also inappropriate management of temporary change. This event involved a temporary bypass that had been installed two months before the incident in order to maintain production while a failed cyclohexane oxidation reactor was repaired. A temporary piping modification – including the pipe, bellows, and pipe support - was designed, installed, and operated without appropriate mechanical review by qualified personnel. The temporary bypass piping failed during a plant start-up that followed unrelated repairs. A vapor cloud explosion followed the release of a large amount of cyclohexane. The explosion killed 28 employees, injured 89 employees and members of the surrounding community, destroyed the plant, and damaged many nearby homes and businesses.

In the intervening years, and especially within the last 25 years since the enactment of the Occupational Safety and Health Administration (OSHA) PSM regulation [1] and similar regulation in other jurisdictions, better PSM practices have become more widely understood and implementation has improved. Most operating companies now routinely apply management of change practices that are intended to address the causes that contributed to the Flixborough incident. Management of Change requirements are widely understood to include appropriate technical and process safety review of the design, drawing and other process safety information (PSI) updates, procedure updates, completion of necessary training and awareness communications, and pre-start-up review to name a few.

Those better practices include demanding equivalent review and approval requirements for all changes – including temporary changes. Where the change is temporary, certain other requirements are stipulated and must be addressed. These additional requirements relate to specifying a duration limit for the temporary change. The temporary change may also include additional safeguards and enhanced monitoring, inspection, and/or maintenance procedures that are specifically applicable for the duration of the temporary change.

This paper examines several myths related to management of temporary changes. While having a limit on the duration for a temporary change is appropriately understood to be an essential requirement for managing any temporary change, this paper addresses several pitfalls where specific interpretations of and practices related to that requirement may actually introduce unintended consequences and risk into the operation. Likewise, the allowance for conversion of temporary MoCs to permanent will be addressed. This practice is often allowed per procedure but is also not advisable or recommended. These practices coupled together can and will often lead to unintended negative consequences. The author offers a preferred approach for defining requirements specifically related to temporary MoCs and outlines a work flow process that places design requirements for the change – including duration and risk assessment - at the front end of the design and approval process.

2 Background

Below is a brief recap of a number of background assumptions and some history regarding management of change and management of temporary changes.

2.1 Knowns

- There is a known history of major, minor, and near-miss incidents and failures associated with improper management of change.
- Temporary changes have been part of that failure history.
    - o As per OSHA Appendix C, "Temporary changes have caused a number of catastrophes over the years" [5].

- Temporary changes often do not meet company specifications, standards, and/or normal operating limits and, therefore, are expected to have shorter service lives than permanent changes and may also require more rigorous monitoring, inspection, and/or maintenance and may include additional safeguards or layers of protection.

Two examples of familiar temporary changes are shown below:

Figure 1. Blue tarps on roofs after hurricane [6].

Figure 2. Piping repair using pipe clamps [7].

  - Due to the above as well as regulatory requirements, approval of temporary MoCs must include a duration limit. Without a duration limit, the operator may lose track of the temporary change and it may effectively become permanent even if not originally intended to be permanent [5].

2.2 Debunked Untruths Some early presumptions are now widely understood to *not* be true.

- It is not true that temporary changes are of such a short duration that review requirements are lesser than standard changes.

  - Review requirements for temporary and permanent MoCs are the same. Procedures must stipulate equivalent review requirements.
    - OSHA 1910.119 (l) provides no allowance for lesser or weaker review requirements for temporary MoCs. No distinctions between temporary and permanent MoCs are mentioned in the PSM regulation [1].
    - As per OSHA Appendix C, "Temporary changes are subject to the management of change provisions [5]."
- It is not true that Process Safety Information updates are waived or are inapplicable for temporary changes [1].
  - Current Process Safety Information (PSI) must be available to personnel involved in the operation regardless of the nature or duration of the change.
    - As per OSHA Appendix C, "Copies of process changes need to be kept in an accessible location to ensure that design changes are available to operating personnel as well as to PHA team members when a PHA is being done or one is being updated [5]."
  - How a company addresses this requirement will vary based on individual company protocols and document management procedures.
    - The regulation does not stipulate the format, such as electronic updates versus hand mark-ups.
    - The regulation does not stipulate the timing for formal updates. Updates for larger studies may require special protocol. Depending on type of study and length of temporary MoC, certain updates may be deferred. Company procedures should ensure that any unincorporated changes are communicated and that a reference to the MoC is made available.

2.3 Myths This paper will address the following myths and propose alternative recommendations.

- Myth 1: Most temporary changes are terminated by returning the system to the prior condition, also known as reversing the change.
- Myth 2: The best way to ensure limits on durations for temporary changes is to have a universal duration limit for temporary MoCs prescribed within the company Management of Change procedure.
- Myth 3: Converting a temporary MoC to a permanent MoC is an acceptable practice especially for longer duration temporary MoCs.

3 Definitions

At this time, it would be useful to ensure that we have consistent definitions for several terms. Typically, changes are defined within Management of Change procedures as: Emergency, Temporary, or Permanent.

> Note: Emergency changes entail some similar concerns as temporary changes. Due to the highly reactive nature of an emergency change, special provisions are frequently considered allowable for emergency changes that are not allowable for any other type of change. These changes require special considerations and management requirements, but detailed discussion of emergency changes is not within the scope of this paper.

3.1 Temporary

Consider the definition of the word *temporary*:

> Merriam-Webster [10] defines temporary as:
> lasting for a limited time

Below are several definitions provided within reference documentation for the term *temporary change:*

> Temporary Change:
> 1) "implemented for a short, predetermined, finite period [2]"
> 2) "a clear time limit for the validity of the change is defined [4]"
> 3) "planned for limited duration [8]"
> 4) "provided for a defined time period [9]"

The theme is very clear. While the term "short" is a relative term that may have varying interpretations, the reference to a finite, limited, or prescribed time period clearly conveys that the length of time that the change is acceptable to be in service must be stipulated for each temporary change.

3.2 Permanent

Conversely, the antonym *permanent* may be more difficult to define. These references provide the following definitions:

> Permanent:
> Merriam-Webster [10] definition:
> continuing or enduring without fundamental or marked change
>
> Cambridge Dictionary [11] definition:
> Lasting for a long time or forever
>
> Dictionary.com [12] definition:
> existing perpetually; everlasting, especially without significant change.

The definition of *permanent* is less clear, because depending on ones' field of view, all things could be considered to be temporary given that equipment, human life, and even our planet have a lifespan, and therefore, do not last forever or perpetually. In the case of equipment, lifespan is affected by rate of wear, availability of replacement components, potential for obsolescence of mechanical or electronic components, maintenance and operational practices, and system changes affecting design requirements.

For our purposes when considering processing equipment and plants, a good working definition of *permanent change* might be:

Permanent Change:
Expected to remain in place indefinitely. Effective lifespan will frequently be in the range of 30 to 50 years assuming normal operating and maintenance practices. Lifespan typically corresponds to life of the field or the facility.

For the purposes of this paper, *permanent change* will reflect the definition given above.

3.3 MoC

The acronym MoC, Management of Change, refers broadly to requirements associated with implementing changes. When used in the context "the MoC," the acronym is sometimes used as a noun meaning: "the change"; "the project"; "the project ID or record from the MoC application representing a given change or project." The acronym MoC may be used in this context in this paper.

Note: Although temporary changes may include temporary organizational changes, organizational changes are not included within the scope of this paper.

4 Regulation and Guidance

4.1 Regulatory requirements (United States)

It would be useful at this time to ensure that we are clear on the language and intent of regulation regarding temporary management of change.

The OSHA PSM regulation, management of change section, reads as follows.

As per OSHA 1910.119 [1]:
1910.119 OSHA Process Safety Management of highly hazardous chemicals
1910.119 (l) *Management of Change*
1910.119 (l)(1) The employer shall establish and implement written procedures to manage changes (except for "replacements in kind") to process chemicals, technology, equipment, and procedures; and, changes to facilities that affect a covered process.
1910.119 (l)(2) *The procedures shall assure that the following considerations are addressed prior to any change:* . . . . (i)technical basis. . . ;(ii) impact . . . on health and safety; (iii) modifications to procedures;
1910.119(l)(2)(iv) *Necessary time period for the change* . . .
(v) authorization requirements . . .
1910.119(l)(3) [Operators, maintenance personnel, and contractors] ". . . shall be informed of, and trained in, the change prior to start-up . . ."
1910.119(l)(4) "process safety information . . . shall be updated accordingly. . ."
1910.119(l)(5) "procedures or practices shall be updated accordingly."

The regulation itself makes no mention of temporary versus permanent, but does indicate that the time period for which the change is approved, installed, and operated must be defined, meaning that the time period must be defined *for all changes*. Given that the period of the change is normally not explicitly defined for each permanent change, companies should formally define the term *permanent* globally within their management of change procedure. This global definition would effectively satisfy the requirement for defining the period of the change for any and all permanent changes as these changes will generally have a consistently understood duration. Given this stipulation in the regulation, it is clear that any change classified as temporary (or emergency) must define the duration that the specific change is allowed to remain in place.

4.2 Guidance

Several clarification and guidance publications support and emphasize the requirement for defining the duration of temporary changes.

> As per OSHA Appendix C:
> It is important that a time limit for temporary changes be established and monitored since otherwise, without control, these changes may tend to become permanent [5].

Likewise, *CCPS Guidelines for the Management of Change for Process Safety* has similar language that refers to the requirement for specifying the allowable duration of the change.

> As per *CCPS Guidelines for the Management of Change for Process Safety*:
> A temporary change is approved with the stipulation that the change last no longer than a specified time and that additional safety features be implemented . . . Temporary changes need prescribed time limits, appropriate authorization, and proper documentation [2].

5 Categories of Temporary MoCs: Why are Temporary MoCs needed?

In order to understand issues, constraints, duration limits, and life cycle requirements for temporary MoCs, it is important to recognize the different reasons for these changes. Temporary MoCs will generally fall into one of six categories:

1) Temporary trial use of new equipment, chemicals, process parameters, process configuration, and/or procedures.
2) Temporary modification following a failure but preceding implementation of the permanent, long-term solution. The pending permanent change may be:
    a. A completely new design OR
    b. Return of the system to original specifications.
3) Temporary bypass or failure of safety critical devices or other components. Device or component is out-of-service due to:
    a. Bypass, override, impairment, or removal for proactive maintenance or inspection OR
    b. Failure or discovered defect. Device or component is not functional or not fully functional. Repairs are pending.
4) Temporary connections and/or temporary operations, such as in advance of shut-downs for purging or other occasional operations.
5) "Temporary" deferral of maintenance or inspection procedure due date including turnaround (TAR) deferrals.
6) "Temporary" installation intended to be followed by a future installation having more capacity or capability.

5.1 Category 1 - Temporary trial or field test

Category 1 Temporary MoCs are proactive and planned in nature. These changes would typically be considered trials or pilot studies performed in order to field-test a new idea before major or permanent modifications are made to accommodate the change. The new idea and change could be related to equipment, instrumentation, controls, chemicals, process parameters, process configuration, or procedures.

5.2 Category 2 – Temporary modification following a failure

Temporary MoCs in Category 2 are generally more reactive in nature. The Category 2 MoC is planned, reviewed, and approved but typically on a much shorter timeline in response to an unplanned failure or event. This type of MoC is always considered an interim step until a final planned modification (another MoC project) or repair may be implemented.

This two-step solution may be due to lack of availability of required materials or resources, additional time needed for design and fabrication, or a need for a larger outage in order to implement the permanent change. The long-term solution will typically require a more lengthy and deliberate project management cycle involving detailed review of the causes of the failure, full assessment of the needs, consideration of alternative solutions, detailed design of the new installation, and execution involving construction activities, installation, and start-up.

In some instances (Category 2a), the long-term solution will be a completely new design or modification addressing the cause of the failure. In other instances (Category 2b), the long-term solution may be to return the system to original specifications if the cause of the failure was related to off-specification materials, installation issues, operational or maintenance issues, or an improperly approved prior change. In those cases, other mitigations unrelated to the original design should be implemented to address the cause(s) of those introduced conditions. Although this type of "change" could represent a return to original prescribed specifications and conditions, the long-term solution to restore the original design is now effectively a change. As such, regulatory requirements as well as good project management practices dictate that all required elements of MoC be applied to this change as well. Some straightforward repairs may be excluded and not require an MoC. The permanent change, long-term solution, should be handled as a separate management of change project but should be related to the temporary MoC record addressing the same failure. Start-up of the long-term solution, a permanent modification or repair, will correspond to the closure date of the temporary modification.

Based on the author's field experience, these types of temporary changes represent a significant percentage (estimated at over half) of the MoC projects that are managed within the core MoC application.

5.3 Category 3 – Temporary bypass or failure of safety critical devices or other components

Category 3 Temporary MoCs involve situations where safety critical devices or components are out-of-service or inoperable for any reason including 3a) being bypassed, jumpered, overridden, or removed or 3b) found failed, impaired, not functional, or not fully functional. Category 3 MoCs fit the definition of temporary management of change; however, these types of changes may be managed under a different procedure or protocol [13] due to the nature of the condition, mitigations, and communication requirements, as well as inherent safety concerns.

Category 3a changes are planned, proactive work based on pre-written procedures, such as routine scheduled inspection, testing, and maintenance of relief devices. Category 3b changes are in response to failures and, therefore, represent reactive changes.

Although potentially handled through alternative procedures, databases, and communication tools, Type 3 Management of Change must address all items required under 1910.119 (l) OSHA PSM Management of Change. These requirements include: defining robust authorization requirements for approval of the change; completing an appropriate hazard and risk assessment to determine safety and health impacts, implementation of required mitigations and special procedures while the device is out-of-service; completion of ongoing communication and training requirements; making current PSI available; and defining a duration limit.

5.4 Category 4 – Temporary connections and/or temporary operations

Category 4 Temporary MoCs involve temporary connections to existing process equipment and/or occasional operations. These temporary changes are procedural. The procedures may involve connecting hoses, portable skids, or trucks to process equipment. These procedures may be related to equipment shut-down and isolation activities such as venting and purging of equipment. Or the procedure may represent a routine or occasional operational activity such as off-loading chemicals.

All standard management of change requirements must be addressed for the initial approval of new procedures. If the identical equipment and procedure is used in the future and the procedure addresses all management of change requirements including safety and hazard review and authorization requirements within the procedure steps, the procedure may be reused without creating another MoC record.

5.5 Category 5 – "Temporary" deferral of maintenance or inspection procedure due date including turnaround (TAR) deferrals

Category 5 temporary MoCs involve deferral of planned maintenance or inspection work beyond the scheduled due date. A temporary MoC could be used for a one-time deferral. A permanent MoC would be used for a permanent change to the maintenance or inspection interval prescribed by a particular maintenance plan. Likewise, deferral of turnarounds should be considered in the same category due to the number of planned maintenance and inspection activities that require an outage in order to be executed.

> Note: Companies may chose to consider deferral of an inspection or maintenance workorder due date as a permanent change associated with the single instance of the procedure, inspection, or turnaround activity. If there is sufficient tracking of the incomplete workorder activity through the maintenance management system, that classification would be sufficient. Identifying the change as temporary would add a secondary due date flag providing additional visibility. An MoC is required whether the single instance due date deferral is considered a temporary or a permanent change.

5.6 Category 6 – "Temporary" installation intended to be followed by a future installation having more capacity or capability*

Category 6 temporary MoCs are included for the purpose of completeness. However, it is recommended that this category not be considered temporary. This category represents a two-stage modification – similar to Category 2. However, in this case, both stages of the project are fully planned and proactive. The installation of the final modification will be driven by business rather than technical needs. The interim modification will typically have less capacity or capability than the planned final modification. However, the interim modification is acceptable from a technical and safety standpoint to remain in place indefinitely. In this scenario, it is recommended that the user classify this change as a permanent MoC. Business drivers and budget availability may change. Changes of this type do not require expiration dates or flagging as overdue for closure. Occasional exception may apply.

> * Note: Category 6 changes will not be discussed further, as these types of changes are recommended to be characterized as permanent.

5.7 Summary of Temporary MoC Categories

Table 1 is a summary of the temporary MoC categories discussed above. The administrative management system commonly used to manage the change category is also shown.

Table 1. Temporary MoC Category Descriptions and MoC Management System

	Temporary MoC Categories							
Category	1	2		3		4	5	6
Sub-Category		2a	2b	3a	3b			
Temporary MoC Category Description	Trials, Field Tests	Interim solution following failure		Bypass or failure of safety device or component		Temp connec-tions / ops	PM/TAR due date deferral	Staged improve-ment*
Sub-Category Description		Long-term: new design	Long-term: original design	Pro-active. Proce-dural	Reactive: Operate without device			
Management (Mgmt) System	MoC System	MoC System	MoC System or Repair	Bypass Mgmt System; Proce-dures	Bypass Mgmt System	MoC System; then Proce-dures	MoC System	MoC System; New Projects
Note:		Interim solution is Temp MoC. Long-term solution is Perm MoC. Both MoCs relate to the failure and the other MoC.		Bypass or failure of safety critical devices could be managed in MoC system, if preferred.			Temp or Perm MoC for single instance deferral.	*Call Perma-nent

6 Dispelling Myths

Section 2.3 summarizes three myths that commonly exist in industry and literature regarding temporary MoCs. Many of these myths, or assumptions, do not accurately reflect the real world usage and service of temporary MoCs. More concerning, many of these assumptions have elicited internal company rules, protocols, and procedures that have a possibility and a tendency to introduce greater risk into the operation.

The author will address each myth with respect to the various categories of temporary MoCs and conclude with preferred recommended practices.

6.1 Myth 1: Most temporary changes are terminated by returning the system to the prior condition, also known as reversing the change.

6.1.1 History/Past Assumptions:
Much literature on the topic speaks to closing temporary changes by returning or reversing the change to its original or prior condition [2][4][5][13][15]. Those references imply a Category 1, 3a, or 4 change is being discussed.

> As per OSHA Appendix C:
> In addition, the management of change procedures are used to [ensure] that the equipment and procedures are returned to their original or designed conditions at the end of the temporary change [5].
>
> As per Marsh:
> For temporary changes (handled under a temporary MoC), one of the key and final actions is the return of the plant to the original design [4].

6.1.2 Alternative Perspective:
Regarding Category 2 or 3b changes - temporary MoCs involving equipment or component failure, reversing the change to the original or prior condition may not be desirable because the original or prior condition may have contributed to the cause of the failure or may not be possible because the failed condition has not yet been corrected.

Consider these examples of temporary changes associated with failures (Category 2):

1. Blue tarps on roofs after a hurricane before roof repair or rebuild.
2. Pipe clamps on pipelines following a leak before permanent pipeline section repair or replacement.
3. De-rating a pipeline to address an inspection finding before permanent section repair or replacement.
4. Removing a pipeline, piece of equipment, or component from service until repair, replacement, or permanent decommissioning.

From the author's experience, Category 2 and 3b, represent the most common types of temporary MoC projects.

Table 2. Estimated Percentage of Temporary MoCs by Category

	Temporary MoC Categories						
Category	**1**	**2**		**3**		**4**	**5**
Sub-Category		**2a**	**2b**	**3a**	**3b**		
Category Description	Trials, Field Tests	Interim solution following failure		Bypass or failure of safety device or component		Temp connec-tions / ops	PM/TAR due date deferral
Sub-Category Description		Long-term: new design	Long-term: original design	Pro-active. Procedural	Reactive. Operate without device		
Management (Mgmt) System	MoC System	MoC System	MoC System	Bypass Mgmt System; Procedures	Bypass Mgmt System	MoC System; then Procedures	MoC System; New Projects
% of Temp MoCs (MoC System) - Based on Count **	**5 - 20%**	**50 - 75%**				Initial approval of procedure: **5 - 20%**	**5 - 20%**
% of Temp MoCs (Bypass Mgmt System) - Based on Count**				**50 - 75%**	**25 - 50%**		
Based on OOS Period***				**10 - 50%**	**50 - 90%**		
Note:	** Estimated based on the author's empirical experience. *** Estimated based on period or number of days out-of-service (OOS) rather than number of instances or MoC count, based on author's empirical experience.						

As per Table 2, the majority of temporary MoCs cannot be reversed to the prior condition at an arbitrary date. Many temporary changes are interim responses to failures that require other repairs or projects to be completed before the temporary change can be removed and replaced with another change.

6.1.3 Truth 1: It is important to recognize the differing reasons for utilizing temporary MoCs. Following a failure – which is one of the primary reasons for creating a temporary MoC – the operating company must complete other work before the temporary MoC may be closed. This other work (e.g. repairs, replacements, modifications, etc.) dictates the timeline for the temporary MoC closure and may result in a new configuration.

6.2 Myth 2: The best way to ensure limits on durations for temporary changes is to have a universal duration limit for temporary MoCs that is not "too long" prescribed within the company Management of Change procedure.

6.2.1 History:
Operating companies have widely, if not universally, taken the approach that the best way to ensure that limits are established on temporary MoC duration is to prescribe a universal "limit" in the company Management of Change procedure or protocol. Globally established procedural limits for temporary MoC duration typically range from 90 days to one year. Some companies indicate two procedural limits: an initial limit that may be followed by extension up to a maximum limit, usually not exceeding one year. Other companies indicate a "limit" in the procedure, generally not exceeding one year, but allow for extensions beyond that limit, making the initially defined period not a true limit but a target and a checkpoint requiring re-authorization. Some companies establish an ultimate limit by using "shall" or "must" statements that preclude approving extensions. Other companies may not prescribe fixed duration limits, but due to discomfort with longer duration temporary MoCs, simply do not allow or approve lengthy temporary MoCs.

Presumably, this approach is seen as the best way to address the following realities. Temporary MoCs:

- ⇨ involve a change that is approved for only a limited duration
- ⇨ are required by regulation or recommended by guidance literature to have a stipulated duration limit and
- ⇨ are often seen as having a risk or tendency to become effectively permanent if they linger or remain in place "too long."

The consensus solution seems to be: apply a universal limit for temporary MoCs and make it not "too long." Interpretations of not "too long" vary but typically land on one year or less. Some companies may interpret the regulation as requiring a single, fixed, defined limit for all temporary MoCs.

6.2.2 Alternative Perspective:
This approach of indicating a single, universal duration limit for all temporary MoCs – while well-intended - is inherently flawed and inherently unsafe. This approach will inevitably lead to unintended consequences that could introduce more risk into the operation. Here's why, . . .

1) First of all, . . .
it is important to recognize the differing reasons for utilizing temporary MoCs as the differing categories of temporary MoCs will have very different drivers for both initiating the temporary MoC and very different time horizons and dependencies for closure of the temporary MoC.

Category 1 changes, trials and science projects, are entirely proactive. Although each temporary change must have a specifically defined period, this category of temporary change could have a wide range of desired and acceptable durations. The change must be designed to safely meet the intended and necessary life of that change. Thus, globally prescribed limits or even targets are not necessary.

The driver for the duration should be defined within the project scope. Similarly to project cost and schedule estimating, duration estimates should incorporate contingency planning so that the design and assessment properly consider the realistic worst case rather than the best case for how long the change could remain in place. The design should be robust so that the risk falls within the company prescribed risk management boundaries for the realistic duration of the temporary change (See Figure 3). The change may be able to be designed so that no incremental risk is incurred during the term of the temporary change as per the process hazard assessment.

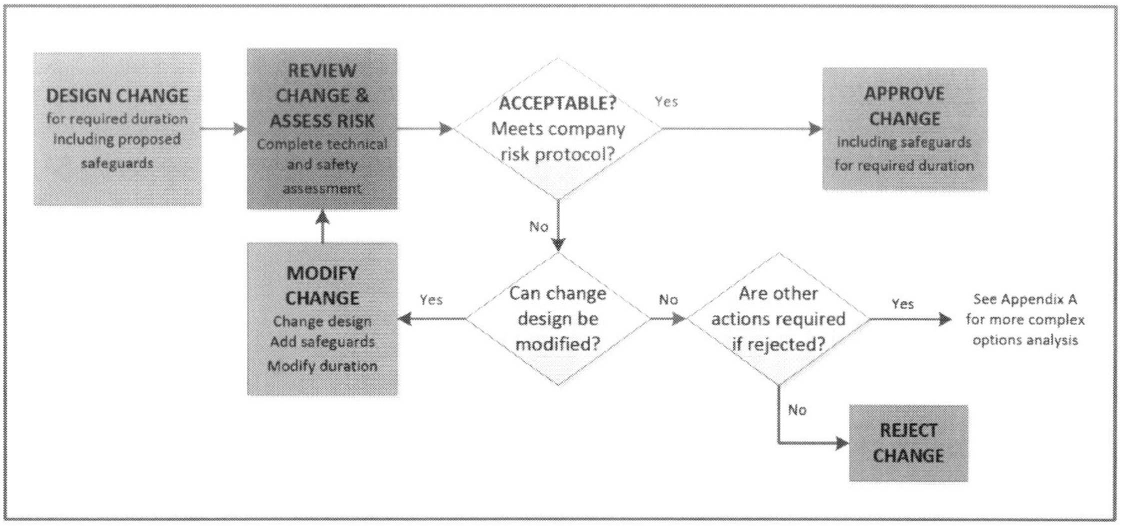

Figure 3. Simplified Work Flow Process: Temporary MoC Review and Approval

As with all temporary MoCs, the expiration date prescribed within the specific MoC should be queryable and monitored so that as the deadline approaches, planning will accommodate closure, which means reversal in this case, by the due date. The contingency built-in up front will aid in on-time closure and make it more likely that the original design and initial technical, safety, and operational reviews consider the full-term of the change.

Category 2 temporary changes are interim solutions in response to failures or unplanned events. The duration limits for Category 2 changes are entirely driven by the timeline and critical path for the implementation of the longer-term solution. This long-term solution is directly related to the failure that drove the interim solution. Duration needs for Category 2 MoCs will also be highly variable. A single prescribed limit is not appropriate for this type of MoC.

As with Category 1 MoCs, when defining duration for the Category 2 MoC, it is important to consider the worst-case timeline that incorporates contingency for the long-term project schedule. The long-term solution and all critical path items associated with the long-term solution may not be fully defined at the time that the temporary change will be initially designed and approved. However, some critical path drivers may be known, such as timing for procuring material and future planned shut-down opportunities if a larger unit or plant shut-down is required.

In the case of Category 3a changes - proactive procedures for performing inspection, testing, and maintenance of safety critical devices, these procedures should have very limited duration allowances, typically in the range of 8 to 12 hours (single shift) or less. However, even these procedures should have individually prescribed duration limits based on the specific application and requirements of that procedure. (Refer to Figure 4, Example 1 for an example of a Category 3a change and associated risk assessment.)

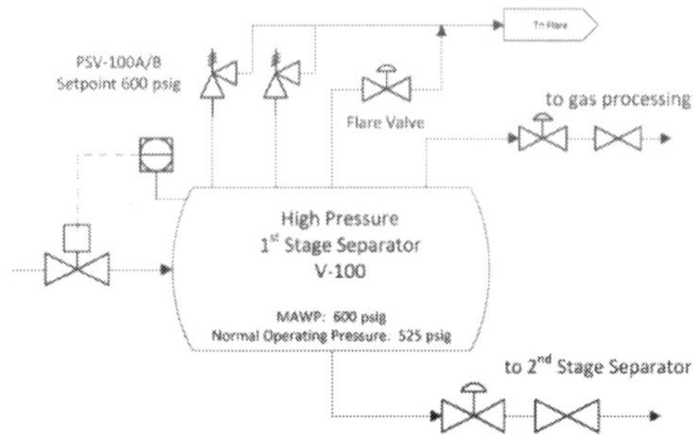

Category 3a: Planned Maintenance Work:
Removing one PSV at a time from service for scheduled inspection, testing, and maintenance is typical Category 3a planned work. Procedures should be prepared in advance which define the hazards (safety risks), required mitigations, and the approved duration for performing this work.

Category 3b: Unplanned Repair Work
In the event that a PSV requires repair or extended removal for any other reasons, the same prepared hazard and risk assessment and mitigation plan may be used as the basis for planning the repair work.

In the case of this PSV removal, the mitigation defined is reducing maximum inflow rate from 200 MMSCFD to 150 MMSCFD. This change will impact production; however, it should be less of an impact than a complete shut-down. The maximum inlet flowrate alarm will be reset to the lower limit. Prescribed operator actions remain the same.

Based on the risk assessment for this PSV setting, the safety risk rank is reduced for the duration of the out-of-service period due to the mitigation of lowering max rate. The flare valve has full relief capacity at the lower rate (could be considered an IPL) and a single relief valve has full relief capacity at the lower rate.

Risk Rank Comparison - PSV-100A or PSV-100B Out-Of-Service (a)

Relief Capacity per PSV	150	MMSCFD		
Relief Capacity - Flare Valve	175	MMSCFD		
	1 **Normal Ops**		2 **1 PSV Out-of-Service**	
Normal Max Inflow Rate	**200**	MMSCFD	**200**	MMSCFD
Mitigation: Reduce Max Rate by	**N/A**	MMSCFD	**50**	MMSCFD
Relief Requirement	**200**	MMSCFD	**150**	MMSCFD
Number of PSVs Available	2		1	
PSV(s)	IPL	Both Rqd	IPL	One Rqd
Shut Down Valve	IPL	SIL 1	IPL	SIL 1
Flare Valve	Not IPL	BPCS	IPL	BPCS
Risk Rank	**B2**		**B1**	
Risk Category	**Med - 6**		**Med - 5**	

Risk Rank Matrix (b)

CONSEQUENCE		LIKELIHOOD (per year)					
		< 1.E-05	1.E-04	1.E-03	1.E-02	1.E-01	1.E+00
SAFETY		1	2	3	4	5	6
multiple fatality	A	M6	H7	H8	VH9	VH10	VH11
fatality onsite	B	(2) M5	(1) M6	H7	H8	VH9	VH10
perm disability	C	L4	M5	M6	H7	H8	VH9
hospital / rehab	D	L3	L4	M5	M6	H7	H8
medical treatment	E	L2	L3	L4	M5	M6	H7
first aid	F	L1	L2	L3	L4	M5	M6

Given the safety risk assessment , the duration limit for this Category 3 temporary change will be driven by commercial rather than safety related concerns.

Figure 4. Example 1: Category 3 Temporary MoC - Separator PSV Out-Of-Service. Risk comparison is shown as (a) with data and independent protection layers (IPL)s. Risk Rank Matrix is shown as (b).

Category 3b situations - out-of-service or failed safety critical devices or other components - will have initial authorization requirements. Then, these instances will typically have reauthorization and communication requirements prescribed on daily, weekly, and monthly intervals, which are similar to duration limits. Re-authorization requirements typically escalate to higher levels of management as the out-of-service duration extends. However, as with all temporary MoCs, authorization or reauthorization of the 3b temporary MoC must consider all elements of the change including current condition, safeguards, mitigations, and risk. Example 1 in Figure 4 also discusses the PSV removal in the context of a Category 3b reactive change.

In certain cases, appropriately addressing the cause of the Category 3b failure could require extended efforts. If modifications are required, a separate MoC is typically written for the long-term solution. In these cases, the resolution timing for the Category 3b failure would resemble the timelines associated with Category 2 temporary changes.

> Note: Category 3 MoCs are sometimes managed through alternative protocols and procedures. If a modification is required to address the cause of a Category 3b failure, the temporary change and the long-term change may now be managed in two systems.

Regardless of the system in which the change is managed and the semantics of duration limits versus reauthorization or extension, the review of the change for authorization, re-authorization or extension must receive detailed technical and safety review, assurance of adequate safeguards and procedures, evaluation of risk, and consideration of alternative options including process shut-down, as needed. Shut-down considerations must weigh risks associated with shut-down and start-up activities against the risk of continued operation in the current state. The realistic timing for the resolution should be understood and be built into subsequent reauthorization or extension requests. Refer to Appendix A for a more detailed version of Figure 3 work flow where more complex options analysis is required. Although it is desirable for Category 3 temporary changes to have very short durations, a failed device or device prone to failure cannot be returned to service.

Category 4 changes – temporary connections and/or temporary operations - will have duration limits defined by the procedural activity to which these changes relate. These limits may be in the range of one shift to 48 hours. This timing will be specifically dependent on and defined by the activity to be completed as well as the risk assessment. These activities might be handled within the MoC system for only the first instance of approving and using a new procedure.

The Table 3 summarizes the typical range of temporary MoC duration by category of temporary MoC. Defining and understanding the change category or type should aid in understanding the duration needs for a given change, which in turn leads to appropriate design and more robust approval protocol.

Table 3. Temporary MoC Category Descriptions and Typical Duration Range

	Temporary MoC Categories						
Category	1	2		3		4	5
Sub-Category		2a	2b	3a	3b		
Category Description	Trials, Field Tests	Interim solution following failure		Bypass or failure of safety device or component		Temp connec-tions / ops	PM/TAR due date deferral
Sub-Category Description		Long-term: new design	Long-term: original design	Pro-active: Procedure	Reactive: Operate without device		
Typical Duration Range	Variable (Months to Several Years)	Variable (Months to Several Years)	Variable (Months to Several Years)	8 to 12 hours (single shift)	Days to Months (or longer); intervals prescribed to reauthorize	Hours to Days	Variable (PM: Days to Months; Defer TAR: 1 -2 years)
Duration Tied to	Temp MoC design intent	Perm MoC timeline	Perm MoC timeline	Procedure guidance	Repair or Perm MoC timing (like Cat 2 Temp MoC)	Procedure guidance	Various: outages, resources
Note:	Process hazard risk assessment is built into all temporary MoC project or procedure design including prescribing safeguards and mitigations required to bring risk to an acceptable level for the required project duration. (See Figure 3 and Appendix A.)						

2) Secondly, . . .
it is important to recognize that the language in these global procedures – specifically the universal duration limits - does not represent truth or reality. Some temporary MoCs will inevitably need to go longer than one year . . . and some should.

By recognizing the driver(s) for the duration and defining a plausible timeframe for the temporary change, the technical, safety, environmental, operational, and process safety reviewers will assess the change with a more critical and realistic eye for actual design needs for the project. Placing arbitrary or unrealistic limits on the duration of the temporary MoC will improperly communicate the true design needs to the technical and operational professionals evaluating the change. If the change needs to remain in place over four years to reach the next plant outage opportunity, the design, operational, and maintenance requirements must reflect the needs to safely operate for that period of time. If the change is unacceptable for that duration, a design change or another alternative option must be considered. (Refer to Figure 3 and Appendix A.)

By placing arbitrary limits taken from global standards on the maximum duration for temporary MoCs and then allowing for extension, the true design requirements of the change are not accurately reflected when the design is reviewed and approved. Documentation and approvals must reflect the true anticipated life. If the documented duration of the change does not reflect the intent or need, the MoC could be subject to audit findings.

Even with specificity and contingency built-in, changes to the long-term project scope or definition may affect the implementation timeline and, therefore, require review and re-evaluation of the temporary MoC to consider extension of that change. However, the number and length of extensions can and should be minimized by avoiding use of arbitrary, globally applied duration limits for temporary changes.

Of note, when extension requests are reviewed and approved, the phenomenon of "normalization of deviation" can lead to weaker scrutiny when perpetuating a temporary change that is already in place. Therefore, limiting the number of extensions is desirable.

It is also important to keep in mind that defining realistic and reasonable schedules rather than forced, aspirational but unrealistic schedules will lead to more robust ultimate solutions. Recall the good-fast-cheap triangle [14] and the truth that it teaches us regarding project management.

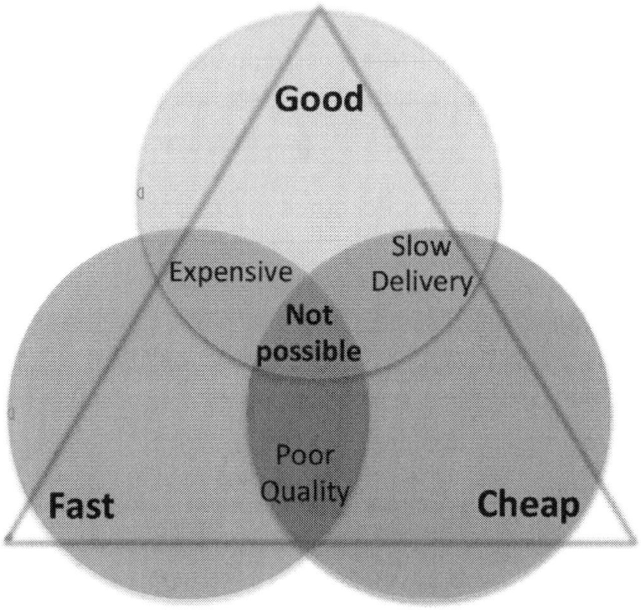

Figure 5. Good-Fast-Cheap project delivery triangle

In project management and delivery, you can only pick two of the three of the corners of the good-fast-cheap triangle. And, if you pull too hard on any one corner, you may only get one of the three.

The OSHA PSM regulation requires duration limits for *any* MoC project. Ironically, it is more appropriate to apply a universal definition to "permanent" than "temporary" when defining change duration, although the reverse is generally done.

3rd, finally and importantly . . .
Many companies may misread or misinterpret the language and intent of regulation and guidance.

The OSHA PSM regulation [1] and well established industry guidance, such as *CCPS Guidelines for the Management of Change for Process Safety* [2] never state that *universal* limits are required for temporary MoCs. The OSHA PSM regulation does *not* define a maximum length for the duration of a temporary change. Likewise, *CCPS Guidelines for the Management of Change for Process Safety* [2] does *not* define a maximum limit. The regulation states and intends that *each* temporary MoC have a duration period stipulated and that the change is appropriately designed, reviewed, and approved considering that duration.

6.2.3 Truth 2: Each Temporary MoC must be designed to meet a specifically stipulated duration limit. Universal caps on duration for temporary MoCs are neither appropriate nor effective.

6.3 Myth 3: Converting a temporary MoC to a permanent MoC is an acceptable practice especially for longer duration temporary MoCs.

6.3.1 History:
Allowance for conversion of temporary MoCs to permanent MoCs is frequently stipulated within company procedures as one of several ways to disposition a temporary MoC that is approaching its deadline.

The following choices are often provided in company MoC procedures as options for addressing the temporary change before the expiration date [4][8][9]:

- At the end of the duration limit, the change shall or must be:
    - Removed or reversed and the plant restored to original condition OR
    - Converted to a permanent change OR
    - Granted an extension

Where extensions are not allowed, the extension option is removed leaving two choices, neither of which may be technically acceptable or feasible.

> Note: Shutting-down the equipment, process, or facility should always be considered as an option, but shut-downs are rarely given as one of the options to consider upon approaching the due date. Of course, shut-downs and start-ups present inherent safety risks as well. But a shut-down may be the only technically acceptable option depending on the options made available.

If a temporary change (of defined limited duration) is converted to a permanent change (of indefinite duration), an essential basis for the prior design and approval has been changed.

Therefore, the original approval is fully invalidated. Any conversion necessarily requires review by any and all prior reviewers and approvers of the temporary MoC (or equivalent discipline representatives).

Regarding conversion administration, in some companies the original MoC record may be redefined as permanent. In other companies, conversions are accomplished by opening and approving a new MoC record describing the same change****. This method of handling the conversion tends to be cleaner administratively allowing for accurate reflection of history, clarity of the scope, insertion of new technical, safety, environmental, operational, and managerial reviews, and documentation of new authorizations. Also, the new permanent MoC will require final updates of all applicable PSI, whereas updates for PSI associated with temporary MoCs may have differing protocol and requirements regarding format and timing. The two MoCs should be related or cross-referenced to each other in the administrative system.

> **** Note: If a new MoC record is opened that describes a new change or condition, then the new MoC is a new permanent change rather than a conversion, although still related to the temporary MoC.

The three disposition options described above, or two where extensions are not allowed, are broadly applied throughout industry. The same choices are also reflected in technical articles [4] [15] on the topic.

6.3.2 Alternative Perspective:

This approach of offering conversion to permanent as a method for managing temporary MoCs approaching their due date – while well-intended - is inherently flawed and inherently unsafe. This approach will inevitably lead to unintended consequences that could introduce more risk into the operation. Here's why, . . .

Conversion of a temporary MoC to a permanent MoC should not be allowed unless the conversion is based on one of a set of defined and defensible reasons. Possible legitimate reasons for conversion of a temporary MoC to permanent could include: 1) there was a technical error in the initial assessment, 2) a global company standard or specification has been relaxed specifically impacting this change, 3) another system change has occurred impacting this change (such as safe operating limits have shifted and those new limits have been separately technically approved), 4) the change was a Category 6 change called temporary for business reasons only, not technical or safety reasons, or 5) the change fell into another category due to original duration intent only but meets all standard technical, safety, and operational protocol for permanent operation and is now intended for indefinite operation in the current configuration. **Bumping up against an arbitrary temporary MoC duration limit is *not* one of those reasons.** If criteria are met that allow for conversion, all process safety information updates must occur as per protocol for permanent changes.

Temporary MoCs are typically defined as temporary because the change is unacceptable technically to remain in place for an indefinite timeframe as a permanent installation. Conversion of a temporary MoC to permanent is, therefore, inherently problematic.

> Note: Closure of temporary MoCs via conversion to permanent MoCs, if allowed by the company, should be a red flag to auditors. If not based one of the reasons given above or other legitimate reasons defined by the company, the approval of the conversion to permanent may not be justifiable or technically sound.

Temporary MoCs that are inappropriately converted to permanent MoCs place the operating company at higher risk, as the flagging system based on the temporary MoC expiration date will be removed. The change will no longer be flagged in the system and will, therefore, be less visible for further review, monitoring, and closure. Additionally, drivers for outages will be masked when the temporary designation and associated due date is removed.

6.3.3 Truth 3: Converting a temporary MoC to permanent MoC is rarely an acceptable practice. Call a spade a spade, or we induce bad behavior and bad outcomes!

7 Two Wrongs Make a Wrong!

Much literature has been written regarding common problems associated with managing temporary MoCs [15]. There seems to be little recognition that the problems may in fact be self-fulfilling prophesies when our management systems steer personnel down paths toward failure. For example, Ozog and Stickles [15], describe common quality "problems" seen with regard to managing temporary MoCs.

> Temporary changes
> The documentation, closure, and communication of temporary changes are often cited as major quality issues in MOC programs. Common issues include:
> a. *The authorized time period expires without removing or making the change permanent.*
> b. *The change is made permanent*, but is not reflected in the piping and instrumentation diagrams (P&IDs) and other process safety information (PSI).
> c. *The change is improperly or incompletely removed* [15].

Causes for these problems are embedded in the problem statements. Conversions to "permanent" are not technically acceptable for a true temporary change, but are frequently inappropriately applied in order to manage deadlines that are neither realistic nor appropriate.

Many of these MoC conversions are being driven by the combination of these two problematic requirements being coupled onto temporary MoCs: arbitrary universal duration limits that are too short *and* allowance for conversion of temporary MoCs to permanent MoCs as a standard, sanctioned method of resolution when MoCs approach these arbitrary deadlines.

In other words, we are fabricating a problem that should not exist by applying an unnecessary constraint, . . . and we are giving people an inappropriate tool to deal with that "problem."

Humans that are forced to make a decision between several untenable choices will inevitably make a poor decision. These types of errors in human decision-making that are propagated by management system flaws are routinely found to be causal factors for major and minor industry incidents [16]. As stated in CCPS, *Recognizing Catastrophic Incident Warning Signs in Process Industries:* "Management system deficiencies are a root cause in all effective incident investigation methodologies [16]."

Insidiously, these MoC procedural constraints and requirements may actually build defects into the operation. Auditing to those procedures can introduce even more defects into an operation, as well.

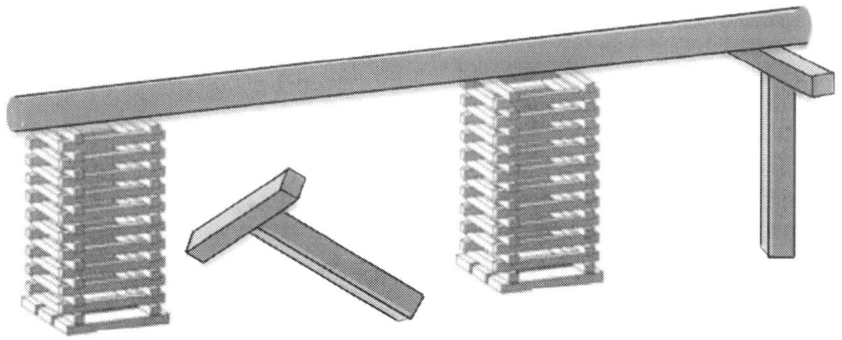

Category 2 Temporary MoC:
Due to soil instability issues, a series of pipe supports servicing a high pressure gas pipeline began falling over.

The timing for implementation of the long-term solution was constrained by seasonal weather windows and a need to consolidate the shut-down with a wider plant outage.

As an interim solution, crib stacks (10' – 15' tall) were specifically designed and engineered by a civil/structural engineer as temporary pipe supports until the new replacement pipe supports could be designed and installed. The crib stacks were reviewed and signed off during the temporary MoC review by all relevant disciplines including mechanical, operations, environmental, and safety. A hazard assessment was conducted.

When the wider plant outage was delayed for one year, the civil/structural and mechanical engineers approved allowing the crib stacks to remain in place for three years with required inspections continuing throughout the period.

However, because of a universal temporary MoC duration limit of one year imposed by the company MoC procedure, the temporary MoC could not be extended. Operations and Engineering Management preference was to convert the MoC to a permanent MoC rationalizing that it would be a "3-year permanent" MoC. This definition of permanent would be in conflict with standard definitions and the intent of permanent installations.

Figure 6. Example 2: Category 2 Temporary MoC – Interim Pipe Supports, Tall Crib Stacks

8 Conclusions/Recommendations

The following recommendations are offered as preferred practices for managing temporary changes:

1. Recognize that there are distinct categories of and reasons for implementing temporary MoCs.
   - Consider defining the temporary change category in the header information in order to better define likely lifecycle range for that change.
   - Recognize that many temporary MoCs are interim solutions associated with failures (Category 2 or 3b). Those MoCs require implementation of another project, the long-term solution, as a prerequisite for closure of the temporary MoC.

  - Cross-reference temporary MoCs that are interim solutions for failures (Category 2 or 3b) with permanent MoCs that are the long-term solutions for the same failure.
  - Define and monitor critical path item(s) that will define closure timing.
2. Do not dictate universal fixed duration limits for temporary MoCs in company management of change procedures. Design the change considering safeguards and including inspection and maintenance requirements in order to meet company risk protocol and safely operate for the intended duration of the change.
  - Ensure that every temporary MoC has a specified duration limit defined for the specific change as a requirement for approval. In electronic systems, expiration date should be a required field.
    - Note: The OSHA PSM regulation does not require a universal fixed limit on temporary MoC durations.
  - Consider real world constraints when setting the temporary MoC duration such as: materials deliveries, resource availability, shut-down opportunities, etc. Incorporate contingency into setting expiration dates.
  - Refer to Figure 3 and Appendix A for temporary MoC review and approval work flow process that considers duration requirements as an input to the design and review process.
  - Develop internal company protocol that addresses PSI management needs for MoCs of varying temporary lengths.
  - Define the meaning and duration of "permanent change" globally within the company management of change procedures.
3. Conversion of temporary MoCs to permanent should rarely be allowed. An inappropriate conversion could create a hidden or latent process safety hazard.
  - Define a listing of acceptable reasons that temporary MoCs may legitimately be converted to permanent. Allow conversion only if one or more of those conditions is met.
    - Note: A temporary MoC that is approaching an expiration date is not an acceptable reason to convert the MoC to permanent.
  - Clearly define what conditions changed from the prior temporary MoC making the conversion allowable for each temporary MoC that is converted.
  - For the rare instances where conversions are allowable, initiate a new MoC and relate the new permanent MoC to the prior temporary MoC.

Many of the errors that have historically been associated with managing temporary MoCs may be directly attributable to constraints and flaws built into the management systems. Addressing these underlying issues will eliminate some of the causal factors for human error when using these procedures. Allowing personnel the flexibility to design, approve, and manage temporary changes based on the system need as well as the assessed risk will ultimately lead to more efficient and resilient [17] work processes and safer, more reliable operations.

9 Acknowledgements

The author wishes to thank Mike Marshall, OSHA; John Wincek, Croda, Inc.; Jan Hayes, RMIT University; Robb Moss, Cal Poly San Luis Obispo; Lance Muniz, Vetergy Group; and David Wilbur, Vetergy Group for their review and comments on this paper.

10 References / Resources

10.1 Cited References

1) U.S. Department of Labor Occupational Safety and Health Administration. Process safety management of highly hazardous chemicals, 29 CFR. 1910.119, Jun 1, 1992, as amended Feb 8, 2013.
2) CCPS. *Guidelines for Management of Change for Process Safety*. Center for Chemical Process Safety. American Institute of Chemical Engineers, New York, NY, 2008.
3) T. Takegawa and H. Kobayashi, Hideo, "Disaster of Chemical Plant at Flixborough," Failure Knowledge Database / 100 Selected Cases, Available at http://www.sozogaku.com/fkd/en/hfen/HB1058048.pdf, Accessed on Jan 15, 2017.
4) Marsh. "Management of Change, Risk Engineering Position Paper – 05," 2015, Available at https://www.marsh.com/uk/insights/research/risk-engineering-position-paper-management-of-change-moc.html, Accessed on Jan 15, 2017.
5) U.S. Department of Labor Occupational Safety and Health Administration. Appendix C to 1910.119 – Compliance Guidelines and Recommendations for Process Safety Management (Nonmandatory), Feb 8, 2013.
6) J. Amy, "After Katrina: Roofs remain source of concern," AL.com, Available at http://blog.al.com/live/2010/08/after_katrina_roofs_remain_sou.html, Accessed on Jan 15, 2017.
7) Dalian ZhuHong Mechanical Co., Ltd, Various piping repair pictures, Available at http://www.dlzhjx.com/zz31.asp, http://www.dlzhjx.com/images/7/17.jpg, Accessed on Jan 15, 2017.
8) BASF, "Management of Change Procedure – Wyandotte," Available at http://www2.basf.us/DocSearchWeb/displaydoc?docbase=Wyandotte_MI_US&cabinet=EHS&folder=Publish/WSTD&docname=WYN032.075&rend=pdf, Accessed on Jan 15, 2017.
9) INEOS Nitriles, Green Lake "SOSO Management of Change (MoC) [Procedure]," Available at http://greenlake.biz/soso/soso/IA0900.DOC, Accessed on Jan 15, 2017.
10) Merriam-Websters Dictionary [online], Available at https://www.merriam-webster.com/dictionary/, Accessed on Jan 15, 2017.
11) Cambridge Dictionary [online], Available at http://dictionary.cambridge.org/us/dictionary/english/, Accessed on Jan 15, 2017.
12) Dictionary.com [online], Available at http://www.dictionary.com/browse/, Accessed on Jan 15, 2017.
13) WorkSafe Victoria, "Management of change at a major hazard facility," Available at http://www.worksafe.vic.gov.au/__data/assets/pdf_file/0018/12393/Management_change.pdf, Accessed on Jan 15, 2017.

14) Wikipedia. "Project management triangle," Available at https://en.wikipedia.org/wiki/Project_management_triangle, Accessed on Jan 15, 2017.
15) Ozog and R.P. Stickles, ioMosaic, "Management of Change or Change of Management?" Available at https://www.chemicalprocessing.com/assets/Media/MediaManager/ManagementOfChange.pdf, Accessed on Jan 15, 2017.
16) CCPS. *Recognizing Catastrophic Incident Warning Signs in Process Industries*. Center for Chemical Process Safety. American Institute of Chemical Engineers, New York, NY, 2011.
17) D. Wilbur, Vetergy. "Admitting Human Error in the Workplace," Jan 2016, Available at http://vetergy.com/, Accessed on Jan 25, 2017.

10.2 Uncited References

1) U.S. Department of Labor Occupational Safety and Health Administration. OSHA 3133 Process Safety Management Guidelines for Compliance, 1994 (Reprinted).
2) U.S. Department of Labor Occupational Safety and Health Administration. OSHA 3132 Process Safety Management, 2000 (Reprinted).
3) Y. Malmen, et al. "Shutdowns and Management of Change." *Hazards XXII* Symposium Series No. 156, iChemE, 2011, Available at https://www.icheme.org/~/media/Documents/Subject%20Groups/Safety_Loss_Prevention/Hazards%20Archive/XXII/XXII-Paper-30.pdf, Accessed on Jan 15, 2017.
4) C. Harrison, "Managing Change in Manufacturing & Production Facilities," ByDesign, 2012, Available at http://www.asse.org/assets/1/7/carl_harrison_article.pdf, Accessed on Jan 15, 2017.
5) Safe Work Australia, Model Work Health and Safety Regulations (28 November 2016). Available at http://www.safeworkaustralia.gov.au/sites/SWA/about/Publications/Documents/616/Model-WHS-Regulations-28Nov2016.pdf, Accessed on Jan 15, 2017.
6) National Offshore Petroleum Safety and Environmental Management Authority (NOPSEMA), Available at https://www.nopsema.gov.au/safety/safety-case/safety-case-guidance-notes/, Accessed on Jan 15, 2017.
7) UK Heath and Safety Executive. "Plant modification / Change procedures," Available at http://www.hse.gov.uk/comah/sragtech/techmeasplantmod.htm, Accessed on Jan 15, 2017.

APPENDIX A

Work Flow Process: Temporary MoC Review and Approval with more complex alternatives analysis

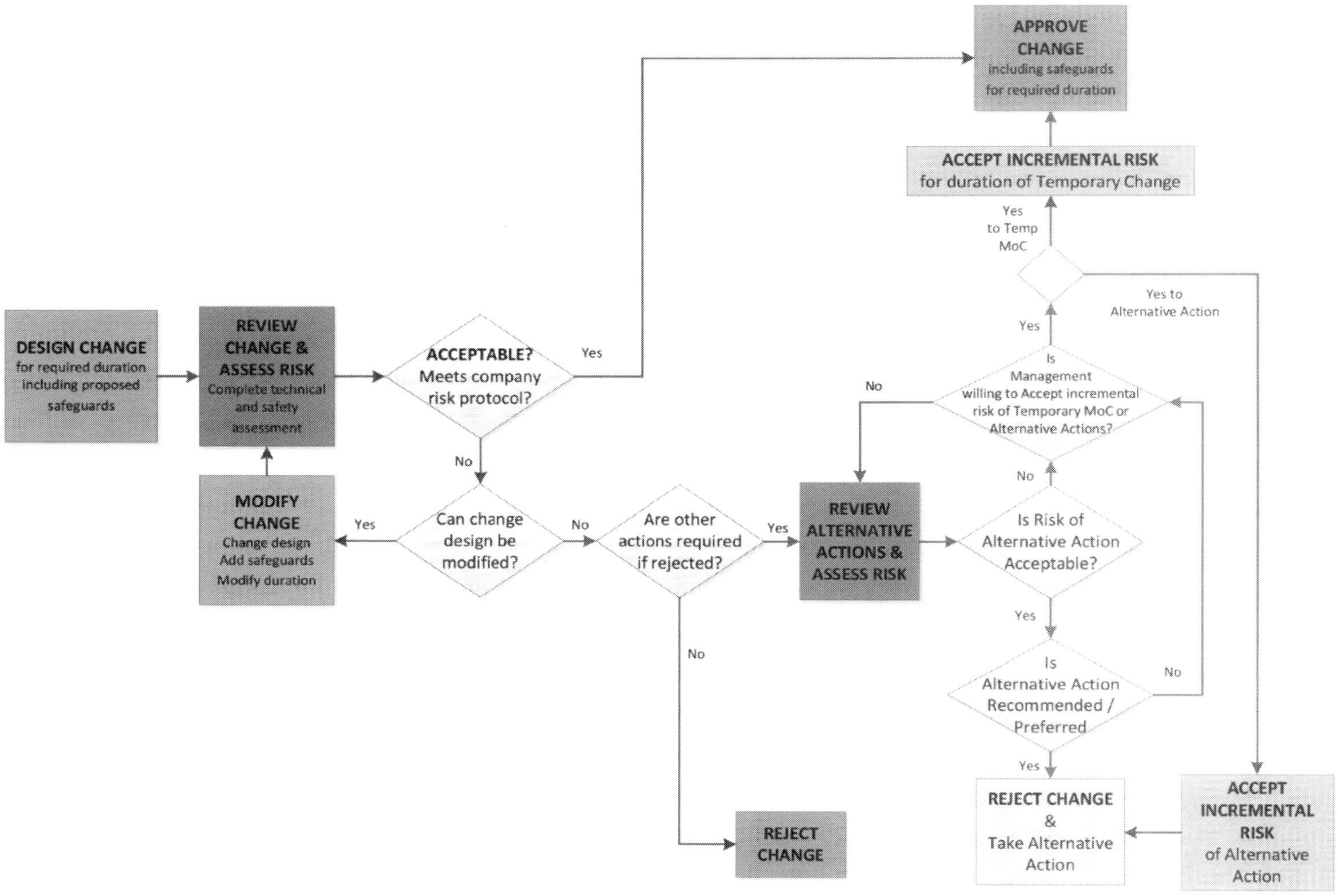

Don't Do What We Did – Stupidity, Luck, and Making Good Choices

Mike Moosemiller
Baker Engineering and Risk Consultants, Inc.
8430 West Bryn Mawr Ave, Suite 720, Chicago, IL 60631
mmoosemiller@bakerrisk.com

Kathleen A. Kas, PE
The Dow Chemical Company
4460 State Hwy 225
Deer Park, TX 77536
kkas@dow.com

Prepared for Presentation at
American Institute of Chemical Engineers
2017 Spring Meeting and 13th Global Congress on Process Safety
San Antonio, TX
March 26 – 29, 2017

Don't Do What We Did – Stupidity, Luck, and Making Good Choices

Mike Moosemiller
Baker Engineering and Risk Consultants, Inc.
8430 West Bryn Mawr Ave, Suite 720, Chicago, IL 60631
mmoosemiller@bakerrisk.com

Kathleen A. Kas, PE
The Dow Chemical Company
4460 State Hwy 225
Deer Park, TX 77536
kkas@dow.com

Keywords: Human Error, Hazard Awareness, Commissioning, Unsafe Act

Abstract

Fortunately, the focus on safe behaviors, hazard recognition, and risk management has greatly improved over the past few decades. However, in the early days of the authors' careers, there wasn't the same level of safety awareness - and much more risk-taking behavior. In this paper, the authors recount events in which they (and others) were ignorant, careless, or intentionally put themselves in dangerous situations early in their careers. These poor choices could have, or did actually result in, chemical exposures, fall hazards, overpressure events, and process upsets. The authors describe the circumstances leading to the decisions to take personal risks, what happened (and what could have happened), and the consequences resulting from those poor choices. As George Santayana said in *The Life of Reason* (1905), "Those who do not remember the past are condemned to repeat it." These same events could happen today to inexperienced (or even experienced) engineers, and it is hoped that the lessons learned by the authors will prevent others from putting themselves in risky situations.

1 Introduction

The early days of chemical processing were fraught with more hazards to workers than we have today. A classic example was the production of nitroglycerin in the 1870s [1]. Control of the reactor was done manually, with frequent adjustment needed by the operator to keep the equipment within safe operating limits so that an explosion would not occur. (Operators were given one-legged stools so that they would not fall asleep and allow the equipment to migrate outside of safe operating limits – with potentially deadly consequences.)

Fortunately, safety performance in the chemical industry has improved dramatically over the past several decades (Figure 1).

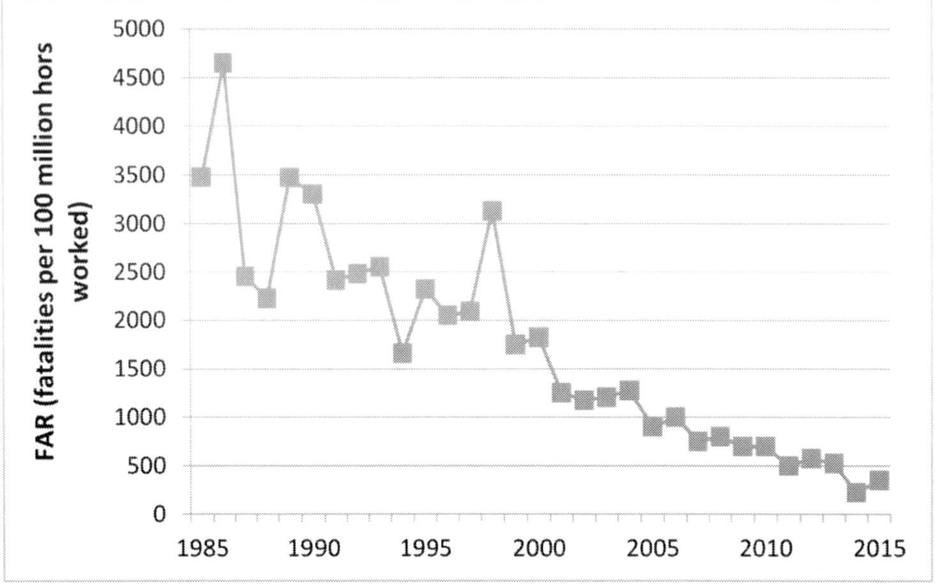

Figure 1: Reported Work Hours and Fatal Accident Rate (1985-2015) [2]

The commitment of employers, as well as government regulations and standards such as the Occupational Safety and Health Administration (OSHA) Process Safety Management (PSM) standard [3], the US Environmental Protection Agency (EPA) Risk Management Program (RMP) rule [4], Directive 82/501/EC (Seveso) [5], the Control of Major Hazards (COMAH) regulation [6], and others have helped to drive improved process safety performance in industry. Certainly, through the learnings from significant industry incidents, chemical companies have recognized the business case – and moral imperative – for eliminating process safety events. Responsible companies no longer tolerate types of behaviors (such as those described in this paper) that may have been practiced 30 or more years ago, and <u>the authors wish to stress that their current and prior employers did not encourage, authorize, or condone the specific activities described in this paper</u>. Although such behaviors would no longer be considered acceptable, the lessons, learned the hard way, should not be lost.

As young engineers, the authors made some classic rookie mistakes. They certainly got themselves into trouble when they did not maintain a keen awareness of their surroundings and the hazards that were present. In some instances, the authors did not recognize the hazards associated with the operation; in other cases, they took it upon themselves to take actions that were either inherently unsafe or for which they had not been properly trained. Poor management of change led to other undesired events, and the authors were also caught in a few traps created by a lack of attention to human factor considerations in equipment design.

The authors (and so presumably, the readers of this paper) are often impacted most strongly by learnings that come through the stories told by those who were personally involved in the incidents. The consequences of the poor choices become so much more "real." From that perspective, the authors are sharing their personal stories regarding mistakes they or others made in their early days and the impact of those poor decisions. It is hoped that the learnings from these mistakes will help newer engineers to have a higher level of awareness of the potential consequences of making dumb decisions as well as a better appreciation for the process safety management systems we have today that help prevent these situations.

2 Learning Opportunities: What Could Go Wrong?

2.1 But we vented the pressure – didn't we?

2.1.1 The Situation

A new emulsion product was being scaled up in a pilot plant facility. One of the authors was working night shift in support of the new product development, wearing normal street clothes. The emulsion was being cooled by passing it through a heat exchanger using an air-operated diaphragm pump. At the end of the transfer, the author turned off the air to the pump and opened a bleed valve on the product discharge hose to remove pressure from the line prior to disconnection, as illustrated in Figure 2. No material came out of the bleed.

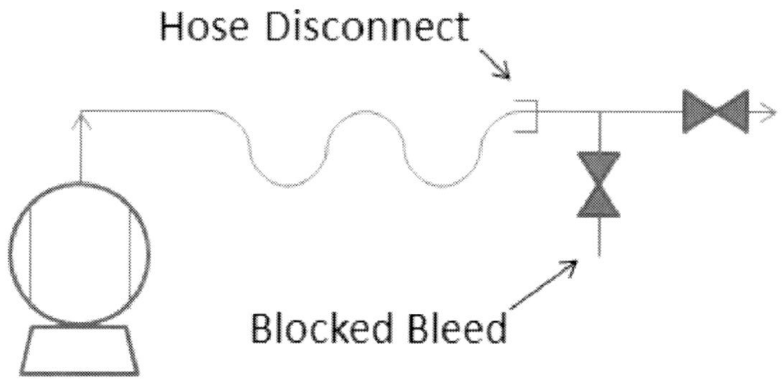

Figure 2: Hose and Bleed Assembly

2.1.2 The Poor Choice

The author believed that, because no material came out of the bleed, the hose was vented. Wearing only goggles and gloves, the author disconnected the hose.

2.1.3 Potential/Actual Consequences

The hose whipped around and doused the author's legs with warm emulsion. Fortunately, the material was not corrosive and had been sufficiently cooled so as not cause a burn. The author immediately went to the closest bathroom with a shower rather than to the safety shower (another poor choice). Fortunately, there was no injury from the exposure – and someone had left spare work clothes in the shower room, sparing the author the embarrassment of wandering in undergarments in search of substitute clothing.

2.1.4 Key Learnings

The author believed, at that time, that, if no material came out of a bleed valve, the line was cleared. However, materials such as emulsions are prone to fouling due to polymerization. The author should have been suspicious when no material was released; surely, there would have been a few drops of liquid, or a slight sound as residual pressure was relieved.

The author also should have worn heat-resistant work clothes and should have used the closest safety shower rather than a bathroom shower further away.

2.1.5 What Should You Do?

1. Maintain a sense of vulnerability. The author was expecting the hose to be vented, and the lack of material exiting the hose seemed to confirm what the author expected. As a result, the author didn't consider whether the interpretation of the observation data (no flow = no material) was correct

2. Do a Job Safety Analysis. If the engineer had considered what could happen if a vent were blocked, a second vent could have also been opened to verify that the pressure was vented. A whip check could have been used to connect the hose to the piping, preventing the hose from whipping around when disconnected.

3. Follow proper safety practices. Wear the Personal Protective Equipment (PPE) that would be required if an exposure was to occur. Use the local safety shower rather than going to the locker room. Wear clothing that is appropriate for the potential hazards that might be encountered (including thermal burns).

2.2 Where common sense did not prevail – or, even if you're the least experienced person, don't keep quiet if you see something that doesn't look right

2.2.1 The Situation

A stab-in reboiler on a distillation column was being removed for maintenance. The removal required two maintenance people to work on an elevated deck to knock out bolts from the reboiler flange with the column in order to free the reboiler for removal. At some point it was possible that the flange could come free and fall onto the workers, and so its weight (several tons) was being supported by a crane (Figure 3).

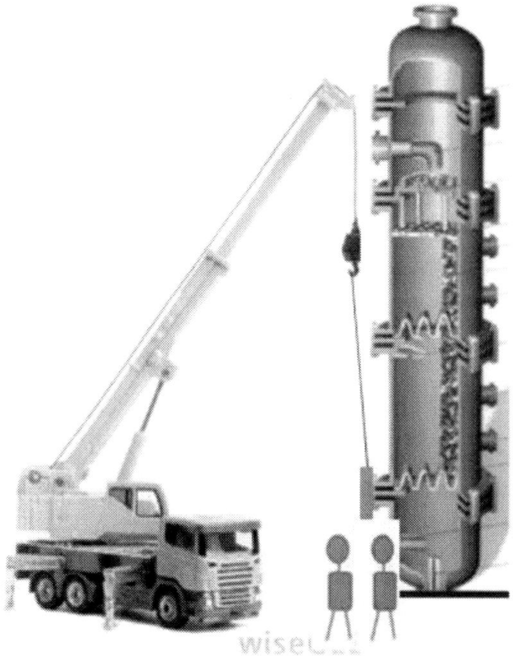

Figure 3: A Crane Under Strain

The author was not associated with this operation, and had only been working in this (the author's first) refinery for two weeks. The author was watching from a distance while several plant supervisors were next to the crane operation. Although the author knew nothing about crane operations, it seemed odd that the crane was pulling so hard on the flange that its rear tires were off the ground. (The nearby plant supervisors were facing away from the wheels and may not have known they were elevated.)

2.2.2 The Poor Choice

Since the author was the least experienced person observing the operation, the author continued to watch but did not engage the plant staff about whether the crane operation was proper.

2.2.3 Potential/Actual Consequences

Once all but two of the bolts holding the flange in place were removed, the final two bolts sheared due to the stress of the over-torqued crane operation. The flange flew up in the air and landed on the backs of the workers who had turned away from it, knocking them to the deck. The flange rebounded and came back down just when one of the workers had staggered back to their feet, hitting him again. The workers were subsequently taken away by ambulance to a hospital, but fortunately did not sustain serious injuries.

2.2.4 Key Learnings

If something doesn't "feel right," then there is a reasonable chance it is not - even if everyone else is comfortable.

2.2.5 What Should You Do?

1. Don't let your seniority blind you to what is wrong and don't let lack of experience keep you from speaking out if something doesn't look right.

2. If something isn't right, it isn't right whether you are an employee of the facility or not. Speak up!

2.3 Nobody else wore goggles…

2.3.1 The Situation

During a summer internship, the author worked in a quality control laboratory of a cosmetic company, which performed quality tests on incoming raw materials. The author was taught to conduct a number of laboratory tests on these materials. No eye protection was required in the lab. In fact, the author wore contact lenses on a daily basis. One day, the author attempted to inject a solvent into a sodium chloride cell for Infrared (IR) analysis (Figure 4). Unfortunately, the cell port was blocked.

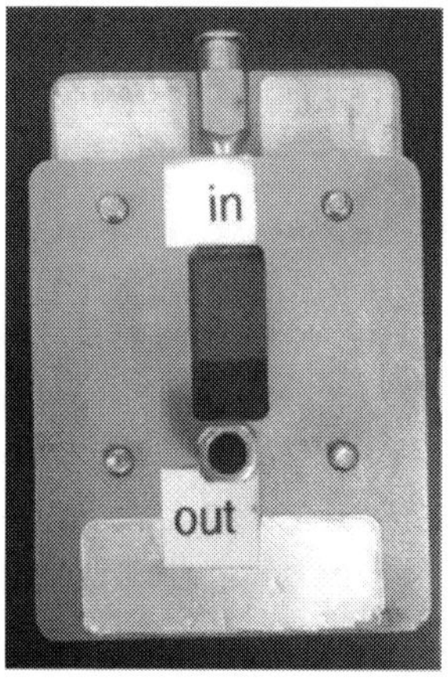

Figure 4: Example of Infrared (IR) Solution Cell [7]

2.3.2 The Poor Choice

The author, not wearing eye protection of any kind, applied excessive force to the syringe. During the injection, material squirted back into one eye.

2.3.3 Potential/Actual Consequences

The author's eye immediately started to sting and water. The author immediately went to the safety shower to start rinsing the affected eye. However, the contact lens was still in place. The author struggled to remove the contact lens so that the eye could be completely flushed. Fortunately, there was no serious injury.

2.3.4 Key Learnings

1. Eye protection should be worn in any situation where it is possible to be exposed to hazardous liquids. (Many plants and labs require that eye protection be worn at all times, since it is not possible to anticipate all spills, leaks, and drips.)
2. The hazards of a blocked IR cell – and of applying excess force to the syringe – were not identified and documented. The need to apply more than normal force to any task is generally a warning signal.
3. The author was a college student who was not properly trained and supervised. Co-ops, interns, and new hires need to be closely monitored. Set a good example and do not assume that they are fully skilled until they have actually demonstrated competency.

2.3.5 What Should You Do?

1. Wear the required personal protective equipment (PPE) at a minimum but think – should I be wearing more? You are the one at risk! Challenge PPE policies where they don't make sense. A general laboratory safety policy may be inadequate to address the hazards of specific/novel equipment.
2. Train your co-ops and new hires. Watch others' co-ops and new hires carefully. Supervise them closely and do not assume competency. Rely on proper demonstrations of their skills.
3. Conduct Job Safety Analyses (JSAs) to identify potential hazards. Capture the potential hazards and precautions in training materials and operating instructions so that the learnings are not lost.

2.4 Use your head – but please, not like that

2.3.1 The Situation

During new equipment commissioning, or turnaround, there is often the opportunity to view the internals on vessels that is not possible during normal operation. Naturally, one wants to look into vessels – sometimes out of curiosity, sometimes to see if someone you are looking for is inside, etc. (Figure 5).

Figure 5: Looking Inside a Vessel

2.3.2 The Poor Choice

There is often a hesitancy to completely climb into the vessel, either because of safety concerns, time concerns ("it's too much trouble to get a confined vessel entry permit or follow those procedures") or tendencies toward claustrophobia. However, surely just sticking one's head in the manway would be all right …wouldn't it?

2.3.3 Potential/Actual Consequences

The authors are aware of multiple occasions where someone put their head through a vessel manway to 'just have a peek,' and died due to asphyxiation – sometimes with most of their body still outside the vessel. Generally, this is because the vessel was purged with nitrogen.

2.3.4 Key Learnings

Just because an environment *looks* safe doesn't mean that it *is* safe. Beware of the hazards you cannot see as well as the ones that you can see.

2.3.5 What Should You Do?

1. Never allow any part of your body to approach the plane of a vessel (or any confined space) without positive proof that the environment is safe (e.g. testing the atmosphere in the vessel). Verbal assurances are not good enough – the authors know of one case where someone was told that the vessel had been purged with air, when in fact it was purged with nitrogen instead.
2. Be cautious when placing your head around potential leak points on inerted equipment. The authors are aware of a colleague who briefly lost consciousness when looking into the sight glass of a nitrogen-inerted vessel. The sight glass had a leak, and the area around the sight glass was oxygen-deficient.
3. Don't assume that you can just hold your breath and take a quick look – an involuntary reflux that causes you to take a breath could be fatal. Even very short exposures to nitrogen can be fatal.

3 Learning Opportunities: Mis-Management of Change

3.1 But it worked in the lab!

3.1.1 The Situation

A new agricultural intermediate process was being operated at a 6,000 gallon scale for the first time. The process included a water wash of the organic phase, followed by a phase separation, to remove salts from the batch. Unfortunately, the batch formed an emulsion and the phases would not separate. This situation had occurred in the lab a few times, and the problem had been remedied by the addition of potassium carbonate salts, which increased the density of the aqueous phase and broke the emulsion. The author took a sample from the commercial vessel, brought it to the plant lab, and added potassium carbonate. The phases then separated.

3.1.2 The Poor Choice

The issue occurred on midnight shift. Experienced technical resources were limited. The author decided to try to break the emulsion so the batch could proceed. The author instructed the operator to open the manway of the vessel and add several bags of potassium carbonate to the vessel.

3.1.3 Potential/Actual Consequences

Shortly after the bags were added, the batch began to swell and material started "burping" out of the top of the vessel. The operating staff retreated to a safe distance. The material oozed along the floor and immediately solidified due to the colder ambient temperature. The crew later spent hours shoveling waste material off of the floor.

3.1.4 Key Learnings

The author did not take into consideration fundamental differences between operating at the lab scale and at the plant scale. In this case, the mixing in the commercial vessel was poor. The bags of salt were dumped and went straight to the bottom of the vessel. Large carbon dioxide bubbles formed at the bottom of the vessel, forcing material from the top. This had not been observed at the lab scale because of better mixing and slower addition of the salt.

3.1.5 What Should You Do?

1. Understand the fundamentals of scale-up. Parameters such as mixing and heat transfer change as a process is scaled. It is important to take those changes into consideration when designing different size processes.

2. Conduct a Management of Change Review. The practice of adding salt to break the emulsion had been practiced in the lab but not reviewed at the plant scale. The event occurred on midnight shift, when technical resources were not readily available. As a result, expert input that would likely have identified the potential for this type of event was not obtained. Avoid troubleshooting and creating procedures "on the fly," as there may be unforeseen consequences.

3.2 It looked like the right stuff…

3.2.1 The Situation

A new process was being started up at an external manufacturer, and the author was responsible for supporting the start-up. The last reaction step of the synthesis process involved the use of a boiling solvent saturated with a toxic acid gas – a highly corrosive environment [7]. Polytetrafluoroethylene (PTFE) lined piping had been specified as the preferred material of construction. A few hours into the first batch, the recirculation pump was turned on to recirculate the batch for sampling and to provide additional mixing.

3.2.2 The Poor Choice

The external manufacturer reused existing pipe. It had a white lining, and it was assumed that the pipe was PTFE-lined. In fact, the pipe was polypropylene-lined, which was not resistant to the process fluid at operating conditions. The Hazard and Operability Study (HAZOP) had not considered the deviation of wrong material of construction.

3.2.3 Potential/Actual Consequences

A few hours after the recirculation loop was turned on, local toxic gas detectors alarmed, signaling the presence of acid gas fumes in the process area. The area was evacuated. A gas cloud formed, filled the building, and left the site. The plant was evacuated, the local fire department was called, and a local highway was shut down for several hours. Painful consequences of the event included media coverage, regulatory visits, an incident investigation, clean-up, and replacing all suspect or compromised piping.

3.2.4 Key Learnings

Although the author knew that existing vessels would be re-used (and had verified that the vessels were of the appropriate material of construction), the re-use of pipe from a "boneyard" (area for storage of equipment not currently in use) had not been considered or included in the Hazard and Operability Study (HAZOP). Although the proper material of construction to be used for the pipe had been communicated to the external manufacturer, they did not have a robust process to ensure that new pipe of the proper specification was used - or that the material of construction of used pipe was verified.

3.2.5 What Should You Do?

1. Determine whether it is appropriate to re-use equipment for a new process. Is it the right material of construction? Has the integrity of the equipment been verified?

2. Consider the use of the wrong material of construction in Process Hazard Reviews (PHAs). Even if the right material of construction is used in initial construction, later maintenance or process changes could result in the inadvertent substitution of the wrong materials of construction. Considering this type of deviation in a PHA would lead to a discussion of the methods to be used to confirm that the right materials are being used and would hopefully lead to the implementation of a work process to validate proper material identification and prevent such incidents. Certainly, positive material identification could be a useful means of verifying the correct material of construction, especially in sensitive applications.

4 Learning Opportunities: "I *Can't* Do It Myself!"

4.1 It will take forever to get that guy back, so I'll do it myself

4.1.1 The Situation

A level instrument on a vessel that was separating 20 Baume' (14 wt%) caustic from hydrocarbon liquid and vapor phases had just been fixed by an instrument technician. However, the technician had left the area without recommissioning the instrument. (He had left its isolation valves closed.) It had taken a long time to get the technician out to fix the instrument, and calling the technician or an operator to the area to open up the isolation valves would have been a nuisance. (Figure 7)

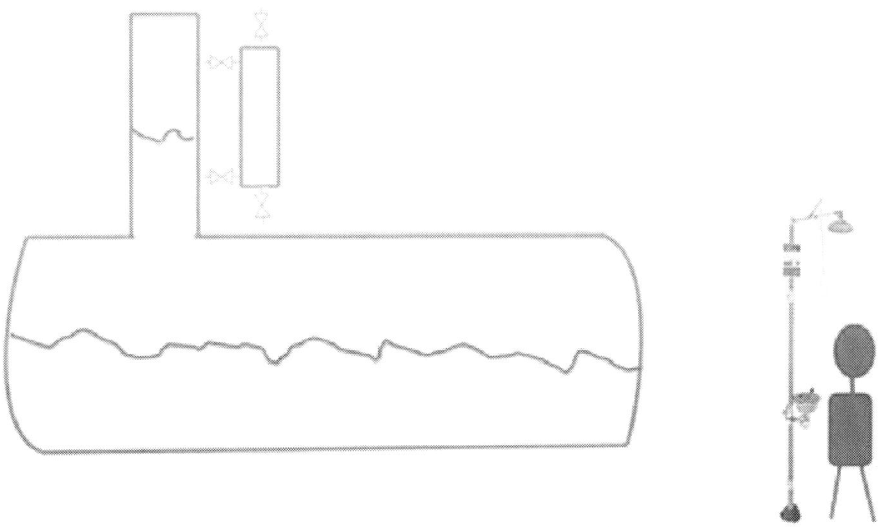

Figure 7: Caustic/Hydrocarbon/Vapor Separator

4.1.2 The Poor Choice

The author believed that, since recommissioning the instrument just involved opening the isolation valves, the author might as well just do it personally rather than try to find someone and persuade them to come back out to the unit. The author checked the instrument drain valve to make sure that it was closed, then opened up the lower valve connecting the process to the instrument.

4.1.3 Potential/Actual Consequences

Unfortunately, the author didn't also check the position of the vent valve from the instrument. About one second after opening the valve, a torrent of caustic solution started raining down on the author's head (covered with a hardhat, at least). The author recognized it as being the caustic since it was slippery. For the same reason, the author decided it was more prudent to jump the six feet from the separator deck to the ground rather than risk grabbing the ladder and slipping off, potentially causing further injury. Soaked to the skin in caustic, the author ran to the nearby safety shower/eyewash station, stripped, and spent the next 15 minutes alternating between the shower and the eyewash. While the author's hardhat changed color from blue to white, the author received only a small minor burn on one arm.

4.1.4 Key Learnings

The author believed that it was appropriate to do another person's work for the sake of expediency. The author also assumed that if the instrument drain valve was closed, the vent valve must also be closed, which turned out to be wrong. On the plus side, the author did some things correctly, which prevented a severe personal injury or fatality, including the following:

- Before the startup, the author had been asked where the safety shower/eyewash station for the unit should be located. As it turned out, the author had felt that this area had the highest incident potential, and as a result picked the perfect location for the subsequent event. (It was not practical to put a shower on the deck where the incident occurred.)
- During the startup phase, the eyewash station nozzle was repeatedly broke by workers using it to wash their hands. The author had refused to enter the unit until the eyewash nozzle was permanently fixed – and it had been repaired before the event.
- Having realized that the spray was caustic, the author correctly chose to jump from the separator deck and risk spraining an ankle rather than using the ladder and risk an uncontrolled fall that had the potential to render the author unconscious with no one around.
- The author correctly realized the gravity of the situation and stripped completely naked in order to wash off the caustic as thoroughly as possible.

4.1.5 What Should You Do?

1. Don't do a job for which you haven't been trained, and do not make assumptions about the status of equipment based on the status of nearby equipment.

2. Think carefully about the siting of safety equipment – the life you save may be your own!

3. Be aware of your surroundings, and know where emergency equipment is located.

4. Wear the Personal Protective Equipment (PPE) that would be required if an exposure was to occur.

5. Use the local safety shower rather than going to the locker room.

6. A chemical exposure event is no time to be modest. If you have to get buck naked to prevent an injury, go ahead! Better for people to have a story to tell about the engineer that was naked than for them to tell the story about the engineer who died or suffered horrible chemical burns.

4.2 Why did the pressure go up so fast?

4.2.1 The Situation

An existing pump had been replaced with a lobe gear pump. The author was monitoring the water testing of the equipment. The next step was to redirect the flow path from the pump to a different location. (Figure 8)

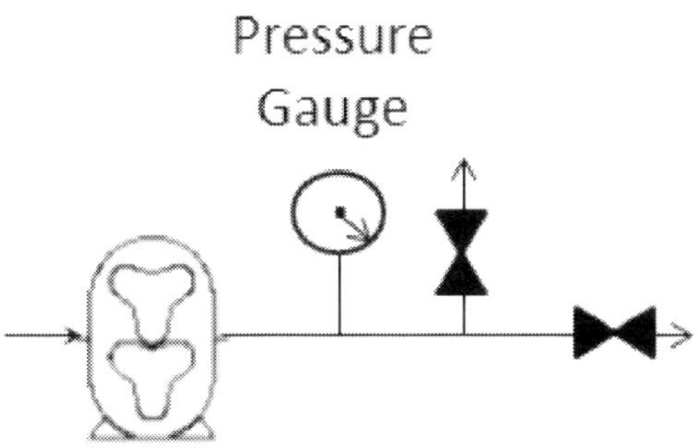

Figure 8: Deadheading a Positive Displacement Pump

4.2.2 The Poor Choice

The author chose not to wait for an operator. Instead of opening the valve to the new flow path first, the operator closed the valve on the discharge of the pump, resulting in "deadheading" the pump.

4.2.3 Potential/Actual Consequences

In an instant, the pressure in the 150 psig line rapidly rose and pegged the pressure gauge. Fortunately, the pressure gauge was clearly visible and author's hand was still on the valve handle. The valve was opened immediately and the line did not fail from the overpressure. (The presence of an internal relief in the pump likely significantly contributed to the successful outcome.)

4.2.4 Key Learnings

The author had worked with operators in the past to troubleshoot centrifugal pumps. Sometimes, when a pump was "air bound," operators would deliberately throttle or briefly close the discharge valve of the pump to remedy the problem. The author did not see the act of closing a pump discharge valve for a few seconds as risky. However, the author did not appreciate that this was a positive displacement pump, which had the capability to quickly generate a pressure sufficient to rupture the pump discharge pipe.

4.2.5 What Should You Do?

1. Understand equipment before you operate it.

2. Be cautious about extrapolating "rules of thumb" and other troubleshooting/operating practices between equipment without first understanding when they are/are not appropriate. Not all equipment is created equal!

4.3 Process control mishap

4.3.1 The Situation

The author was working in a batch processing plant when an operator asked for assistance in troubleshooting a problem on another process. The Programmable Logic Controller (PLC) would not progress to the next step because it was not receiving a signal from a switch in the field. An instrument technician was not available to fix the issue and product was needed for a customer order.

4.3.2 The Poor Choice

The author had recently received training on PLCs and wanted to use that knowledge. After determining that the field switch was faulty, the author (with limited knowledge of the workings of the PLC and its programming), forced the process control sequence into the next step.

4.3.3 Potential/Actual Consequences

The control sequence did proceed to the next step and started to charge material to the reactor; however, the bottom valve of the reactor was still open! Fortunately, a manual valve was closed, preventing the release of material. The operation was stopped until the valves were aligned in the correct position and a release was averted.

4.3.4 Key Learnings

The author did not recognize that, by advancing the sequence, the ladder logic that would have sent a signal to close the reactor bottom valve was bypassed. Clearly, the author chose to take action without possessing the necessary process control expertise.

4.3.5 What Should You Do?

1. Follow established review and approval protocols when bypassing control systems, advancing sequences, operating manually, or downloading code while equipment is in operation. Equipment might be placed in an improper configuration or safety systems may be bypassed. The author is aware of an incident in another plant, where an engineer attempted to download a new sequence while the equipment was in operation. This resulted in the loss of control of the charging sequence and a deactivation of the high level alarm. A flammable spill resulted; fortunately, it did not ignite.

2. Ensure that you are properly trained on the control system platform and logic, and have a good understanding of the chemical process, before making any changes to the operation of a control system. There may be unintended consequences!

5 Learning Opportunities: Human Factors Traps

5.1 The pros and cons of ball valves

5.1.1 The Situation

Ball valves are often selected for use in chemical facilities. Ball valves can be very convenient – a simple quarter-turn and they open or close; it doesn't get much easier than that (Figure 9). However, the ease of opening or closing ball valves intentionally also implies that they can be opened or closed not only easily, but *accidentally*.

Figure 9: A Typical Ball Valve

5.1.2 The Poor Choice

In one unit, the engineers specified the use of ball valves in hazardous service. In one recent incident, a newly-hired field operator was going through a unit tour when the sleeve of his Nomex suit caught on the handle of a ball valve, causing the valve to open.

5.1.3 Potential/Actual Consequences

The valve was on a high pressure hydrogen line, which then discharged up the sleeve of the new-hire. The valve was quickly closed, but he had quite a shocking experience and his arm was badly bruised. He was taken to a local hospital as a precaution; fortunately, there were no other injuries.

5.1.4 Key Learnings

The event was a wake-up call that the ease of use of ball valves has a down side, and staff and especially new-hires should be on the lookout for this hazard and consider the potential hazards of specifying ball valves in piping design.

5.1.5 What Should You Do?

1. When determining whether to install a ball valve or other type of valve in a given application, consider the implications of both the ease of use and ease of misuse of the valve. In some cases, having a ball valve that can be easily turned could result in unsafe consequences. Some companies restrict the use of ¼ turn ball valves in hazardous service, for example.

2. Orient operators and visitors to a unit that has ball valves to the potential hazards of accidental operation.

3. Be mindful that loose clothing or jewelry can catch on ball valve handles. Not only could this inadvertently operate the valve, but it could also cause an injury. An author observed an operator wearing large hoop earrings bend down and catch one earring on a ball valve handle. Fortunately, the operator recognized the issue and detached the earring before standing up, preventing an injury to the ear or turning the ball valve.

4. The risk of ball valves can be mitigated by using oval handles, spring-loaded handles and plugs/caps in terminal connections Some ball valves also have locking handles, which keep the valve from being inadvertently opened.

5.2 The ergonomics of an emergency shutdown (ESD) button

5.2.1 The Situation

A new unit was built, along with a new control room. The control room had a typical layout for its day, with computer control panels and instrumentation along a long wall that had largely open walk-around space in front of it. The unit had been operational for a week or so.

5.2.2 The Poor Choice

A colleague was chatting with one of the regular plant staff one shift and casually leaned against the control panel wall. Unfortunately, she happened to place her hand against the wall precisely where the unit emergency shutdown button was located. (Figure 10)

Figure 10: Sometimes, Talk Isn't Cheap

5.2.3 Potential/Actual Consequences

The button depressed, resulting in an unplanned unit shutdown. The outage caused approximately $250,000 in losses.

5.2.4 Key Learnings

1. During the design phase, it is important to give sufficient consideration to the potential for inadvertent activation of safety devices.

5.2.5 What Should You Do?

Design for ergonomics and the potential for human error. After the event occurred, the plant purchased a plastic cover for about $1 and placed it over the ESD button. A better design that would have prevented the event entirely is if the ESD was operated as a pull button rather than a push button, or locating the ESD at an easily reached position not at convenient "accidental" activation locations.

6 Learning Opportunities: Situational Unawareness

6.1 Watch your step, even if you're tired

6.1.1 The Situation

A new distillation tower had been installed but not yet commissioned. Part of pre-commissioning activities included climbing the tower to make sure that all equipment was in place. The tower was approximately 230 feet tall. Reaching the top of the tower involved climbing a series of caged ladders that went from one deck to another in about 20 foot intervals.

6.1.2 The Poor Choice

The author had many activities to check on, so (being in much better shape in those days) the author climbed the tower fairly fast, to the point where the author was winded after reaching the deck at the 175-foot level.

Upon stepping off the top of one ladder onto the deck, the author almost placed a foot through a section of missing grating adjacent to the ladder, which would have resulted in a free-fall of 175 feet. (Figure 11)

Figure 11: Getting Fresh Air Without Benefit of a Parachute

6.2.3 Potential/Actual Consequences

Fortunately the author managed to stop just in time, so nothing bad happened (except for a rapid acceleration of his heartbeat for a minute or two, and experiencing the sensation of having his life pass before his eyes). The author provided an emphatic description of the safety issue and pushed for prompt resolution.

6.2.4 Key Learnings

The author allowed himself to become winded, increasing the chance of falling through the hole.

6.2.5 What Should You Do?

1. Do not allow yourself to become winded – it leads to inattention, and inattention can lead to a bad outcome in a hazardous situation.

2. If you see an unsafe situation, do what you can to mitigate the problem. In this case, the author found some red plastic ribbon trash that was stuck at that deck, and wrapped the ribbon around the grating at the gap to try to warn someone else of the situation if they happened to enter the area before the problem was fixed.

3. If you see an unsafe situation, immediately notify the appropriate people in the unit. Continue to monitor the situation occasionally, and remind people of the issue until it is fixed.

6.3 The hazards of following piping in a pipe rack

6.3.1 The Situation

It is common for anyone who is trying to learn a new process unit to walk down the equipment in the field. This includes following runs of piping in pipe racks that may swerve back and forth in the rack, or at times might be partially obscured by other piping in the pipe rack. (Figure 12) Sometimes it requires a great deal of focusing on the line to not lose track of its route.

Figure 12: A Typical Pipe Rack

6.3.2 The Poor Choice

A co-worker of the author was undertaking such an activity and walked for an extended distance while continually looking up in the pipe rack.

6.3.3 Potential/Actual Consequences

Unbeknownst to this co-worker, some distance from the start of this pipe run was an unmarked open manhole below the pipe run. The co-worked walked into the manhole and fell into the sewer. Fortunately, there was no injury from the fall. However the person's smell was so offensive that his carpool colleague refused to allow him in the car at the end of his shift – resulting in a 10-mile walk home for the unfortunate sewage-drenched co-worker after the shift.

6.3.4 Key Learnings

The person incorrectly assumed that there would be no obstructions or dangers along their path.

6.3.5 What Should You Do?

Before tracing long runs of pipe, view the entire path to make sure there are no hazards along the path. One can also stop periodically and check one's surroundings.

7 Conclusion

The authors and others clearly made poor choices as a result of poor situational awareness, lack of hazard recognition, inadequate management of change, inadequate human factors design, and taking it upon themselves to do tasks which were inherently unsafe or for which they had been inadequately trained. As the authors reflected on this collection of cautionary examples, they feel lucky to have survived the stupidity and poor choices of their formative engineering years largely intact (except perhaps for their collective pride)! Fortunately, the authors are somewhat older and wiser from these experiences. More importantly, the authors are grateful that process safety behaviors, management systems, and equipment design have evolved to the point where these activities would not be necessary nor tolerated today.

However, it should be noted that, although processes have become better designed and automated, people have become less hands-on with the process and equipment. There is a risk that operations personnel (including engineers) will lose the deep level of process knowledge that hands-on operation once enabled. Although it is quite fortunate that today's engineers are not placed in a position to take the risks that were taken several decades ago, it is important that they do not lose their appreciation for what can happen when these types of poor decisions are made. As George Santayana said, "Those who do not remember the past are condemned to repeat it."[9] It is hoped that the lessons learned by the authors will heighten the awareness of newer engineers regarding the potential consequences of putting themselves into risky situations such as these, and perhaps inspire the next generation of engineers to make better, smarter choices than the authors did.

8 References

[1] "Alfred Nobel in Scotland," The Official Website of the Nobel Prize, Available at: https://www.nobelprize.org/alfred_nobel/biographical/articles/dolan/. Accessed on Dec. 22, 2016.

[2] "Safety performance indicators - 2015 data," International Association of Oil & Gas Producers, June 2016.

[3] US Department of Labor Occupational Safety and Health Administration (OSHA) Standard 29 CFR 1910.119, "Process Safety Management of Highly Hazardous Chemicals," 1991.

[4] US Environmental Protection Agency (EPA) Rule 40 CFR Part 68, "Risk Management Program," first promulgated in 1991.

[5] European Union Directive 82/501/EC, "Directive on the control of major-accident hazards involving dangerous substances," first promulgated in 1982.

[6] UK Health and Safety Executive (HSE), "Control of Major Accident Hazards," 1999.

[7] Infrared Spectroscopy. (n.d.) In Wikipedia. Photo retrieved from https://commons.wikimedia.org/wiki/File:SolnIRcell.jpg#filelinks and modified on January 15, 2017.

[8] Hendershot, D., Johnson, S., and Kas, K., "What You Ask For Isn't Always What You Get," presented at the Mary Kay O'Connor Process Safety Center Symposium at Texas A&M University, 2000.

[9] Santayana, George, *Reason in Common Sense*, P. 284, volume 1, of *The Life of Reason*, 1905.

Inherently safer automation

Angela Summers, PhD. President
SIS-TECH Solutions
12621 Featherwood Dr. Suite 120, Houston, TX 77034
asummers@sis-tech.com

Prepared for Presentation at
American Institute of Chemical Engineers
2017 Spring Meeting and 13th Global Congress on Process Safety
San Antonio, TX
March 26 – 29, 2017

Inherently safer automation

Angela Summers, PhD. President
SIS-TECH Solutions
12621 Featherwood Dr. Suite 120, Houston, TX 77034
asummers@sis-tech.com

Keywords: inherently safer, automation, engineered systems, administrative controls, functional safety

Abstract

Inherently safer design and functionally safe design go hand-in-hand in assuring safe operation. As processes go from conceptual to existing installations, the opportunities to apply the inherently safer strategies to the process design become constrained by cost and practicality. A wider perspective shows that the inherently safer strategies can be used to create automation systems that have less potential for dangerous failure, whether the failure occurs due to safeguard design, to a support system disruption or to human error. One might argue that the application of these strategies to automation can only make a process safer, rather than inherently safer. However, when such strategies are applied systematically across the site, the resulting design and management practices become part of "*the way things are done*" and result in an inherently safer process operation.

This paper will discuss the application of the inherently safer strategies to automation systems. The cost and sustainability of different risk reduction options are presented. Then, guidance on and practical examples of the application of the inherently safer strategies for automation are provided.

1 Inherently Safer and Functionally Safe

Chemical processing is an industrial activity that involves using, storing, manufacturing, handling, or moving chemicals. Chemical processing may be accomplished in a single vessel or a group of interconnected vessels and process equipment. Process operation poses different types of risk, dependent on the hazardous nature of the chemicals, the quantity of chemicals processed, and the process operating conditions.

Inherently safer design involves making conscious choices to design and operate the process in a manner that avoids the hazard or minimizes the likelihood and consequence of the loss events. The word inherent means that the design feature is an essential constituent or characteristic of the process design. In contrast, functional safety management involves the addition of safeguards that act to achieve or maintain a safe state of the process when abnormal conditions occur. These safeguards are implemented in protection layers that are not inherent to the process design; they are added to the process to ensure functional safety. *Guidelines for Safe Automation of Chemical Processes 2nd Edition* [1] states that functional safety is part of the overall safety plan relating to the process and its control system, which depends on the correct functioning of the safety controls, alarms, and interlocks (SCAI) and other protection layers.

Safe operation of chemical processes is achieved through a process safety management program supported by the twin pillars of inherently safer design and functional safety management (Figure 1). Most process designs incorporate aspects of both inherently safer design and functional safety management. Ultimately, it is the owner/operator's responsibility to determine and document that the equipment is designed, maintained, inspected, tested, and operating in a safe manner, regardless of the means used to achieve this objective. Moving forward, industry will be increasingly challenged to seek a balance of inherently safer design and functionally safe design.

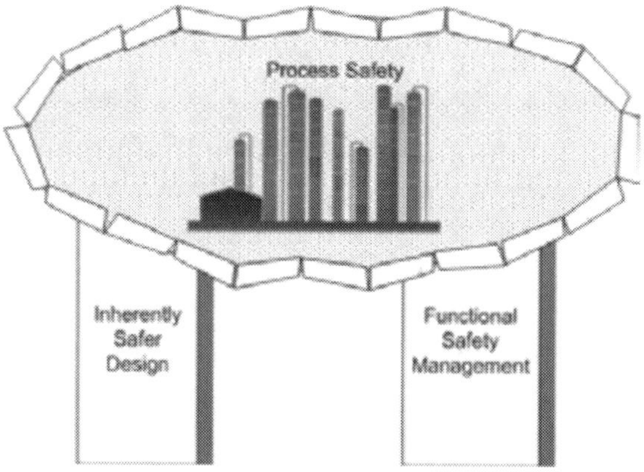

Figure 1. Process Safety Supported by Inherently Safer Design and Functional Safety Management [1]

2 Inherently Safer Challenges to Process Design

Process design typically starts with the definition of production and product quality requirements. Next, competing production technologies are identified and evaluated with respect to selected criteria, such as economics, reliability, regulatory and corporate requirements, tolerable risks, etc. Selecting a preferred process technology includes identifying hazards and making risk judgments. Many processes contain energy sources and process materials that under abnormal conditions can be harmful.

One means to achieve safe operation is to create inherently safer processes, where the process and its equipment are designed and maintained to eliminate or minimize risk. A report issued by CCPS [2] to the Department of Homeland Security stated, "A technology can only be described as inherently safer when compared to a different technology, including a description of the hazard or set of hazards being considered, their location, and the potentially affected population." The CCPS book, *Inherently Safer Chemical Processes: A Life Cycle Approach* [3], discusses various means to reduce process risk through facility siting, process chemistry, unit operations, control system design, operating plan, inventories, etc. The inherently safer strategies of minimize, substitute, moderate, and simplify can be used to eliminate or reduce the potential for harm.

Inherently safer design involves design choices that make the process and its equipment less susceptible to human error, support system failure, and dangerous failure during the facility life, but the installed equipment is still subject to degradation mechanisms that over time can erode the inherently safer assumptions. For example, what was an inherently safer design for the process equipment 30 years ago could now be a degraded foundation, vessel, or piping network in need of replacement.

Risk management and the inherently safer strategies must be fully integrated into the operating objectives for a particular process to ensure that its operating plan is in alignment with safe operation (Figure 2). Industry uses a mixture of prescriptive and performance-based practices to ensure that the process is designed for process safety [4, 5]. Inherently safer design becomes integral to the operating plan and process design basis. The design strategies typically are incorporated into customary practices, or "*the way things are done*," at a site, so people come to expect certain types of design and management depending on the equipment classification. Experienced and trained personnel use their knowledge, backed with documented practices and standardized templates, to design systems that maximize human, equipment, and process performance. Actual operating and maintenance data is needed to evaluate system performance and to initiate change when needed.

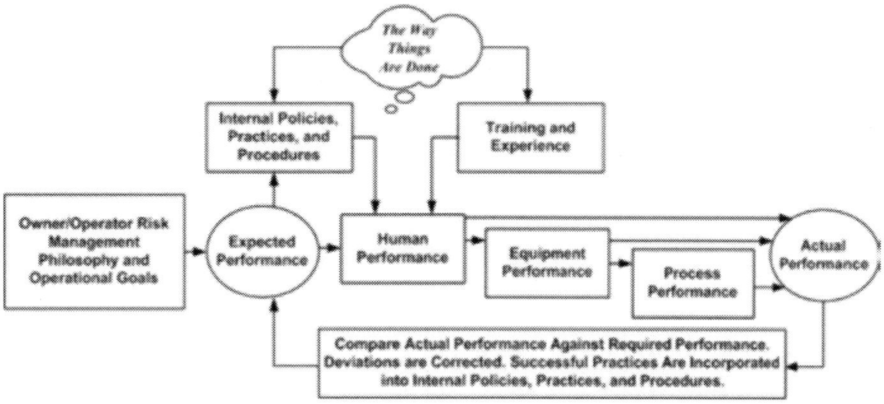

Figure 2. Feed Forward and Feed Back Work Processes for Quality Assurance [6]

Consideration for inherently safer design should be fully embedded in new projects and in the cyclic process hazards analysis as a means to address identified risk gaps [7]. Functional safety relying on protection layers should only be applied after first considering and ruling out inherently safer process design options [7]. Although the opportunities for inherently safer process design diminish over the process life, there are often ways that risk can be minimized through the inherently safer strategies (Figure 3).

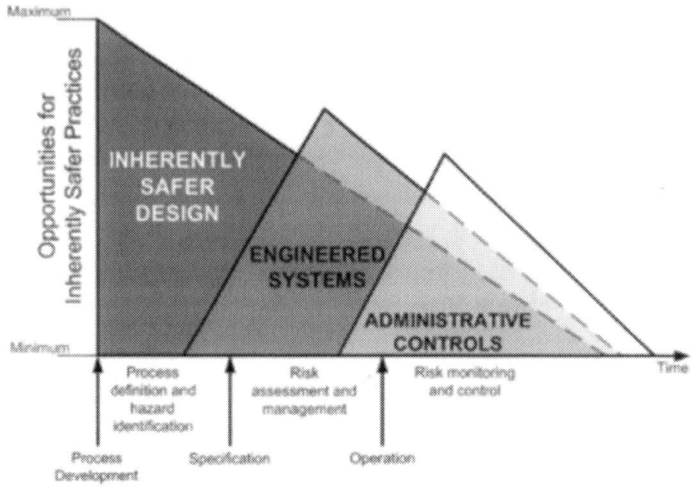

Figure 3. Opportunities for Inherent Safety Diminish Over Time [1]

3 Cost versus Value Trends

Process hazards can sometimes be reduced, or perhaps eliminated, during the design phase through inherently safer choices in process technology, equipment design, and operating parameters. When practicable, inherently safer design can minimize or eliminate the need for safeguards. Changes to the process design and operating plan should be considered as early as possible during the project life, since the relative cost of these changes typically escalates as the project progresses towards maturity (Figure 4). The particular means used to address risk is often influenced by the perceived effectiveness, availability, reliability, and sustainability of the protection relative to its lifecycle costs.

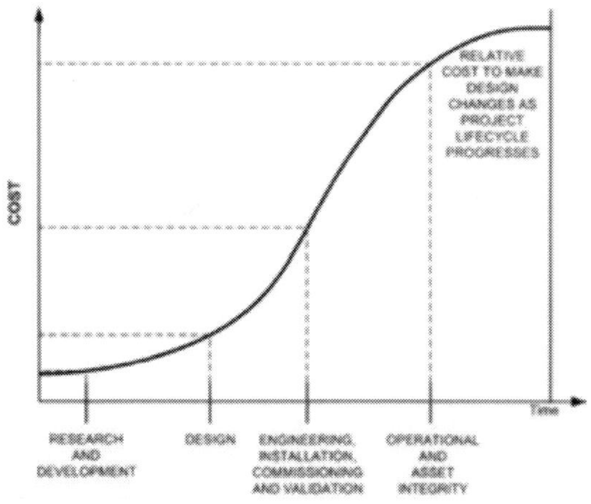

Figure 4. Relative Cost to Make Design Changes as a Function of Project Phase [6]

Inherently safer design, engineered systems, and administrative controls affect the capital and operating costs of the process differently (Figure 5). Historically, the capital project process has emphasized the need to minimize capital costs and reduce engineering time. This emphasis tends to favor a functional safety strategy relying on engineered systems and administrative controls rather than inherently safer design. However, inherently safer design applied in early phases of the process design can be a simple change in the equipment specification with no capital cost impact. Inherently safer design typically results in a less complex solution with lower operating and maintenance costs. Less automation complexity generally results in less potential for unnecessary process disruption due to automation failures.

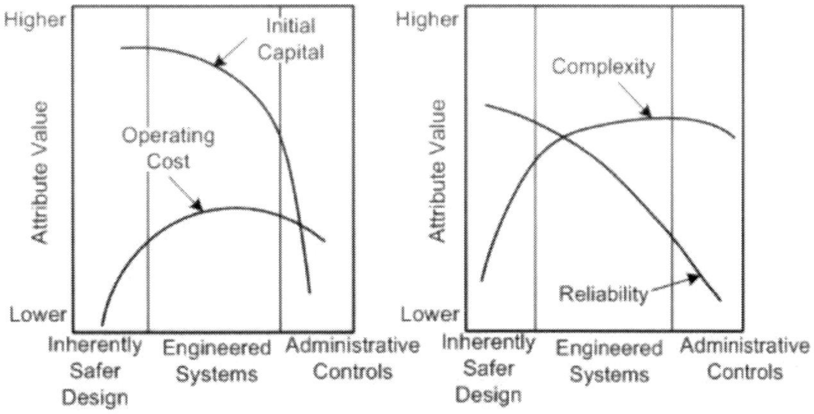

Figure 5. Typical Cost Versus Value Trends for Risk Reduction Means [3]

The impact of inherently safer design on operational flexibility should be considered, as inherently safer processes can sometimes be more difficult to operate (e.g. reduced or eliminated inventories between unit operations removes operational flexibility and may transfer disturbances downstream). So an inherently safer process might be less prone to process safety concerns, but it might actually be more prone to upsets due to operational difficulties.

Inherently safer process design reduces the process risk through process design and operating constraints rather than controlling it using safety systems and other protection layers. Administrative controls need the least capital, but ensuring that humans consistently do the same thing repeatedly can be difficult. The complexity of the necessary procedures and the rigor of the required administrative processes are substantially higher, but the resultant reliability is lower. Engineered systems have the highest operating costs because effort must be expended to sustain the equipment performance, but engineered systems have very predictable responses to abnormal operation and can achieve high integrity and reliability.

Examples of engineered systems often identified as safeguards are illustrated in Figure 6. The size of each bubble represents the relative risk reduction provided by the system. The bubble location is related to the relative ease of sustaining the system's risk reduction and reliability. Sustainability of these systems can be significantly different even when they are designed and managed to provide similar risk reduction. The process control system, safety alarm system, and SIL 1 SIS may achieve similar risk reduction from a hardware integrity standpoint, but the resilience of the SIS to systematic failure is higher due to its more rigorous design, verification, and validation processes. This makes the SIS performance more sustainable long-term. A pressure relief valve and a check valve are both mechanical devices, yet the pressure relief valve achieves much higher risk reduction with greater sustainability. *Choosing protection layers that are more resilient to systematic failures is an inherently safer practice.*

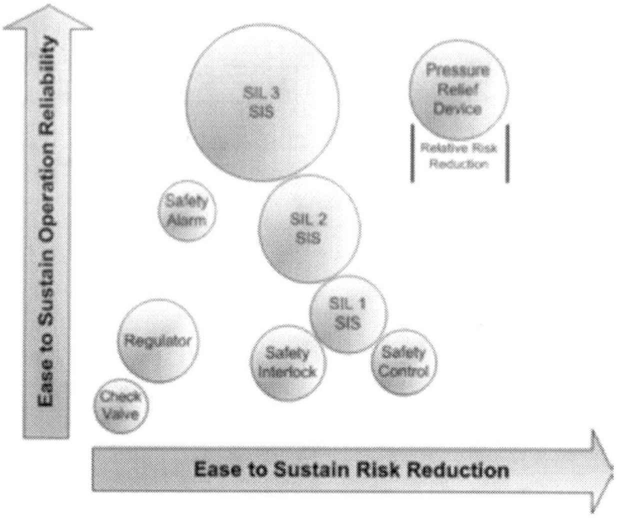

Figure 6. Protection Layers Showing Relative Risk Reduction, Reliability and Sustainability [1]

4 Functional Safety Through Effective Safeguards

Safeguards are an essential part of the functional safety management plan. Once the process design is complete, the risk of process operation can be further reduced through the implementation safeguards. Instrumentation in the form of process control and safety systems plays an important role in reducing the frequency of loss events and subsequent harm. *Guidelines for Safe Automation of Chemical Processes 2nd Edition* [1] provides guidance on the design and operation of process control and safety systems, which are the most common safeguards used in the process industry. A properly designed process control system, typically monitored by trained and alert operators, is the first line of defense, beyond sound process design, in preventing loss events.

Safeguards achieve different outcomes dependent on when and how they act during the event propagation [8]. Proactive safeguards act to regain control of the process or to shutdown the process. Reactive safeguards take action after the loss event occurrence, acting to stop loss event escalation and to minimize the harm. The number of safeguards and the overall risk reduction required from these safeguards is determined by the severity of the loss event and the frequency of loss of control. The safeguards are implemented in a defined order to achieve a specified outcome. This sequence of safeguard operation may be viewed as in Figure 7, which represents an onion where each safeguard is deployed by a layer of the onion [6]. These layers consist of engineered systems and administrative controls that work in concert to reduce process risk. As the event propagates through the onion, the impact on the process operation becomes greater as does the uncertainty of the final outcome.

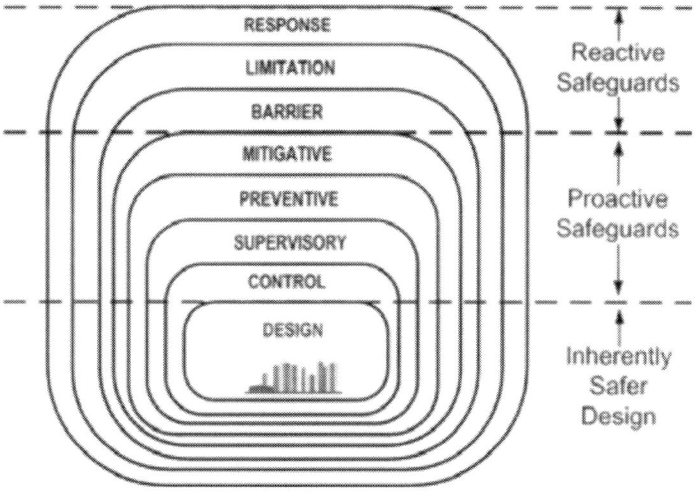

Figure 7. Typical Layers of Protection [Adapted from 6]

The highest priority for reducing the risk as low as reasonably practicable (ALARP) is to employ the inherently safer strategies during the process design (Figure 8). For example, equipment segregation is a proven strategy for minimizing common cause and reducing the potential for failure escalation that impacts multiple systems. It is often impractical to design the process to be inherently safe for all possible loss events. Once the process design is complete, functional safety is achieved using safeguards, such as engineered systems and administrative controls. Events with significant severity outcomes may require more independent safeguards than events posing less harm.

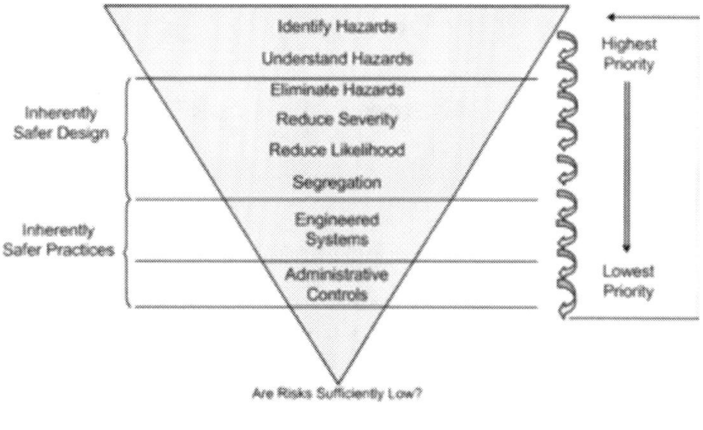

Figure 8. Priority of Inherently Safer Design and Protection Layers in Risk Management (Broadribb 2010)

Lower priority is placed on implementation of engineered systems and administrative controls (Figure 2.3), since the performance of these risk reduction measures are highly dependent on the rigor of the functional safety management plan. In contrast, a process designed to be inherently safer will remain so as long as the inherently safer feature is sustained. Engineered systems, particularly those relying on automation, are critical to achieving ALARP for many loss events in the process industry. For example, safety alarms are needed to direct the operator's attention to process deviations with safety consequences. For the operator to be effective, many performance shaping factors must be addressed as part of the alarm system design and during operator procedure development.

Administrative controls, while the least priority from a risk reduction strategy standpoint, are necessary to ensure that engineered systems are functional; operators know how the system works and how they interact with it; maintenance knows how to keep all of the equipment in its "as good as new" condition; and management has metrics to measure performance.

Engineered systems that act upon control system failure include many types of systems, such as relief systems, de-inventory systems, safety alarms and safety instrumented systems. Taking action at the earliest step in the propagation of a loss event minimizes the impact of abnormal operation (Figure 9). Reliable safeguards act when and as required to achieve or maintain a safe state.

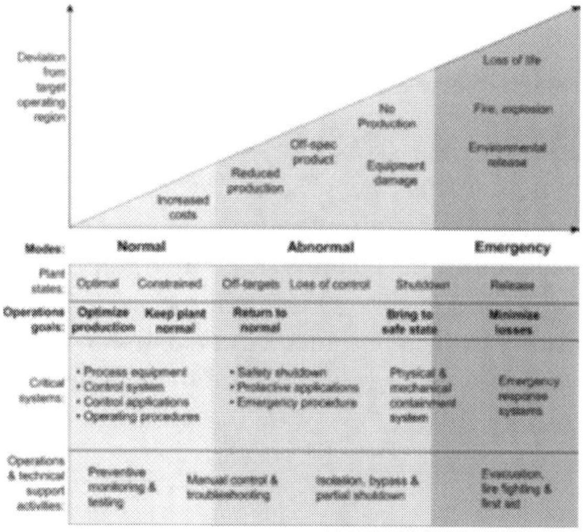

Figure 9. Anatomy of a Loss Event [9]

5 Inherently Safer Automation

Automated systems, whether in manual or automatic mode, are complex systems where many different devices must work successfully to achieve the desired functionality. Sustaining automated system performance requires many different skill sets and planned activities to ensure that the systems work as desired when required. In general, inherently safer practices can create:

- *Reliable control systems that place fewer demands on the safeguards*
- *Safeguards that have less potential for dangerous failure*

These benefits are realized whether the failure occurs due to the automation design, to support system disruption or to human error. Some inherently safer practices bring a higher potential for spurious, or unnecessary, activation of the safety systems. If spurious operation causes intolerable losses, the functional specification should state a target spurious trip rate.

Inherently safer practices can significantly influence the automation equipment selection, normal operating limits, fault tolerance, and response to detected failure. It is not possible to create a complete list of the automation features that could be considered inherently safer than alternative choices. Instead, each inherently safer strategy is defined below as it applies to automation. Then, a detailed list of inherently safer practices is provided to illustrate the strategy.

Minimize applied to automation - reducing the use of automation features that tend to increase the failure mechanisms that result in system failure.

- Avoid using materials of construction that are incompatible with normal or abnormal operating conditions
- Select devices that do not require additional equipment for protection from the operating environment conditions, such as climate controls
- Select devices that do not require additional instrumentation in order to make them function properly in the given process; for example, using a remote sealed level in plugging services instead of using a device that requires process connection purging, and using a mass flow meter instead of using pressure, temperature, and density to compensate a volumetric flow reading
- Minimize the number of devices using a justification process that ensures every installed device is needed for a particular purpose
- Minimize the number of alarms using a rationalization process that enhances operator effectiveness
- Minimize blind spots in measurement by using devices that are applicable over the full range of process operating modes
- Where possible, eliminate inherently weak components, such as sight glasses, hoses, rotameters, bellows, and plastic components
- Minimize the length of impulse lines
- Minimize common cause by segregating redundant signals onto separate input/output modules
- Minimize common cause by not the sharing devices across the initiating event and protection layers for a particular loss event
- Use air-gapping to protect the instrumented safeguards from external cybersecurity attacks
- Minimize hazardous testing by installing facilities that allow the system to be tested without creating an abnormal process condition
- Use mechanical connection for blanking, draining, cleaning, and purging connections so that maintenance activities cannot be started without first disconnecting the lines
- Install devices to be accessible for inspections, tests, and preventive maintenance tasks

- Apply advanced alarm management techniques to reduce the number of alarms to a rate that can be handled by an operator, particularly during abnormal operating conditions

- Avoid co-mingling non-safety and safety logic in the same logic solver, unless required to support instrumented safeguards

Substitute applied to automation- replacing an automation feature with an alternative that reduces or eliminates the frequency of dangerous failure.

- Replace a device with a lower failure rate device

- Use materials of construction with lower corrosion or erosion rates

- Use a device that provides a direct measurement of the process parameter being controlled rather than using an indirect measurement

- Select devices that fail to a specified safe state on loss of any utility, such as power or instrument air, instead of devices which require energy to take action

- Configure devices to fail to a specified safe state on detected failure

- Select the process variable for the alarm system that gives the operator more time to correct abnormal operation

- Use safety interlocks instead of safety alarms to reduce the potential for operator error to allow abnormal events events to propagate

Moderate applied to automation- using automation features to facilitate operating the facility under less hazardous conditions; using automation features to limit the impact of dangerous failure of the automation system on the process operation.

- Use normal operating limits that are well within the safe operating limits

- Provide operator with redundant indication of safety variables using simple graphical displays that build trust in the automation system

- Select setpoints that provide enough time for the system to sense an upset and complete corrective action prior to crossing never exceed limits

- Consider minimum flow stops to prevent loss of flow in sensitive services

- Configure limits on manual entry of process control inputs and setpoint changes

- Limit the maximum step changes to setpoints to prevent upsetting process

- Use confirmation of change prior to taking action on operator commands

- Provide bumpless transfer and setpoint tracking for switching among automatic, manual and cascade control

- Provide first out indication and sufficient additional information to allow the operator to quickly diagnose and respond to the causes of process deviation

- Provide the operator with detailed guidance on the expected process response to a specific action, e.g., how much lead or lag time to expect or what percentage should a valve be open to achieve the desired result

- Use closed coupled connections to reduce the chances of plugging in applications where solids formation or deposition can be a problem

- Use equipment that are sufficiently mature that embedded and operating software updates are infrequent.

Simplify applied to automation– specifying automation features in a manner that eliminates unnecessary complexity and makes operating and maintenance errors less likely, and which is forgiving of errors.

- Reduce complexity; keep the functions simple

- Select devices that are less complex, e.g., have fewer configuration options that could be set incorrectly

- Use device redundancy and simple redundancy schemes to provide fault tolerance to manufacturing errors, maintenance errors, or unexpected hardware failures instead of using complex configurations or relying on theoretical analysis

- Configure devices such that loss of communication or loss of signal results in a specified safe state

- Implement a fully independent and separate process control system from the safety system to simplify cybersecurity requirements, auditing practices, controlling bypasses, and managing change

- Display safety alarms so that they are easily distinguishable from other alarms [10, 12, 13]; Provide an independent safety alarm interface where possible (ISA-TR84.00.04 Appendix B [10])

- Apply alarm management to establish a clear priority for the operator's attention (IEC 62682 [11]);

- Make the navigation of the operator HMI and safety HMI intuitive and user-friendly

- Present information to the operator in a logical manner that provides situational awareness (e.g., process overviews, equipment overviews)

- Display redundant measurements in a manner that supports operator confirmation of the process condition

- Use distinctive labeling in plant documentation, the operator HMI, and on the components in the field for safety devices: use logical numbering for device groupings

- Use valve designs that offer a visual indication of actual position
- Write user-friendly operating and maintenance procedures that provide detailed task descriptions in order to reduce human error
- Repair and spare parts management procedures should include part number of replacement equipment. This same number should be used to label the part and its storage location in the warehouse
- Design maintenance facilities and parts warehouses to minimize the potential for mistakes

These inherently safer practices should be implemented as part of the design, operation, maintenance, and testing of the control and safety systems, where practicable.

6 Conclusion

The process and its equipment can be designed using inherently safer strategies to ensure safe operation during foreseen process upsets. An inherently safer process is designed to eliminate the potential for loss events with features that are inseparable from the process equipment. When process equipment is not designed to inherently withstand abnormal operation, process safety is achieved through functional safety management. Automated systems are specified to control the process within the normal operating limits and to act when there is a loss of control. These systems reduce operational risk to an acceptable level. Sustainability of automated system performance can be significantly enhanced through the application of the inherently safer strategies.

7 References

[1] CCPS. *Guidelines for Safe Automation of Chemical Processes, 2nd Edition,* New York: AIChE. 2016.

[2] CCPS. "Final Report: Definition for Inherently Safer Technology in Production, Transportation, Storage, and Use," New York: AIChE. 2010.

[3] CCPS. *Inherently Safer Chemical Processes: A Life Cycle Approach,* New York: AIChE. 2009.

[4] CCPS. *Guidelines for Engineering Design for Process Safety, 2nd Edition,* New York: AIChE. 2012.

[5] CCPS. *Guidelines for Risk Based Process Safety,* New York: AIChE. 2007.

[6] CCPS. *Guidelines for Safe and Reliable Instrumented Protective Systems*, New York: AIChE. 2007.

[7] Broadribb, M., M. R. Curry. "HAZOP/LOPA/SIL Be Careful What You Ask For!," Paper presented at 6th Global Congress on Process Safety, San Antonio, TX, March 22-24. 2010.

[8] Summers, Angela E., and William H. Hearn. "Risk Criteria, Protection Layers, and Conditional Modifiers," Process Safety Progress. June, Volume 31(2). Hoboken: AIChE. 2012.

[9] CCPS. *Guidelines for Hazard Evaluation Procedures, 3rd Edition*, New York: AIChE. 2008.

[10] ISA. *Guidelines for the Implementation of ANSI/ISA 84.00.01- Part 1*, TR84.00.04-2015, Research Triangle Park: ISA. 2015.

[11] IEC. *Management of Alarm Systems for the Process Industries, IEC 62682,* Geneva: IEC. 2014.

[12] CCPS. *Human Factors Methods for Improving Performance in the Process Industries,* New York: AIChE. 2007.

[13] CCPS. *Guidelines for Initiating Events and Independent Protection Layers in Layer of Protection Analysis,* New York: AIChE. 2014.

CCPS Project 246: Second Edition of the Guidelines for Siting and Layout of Facilities

Bruce K. Vaughen
Baker Engineering and Risk Consultants, Inc.,
Chicago, Illinois
bvaughen@BakerRisk.com

Martin Timm
Praxair Inc.
Buffalo, New York
Martin_Timm@praxair.com

Donald Connolley
BP Corporation North America Inc.
Aiken, South Carolina
Don.Connolley@bp.com

Charles Cowley
CCPS Staff Consultant
London, UK
charlescowley@btinternet.com

Prepared for Presentation at
American Institute of Chemical Engineers
2017 Spring Meeting and 13th Global Congress on Process Safety
San Antonio, TX
March 26 – 29, 2017

CCPS Project 246: Second Edition of the Guidelines for Siting and Layout of Facilities

Bruce K. Vaughen
Baker Engineering and Risk Consultants, Inc.,
Chicago, Illinois
bvaughen@BakerRisk.com

Martin Timm
Praxair Inc.
Buffalo, New York
Martin_Timm@praxair.com

Donald Connolley
BP Corporation North America Inc.
Aiken, South Carolina
Don.Connolley@bp.com

Charles Cowley
CCPS Staff Consultant
London, UK
charlescowley@btinternet.com

Keywords: facility selection, facility siting, facility layout, equipment layout, spacing, distance

Abstract

The second edition of the CCPS *Guidelines for Siting and Layout of Facilities* (2017) has been written to address the many developments in the last decade which have improved how companies survey and select new sites, evaluate acquisitions, or expand their existing facilities. By updating the title of the first edition issued in 2003, *Guidelines for Facility Siting and Layout*, it has been emphasized that this book focuses not only on siting of buildings and unit operations within a facility, but also on siting of facilities within a community. This guideline addresses issues for those involved in business and project development of processes with hazardous materials and energies, provides guidance for facility layout experts, and provides information for authorities and the public involved in the project's approval. In addition, this edition adds discussions on how the life cycle and long-term risks of the facility are addressed, how toxic dispersions and the effects of explosions are modelled, how changes can be more effectively managed, and how to select the optimal distances between the equipment. This paper will provide an overview of the changes to the first edition, as well as provide an overview of the significant reviews and updates to the appendices containing the recommended separation distances and checklists that will help the teams obtain the information they need when locating the project, when arranging the processes, and when arranging the equipment.

1 Introduction

This book provides guidance on how to select a facility's location, how to identify and assess potential long-term risks, and how to layout the process units and equipment within the new facility location. It addresses the location of - the siting of - facilities within a community, as well as how best to design the locations of and safely arrange buildings, unit operations and process equipment within the facility. The location of a new facility, an acquisition, or expansion project will impact the surrounding community in positive ways, such as increased economic activity and wages, and possibly in negative ways, such as increased noise or vehicle traffic. The potential for negative consequences increases if the process must handle hazardous materials and energies. To ensure that these process hazards and potential consequences are evaluated at the very beginning of the project, a company should follow a basic, time-proven approach to help reduce the potential consequences and, thus, the overall operational risk for the life of the facility. This book has been written to provide this fundamental guidance and help the venture succeed.

2 The Guideline's Framework

The chapters in this guideline are arranged in sequence for the step-wise approach when siting and locating process units and equipment within a facility. The steps are illustrated in **Figure 1**. The hazards and their consequences, evaluated in Step 1, are used by the location selection team in Step 2, the process unit layout team in Step 3, and the equipment layout team in Step 4. It is noted that effective layout teams consider the unit and equipment life cycle and integrity programs as they arrange the equipment within the facility, as well.

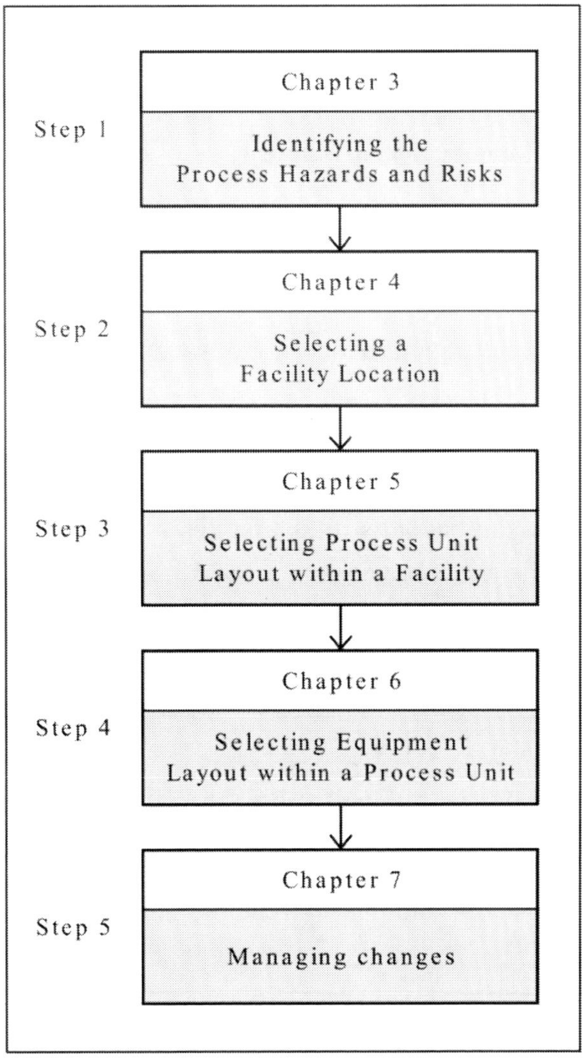

Figure 1.

The Chapters Corresponding to the Steps for Siting and Laying out of Facilities

Additional guidance and insight for the process unit and layout teams for effectively managing future changes are provided in this approach as Step 5. Although this may seem obvious for Step 2, experience shows that the effort spent in selecting a facility location team with members who understand the process hazards and potential consequences for a specific project ensures a more beneficial and thorough assessment of the locations under evaluation. The chapters provide an overview of the benefits for selecting proper facility, process unit and equipment locations, information on how to identify the process hazards and consequences, how to select a facility location, how to select the process unit layout within the facility's location, how to select the equipment layout within a process unit, and how to manage changes. New case histories are described to support the need for understanding and addressing potential issues that arise when locating processes with hazardous materials and energies. The objectives for each of the guideline's chapters are summarized in **Table 1**.

The second edition begins by describing the benefits of combining and applying inherently safer principles and protection layer concepts when addressing the siting and layout of facilities. There are five detailed chapters focusing on how to:

- identify the process hazards, potential consequences and risks
- select a facility location
- select the process unit layout within a facility
- select the equipment layout within a process unit
- manage changes.

The project's hazards analysis helps guide the location selection team, the process unit layout team, and the equipment layout team when they are deciding where to locate the facility, where to locate the process units within the facility, and where to the locate the process equipment within each process unit, respectively. Successful projects have multi-disciplined teams, and depending on the scope and phase of the project, may have the same people on all three teams.

The 2017 guideline has enhanced the previous edition with a new discussion on the facility's life cycle (addressing the long-term risks of the facility, including those off-site), and when selecting optimal distances between processes and equipment (i.e., arranging them within the facility). The book provides guidance on the issues which impact the ultimate location of the equipment in the process unit, including the:

- location's topography, weather, and surrounding environment or communities
- accessibility of the operating and maintenance personnel to the process
- accessibility of emergency responders, when needed.

The book provides guidance on how to manage continuous improvement efforts using a risk based approach to better understand the change's potential effects on the operating risks. Case studies have been enhanced or added in each chapter, helping the reader visualize the approaches used to effectively site and layout facilities.

Table 1.

Objectives for the Guideline Chapters

Guideline Chapter		Chapter Objective
1	Introduction	Describes the scope of this guideline: to provide information on how to select a facility's location, how to identify and assess potential long-term risks, and how to layout the process units and equipment within the new facility location.
2	Benefits Overview	Describes the benefits of combining and applying inherently safer principles and the barrier concepts when addressing the siting and layout of facilities.
3	Identifying the Process Hazards and Risks	Describes what process safety information is needed when performing preliminary hazards analyses of the new process or processes. The hazards and their consequences are then used by the location selection team (Chapter 4), the process unit layout team (Chapter 5), and the equipment layout team (Chapter 6).
4	Selecting a Facility Location	Describes who should be included on the location selection team, some potential project-related issues, and how to select a facility location when evaluating and comparing both the pros and cons of the potential location and its surroundings.
5	Selecting Process Unit Layout within a Facility	Describes how the process hazards and consequences and the proposed facility's topography, environment and surroundings influence the layout of the process units within the facility.
6	Selecting Equipment Layout within a Process Unit	Describes how the process hazards and consequences and the potential operating and maintenance accessibility issues influence the layout of the equipment within the process units.
7	Managing Changes	Describes how to manage process and equipment layout changes.
8	Case Histories	Shares examples of incidents which had severe consequences due, in part, to inadequate consideration and handling of siting and layout issues.

Similar to the first edition, a simplified siting and layout project is used as an example case study, specifically describing an expansion at an existing petrochemical facility. This example is structured to illustrate the progression of a project through its location selection (siting as described in Chapter 4), its process unit layout (Chapter 5), and then its equipment layout (Chapter 6) using the fire consequence-reducing separation distances provided in Appendix B (see Section 2.1 below). This example is also based on the protection layer / barrier risk reduction approach, considering how best to provide the new facility with adequate distances that reduce the potential consequences and, thus, the overall risk. The "conclusion" of this case study for equipment arrangement and spacing, based only on the potential for fire consequences, is presented in **Figure 2**.

The second edition enhances three major aspects in the foundation established by first edition: 1) developing and adding a chapter on effectively managing changes, 2) adding case histories from process safety incidents which have occurred since 2003, and 3) restructuring and updating the tables and checklists in the appendices. Many incidents which have resulted in significant impact, such as fatalities, severe injuries, environmental damage, and business interruption were caused, in part, by ineffective management of change to personnel or technologies. The new chapter on managing changes has been written to provide overview guidance on how change can be effectively managed. The new incident case histories were selected because there were severe consequences due, in part, to inadequate consideration and handling of siting and layout issues. It is hoped that the reader will review these case histories, recognize, evaluate, and then address potential siting and layout issues early in their personal project.

The appendices have been extensively reviewed and updated, providing both updated, industry-wide recommended separation distances (aka, the "spacing" between process equipment) and thorough team-specific checklists to help the location selection, process unit location, and equipment arrangement teams identify the information they need for their efforts. It is hoped that the reader will be able to use these tools in the appendices to effectively manage the potential consequences and, thus, the operational risks by selecting adequate separation distances early in their projects with processes handling hazardous materials and energies.

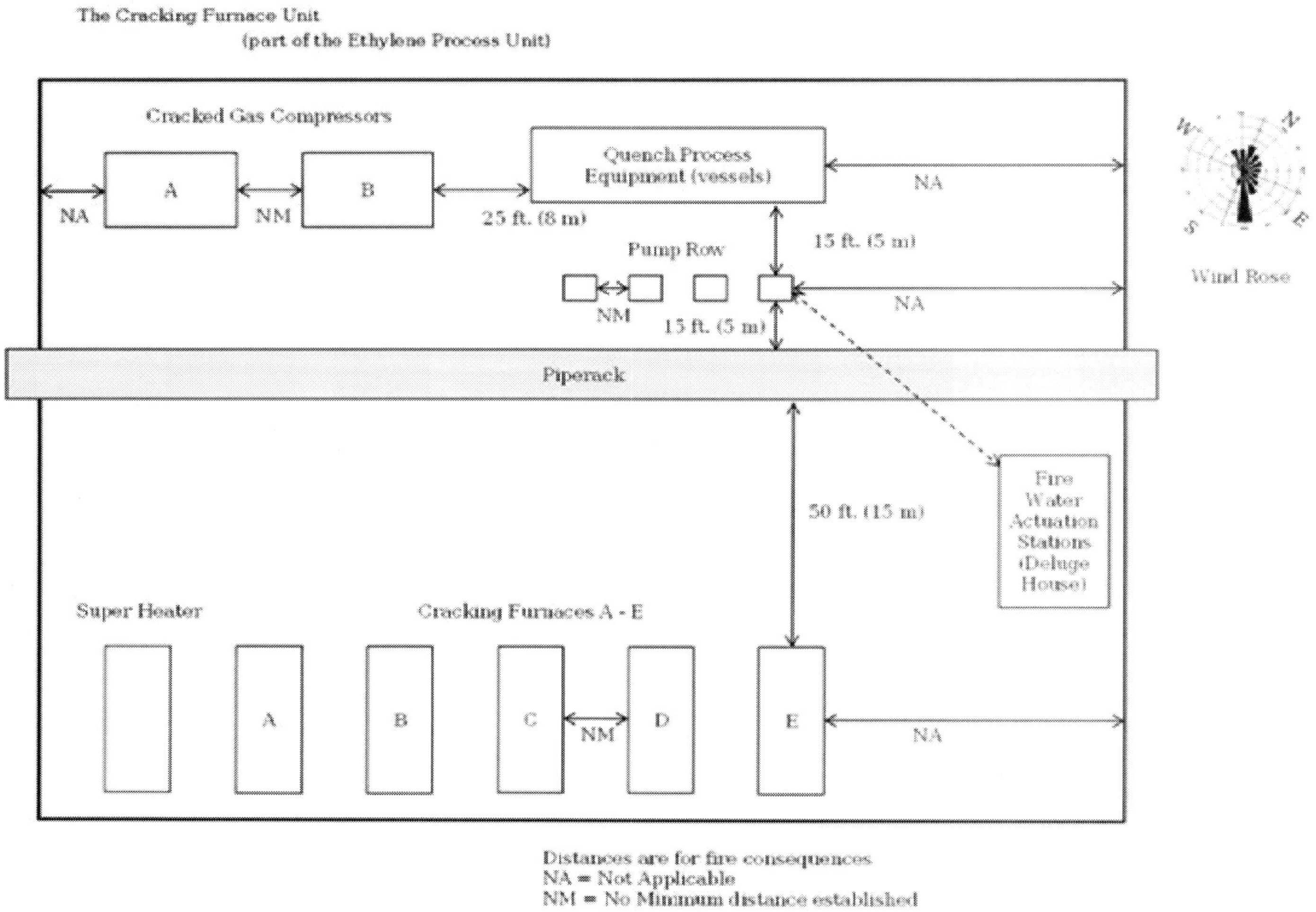

Figure 2.

Proposed Layout of the Equipment for the New Petrochemical Facility Illustration

2.1 Updated Equipment Separation Distances for Fire Consequences – Appendix B

This expanded and updated appendix provides industry-recommended separation distance guidance for fire consequence scenarios. These tables are based on compiled industry data over the decades and help reduce the consequences during fires in outdoor processes. Historical incident-related experience and learnings, additional engineering experience, industry consensus guidelines and standards, regulatory requirements, insurance loss reviews, and insurance guidelines have been used to develop and update these tables. Overall, these distances did not change significantly, with the changes noted using a ***bold italicized*** font for those that were updated in the second edition. These tables provide specific guidance for distances between process unit equipment, between tanks and process unit equipment, between tanks of hazardous materials, and between on-site buildings. Recommended layout distances are included for flare systems and for emergency response and operations accessibility. The general process hazards are based on moderate, intermediate and high hazard designations [1]. Although these tables may not provide an exact, analytical answer, they can be used to help with preliminary process unit lay out design and then with preliminary equipment lay out designs, taking advantage of industrial experience. The summary of this enhanced appendix is provided in **Table 2**.

Table 2.

Summary of the Updated Separation Distance Tables for Fire Consequences

Table in Appendix B	Provides Guidance for Typical Facility and Layout Distances (Fire consequences only)
1	Between Process Unit Equipment
2	Between Tanks and Process Unit Equipment
3	Between Tanks of Hazardous Materials
4	Between On-Site Buildings
5	Between Other Types of Equipment and Operations
6	For Emergency Response and Operations Accessibility
7	For Flare Systems

Notes:

1) Tables are provided in both English and Metric units.

2) The general hazardous process designations used when selecting these separation distances are as follows [1]:

Moderate hazard: This category includes processes, operations or materials having a limited explosion hazard and a moderate fire hazard. This class generally involves endothermic reactions and nonreactive operations, such as distillation, absorption, mixing and blending of flammable liquids. Exothermic reactions with no flammable liquids or gases also fit in this hazard group.

Intermediate hazard: This category includes processes, operations or materials having an appreciable explosion hazard and a moderate fire hazard. This class generally involves mildly exothermic reactions.

High hazard: This category includes processes, operations or materials having a high explosion hazard and moderate to heavy fire hazard. This class involves highly exothermic or potential runaway reactions and high hazard products handling.

2.2 Updated Case Histories

The chapter containing case histories has been enhanced with descriptions of over ten new process safety incidents which had, in part, issues associated with poor siting and layout of the facility. The new case histories that resulted in fatalities, severe injuries, environmental damage, and business interruption were selected from process safety incidents which occurred since publication of the first edition in 2003. Each of these incidents resulted in the severe consequences due, in part, to inadequate consideration and handling of siting and layout issues. The case histories illustrate the need to proactively address potential siting and layout risks as a part of a company's risk reduction and management effort. An overview of the case histories included in the second edition is provided in **Table 3**. Additional hydrocarbon industry case histories reflecting both site and community evacuations from incidents with large property damage losses are also provided by Marsh [2]. As a part of their review, Marsh notes that companies should apply siting and layout principles to help minimize the consequences, such as good layout, good separation, and process isolation.

3 The Step-Wise Approaches Described in the Guideline

The guideline has been enhanced with more detailed discussions on the approaches for locating the facility, locating the process units and associated personnel and utilities, and locating the process equipment within the units. Before these facility, process unit and equipment location approaches can be used, a thorough understanding of the process hazards and their potential off-site and on-site impacts is essential to effectively manage the risks. This section of the paper provides an overview of the approaches that can be used by the different siting and layout teams. The approach in the guideline focuses on guidance for the location selection team (Chapter 4), the process unit layout team (Chapter 5), and the equipment layout team (Chapter 6). Although these teams may have different members, schedules and goals, effective siting (i.e., locating the future facility) and locating the process units and associated equipment depends on both understanding and considering the technical aspects of the process design, its hazards, and its potential risks, both off-site and on-site. The checklist and steps essential for effectively evaluating the process hazards and their associated consequences are discussed below in Section 4.1.

Table 3.

Siting and Layout Incident Case Histories in the Two Editions

			Edition	
Incident Case History		**Siting and Layout Issue**	**1st**	**2nd**
1984	PEMEX LPG, Mexico	Location of LPG Storage: Effects of BLEVEs	X	X
1984	Bhopal, India	Community encroachment	X	X
1988	PEPCON, USA	Location of large quantities of reactive chemicals	X	X
1999	Concept Sciences, Inc., USA	Location of equipment with highly reactive chemicals	X	X
2005	BP Texas City, USA	Staging of non-operating personnel		X
2005	Praxair, USA	Location of LPG Storage: Effects of ambient conditions		X
2006	Danvers, USA	Location of confined vapors; community impact		X
2007	Little General Store, USA	Location of propane storage tanks		X
2007	Valero McKee Refinery, USA	Location of equipment: Effects of jet fires		X
2008	Imperial Sugar, USA	Location of dust handling equipment		X
2010	T2 Laboratories, Inc., USA	Location of equipment with highly reactive chemicals		X
2011	Mont Belvieu, USA	Locating new facilities in existing communities		X
2012	Amuay, Venezuela	Community encroachment		X
2013	West Fertilizer, USA	Community encroachment		X

3.1 Locating the facility

A time-tested approach is discussed first to ensure that new or modified facilities consider essential issues early when searching for the facility's location. This approach helps anticipate and address issues that may become costly short-term project-related design changes, may involve costly design changes during construction, or may become costly long term operating and maintenance protocols once the facility is built. This siting philosophy begins with a review of the material and processing hazards, such as toxicity, flammability, explosivity, reactivity, or a combination of these hazards (see Section 4.1 below). Other potential hazards should also be considered since they may be unacceptable to the surrounding community, such as odors, loud noises, or the light from flares.

Once these hazards have been identified and their potential off-site and on-site impacts are understood, the next step includes determining how the local terrain affects the unlikely but credible release scenarios (the ultimate impact on the surrounding community), emergency responder accessibility, and security accessibility (the risks at the facility's boundary). At the same time, the layout of the process units and associated areas within the facility, such as storage tank areas or flares, should be arranged to reduce the potential consequences. The layout of the equipment, including both their orientation and the separation distances between them, may affect both off-site and on-site consequences, as well. Since the layout of equipment can affect the efficiency of day-to-day operations, it is important to address the balance between reduced or increased distances and the impact on accessibility when evaluating the on-site consequences. The illustration in **Figure 3** provides a high level view of this approach, beginning with understanding the hazards and potential consequences, understanding the effects of the location's terrain, and then understanding both the potential off-site and on-site impacts due to the potential hazards.

Since proper location of the process unit equipment is an important aspect of risk management, the overall business risk depends on a suitable facility location and its equipment layout for the life of the facility. The overall risk to people, the environment, and property both internal and external to the facility must be managed from beginning to end, with the best time for reducing life cycle costs being early in the project's development phase. As is well-known in the industry, the most effective strategy for reducing the facility risks and its life cycle costs is to evaluate and implement Inherently Safer Design (ISD) principles early in the project [3, 4, and 5]. This guideline emphasizes that these ISD principles should be applied early on when locating the new facility, locating the process units within the new facility, and then arranging the equipment layout within each process unit.

Once these ISD principles have been applied and the facility is constructed and operating, the integrity of the process units and their associated equipment has to be sustained for effective management of the process safety risks. The eight distinct phases for maintaining the facility's equipment integrity are as follows: 1) design; 2) fabricate; 3) install; 4) commission; 5) operate; 6) maintain; 7) change; and 8) decommission. These typical life cycle phases are depicted in **Figure 4**. The guideline provides additional discussions on maintaining the integrity of the facility to ensure safe and reliable operations during its life cycle.

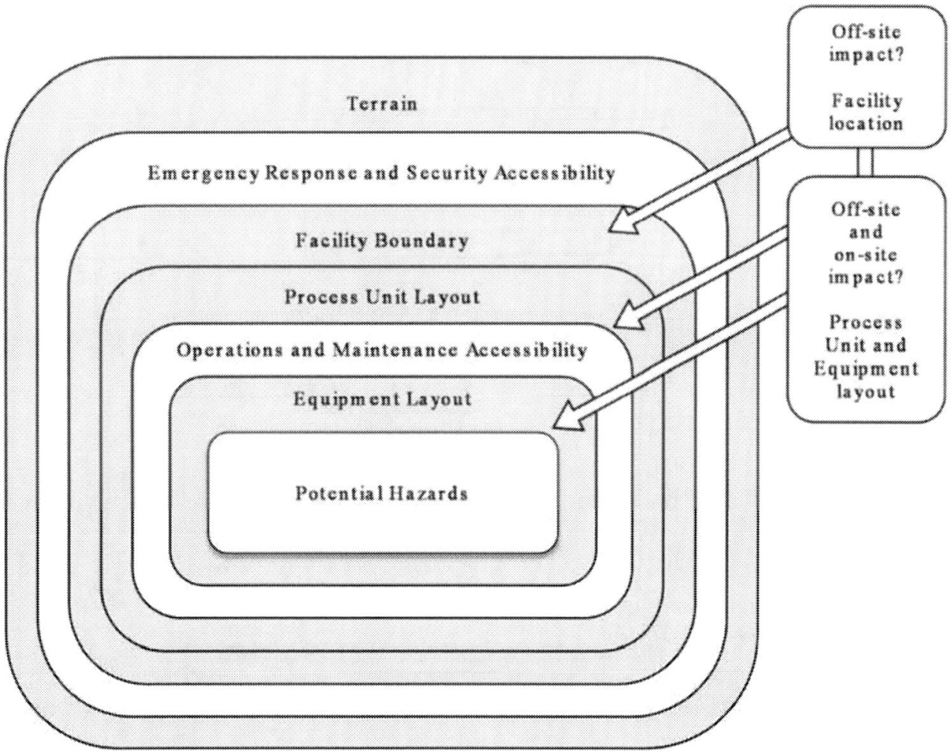

Figure 3.

A Siting and Layout Approach

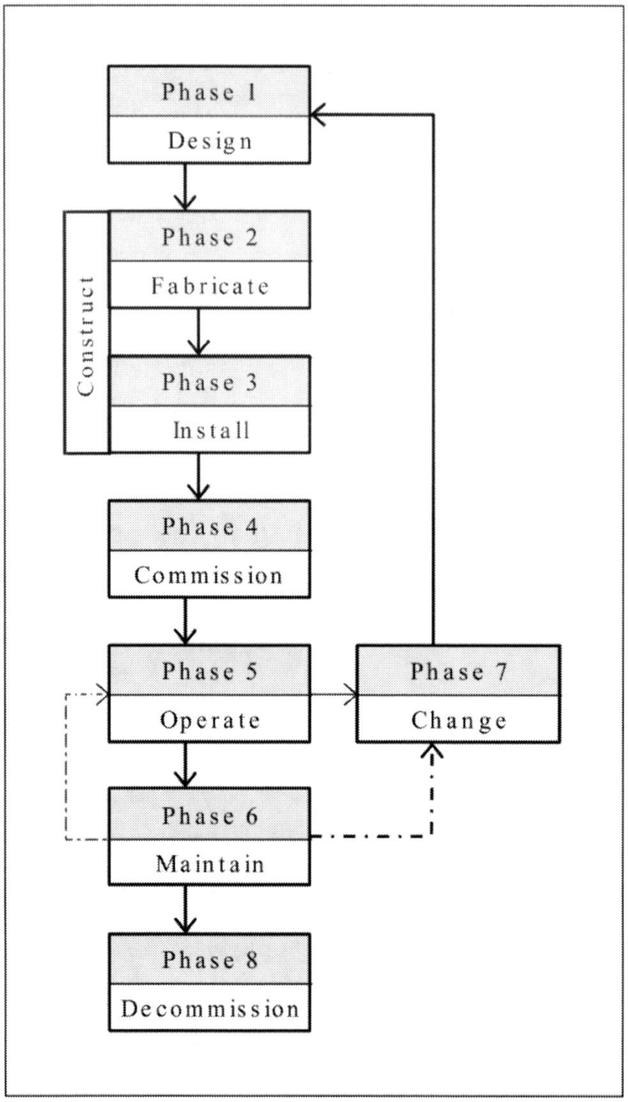

[Adapted from 14]

Figure 4.

Life-cycle Phases for a Facility, a Process Unit, or the Equipment

3.2 Locating the process units

The next approach, applied once the facility's location and size has been identified, uses an iterative approach of a "block" layout methodology. These steps help arrange the process unit blocks within the facility by identifying the process units that have hazardous consequences, determining the off-site and on-site consequences, determining the separation distances needed to reduce these consequences, and verifying the distances between process units and the types of structures. When the separation distances are not acceptable, the process unit location team must re-evaluate the locations for other potentially inherently safer designs, rearrange process units, or identify other potential engineering and administrative safeguards to help prevent or reduce the consequences. The final step is to verify that the distances between the process units and types of structure construction are acceptable, and if possible, optimize the distances. These steps are depicted in the flowchart in **Figure 5**, which shows the potential for reiterations and refers to Appendix B, the CCPS recommended distance tables for the siting and layout of facilities.

3.3 Locating the process equipment

The third approach, applied once the process unit block locations have been identified, uses an iterative approach to identify where the process equipment handling the hazardous materials should be located within each of the process unit blocks. The equipment layout methodology uses a method similar to the method used when arranging the blocks within the facility. The steps in this approach help arrange the equipment by: 1) reviewing the process unit's overall hazards and block location-related issues (the information from the process unit layout team); and then 2) evaluating each of the specific process unit's equipment-related hazards and consequences. The equipment is then arranged within the process unit's boundaries based on acceptable separation distances. When the separation distances are not acceptable, the equipment location team must re-evaluate the equipment for other potentially inherently safer designs, rearrange the equipment, or identify other potential engineering and administrative safeguards to help prevent or reduce the consequences. The final step is to verify that the distances between the equipment are acceptable, and if possible, optimize these separation distances. These steps, including potential reiterations, are depicted in the flowchart in **Figure 6** (note the similarity to the flowchart presented in **Figure 5**). Essentially, the hazards information for both on-site and off-site consequences are used for both the block layout and the equipment layout teams. The distance tables for fire consequences provided in the guideline's Appendix B (described in Section 2.1 above) are used to help with the layout of the equipment within the unit. Other methods, including consequence-based assessments, are used to determine the areas impacted from potential explosions or toxic releases.

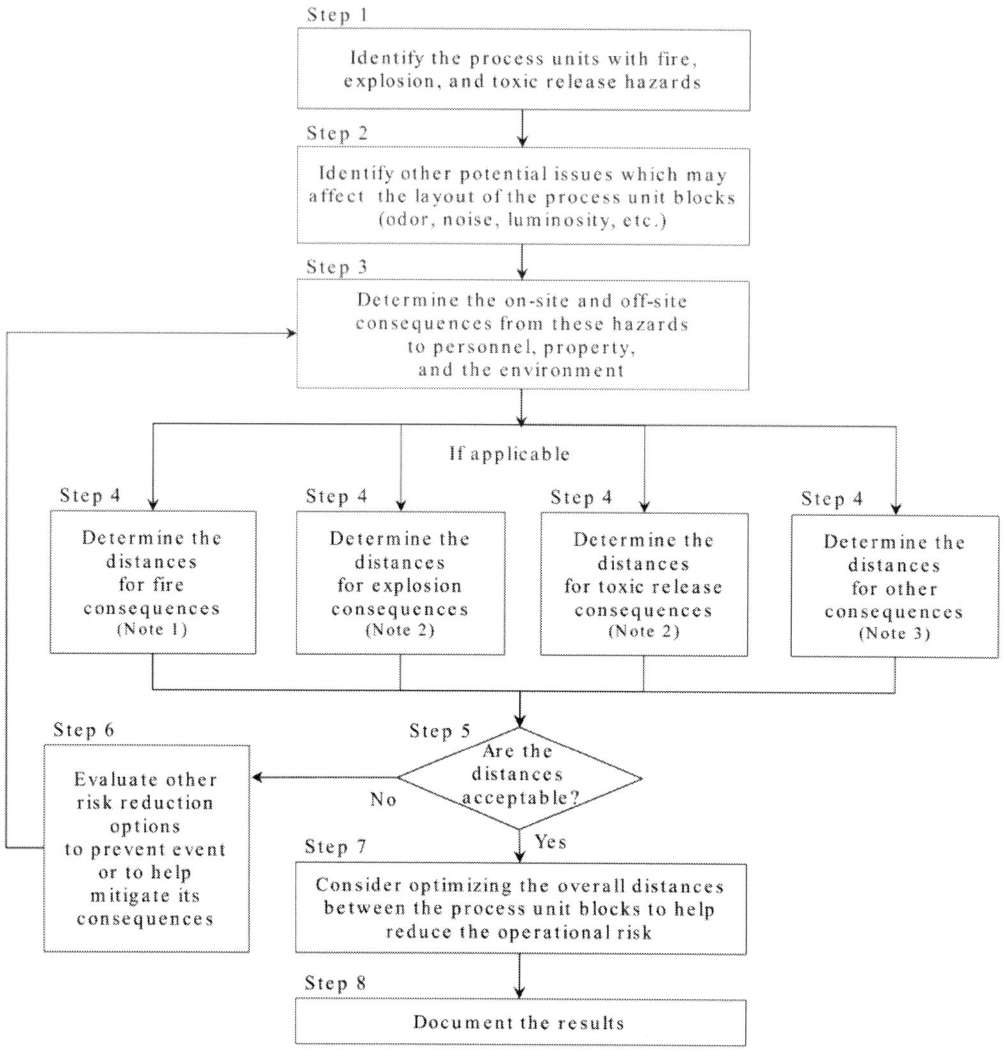

Figure 5.

Flowchart used to Determine the Process Unit Layout Distances at a Facility

Notes for Figure 5:

1) Appendix B, CCPS recommended distance tables for siting and layout of facilities, contains distances primarily based on fire consequences. Additional thermal radiation modeling can be performed, as needed.
2) For explosion or toxic release consequences, distances should be determined using explosion contour diagrams or toxic contour diagrams modeled with a consequence-based or risk based approach when siting structures (see additional discussion in guideline). These distances may differ to those determined using Appendix B, which are for primarily for fire consequences.
3) For other consequences, use the local area zoning requirements to establish the distances (e.g., for odors, noise levels, luminosity levels, flare heights, etc.)

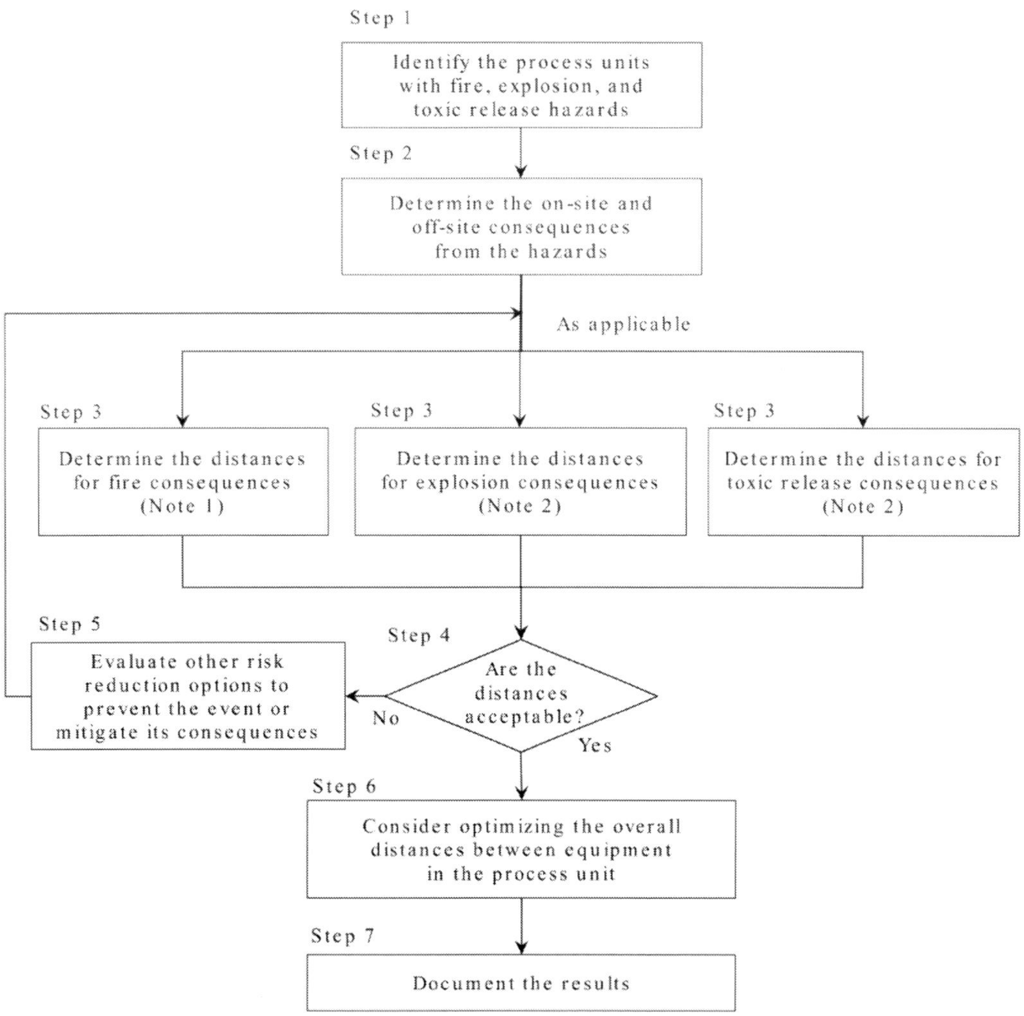

Figure 6.

Flowchart used to Determine Equipment Layout Distances Within a Process Unit

Notes for Figure 6:

1) Appendix B, CCPS recommended distance tables for siting and layout of facilities, contains distances primarily based on fire consequences. Additional thermal radiation modeling can be performed, as needed.
2) For explosion or toxic release consequences, determine distances using explosion contour diagrams or toxic contour diagrams modeled with a consequence-based or risk based approach to siting buildings, such as that described in API RP 752 (e.g., through a QRA; note additional discussion on structure and building locations is included in Chapter 5). These distances may differ than those determined using Appendix B, which are primarily for fire consequences.

3.4 Managing changes

If you don't like change, you are going to like irrelevance even less.
Eric Shinseki, military leader

Among the many enhancements in the second edition, the guideline provides a new chapter on effectively managing change when siting and laying out facilities. Change management systems which did not adequately address potential siting and layout issues have been responsible, in part, for exacerbating the consequences of significant process safety incidents, including Bhopal, Buncefield and Texas City [6, 7, 8, 9, 10, and 11]. The new chapter focuses on but one portion of a facility's MOC system: evaluating potential impacts on siting and layout issues due to new, temporary, or expanded processes at an existing facility. An approach for managing these changes and understanding how they may impact siting and layout issues is illustrated with **Figure 7**. Both on-site and off-site changes might affect the acceptable locations of, and separation distances between, the equipment within the facility's process units and between the facility and its neighbors.

Although temporary activities have a defined duration and are agreed upon between the affected groups (e.g., engineering, operations and maintenance), one of the most high risk times to the workforce is during simultaneous operations (SimOps). The pre-, during- and post-SimOps-related turnaround activities are noted in this chapter and more fully developed discussion in Chapter 5 and the new Case History 8-1 describing the BP Texas City incident in which temporary trailers containing personnel were staged in a hazardous area during the start-up of a hazardous process after a turnaround.

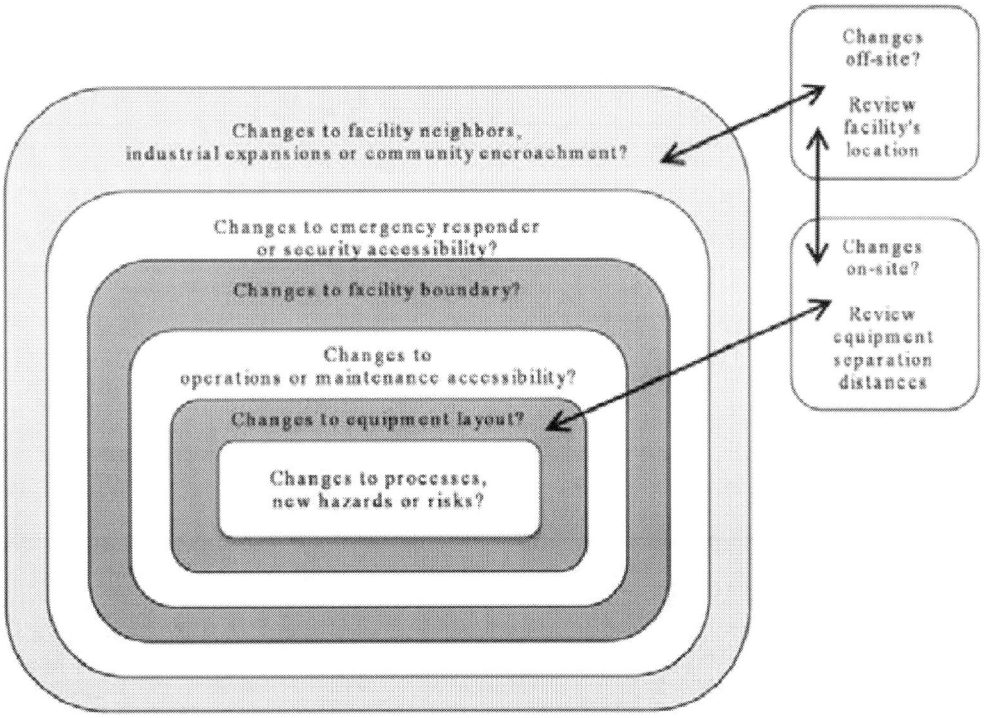

Figure 7.

A Siting and Layout Approach when Managing Changes

4 The Checklists for Siting and Layout Teams

The second edition significantly restructured, expanded and updated the tables and appendices created for the first edition. The appendices were reviewed and updated, providing both updated, industry-wide recommended separation distances (aka, the "spacing" between process equipment) and enhanced, thorough team-specific checklists to help the location selection, process unit location, and equipment arrangement teams identify the information they need for their efforts. The updated separation distance tables for fire consequences, Appendix B, was discussed in Section 2.1 above. The appendices for the four updated checklists are described in more detail below.

4.1 When identifying the process hazards and risks

The CCPS Risk Based Process Safety (RPBS) approach is used in this guideline when identifying the process hazards, consequences and risks [12]. The RBPS discussion in Chapter 3 is new for the second edition. Additional enhancements and details are provided in the guideline for assessing:

- fire scenarios (i.e., thermal radiation impacts from jet fires, pool fires, and flares)
- explosion scenarios (i.e., damage from vapor cloud explosions – VCEs, such as blast waves and flying debris)
- toxic release scenarios (i.e., personnel impact from vapor dispersions of toxic gases or evaporating toxic liquids).

The objective of the checklist provided in Appendix C is to help the selection and layout teams identify the types of process hazards, potential consequences and risks associated with the new or expanded facility.

It is important to note at this point that the separation distances noted in Appendix B focus on those for fire consequences only (see Section 2.1 above). The checklist in Appendix C provided in the second edition also refers to the changes and enhancements to toxic dispersion and explosion overpressure modelling efforts (i.e., a Quantitative Risk Analysis, QRA). The QRA can provide siting and layout teams with additional information when toxic materials are released and dispersed and when explosions cause damaging blast overpressures. These modelling efforts can be used to show siting and layout teams potential impact areas. Impact areas are specific to the release locations and the loss of containment scenarios, and may consider impact on people (including those inside buildings), to the environment (including potential off-site consequences) and to adjacent property and structures (such as building damage levels). **Figure 8** depicts one of the illustrations describing current modelling efforts for potential blast overpressures from materials with explosive hazards. These modelled scenarios provide more realistic distances to help siting and layout teams better evaluate the potential risks.

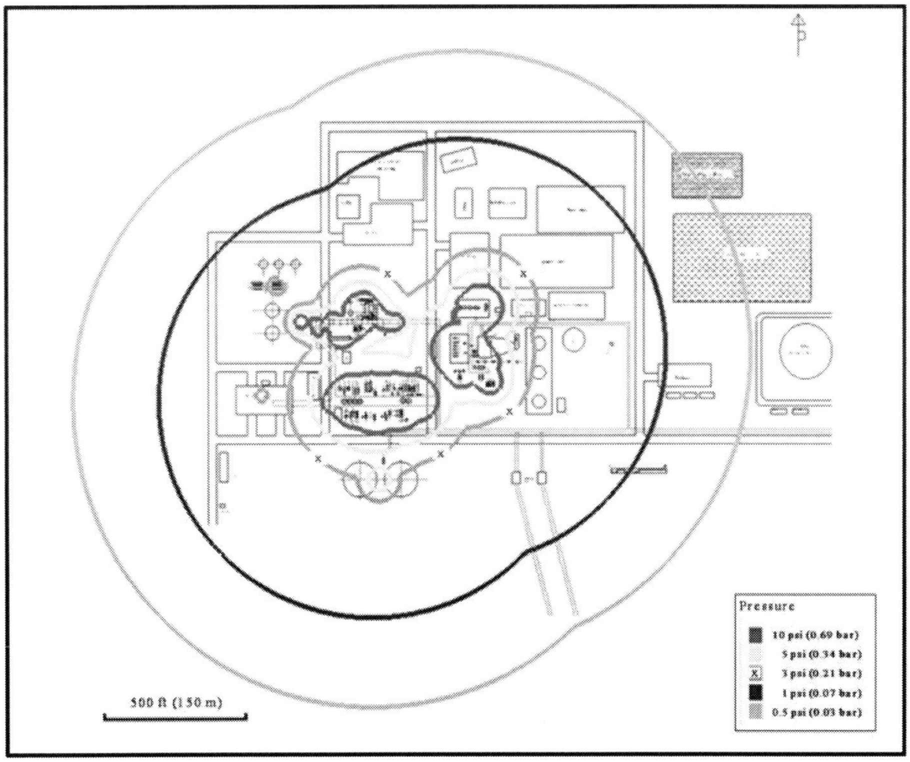

Courtesy: BakerRisk®

Figure 8.

An illustration of potential blast overpressure contours

4.2 When selecting a facility location

The objective of this checklist is to help the location selection team assess and compare both the pros and cons of the potential location and its surroundings. The terrain properties, provided by a topographic map (a detailed graphical representation of the terrain with both man-made and natural ground characteristics), shows different elevations (such as hills and valleys), and may include highways, railroads, nearby industries or communities, rivers, shoreline, sand, rock and vegetation. Some of the characteristics which need to be considered when selecting a facility location are depicted in **Figure 9**.

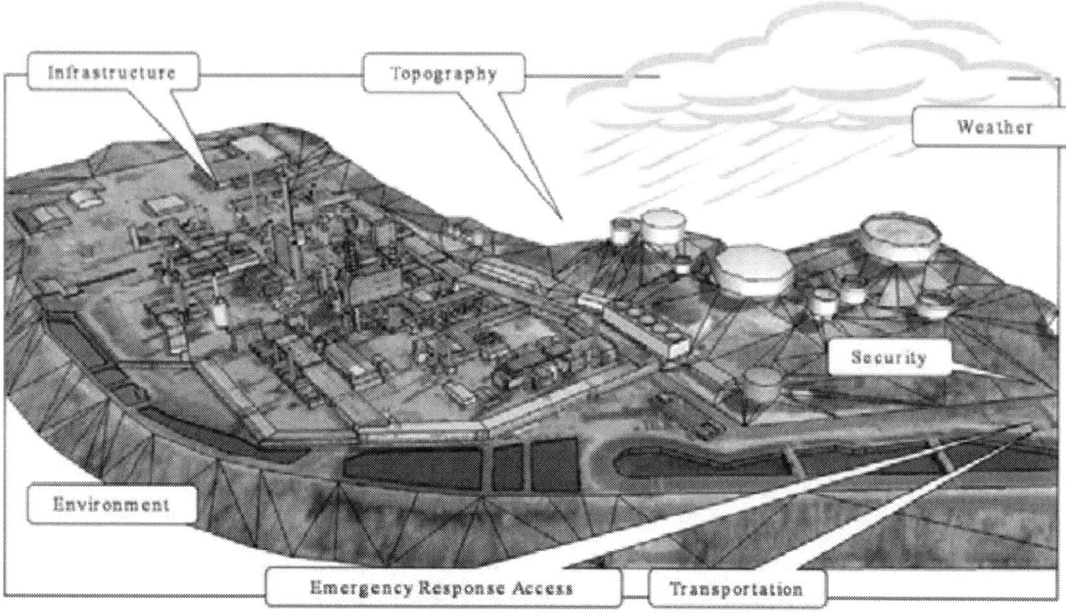

Figure 9.

Characteristics Considered when Selecting a Facility Location

Note that the hazards, consequences and risks associated with the new facility, identified with the checklist provided in Appendix C, will influence the final location, whether it is a greenfield or a brownfield site, a grassroots facility, an acquisition, or an expansion to an existing facility. The facility location selection team can choose from a variety of topics in Appendix D that apply to their specific project. One goal of the location survey effort is to obtain sufficient location-specific information which is then used to help bridge the hand-off from the location selection team to the process unit and equipment layout teams.

Since the property line of a facility separates the ownership of a company's land to its neighbors, there are three conservation balances which apply across the facility's property line:

- Materials in and materials out (e.g., feedstocks, equipment, materials, products, etc.)
- Energy in and energy out (e.g., electricity, steam, fuels, etc.)
- People in and people out (e.g., security, employees, visitors, emergency response, etc.).

The checklist used by the location selection team has been structured to address this overall "conservation balance" across a facility's boundary (property line), as is depicted in **Figure 10** [adapted from 13], with the general balances form the foundation of the information needed to address the facility's location (e.g., its topography, weather, local resources), and its infrastructure (e.g., accessibility), transportation, utilities, emergency response, and local environmental and permitting issues.

4.3 When selecting the process unit layout

The objective of this checklist, Appendix E, is to help the process unit layout team assess potential geographical and environmental issues at the new location. The checklist is structured to build on the efforts of the facility location team. At the time of publication, the checklists in Appendix E reflected the current information available to the CCPS. Since updates and changes are expected in the future, the reader should access the latest version of this appendix which will be maintained on the CCPS Website.

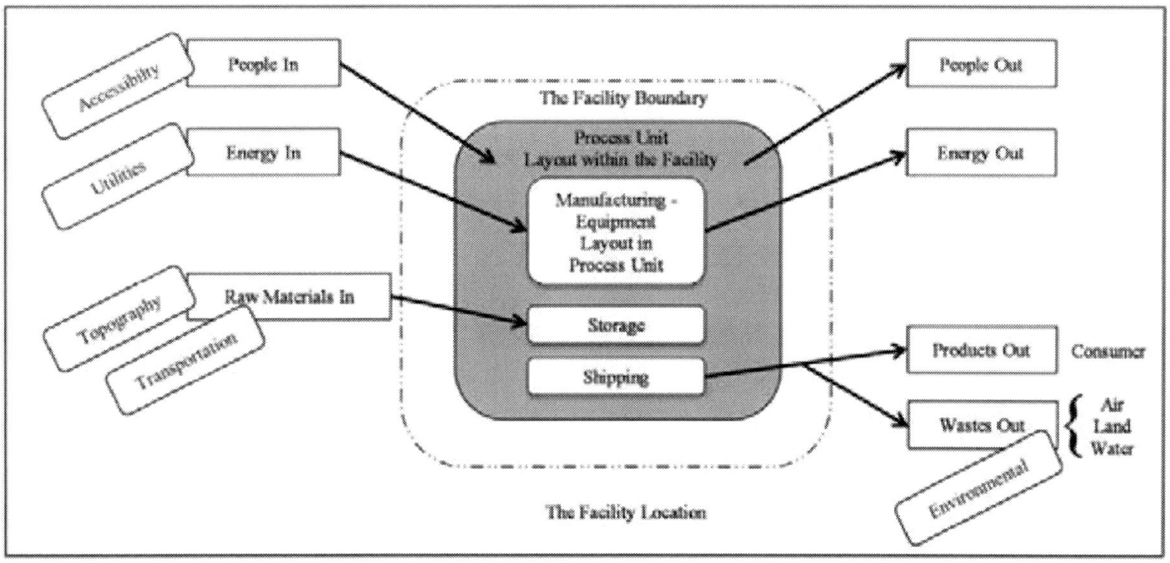

Figure 10.

The Conservation Balance Across a Facility's Property Line

4.4 When selecting the equipment layout

The objective of this checklist, Appendix F, is to help the equipment layout team assess potential operating and maintenance accessibility issues within the process unit. As with Appendix E, the checklist is structured to build on the previous efforts of the facility location and process layout teams, and, at the time of publication, the checklists in Appendix F reflected the current information available to the CCPS. Again, since updates and changes are expected in the future, the reader should access the latest version of this appendix which will be maintained on the CCPS Website.

5 Summary

The second edition addresses the many developments in the last decade which have improved how companies survey and select new sites, evaluate acquisitions, or expand their existing facilities. The changes and revisions in the second edition will help siting and layout teams better prepare for, anticipate, and propose locations for new projects and help reduce the potential consequences and, thus, the overall operational risks during the facility's lifetime. Adequate separation distances between process equipment handling hazardous materials are used to help reduce the consequences to other processes and the surrounding community. The approaches described in the updated guideline area should help reduce potential on-site consequences due to the equipment layout, which in turn, helps reduce potential off-site consequences.

The major enhancements in the second edition include updating the:

- evaluations of the process hazards, consequences and risks, both on-site and off-site
- discussions on the project life cycle
- discussions on modelling fires, explosions and toxic releases
- discussions on managing changes
- tables listing the separation distances (the "spacing") for scenarios with fire consequences
- the case studies within each chapter and the case histories from process safety incidents which have occurred since 2003
- checklists for the information needed by the site, process unit and equipment location selection teams.

This guideline addresses hazardous materials and process issues for those involved in business and project development, provides guidance for facility layout experts, and provides information for authorities and the public involved in any planning approval consultation process. The second edition complements the existing process safety literature by providing information and approaches on how best to address potential siting and layout issues when protecting people, the environment and the business from potential adverse effects with the loss of containment of hazardous materials and energies.

6 Acknowledgements

The guideline's primary author and CCPS Subcommittee chairs express their appreciation and gratitude to all members of the CCPS Subcommittee and their CCPS member companies for their generous support and technical contributions in the preparation of this guideline (Project 246). This appreciation is extended to the team of BakerRisk® authors who contributed to specific parts of this second edition, and to CCPS staff and consultants who supported this work. In addition, before publication, all CCPS books are subjected to a thorough peer review process. Thus, the insightful suggestions and comments of these peer reviewers were reviewed and incorporated into this second edition, as well. The authors and chairs of this CCPS project extend their appreciation to these peer reviewers. We sincerely hope that the information presented in this updated guideline will complement the existing process safety literature to help protect people, the environment and the business by effectively managing processes with hazardous materials and energies.

7 References

1 Global Asset Protection Services, LLC, (XL-Catlin), GAP Guidelines 2.5.2.A, Hazard Classification of Process Operations for Spacing Requirements, September 1, 2007.

2 Marsh, *The 100 Largest Losses, 1974-2013, Large Property Damage Losses in the Hydrocarbon Industry, 23rd Edition*, 2014. www.uk.marsh.com

3 The Center for Chemical Process Safety (CCPS), *Inherently Safer Chemical Processes: A Life Cycle Approach, 2nd Edition*, AIChE and John Wiley & Sons, Inc., Hoboken, New Jersey, 2008.

4 Kletz, T. A., and P. Amyotte, *Process Plants: A Handbook for Inherently Safer Design, Second Edition*, CRCPress, Taylor and Francis, Boca Raton, FL, 2010.

5 The UK Health and Safety Executive (UK HSE), Plant Layout reference. www.hse.gov.uk

6 Kletz, T., *What Went Wrong? Fifth Edition,* Gulf Professional Publishing, Elsevier, Burlington, MA, 2009.

7 Buncefield Major Incident Investigation Board (MIIB), *The Buncefield Incident 11 December 2005: The final report of the Major Incident Investigation Board, Volume 1,* UK HSE Books, 2008.

8 Buncefield Major Incident Investigation Board (MIIB), *The Buncefield Incident 11 December 2005: The final report of the Major Incident Investigation Board, Volume 2,* UK HSE Books, 2008.

9 United States Chemical Safety and Hazard Investigation Board, Investigation Report, "*Refinery Explosion and Fire, BP Texas City, Texas, March 23, 2005*," Report No. 2005-04-I-TX, issued March 2007.

10 Atherton, J., and F. Gil, Incidents that Define Process Safety, CCPS and John Wiley & Sons, Hoboken, NJ, 2008.

11 Bloch, Kenneth, *Rethinking Bhopal: A Definitive Guide to Investigating, Preventing, and Learning from Industrial Disasters*, IChemE, Elsevier, Amsterdam, Netherlands, 2016.

12 The Center for Chemical Process Safety (CCPS)/American Institute of Chemical Engineers, *Guidelines for Risk Based Process Safety (RBPS),* John Wiley & Sons, Inc., Hoboken, New Jersey, 2007.

13 Klein, J. A. and R. A. Davis. 2011, "Conservation of Life as a Unifying Theme for Process Safety in Chemical Engineering Education," *Chemical Engineering Education,* **45**:126-130.

14 Klein, J. A., and B. K. Vaughen, *Process Safety: Key Concepts and Practical Applications,* Boca Raton: Taylor & Francis, 2017.

CCPS Vision 20/20: What and how are WE doing?

Cheryl Grounds
cheryl.grounds@att.net

Jack McCavit, CCPS Staff Consultant
jlmconsulting@sbcglobal.net

Rukyah Hennessey, CCPS Engineering Specialist
rukyh@aiche.org

Prepared for Presentation at
American Institute of Chemical Engineers
2017 Spring Meeting and 13th Global Congress on Process Safety
San Antonio, TX
March 26 – 29, 2017

CCPS Vision 20/20: What and how are WE doing?

Cheryl Grounds
cheryl.grounds@att.net

Jack McCavit, CCPS Staff Consultant
jlmconsulting@sbcglobal.net

Rukyah Hennessey, CCPS Engineering Specialist
rukyh@aiche.org

Keywords: CCPS Vision 20/20, process safety performance, benchmarking

Abstract

For the past couple of years, the Center for Chemical Process Safety (CCPS) has been promoting Vision 20/20 in hopes that industry, regulators, academia, and the community worldwide will collectively and fervently drive great process safety performance through use of the five industry tenets and four global societal themes. CCPS has developed an assessment tool that is intended to help a company assess its current process safety implementation against Vision 20/20 tenets/themes. Although the tool is intended to be used internally, we continue to receive requests for benchmark data to allow companies to compare their results with their peers. This presentation will share examples of how CCPS Vision 20/20 tenets and themes appear in action in various companies as well as a new set of benchmark data generated from CCPS member companies this year.

1 A brief reminder

CCPS's Vision 20/20, looks into the not-too-distant future to describe how great process safety is delivered when it is collectively and fervently supported by industry, regulators, academia, and the community worldwide. The Vision 20/20 project focuses on defining the characteristics of companies that have great process safety performance.

CCPS Vision 20/20 is comprised of five industry tenets and four societal themes as shown in Figure 1. The fact that there are both industry and societal aspects to the vision underscores that we should all contribute to this vision - we all have a role to play.

The industry tenets describe the characteristics of companies with great process safety performance.

- Committed Culture
- Vibrant Management Systems
- Disciplined Adherence to Standards
- Intentional Competency Development
- Enhanced Application & Sharing of Lessons Learned

The societal themes are important for industry to achieve great process safety performance.

- Enhanced Stakeholder Knowledge
- Responsible Collaboration
- Harmonization of Standards
- Meticulous Verification

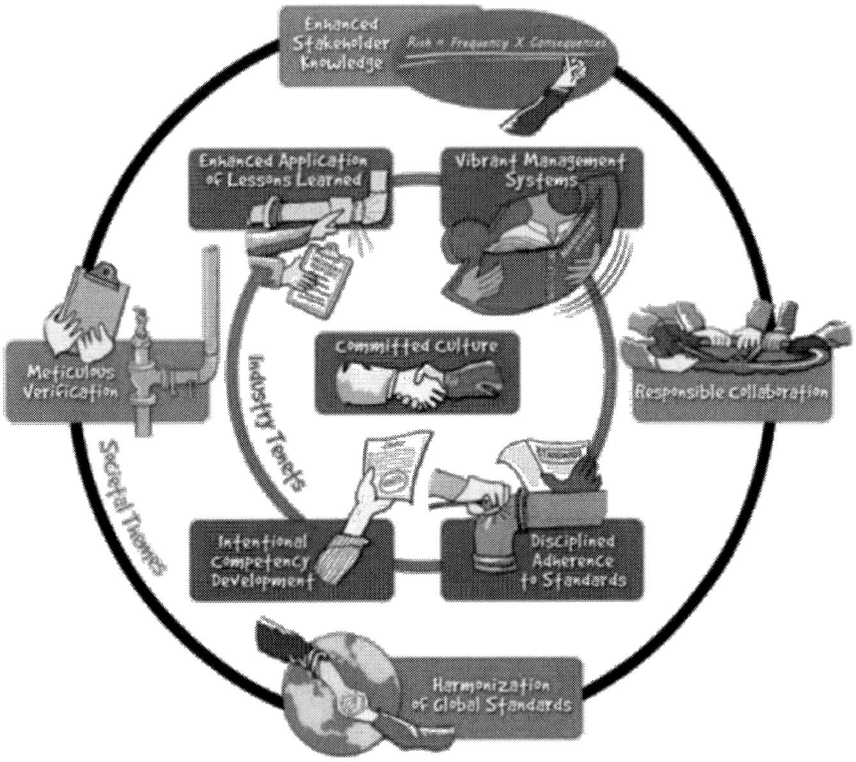

Figure 1. CCPS Vision 20/20

2 How the tenets and themes appear in action

Operating companies and other entities have used CCPS Vision 20/20 to drive process safety improvement in various ways. Some companies have used the Vision 20/20 materials directly from the CCPS website http://www.aiche.org/ccps/resources/vision-2020 to communicate the Vision and prompt activities in their company aimed at process safety performance.

In CCPS

Over the past couple of years, the CCPS Vision 20/20 journey gained significant momentum. The Vision 20/20 Call to Action has been a key element of all of the CCPS meetings, conferences, and workshops, held around the world. These include the Global Congress on Process Safety held in the USA, the CCPS Global Summit in South East Asia, and the Latin American Conference on Process Safety held in Latin America. Vision 20/20 is the platform on which all of CCPS growth and efforts are based.

In BP

Some companies embody the Vision 20/20 concepts in their company processes as opposed to using the Vision 20/20 tenets and themes directly A paper in 2016 MKOPSC Symposium shared how the CCPS Vision 20/20 tenets and themes could be seen in BP. Excerpts from this paper are highlighted below.

- Committed Culture – In BP, culture starts with the values: safety, respect, excellence, courage and one team. The tone is set at the top through BP leadership being personally involved in process safety including reviewing performance, asking questions in the field and reviewing risk action plans.

- Vibrant Management Systems - The Operating Management System (OMS) is the Group-wide framework designed to help BP manage risks in their operating activities. The elements of the CCPS Risk Based Process Safety can be seen in the BP OMS.

- Disciplined Adherence to Standards – Engineering practice control authorities decide which engineering practices are required in the projects. For existing equipment, BP Upstream has implemented an approach called fit for service.

- Intentional Competency Development – Learning offers are organized in the BP Academies. There is an intent to reach two audiences; the Process Safety Engineers who need a deep technical expertise and the broader BP population that needs to understand process safety fundamentals.

- Enhanced Applications & Sharing of Lessons Learned – A new record/report/learn tool called IRIS will help BP analyze and learn from both leading and lagging data.

- Enhanced Stakeholder Knowledge – BP is involved in numerous activities to support STEM programs. Many BP process safety professionals have lectured at the university level on process safety topics.

- Responsible Collaboration – Like many companies, BP is involved in joint industry committees that create codes, standards and guidelines as well as professional institutes that focus on furthering capability and impact.

- Harmonization of Global Standards – BP cited an example of active involvement in API RP 754 and IOGP 456 on process safety indicators. These documents are purposefully harmonized.

- Meticulous Verification – BP implements 3 lines of defense that work together to self-verify, assure and audit the health of the barriers managing major accident risks.

In LANXESS

The specialty chemical company LANXESS has several examples of how they are embracing Vision 20/20 within their NAFTA region operations:

- Starting in 2013, LANXESS began introducing the Vision 20/20 tenets and themes during quarterly internal process safety web meetings with NAFTA sites.

- They have incorporated core expectations of the Vision 20/20 tenets into their NAFTA regional process safety requirements. Critical enhancements included greater clarity on senior management roles and expectations, increased focus on applying and adhering to standards, and including competency requirements for critical roles.

- As part of a three-day process safety internal workshop for NAFTA sites in 2016, the participants divided into five subgroups and completed the Vision 20/20 Assessment Tool, then reconvened and discussed the results.

- The NAFTA Manager of Process Safety presented the following topics to the NAFTA Operating Committee:

    - Senior management roles and expectations regarding process safety
    - The Vision 20/20 goal, background, and tenets
    - Applicable process safety regulations and industry expectations
    - The structure of the LANXESS process safety program

In Huntsman

Huntsman, a global manufacturer and marketer of differentiated chemicals, is applying the CCPS Vision 20/20 in the following ways.

- A series of workshops building from Vision 2020 concepts also utilizing the Vision 2020 assessment tool (Culture, Competency)

    - Executive
    - Site Level Leadership
    - Operators and Supervisors

- Development of a consolidated standards set dedicated to Project Management which intentionally integrates process safety (Competency, Culture, Management Systems, Standards)

- Site Leadership technical competency review focusing on hazard recognition, knowledge, and risk management (Competency)

    - Site level requirement for assessment and appointment of designated "Qualified Persons" across process safety relevant technical areas (Competency)
    - "Red Flyer" program which communicates potentially significant internal incidents and issues action items for relevant sites/locations (Lessons Learned)

In NOVA Chemicals

NOVA Chemicals, a leading producer of plastics and chemicals that make everyday life safer, healthier and easier, utilized the Vision 2020 framework as a way to evaluate the completeness of its Process Safety programs.

- Preliminary benchmarking from the 2015 Global Congress on Process Safety was used to assess NOVA's focus areas, relative to the benchmarking data, and aided in the setting of priorities.

- The model within the Committed Culture Industry Tenet was utilized as a starting point to build a broad model to address Culture, as related to not just Process Safety, but to all disciplines within Health, Safety, Environment and Sustainability.

- Elements related to Enhanced Stakeholder Knowledge were enacted through further development of relationships and volunteering efforts with major Universities within the communities where NOVA Chemicals operates.

- NOVA Chemicals took an active role supporting the development of the recently published CSA Standard Z767 - Process Safety Management, as an example of their commitment to Harmonization of Standards.

- In addition, earlier initiatives from the Vision 2020 project, such as the "Day in the Life" stories have been integrated into leadership presentation and discussions helping to shape further Process Safety dialogue.

Chemical Engineering Progress (CEP) Series

CEP is running a series of one-page feature articles on Vision 20/20. With the broad reach of the CEP readership of 45,000 AIChE members, this will share the CCPS Vision 20/20 message to an even greater audience, worldwide. This series started in November 2016 and will continue through December 2017. Each tenet and theme will be highlighted with content included on:

- What does it mean?,
- What is the value?,
- What can I do?, and
- What does it look like?

The objective of this column is to deliver a message to the reader that Vision 20/20 is within reach for everyone. These articles provide the reader with steps that can be taken to implement Vision 20/20 where they work.

Risk Not Chance

An exciting new example of how the CCPS Vision 20/20 tenets and themes appear in action is in a United Engineering Fund Grant titled "Risk Not Chance". This effort will develop and distribute a series of three educational outreach videos to increase risk awareness in key stakeholder groups: high school/STEM students; general public/community, and first responders. The vides will use stakeholder-specific examples which will introduce risk concepts and barriers in place to prevent or minimize impacts. These educational videos support the CCPS Vision 20/20 societal theme of Enhanced Stakeholder Knowledge and Responsible Collaboration.

3 Benchmarking using the Assessment Tool

The **CCPS Vision 20/20 Assessment Tool** focuses on the five industry tenets associated with Vision 20/20. It is intended to help a company assess its current implementation against Vision 20/20 elements and identify areas for further emphasis. The tool contains a series of questions for each of the five industry tenets. Each question is answered with: *always, most of the time, some of the time, or infrequent or never.* The assessment tool, and instructions, are available on the CCPS Vision 20/20 webpage. http://www.aiche.org/ccps/resources/vision-2020

CCPS received feedback following the launch of this assessment tool that some of the terminology and references to organizational structure might make it challenging for smaller to mid-size companies to use the tool. A second version of the tool has been developed for this small to mid-size company audience and is available at the same webpage.

The assessment results are intended to be used internally by a company, to identify opportunities to better deliver great process safety performance through Vision 20/20 tenets. However, CCPS continues to receive requests for benchmark data to allow companies to compare their results with their peers. This paper summarizes the original 2015 Global Congress on Process Safety (GCPS) benchmarking data and outlines a plan to gather future data.

2015 Global Congress on Process Safety Benchmarking

One set of benchmarking data was developed at the 2015 Global Congress on Process Safety and included 68 complete surveys. Key excerpts from that benchmarking effort are included below. The full presentation can be found at http://www.aiche.org/sites/default/files/docs/pages/gcps2015vision2020.pdf The benchmarking conducted at the 2015 GCPS indicated a few points.

As seen in Figure 2, there was room for improvement in each of the five tenets. The red dashed line is the highest single average score from a respondent. Green is the lowest single average score. Black is the average of all the respondents for a specific survey.

As seen in Figure 3, the CCPS Vision 20/20 was experiencing good exposure globally and across different types and sizes of companies.

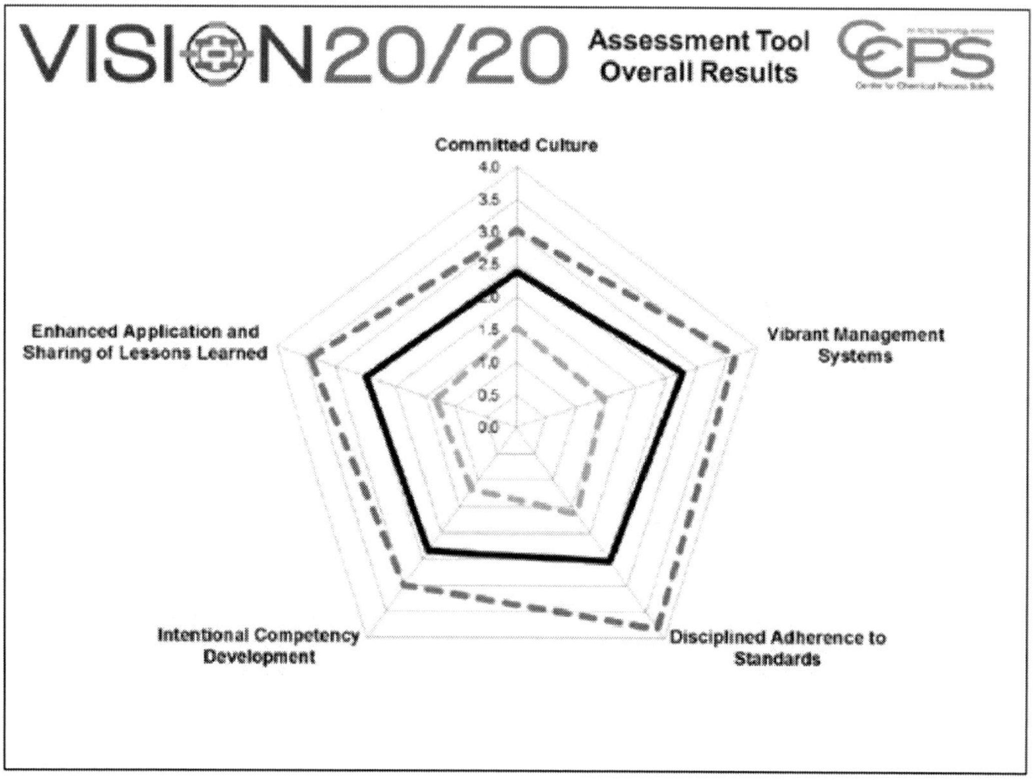

Figure 2. Overall Results from 2015 GCPS Benchmarking

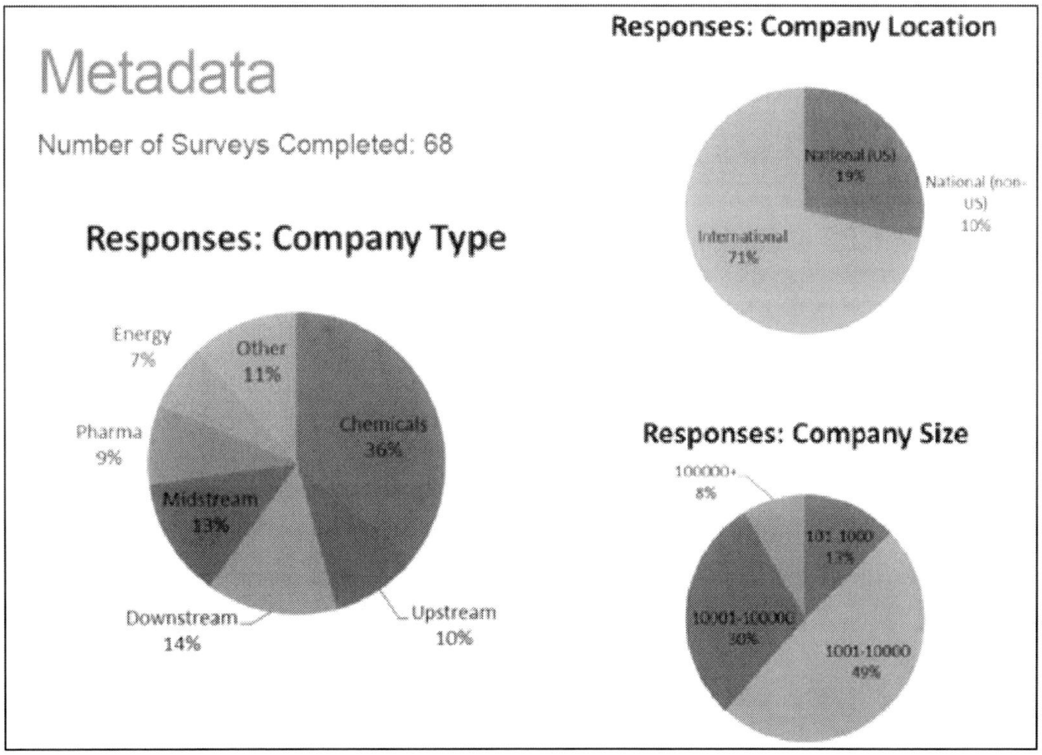

Figure 3. Metadata from 2015 GCPS Benchmarking

2017 Benchmarking

In the January and February of 2017, a CCPS Business Blast and an email to the TSC requested participation in a CCPS Vision 20/20 benchmarking exercise. Thank you to those who responded. Unfortunately, there were very few responses and so we are unable to share results in this paper.

As we have repeatedly heard the request for benchmarking data, the CCPS Vision 20/20 committee would like to extend the benchmarking participation request to the 2017 GCPS audience. Both during the 2017 GCPS through hard copy surveys and following the conference via email, the 2017 GCPS attendees will be encouraged to participate in a benchmarking exercise by completing the survey and submitting (anonymously) their data. These benchmarking results will be shared on the CCPS Vision 20/20 webpage.

This benchmarking will use the following metadata which will enable comparison with the benchmarking results from 2015.

- Company type: Refining, Chemical, Upstream, Midstream, Biofuels, Wind, Integrated Energy Company, Power Generation, Pharmaceutical

- Operating region: North America, Central & South America, Europe, Asia, Australia, Global

- Company size: 101 to 1000 employees, 1001 to 5,000 employees, 5,001 to 10,000 employees, 10,001 to 100,000 employees, greater than 100,000 employees

4 Conclusion

Vision 20/20 provides an opportunity for companies to reflect on the characteristics of those companies with great process safety performance. It encourages us all to consider how WE can improve our process safety performance. We have seen that companies have opportunities for improvement. Many companies still struggle with learning from incidents and many companies have opportunities for improvement on standards for existing equipment.

The assessment tools support identification of areas for potential improvement through self-evaluation. Many have asked for benchmarking data from the assessment tool to allow comparison with other companies. We encourage the 2017 GCPS attendees to take the opportunity to practice the CCPS Vision 20/20 theme of ***responsible collaboration*** by participating in the current benchmarking request. Results will be shared on the CCPS Vision 20/20 web page.

5 References

[1] Cheryl Grounds, Samantha Scruggs, and Jack McCavit. "CCPS's Vision 20/20: Improving YOUR vision so WE ALL can see better", 2016 Mary Kay O'Connor Process Safety Center International Symposium, October 25 - 27, 2016

[2] Cheryl Grounds, Jack McCavit, and Samantha Scruggs. "Vision 20/20", GCPS 2015

[3] Jack McCavit, Todd Aukerman, Jeff Fox, and Rukyah Hennessey. "Used, Classic, or Antique; Solving the Enigma of Disciplined Adherence to Standards for Existing Equipment", GCPS 2016

6 Acknowledgements

[1] Todd Aukerman, Senior Process Safety Engineer, LANXESS

[2] Shannon Ross, Process Safety Technologist, Huntsman

[3] Fred Henselwood, Process Safety Leader, NOVA Chemicals

Waste Solvents to Trash Haulers: Lessons Learned from Hazardous Waste Accidents

Delmar "Trey" Morrison, Ph.D., P.E.
Exponent, Inc.
4580 Weaver Parkway, Suite 100, Warrenville, IL 60555, USA
tmorrison@exponent.com

Michael Stern, Ph.D., P.E.
Exponent, Inc.
Natick, MA, USA
mstern@exponent.com

Carmen H. Osorio-Amado, Ph.D.
Exponent, Inc.
Los Angeles, CA, USA
cosorio@exponent.com

Prepared for Presentation at
American Institute of Chemical Engineers
2017 Spring Meeting and 13th Global Congress on Process Safety
San Antonio, TX
March 26 – 29, 2017

Waste Solvents to Trash Haulers: Lessons Learned from Hazardous Waste Accidents

Delmar "Trey" Morrison, Ph.D., P.E.
Exponent, Inc.
4580 Weaver Parkway, Suite 100, Warrenville, IL 60555, USA
tmorrison@exponent.com

Michael Stern, Ph.D., P.E.
Exponent, Inc.

Carmen H. Osorio, Ph.D
Exponent, Inc.

Keywords: Hazardous waste, incident investigation, chemical reactivity

Abstract

Process safety management (PSM) of hazardous chemicals is a core tenet of the chemical, oil, and gas industries. Organizations in these industries have increasingly woven the elements of PSM into their strategic vision, day-to-day operations, and interactions with contractors and vendors. This cultural evolution has, in part, been driven by major accidents in these fields. In these industries, the chemicals, their hazardous properties, and their potential interactions, are usually either well-defined or capable of being well-defined. Control and mitigation strategies for those chemicals can be developed based on this available knowledge. In contrast, companies that handle wastes have experienced catastrophic accidents due to the ill-defined nature of the hazards. This industry handles the cast-offs, dregs, and refuse, which makes identification of the hazards and prevention of incidents more challenging. How do you implement a PSM program when your input streams can vary due to processes at a plant you do not control or monitor? Does the EPA (Environmental Protection Agency) Resource Conservation and Recovery Act (RCRA) provide sufficient guidance of safely handling of wastes? The overriding theme in the incidents is that PSM is more complex to apply to unknown, mixed, and uncontrolled chemicals, both solid and liquid. To assist companies that handle wastes, this paper provides several case studies of fires and explosions involving multiple waste material handling processes at different facilities. The incidents involved numerous matrices from solid industrial wastes, EPA RCRA hazardous wastes, to waste solvents. An analysis of the lessons learned will be presented and used to provide realistic guidelines to reduce safety risks based on a pragmatic implementation of process safety management principles. The paper will also compare OSHA (Occupation Safety and Health Administration) PSM with EPA RCRA approaches to characterizing wastes versus managing the hazards, and will emphasize those elements that played critical roles in the accident case studies.

1 Introduction

This paper discusses several case histories of process incidents occurring at waste handling or processing facilities and compiles lessons learned from those accidents. Depending upon the source, the industry, and the regulatory classification of the waste, industrial waste may pass through a municipal waste transfer station to a landfill or be collected, aggregated and then incinerated by a RCRA (Resource Conservation and Recovery Act) [1] permitted hazardous waste Treatment, Storage, and Disposal Facility (TSDF). Managing waste is inherently more challenging than managing typical chemical process hazards where the identity and properties of the process chemicals can be effectively defined. While many types of hazardous waste facilities test for the properties of incoming waste streams, these tests can be challenging if the waste is phase separated, if the waste is still reacting and changing over time, or if the waste is incompletely identified by the generator. Similarly, investigations of accidents are complicated by the fact that wastes involved in the accident could have contained unknown chemicals with unknown properties.

Several case studies are presented in this paper. The accident locations range from waste transfer stations, waste storage facilities, waste treatment facilities, and a cement manufacturing facility using waste as a fuel source. The wastes involved in the accidents ranged from municipal waste, unregulated solid and liquid wastes, to regulated hazardous solid and liquid wastes. Significant fires and explosions occurred in these incidents. The overarching causes of these incidents were due to the gap between the true properties and the recognized properties of the wastes. While the task of characterizing waste streams may be challenging, the overall process safety and incident investigation principles remain similar. In this paper, we start with a typical decision tree analysis for screening chemical reactivity hazards and modify it for use for waste processing facilities. We will then discuss several case studies involving fires and explosions in waste handling and treatment and provide lessons learned.

2 Background

Application of Process Safety Management (PSM) principles in the chemical industry has been codified for highly hazardous chemicals above threshold quantities through Title 29 Part 1910.119 of the Code of Federal Regulation since 1992 [2]. The Center for Chemical Process Safety (CCPS) subsequently expanded on the process safety model presented by the regulations to develop twenty elements of risk-based process safety [3]. The overlap between the CCPS process safety model and the OSHA PSM regulatory elements (bolded) is represented in the list below (Table 1).

Table 1. Overlap between OSHA (bold) and CCPS PSM elements

1. Process Safety Culture	2. Contractor Management **(h) Contractors**
3. Compliance with Standards	4. Training and Performance Assurance **(g) Training**
5. Process Safety Competency	6. Management of Change **(l) Management of change**
7. Workforce Involvement **(c) Employee participation**	8. Operational Readiness
9. Stakeholder Outreach	10. Conduct of Operations
11. Process Knowledge Management **(d) Process safety information**	12. Emergency Management **(n) Emergency planning and response**
13. Hazard Identification and Risk Analysis **(e) Process hazard analysis**	14. Incident Investigation **(m) Incident investigation**
15. Operating Procedures **(f) Operating procedures**	16. Measurement and Metrics
17. Safe Work Practices **(k) Hot work permit**	18. Auditing
19. Asset Integrity and Reliability **(j) Mechanical integrity**	20. Management Review & Continuous Improvement

Deficiencies in many of these elements can be traced back to the root causes of hazardous materials incidents, as will be illustrated in the case studies in this paper. However, it is uncommon for TSDFs and other waste handling facilities to formally apply PSM to their processes. Usually, facilities follow a prescriptive approach to waste characterization and handling that is based upon the EPA (Environmental Protection Agency) regulations and relies heavily upon the waste generator's characterization of the waste properties [4]. This reliance is not unusual when considering the quantity of distinct and unique waste streams that an individual municipal facility or TSDF may handle and dispose.

The primary hazard identified in the case studies is delayed reaction leading to pressure generation, heat generation, and unintended ignition of packaging or the waste material itself. This is commonly referred to as spontaneous ignition or spontaneous combustion since a piloted ignition source is not necessary. Trash and municipal waste handling facilities are frequently susceptible to delayed fires due to spontaneous combustion. Such ignition events may be caused by unintentional mixing of incompatible chemicals or exothermic oxidation of self-heating materials, such as vegetable-derived oils. For example, the National Fire Protection Association (NFPA) estimated that 4,460 annual fires involving spontaneous combustion of trash occurred between 2005 and 2009 [5].

Recent studies have identified several trends in hazardous waste TSDF fires and explosions [6], [7], [8]. Incidents can occur in any stage of the process from initial storage, waste repackaging, or transfer to the final treatment stage (i.e., incinerator, calciner, or kiln). The consequences of a fire or explosion can accelerate rapidly, depending upon the nature of the initial waste and its proximity to surrounding waste materials, surrounding waste materials' properties, and the adequacy of fire protection and emergency response actions. The causes of fires and explosions typically derive from unknown or unrecognized chemical reactivity issues with the waste material in question.

2.1 Waste generation in the United States

According to the US EPA, 26,237 generators produced 33.6 million tons of hazardous waste, of which over 6 million tons were non-wastewater [9]. Of the 6 million tons, 26% is combusted for energy recovery, 19% is put into landfills, and 13% is incinerated. Over 76% of the non-wastewater hazardous waste is treated off-site by a TSDF.

The EPA statistics also show that 258.5 million tons of non-hazardous Municipal Solid Waste (MSW) was generated in the United States in 2013. While the per capita waste has slightly decreased since the year 2000, the total annual generation continues to climb due to population growth [10]. Approximately 89.4 million tons (35%) of the MSW were recycled in 2013. Another 13% of the total MSW was combusted with some amount of energy recovery. The remaining 52.6% was placed into landfills.

2.2 Resource Conservation and Recovery Act

A hazardous waste can be any waste material with a potential to harm human health or the environment, either by itself or through interaction with other factors. However, for regulatory purposes in the U.S., a waste material must meet specific criteria to be considered as a hazardous waste. These conditions, summarized in Figure 1, are defined by the EPA under the RCRA in Title 40 of the Code of Federal Regulations [1].

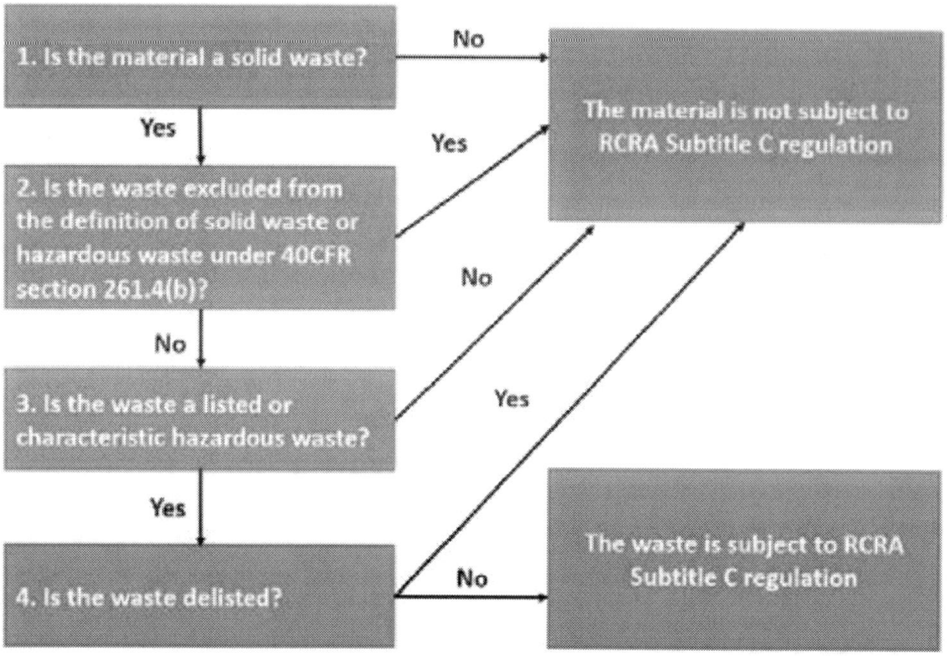

Figure 1. EPA's Hazardous waste identification process [11]

Per the EPA, the first condition of a hazardous waste is to be a solid waste, which paradoxically includes not only wastes that are physically solid, but also wastes that are liquid, semi-solid, or contained gaseous material. A hazardous waste is also required to meet certain characteristics based on its toxicity, ignitability, reactivity, and corrosivity, and needs to be listed in either the F, K, P, or U lists found at the 40CFR section 261 [12].

Some solid wastes are specifically excluded from RCRA regulatory compliance due to different reasons such as impracticability, economy impact, lack of information to justify applying the regulation, regulation by other laws, or they were removed (delisted) from this category. These exclusions refer to conditions (2) and (4) from the hazardous waste identification process shown in Figure 1. For example 40 CFR 261.4(b) excludes household waste and solid waste from commercial or industrial sources (that don't contain solid hazardous waste). It also excludes fly ash waste. These types of waste material may pose physical and chemical hazards, but they are not regulated under RCRA.

2.3 RCRA chemical compatibility testing

As part of the RCRA regulations, hazardous waste generators are required to classify their waste. When TSDFs receive waste, they must perform pre-acceptance analysis to confirm the waste's characteristics and identify incompatible wastes [13]. The EPA states "Compatibility testing may include determining water reactivity (e.g., foaming, heat generating, explosivity), oxidation-reduction potential, spot tests for cyanide and sulfide, among others" [14]. The EPA provides a list of some common potentially incompatible wastes in Appendix V to Part 264. When analysis is required, the EPA has published guidance on how to qualitatively evaluate the compatibility of wastes [15]. The guidance document uses a two-part approach: step-by-step compatibility analysis procedures and then a compatibility chart that considers 41 reactivity groupings. To simplify this analysis when chemical testing is not required, free computer software packages are available for performing quick assessments of chemical compatibility [16].

2.4 Guidelines for evaluating chemical reactivity hazards

According to the U.S. Chemical Safety Board (CSB), a chemical reactive hazard can be defined as "*a sudden event involving an uncontrolled chemical reaction with significant increases in temperature, pressure, or gas evolution that has the potential to, or has caused serious harm to people, property or the environment*" [17]. In order to decide whether a chemical can be considered a reactive hazard or not, the CCPS published a preliminary screening test based on a series of questions summarized in Figure 2 [18].

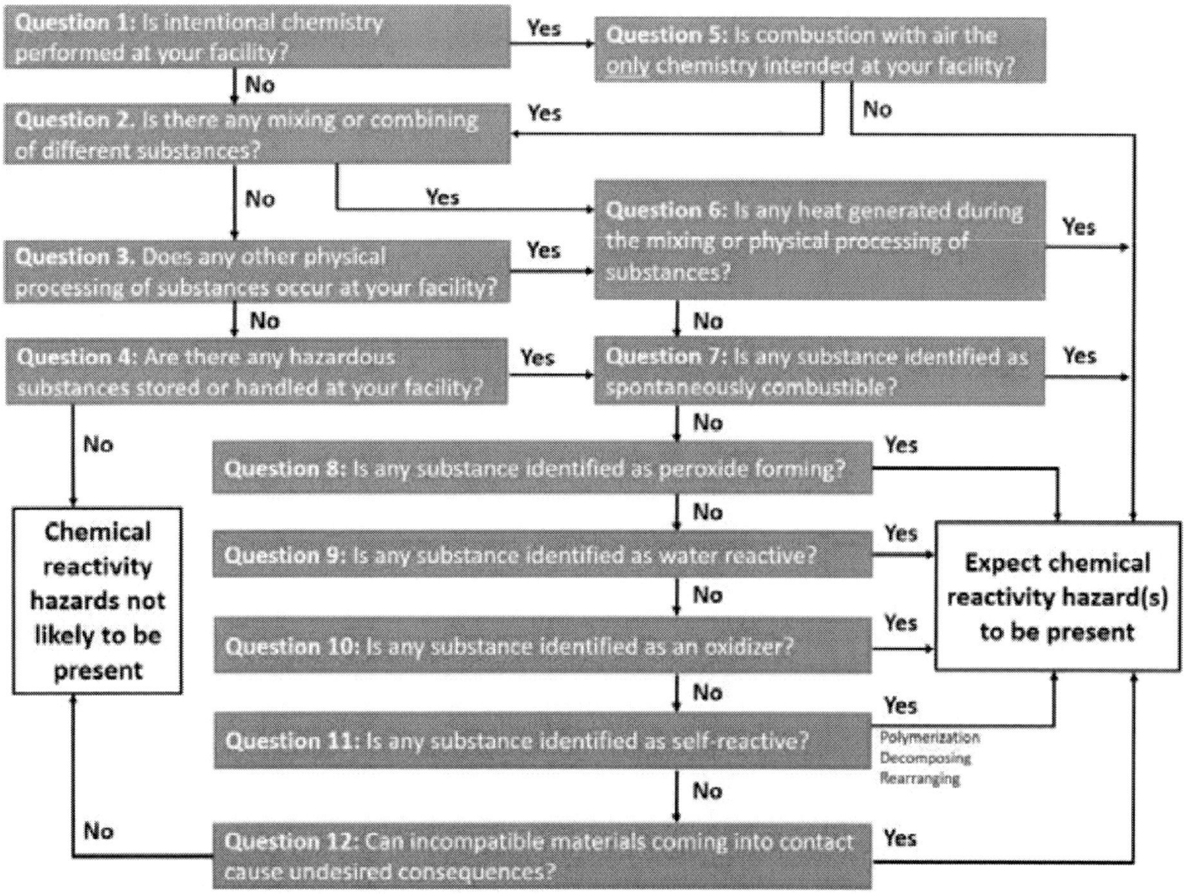

Figure 2. Preliminary screening for chemical reactivity hazards [18].

This test should be conducted on an ongoing basis, in particular when there have been changes in the process/conditions or a new chemical is introduced into the facility. If a chemical reactive hazard is expected to be present at the facility, the CCPS also suggests the implementation of a chemical reactivity management program structured consistent in Figure 3.

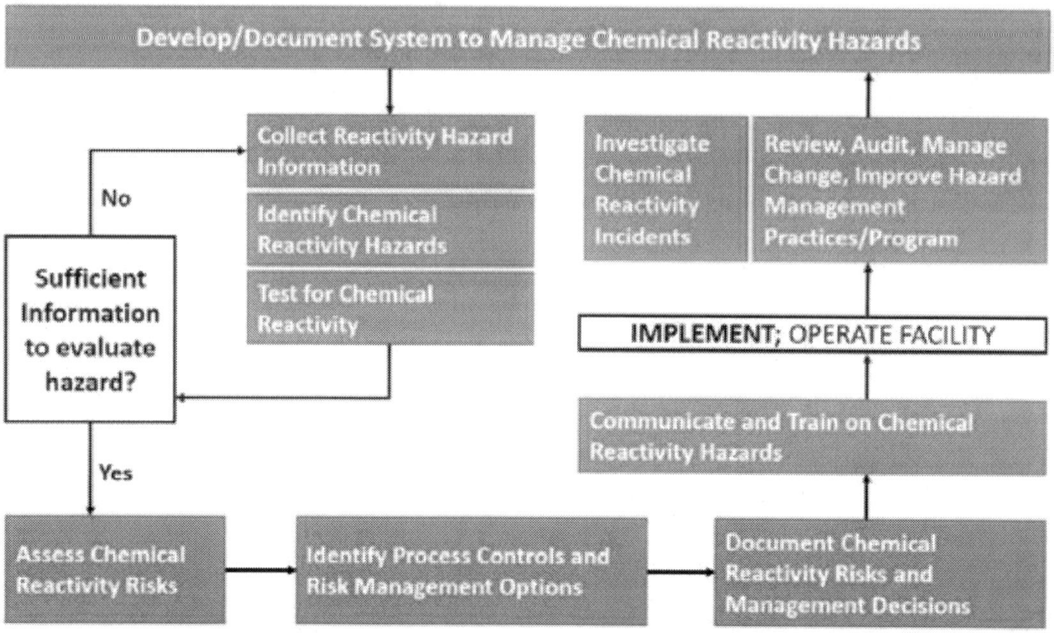

Figure 3. Flowchart for implementing chemical reactivity hazard management [18].

2.5 Why and how an event happened? Protocols for investigating reactive chemical incidents

Given the nature of fires and explosions, especially when involving hazardous waste materials, necessary evidence may be destroyed or lost during the incident or subsequent emergency response. Thus, conclusive determination of the chemistry and proximate causes may be impossible. The investigation process must rely upon the scientific method to develop hypothetical causes, test those causes, and determine the most likely causes and contributing factors.

Two commonly relied upon resources for performing incident investigations per the scientific method are NFPA 921 *Guide for Fire and Explosion* Investigations [19] and the CCPS publication Guidelines for Investigating Chemical Process Incidents [20]. Figure 4 shows the progression of an investigation as described in Chapter 4 of NFPA 921. Two essential concepts, inductive reasoning and deductive reasoning, are used to develop and test hypotheses, respectively. Inductive reasoning is using the analyzed data to develop a consistent hypothesis. Deductive reasoning is used to test the hypothesis by comparing it to all known facts of the accident and scientific literature. NFPA 921 cautions that "[t]he investigator does not have a valid or reliable conclusion unless the hypothesis can stand the test of a careful and serious challenge." Recent editions of NFPA 921 have also added cautionary language regarding the use of *negative corpus* also known as the process of elimination. Investigators should not confuse the lack of an alternative hypothesis for affirmative proof of a hypothesis with no supporting evidence (but that cannot be ruled out with the available information).

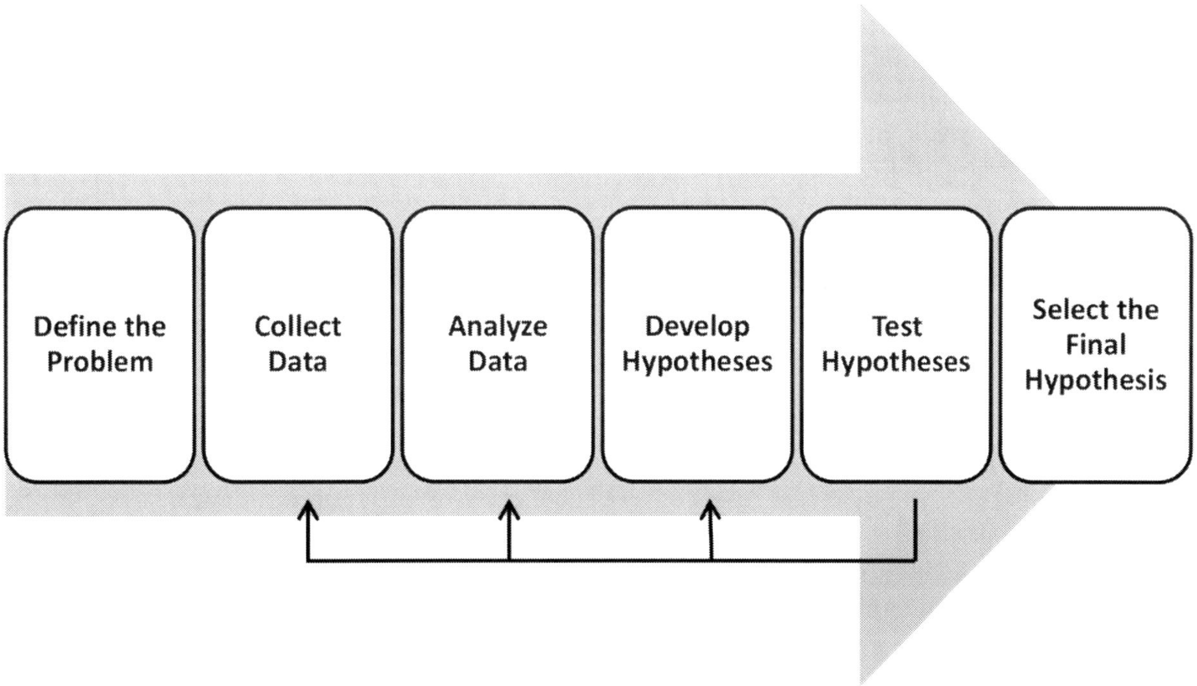

Figure 4. The scientific method of investigation.

The CCPS Guidelines text provides a similar scientific methodology for accident investigation but also offers information specific to industrial accident investigation. For example, information for designing incident management systems within a corporation is provided. Specific investigation and root cause analysis tools commonly employed for investigations, organized by the level of structure, are shown in Figure 5.

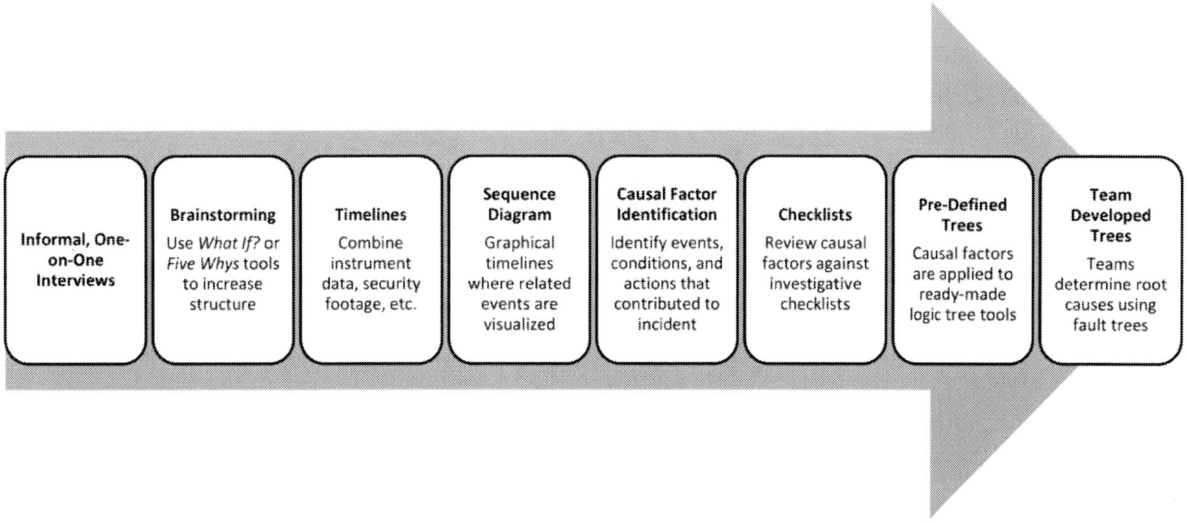

Figure 5. Investigation tools for process incident investigations.

The case studies below describe the causes and lessons learned from MSW handling, solvent reclamation, and TSDF incidents. All of the incidents were characterized by fires or explosions, which caused significant damage and posed potential for catastrophic injuries or fatalities to onsite personnel. Many of the incidents were investigated by external public agencies in addition to the company investigations. The authors also participated firsthand in the investigations of the incidents. Where possible, lessons learned from the investigation process as one element of PSM are shared.

3 MSW Transfer Facility Fires

Two fires, separated by three days, occurred at a MSW transfer station. The first fire involved a trailer of waste awaiting transfer to the landfill, and the second fire involved a different trailer and a large portion of the building and waste contained within. The facility was unsophisticated when considering the element of process safety management, which is understandable because it did not handle reactive materials. As a consequence, early investigation of the causes of the fires relied upon supposition and poor science to conclude that one waste stream was the cause of the fire. The incorrect conclusion adopted by the facility meant that the real cause went unaddressed until a later, thorough engineering investigation occurred.

The waste transfer station was a large, open warehouse structure with two sunken loading bays. The facility handled trash classified as municipal solid waste (MSW) and construction and demolition (C&D) debris. Waste materials were piled on the concrete and moved around using a front end loader. Different materials were commingled by the facility operators, both in piles and in the open trailers located in the loading bays. The commingled waste was periodically hauled to a municipal landfill. The facility did not accept RCRA hazardous waste or liquid wastes. The local regulations and the permit for the facility both acknowledged the reasonable possibility of "hot loads" - in other words, the receipt of loads of waste material that are undergoing self-heating or are already smoldering.

The first fire involved the contents and the rear of a trailer (Figure 6), which was located in a loading bay overnight. The trailer contained a mixture of bottom ash from a fluidized bed coal combustor and other unknown MSW and C&D materials. No formal investigation was conducted to attempt to determine the cause of the trailer fire. The waste material in the trailer was not fully extinguished or overhauled by the fire department, and the facility located the trailer away from the building so that it could eventually burn itself out. The fire witness statements indicated that the contents of the trailer continued to burn for the next several days. The fire damage patterns to the trailer are consistent with a fire originating at the rear wheels, perhaps due to overheated brakes. Three days after the initial fire, a second fire was discovered after a different trailer was loaded with similar waste materials from within the building and held overnight in the undamaged loading bay. The second fire burned to completion and destroyed the building.

Figure 6. Photograph of the trailer damage resulting from the first fire.

An early investigation of the second fire concluded that the combustor bottom ash was the equivalent of pure lime (CaO), caused both fires due to exothermic heat of hydration, and should have been classified as a RCRA hazardous waste. Pure CaO is referred to as "lime" or "quick lime." The reaction of CaO with water vapor results in solid calcium hydroxide, $Ca(OH)_2$, while mixing with excess water creates an aqueous solution, according to the chemical reaction:

$$CaO_{(s)} + H_2O \rightarrow Ca(OH)_2$$

This reaction is exothermic, meaning that it generates heat, with a heat of reaction of -65.27 kJ/mole of CaO reacted [21]. The quantity of heat released is directly proportional to the quantity of CaO and H_2O undergoing the reaction.

The transfer station's incorrect initial conclusion was based on the description of the waste by the generator as "agricultural lime," which is commonly used to describe calcium carbonate ($CaCO_3$), hydroxide ($Ca(OH)_2$), and oxide (CaO) used in farming. Further, the early investigation concluded that the mischaracterization of the waste was a direct cause of the fires. A subsequent engineering investigation analyzed that conclusion and also identified several plausible causes including an electrical fire, spontaneous combustion of MSW or C&D, or re-kindling of the smoldering remains of the first fire after being transported across the facility.

Chemical analysis of the combustor bottom ash and a review of the literature indicated that the ash did not contain sufficient CaO to cause spontaneous combustion through hydration and was excluded under 40 CFR 261.4(b). The residual CaO in combustor tram will undergo hydration by reacting with water (H_2O) in either a vapor (e.g., atmospheric) or liquid form. The hydration reaction for residual CaO would run largely to completion within a few days when exposed to the outdoor atmosphere [22]. Comparison samples of fresh combustor bottom ash revealed that it contained approximately 10% CaO by weight.

If concentrated CaO is mixed with liquid water, the hydration reaction can lead to very high temperatures capable of igniting combustible materials. However, it is unlikely that these conditions existed in the current matter. The maximum temperature reached by the combustor ash via this reaction, if any unreacted CaO is present, will be tempered by the presence of liquid water and surrounding solid material (i.e., $CaCO_3$, $Ca(OH)_2$, $CaSO_4$, sand, and onsite waste materials). The water will absorb the released heat and act to cool the material as the water evaporates; the other solid species will act to absorb the liberated heat and increase in temperature.

The false conclusion that the ash waste was CaO, thus causing the fires, highlights the importance of precise descriptions for waste streams even if they are not bound to the RCRA regulation. Effective description can help investigators, who may be unaware of the precise waste compositions, more effectively assess the important properties of the waste stream after an accident. Similarly, investigators must be careful when examining the properties of non-RCRA waste streams as their descriptions may be more colloquial. Conclusions should not be drawn until a more detailed composition of the waste can be obtained through chemical analysis or discussion with the generator.

4 Solvent Recovery Plant Vapor Cloud Explosion

In 2009, a release of ignitable liquid and vapors, followed by catastrophic vapor cloud explosion and subsequent fires, took place in a solvent recovery plant. The explosion severely injured two employees, destroyed part of the plant (see Figure 7), and broke windows in neighboring facilities and residences. The explosion was investigated by the Chemical Safety and Hazard Investigation Board (CSB) [23].

The solvent recovery plant (Figure 8) was located at a facility that handled hazardous and non-hazardous waste materials that were received via tanker truck, drums, totes, and lab packs. The facility performed fuel blending, solvent recovery, and waste consolidation. Incoming dirty solvents were unloaded into tanks in the dirty tank farm. Clean solvent was recovered from the dirty solvents via one of three distillation units (Unit 1, Unit 3, and Unit 4). The recovery process involved the circulation of distillation bottoms back into the source tank in the dirty tank farm, while purified "clean" solvent was condensed and transferred to the clean solvent tank farm. Final waste liquids and sludge were shipped offsite to a waste incinerator as blended fuels.

At the time of the incident, the three Processing Units were being run by a single outside operator. Unit 1 was used to recycle a paint thinner material whose primary compound was acetone, while Unit 3 was processing a tetrahydrofuran (THF)/water mixture. Unit 4 worked as a closed circuit reflux-distillation unit where a mixture of THF and water near the azeotropic concentration was processed.

Figure 7. Aerial view of the solvent recovery plant after the incident from the CSB report [23].

Figure 8. Facility layout (not to scale) [23].

Three of the six workers onsite reported hearing a loud noise like a pressurized release, accompanied by a strong solvent odor and a visible aerosol/or vapor cloud coming from the Dirty Tank Farm prior to ignition. Security camera footage from onsite and a nearby business supported the conclusion of a pressurized release, likely of Tetrahydrofuran (THF) solvent mixture; however, the exact composition and source of the flammable material released could not be determined.

The incident investigation concluded that the vapor cloud spread out from the Dirty Tank Farm, entered the laboratory/Operations Building, and ignited inside the Boiler Room causing, an initial explosion. Consequently, additional flammable material within the Dirty Tank Farm was also ignited causing a blast pressure wave (vapor cloud explosion) and subsequent pool fire that burned for several hours.

The CSB's findings from this incident included the following facility siting concerns:

- Lack of hazard analysis to determine the risks associated to the proximity of occupied buildings to processes: The most critical injured employees were at the Laboratory/operations building at the time of the incident. The nearby building was occupied with non-essential personnel during the day shift.
- Inappropriate area classification: Two boilers were in service in the Laboratory/Operation Building, where the ignition source was identified. The flammable vapor cloud was ingested and ignited by one of the boilers.

Although the facility was covered by the OSHA PSM standard, the CSB found that the TSDFs did not have industry-specific guidance for fire safety and facility siting.

5 Fires and Explosions in Storage Operations at TSDFs

Several fire incidents have occurred during storage of hazardous materials at TSDFs. The ultimate fate of hazardous materials is often destruction through incineration. The type of incinerator and local environmental emission limits dictate handling activities with the hazardous waste. To manage the limits on the heat of combustion for the incinerator, heavy metal emissions, and other factors, TSDFs commonly repackage wastes by either combining similar waste streams in the same container or by breaking up larger containers into smaller container sizes. For example, if the bulk heat of combustion of the 55-gallon drum of waste received by a TSDF is too high for the incinerator's operating limit, then the 55-gallon drum must either be metered (if a liquid) into the incinerator, or the drum must be subdivided into smaller containers (e.g., 30-gallon drums).

Thus, hazardous materials often pass through the following stages at a TSDF:

1. Receipt and characterization
2. Temporary storage
3. Repackaging – either bulk up or break down
4. Storage awaiting disposal
5. Final disposal

The waste generator must provide specific information on the hazardous waste that they are providing to the TSDF. The TSDF oftentimes works with the generator to help characterize the waste stream as part of the initial contracting process. The TSDF typically samples and performs basic characterization of individual wastes as they are received. However, this process is imperfect because the TSDF cannot be guaranteed that a set of samples from distinct locations in a waste container represent the bulk of the waste in that container. The potential deviation of actual properties from the results of this sample may grow significantly for larger-sized containers (e.g., a liquid tote or a 1 m^3 box), wastes that are heterogeneous, wastes with layers or pockets of different composition, or large volume wastes that are not completely sampled (e.g., sampling a subset of containers).

After initial contracting and waste specification, a TSDF will develop a plan for handling that waste stream, which will include sampling, characterization, and verification that as-received waste meets the contract specification. This process falls under Process Knowledge Management in the 20 elements of CCPS PSM. Deviations of as-received waste from the specification should be addressed under a Management of Change element, but the TSDF may not have a robust change management policy. Additionally, changes from the specification may not be identified during the characterization process as discussed above.

Incident investigation after a hazardous materials incident at a TSDF can be very challenging. A TSDF storage area may contain hundreds or even thousands of individual waste containers (see Figure 9). A fire or explosion in the storage area will destroy or damage the waste containers, displace waste containers, and mix the residue from involved containers. All these aspects make determination of the cause of the incident difficult. The likelihood of determination is usually directly related to whether the incipient stages of the fire or explosion were witnessed. The following three case studies provide incident summary and investigation results for a progressively larger TSDF fire incidents in order to characterize the difficulty in cause determination and lessons learned.

Figure 9. Example of TSDF storage warehouse.

5.1 Delayed Ignition - No Repackaging – Cause Undetermined

An investigation was performed on a fire that occurred in a small TSDF's storage bay. The facility handled a wide variety of wastes including flammables materials and acids. The facility included storage bays, laboratories, treatment equipment and office areas. In the storage bay where the fire occurred, several different types of containers were present including:

- Plastic buckets
- Plastic, steel, and paper fiber drums
- Steel encaged plastic totes known as intermediate bulk containers (IBCs)

The 55-gallon drums were placed on wooden pallets in sets of four. The pallets of 55-gallon drums and IBCs were stacked two or three levels high. A photograph of the bay after the fire is shown in Figure 10.

Figure 10. Photograph of storage area after fire.

Based on witness statements, the fire in the storage bay was preceded by a loud hissing sound and a pressurized release of a reddish dust/mist from a container near the center row of the image. The witness did not see the exact container where the release originated, but was able to place it within a few pallet stacks in the area. Immediately after the pressurized release, a fire ensued in the area. Due to fire suppression efforts, the fire damage was contained to the storage bay where it originated. The investigation focused on the adjacent rows containing the waste drums that were damaged. The exact location of each waste container in the neighboring pallet stacks was reconstructed based upon storage records, remaining containers, damage patterns, and witness statements about container placement. The origin was confirmed to be one pallet stack in the center row.

The containers in the center row had not been repackaged or disturbed from the time of receipt until the incident; therefore, commingling of incompatible wastes or mixing of a reactive pocket within the source drum were refuted as potential causes. Delayed decomposition of waste material in one of the 55-gallon drums is the likely source of the pressurization. Delayed decomposition may be the result of chemical incompatibility or thermal instability leading to a temperature rise, gas generating reaction, or both. Eventually, the pressure increased to the level that the source container ruptured. It was most likely that this release occurred from either a steel drum or a plastic drum given their ability to contain some degree of pressurization compared to a fiber drum. After the initial pressurized release and fire, several other containers were damaged and ruptured as a result of the growing fire.

Based on the photo-documentation of the scene, it was possible to identify several potential drums that showed evidence of pressurization and/or rupture. However, due to the extent of the damage, it was not possible to identify the serial number on all of the containers, and therefore, their origin and contents. Examination of the fire damage showed evidence of burned away paint, burned pallets, and melting of non-metal containers. Based on the damage patterns, it was possible to rule out some waste containers including a set of drums containing phosphoric acid. A ground-level area containing four drums was identified as the likely area of origin, but further specificity was not possible due to the extent of the damage and several other drums that could not be conclusively eliminated. In a regular facility, the containers of all the stored containers would be known, providing valuable information to incident investigators. In a waste handling facility, significantly less certainty is available about the exact contents of each container. The ability for many of the waste container to have been the source of the pressure release significantly complicated this specific investigation. The final determination as to which container was the source could not be determined given the degree of damage.

The extent of damage does not always preclude determination of the likely cause of a fire in storage. A similar TSDF facility experienced a fire in a storage bay, which consumed all of the hazardous waste and combustible containers in the area (see Figure 11). The point of origin of the initial fire could be more precisely placed than the incident just discussed because a nearby operator observed the location of the incipient fire at ground level at a specific pallet stack. The source of the fire was most likely a leak of fuming red nitric acid that spilled onto a wood pallet holding the drums. The drums were stacked on combustible wood pallets, and fuming nitric acid is a strong acid capable of igniting such materials. The facility did not identify the potential consequences of a spill of the acid; thus, a noncombustible pallet (i.e., steel) was not selected. Use of a what-if strategy for packaging and storage selection may have prevented the fire (but not necessarily the leak).

Figure 11. Another TSDF storage bay after extensive fire damage.

5.2 EQ Industrial Services Warehouse

A hazardous waste TSDF was complete destroyed due to a fire for which the initial cause was undetermined (see Figure 12). The magnitude of this incident forced thousands of local residents to be evacuated in two days. The CSB investigated the cause of the incident, and identified several hypothetical causes. Their recommendations focused on eliminating those hypothetical causes.

Figure 12. Facility after the incident [24].

Based on witness statements, a strong chlorine odor and plume of smoke came from the TSDF on the night of the incident. Initial firefighters responding to the scene reported that a small fire started in one of the hazardous waste storage bays and rapidly spread to the other locations of the facility, including a bay containing many 55-gallon drums of flammable solvents. The incident investigation revealed that the fire originated in a storage bay where containers of solid chlorine-based pool chemicals and unspent aircraft oxygen generators awaiting for final disposal.

Many incidents have occurred involving storage of chlorine-based pool chemicals (see for example [25]-[28]), and around the time of the incident, the NFPA was pursuing research into fire protection considerations for pool chemicals in warehouse settings [29]. Concentrated calcium hypochlorite or sodium dichloroisocyanurate containing pool chemicals pose potential thermal runaway concerns if contaminated with organic material or if stored at elevated temperature. Given their nature as solid oxidizers, pool chemicals would also serve to accelerate fire growth. The witness observations of a chlorine-odor associated with the smoke likely indicate that these chemicals were present near the fire origin, but the investigators could not conclude that these were cause of the fire.

Although the first material involved could not be determined due to the extent of damage, the CSB investigation findings focused on chemical oxygen generators [30]. The CSB concluded that chemical oxygen generators most likely contributed to the intensification of the fire, which quickly spread to an adjacent bay where flammable materials were stored. The investigation found that the unspent oxygen generators, coming from an aircraft maintenance facility, were not properly packaged by the aircraft maintenance personnel, and were misidentified as general oxidizer waste on shipping paperwork. A lesson learned from this incident was to consider oxygen generators that have exceed their service life as still posing a risk of heat and oxygen generation if accidentally actuated. Therefore, the CSB recommended that unspent oxygen generators should be actuated before they are transported [30].

5.3 Teris El Dorado Warehouse

This incident occurred in a warehouse at a TSDF that contained 4000 to 5000 drums of different wastes. The facility operated two rotary kiln incinerators for solid disposal and thermal oxidation for liquid incineration [31]. The incident was witnessed by an operator who saw a drum rupture, which quickly spread fire throughout the warehouse contents (see Figure 13). The entire contents of the warehouse were consumed, damaged, and/or released during the incident. No injuries or fatalities were reported; however, nearby residents were evacuated during the emergency response.

The investigation found that the fire was caused by spontaneous combustion of sodium chlorate solid waste recovered from a rail car accident site. This material was brought to the facility in 1m^3 boxes to be incinerated shortly before the incident. However, the waste generator incorrectly profiled this material as containing only sodium chlorate in soil; the waste actually contained vegetative matter, rail car components, and portions of railroad ties. The organic contaminants are incompatible with the strong oxidizer and would act as fuel sources, allowing spontaneous ignition of the mixture. The material was repackaged into smaller drums and a portion was moved into the warehouse prior to the fire. The remainder of the material had not been relocated at the time of the fire. This residual material was isolated in a trailer onsite. The trailer containing the residual material underwent spontaneous ignition several days after the original fire incident, which supported the conclusion that the sodium chlorate waste material was the cause of the fire.

Figure 13. Aerial image during fire event [31].

Given the non-homogeneous nature of the waste material received, grab sampling of a single container was insufficient to fully characterize the mixed material. In contrast to liquid sampling, a thorough visual examination of the mixed, non-homogeneous waste material may have identified the presence of organic materials in the oxidizer and soil waste. Further, the use of "soil" in the waste analysis plan submitted by the generator may have triggered a further review by the TSDF since soil typically contains a significant organic component. It is likely that the repackaging activity mixed reactive sodium chlorate with fresh organic fuel materials that led to the delayed spontaneous combustion event. After the incident, the facility enhanced their material analysis and handling plans for wastes containing solid oxidizers.

6 Fires & Explosions during Incineration Processes

6.1 Explosion during Solid Waste Homogenization

An explosion occurred at an incineration facility that processed hazardous and non-hazardous solid and liquid wastes through thermal destruction in a rotary kiln. Drums of waste were homogenized in a nitrogen-inerted shredder. Drums were scheduled to be incinerated and assembled onto pallets based upon the waste characteristics. The system automatically moved individual pallets into an airlock, purged the airlock with nitrogen, then dumped the pallets into the already inerted shredder chamber. The shredder was used to shred steel drums as well as plastic drums. After shredding, the homogenized waste was conveyed to the rotary kiln via a screw conveyor.

The pallets, packaging materials, and waste materials are combustible materials, and flammable materials may also be processed based on the incinerator waste schedule. Due to the potentially flammable nature of the waste and the likelihood of mechanical sparks and heat generation from the shredder, the system was inerted with nitrogen gas. The nitrogen inerting function was performed by using a fixed timing of nitrogen addition. Although the system did not use an oxygen measurement control loop to dictate the duration of purging, it did use oxygen measurement as a permissive control for the automated conveyor movements and door opening between chambers in the system.

A central incinerator control room at the facility monitored several operating parameters, including the oxygen concentration and temperature within the inerted system. Since one of the primary safety mechanisms for the unit was a reduced oxygen atmosphere, the unit was not intended for processing oxidizing materials including solid oxidizers, oxygen containers/cylinders, explosives, and/or water reactive materials. During one waste load, an explosion occurred in the shredder chamber, which damaged the unit and caused explosion panels to open on the shredder chamber and airlock.

An approximation of the oxygen concentration data and shredder chamber temperature is provided in Figure 14. There was a lag noted in the oxygen sensor response due to the delay for gas to traverse the sampling leg to the sensor. The temperature measurement had less of a response lag; thus, the sudden temperature rise in the shredder chamber actually occurred coincidentally with the explosion and the opening of the explosion vents. On the night of the incident, an additional waste load was fed into the unit, and shredding commenced less than one minute after it was added. Approximately one minute after shredding began, the oxygen concentration started to increase from 2% to over 13% over a period of 20 seconds. According to the recorded data, the temperature in the shredder chamber suddenly increased at approximately the same time as the oxygen concentration increased. This temperature increase is consistent with the heat generated by the explosion within the shredder chamber. An engineering analysis concluded that the most likely source of the rapid oxygen influx was from an item in the waste stream that released oxygen when shredded. All potential sources of air via external sources were examined and rejected.

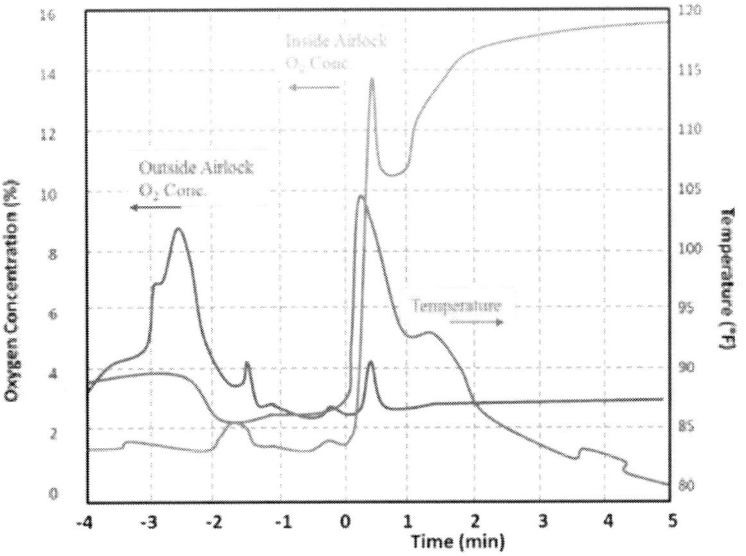

Figure 14. Process oxygen concentration and temperature during incident.

The explosion in the shredding system demonstrated the serious consequences of processing waste with unintended properties. In this case, a system that was intended to operate while inerted inadvertently contained compressed oxygen, resulting in a damaging explosion. While the waste was screened prior being fed into the shredder, the screening process did not identify and divert the oxygen-containing waste. In this case, the explosion mitigation system worked as intended by relieving the explosion pressure through explosion vents. The system was damaged, but the damage was not catastrophic, and there were no injuries.

6.2 Spontaneous Combustion at a Calciner

A fire occurred at a cement production facility that used an engineered solid fuel composited from MSW and other industrial wastes to co-fire in the calciner. The engineered fuel was a blended mixture of solid wastes including paper, plastic, and sawdust as well as non-hazardous waste liquids, slurries, and sludge. Because the waste streams possessed variable energy content, a qualifications process (including laboratory analysis) was used to determine the appropriate mixture for creating a consistent engineered feed. A block flow diagram of the shredding, blending, and mixing process is shown in Figure 15. The engineered fuel had the potential for self-heating due to containing oxidizing materials such as polymerizing agents and vegetable-derived oils. Material handlers inspected the stockpiles for external signs of high temperature due to self-heating and excavated the piles to identify internal high temperatures. Infrared pyrometers were used for surface temperature measurements. The temperatures were monitored with the handheld pyrometers and logged prior to the loading the material into trucks for transport to the cement production facility.

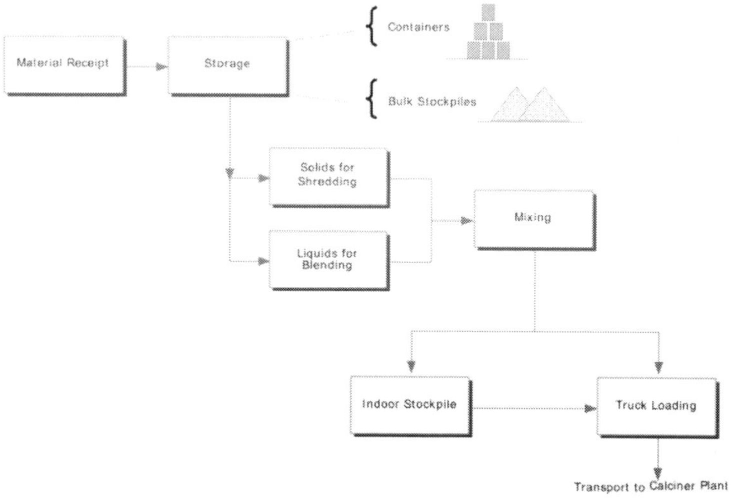

Figure 15. Block flow diagram of the engineered solids feed preparation process.

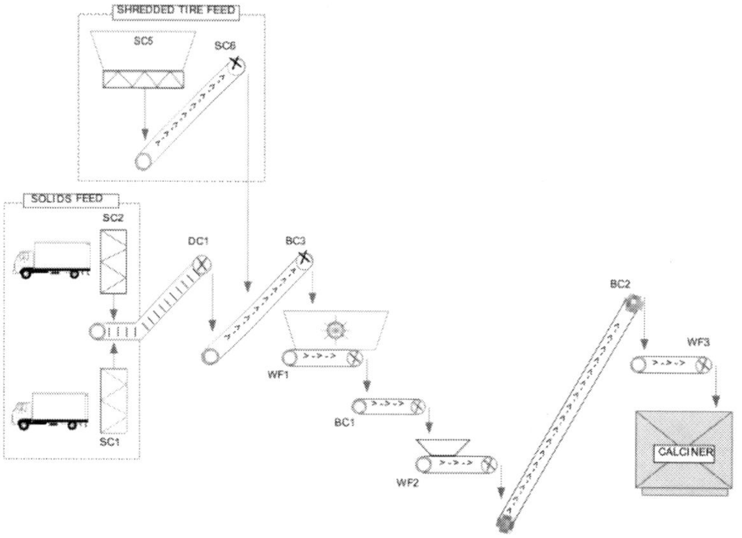

Figure 16. Conveyor flow diagram between truck delivery and the calciner.

The engineered fuel was transported to the cement production facility in tractor trailers for eventual feed into the calciner unit. A conveyor flow diagram is provided in Figure 16. The engineered fuel was unloaded into enclosed hoppers that discharged into horizontal screw conveyors. The horizontal screw conveyors discharged into a central drag conveyor that delivers the fuel to a belt conveyor which also received a separate shredded tire feed. The combined feed then traveled through an additional series of hoppers and conveyors before being deposited on a 1,600 feet long enclosed belt conveyor that climbed from ground level to the upper levels of the calciner structure. The belt of the long conveyor was originally a flame-resistant belt material but had been subsequently patched with non-flame resistant material. The total travel time of the engineered fuel was estimated to be 20-30 minutes by the cement facility's personnel.

A fire was discovered overnight soon after the conveyor system was restarted after a two-day outage. During that two-day period, the belts were left loaded with the engineered fuel and combined feed. When the kiln was restarted, the system was operated for approximately one hour then shut down due to an upstream clog during which a fire was discovered near the head section of the long conveyor. The belt failed and broke as a result of the fire.

An investigation into the fire cause determined that, while various friction-based ignition sources (belts, bearing, etc.) did exist within the conveyor system, the most likely cause of the fire was spontaneous ignition of the engineered feed. This was substantiated because three days after the fire, a stockpile of the engineered feed that had been fed to conveyor prior to the fire enflamed at the storage warehouse. Although staff had monitored stockpiles at the feed mixing facility for external signs of ignition, external methods can be insufficient to detect self-heating inside the pile. After exposing the self-heating pile to oxygen during transport and handling, self-heating may accelerate and lead to spontaneous ignition. The facility implemented a strategy that avoided leaving materials on conveyors for long periods of time. Through the use of conveyor interlock strategies and external dump locations, the system can be operated to empty conveyors in the event of an unintended shutdown.

6.3 TDI Waste Rail Tank Car

An explosion occurred at another TSDF cement plant in the liquid fuels rail tank car unloading area [6], [32]. As consequence, the tank ruptured and was propelled about 750 feet away. Other tank cars, as well as storage tanks located nearby, were also heavily damaged (Figure 17). No injuries or fatalities were reported; however, total property losses were estimated at nearly $8.2 million.

This facility incinerated liquid hazardous waste streams as supplementary fuel in rotary kilns, and used to receive several tank cars containing column bottoms waste from a toluene diisocyanate (TDI) manufacturing process. TDI is a monomer feedstock used in the manufacture of polyurethane polymers. The waste generator had stored tank cars containing this type of waste material for several years, and reprocessed them as the opportunity arose. Prior to the incident, several tank cars were shipped to the TSDF for incineration, and had been successfully emptied, with contents incinerated in the kilns. Despite the TDI waste was partially polymerized in most tank cars, the car's contents were non-homogeneous and did not have identical properties between individual tank cars.

Figure 17. Locations (1) and (2) indicate positions of the tank car's vessel before and after the explosion, respectively [33].

In order to decrease the viscosity of the polymerized TDI, so that it was pumpable to the kilns, the facility had successfully used external steam heating through the built-in steam jackets on the tank cars.. A schematic of the approximately 30,000 gallon tank car is provided in Figure 18. While attempting to unload the incident TDI material, the tank car vessel was steam-heated in two opportunities to ensure the waste mixture was sufficiently fluid to be pumped. After three unsuccessful unloading attempts, the shipment was rejected, but the tank car was left in the unloading location. The explosion incident took place two days after the last heating attempt.

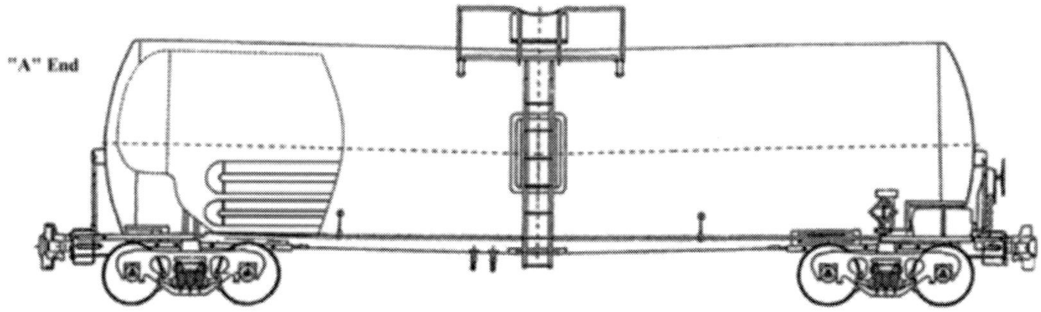

Figure 18. TDI tank car schematic with cutaway showing the external steam heating coils. [20]

Retained samples of the original material and residue ejected during the incident were analyzed, and the presence of self-polymerized TDI, TDI monomer, and ureas were determined. Thermal analysis of the materials indicated that the polymerized TDI and TDI monomer exhibited an onset temperature for thermal runaway that was much higher than possible with the steam heating system. A chemical hazard assessment revealed that the cause of the explosion was the runaway thermal decomposition of a contaminant mixed with the TDI waste. The TSDF operator relied on the waste generator for information regarding the waste composition and safe heating procedures. In this case, the waste generator supplied incomplete waste characterization data and inadequate instructions and warnings for heating and handling the TDI waste.

Four critical factors and lessons learned were identified from this incident [6] [20]:

(1) Use of inappropriate tank car design that was unsuitable for this service, creating difficulties for unloading of TDI solid waste.

(2) Storing of TDI waste for several years in a warm climate

(3) Providing inadequate instructions and warnings for unloading the TDI waste.

(4) Commingling an unknown contaminant with the TDI waste; this last factor was the most serious error in this incident.

7 Discussion and Conclusions

In the case studies presented above, hazardous properties that were not known to exist led to accidents. In several cases, the accidents were the result of a chemical reaction, leading to a release of heat or pressure that caused a storage vessel to fail. Chemical reactions due to the mixing of incompatible chemicals, storage of unstable chemicals, or heating of thermally unstable chemicals occurred in several of the case studies described. While characterization is a significant preventative safeguard for these types of failures, it may not always be possible to determine if a waste stream is completely stable based on testing; therefore, facilities may still need to consider the unlikely event of an exothermic or pressure-generating reaction within their waste streams. This may be accomplished by using separation distance or barriers between certain waste streams, fire protection systems, or fast turnaround from receipt through final processing.

In several incidents, oxidizers including oxygen tank, oxygen generators/tanks and sodium chlorate, are involved in the initiation of the incident or contribute to the severity of the incident. Because many waste streams contain flammable or combustible materials, enriched oxygen atmospheres or the presence of chemical oxidizers increases the likelihood and severity of a fire or explosion. While in many process operations, inerting atmospheres may be an effective and robust protective measure, due to the possibility of oxidizer ingress in mis-identified waste, additional protections such as explosion vents may be required. The following flowchart was created based on the cases presented in this paper, and may serve as a guideline for investigating suspected waste streams and their role on initiating a fire or explosion event.

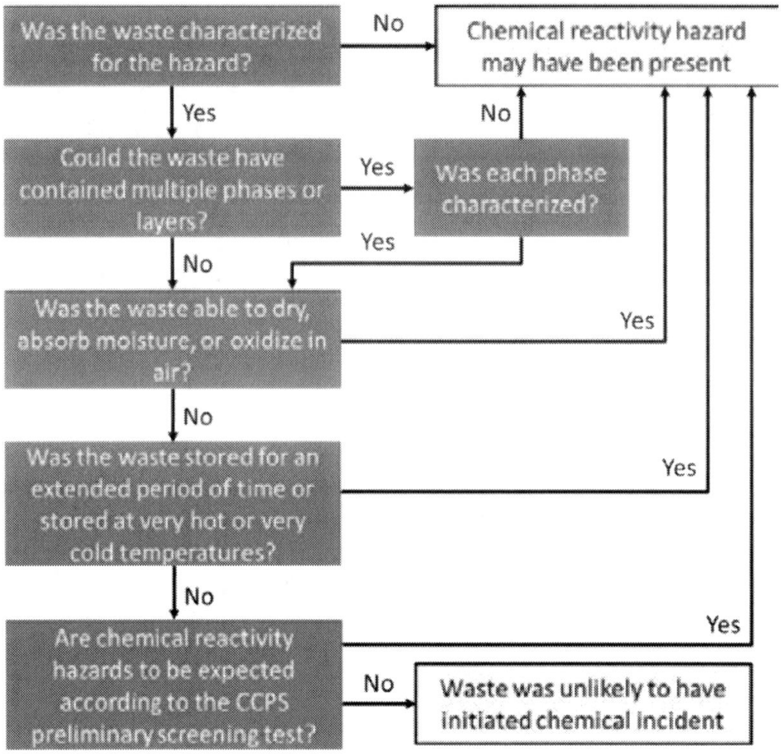

Figure 19. Role of suspected waste stream on initiating a fire or explosion.

8 References

[1] U.S. Environmental Protection Agency (EPA) Resource Conservation and Recovery Act, Title 40 of the Code of Federal Regulations.

[2] 29 CFR §§1910.119

[3] Center for Chemical Process Safety (CCPS). Guidelines for Risk Based Process Safety. Center for Chemical Process Safety, American Institute of Chemical Engineers, New York, NY, 2007.

[4] U.S. Environmental Protection Agency: Resource Conservation and Recovery Act (RCRA) Laws and Regulations, available at https://www.epa.gov/rcra. Accessed on February 15, 2017.

[5] Evarts, B. "Fires Caused by Spontaneous Combustion or Chemical Reactions," National Fire Protection Association, November 2011.

[6] Cox BL, Carpenter AR, Ogle RA. Lessons learned from case studies of hazardous waste/chemical reactivity incidents. Process Safety Progress 2014; 33(4):395–398.

[7] E.C. Buc. "Hazardous Waste Treatment, Storage, and Disposal Facility Fire Code Gap Analysis, Final Report," The Fire Protection Research Foundation, 2015.

[8] Environmental Protection Agency. Report on Emergency Incidents at Hazardous Waste Combustion Facilities and Other Treatment, Storage, and Disposal Facilities (TSDFs). EPA 1999

[9] U.S. Environmental Protection Agency (EPA): 2015 Biennial Report Summary, available at: https://rcrainfo.epa.gov/rcrainfoweb/action/modules/br/national/view. Accessed January 15, 2017.

[10] U.S. Environmental Protection Agency: Advancing Sustainable Materials Management: 2014 Fact Sheet. US EPA, November 2016.

[11] U.S. Environmental Protection Agency, Available at https://www.epa.gov/hw. Accessed on February 15, 2017.

[12] 40 CFR §§261

[13] 40 CFR §§264/265.17

[14] U.S. Environmental Protection Agency. Waste Analysis at Facilities that Generate, Treat, Store, and Dispose of Hazardous Wastes – Final: Guidance Manual (EPA 530-R-12-001). April 2015.

[15] Hatayama H.K., J. J. Chen, E. R. de Vera, R. D. Stephens, and D. L. Storm. A Method for Determining the Compatibility of Hazardous Wastes. U.S. Environmental Protection Agency, Washington, D.C., EPA/600/2-80/076 (NTIS PB80221005), 1980.

[16] National Oceanic and Atmospheric Administration (NOAA), "Chemical Reactive Worksheet," NOAA, available at http://www.aiche.org/ccps/resources/chemical-reactivity-worksheet-40. Accessed January 15, 2017.

[17] Davis E.M., D. Silva, and J. Murphy. The CCPS Chemical Reactivity Evaluation Tool. 8th Global Congress on Process Safety. April 2012.

[18] Center for Chemical Process Safety (CCPS). Essential Practices for Managing Chemical Reactivity Hazards. Center for Chemical Process Safety, American Institute of Chemical Engineers, New York, NY, 2013.

[19] National Fire Protection Agency. NFPA 921: Guide for Fire and Explosion Investigations

[20] Center for Chemical Process Safety (CCPS). Guidelines for Investigating Chemical Process Incidents. 2nd Ed. 2003

[21] Anthony, EJ, L Jia, M Caris, F Preto, and S Burwell, "An examination of the exothermic nature of fluidized bed combustion (FBC) residues," Waste Management Journal, 19 (1999).

[22] Serris, E, L Favergeon, M Pijolat, M Soustelle, P. Nortier, RS Gartner, T Chopin. "Study of the hydration of CaO powder by gas–solid reaction," Cement and Concrete Research 41(2011) 1078-1084.

[23] U.S. Chemical Safety Board (CSB). Veolia ES Technical Solutions, L.L.C. Case Study. Report 2009-10-I-OH. Washington, D.C. 2010

[24] U.S. Chemical Safety Board (CSB). EQ North Carolina Case Study. Report 2007-01-I-NC. Washington, D. C. 2008.

[25] Gray B.F. and B.W. Halliburton. 2000. "The thermal decomposition of hydrated calcium hypochlorite (UN 2880)." *Fire Safety Journal*. 35.3: 223-239.

[26] Gray B.F. 2002. Spontaneous Combustion and Self-Heating. In SFPE Handbook of Fire Protection Engineering, eds. P.J. DiNenno, D. Drysdale, C.L. Beyler, W.D. Walton, R.L.P. Custer, J.R. Hall and J.M. Watts, Quincy, Massachusetts: National Fire Protection Association, 211-228.

[27] Ogle R. A, Carpenter A. R. Toxic gas release caused by the thermal decomposition of a bulk powder blend containing sodium dichloroisocyanurate. Process Safety Progress 2003; 22:75–82.

[28] Ogle RA, Morrison DR, Carpenter AR, Su YS. Missed opportunities in reactive chemical hazard evaluations. Process Safety Progress, March, 2006, 25(1):2-7.

[29] Buc E.C. and D.J. Hoffman, "Development of an Enhanced Hazard Classification System for Oxidizers Research Project," The Fire Protection Research Foundation, One Batterymarch Park, Quincy, Massachusetts, April, 2006

[30] U.S. Chemical Safety Board (CSB). Safety Advisory Dangers of Unspent Aircraft Oxygen Generators (2007-I-NC-01-SA). 2007

[31] U.S. EPA, "Teris El Dorado Explosion," downloaded on March 13, 2017, https://response.epa.gov/site/site_profile.aspx?site_id=NRC745957.

[32] Ogle R.A, Carpenter AR, Morrison DR. Explosion of a railcar containing toluene diisocyanate waste. Process Safety Progress 2004 Dec; 23(4):316-320.

[33] National Transportation Safety Board. Rupture of a Railroad Tank Car Containing Hazardous Waste Near Clymers, Indiana. Hazardous Materials Accident Report NTSB/HZM-01/01. Washington, DC. 2001

479039 ExxonMobil Torrance Refinery FCC ESP Explosion

Abstract:

On February 18, 2015, an explosion occurred in the electrostatic precipitator (ESP) at the ExxonMobil Refinery in Torrance, California near Los Angeles. The explosion injured four workers caused significant property damage to multiple process units within the refinery, and resulted in an off-site accidental release of catalyst dust. As a result of the incident, the FCC unit was down for over one year. The explosion occurred when hydrocarbons from the FCC unit reached the energized ESP while the unit was idling to perform maintenance on equipment that had failed. Multiple failures contributed to the incident. During the ESP explosion, there was also a near miss release of hydrofluoric acid when a large piece of ESP debris fell within feet of a storage vessel storing thousands of gallons of modified hydrofluoric acid in the neighboring alkylation unit. If the storage vessel had failed due to impact from ESP debris following the explosion, hydrofluoric acid would have been released. This presentation will detail key findings from the CSB investigation into the incident.

13TH GLOBAL CONGRESS ON PROCESS SAFETY

Overfill to Overpressure – Investigation Findings and Contributing Events to the CAPECO Explosion

Scott Davis
GexCon US
Bethesda, MD
sgdavis@gexcon.com

Tom DeBold
Claudio Marsegan
Bryant Hendrickson
GexCon US
Bethesda, MD

Prepared for Presentation at
American Institute of Chemical Engineers
2017 Spring Meeting
13th Global Congress on Process Safety
San Antonio, Texas
March 26-29, 2017

Overfill to Overpressure – Investigation Findings and Contributing Events to the CAPECO Explosion

Scott Davis
GexCon US
Bethesda, MD
sgdavis@gexcon.com

Tom DeBold
Claudio Marsegan
Bryant Hendrickson
GexCon US
Bethesda, MD

Keywords: CAPECO, petroleum, overfill, bulk storage, explosion

Abstract

On October 23, 2009, a vapor cloud explosion occurred after an above ground gasoline storage tank was overfilled at the Caribbean Petroleum Corporation (CAPECO) tank farm facility in Bayamón, Puerto Rico. The blast registered a 2.9 on the Richter scale and sparked several fires throughout the facility that burned for another three days and damaged a total of 21 tanks before emergency responders were able to extinguish the flames.

Operations on the day of the incident included a ship transfer of unleaded gasoline to the CAPECO facility and the transfer was scheduled for four smaller capacity tanks. At approximately 10:00 pm, all of the unleaded gasoline was being transferred to tank 409 and a visible vapor cloud was noted during the walk around at 12:00 am on October 23, 2009. It was determined that tank 409 was overfilled by approximately 5,000 barrels (210,000 gallons) of gasoline, which caused gasoline to flow out of the tank vents near the top of the tank and cascade down the side forming a vapor cloud and liquid pool. This is equivalent to overfilling the tank by over 20 gasoline tanker trucks, which if 10% of the liquid formed the vapor cloud it would cover an area larger than 30 American football fields of a rich flammable mixture. The gasoline vapors mixed with the air forming an ignitable vapor cloud spanning a vast portion of the facility, which subsequently ignited neighboring tanks that burned for three days.

US regulations do not consider CAPECO and other bulk petroleum storage terminals as high hazard facilities, yet a large vapor cloud formed. This paper will present investigation findings and key lessons learned into the massive explosion and fire. This includes: effects of confinement and congestion in a seemingly open space that contributed to the incident; ship offloading and tank filling procedures that revealed significant operational deficiencies; failures to secure the confined drainage system; and level controls.

1 Introduction

On October 23, 2009, a vapor cloud explosion occurred after an above ground gasoline storage tank was overfilled at the Caribbean Petroleum Corporation (CAPECO) tank farm facility in Bayamón, Puerto Rico. The blast registered a 2.9 on the Richter scale and sparked several fires throughout the facility that burned for another three days and damaged a total of 21 tanks before emergency responders were able to extinguish the flames. The explosion and ensuing fires did not cause any fatalities; only a few minor injuries were reported. However, approximately 300 homes and business within 1.25 miles of the site reported damages [1].

GexCon was retained to perform an independent and objective investigation into the cause and origin of the explosion. Due to the nature of the investigation, some information provided to GexCon was confidential and could not be provided. Therefore, information provided in the reports of other agencies were used in the present analysis, and include: the ATF Investigation 764045-10-0011 Report [1]; U.S. Chemical Safety and Hazard Investigation Board Final Investigation Report – Caribbean Petroleum Tank Terminal Explosion and Multiple Tank Fires [2]; and the Puerto Rico Fire Department Fire Marshal Report [3] . This paper will present investigation findings and key lessons learned into the massive explosion and fire that occurred at the CAPECO facility. This includes: effects of confinement and congestion in a seemingly open space that contributed to the incident; ship offloading and tank filling procedures that revealed significant operational deficiencies; failures to secure the confined drainage system; and level controls.

We will first present a brief description of the facility, events leading up to the incident, and details of the explosion. Next, we will evaluate the conditions and contributing factors that lead to the explosion.

CAPECO Facility

The Caribbean Petroleum Corporation (CAPECO) facility was a petroleum storage terminal located in Bayamón, Puerto Rico. At the time of the incident, the petroleum storage terminal employed approximately 65 people. Prior to the gasoline overfill and ensuing explosion, the CAPECO facility (Figure 1.1) consisted of a tank farm of above ground storage tanks, offices, decommissioned refinery, waste water treatment plant and distribution piping that serviced the San Juan Bay. The CAPECO facility stored gasoline, diesel, fuel oil, and jet fuel at the time of the incident. The tank locations along with what fuel was stored in each can be found in Figure 1.2.

The tank farm was split by roads into three areas; the northern, central, and southern tank farms. The tanks were separated by dikes or berms designed to hold 110% of the largest tank capacity in that berm (see Figure 1.2) [3]. The southern tank farm contained eleven tanks, which included diesel and fuel oil tanks, divided by six berms. Jet fuel, diesel, and fuel oil were contained in the central tank farm. The northern tank farm held gasoline, diesel, and fuel oil. CAPECO estimated that 613,816 barrels (25,780,272 gallons) of gasoline, 428,260 barrels (17,986,920 gallons) of diesel, 360,496 barrels (15,140,832 gallons) of fuel oil, and 132,724 barrels (5,574,408) of jet fuel were present at the time of the incident [4].

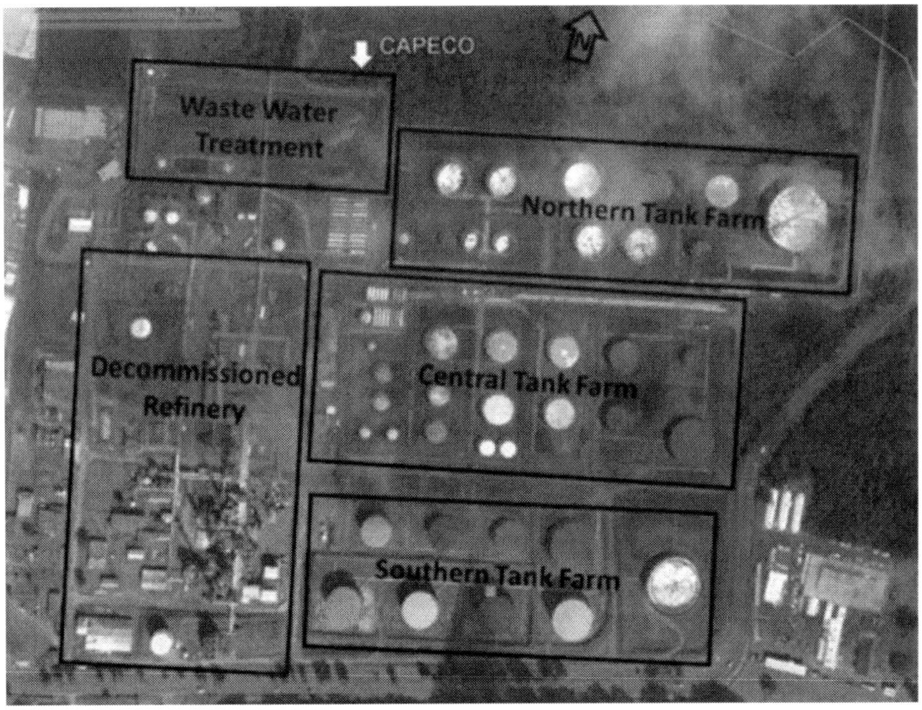

Figure 1.1: Facility Layout.

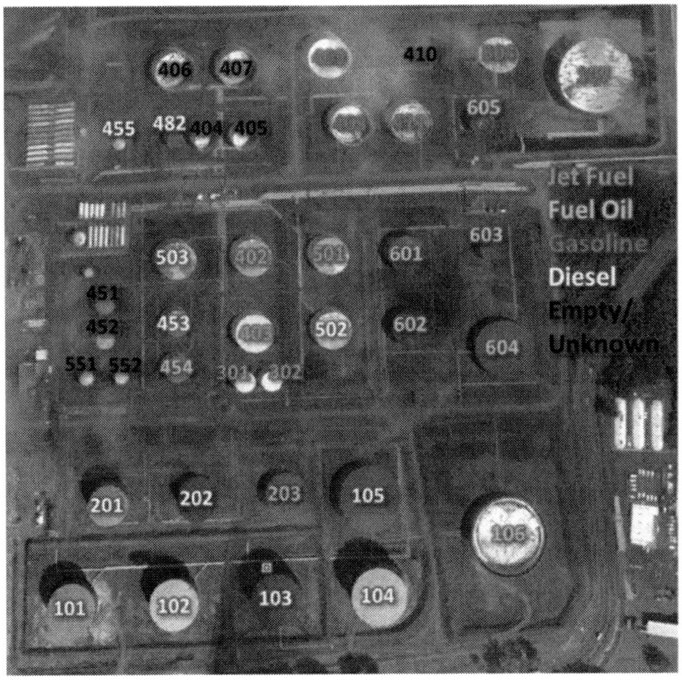

Figure 1.2: CAPECO tank location and inventory.

During fuel transfer operations, vessels would load/unload through the facility's 2 ½ mile pipeline which ended at a loading dock located in the San Juan Bay [2]. The terminal dock could accommodate one tanker of up to 75,000 tons. The transfers were made through six pipelines that connected to the facility transferring crude oil, FCC stock, fuel oil, gasoline, diesel, jet fuel and kerosene. The pipelines were equipped with remote operating valves that shutdown through: an electrical actuator; manually closing the valve; thermal shut-off switch set to 221°F (105°C); five push buttons in the operator's office, guards office, stairway to the main platform, fire pump building and Chiksan platform [5].

Events Prior to the Incident

Operations the day of the incident included: 1) A transfer of diesel fuel to the Puerto Rico Electric Power Authority Plant in Puerto Nuevo from tank 202; 2) A transfer of fuel oil to the Puerto Rico Electric Power Authority Plant in Palo Seco from tank 105; and 3) A transfer of gasoline from the Cape Bruny vessel to the CAPECO facility. Due to CAPECO's largest tank 107 (capacity 500,000 bbls – 21,000,000 gal) being used for a different quality gasoline, the transfer was scheduled for four smaller capacity tanks (see Figure 1.3) [6] .

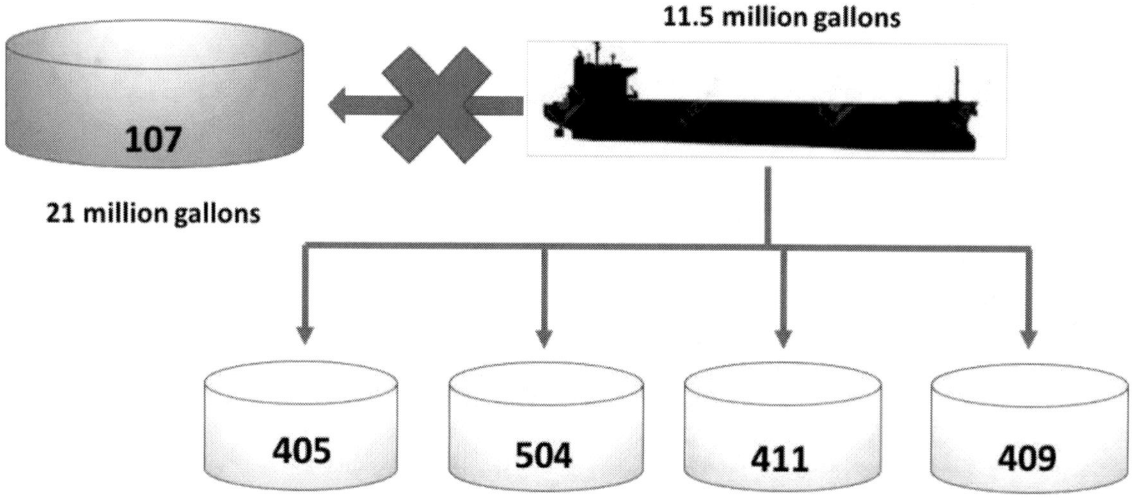

Figure 1.3: Cape Bruny transfer plan.

Pumping from Cape Bruny to the CAPECO facility began on October 21, 2009. Four tanks (Tank 405, 504. 409, and 411) in the northern tank farm received gasoline during the transfer. Initial pumping started at approximately 8:47 p.m. A sequence of operations can be found in Figure 1.4 [2].

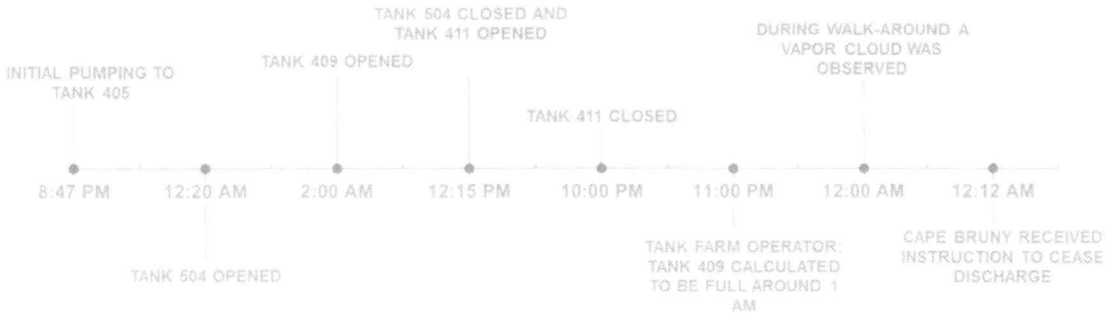

Figure 1.4: Transfer Operations [6].

Shortly after 11:00 p.m., the tank farm operator advised the supervisor in charge of the operations that night that the inspector found a difference between the side gauge, monitor console reading and the physical status of Tank 411 (already closed), and advised that the discrepancy was to be investigated [1].

ATF interviews determined that numerous operators and employees observed a vapor cloud starting at midnight. Tank 409 began to overfill between the 11:00 p.m. and 12:00 a.m. tank checks while the average pumping rate was approximately 11,134 bbls/hr (467,628 gal/hr) [7]. The overfilling of tank 409 caused gasoline to flow out of the tank vents near the top of the tank and cascade down the side forming a vapor cloud and liquid pool (as will be discussed in detail later). Tank 409 was contained in a berm with Tank 410 having a capacity of approximately 132,000 barrels (5,544,000 gallons) based on the 110% capacity requirement [3] . As the overfilling continued, gasoline (liquid and vapor) began to fill the containment area and expand towards the waste water treatment facility resulting in the white mist or vapor cloud that was seen in the facility by employees [6]. It was determined that approximately 5,082 barrels (213,444 gallons) of gasoline overflowed from tank 409 [7]. This is equivalent to overfilling the tank by over 20 gasoline tanker trucks, which if 10% of the liquid formed the vapor cloud it would cover an area larger than 30 American football fields of a rich flammable mixture.

The employees saw a white cloud extending approximately two to three feet above the ground by tank 404, and the white vapor cloud could be seen throughout the area between tanks 453 and 301, and near tanks 201 and 409 (Figure 1.5). The vapor cloud could only be seen in areas that were illuminated and extended approximately 200 yards wide by 400 yards long (183 m by 366 m) [1].

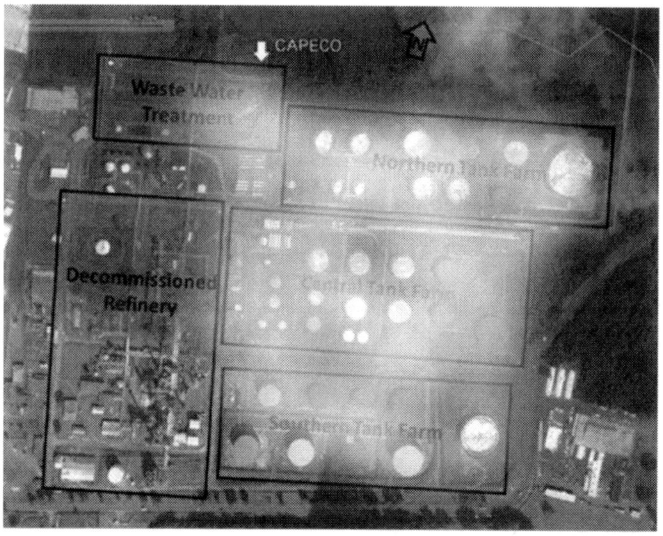

Figure 1.5: Approximate migration area of the vapor cloud.

Incident

At approximately 12:23 a.m. on October 23, 2009, the resulting vapor cloud from the overfill of gasoline ignited on the western side of the facility, away from the tank farm area. As the flame propagated (Figure 1.6 and Figure 1.8) towards the above ground storage tanks, a large explosion (Figure 1.8) occurred in the tank farm area that registered 2.9 on the Richter scale. Due to ignition of the large vapor cloud and ensuing explosion, many of the facility storage tanks were consumed by fire, which burned for another three days before emergency responders extinguished the flames [3].

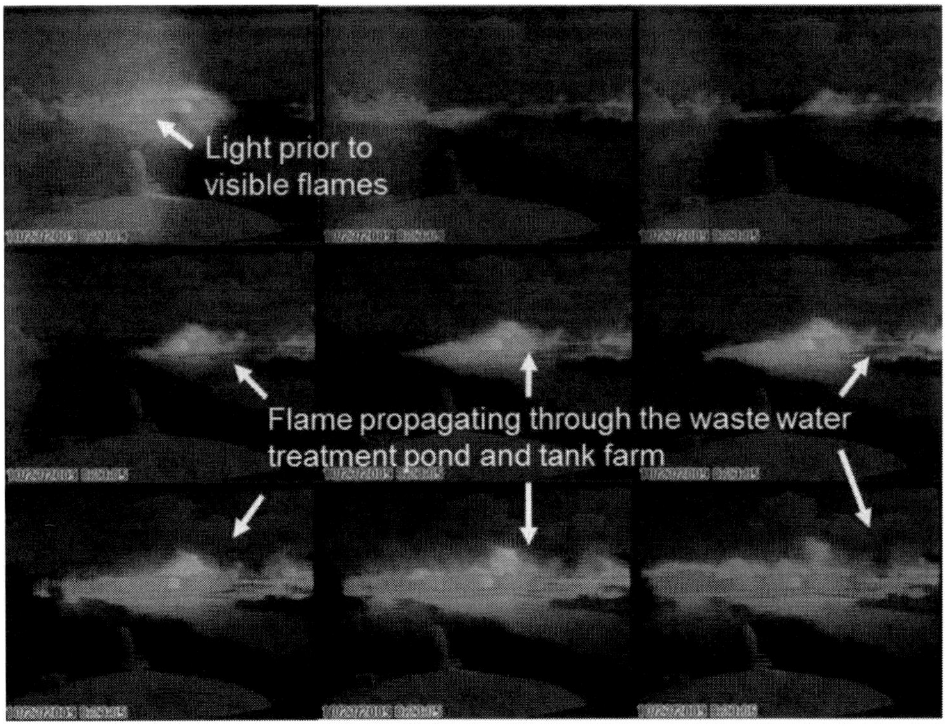

Figure 1.6: CAPECO security footage from camera 4. Visible light is seen prior to the flame propagation through the waste water treatment pond and tank farm.

Figure 1.7: Approximate propagation of the flame front prior to the explosion in the underground drainage as indicated by arrows. Camera locations and angles (red-camera 3 and blue-camera 4) at the first sign of ignition.

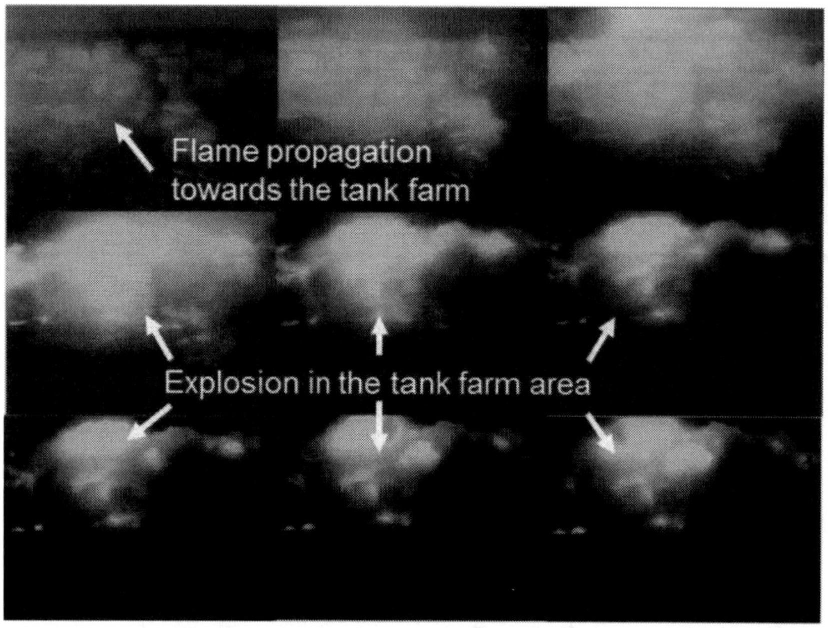

Figure 1.8: CAPECO security footage from camera 3 as the flash fire approached the tank farm from the west, leading to an explosion within the tank farm area (top-left to bottom-right).

Post Incident and Blast Damage

Emergency crews were alerted to the incident when a nearby citizen heard a large explosion and saw flames at the CAPECO facility [1]. At approximately 12:30 a.m. on October 23, 2009 the Bayamón and Cataño Fire Departments arrived at the CAPECO facility where they waited for the site fire chief to arrive approximately 45 minutes later. Eleven of the CAPECO tanks were damaged by the initial fire and explosion.

Firefighting efforts, as shown in Figure 1.9, lasted almost three days until approximately October 25, 2009 at 11:40 a.m. when the fires were extinguished [6]. Additional tanks were damaged by ensuing fires (see Figure 1.10 and Figure 1.11) [3] and when the fire was completely extinguished, the damage caused by the large gasoline vapor cloud explosion and fire included 21 tanks onsite. Both blast and thermal damage were observed on the tanks at the CAPECO facility (see Figure 1.12), consistent with the 11 tanks initially damaged by the large vapor cloud explosion and ensuing gasoline fire, and the others falling victim to the resultant fire [1].

Figure 1.9: Firefighting efforts after the incident [8].

Figure 1.10: Aerial view of the CAPECO facility after the initial gasoline vapor cloud explosion.

Figure 1.11: Secondary tank fires caused by the initial vapor cloud explosion and fire from the gasoline vapors.

Figure 1.12: Post incident damage to the tank farm.

2 Physics of vapor cloud explosions

Prior to the explosion at the CAPECO facility, tank 409 was overfilled and gasoline was being discharged from the overflow vents at the top of the tank. Recall that ignition and burning of liquid hydrocarbons occurs when the hydrocarbon is in the vapor state. A large vapor cloud developed resulting in the 'white cloud' described by CAPECO employees. As the vapor cloud grew from the continuous overflow, the gasoline vapors continued to mix with air forming a flammable mixture throughout the facility. This cloud found a competent ignition source in the western side of the facility and flames propagated towards the tank farm leading up to the main explosion event. This phenomenon is referred to as a vapor cloud explosion (VCE). The specifics on how the vapor cloud was formed and on the VCE will be discussed next.

Formation of vapor clouds

When evaluating the potential for large vapor cloud formation, which may lead to a massive explosion event, it is important to consider the mechanism by which a vapor and vapor clouds are formed. Recent experimental and modeling research has been performed to explore the generation of vapor clouds from overfilling gasoline storage tanks [9], [10]. The research determined that during a tank overfill, the gasoline would flow out of the ventilation holes at the top of the tank and over the deflector plate. The deflector plate channeled some of the escaped fuel onto the tank wall, but the rest ran over the top of the plate, fragmenting into droplets that cascaded through the air. Most of the fuel running down the wall hit a wind girder (a structural stiffening ring) and detached from the tank wall, creating a second cascade of droplets.

These conditions promoted the evaporation of the lighter components of gasoline (e.g., butanes, pentanes and hexanes). The free-fall of droplets leads to entrainment of air and mixing between the air and fuel vapor, and the formation of a rich fuel/air mixture. In addition, the wind conditions at the time of the overfill were calm, and had the wind speeds been higher, the evaporating gasoline vapor would be quickly diluted as it escaped the berm. Therefore, the overfill led to conditions that increased the break-up of gasoline into droplets, which significantly increased the evaporation rate of the lighter components of gasoline as compared to that from the pooled gasoline. Therefore, the overfill rate coupled with low wind speeds generated very large vapor clouds capable of filling a significant region of the CAPECO tank farm as seen in modeling shown in Figure 2.1.

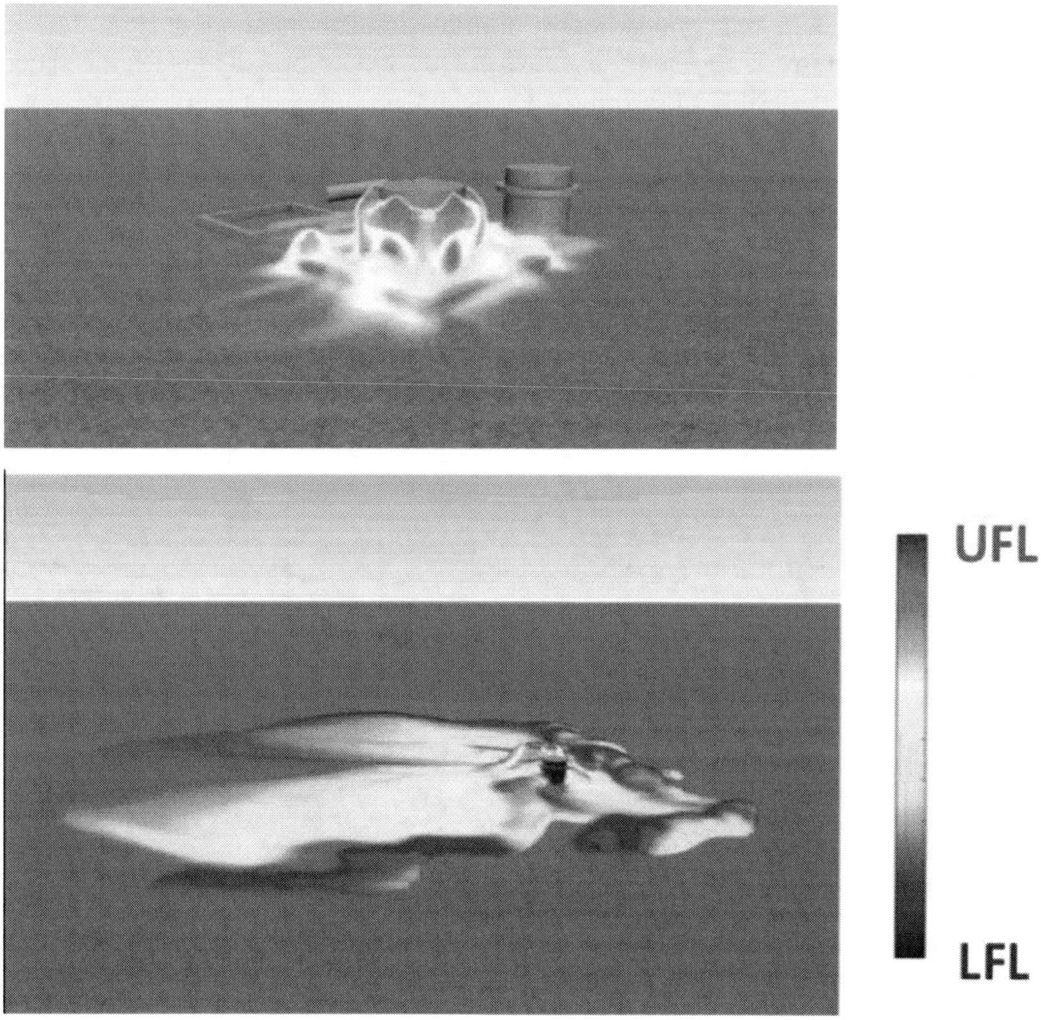

Figure 2.1: Vapor cloud formation and growth as gasoline initially cascades down the side of a storage tank and is dispersed.

Vapor cloud explosions

Vapor cloud explosions (VCEs) are oftentimes understood as "confined explosions", where a flammable mixture of gas (or vapor) and air is ignited within a confined structure or vessel. The heat released from the burning mixture causes the gases to expand and the pressure within the confinement increases above normal atmospheric pressure; this is referred to as "overpressure". This occurs until the envelope of the structure ultimately fails, similar to an ignition of a flammable mixture within an empty tank causing the roof to blow off. The sudden failure of the structure/vessel causes a shock or blast wave to form. However, conditions at CAPECO were mostly unconfined and open, except for the underground drainage system. An open and unconfined environment presents little to no resistance of an expanding flame front when a mixture is ignited; therefore, an overpressure is not expected if the gases are allowed to freely expand. The formation of a shock or blast wave in an open area cannot be explained in the same manner, and in fact, requires a unique set of circumstances.

Overpressures can indeed be created for unconfined vapor cloud explosions and require mechanisms that accelerate the combustion or burning process. The combustion rate for burning vapor/air mixtures is measured by how fast the flame propagates (flame speed) relative to the unburned mixture. For quiescent fuel/air mixtures the flame propagates at approximately 0.5 m/s and minimal overpressure is observed as the hot gases can expand in the unconfined environment. However, the flame propagation rate is extremely sensitive to turbulence in the flammable mixture into which the flame is burning. This turbulence can be caused by repetitive obstacles, (e.g. pipes, equipment, trees) within the vapor cloud itself. This results in a feedback loop, which has been extensively studied by Gexcon (see Figure 2.2), whereby the turbulence causes the flame to burn faster, which in turn pushes the unburned gases past the obstacles even faster creating higher turbulence levels, which in turn causes the flame to burn even faster. This results in the flame accelerating and can lead to very high propagation rates.

Figure 2.2: Flammable fuel air experiments for "unconfined" vapor clouds with no obstacles (top) and with trees (bottom). Clouds ignited on the left and maximum flame speeds with trees are six times larger than without.

The flame may propagate at such high speeds that the expanding gas ahead of the flame is actually compressed, which can ultimately result in a shock or blast wave. The strength of the blast wave will depend on the flame speeds achieved during the acceleration process.

If the congestion in the cloud is low, the flame speeds will remain low and will not generate a shock or blast wave (i.e., a "flash" or "whoosh" type sound), while relatively higher flame speeds can occur in high congestion areas, resulting in shock wave and "boom" type sound. Flame speeds between 100-150 m/s can generate significant overpressures, and those above 400 m/s can generate pressures well in excess of 1 barg, and can even transition to detonations. This phenomenon was precisely what occurred at the Buncefield explosion in 2005, whereby the flame accelerated into highly congested tree lines that extended over 100 meters. No confinement was found at the Buncefield site and the flame accelerations, and extremely high overpressures, were exclusively caused by the tree density [11].

In order to generate the massive explosion observed at the CAPECO facility, the resulting large vapor cloud needs to migrate into a congested (as described above) or confined region to accelerate the flame. Similar to Buncefield, the CAPECO facility had congestion that included numerous low-lying pipelines connecting the storage tanks, and vegetation to the north and east of the facility. During the post-incident scene investigation, thermal damage was found to extend into the tree line approximately 50 yards past the east property fence [1]. Unlike Buncefield, the CAPECO facility had additional long channels of confined drainage pipes that were not properly closed. Recall that the vapor cloud was ignited western side of the plant propagating towards the waste water treatment (see Figure 1.7) and it propagated east towards the tank farm and vegetation. Video analysis confirmed (see Figure 2.3) that while the main flame propagated above ground through the tank farm, a faster flame front traveled via the underground drainage pipes where a fireball exited the sump pump encasement at the northeast corner of the central tank farm well before the above ground flame.

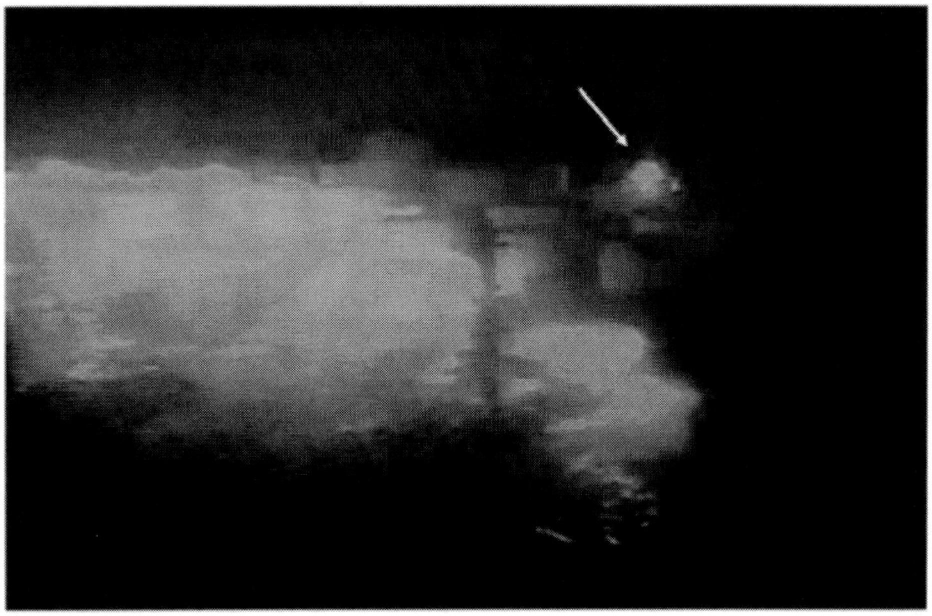

Figure 2.3: View from camera 3 showing the flame front in the underground drainage pipes reaching the northeast corner of the central tank farm before the above ground flame front. Ignition occurred on the western side of the facility resulting in flame propagation from left to right in the image above.

In fact, the 10 ft x 10 ft concrete lid on the sump pump, weighing about 12,000 pounds, was found over five feet from its original location [1]. Furthermore, blast damage to the above ground storage tanks was generally in the direction of the sump pump encasement and eastern vegetation and not the western side of the facility where the cloud was first ignited [1]. This confirms that the explosion pressures were exacerbated due to the confinement caused by the underground drainage pipes, which resulted in higher overpressures and faster accelerating flames that also entered the congested region. To evaluate the effect of the underground drainage system, FLACS simulation were performed to evaluate the resulting overpressures and blast damage both with and without the presence of the drainage system.

The CAPECO tank farm area (both with and without the drainage system) was modeled in FLACS, a computational fluid dynamics (CFD) tool predominantly used in the Oil & Gas and chemical industries to predict the consequences of hydrocarbon deflagrations. A large vapor cloud (e.g., stoichiometric cloud 4 meters tall) was placed throughout the tank farm and water treatment areas, and ignited in the water treatment area. Figure 2.4 shows the tank farm with the large vapor cloud and drainage system.

The flame propagation rates were evaluated both through the above ground cloud and through the drainage system. Figure 2.5 shows that the vapor cloud entering the drainage system resulted in flame speeds of over 400 m/s, significantly higher than above ground flame propagation. In fact, the confinement and acceleration of the flame in the drainage system resulted in overpressures that exceeded 2 barg (see Figure 2.6), which could possibly result in condition where the exiting deflagration transitions to a detonation. In contrast, overpressures generated by the flame burning through the above ground portion of the cloud did not exceed 0.03 barg. Therefore, the confined underground drainage system resulted in flames burning faster than the speed of sound and with pressures exceeding 2 barg, which ignited the remaining unburned vapor cloud in the tank farm. This high-pressure, supersonic flame front resulted in a significantly more devastating explosion than the "flash" or "whoosh" that would have been observed had the drainage system been properly closed prior to the overfill.

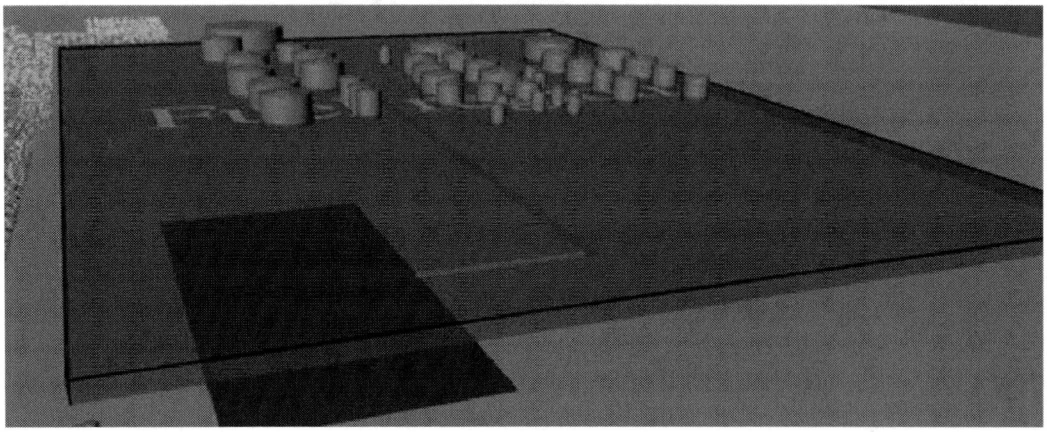

Figure 2.4: FLACS model of the CAPECO underground drainage system.

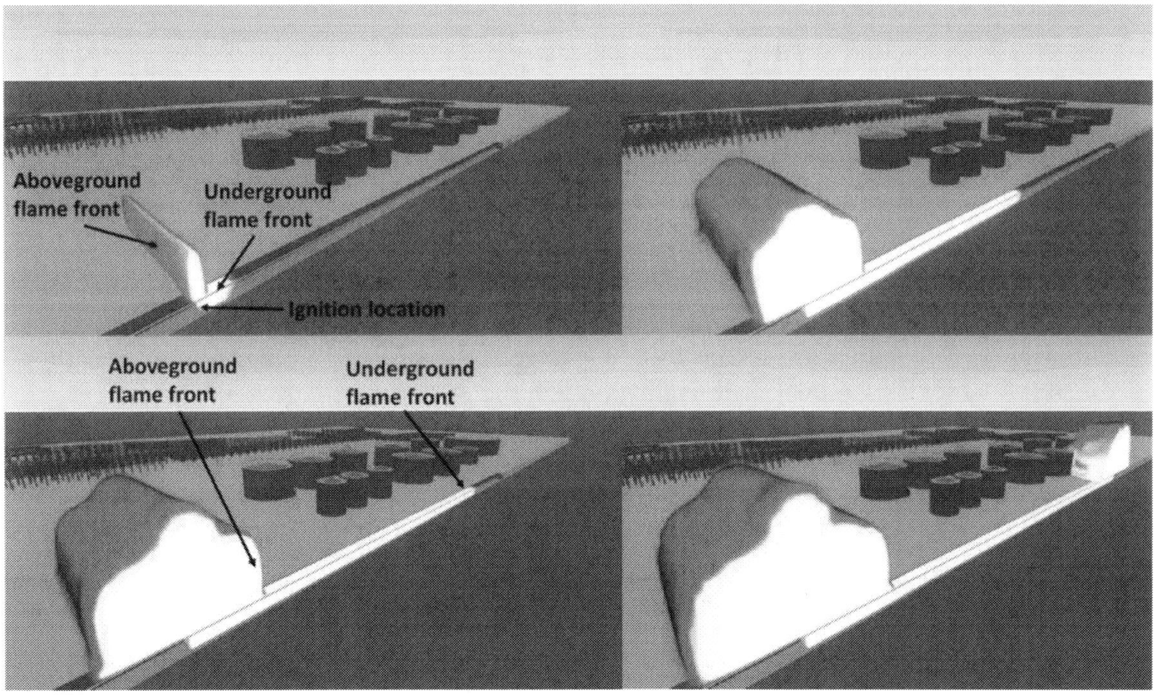

Figure 2.5: Flame front above and underground.

Figure 2.6: Pressure generated from the explosion in the underground drainage system.

3 Investigation findings and key lessons learned

Many of the contributors to the explosive atmosphere present at the CAPECO facility during the early morning hours on October 23, 2009 could have been avoided if the proper precautions were followed. While previous investigation reports have outlined certain contributing factors related to the incident, other major findings and lessons learned were not identified. For example, other groups have determined that the facility did not have automatic level alarms and that the automatic tank gauge system at the facility was not operational, which contributed to the event by not providing key indicators automatically and relying solely on manual measurements. However, key contributors to the incident (in addition to the automatic gauge system) were not identified and included: a failure to accurately monitor and verify the rate at which the gasoline storage tanks were being filled during the vessel offloading; and a failure to ensure "inherent safety" at the facility by not securing the dike drainage systems, which resulted high-pressure, supersonic flame fronts capable of causing the massive explosion as opposed to a "flash" or "whoosh" had the vapor cloud remained above ground. These factors will be discussed next.

Operational deficiencies that led to the Tank 409 Overfill

Recall that automatic controls were down and operations were manually monitored. Using the vessel discharge rates and tank strapping tables, it was estimated that when tanks 504 and 409 were being filled from 4:00 a.m. to approximately 12:00 noon on October 23, 2009, the transfer to each tank was approximately equal (see Figure 3.1). However, when 411 was brought online from 12:00 noon to 10:00 p.m. on October 23, 2009, it was determined that approximately 66% of the fuel was being transferred to tank 411 (see Figure 3.2), with the remaining fuel going to tank 409. This is consistent with operators' observations when preparing to take tank 411 offline: the valve to tank 409 was noted to be only open a little, confirming that tank 411 was receiving a much higher flow rate than tank 409 while they were being filled simultaneously [1].

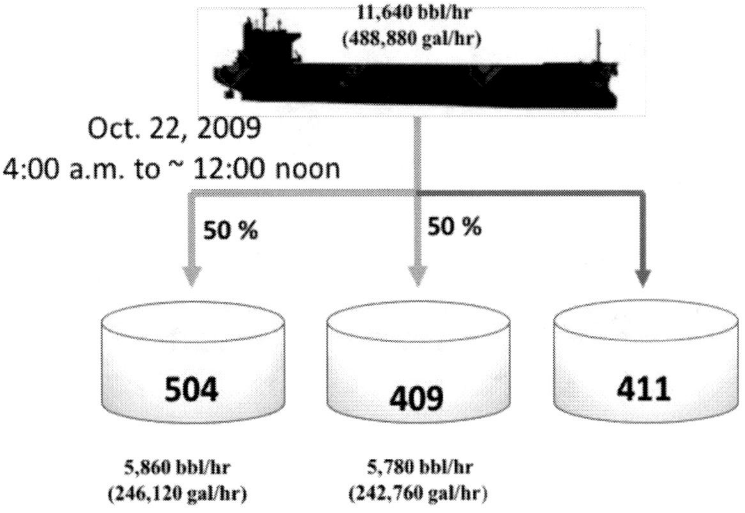

Figure 3.1: Transfer from the Cape Bruny to tanks 504 and 409 from 4:00 a.m. to approximately 12:00 noon on October 22, 2009.

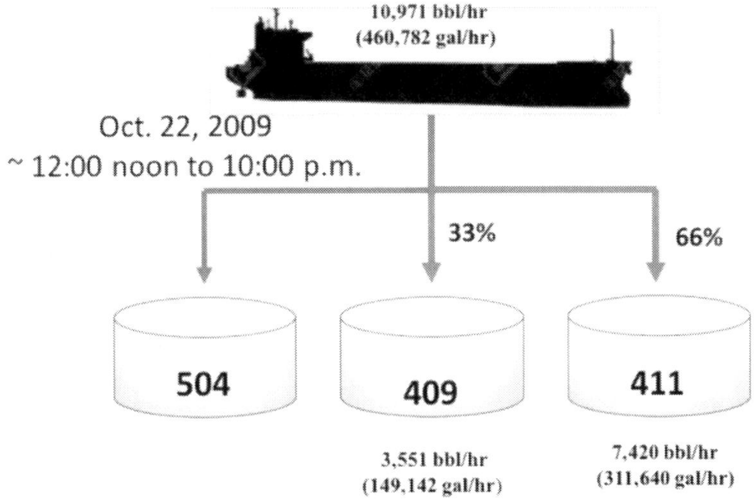

Figure 3.2: .Transfer from the Cape Bruny to tanks 411 and 409 from approximately 12:00 noon to 10:00 p.m. on October 22, 2009.

At 10:00 p.m. on October 22nd, tank 411 was closed and tank 409 was fully opened. Tank 409 contained approximately 98,716 gross barrels (4,146,072 gallons) of gasoline (overfill occurs at approximately 121,062 barrels).[1] The supervisor incorrectly estimated that the tank would be full at 1:00 a.m. by assuming only 33% of the discharge rate was going to tank 409. However, with the full vessel discharge of approximately 11,134 bbl/hr [7] going into tank 409, approximately 5,082 barrels (213,444 gallons) of gasoline overflowed from tank 409 contributing to the generation of a large flammable vapor cloud. This means that overflow occurred for approximately 27 minutes, beginning at approximately 11:45 p.m. on October 22nd until 12:12 a.m. the next morning.

Table 3.1: Estimated amounts of gasoline added to each tank and overflown at the CAPECO facility.

Tank	Estimated gasoline added Gross Barrels (Gallons)
405	4,411 (185,262)
504	62,984 (2,645,328)
411	74,198 (3,116,316)
409	114,325 (4,801,650)
Total	255,918 (255,918)
Total from Cape Bruny	261,000 (10,962,000)
Overfill	**5,082 (213,444)**

Therefore, one major cause of the incident was due to an incorrectly estimated fill time of 1:00 a.m. for tank 409, likely caused by ignorance of the increased flow rate to tank 409 after closing tank 411. Therefore, failure to compare vessel discharge rates to tank fill rates and verify that tank 409 had sufficient space to receive the scheduled amount of product resulted in tank overflow which generated a large flammable vapor cloud [12].

Drainage system

In the post incident investigation, the drainage valve in the berm surrounding tank 409 was found to be in the open position. The open valve drained to the waste water treatment facility, creating a pathway for the liquid gasoline to spread, increasing vapor generation and allowing for flammable gasoline vapor to fill the underground drainage system. The dike drainage valves should be closed during transfer operations so that in the event of an overfill, pool spread is limited. In this case, failure to close the drainage valve led to flammable vapors building up within the underground drainage system, and after ignition, resulted in a deflagration propagating within a completely confined drain and causing significant flame accelerations and overpressures. CFD analyses confirmed that these flame accelerations led to high pressure, supersonic flame fronts capable of causing the massive explosion as opposed to a "flash" or "whoosh" had the vapor cloud remained above ground. Keeping the drainage valves closed during a fuel transfer would have vastly limited the consequences and ensured inherent safety at the facility.

Level controls

The CAPECO facility has an automatic tank gauge system that transmitted the tank liquid depth to the operator's office. It would calculate the fill rate, as well as plot the flow rate and tank liquid depth versus time. Due to issues with the automatic tank gauge system (typical float and tape gauge), manual inspection of the tanks and calculation of the fill time had to occur through hourly readings.

4 Conclusions

Although US regulations do not consider bulk petroleum storage terminals as high hazards, an explosion at the CAPECO facility and ensuing fire destroyed a vast majority of the tank farm facility. This explosion was caused by a vapor cloud resulting from the overfill of a gasoline bulk storage tank, and this vapor cloud entering an improperly closed drainage system. The two main lessons learned from this incident (and previous incidents like Buncefield) include: (1) having systems and procedures in place to limit the likelihood of having such an overfill condition, and (2) providing inherent safety to such facilities to vastly limit the explosion consequences in the event such an overfill occurs (e.g. limiting confinement and congestion).

Regarding reducing the likelihood of such an event, the CAPECO facility was operating as a fully-attended facility and was in compliance with federal regulations at the time of the incident. As of today, fully-attended tank farm facilities do not require independent liquid level alarms and automatic tank gauging which allows other facilities to operate in the same fashion as CAPECO when transferring and storing flammable liquids such as gasoline. Therefore, such facilities should consider redundant controls and procedures to ensure: fill and transfer rates are constantly updated and verified throughout the transfer; and that the automatic tank gauge system is operational and complements manual measurements made at the facilities; and areas of high confinement like dike drainage systems are secured and closed prior to transfer of volatile hydrocarbons.

Inherent safety should also be evaluated at such facilities to ensure significantly lower consequences, even in the event of a significant release. CFD simulations have confirmed that physical layout of both Buncefield and CAPECO resulted in significantly higher consequences that could have resulted for identical overfill conditions. In fact, the densely coppiced tree lines at Buncefield and high confinement of the improperly opened drainage system at CAPECO resulted in explosion pressures and consequences orders of magnitude higher had such congestion/confinement not been present. Inherent safety of such facilities can be enhanced if areas of confinement and congestion are identified and mitigated in standard risk assessments. In fact, the CAPECO and Buncefield incidents would have likely resulted in large flash fires instead of the more devastating VCEs had these congestion and confinement areas been mitigated.

References

[1] ATF Investigation 764045-10-0011 Report.

[2] U.S. Chemical Safety and Hazard Investigation Board Final Investigation Report – Caribbean Petroleum Tank Terminal Explosion and Multiple Tank Fires.

[3] Puerto Rico Fire Department Fire Marshal Report.

[4] Inventory Logs.

[5] Dock Operating Manual.

[6] ATF Investigation 764045-10-0011 Interviews.

[7] Ship Transfer Logs.

[8] Caribnewspr https://www.flickr.com/photos/43578995@N03/.

[9] Davis, SG, Hinze, PC, Hansen, OR, van Wingerden, K (2010). 2005 Buncefield Vapor Cloud Explosion: Unraveling the Mystery of the Blast. Interflame 2010, Nottingham, UK..

[10] Coldrick, S., Atkinson, G.T., and Gant, S.E., Health and Safety Laboratory, Hazards XXII, 2011, pp. 560- 569..

[11] Van Wingerden, M. and Wilkins, B. (2008) Experimental investigation of the effect of flexible obstructions on flame propagation in vapour cloud explosions, GexCon report no. GexCon-08-F44110-RA-1, rev 01.

[12] API RP 2350, Appdx B.3.3 (2005).

Produced Water Tank Explosion Case Study

Presented by:
J. Kelly Thomas, Ph.D.
Baker Engineering and Risk Consultants, Inc.

2017 AIChE
Global Congress on Process Safety

BAKERRISK

BAKERRISK

Outline

- Purpose
- Background
- Facility Description
- Incident
  - Overview
  - Fuel
  - Oxidant
  - Ignition
- Related Incidents
- Lessons Learned
- Potential Prevention and Mitigation Approaches

2

Purpose

- Highlight commonly encountered hazard
  - Produced water tank with flammable atmosphere
  - Similar hazard associated with waste (slop) water tanks
    - Incidents involving waste water tanks in industry have occurred
- Identify lessons learned
- Identify potential prevention and mitigation approaches
  - Examples, not exhaustive compilation
  - Other approaches may provide similar level of risk reduction

3

Background

- Produced water (brine water)
  - Water separated from oil / condensate / gas streams
  - Collected in tanks and treated as waste
  - Head space may be in flammable range depending on operating parameters and separator performance (e.g., condensate carryover)
- Not typically viewed as posing significant explosion risk
  - Reasonable on a relative basis if other explosion hazards present which pose much more severe consequences and/or risk
  - Reasonable from a building siting perspective
  - Hazard is local, although debris may be thrown considerable distance
  - However, produced water tanks can, and do, explode
- Should be considered in terms of hazard/risk to personnel

4

Facility Description (1 of 6)

- Incident occurred at gas production site (West Virginia, 2013)
- 6 gas wells (no oil or condensate produced), each with own separator
  - Separators are 2.5 ft diameter by 8 ft long, and operate at 675 psig
  - Gas composition (97.2% CH_4, 2.1% C_2H_6, 0.1% C_3H_8, 0.3 CO_2, 0.3% N_2)
- Produced water tanks
  - Tanks in two groups of 3 (i.e., 6 tanks)
  - Group of 3 tanks share common overflow (3 in.) and vent (4 in.) lines
    - Common vent line not supplied by tank manufacturer
    - No flame arresters on these lines
  - Fixed roof API 12F style, 12 ft. (3.7 m) diameter, 20 feet (6.1 m) tall
  - 400 barrel (16,800 gallons, 64 m^3) & hydrotested at 1.5 psig (0.10 barg)
  - 2 access hatches (Enardo ES-660 and Jayco JT-8); relieve at 0.5 psig

5

Facility Description (2 of 6)

6

Facility Description (3 of 6)

- Fill line and access hatches

7

Facility Description (4 of 6)

- Common vent and overflow lines

- Common vent line terminates with open end at opposite end of 3-tank group

Facility Description (5 of 6)

- Sign at entrance to pad

9

Facility Description (6 of 6)

- Warning sign on each produced water tank
- Flammability rating of 3 would be indicative of gasoline

10

Incident (Overview, 1 of 7)

- Approx. 100 bbl removed from each tank every 3 or 4 days
  - Driver goes up to top of tank via catwalk, opens hatch, and gauges using wooden rod before pumping out desired liquid volume
- Driver arrived between 5:30 and 5:45 am, hooked truck up to tank manifold, and (at some point) ascended catwalk to tanks
- Explosion occurs prior to 6:45 am (February, sunrise at 7:17)
- One tank fails at shell-floor joint, ejected, and thrown 110 feet (34 m)
- One tank fails at shell-roof joint, roof thrown 165 feet (50 m)
- Catwalk damaged and displaced, throwing driver from catwalk and outside tank battery, approximately 30 feet (9 m); fatality.

11

Incident (Overview, 2 of 7)

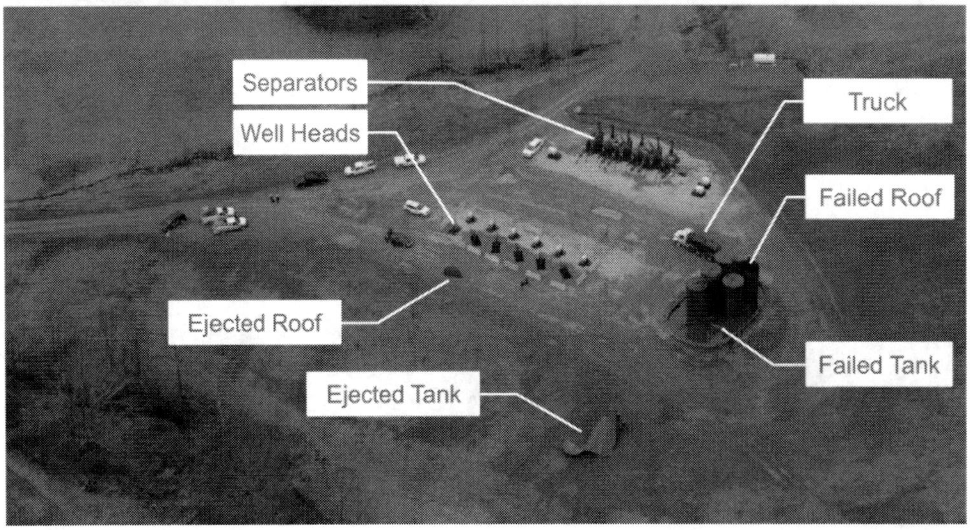

12

Incident (Overview, 3 of 7)

13

Incident (Overview, 4 of 7)

- Ejected tank

14

BAKERRISK

Incident (Overview, 5 of 7)

- Failed tank roof
- Shell-floor joint shows uplift (was approaching joint failure)

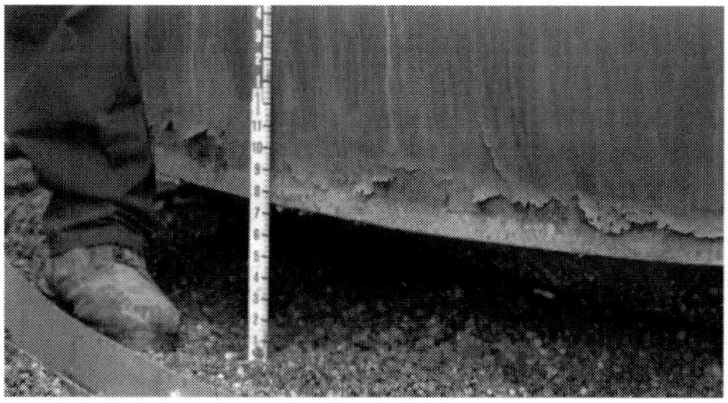

15

BAKERRISK

Incident (Overview, 6 of 7)

- Catwalk displacement (catwalk supported by tank attachment)

16

Incident (Overview, 7 of 7)

- Tank structural capacity roughly 10 to 15 psig (gradually applied internal pressurization) based on past experience
- Trajectory analysis indicates in range of 5 to 15 psig required to achieve observed tank and tank roof throw distances
- Indicates weakly flammable mixture was present in two tanks which failed, and mixture in "middle" tank was either not flammable or barely flammable
- Post-incident flammable gas measurements
  - Tank in adjacent set ≈ 65% LFL days after incident
  - Truck ≈ 15% LFL 2 months after incident
  - Consistent with elevated fuel concentration in tanks at time of incident

17

Incident (Fuel, 1 of 3)

- Separator does not remove "all gas" from the fluid, even when functioning properly and as designed
- Produced water contains gas since at elevated pressure (675 psig), which is released when pressure dropped to atmospheric
  - Gas content function of water pressure, temperature and composition
- Can have gas "carry" over or with stuck/jammed valve in separator
  - Separators were tested post-event and found to work properly
  - Concluded gas source was solely content of produced water transferred to storage tanks from separators

18

Incident (Fuel, 2 of 3)

- Methane solubility data from Duan & Mao (2006) Geochemica, 70: 2269-86.
  - Approximately 0.043 moles/kg at relevant conditions
- See vapor space fuel concentration plot on following slide:
  - Low level to high level of 50 to 150 barrels (1/4 tank capacity)
  - Conditions present (separator pressure, fluid composition, etc.)
  - No air in-leakage (only draw in air when drop liquid level)
- Can easily exceed UFL
  - Fuel concentration would be lower at lower separator pressure
  - Minimum fuel concentration would decrease if either high level or liquid volume extracted were increased (vs. values noted above)
  - > UFL is not a "universal condition" for produced water tanks

19

Incident (Fuel, 3 of 3)

- Fuel concentration for series of 100 bbl fill / extract cycles

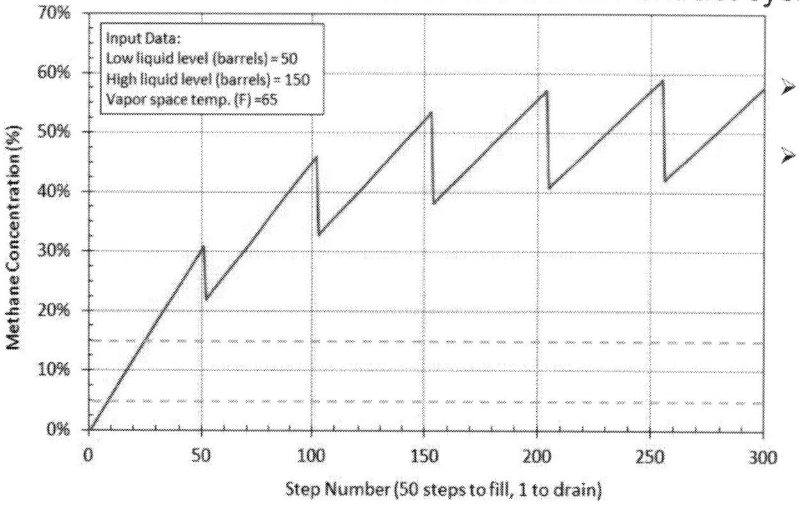

- Can easily exceed the UFL
- Issue is how get enough oxidant to be in flammable range

20

Incident (Oxidant, 1 of 4)

- Need additional air to enter flammable range (i.e., < UFL)
- Additional air can be drawn into tank via natural convection
  - Gas within tank lighter than outside air (i.e., unless outside air is hotter)
  - Light gas can flow out common vent line, pulling air in through leak points (e.g., access hatch seals)
    - Analysis subject to uncertainty since inlet flow path not well defined, but leakage rate in range of 125 scfd (2.5 lpm, 22 bbl/day) reasonable
    - Acts to limit maximum fuel concentration
    - Steady decrease in fuel concentration when tank is "stagnant" (i.e., no produced water additions or removals)
  - Much higher air in-leakage rate when hatch is open (2 to 5 scfm)
- Tanks stagnant for period before explosion & hatched opened
- See illustrate plots on following page

21

Incident (Oxidant, 2 of 4)

- Fuel concentration for series of 100 bbl fill / extract cycles, followed by stagnant period (125 scfd air in-leakage)

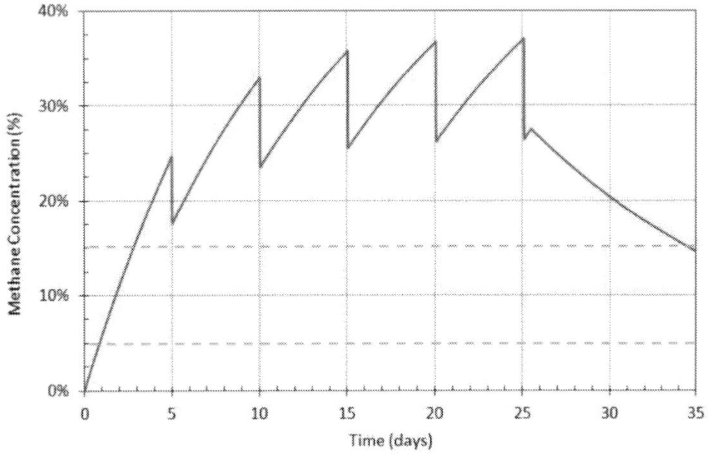

- Above UFL until enter stagnant period
- Air in-leakage during stagnant period allows fuel concentration to drop below UFL and enter flammable range
- Reach flammable condition in about 2 weeks

22

Incident (Oxidant, 3 of 4)

- Fuel concentration for series of 100 bbl fill / extract cycles, followed by stagnant period with hatch open (2 scfm air in-leakage)

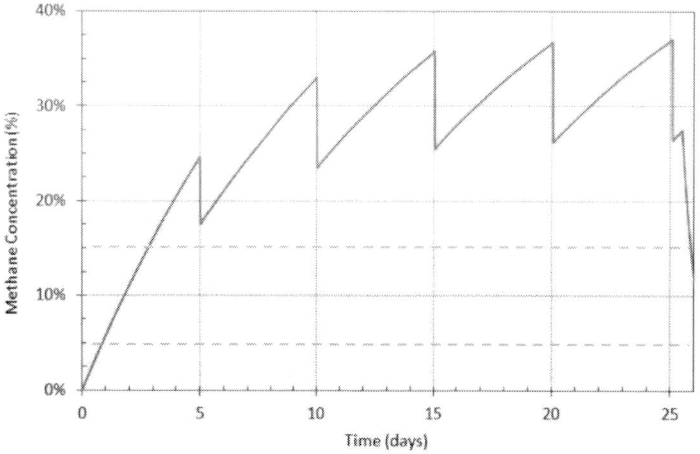

- Above UFL until hatch open for extended period
- Air in-leakage with hatch open for extended period allows fuel concentration to drop below UFL and enter flammable range
- Reach flammable condition in about 8 hours

23

Incident (Oxidant, 4 of 4)

- Natural convection limits maximum fuel concentration and can bring tank vapor space into flammable range, particularly with a tank which is stagnant for an extended period or with a hatch left open
- Note should not conclude that require natural convection in order to form a flammable mixture with a generic produced water tank
  - Lower separator pressure would mean less gas and hence tank could be flammable for portion of time regardless of natural convection
  - Larger volume of produced water withdrawn from tank or higher tank levels could put the tank vapor space into the flammable range
- Should assume a flammable mixture can result if the fuel concentration can exceed the LFL

24

Incident (Ignition Source)

- Driver was performing operation before sunrise
  - No headlamp or flashlight recovered from scene
  - Lighter found 25 feet from body (in same direction as thrown)
  - Driver was a smoker and one glove was off
  - Opened cigarette pack present in truck
  - Reasonable to assume driver used lighter to check liquid level
  - Driver not equipped with an LFL meter
- Static charge cannot be completely ruled out
- Flame propagation to adjacent tanks via connecting lines
- Ignition potential recognized (e.g., API RP 505)

25

Related Incidents

- Produced water tank explosions have happened in the past
- Cursory literature search
  - Texas (2013): Set of six 200 bbl produced water tanks, 2 people smoking on top of tank, tank exploded
  - Canada: Welding on tank which had contained produced water resulted in explosion
- CSB investigation report (2011-H-1):
  - "Public Safety at Oil & Gas Storage Facilities"
  - 26 incidents over 27 year period (44 fatalities and 25 injuries)
  - Includes at least one produced water tank
- Produced water SDS packages vary in flammability warnings (as does the signage on the produced water trucks)

26

Lessons Learned

- Produced water tank explosions can and do happen
  - Lesson already learned in prior incidents .....
- Hazard awareness may not be adequate among all personnel working at oil and gas production sites
- Development of flammable atmosphere in tank depends on separator pressure, tank arrangement and tank ventilation
  - Should assume tank vapor space may be flammable
  - Can also have separator malfunction leading to increased fuel concentration

27

Potential Prevention and Mitigation Approaches

- Increase hazard awareness for all personnel working at site
- Measure level at grade to eliminate routine access to most hazardous location (i.e., tank roof)
- Inert tanks using natural gas
  - Take feed from product gas line
  - Eliminate open vent paths and institute pressure control system (flow as required to maintain positive pressure and preclude air in-leakage),
  - Will result in increased emission of natural gas from facility
- Use a "lead tank" as a secondary separator
  - Inert only this tank with natural gas (limit use of natural gas)
  - Pump or gravity flow from "lead tank" to other produced water tanks
- Monitor tank vapor space

28

Questions

?

29

481704 Milestone Company Incidents and Their Impact on Air Products Process Safety Journey

Abstract:

Air Products has experienced several significant Process Safety events in our history that have helped define the Process Safety culture that exists at the company today. This presentation describes these events and their root causes, as well as the changes that they launched in the company's approach to Process Safety. The first incident was an explosion at a Nitrous Oxide plant in 1976 that resulted in four on-site fatalities. In the aftermath of this event the need for a formal Hazard Review Process was identified. The second event occurred in 1978 at an air separation plant in which a cryogenic Liquid Oxygen storage tank was ruptured and 800 tons of Liquid Oxygen was released. The Liquid Oxygen vaporized and drifted off-site, resulting in five off-site fatalities. This incident highlighted the necessity of Risk Quantification. While these milestone events served as the impetus for Air Products Process Safety journey, that journey continues to current day, and the presentation will close with a brief update on future direction.

481925 Plymouth Incident

Abstract:

The Williams Company LNG facility on their Northwest Pipeline system located in the state of Washington experience an explosion March 30th 2014. The incident occurred during a routine start-up of the liquefaction process. Directly after the facility was deemed safe the site preservation process was initiated and the investigation began. The various regulatory agencies were involved during the investigation to help determine the cause and to provide recommendations to prevent like kind occurrences elsewhere within the industry. Additional 3rd party firms were called in to review potential structural deficiencies, design flaws, metallurgy testing, and auto ignition testing.

AUTHOR INDEX

AUTHOR INDEX